Learning Autodesk® Inventor® 2013

Randy H. Shih
Oregon Institute of Technology

ISBN: 978-1-58503-727-8

Mission, Kansas

Schroff Development Corporation
P.O. Box 1334
Mission KS 66222
(913) 262-2664
www.SDCpublications.com

Publisher: Stephen Schroff

Shih, Randy H.
Learning Autodesk Inventor 2013/
Randy H. Shih

ISBN 978-1-58503-727-8

Printed and bound in the United States of America.

Preface

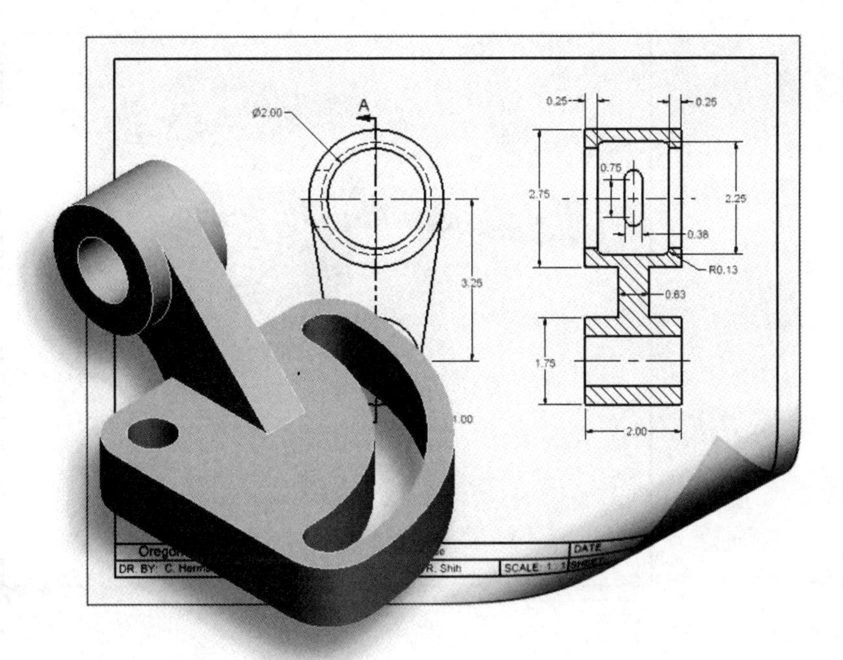

The primary goal of ***Learning Autodesk Inventor 2013*** is to introduce the aspects of designing with **Solid Modeling** and **Parametric Modeling**. This text is intended to be used as a practical training guide for students and professionals. This text uses *Autodesk Inventor 2013* as the modeling tool and the chapters proceed in a pedagogical fashion to guide you from constructing basic solid models to building intelligent mechanical designs, creating multi-view drawings and assembly models. This text takes a hands-on, project-based approach to all the important *Parametric Modeling* techniques and concepts. This textbook contains a series of twelve tutorial style lessons designed to introduce beginning CAD users to ***Autodesk Inventor***. This text is also helpful to *Autodesk Inventor* users upgrading from a previous release of the software. The solid modeling techniques and concepts discussed in this text are also applicable to other parametric feature-based CAD packages. The basic premise of this book is that the more designs you create using *Autodesk Inventor*, the better you learn the software. The book presents the basics to parametric modeling by modeling the *Tamiya Mechanical Tiger* kit. Additional engineering analyses are also performed in both *Autodesk Inventor* and the dynamic geometry software, *GeoGebra*. With this in mind, each lesson introduces a new set of commands and concepts, building on previous lessons. Majority of the main parts of the *Tiger* design are modeled in the chapters. This book does not attempt to cover all of the *Autodesk Inventor's* features, only to provide an introduction to the software. It is intended to help you establish a good basis for exploring and growing in the exciting field of **Computer Aided Engineering**.

Acknowledgments

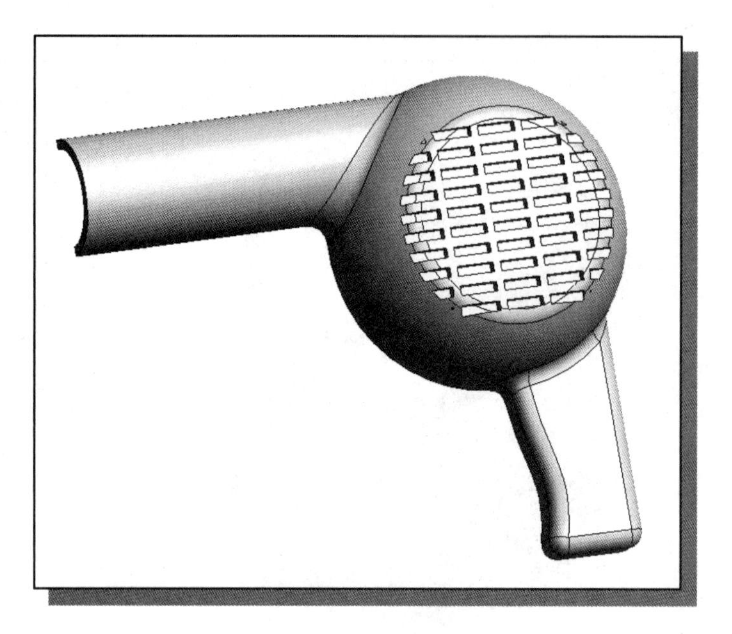

This book would not have been possible without a great deal of support. First, special thanks to two great teachers, Prof. George R. Schade of University of Nebraska-Lincoln and Mr. Denwu Lee, who showed me the fundamentals, the intrigue, and the sheer fun of Computer Aided Engineering.

The effort and support of the editorial and production staff of Schroff Development Corporation is gratefully acknowledged. I would especially like to thank Stephen Schroff for the support and helpful suggestions during this project.

I am grateful that the Mechanical Engineering Technology Department of Oregon Institute of Technology has provided me with an excellent environment in which to pursue my interests in teaching and research.

Finally, truly unbounded thanks are due to my wife Hsiu-Ling and our daughter Casandra for their understanding and encouragement throughout this project.

Randy H. Shih
Klamath Falls, Oregon
Spring, 2012

Table of Contents

Chapter 4
Parametric Constraints Fundamentals

Chapter 5
Pictorials and Sketching

Chapter 6
Symmetrical Features and Part Drawings

Chapter 7
Datum Features in Designs

Chapter 8
Content Center and Gear Generator

Chapter 9
Advanced 3D Construction Tools

Chapter 10
Planar Linkage Analysis using GeoGebra

Chapter 11
Design Makes the Difference

Chapter 12
Assembly Modeling and Motion Analysis

Index

Notes:

Chapter 1
Introduction – Getting Started

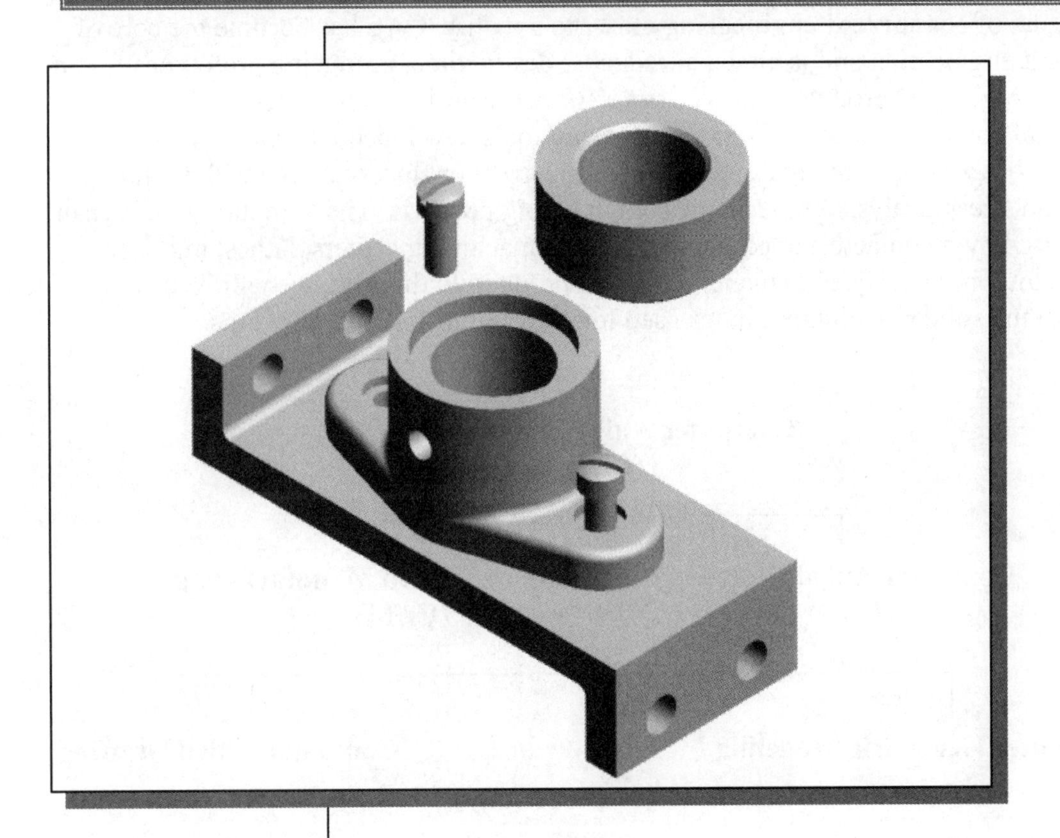

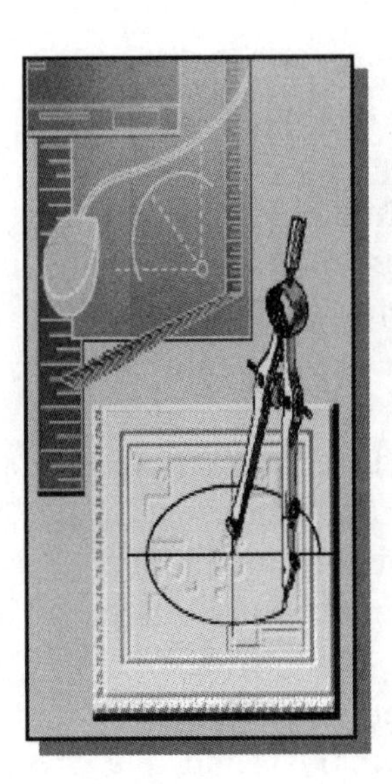

Learning Objectives

- ♦ Development of Computer Geometric Modeling
- ♦ Feature-Based Parametric Modeling
- ♦ Startup Options and Units Setup
- ♦ Autodesk Inventor Screen Layout
- ♦ User Interface and Mouse Buttons
- ♦ Autodesk Inventor On-Line Help

Introduction

The rapid changes in the field of **Computer Aided Engineering** (CAE) have brought exciting advances in the engineering community. Recent advances have made the long-sought goal of **concurrent engineering** closer to a reality. CAE has become the core of concurrent engineering and is aimed at reducing design time, producing prototypes faster, and achieving higher product quality. ***Autodesk Inventor*** is an integrated package of mechanical computer aided engineering software tools developed by *Autodesk, Inc.* ***Autodesk Inventor*** is a tool that facilitates a concurrent engineering approach to the design and stress-analysis of mechanical engineering products. The computer models can also be used by manufacturing equipment such as machining centers, lathes, mills, or rapid prototyping machines to manufacture the product. In this text, we will be dealing only with the solid modeling modules used for part design and part drawings.

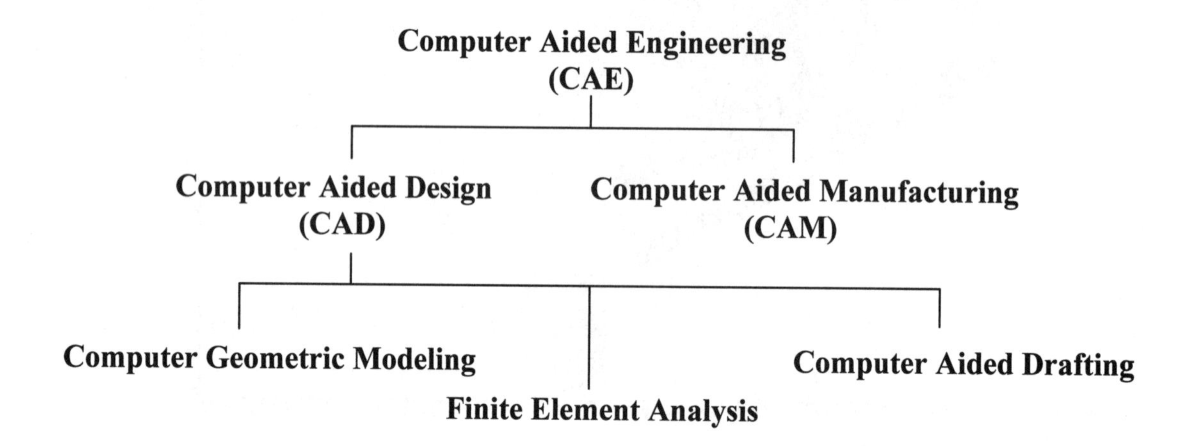

Development of Computer Geometric Modeling

Computer geometric modeling is a relatively new technology, and its rapid expansion in the last fifty years is truly amazing. Computer-modeling technology has advanced along with the development of computer hardware. The first generation CAD programs, developed in the 1950s, were mostly non-interactive; CAD users were required to create program-codes to generate the desired two-dimensional (2D) geometric shapes. Initially, the development of CAD technology occurred mostly in academic research facilities. The Massachusetts Institute of Technology, Carnegie-Mellon University, and Cambridge University were the leading pioneers at that time. The interest in CAD technology spread quickly and several major industry companies, such as General Motors, Lockheed, McDonnell, IBM, and Ford Motor Co., participated in the development of interactive CAD programs in the 1960s. Usage of CAD systems was primarily in the automotive industry, aerospace industry, and government agencies that developed their own programs for their specific needs. The 1960s also marked the beginning of the development of finite element analysis methods for computer stress analysis and computer aided manufacturing for generating machine tool paths.

The 1970s are generally viewed as the years of the most significant progress in the development of computer hardware, namely the invention and development of **microprocessors**. With the improvement in computing power, new types of 3D CAD programs that were user-friendly and interactive became reality. CAD technology quickly expanded from very simple **computer aided drafting** to very complex **computer aided design**. The use of 2D and 3D wireframe modelers was accepted as the leading edge technology that could increase productivity in industry. The developments of surface modeling and solid modeling technologies were taking shape by the late 1970s, but the high cost of computer hardware and programming slowed the development of such technology. During this period, the available CAD systems all required room-sized mainframe computers that were extremely expensive.

In the 1980s, improvements in computer hardware brought the power of mainframes to the desktop at less cost and with more accessibility to the general public. By the mid-1980s, CAD technology had become the main focus of a variety of manufacturing industries and was very competitive with traditional design/drafting methods. It was during this period of time that 3D solid modeling technology had major advancements, which boosted the usage of CAE technology in industry.

The introduction of the *feature-based parametric solid modeling* approach, at the end of the 1980s, elevated CAD/CAM/CAE technology to a new level. In the 1990s, CAD programs evolved into powerful design/manufacturing/management tools. CAD technology has come a long way, and during these years of development, modeling schemes progressed from two-dimensional (2D) wireframe to three-dimensional (3D) wireframe, to surface modeling, to solid modeling and, finally, to feature-based parametric solid modeling.

The first generation CAD packages were simply 2D **computer aided drafting** programs, basically the electronic equivalents of the drafting board. For typical models, the use of this type of program would require that several views of the objects be created individually as they would be on the drafting board. The 3D designs remained in the designer's mind, not in the computer database. Mental translations of 3D objects to 2D views are required throughout the use of these packages. Although such systems have some advantages over traditional board drafting, they are still tedious and labor intensive. The need for the development of 3D modelers came quite naturally, given the limitations of the 2D drafting packages.

The development of three-dimensional modeling schemes started with three-dimensional (3D) wireframes. Wireframe models are models consisting of points and edges, which are straight lines connecting between appropriate points. The edges of wireframe models are used, similar to lines in 2D drawings, to represent transitions of surfaces and features. The use of lines and points is also a very economical way to represent 3D designs.

The development of the 3D wireframe modeler was a major leap in the area of computer geometric modeling. The computer database in the 3D wireframe modeler contains the locations of all the points in space coordinates, and it is typically sufficient to create just one model rather than multiple views of the same model. This single 3D model can then be viewed from any direction as needed. Most 3D wireframe modelers allow the user to create projected lines/edges of 3D wireframe models. In comparison to other types of 3D modelers, the 3D wireframe modelers require very little computing power and generally can be used to achieve reasonably good representations of 3D models. However, because surface definition is not part of a wireframe model, all wireframe images have the inherent problem of ambiguity. Two examples of such ambiguity are illustrated.

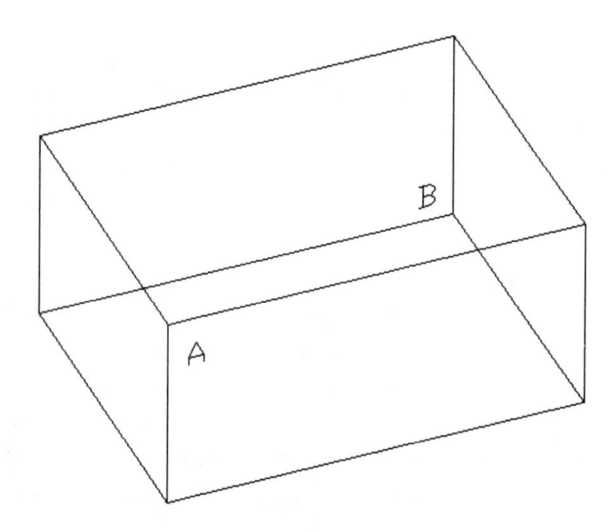

Wireframe Ambiguity: Which corner is in front, A or B?

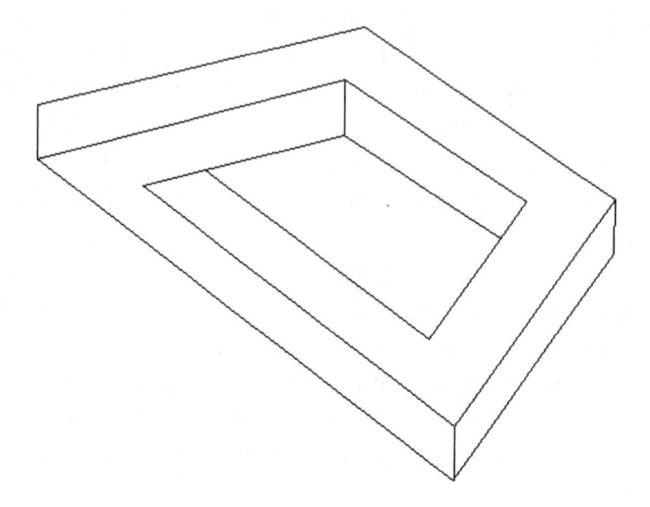

A non-realizable object: Wireframe models contain no surface definitions.

Surface modeling is the logical development in computer geometry modeling to follow the 3D wireframe modeling scheme by organizing and grouping edges that define polygonal surfaces. Surface modeling describes the part's surfaces but not its interiors. Designers are still required to interactively examine surface models to insure that the various surfaces on a model are contiguous throughout. Many of the concepts used in 3D wireframe and surface modelers are incorporated in the solid modeling scheme, but it is solid modeling that offers the most advantages as a design tool.

In the solid modeling presentation scheme, the solid definition includes nodes, edges, and surfaces, and it is a complete and unambiguous mathematical representation of a precisely enclosed and filled volume. Unlike the surface modeling method, solid modelers start with a solid or use topology rules to guarantee that all of the surfaces are stitched together properly. Two predominant methods for representing solid models are **constructive solid geometry** (CSG) representation and **boundary representation** (B-rep).

The CSG representation method can be defined as the combination of 3D solid primitives. What constitutes a "primitive" varies somewhat with the software but typically includes a rectangular prism, a cylinder, a cone, a wedge, and a sphere. Most solid modelers also allow the user to define additional primitives, which are shapes typically formed by the basic shapes. The underlying concept of the CSG representation method is very straightforward; we simply **add** or **subtract** one primitive from another. The CSG approach is also known as the machinist's approach, as it can be used to simulate the manufacturing procedures for creating the 3D object.

In the B-rep representation method, objects are represented in terms of their spatial boundaries. This method defines the points, edges, and surfaces of a volume, and/or issues commands that sweep or rotate a defined face into a third dimension to form a solid. The object is then made up of the unions of these surfaces that completely and precisely enclose a volume.

By the 1980s, a new paradigm called *concurrent engineering* had emerged. With concurrent engineering, designers, design engineers, analysts, manufacturing engineers, and management engineers all work together closely right from the initial stages of the design. In this way, all aspects of the design can be evaluated and any potential problems can be identified right from the start and throughout the design process. Using the principles of concurrent engineering, a new type of computer modeling technique appeared. The technique is known as the *feature-based parametric modeling technique*. The key advantage of the *feature-based parametric modeling technique* is its capability to produce very flexible designs. Changes can be made easily and design alternatives can be evaluated with minimum effort. Various software packages offer different approaches to feature-based parametric modeling, yet the end result is a flexible design defined by its design variables and parametric features.

Feature-Based Parametric Modeling

One of the key elements in the *Autodesk Inventor* solid modeling software is its use of the **feature-based parametric modeling technique**. The feature-based parametric modeling approach has elevated solid modeling technology to the level of a very powerful design tool. Parametric modeling automates the design and revision procedures by the use of parametric features. Parametric features control the model geometry by the use of design variables. The word ***parametric*** means that the geometric definitions of the design, such as dimensions, can be varied at any time during the design process. Features are predefined parts or construction tools for which users define the key parameters. A part is described as a sequence of engineering features, which can be modified and/or changed at any time. The concept of parametric features makes modeling more closely match the actual design-manufacturing process than the mathematics of a solid modeling program. In parametric modeling, models and drawings are updated automatically when the design is refined.

Parametric modeling offers many benefits:

- **We begin with simple, conceptual models with minimal detail; this approach conforms to the design philosophy of "shape before size."**
- **Geometric constraints, dimensional constraints, and relational parametric equations can be used to capture design intent.**
- **We have the ability to update an entire system, including parts, assemblies and drawings, after changing one parameter of complex designs.**
- **We can quickly explore and evaluate different design variations and alternatives to determine the best design.**
- **Existing design data can be reused to create new designs.**
- **Quick design turn-around.**

One of the key features of *Autodesk Inventor* is the use of an assembly-centric paradigm, which enables users to concentrate on the design without depending on the associated parameters or constraints. Users can specify how parts fit together and the *Autodesk Inventor assembly-based fit function* automatically determines the parts' sizes and positions. This unique approach is known as the **Direct Adaptive Assembly approach**, which defines part relationships directly with no order dependency.

The *Adaptive Assembly approach* is a unique design methodology that can only be found in *Autodesk Inventor.* The goal of this methodology is to improve the design process and allows you, the designer, to **Design the Way You Think**.

Getting Started with *Autodesk Inventor*

- *Autodesk Inventor* is composed of several application software modules (these modules are called *applications*), all sharing a common database. In this text, the main concentration is placed on the solid modeling modules used for part design. The general procedures required in creating solid models, engineering drawings, and assemblies are illustrated.

How to start *Autodesk Inventor* depends on the type of workstation and the particular software configuration you are using. With most *Windows* systems, you may select **Autodesk Inventor** on the *Start* menu or select the **Autodesk Inventor** icon on the desktop. Consult your instructor or technical support personnel if you have difficulty starting the software. The program takes a while to load, so be patient.

The tutorials in this text are based on the assumption that you are using *Autodesk Inventor's* default settings. If your system has been customized for other uses, contact your technical support personnel to restore the default software configuration.

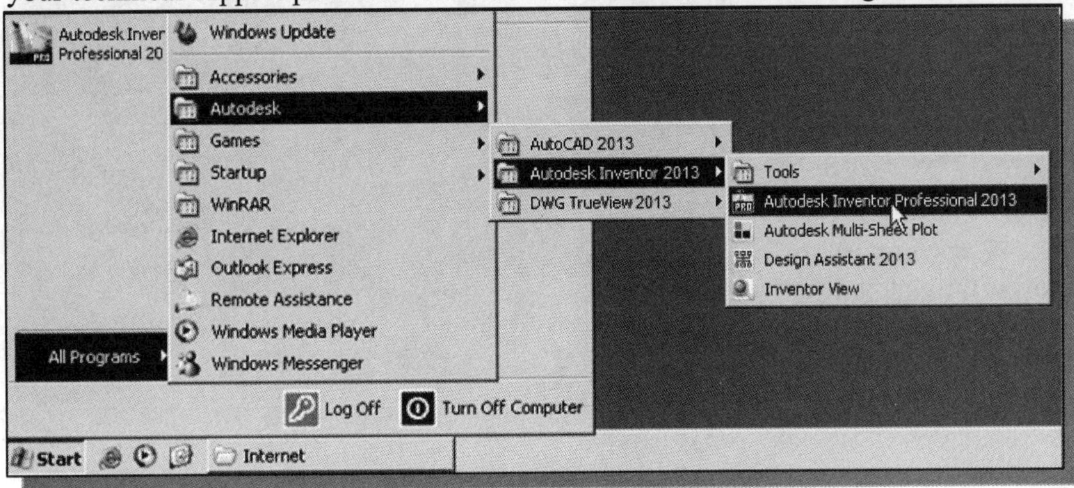

The Screen Layout and Getting Started Toolbar

Once the program is loaded into the memory, the *Inventor* window appears on the screen with the *Get Started* toolbar options activated.

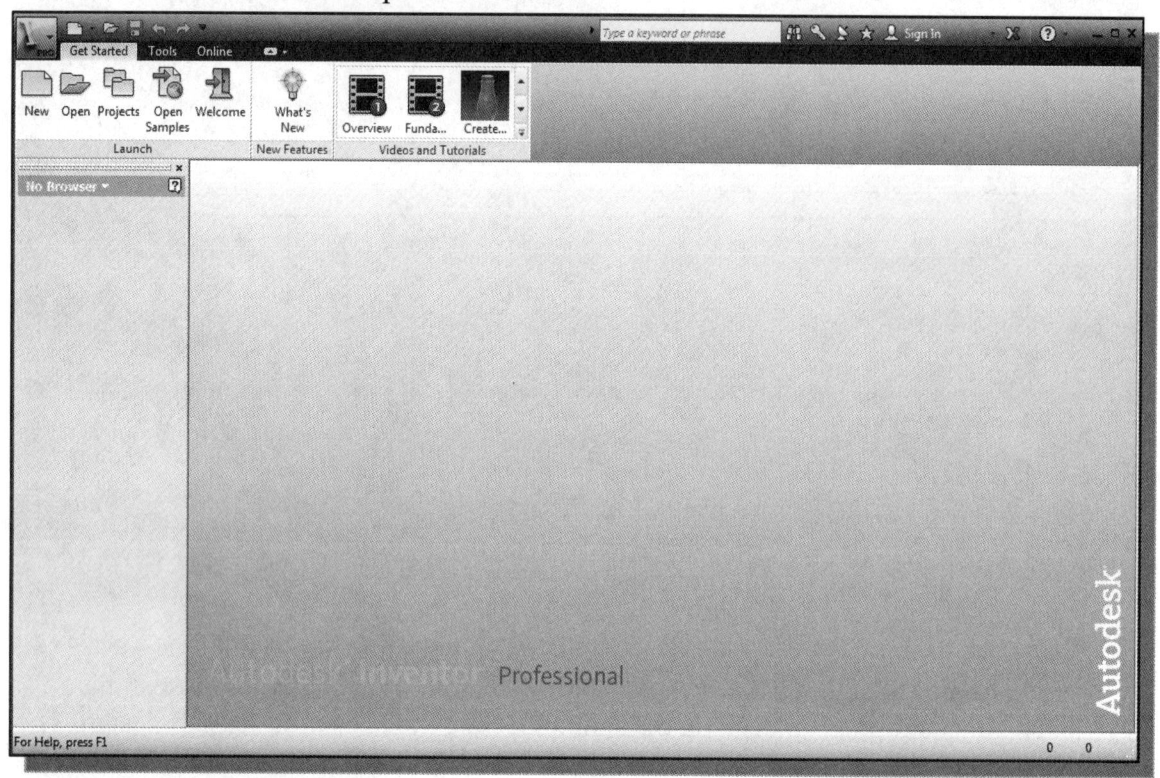

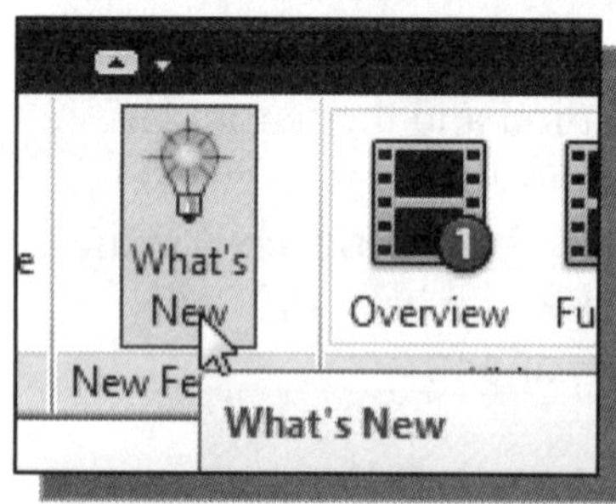

❖ Note that the *Get Started* toolbar contains helpful information in regards to using the *Inventor* software. For example, clicking the **What's New** option will bring up the *Internet Browser*, which contains the list of new features that are included in this release of *Autodesk Inventor*.

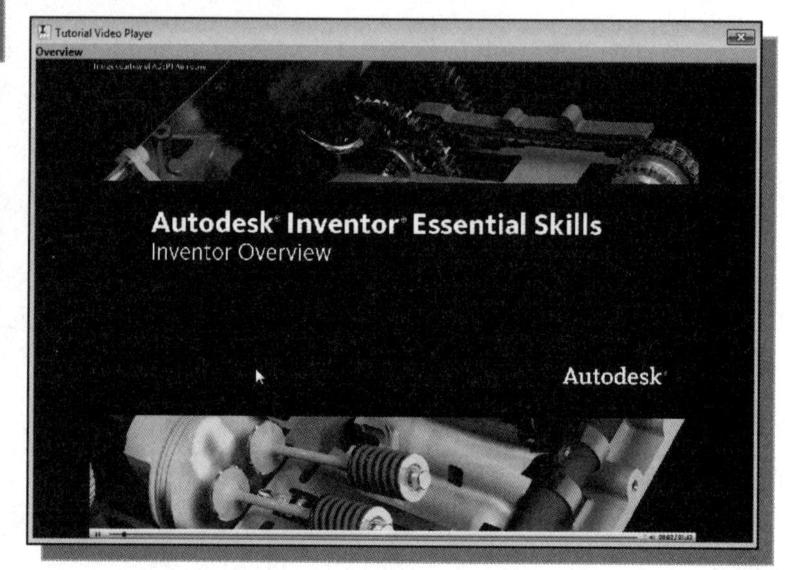

❖ You are encouraged to browse through the general introduction available in the *Videos and Tutorials* section.

The New File Dialog Box and Units Setup

When starting a new CAD file, the first thing we should do is to choose the units we would like to use. We will use the English (feet and inches) setting for this example.

- Select the **New** icon with a single click of the left-mouse-button in the *Launch* toolbar.

❖ Note that the **New** option allows us to start a new modeling task, which can be creating a new model or several other modeling tasks.

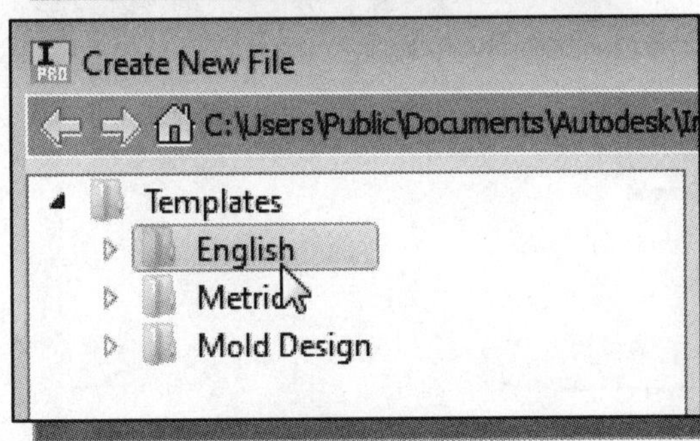

- Select the **English** tab in the *New File* dialog box as shown. Note the default tab contains the file options which are based on the default units chosen during installation.

- Select the **Standard(in).ipt** icon as shown. The different icons are templates for the different modeling tasks. The **idw** file type stands for drawing file, the **iam** file type stands for assembly file, and the **ipt** file type stands for part file. The **ipn** file type stands for assembly presentation.

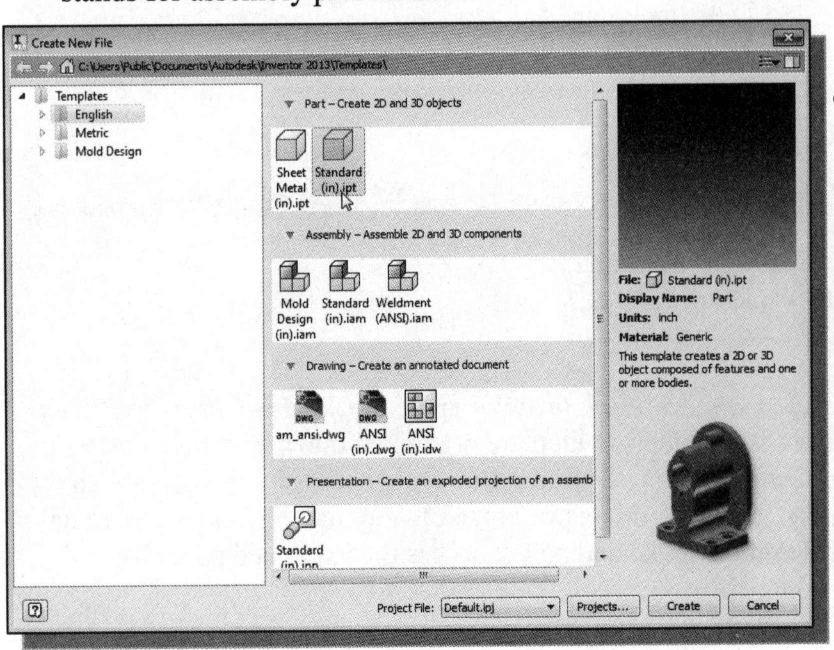

- Click **Create** in the *Create New File* dialog box to accept the selected settings.

The *Default Autodesk Inventor* Screen Layout

The default *Autodesk Inventor* drawing screen contains the *pull-down* menus, the *Standard* toolbar, the *Features* toolbar, the *Sketch* toolbar, the *drawing* area, the *browser* area, and the *Status Bar*. A line of quick text appears next to the icon as you move the *mouse cursor* over different icons. You may resize the *Autodesk Inventor* drawing window by clicking and dragging the edges of the window, or relocate the window by clicking dragging the window title area.

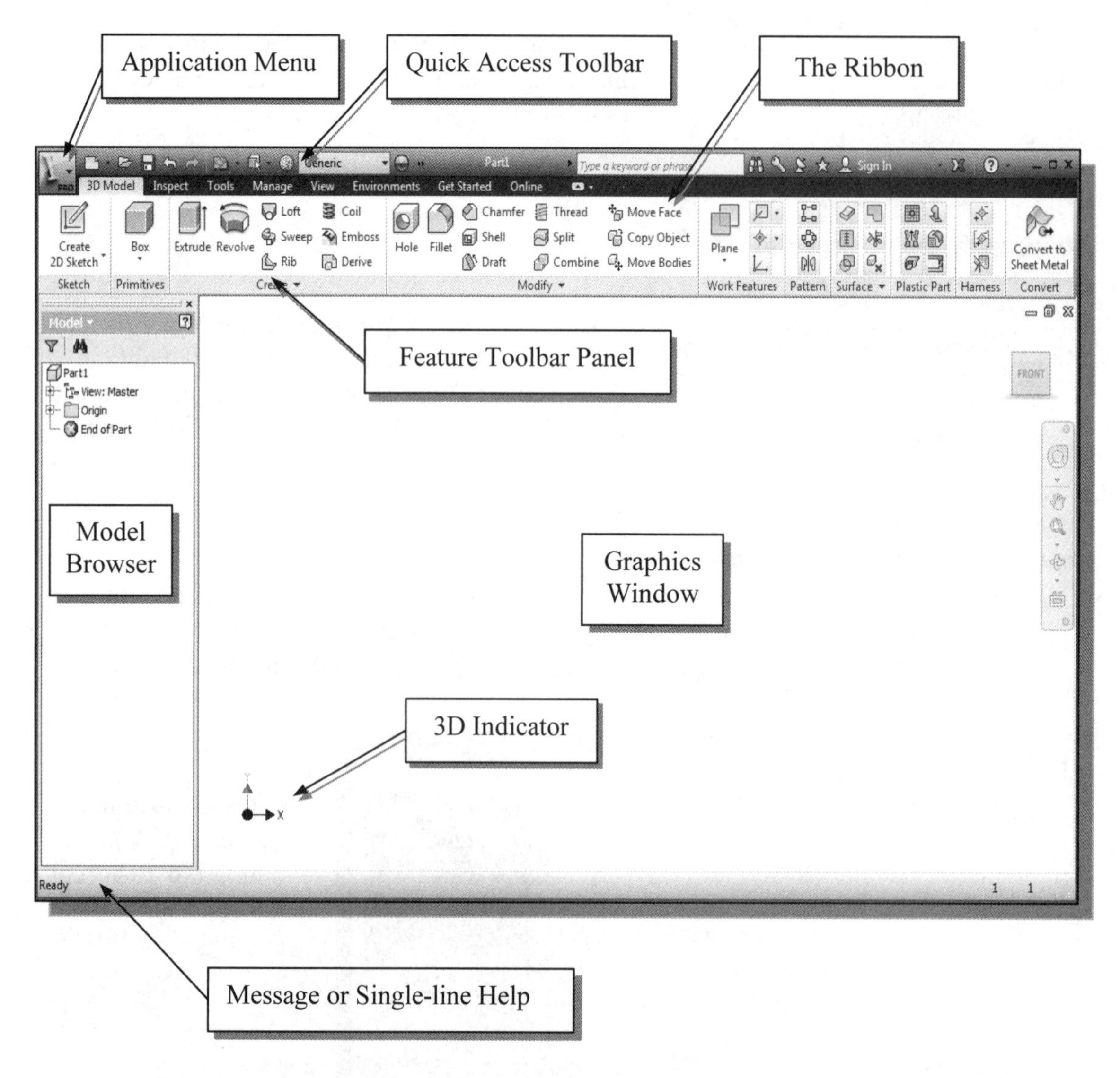

- The ***Ribbon*** is a new feature in *Autodesk Inventor* since the *2011 release*; the *Ribbon* is composed of a series of tool panels, which are organized into tabs labeled by task. The *Ribbon* provides a compact palette of all of the tools necessary to accomplish the different modeling tasks. The drop-down arrow next to any icon indicates additional commands are available on the expanded panel; access the expanded panel by clicking on the drop-down arrow.

- **Application Menu**

The *Application* menu at the upper left corner of the main window contains tools for all file-related operations, such as Open, Save, Export, etc.

- **Quick Access Toolbar**

The *Quick Access* toolbar at the top of the *Inventor* window allows us quick access to file-related commands and to Undo/Redo the last operations.

- **Ribbon Tabs**

The *Ribbon* is composed of a series of tool panels, which are organized into tabs labeled by task. The assortments of tool panels can be accessed by clicking on the tabs.

- **Online Help Panel**

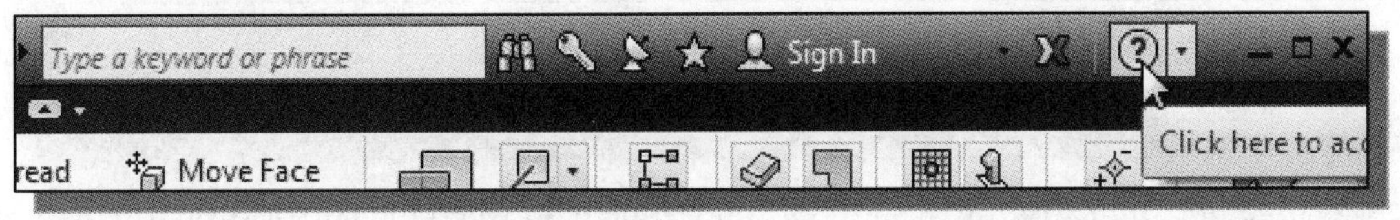

The *Help* options panel provides us with multiple options to access on-line help for *Autodesk Inventor*. The ***Online Help system*** provides general help information, such as command options and command references.

- **2D Sketch Toolbar**

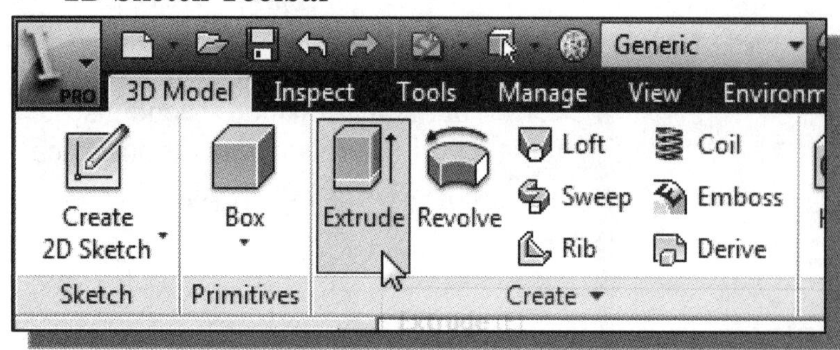

The *Create* toolbar provides tools for creating the different types of 3D features, such as Extrude, Revolve, Sweep, etc.

- **Graphics Window**

The *graphics window* is the area where models and drawings are displayed.

- **Message and Status Bar**

The *Message and Status Bar* area shows a single-line help when the cursor is on top of an icon. This area also displays information pertinent to the active operation. For example, in the figure above, the coordinates and length information of a line are displayed while the *Line* command is activated.

Mouse Buttons

Autodesk Inventor utilizes the mouse buttons extensively. In learning *Autodesk Inventor*'s interactive environment, it is important to understand the basic functions of the mouse buttons. It is highly recommended that you use a mouse or a tablet with *Autodesk Inventor* since the package uses the buttons for various functions.

- **Left mouse button**
  The **left-mouse-button** is used for most operations, such as selecting menus and icons, or picking graphic entities. One click of the button is used to select icons, menus and form entries, and to pick graphic items.

- **Right mouse button**
  The **right-mouse-button** is used to bring up additional available options. The software also utilizes the **right-mouse-button** as the same as the **ENTER** key, and is often used to accept the default setting to a prompt or to end a process.

- **Middle mouse button/wheel**
  The middle mouse button/wheel can be used to **Pan** (hold down the wheel button and drag the mouse) or **Zoom** (turn the wheel) realtime.

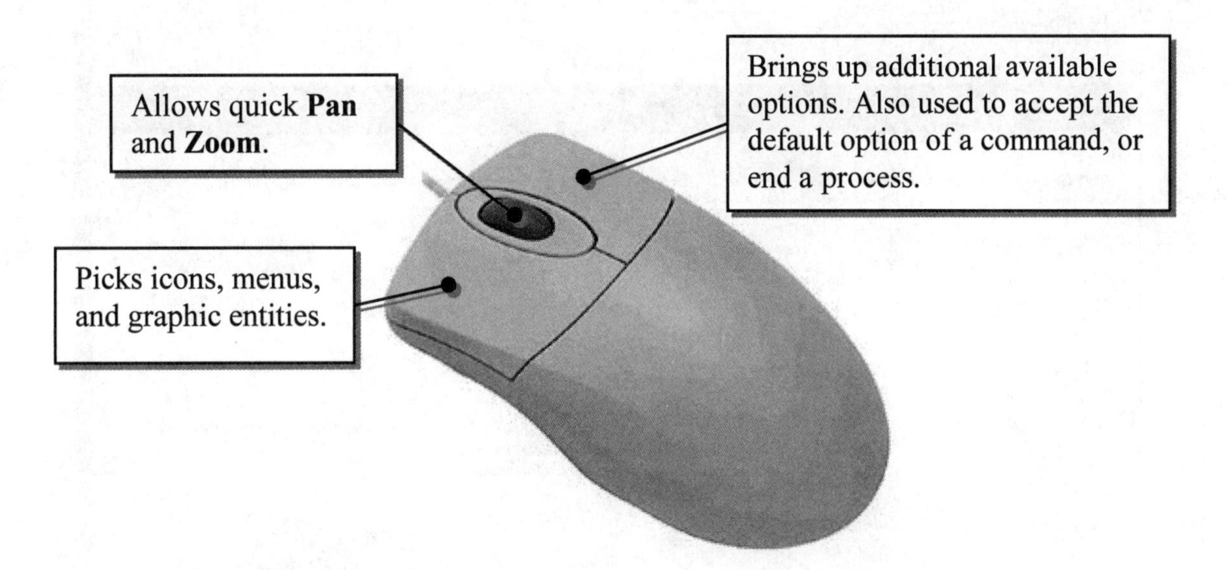

[Esc] – Canceling Commands

The [**Esc**] key is used to cancel a command in *Autodesk Inventor*. The [**Esc**] key is located near the top-left corner of the keyboard. Sometimes, it may be necessary to press the [**Esc**] key twice to cancel a command; it depends on where we are in the command sequence. For some commands, the [**Esc**] key is used to exit the command.

Autodesk Inventor Help System

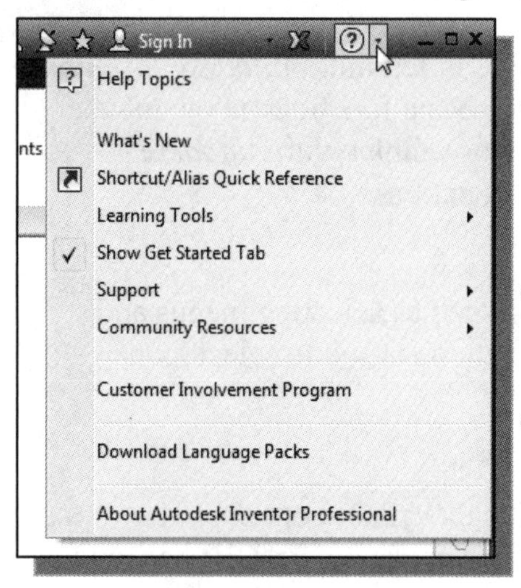

❖ Several types of help are available at any time during an *Autodesk Inventor* session. *Autodesk Inventor* provides many help functions, such as:

- Use the ***Help*** button near the upper right corner of the *Inventor* window.

- Help quick-key: Press the [**F1**] key to access the ***Inventor Help* system**.

- Use the ***Info Center*** to get information on a specific topic.

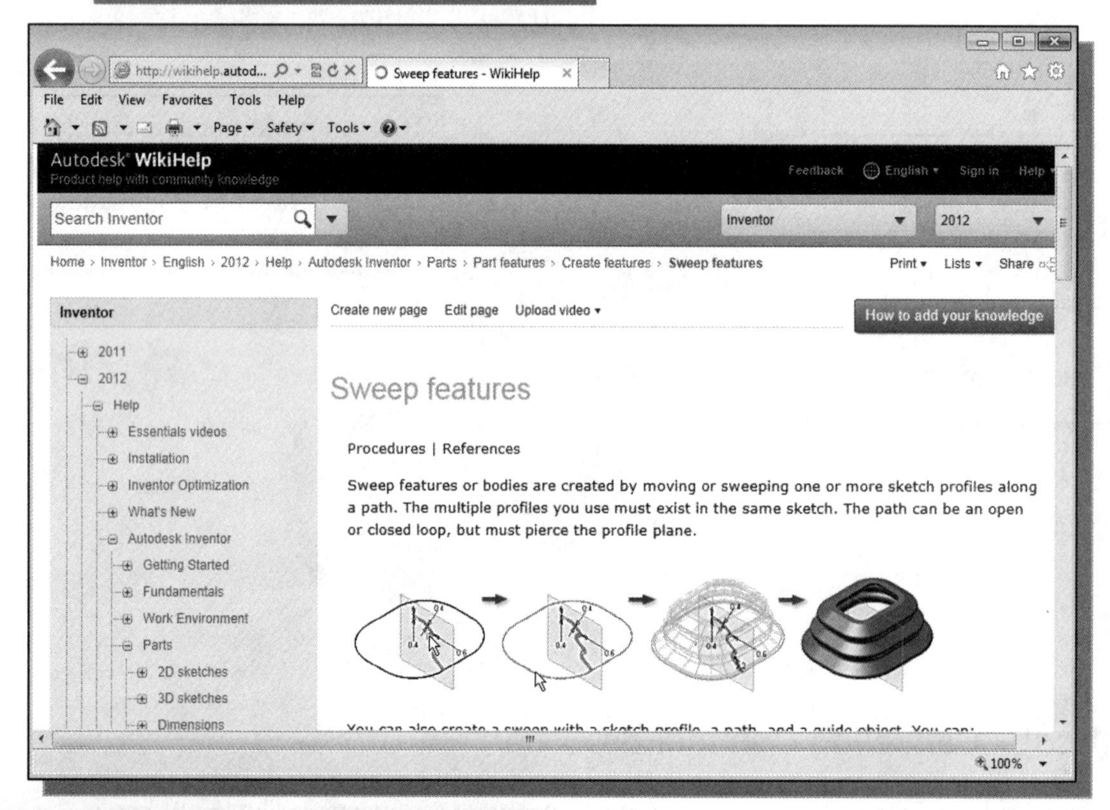

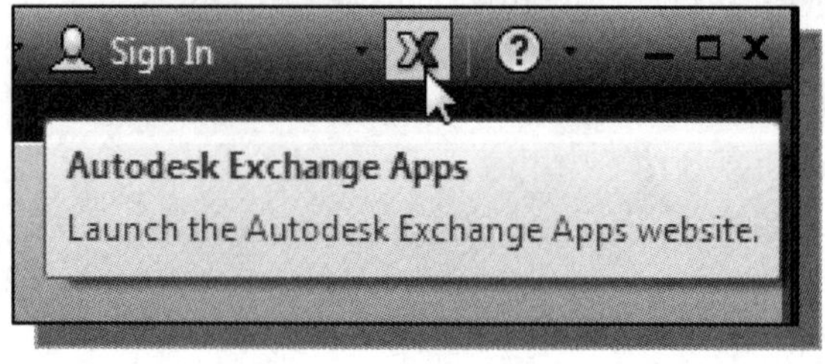

- Use the ***Autodesk Exchange*** to access information on the Autodesk community website.

Data Management using Inventor Project Files

With *Autodesk Inventor*, it is quite feasible to create designs without any regard to using the *Autodesk Inventor* data management system. Data management becomes much more critical for projects involving complex designs, especially when multiple team members are involved, or when we are working on integrating multiple design projects, or when it is necessary to share files among the design projects. *Autodesk Inventor* provides a fairly flexible data management system, so that one person may want to just use the basic option to help manage the locations of the different design files, or a team of designers can use the data management system to manage their projects stored on a networked computer system.

The *Autodesk Inventor* data management system organizes files based on **projects**. Each project is identified with a main folder that can contain files and folders associated to the design. In *Autodesk Inventor*, a *project* file (.ipj) defines the locations of all files associated with the project, including templates and library files.

The *Autodesk Inventor* data management system uses two types of projects:
- **Single-user Project**
- **Autodesk Vault Project** (installation of ***Autodesk Vault*** software is required)

The ***single-user project*** is for simpler projects where all project files are located on the same computer. The ***Autodesk Vault project*** is more suitable for projects requiring multiple users using a networked computer system.

- Click on the **Get Started** tab, with the left-mouse-button, in the *Ribbon* toolbar.

❖ Note the **Get Started** tab is the default panel displayed during startup.

- Click on the **Projects** icon with a single click of the left-mouse-button in the *Launch* toolbar.

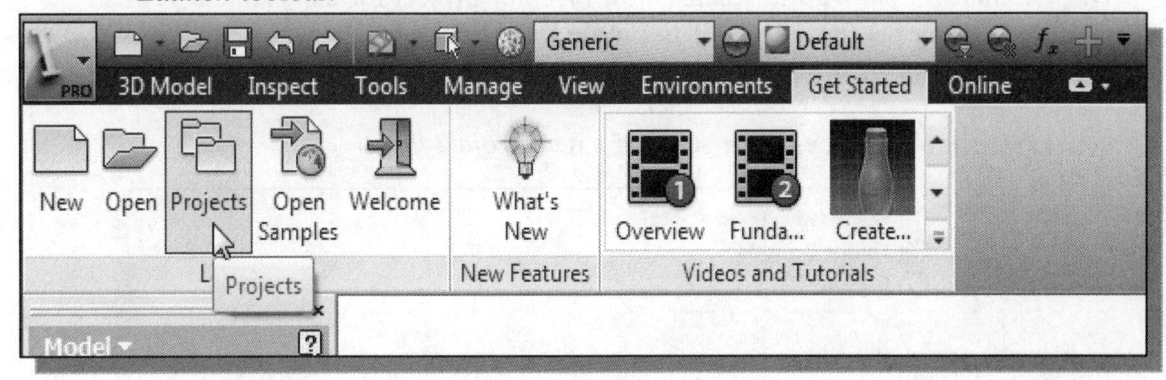

❖ The **Projects Editor** appears on the screen. Note that several options are available to access the *Editor*; it can also be accessed through the **Open file** command.

❖ In the *Projects Editor*: the **Default** project is available.

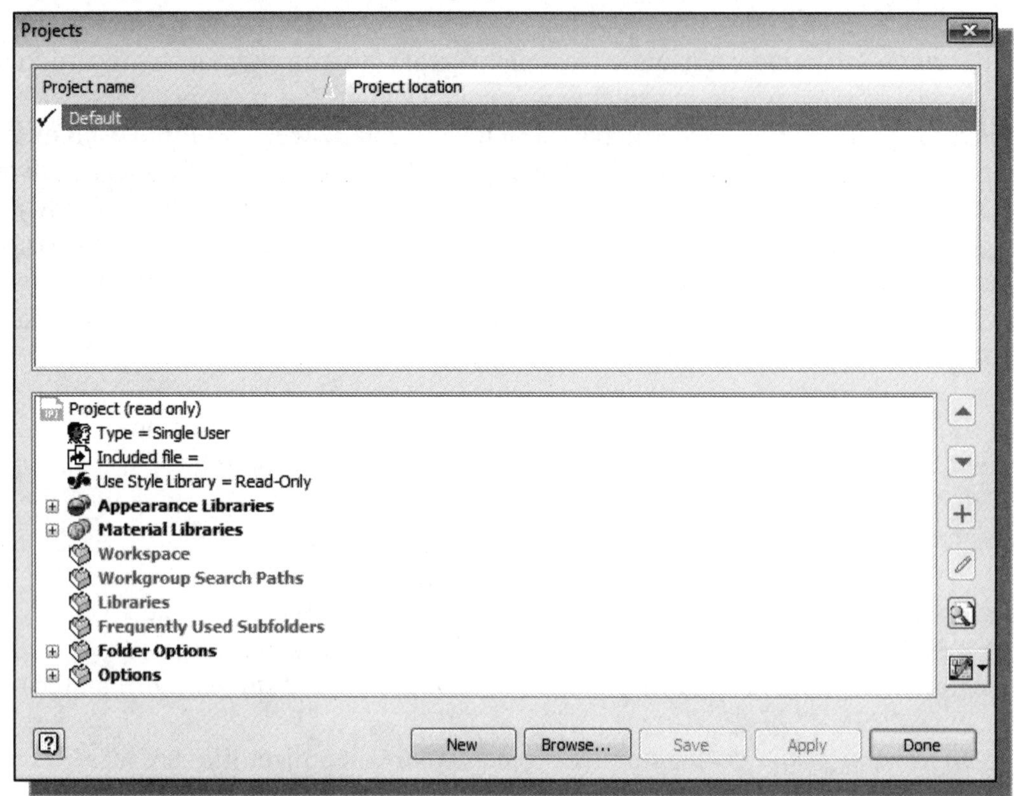

The **Default** project is automatically active by default; and the *default project* does not define any location for files. In other words, the data management system is not used. Using the *default project*, designs can still be created and modified, and any model file can be opened and saved anywhere without regard to project and file management.

Setup of a New Inventor Project

- In this section, we will create a new *Inventor* project for the chapters of this book, using the *Inventor* built-in **Single User Project** option. Note that it is also feasible to create a separate project for each chapter.

1. Click **New** to begin the setup of a new *project file*.

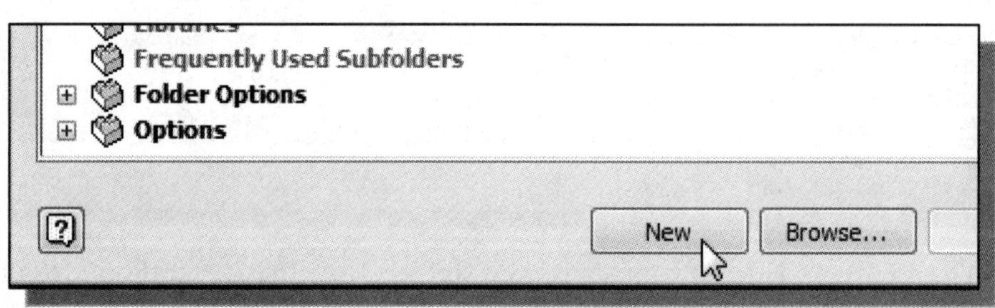

2. The *Inventor project wizard* appears on the screen; select the **New Single User Project** option as shown.

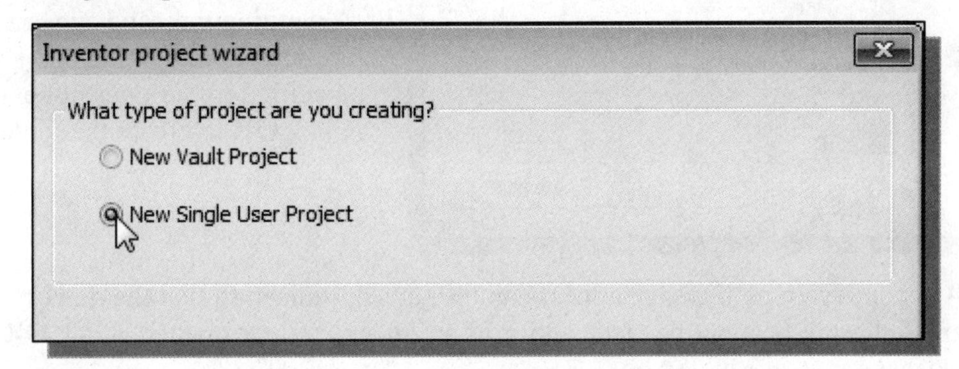

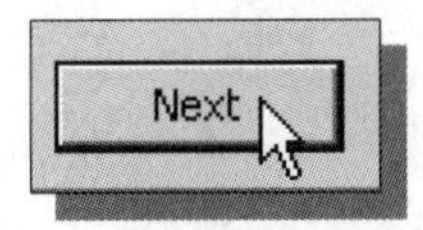

3. Click **Next** to proceed with the next setup option.

4. In the *Project File Name* input box, enter **Parametric-Modeling** as the name of the new project.

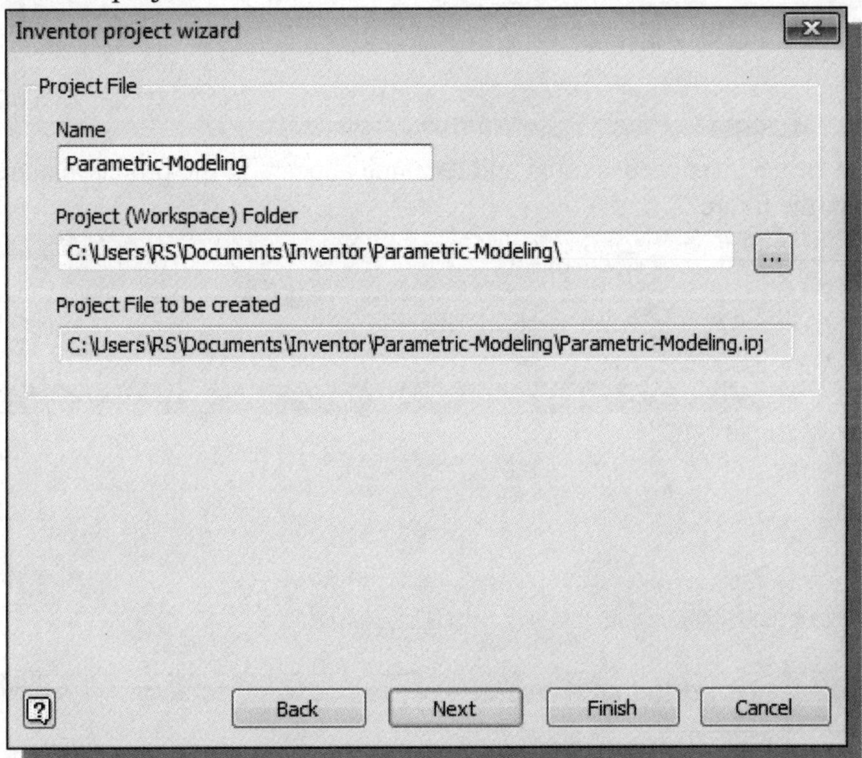

5. In the *Project Folder* input box, note the default folder location, such as **C:\Users\Docuements\Inventor-Data\Parametric Modeling** is displayed; choose a preferred folder name as the folder name of the new project.

6. Click **Finish** to proceed with the creation of the new project.

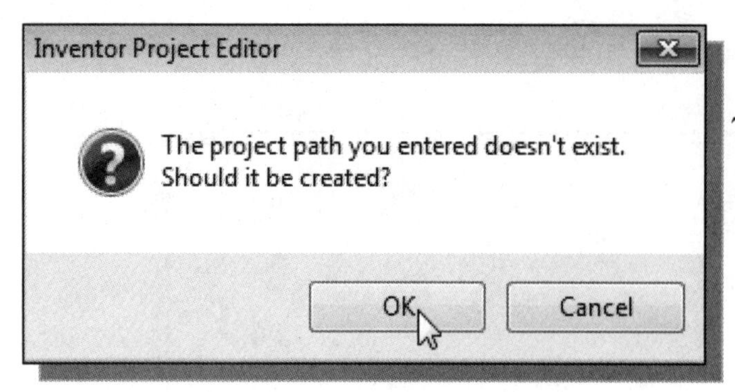

7. A warning message appears on the screen, indicating the specified folder does not exist. Click **OK** to create the folder.

8. A second warning message appears on the screen, indicating that the newly created project cannot be made active as an *Inventor* file is opened. Click **OK** to close the message dialog box.

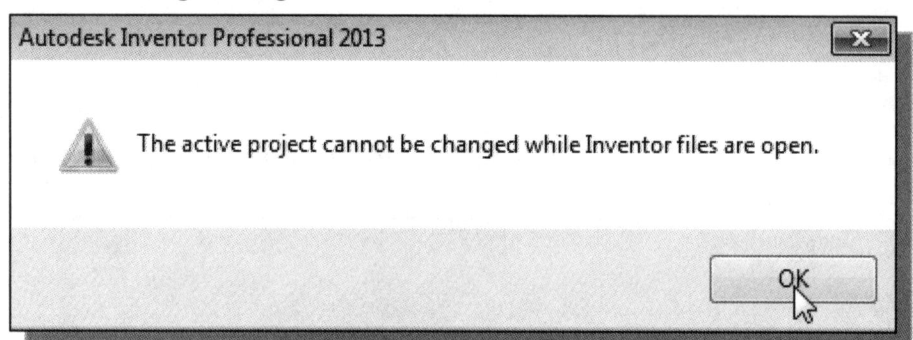

- The new project has been created and its name appears in the project list area as shown in the figure.

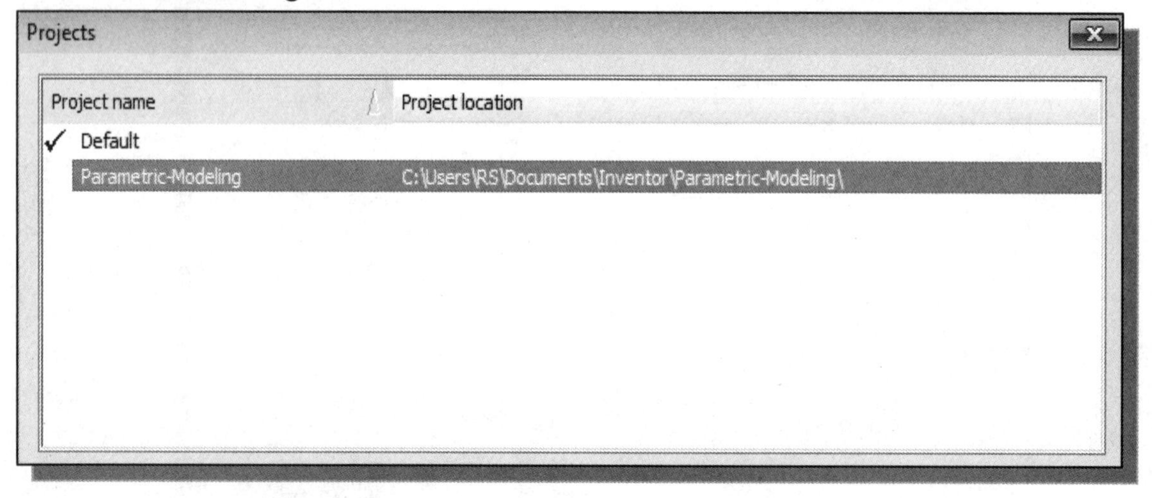

9. Click **Done** to exit the *Inventor Projects Editor* option.

The Content of the Inventor Project File

➤ An *Inventor* project file is actually a text file in .xml format with an .ipj extension. The file specifies the paths to the folder containing the files in the project. To assure that links between files work properly, it is advised to add the locations for folders to the project file before working on model files.

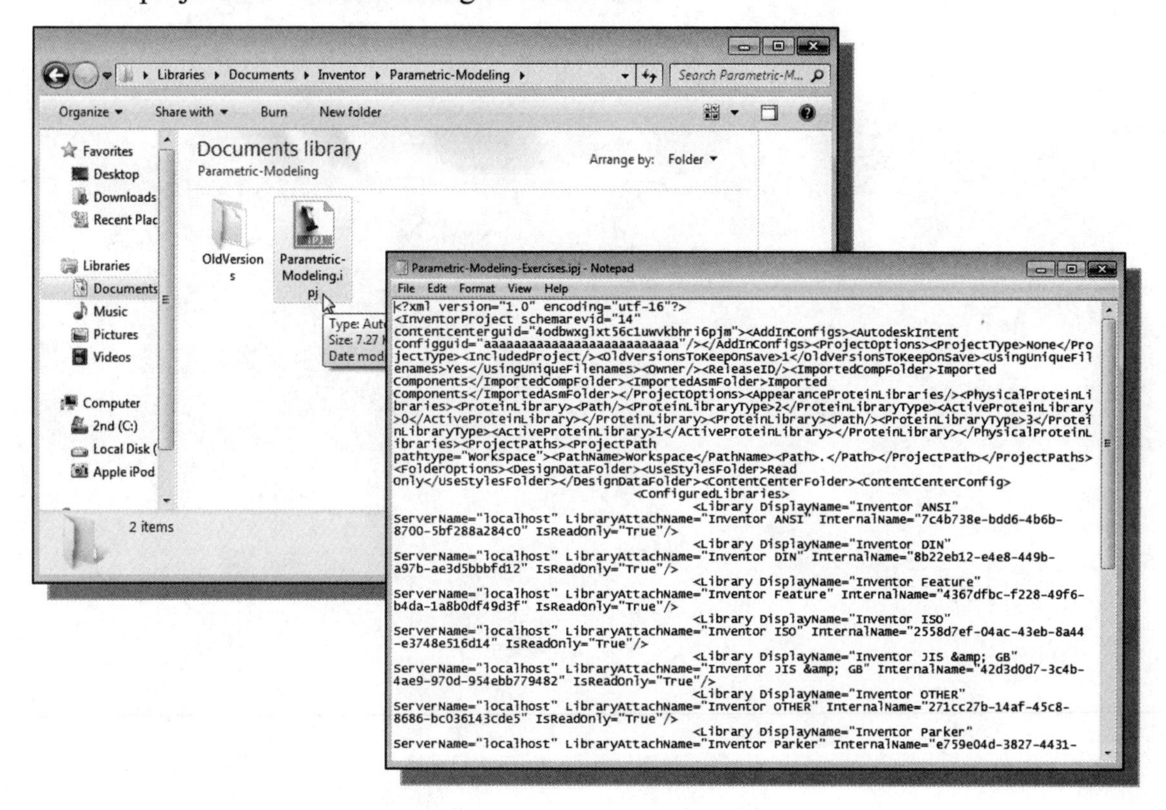

Leaving Autodesk Inventor

➤ To leave the *Application* menu, use the left-mouse-button and click on **Exit Inventor** from the pull-down menu.

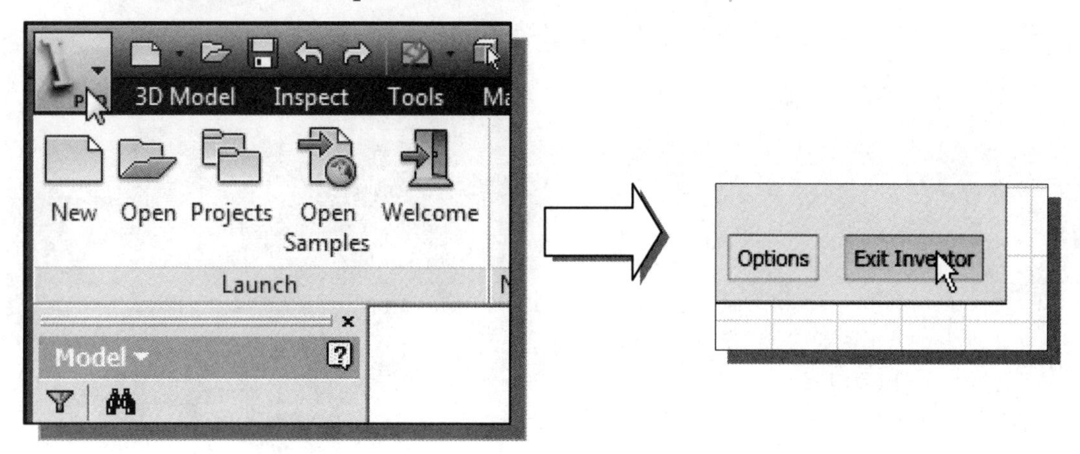

Notes:

Chapter 2
Parametric Modeling Fundamentals

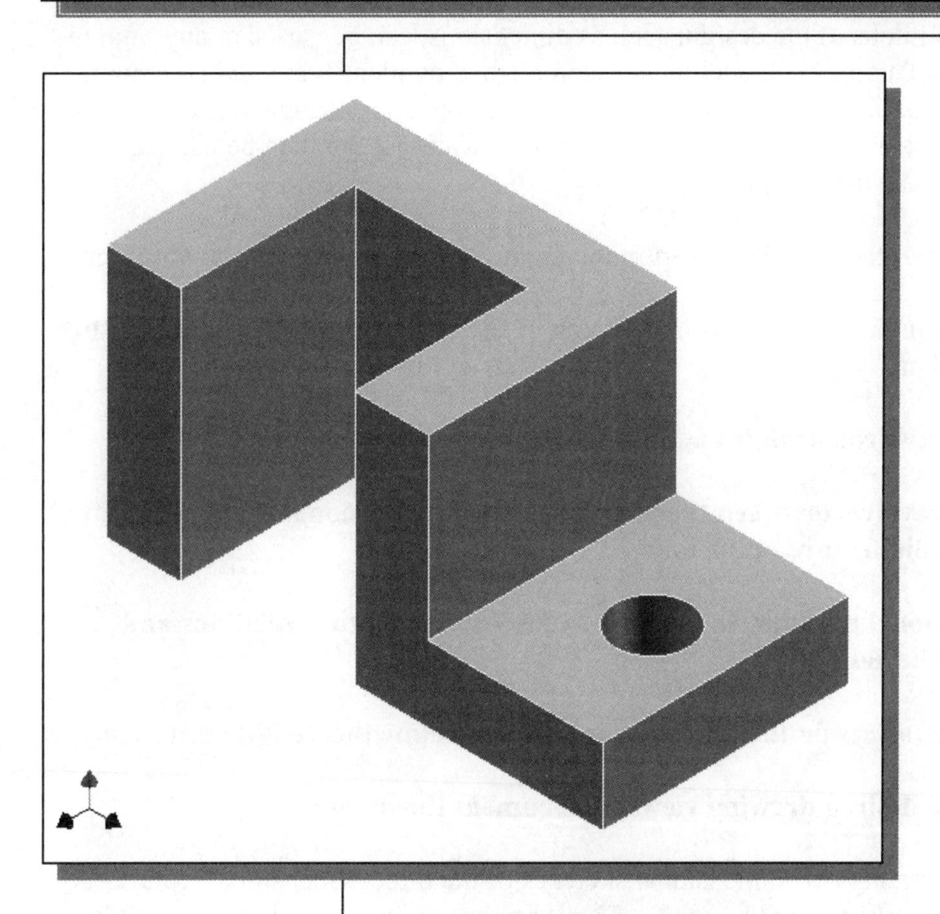

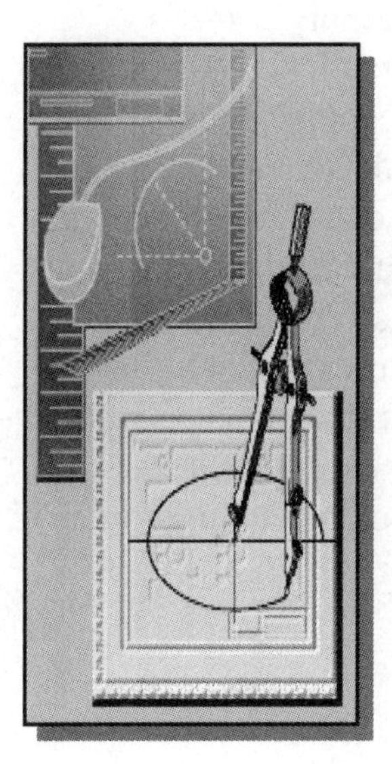

Learning Objectives

- **Create Simple Extruded Solid Models**
- **Understand the Basic Parametric Modeling Procedure**
- **Create 2-D Sketches**
- **Understand the "Shape before Size" Design Approach**
- **Use the Dynamic Viewing Commands**
- **Create and Edit Parametric Dimensions**

Introduction

The **feature-based parametric modeling** technique enables the designer to incorporate the original **design intent** into the construction of the model. The word ***parametric*** means the geometric definitions of the design, such as dimensions, can be varied at any time in the design process. Parametric modeling is accomplished by identifying and creating the key features of the design with the aid of computer software. The design variables, described in the sketches and described as parametric relations, can then be used to quickly modify/update the design.

In *Autodesk Inventor*, the parametric part modeling process involves the following steps:

1. **Create a rough two-dimensional sketch of the basic shape of the base feature of the design.**

2. **Apply/modify constraints and dimensions to the two-dimensional sketch.**

3. **Extrude, revolve, or sweep the parametric two-dimensional sketch to create the base solid feature of the design.**

4. **Add additional parametric features by identifying feature relations and complete the design.**

5. **Perform analyses on the computer model and refine the design as needed.**

6. **Create the desired drawing views to document the design.**

The approach of creating two-dimensional sketches of the three-dimensional features is an effective way to construct solid models. Many designs are in fact the same shape in one direction. Computer input and output devices we use today are largely two-dimensional in nature, which makes this modeling technique quite practical. This method also conforms to the design process that helps the designer with conceptual design along with the capability to capture the ***design intent***. Most engineers and designers can relate to the experience of making rough sketches on restaurant napkins to convey conceptual design ideas. *Autodesk Inventor* provides many powerful modeling and design-tools, and there are many different approaches to accomplishing modeling tasks. The basic principle of **feature-based modeling** is to build models by adding simple features one at a time. In this chapter, the general parametric part modeling procedure is illustrated; a very simple solid model with extruded features is used to introduce the *Autodesk Inventor* user interface. The display viewing functions and the basic two-dimensional sketching tools are also demonstrated.

The *Tiger Head* Design

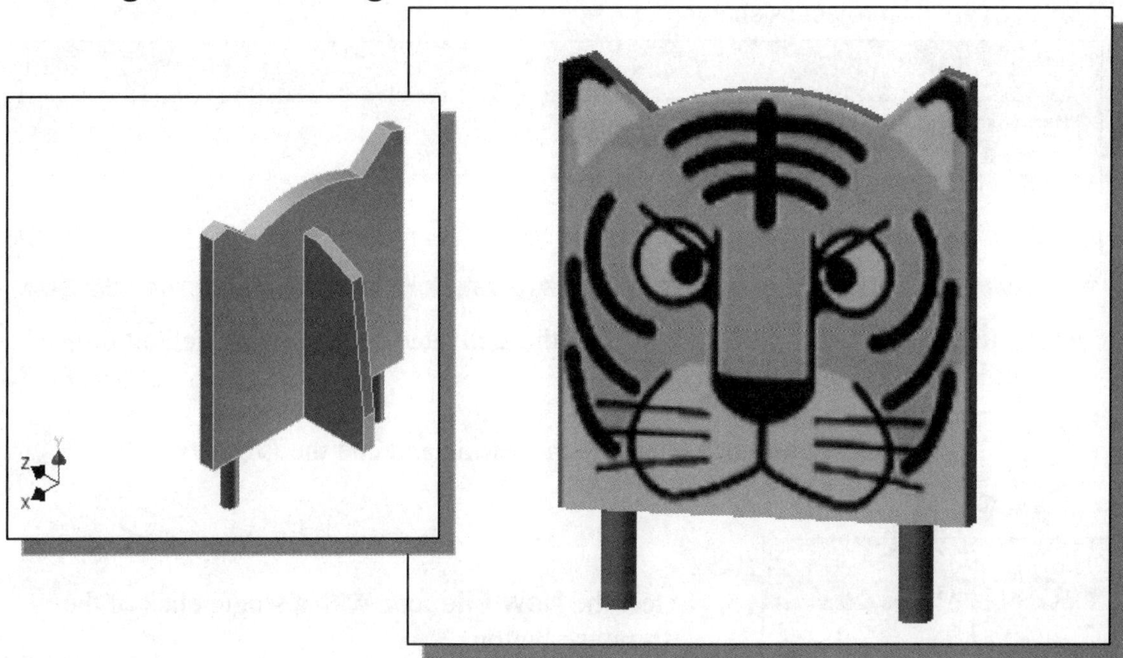

Starting *Autodesk Inventor*

1. Select the **Autodesk Inventor** option on the *Start* menu or select the **Autodesk Inventor** icon on the desktop to start *Autodesk Inventor*. The *Autodesk Inventor* main window will appear on the screen.

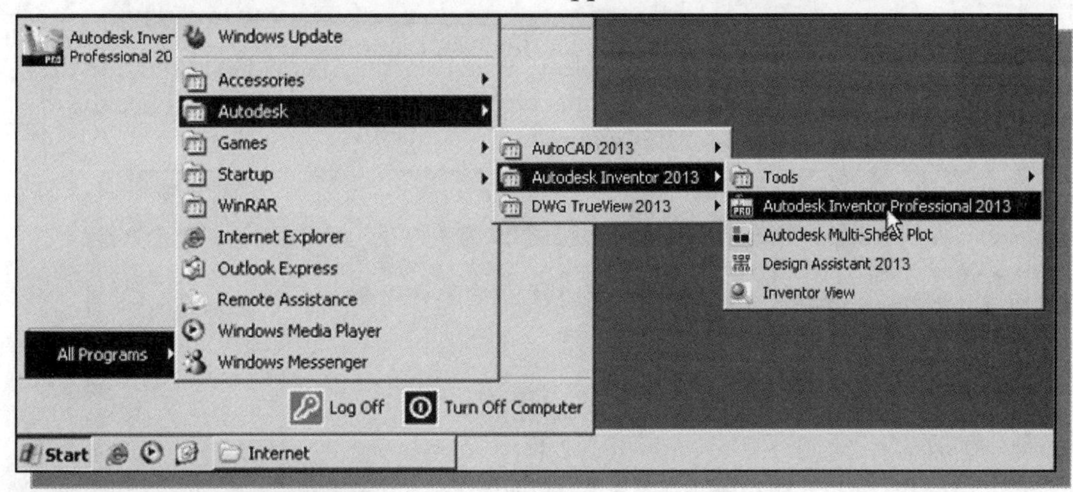

2. Select the **Projects** icon with a single click of the left-mouse-button.

3. In the *Projects* list, **double-click** on the ***Parametric-Modeling*** project name to activate the project as shown.

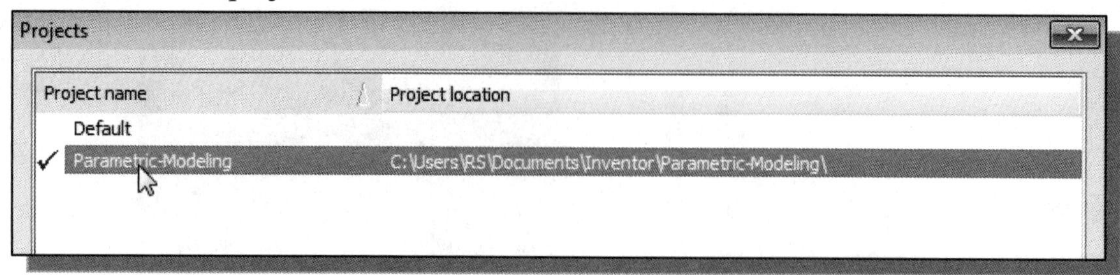

- Note that *Autodesk Inventor* will keep this activated project as the default project until another project is activated.

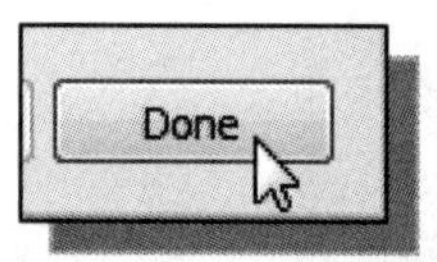

4. Click **Done** to accept the setting and end the *Projects Editor*.

5. Select the **New File** icon with a single click of the left-mouse-button.

- Notice the ***Parametric-Modeling*** project name is displayed as the active project.

6. Select the **English** tab as shown below. When starting a new CAD file, the first thing we should do is choose the units we would like to use. We will use the English setting (inches) for this example.

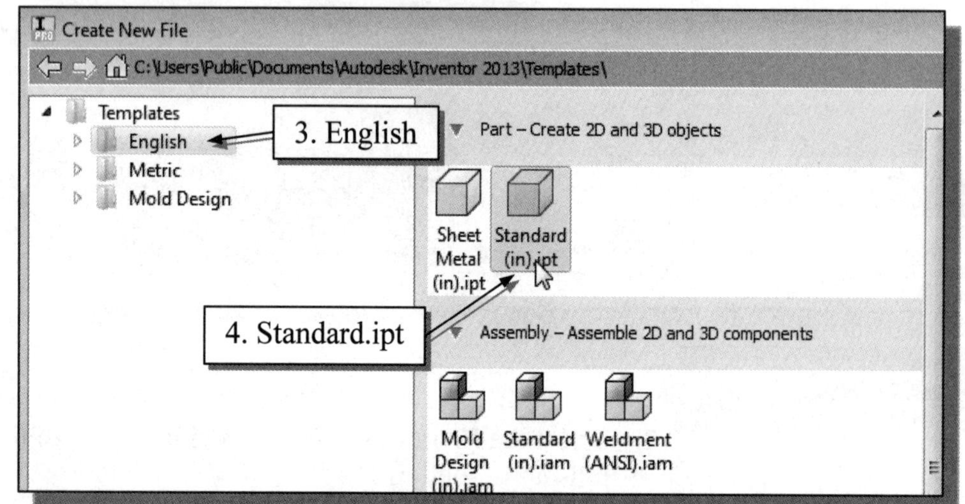

7. Select the **Standard(in).ipt** icon as shown.

8. Pick **Create** in the *New File* dialog box to accept the selected settings.

The Default *Autodesk Inventor* Screen Layout

The default *Autodesk Inventor* drawing screen contains the *pull-down* menus, the *Standard* toolbar, the *Features* toolbar, the *Sketch* toolbar, the *drawing* area, the *browser* area, and the *Status Bar*. A line of quick text appears next to the icon as you move the *mouse cursor* over different icons. You may resize the *Autodesk Inventor* drawing window by clicking and dragging the edges of the window, or relocate the window by clicking dragging the window title area.

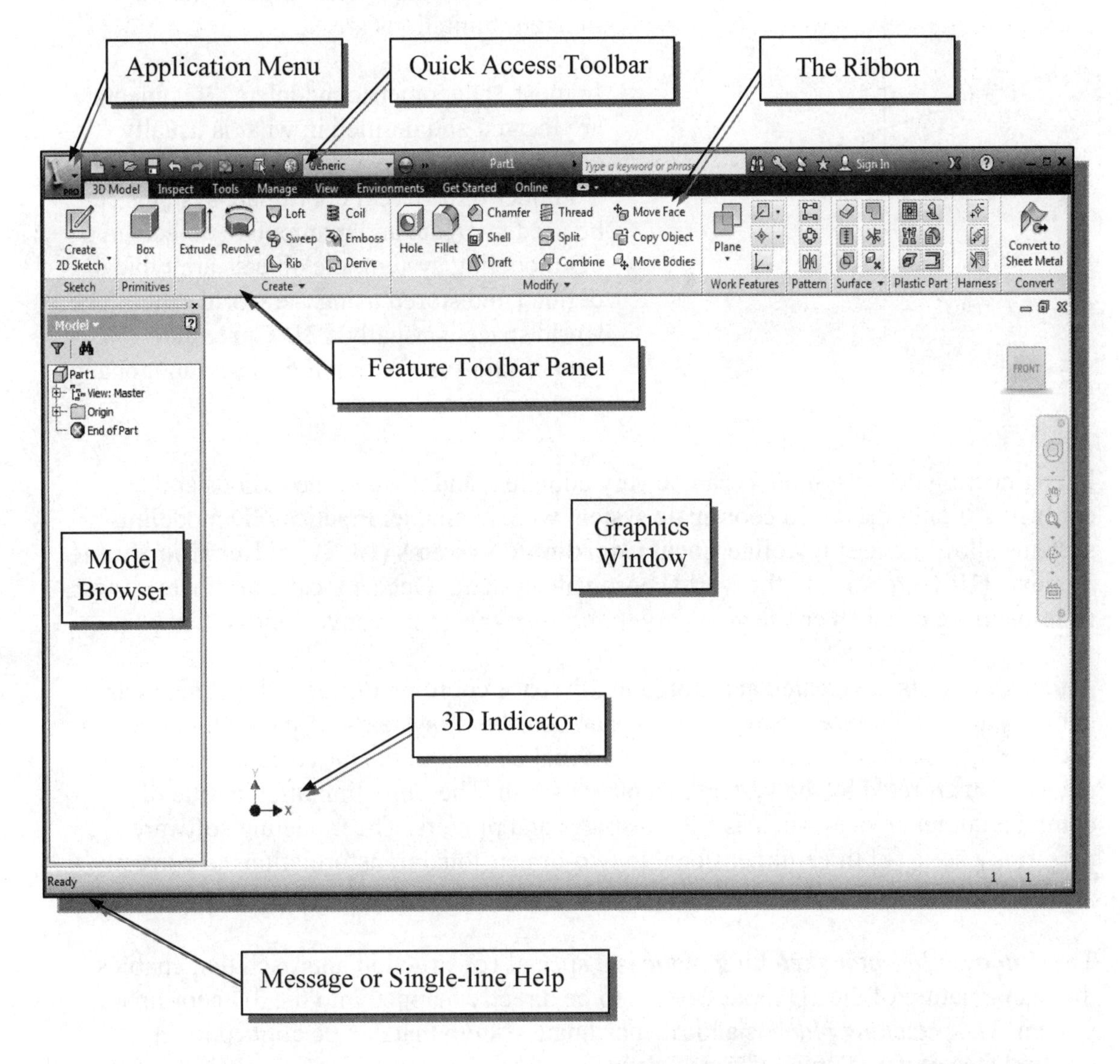

- The ***Ribbon*** is a new feature in *Autodesk Inventor* since the 2011 release. The *Ribbon* is composed of a series of tool panels, which are organized into tabs labeled by task. The *Ribbon* provides a compact palette of all of the tools necessary to accomplish the different modeling tasks. The drop-down arrow next to any icon indicates additional commands are available on the expanded panel; access the expanded panel by clicking on the drop-down arrow.

Sketch plane – It is an XY monitor, but an XYZ World

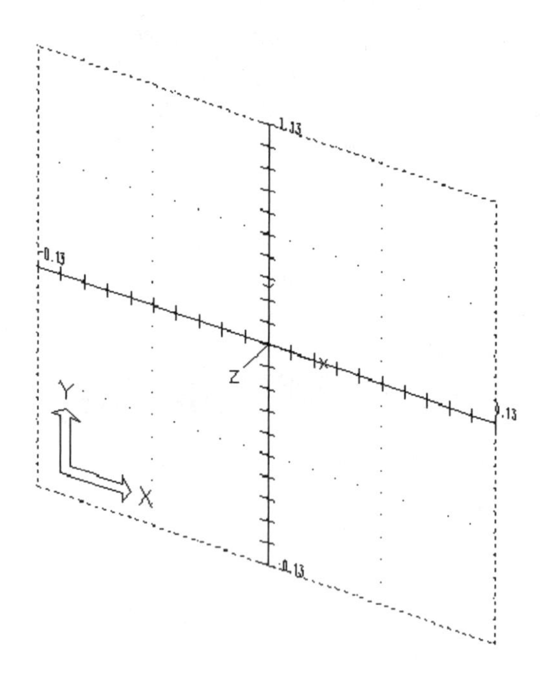

Design modeling software is becoming more powerful and user friendly, yet the system still does only what the user tells it to do. When using a geometric modeler, we therefore need to have a good understanding of what its inherent limitations are.

In most 3D geometric modelers, 3D objects are located and defined in what is usually called **world space** or **global space**. Although a number of different coordinate systems can be used to create and manipulate objects in a 3D modeling system, the objects are typically defined and stored using the world space. The world space is usually a **3D Cartesian coordinate system** that the user cannot change or manipulate.

In engineering designs, models can be very complex, and it would be tedious and confusing if only the world coordinate system were available. Practical 3D modeling systems allow the user to define **Local Coordinate Systems (LCS)** or **User Coordinate Systems (UCS)** relative to the world coordinate system. Once a local coordinate system is defined, we can then create geometry in terms of this more convenient system.

Although objects are created and stored in 3D space coordinates, most of the geometric entities can be referenced using 2D Cartesian coordinate systems. Typical input devices such as a mouse or digitizer are two-dimensional by nature; the movement of the input device is interpreted by the system in a planar sense. The same limitation is true of common output devices, such as CRT displays and plotters. The modeling software performs a series of three-dimensional to two-dimensional transformations to correctly project 3D objects onto the 2D display plane.

The *Autodesk Inventor* ***sketching plane*** is a special construction approach that enables the planar nature of the 2D input devices to be directly mapped into the 3D coordinate system. The *sketching plane* is a local coordinate system that can be aligned to an existing face of a part, or a reference plane.

Think of the sketching plane as the surface on which we can sketch the 2D sections of the parts. It is similar to a piece of paper, a white board, or a chalkboard that can be attached to any planar surface. The first sketch we create is usually drawn on one of the established datum planes. Subsequent sketches/features can then be created on sketching planes that are aligned to existing **planar faces of the solid part** or **datum planes.**

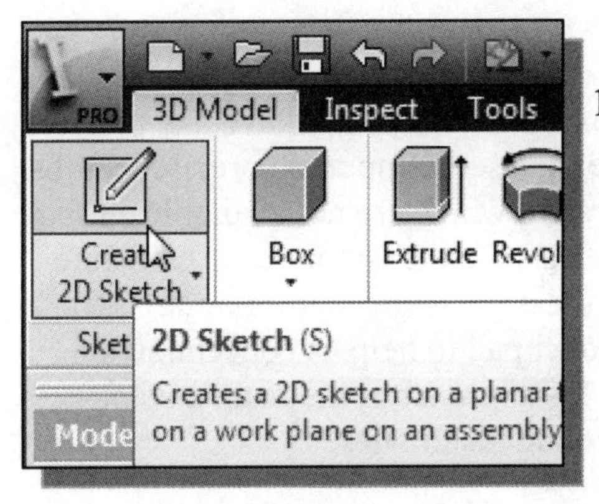

1. Activate the **Create 2D Sketch** icon with a single click of the left-mouse-button.

2. Move the cursor over the edge of the *XY Plane* in the graphics area. When the *XY Plane* is highlighted, click once with the **left-mouse-button** to select the *Plane* as the sketch plane for the new sketch.

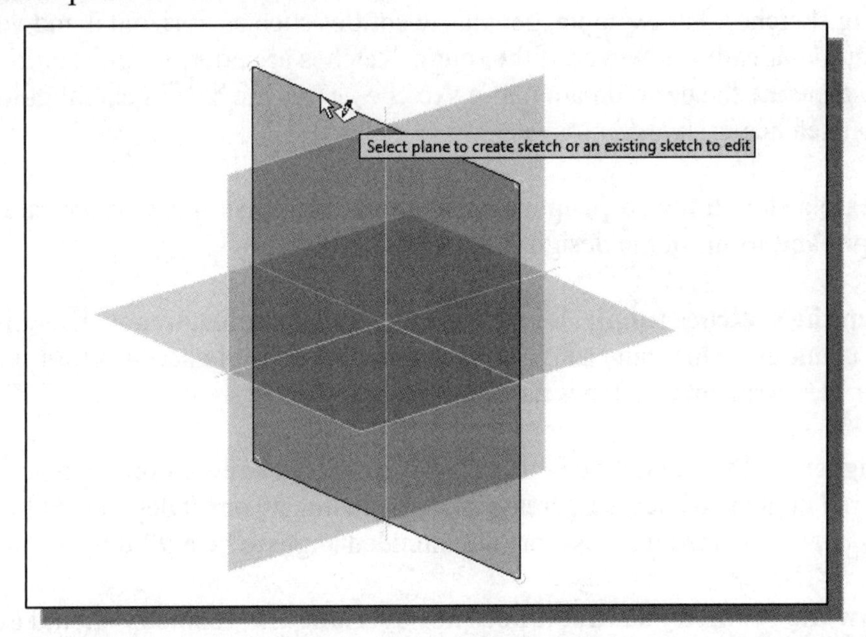

- The *sketching plane* is a reference location where two-dimensional sketches are created. Note that the *sketching plane* can be any planar part surface or datum plane.

3. Confirm the main *Ribbon* area is switched to the **Sketch** toolbars; this indicates we have entered the 2D Sketching mode.

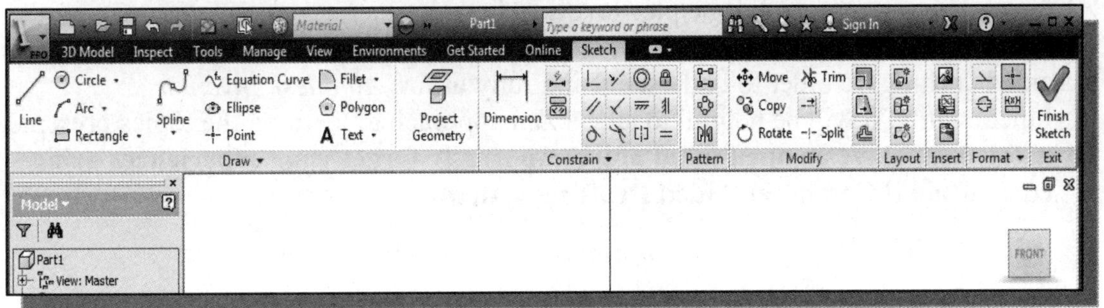

Creating Rough Sketches

Quite often during the early design stage, the shape of a design may not have any precise dimensions. Most conventional CAD systems require the user to input the precise lengths and locations of all geometric entities defining the design, which are not available during the early design stage. With *parametric modeling*, we can use the computer to elaborate and formulate the design idea further during the initial design stage. With *Autodesk Inventor*, we can use the computer as an electronic sketchpad to help us concentrate on the formulation of forms and shapes for the design. This approach is the main advantage of *parametric modeling* over conventional solid-modeling techniques.

As the name implies, a ***rough sketch*** is not precise at all. When sketching, we simply sketch the geometry so that it closely resembles the desired shape. Precise scale or lengths are not needed. *Autodesk Inventor* provides us with many tools to assist us in finalizing sketches. For example, geometric entities such as horizontal and vertical lines are set automatically. However, if the rough sketches are poor, it will require much more work to generate the desired parametric sketches. Here are some general guidelines for creating sketches in *Autodesk Inventor*:

- **Create a sketch that is proportional to the desired shape.** Concentrate on the shapes and forms of the design.

- **Keep the sketches simple.** Leave out small geometry features such as fillets, rounds and chamfers. They can easily be placed using the Fillet and Chamfer commands after the parametric sketches have been established.

- **Exaggerate the geometric features of the desired shape.** For example, if the desired angle is 85 degrees, create an angle that is 50 or 60 degrees. Otherwise, *Autodesk Inventor* might assume the intended angle to be a 90-degree angle.

- **Draw the geometry so that it does not overlap.** The geometry should eventually form a closed region. *Self-intersecting* geometry shapes are not allowed.

- **The sketched geometric entities should form a closed region.** To create a solid feature, such as an extruded solid, a closed region is required so that the extruded solid forms a 3D volume.

➢ **Note:** The concepts and principles involved in *parametric modeling* are very different, and sometimes they are totally opposite, to those of conventional computer aided drafting. In order to understand and fully utilize *Autodesk Inventor's* functionality, it will be helpful to take a *Zen* approach to learning the topics presented in this text: **Have an open mind and temporarily forget your experiences using conventional Computer Aided Drafting systems.**

Step 1: Creating a Rough Sketch

➢ The *Sketch* toolbar provides tools for creating the basic geometry that can be used to create features and parts.

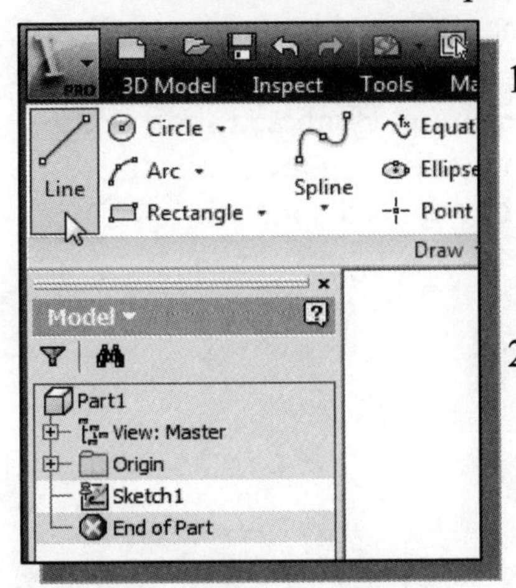

1. Move the graphics cursor to the **Line** icon in the *Draw* toolbar. A *Help tip* box appears next to the cursor and a brief description of the command is displayed at the bottom of the drawing screen: "*Creates Straight line segments and tangent arcs.*"

2. Select the icon by clicking once with the **left-mouse-button**; this will activate the **Line** command. *Autodesk Inventor* expects us to identify the starting location of a straight line.

Graphics Cursors

➢ Notice the cursor changes from an arrow to a crosshair when graphical input is expected.

1. Move the cursor inside the graphics window, left-click a starting point for the shape, roughly to the left side of the screen as shown.

2. As you move the graphics cursor, you will see a digital readout next to the cursor and also in the *Status Bar* area at the bottom of the window. The readout gives you the cursor location, the line length, and the angle of the line measured from horizontal. Move the cursor around and you will notice different symbols appear at different locations.

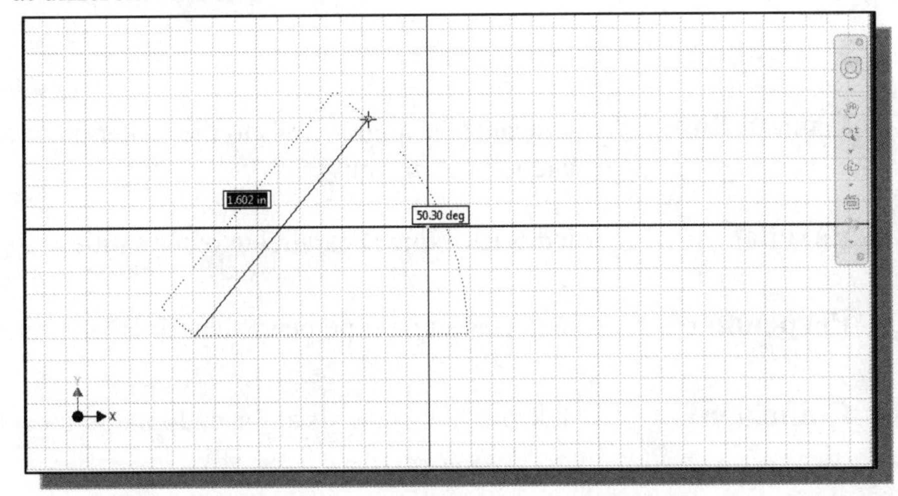

➢ The readout displayed next to the cursor is called the ***Dynamic Input***. This option is part of the **Heads-Up Display** option that is new in *Inventor*. *Dynamic Input* can be used for entering precise values, but its usage is somewhat limited in *parametric modeling*.

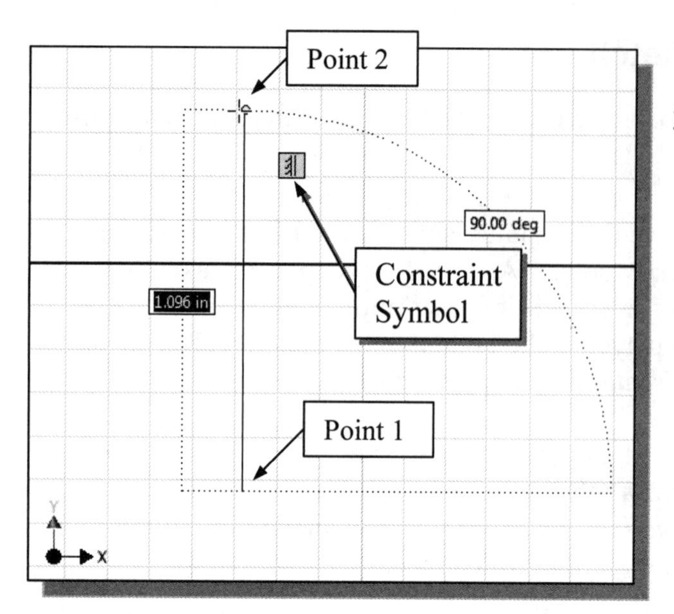

3. Move the graphics cursor above the last point and create a vertical line as shown in the figure (**Point 2**). Notice the geometric constraint symbol, a short vertical line indicating the geometric property, is displayed.

Geometric Constraint Symbols

Autodesk Inventor displays different visual clues, or symbols, to show you alignments, perpendicularities, tangencies, etc. These constraints are used to capture the *design intent* by creating constraints where they are recognized. *Autodesk Inventor* displays the governing geometric rules as models are built. To prevent constraints from forming, hold down the [**Ctrl**] key while creating an individual sketch curve. For example, while sketching line segments with the Line command, endpoints are joined with a Coincident constraint, but when the [**Ctrl**] key is pressed and held, the inferred constraint will not be created.

	Vertical	indicates a line is vertical
	Horizontal	indicates a line is horizontal
- - -	**Dashed line**	indicates the alignment is to the center point or endpoint of an entity
	Parallel	indicates a line is parallel to other entities
	Perpendicular	indicates a line is perpendicular to other entities
	Coincident	indicates the cursor is at the endpoint of an entity
	Concentric	indicates the cursor is at the center of an entity
	Tangent	indicates the cursor is at tangency points to curves

1. Complete the sketch as shown below, creating a closed region ending at the starting point (**Point 1**). Do not be overly concerned with the actual size of the sketch. Note that the **four inclined lines** are sketched **not perpendicular or parallel** to each other.

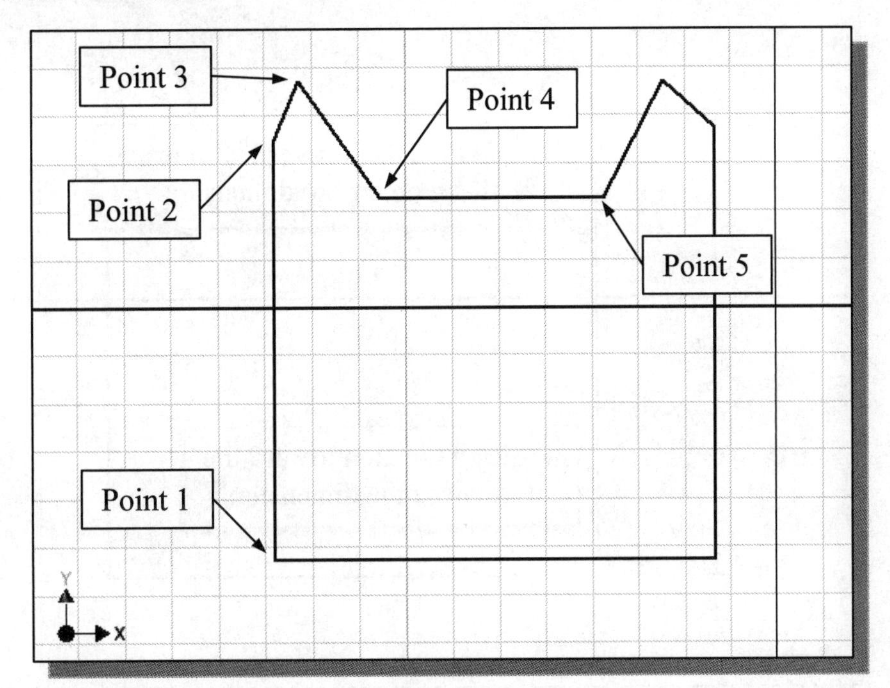

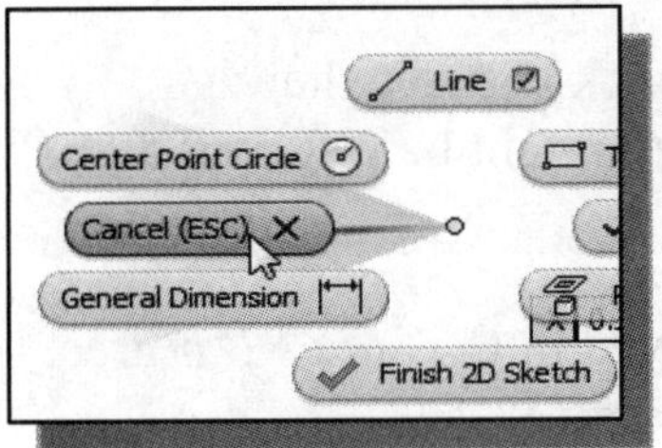

2. Inside the graphics window, click once with the **right-mouse-button** to display the option menu. Select **Cancel [Esc]** in the popup menu, or hit the [**Esc**] key once, to end the Sketch Line command.

Step 2: Apply/Modify Constraints and Dimensions

➤ As the sketch is made, *Autodesk Inventor* automatically applies some of the geometric constraints (such as horizontal, parallel, and perpendicular) to the sketched geometry. We can continue to modify the geometry, apply additional constraints, and/or define the size of the existing geometry. In this example, we will illustrate adding dimensions to describe the sketched entities.

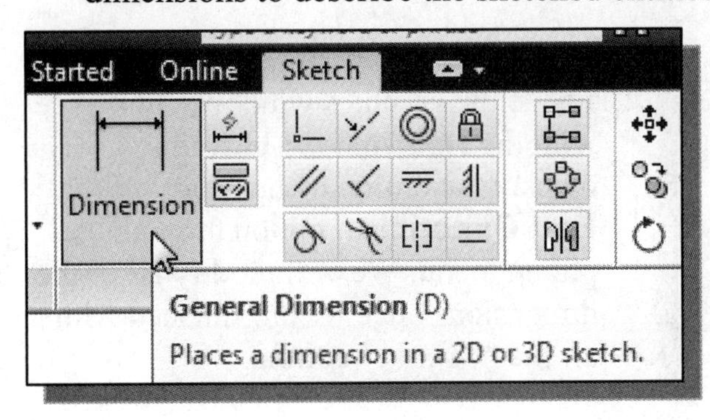

1. Move the cursor to the *Constrain* toolbar area; it is the toolbar next to the 2D *Draw* toolbar. Note the first icon in this toolbar is the General Dimension icon. Left-click once on the icon to activate the Dimension command.

2. The message "*Select Geometry to Dimension*" is displayed in the *Status Bar* area at the bottom of the *Inventor* window. Select the left vertical line by left-clicking once on the line.

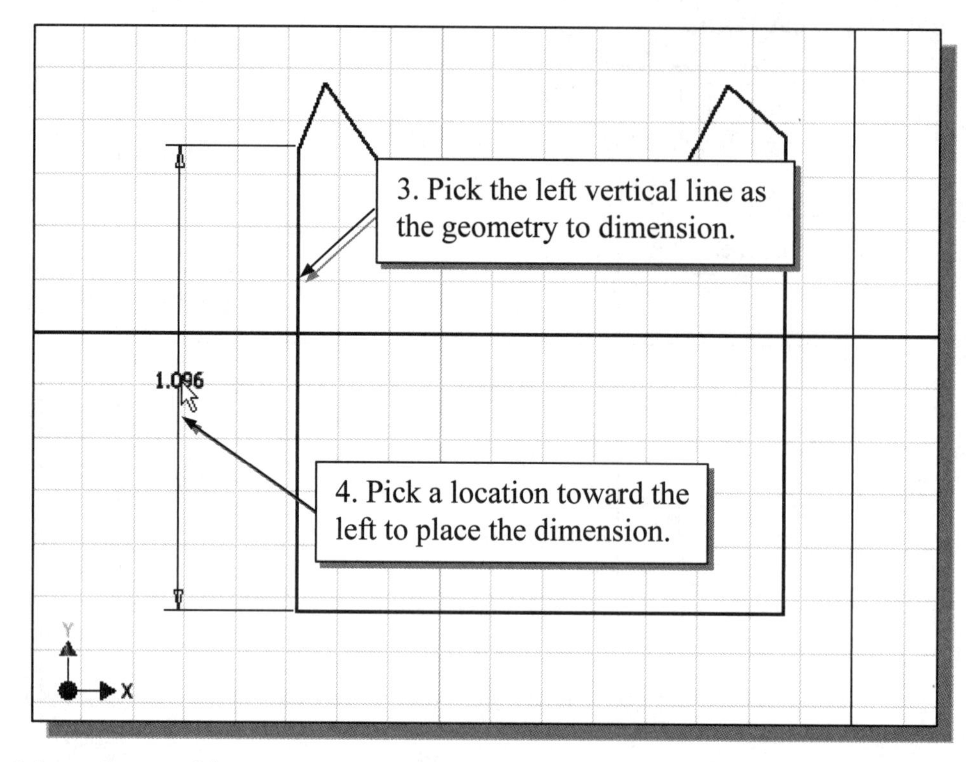

3. Move the graphics cursor toward the left side and left-click to place the dimension. (Note that the value displayed on your screen might be different than what is shown in the figure above.)

4. The message "*Select Geometry to Dimension*" is displayed in the *Status Bar* area, at the bottom of the *Inventor* window. Select the right-vertical line.

5. Pick a location toward the right of the sketch to place the dimension.

6. Click **OK** to accept the default value for the dimension.

- The **General Dimension** command will create a length dimension if a single line is selected.

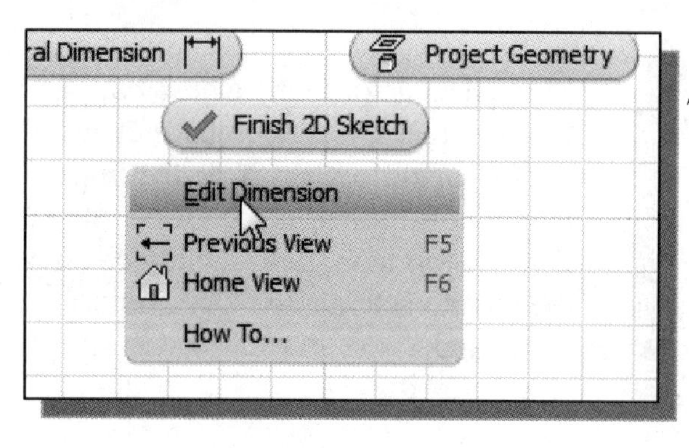

7. Inside the graphics window, click once with the **right-mouse-button** to display the option menu. Turn off the **Edit Dimension** option through the popup menu. We will modify all of the dimensions once we are finished with all the necessary editing.

8. The message "*Select Geometry to Dimension*" is displayed in the *Status Bar* area, located at the bottom of the *Inventor* window. Select the left vertical line as shown below.

9. Select the right vertical line as shown below.

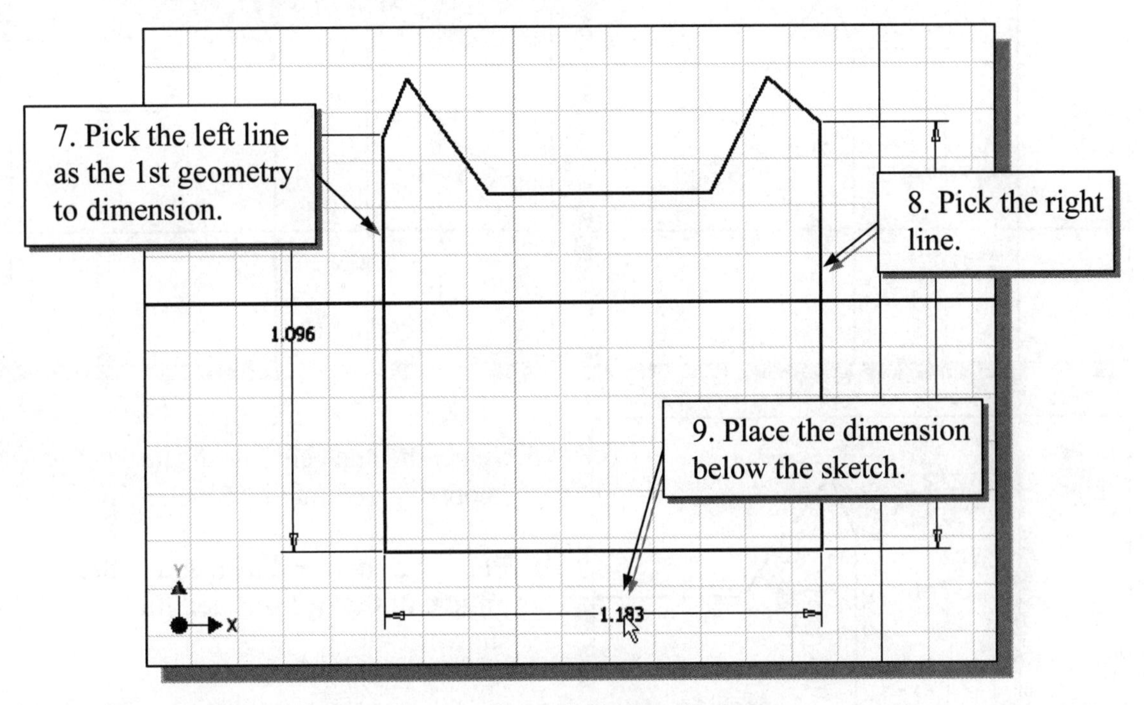

10. Pick a location below the sketch to place the dimension.

11. The message "*Select Geometry to Dimension*" is displayed in the *Status Bar* area, located at the bottom of the *Inventor* window. Select the top left corner as shown.

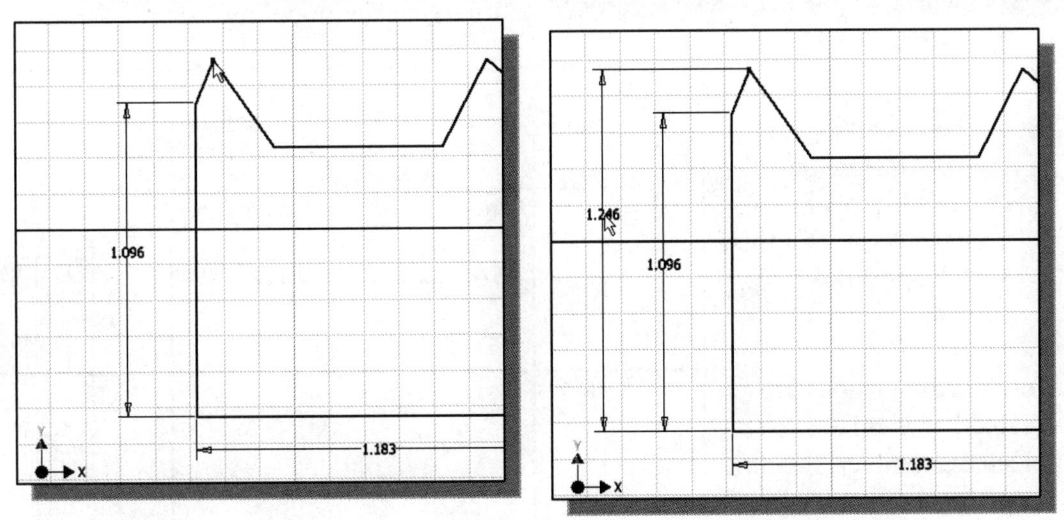

12. Select the bottom horizontal line as the 2nd geometry to dimension.

13. Place the dimension toward the left of the sketch as shown.

14. The message "*Select Geometry to Dimension*" is displayed in the *Status Bar* area, located at the bottom of the *Inventor* window. Select the left vertical line as shown.

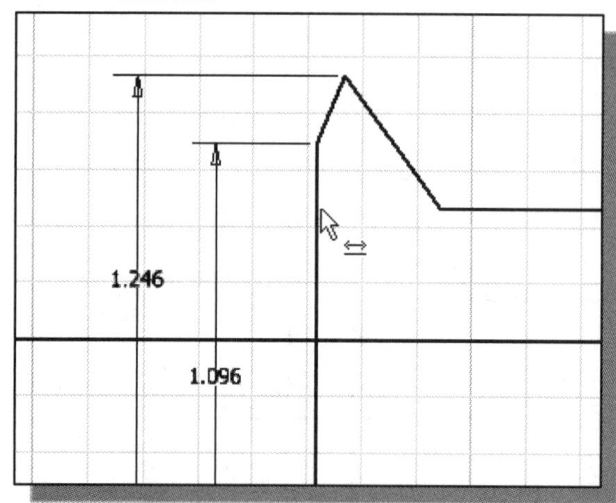

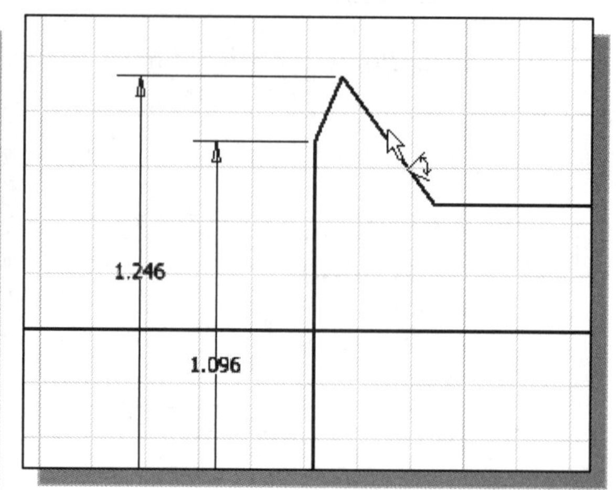

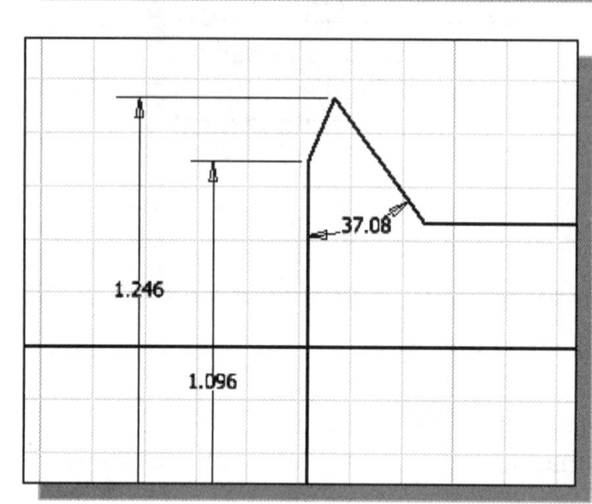

15. Select the adjacent line as the 2nd geometry to dimension.

16. Place the angular dimension in the middle of the two selected lines as shown

❖ Based on selected entities, the **General Dimension** command will create associated dimensions; this is also known as **Smart Dimensioning** in parametric modeling.

16. On you own, repeat the above steps and create additional dimensions so that the sketch appears as shown.

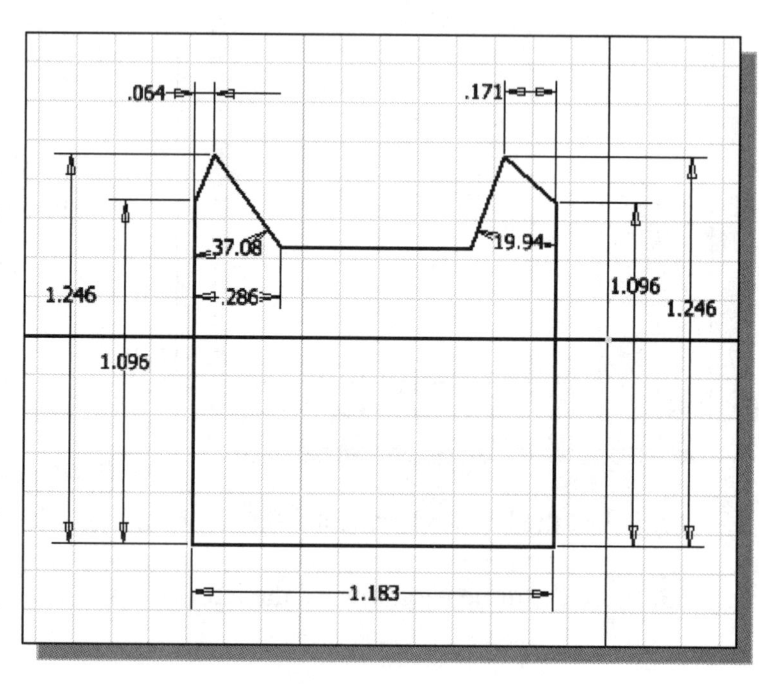

- Note the dimensions are created based on the selected geometry, this is known as the **Smart Dimensioning** feature in *parametric modeling*.

Dynamic Viewing Functions – *Zoom* and *Pan*

- *Autodesk Inventor* provides a special user interface called *Dynamic Viewing* that enables convenient viewing of the entities in the graphics window.

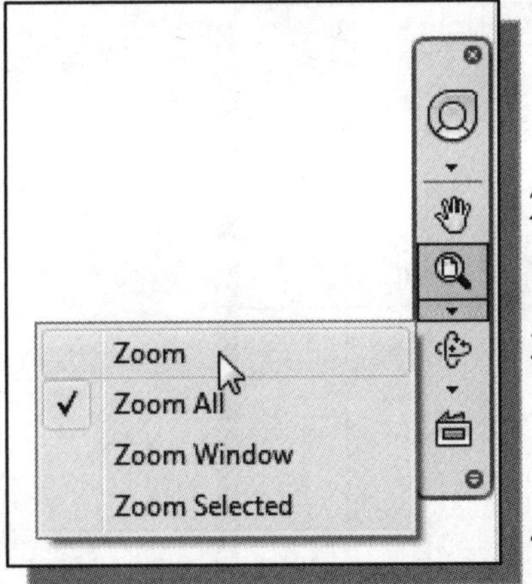

1. Click on the **Zoom** icon, located in the *Navigation* bar as shown.

2. Move the cursor near the center of the graphics window.

3. Inside the graphics window, **press and hold down the left-mouse-button**, then move downward to enlarge the current display scale factor.

4. Press the [**Esc**] key once to exit the Zoom command.

5. Click on the **Pan** icon, located above the Zoom command in the *Navigation* bar. The icon is the picture of a hand.

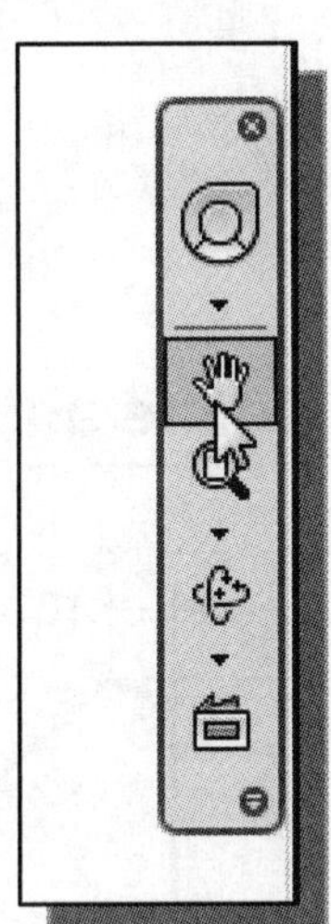

➢ The Pan command enables us to move the view to a different position. This function acts as if you are using a video camera.

6. On your own, use the Zoom and Pan options to reposition the sketch near the center of the screen.

Modifying the Dimensions of the Sketch

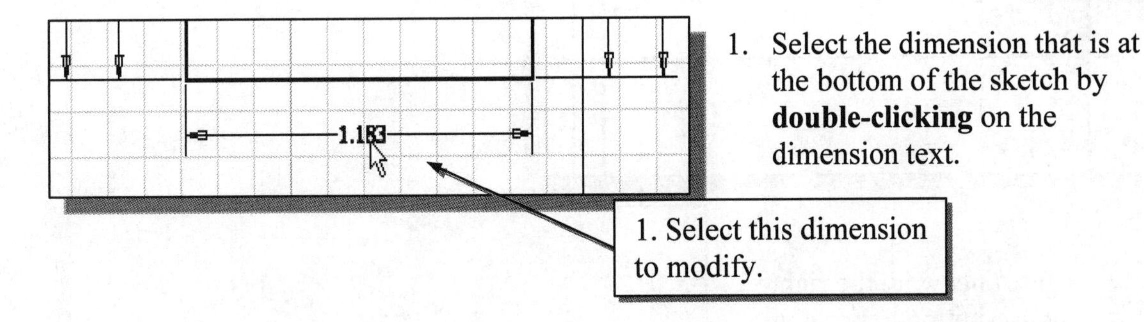

1. Select the dimension that is at the bottom of the sketch by **double-clicking** on the dimension text.

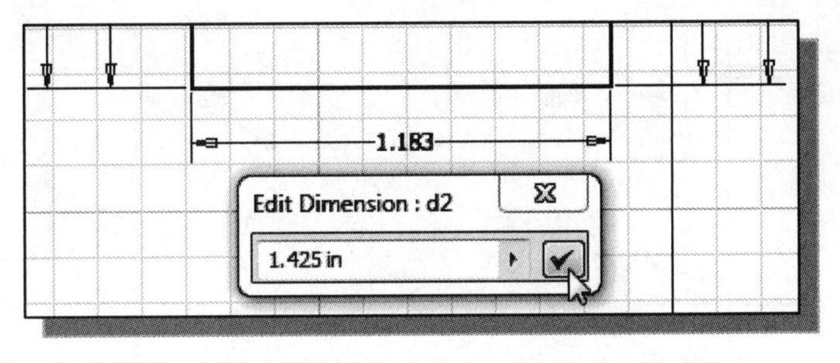

2. In the *Edit Dimension* window, the current length of the line is displayed. Enter **1.425** to set the length of the line.

3. Click on the **Accept** icon to accept the entered value.

➢ *Autodesk Inventor* will now update the profile with the new dimension value.

4. On you own, repeat the above steps and adjust the dimensions so that the sketch appears as shown.

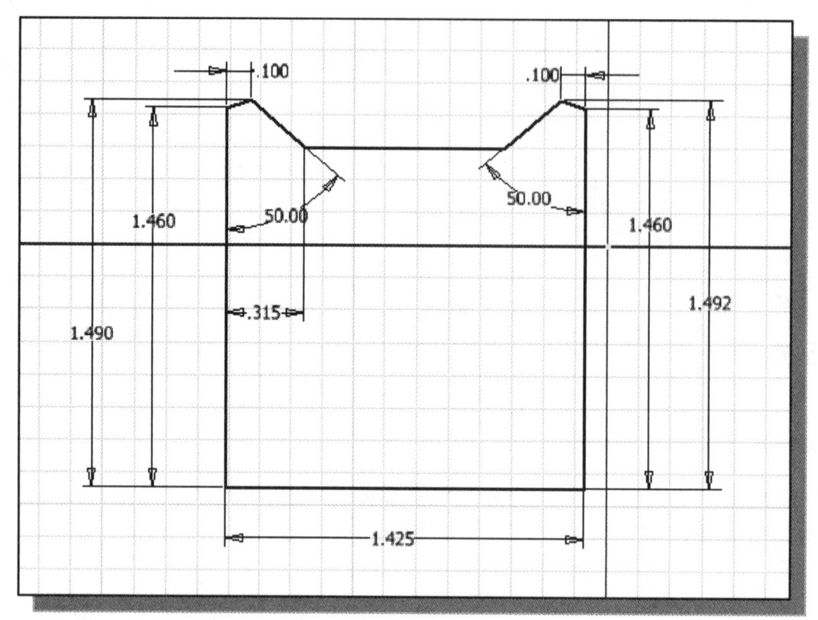

Delete an Existing Geometry of the Sketch

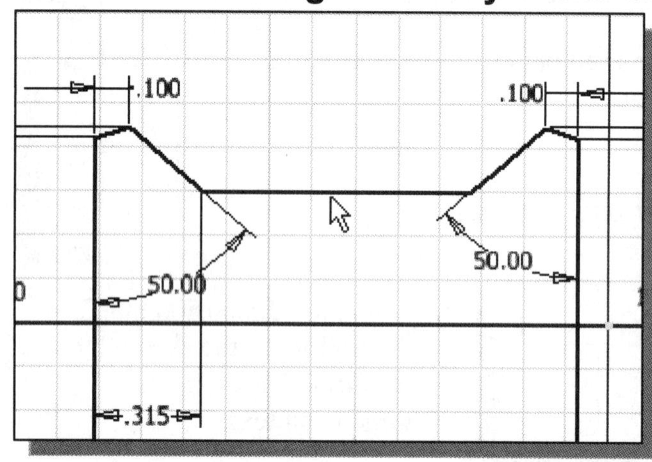

1. Select the top horizontal line by **left-clicking** once on the line.

2. Click once with the right-mouse-button to bring up the **option menu**.

3. Select **Delete** from the *option list* as shown.

➢ Note that any dimension attached to the geometry will be also deleted.

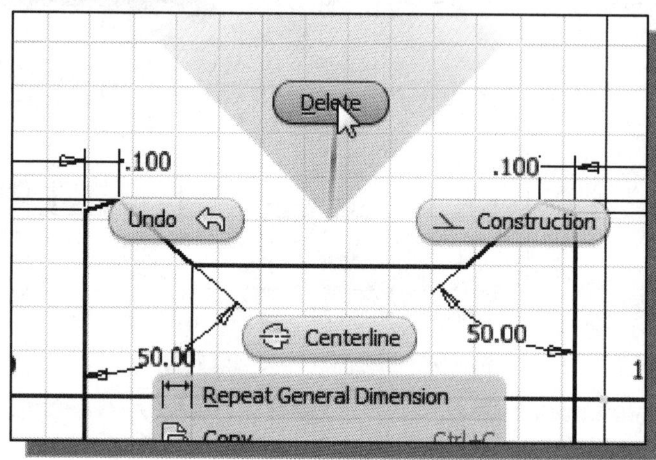

Using the 3-Point Arc Command

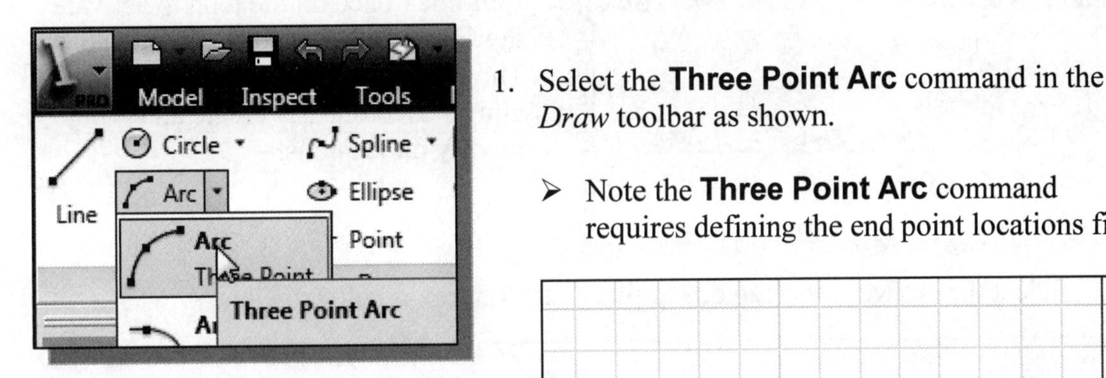

1. Select the **Three Point Arc** command in the *Draw* toolbar as shown.

➢ Note the **Three Point Arc** command requires defining the end point locations first.

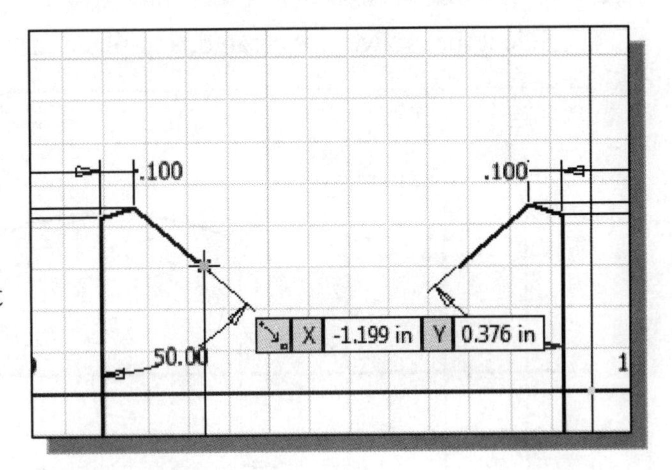

2. Select the endpoint of the longer inclined line on the left as shown.

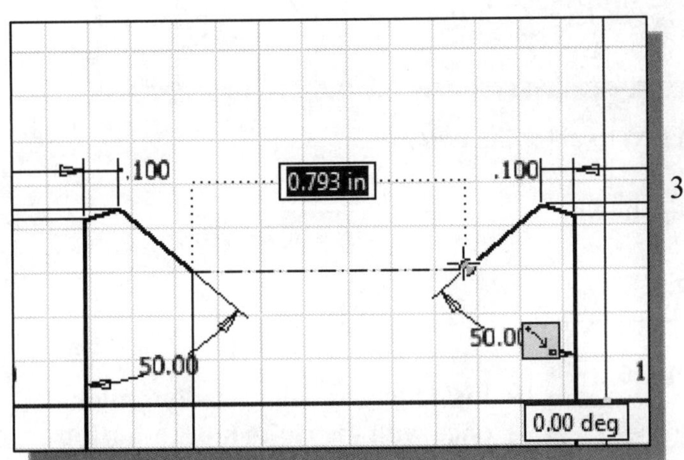

3. Select the endpoint of the longer inclined line on the right as shown.

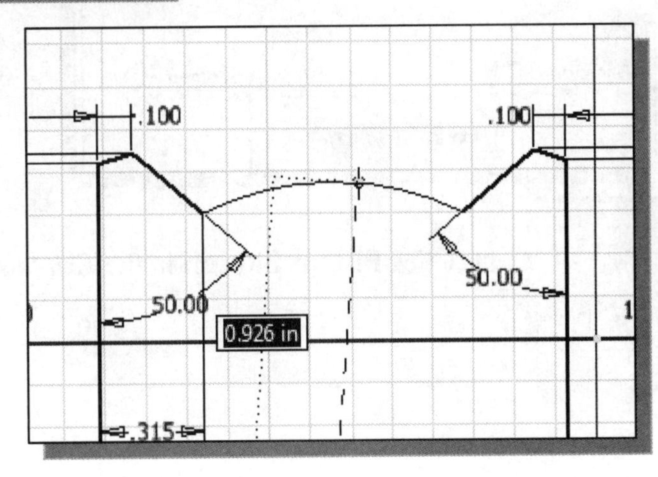

4. Move the cursor above the two selected points to set the curvature of the arc, **left-clicking** once when the radius is roughly 0.9 inch as shown.

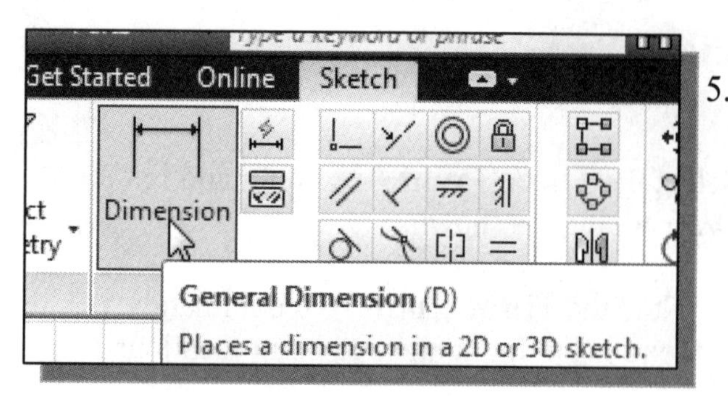

5. Left-click once on the icon to activate the **General Dimension** command. The **General Dimension** command allows us to quickly create and modify dimensions.

6. Select the arc we just created, and place the dimension below the arc.

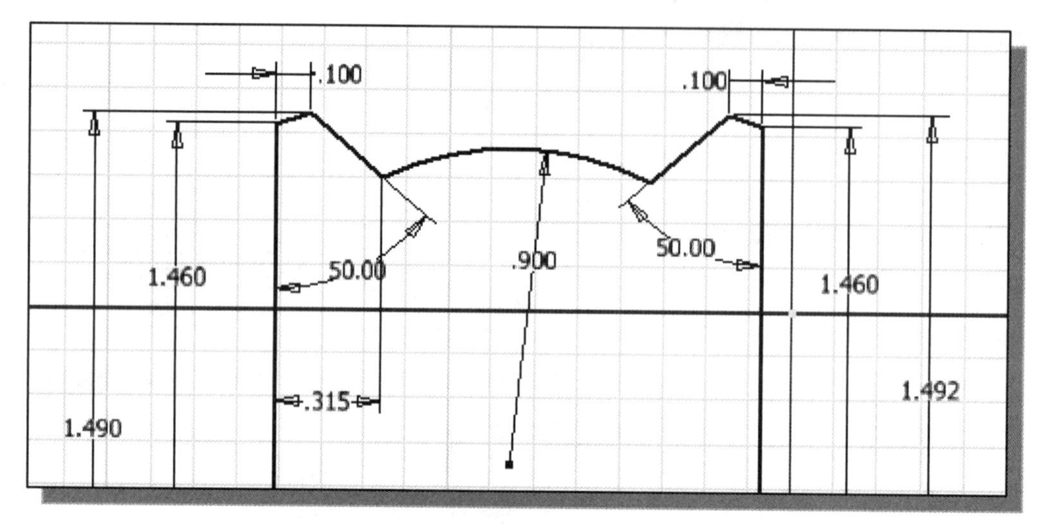

7. Click on the radius dimension text to enter the ***edit mode***.

8. Enter **0.90** as the new radius as shown.

9. Click **OK** to accept the setting.

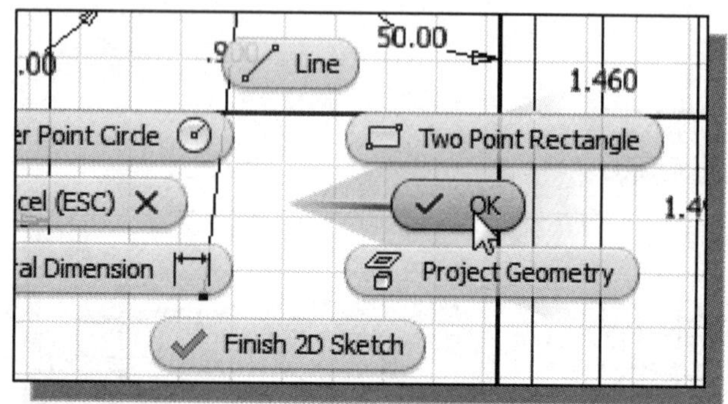

10. Inside the graphics window, click once with the **right-mouse-button** to display the option menu. Select **OK** in the popup menu to end the **Sketch Line** command.

11. Click **Finish Sketch** in the *Exit* toolbar to end the **Sketch** option.

Step 3: Completing the Base Solid Feature

Now that the 2D sketch is completed, we will proceed to the next step: creating a 3D part from the 2D profile. Extruding a 2D profile is one of the common methods that can be used to create 3D parts. We can extrude planar faces along a path. We can also specify a height value and a tapered angle. In *Autodesk Inventor*, each face has a positive side and a negative side, the current face we're working on is set as the default positive side. This positive side identifies the positive extrusion direction and it is referred to as the face's ***normal***.

1. In the **Model** tab (the tab is located in the ***Ribbon***), select the **Extrude** command by releasing the left-mouse-button on the icon.

2. In the *Extrude* popup window, expand the window by clicking on the down arrow; enter **0.1** as the extrusion distance. Notice that the sketch region is automatically selected as the extrusion profile.

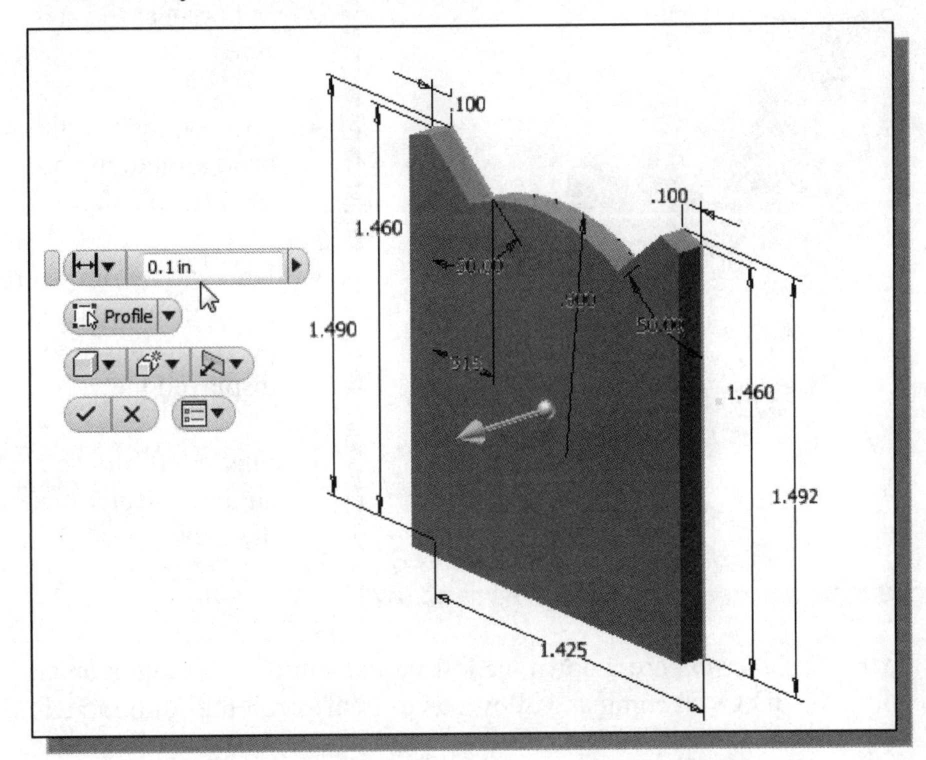

3. Click on the **OK** button to proceed with creating the 3D part.

➤ Note that all dimensions disappeared from the screen. All parametric definitions are stored in the ***Autodesk Inventor*** **database** and any of the parametric definitions can be redisplayed and edited at any time.

Dynamic Rotation of the 3D Block – *Free Orbit*

The Free Orbit command allows us to:

- Orbit a part or assembly in the graphics window. Rotation can be around the center mark, free in all directions, or around the X/Y-axes in the *3D-Orbit* display.
- Reposition the part or assembly in the graphics window.
- Display isometric or standard orthographic views of a part or assembly.
- The Free Orbit tool is accessible while other tools are active. *Autodesk Inventor* remembers the last used mode when you exit the Orbit command.

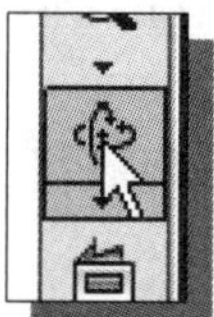

1. Click on the **Free Orbit** icon in the *Navigation* bar.

➢ The *3D Orbit* display is a circular rim with four handles and a center mark. *3D Orbit* enables us to manipulate the view of 3D objects by clicking and dragging with the left-mouse-button:

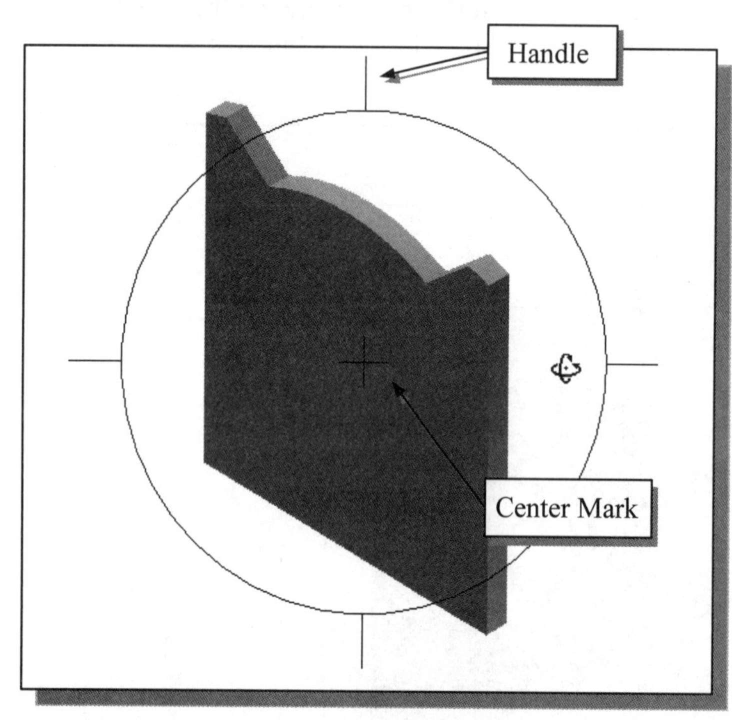

- Drag with the left-mouse-button near the center for free rotation.
- Drag on the handles to orbit around the horizontal or vertical axis.
- Drag on the rim to orbit about an axis that is perpendicular to the displayed view.
- Single left-mouse-click to align the center mark of the view.

2. Inside the *circular rim*, press down the left-mouse-button and drag in an arbitrary direction; the 3D Orbit command allows us to freely orbit the solid model.

3. Move the cursor near the circular rim and notice the cursor symbol changes to a single circle. Drag with the left-mouse-button to orbit about an axis that is perpendicular to the displayed view.

4. Single left-mouse-click near the top handle to align the selected location to the center mark in the graphics window.

5. Activate the **Constrained Orbit** option by clicking on the associated icon as shown.

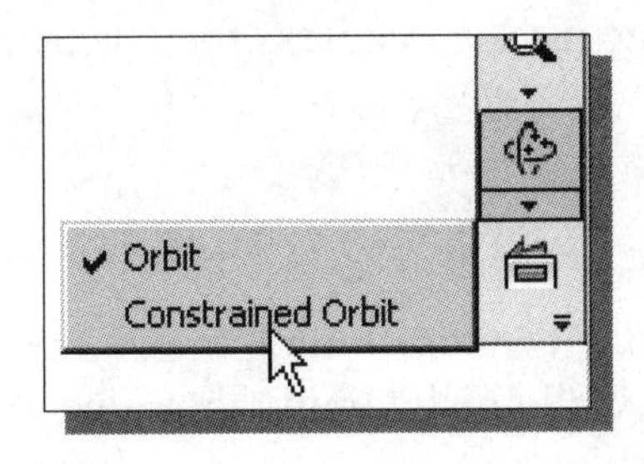

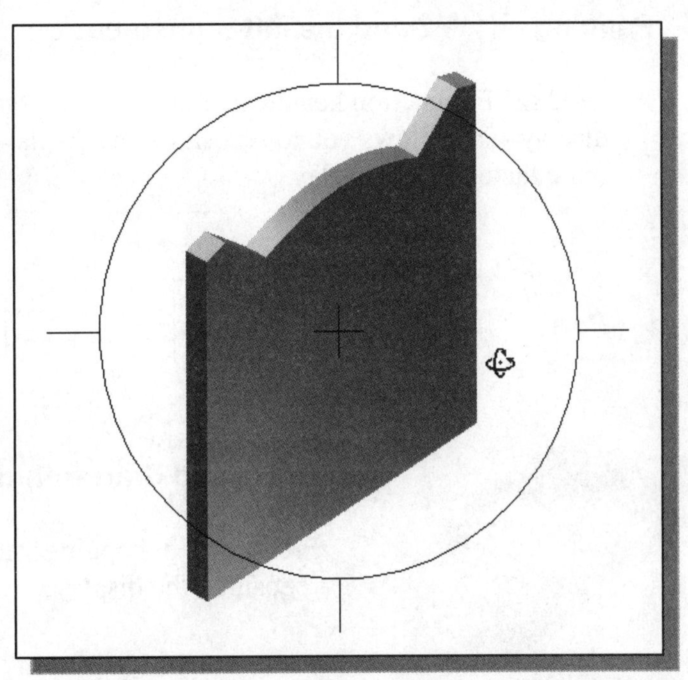

❖ *The* **Constrained Orbit** *can be used to rotate the model about axes in Model Space, equivalent to moving the eye position about the model in latitude and longitude.*

6. On your own, use the different options described in the above steps and familiarize yourself with both of the **3D Orbit** commands. Reset the display to the *Isometric* view as shown in the figure above before continuing to the next section.

❖ Note that while in the **3D Orbit** mode, a horizontal marker will be displayed next to the cursor if the cursor is away from the circular rim. This is the **exit marker**. Left-clicking once will allow you to exit the **3D Orbit** command.

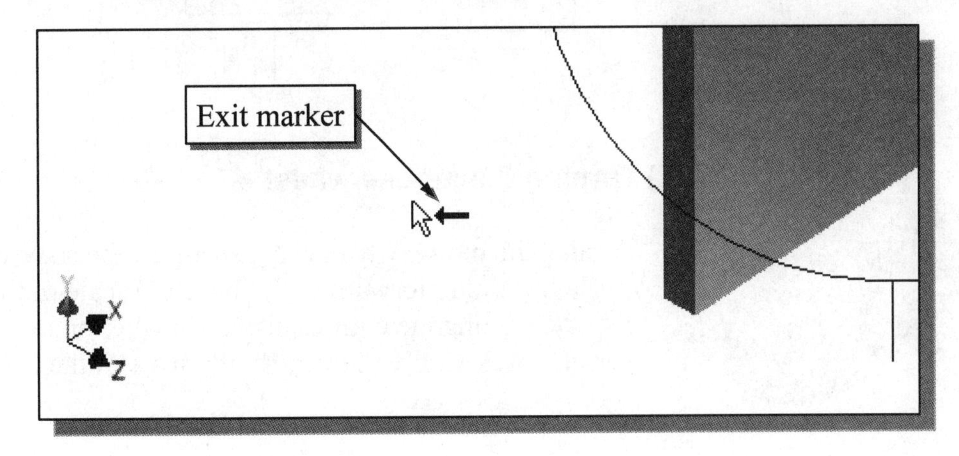

Dynamic Viewing – Quick Keys

We can also use the function keys on the keyboard and the mouse to access the *Dynamic Viewing* functions.

❖ Panning – (1) F2 and the left-mouse-button

Hold the **F2** function key down, and drag with the left-mouse-button to pan the display. This allows you to reposition the display while maintaining the same scale factor of the display.

Pan

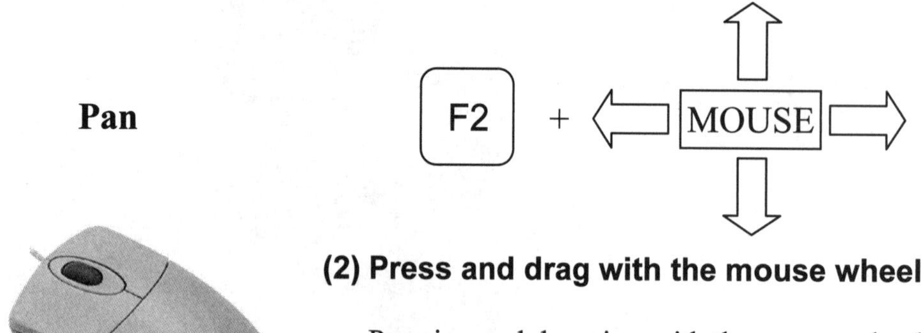

(2) Press and drag with the mouse wheel

Pressing and dragging with the mouse wheel can also reposition the display.

❖ Zooming – (1) F3 and drag with the left-mouse-button

Hold the **F3** function key down, and drag with the left-mouse-button vertically on the screen to adjust the scale of the display. Moving upward will reduce the scale of the display, making the entities display smaller on the screen. Moving downward will magnify the scale of the display.

Zoom

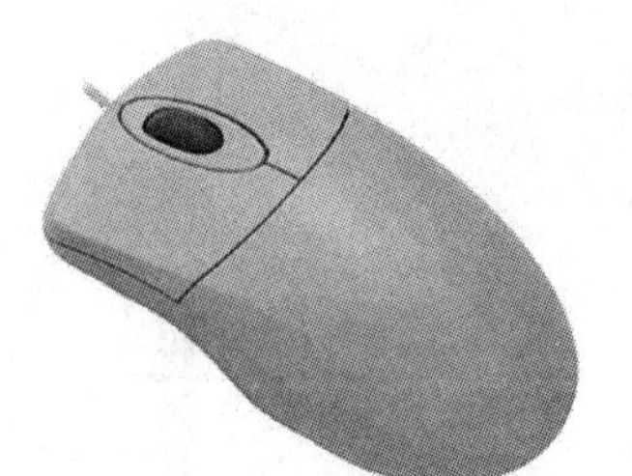

(2) Turning the mouse wheel

Turning the mouse wheel can also adjust the scale of the display. Turning forward will reduce the scale of the display, making the entities display smaller on the screen. Turning backward will magnify the scale of the display.

❖ 3D Dynamic Rotation – F4 and the left-mouse-button

Hold the **F4** function key down and drag with the left-mouse-button to orbit the display. The 3D Orbit rim with four handles and the center mark appear on the screen. Note that the Common View option is not available when using the **F4** quick key.

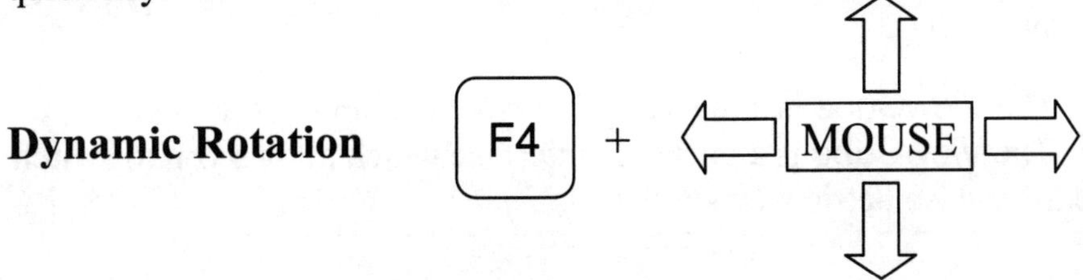

Viewing Tools – Standard Toolbar

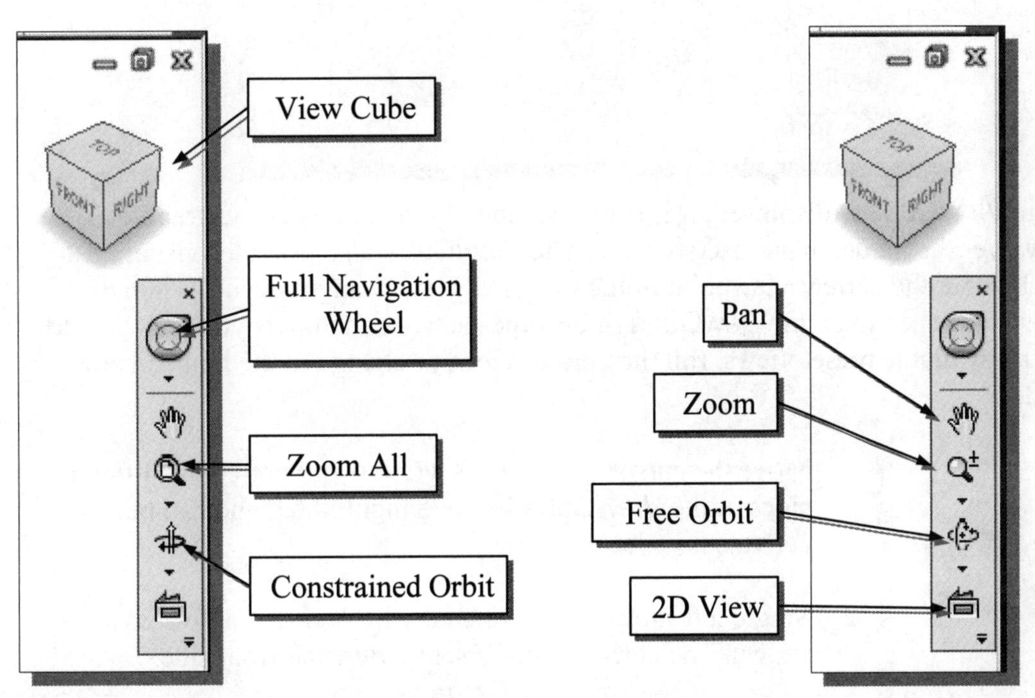

Zoom All – Adjusts the view so that all the items on the screen fit inside the graphics window.

Zoom Window – Use the cursor to define a region for the view; the defined region is zoomed to fill the graphics window.

Zoom – Moving upward will reduce the scale of the display, making the entities display smaller on the screen. Moving downward will magnify the scale of the display.

Pan – This allows you to reposition the display while maintaining the same scale factor of the display.

Zoom Selected – In a part or assembly, zooms the selected edge, feature, line, or other element to fill the graphics window. You can select the element either before or after clicking the Zoom button. (Not used in drawings.)

Orbit – In a part or assembly, adds an orbit symbol and cursor to the view. You can orbit the view planar to the screen around the center mark, around a horizontal or vertical axis, or around the X and Y axes. (Not used in drawings.)

2D View/View Face – In a part or assembly, zooms and orbits the model to display the selected element planar to the screen or a selected edge or line horizontal to the screen. (Not used in drawings.)

View Cube – The ViewCube is a 3D navigation tool that appears, by default, when you enter *Inventor*. The ViewCube is a clickable interface which allows you to switch between standard and isometric views.

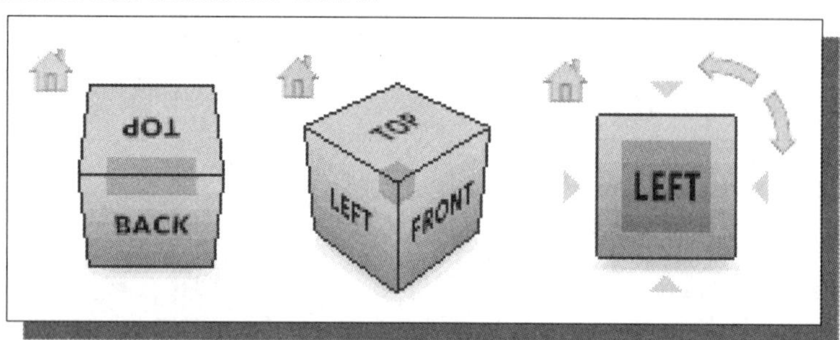

Once the ViewCube is displayed, it is shown in one of the corners of the graphics window over the model in an inactive state. The ViewCube also provides visual feedback about the current viewpoint of the model as view changes occur. When the cursor is positioned over the ViewCube, it becomes active and allows you to switch to one of the available preset views, roll the current view, or change to the Home view of the model.

1. Move the cursor over the ViewCube and notice the different sides of the ViewCube become highlighted and can be activated.

2. Single left-mouse-click when the front side is activated as shown. The current view is set to view the front side.

3. Move the cursor over the counter-clockwise arrow of the ViewCube and notice the orbit option becomes highlighted.

4. Single left-mouse-click to activate the counter-clockwise option as shown. The current view is orbited 90 degrees; we are still viewing the front side.

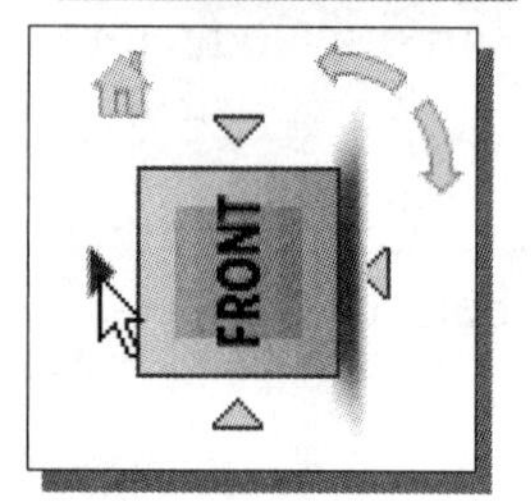

5. Move the cursor over the left arrow of the ViewCube and notice the orbit option becomes highlighted.

6. Single left-mouse-click to activate the left arrow option as shown. The current view is now set to view the top side.

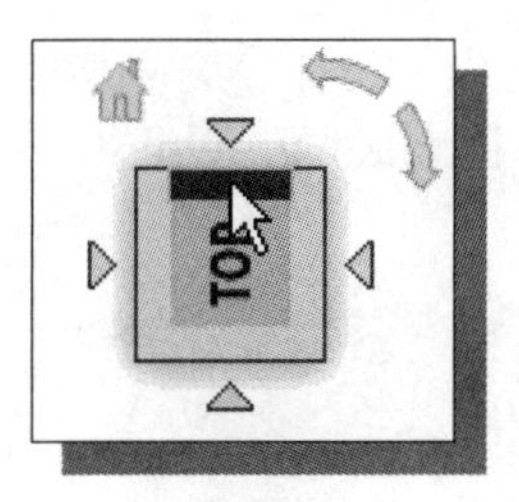

7. Move the cursor over the top edge of the ViewCube and notice the roll option becomes highlighted.

8. Single left-mouse-click to activate the roll option as shown. The view will be adjusted to roll 45 degrees.

9. Move the cursor over the ViewCube and drag with the left-mouse-button to activate the **Free Rotation** option.

10. Move the cursor over the home icon of the ViewCube and notice the Home View option becomes highlighted.

11. Single left-mouse-click to activate the **Home View** option as shown. The view will be adjusted back to the default ***isometric view***.

Full Navigation Wheel – The Navigation Wheel contains tracking menus that are divided into different sections known as wedges. Each wedge on the wheel represents a single navigation tool. You can pan, zoom, or manipulate the current view of a model in different ways. The 3D Navigation Wheel and 2D Navigation Wheel (mostly used in the 2D drawing mode) have some or all of the following options:

Zoom – Adjusts the magnification of the view.
Center – Centers the view based on the position of the cursor over the wheel.
Rewind – Restores the previous view.
Forward – Increases the magnification of the view.
Orbit – Allows 3D free rotation with the left-mouse-button.
Pan – Allows panning by dragging with the left-mouse-button.
Up/Down – Allows panning with the use of a scroll control.
Walk – Allows *walking*, with linear motion perpendicular to the screen, through the model space.
Look – Allows rotation of the current view vertically and horizontally

3D Full Navigation Wheel

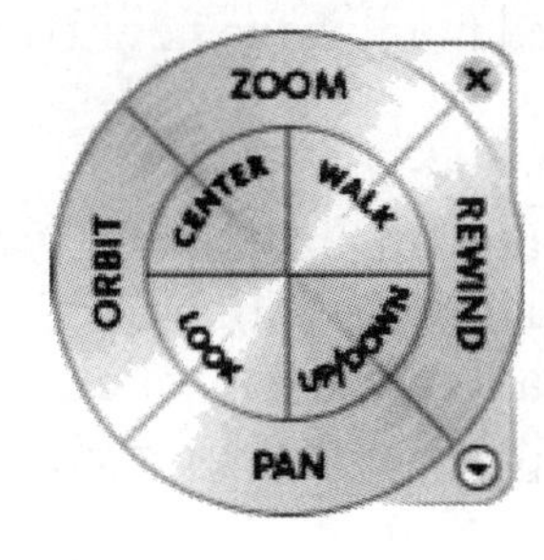

2D Full Navigation Wheel

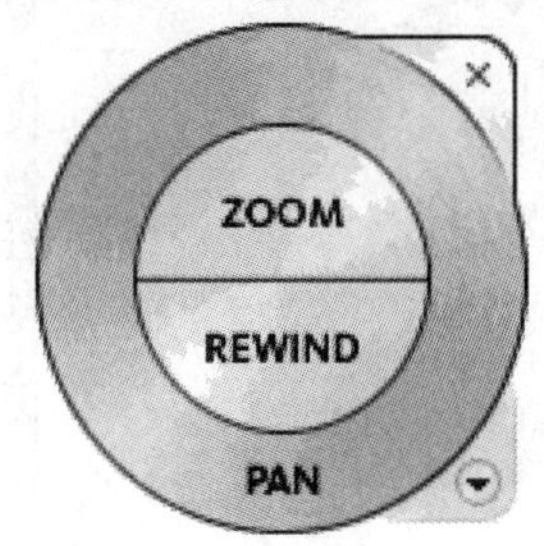

1. Activate the **Full Navigation Wheel**, by clicking on the icon as shown.

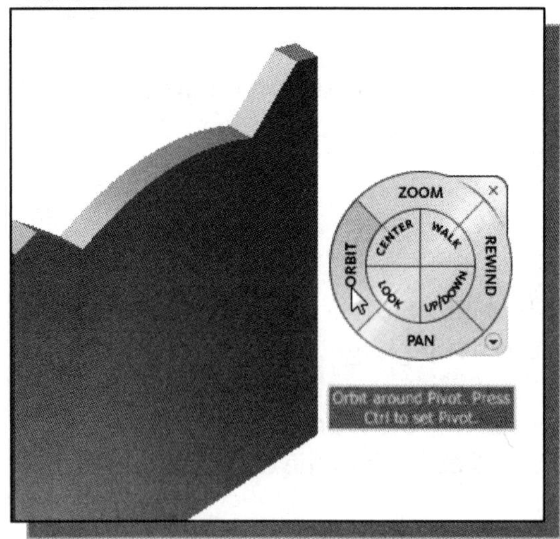

2. Move the cursor in the graphics window and notice the **Full Navigation Wheel** menu follows the cursor on the screen.

3. Move the cursor on the **Orbit** option to highlight the option.

4. Click and drag with the left-mouse-button to activate the **Free Rotation** option.

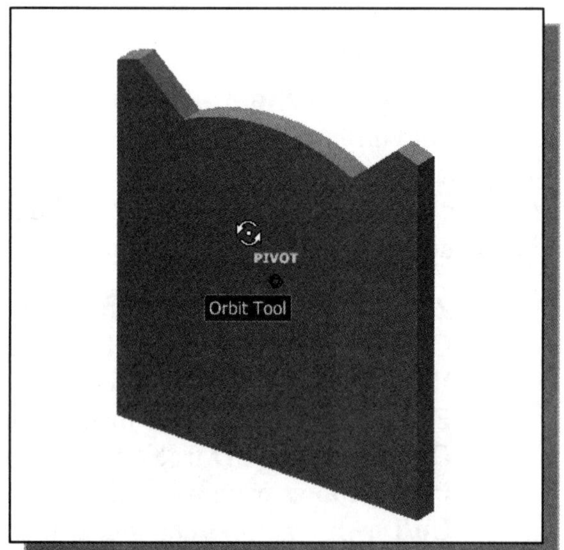

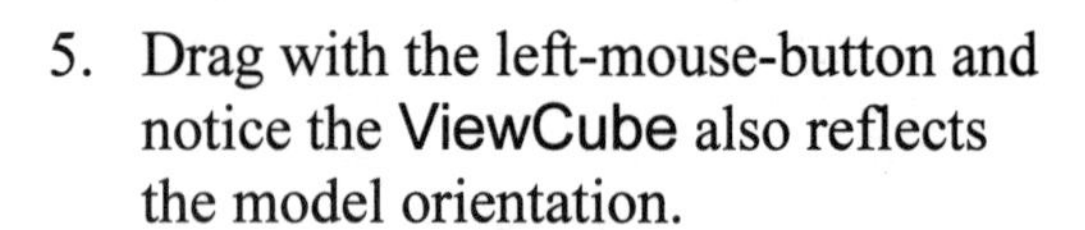

5. Drag with the left-mouse-button and notice the **ViewCube** also reflects the model orientation.

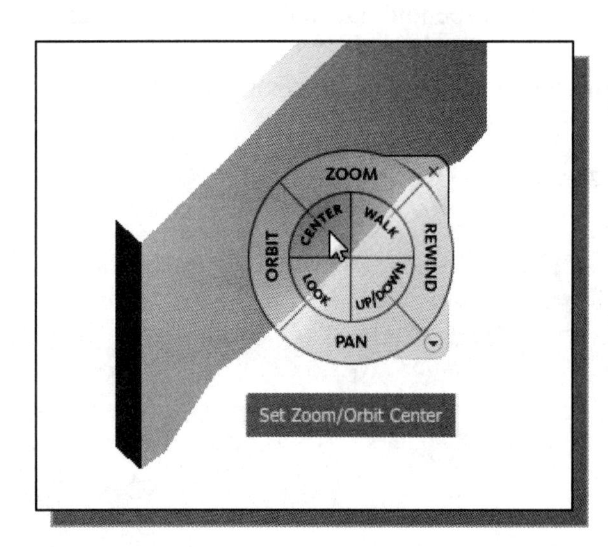

6. Move the cursor to the left side of the model and click the **Center** option as shown. The display is adjusted so the selected point is the new **Zoom/Orbit** center.

7. On your own, experiment with the other available options. Exit the Navigation wheel by clicking on the small [x] near the upper right corner.

Display Modes

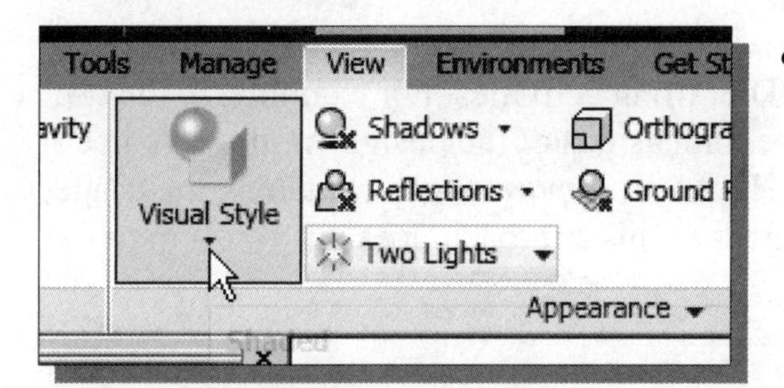

- The **Visual Style** command in the *View* tab has ten display modes, ranging from very realistic renderings of the model to very artistic representations of the model. The most commonly used modes are as follows:

❖ Realistic Shaded Solid:

The *Realistic Shaded Solid* display mode generates a high quality shaded image of the 3D object.

❖ Standard Shaded Solid:

The *Standard Shaded Solid* display option generates a shaded image of the 3D object that requires fewer computer resources compared to the realistic rendering.

❖ Wireframe Image:

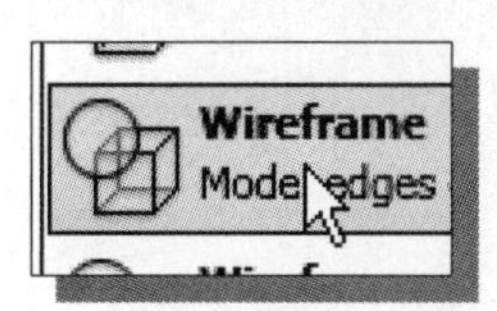

The *Wireframe Image* display option allows the display of the 3D objects using the basic wireframe representation scheme.

❖ Wireframe with Hidden-Edge Display:

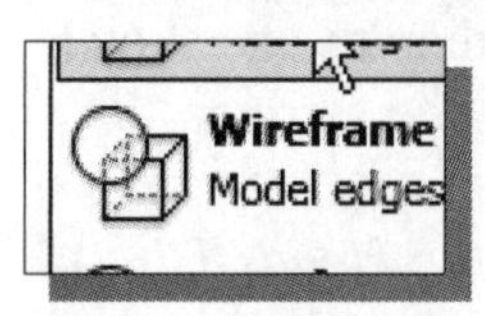

The *Wireframe with Hidden-Edge Display* option can be used to generate an image of the 3D object with all the back lines hidden.

Orthographic vs. Perspective

- Besides the three basic display modes, we can also choose orthographic view or perspective view of the display. Click on the icon next to the display mode button on the *Standard* toolbar, as shown in the figure.

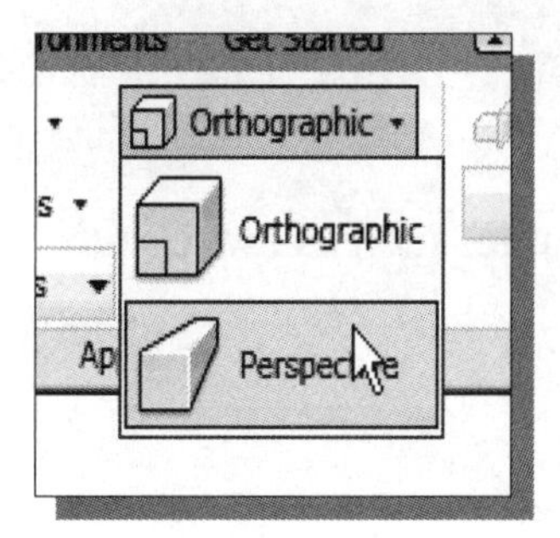

❖ Orthographic

The first icon allows the display of the 3D object using the parallel edges representation scheme.

❖ Perspective

The second icon allows the display of the 3D object using the perspective, nonparallel edges representation scheme.

Disable the Heads-Up Display Option

- The **Heads-Up Display** option in *Inventor* provides mainly the **Dynamic Input** function, which can be quite useful for 2D drafting activities. For example, in the use of a 2D drafting CAD system, most of the dimensions of the design would have been determined by the documentation stage. However, in *parametric modeling*, the usage of the *Dynamic Input* option is quite limited, as this approach does not conform to the "**shape before size**" design philosophy.

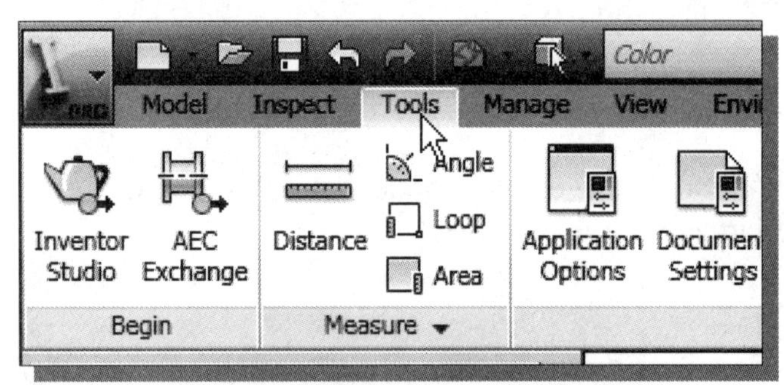

1. Select the **Tools** tab in the *Ribbon* as shown.

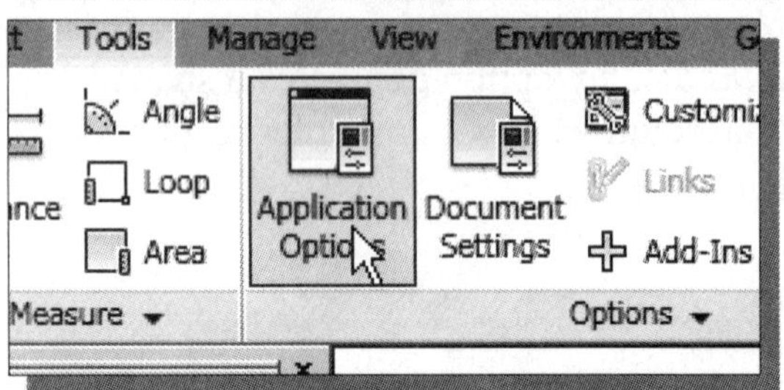

2. Select **Application Options** in the *Options* toolbar as shown.

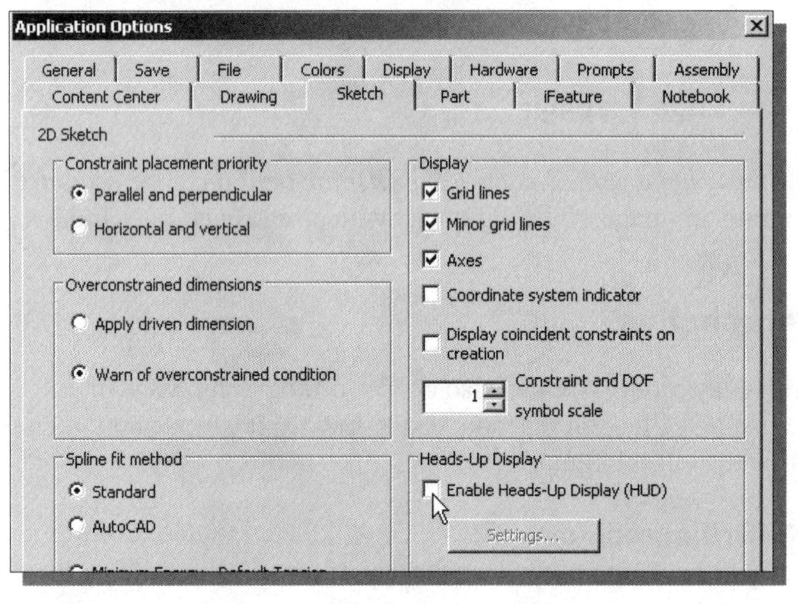

3. Pick the **Sketch** tab to display the sketch related settings.

4. In the *Heads-Up Display* section, turn ***OFF*** the ***Enable Heads-Up Display*** option as shown.

5. On your own, examine the other sketch settings that are available, such as switch **ON** the *Grid lines* in the *Display* section.

6. Click **OK** to accept the settings.

Step 4-1: Adding an Extruded Feature

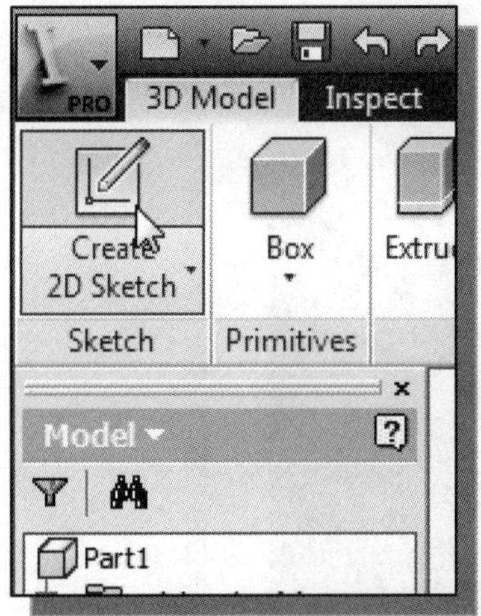

1. Activate the **Model** tab and select the **Create 2D Sketch** command by left-clicking once on the icon.

2. In the *Status Bar* area, the message: "*Select face, workplane, sketch or sketch geometry*" is displayed. *Autodesk Inventor* expects us to identify a planar surface where the 2D sketch of the next feature is to be created. Move the graphics cursor on the 3D part and notice that *Autodesk Inventor* will automatically highlight feasible planes and surfaces as the cursor is on top of the different surfaces. Pick the back face of the 3D solid object.

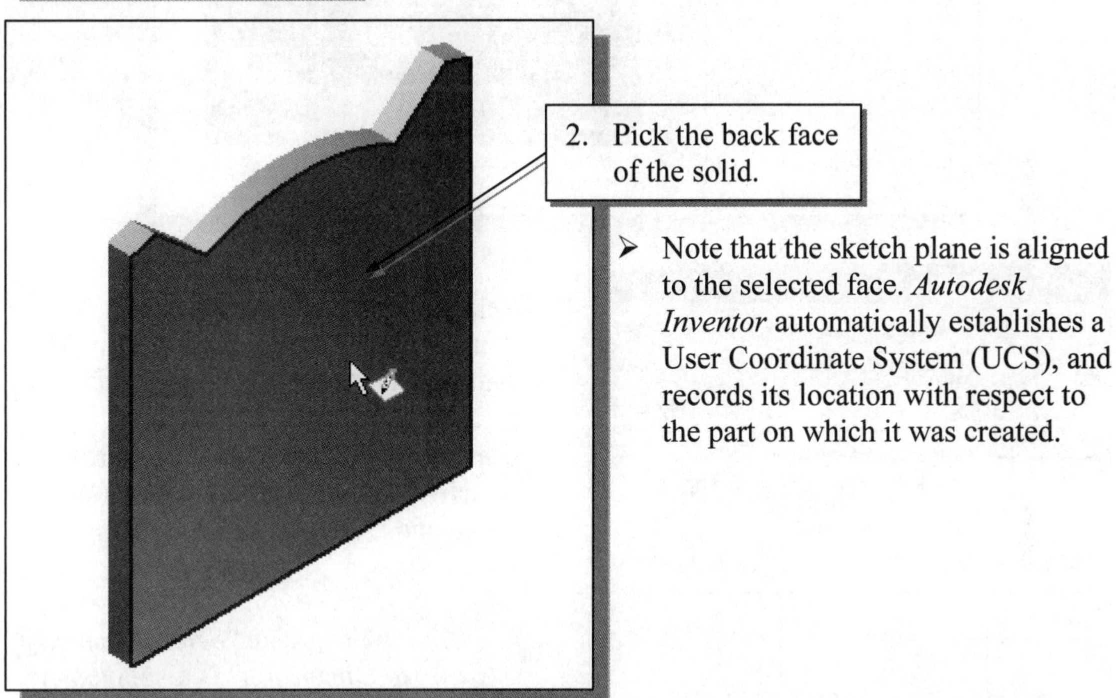

➢ Note that the sketch plane is aligned to the selected face. *Autodesk Inventor* automatically establishes a User Coordinate System (UCS), and records its location with respect to the part on which it was created.

- Next, we will create and profile another sketch, a rectangle, which will be used to create another extrusion feature that will be added to the existing solid object.

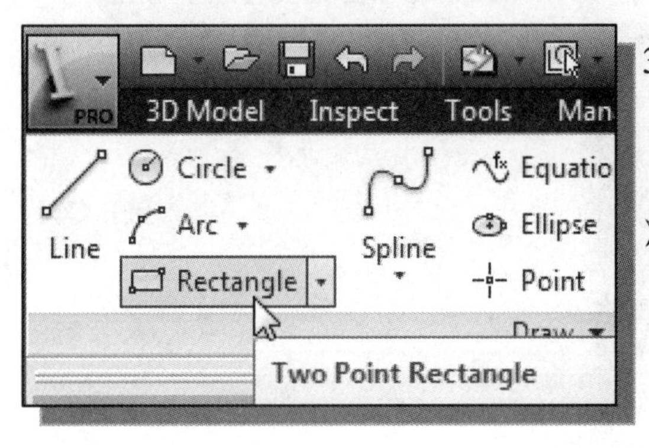

3. Select the **Rectangle** command by clicking once with the **left-mouse-button** on the icon in the *Draw* toolbar.

➢ To illustrate the usage of dimensions in parametric sketches, we will intentionally create a rectangle away from the desired location.

4. Create a sketch with segments perpendicular/parallel to the bottom edges of the solid model as shown below.

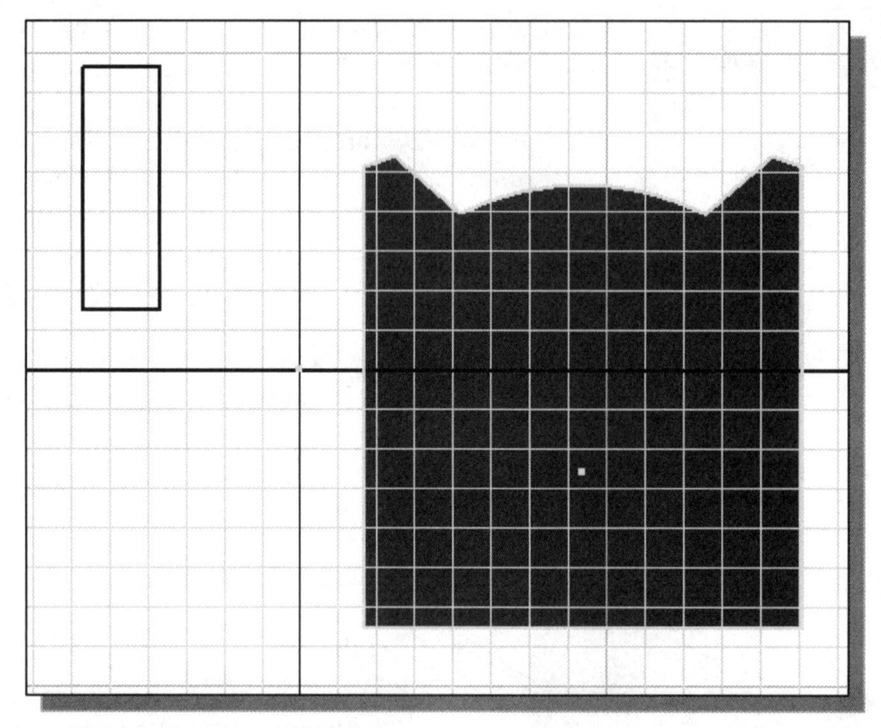

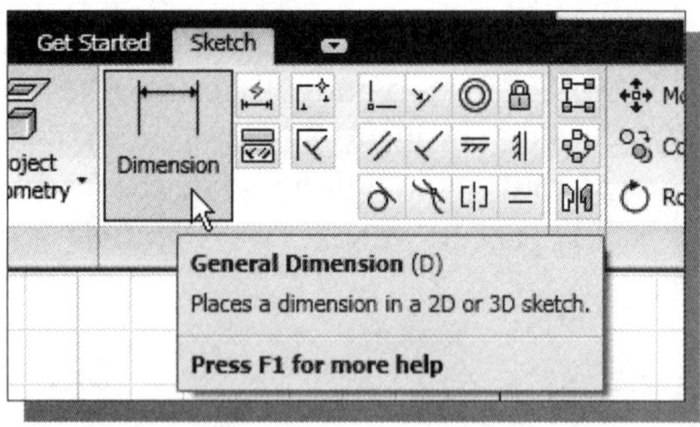

5. Select the **General Dimension** command in the *Constrain* toolbar. The General Dimension command allows us to quickly create and modify dimensions. Left-click once on the icon to activate the General Dimension command.

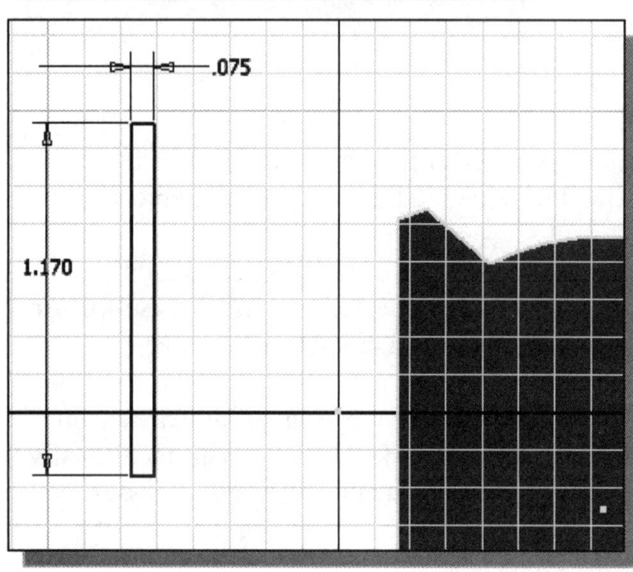

6. The message "*Select Geometry to Dimension*" is displayed in the *Status Bar* area, at the bottom of the *Inventor* window. Create and modify the size dimensions to describe the size of the sketch as shown in the figure.

7. Create the two location dimensions to describe the position of the sketch relative to the top corner of the solid model as shown.

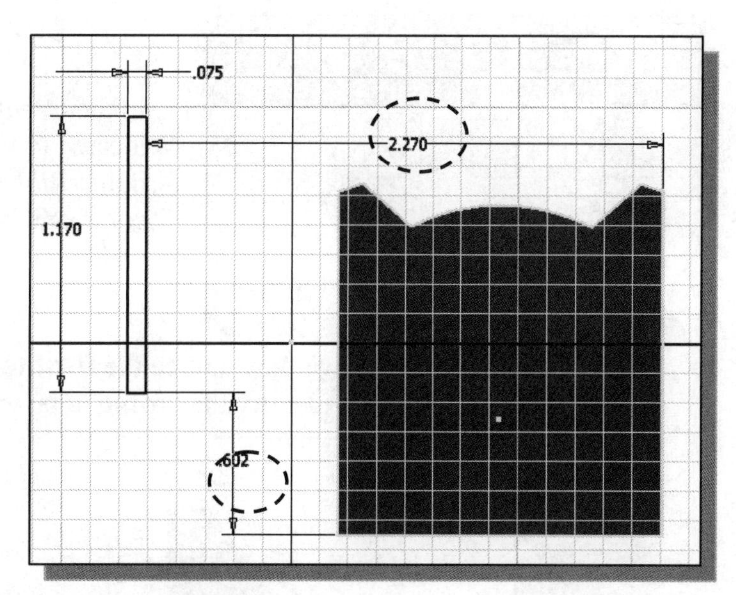

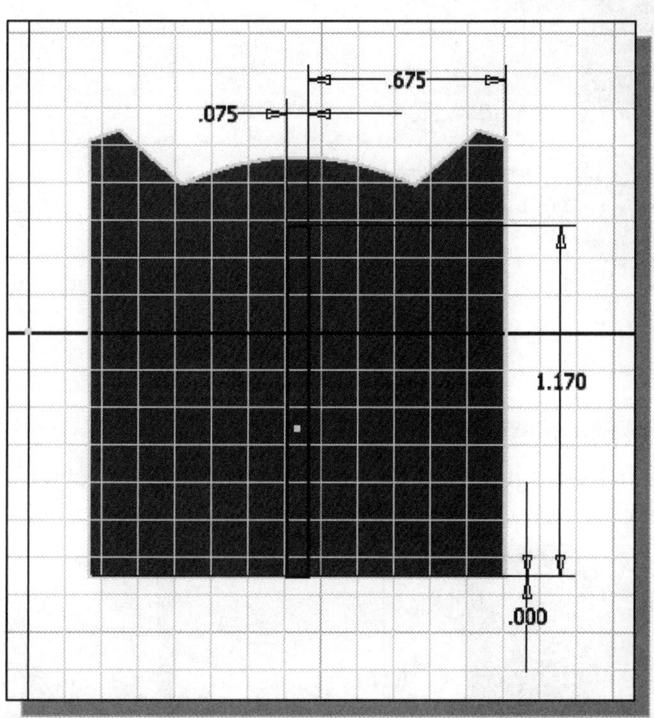

8. On your own, click on the associated location dimensions and modify them to **0.675** and **0.0** as shown in the figure.

➢ In parametric modeling, the dimensions can be used to quickly control the size and location of the defined geometry.

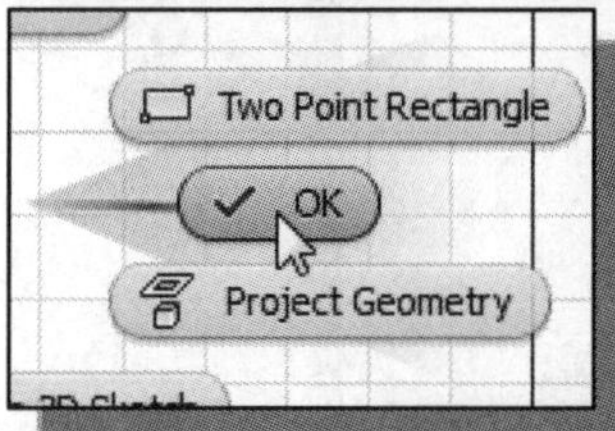

9. Inside the graphics window, click once with the **right-mouse-button** to display the option menu. Select **OK** in the popup menu to end the General Dimension command.

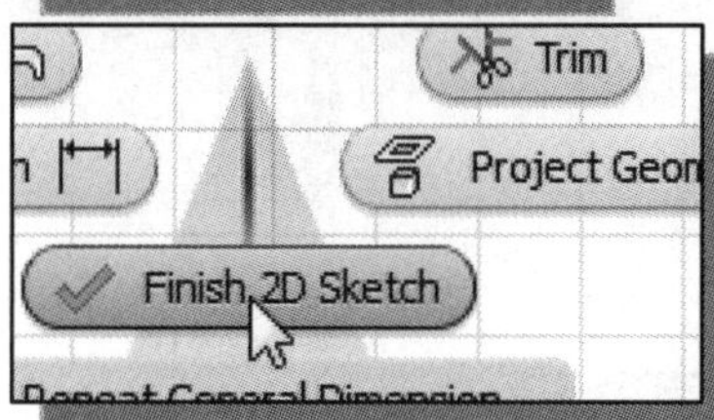

10. Inside the graphics window, click once with the **right-mouse-button** to display the option menu. Select **Finish 2D Sketch** in the popup menu to end the Sketch option.

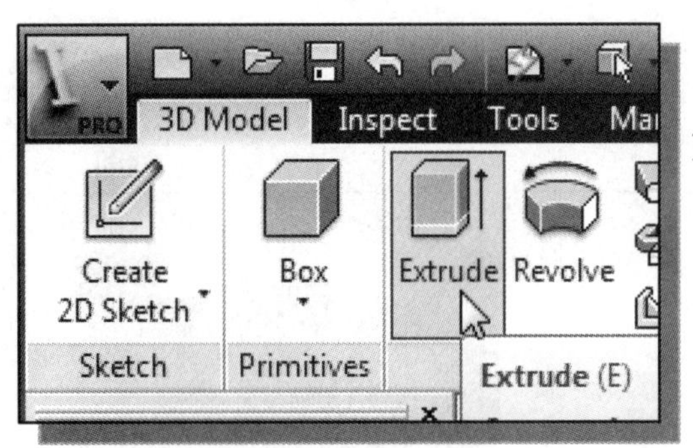

11. In the *Create* toolbar (the toolbar that is located to the right side of the *Sketch* toolbar in the *Ribbon*), select the **Extrude** command by releasing the left-mouse-button on the icon.

12. In the *Extrude* popup window, notice the **Profile** button is activated; *Autodesk Inventor* expects us to identify the profile to be extruded.

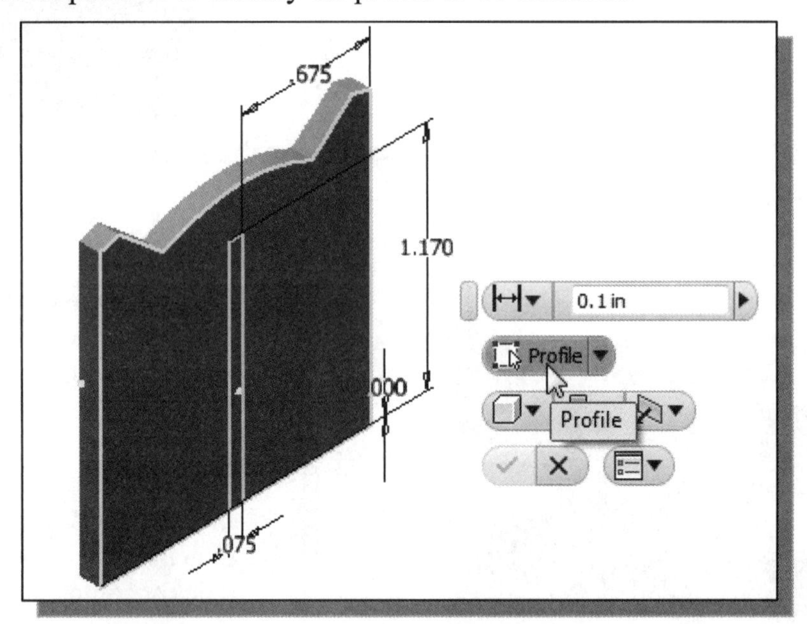

13. Move the cursor inside the rectangle we just created and left-click once to select the inside region as the profile to be extruded.

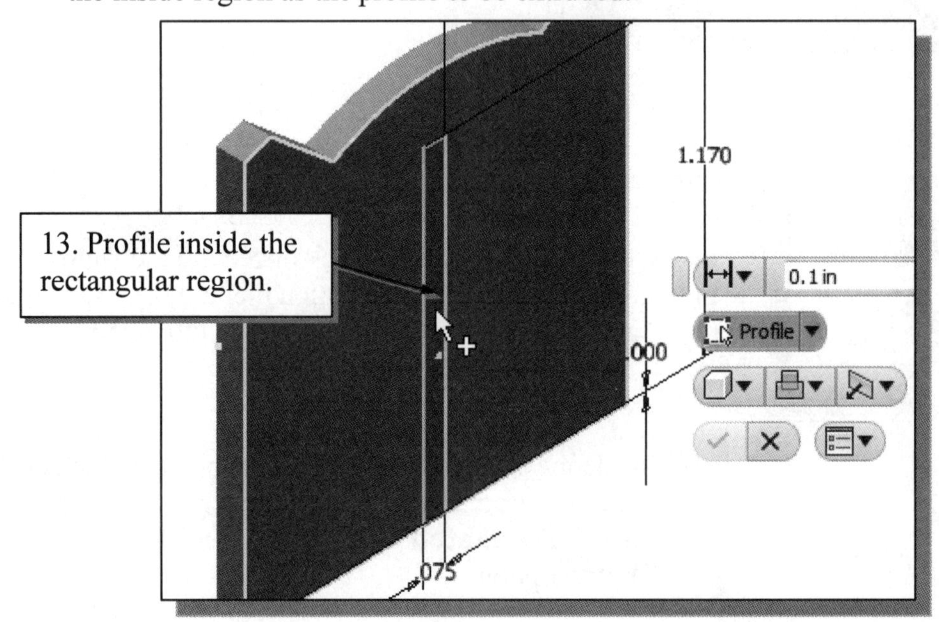

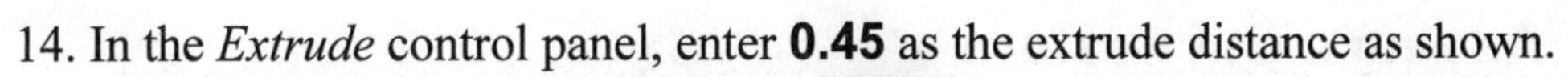

14. In the *Extrude* control panel, enter **0.45** as the extrude distance as shown.

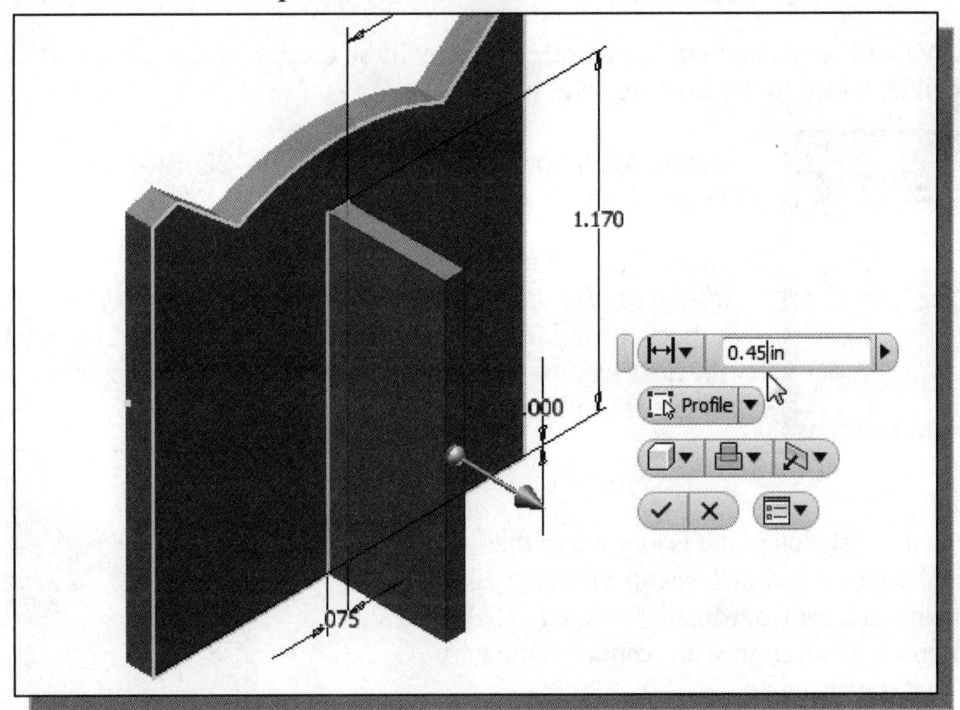

15. On your own, confirm the preview of the extruded feature appears as shown in the above figure.

16. Click on the **OK** button to proceed with creating the extruded feature.

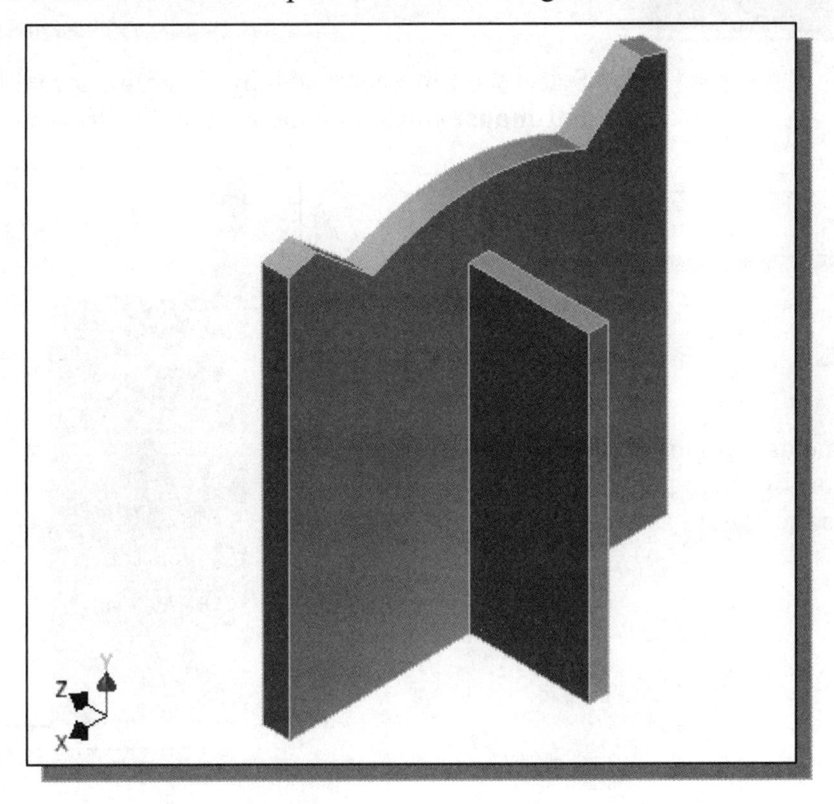

Step 4-2: Adding a Cut Feature

- Next, we will create and profile a circle, which will be used to create a **cut** feature that will be added to the existing solid object.

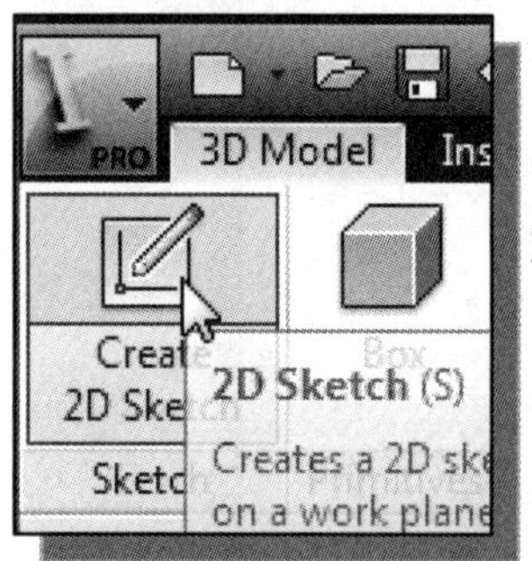

1. In the *Sketch* toolbar select the **Create 2D Sketch** command by left-clicking once on the icon.

2. In the *Status Bar* area, the message: "*Select face, workplane, sketch or sketch geometry.*" is displayed. *Autodesk Inventor* expects us to identify a planar surface where the 2D sketch of the next feature is to be created. Pick the top horizontal face of the 3D solid model as shown.

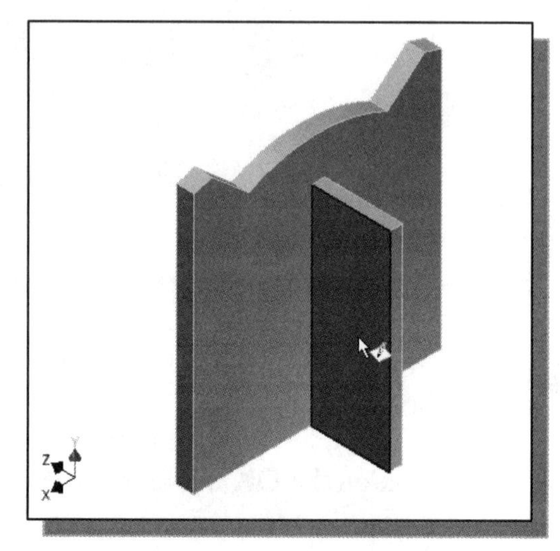

➢ Note that the sketch plane is aligned to the selected face. *Autodesk Inventor* automatically establishes a User Coordinate System (UCS), and records its location with respect to the part on which it was created.

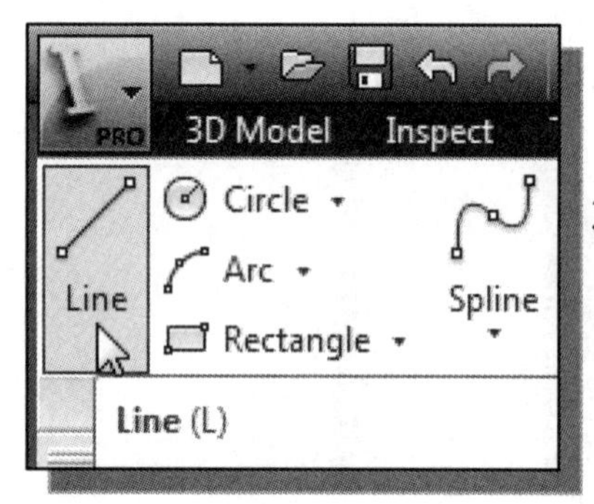

3. Select the **Line** command by clicking once with the **left-mouse-button** on the icon in the *Draw* toolbar.

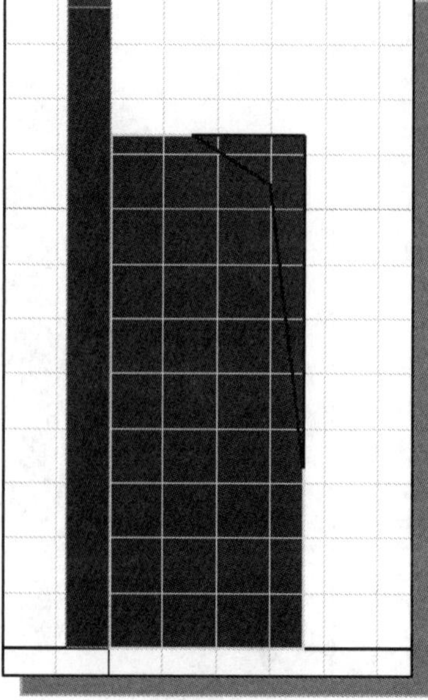

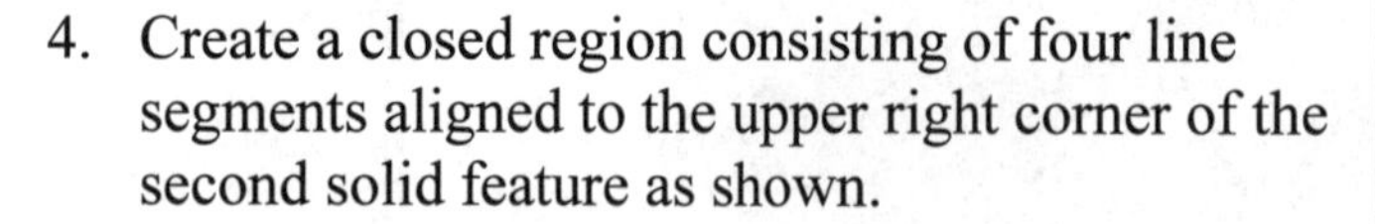

4. Create a closed region consisting of four line segments aligned to the upper right corner of the second solid feature as shown.

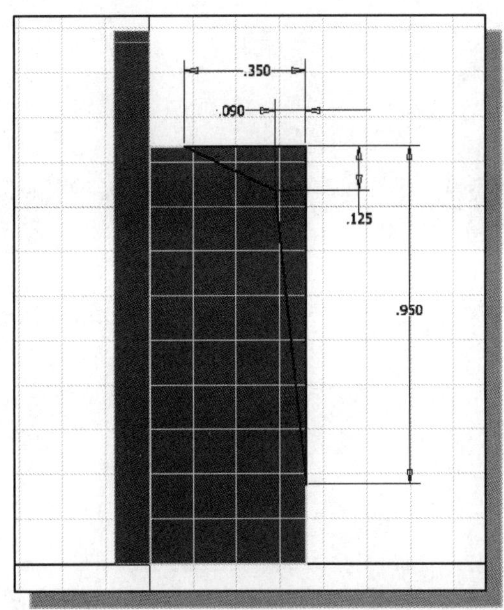

5. On your own, create and modify the dimensions of the sketch as shown in the figure.

6. Inside the graphics window, click once with the **right-mouse-button** to display the option menu. Select **OK** in the popup menu to end the General Dimension command.

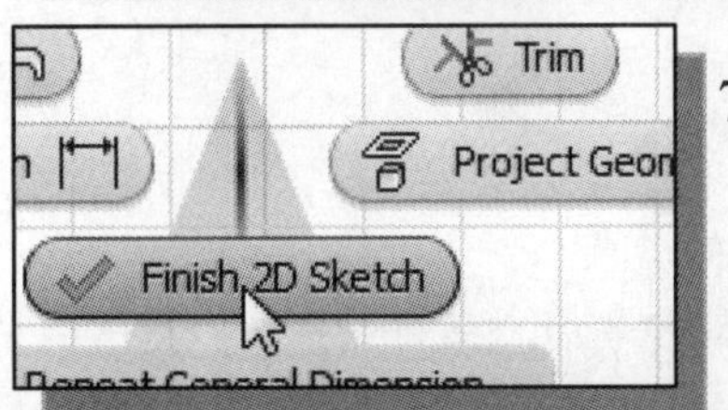

7. Inside the graphics window, click once with the **right-mouse-button** to display the option menu. Select **Finish Sketch** in the popup menu to end the Sketch option.

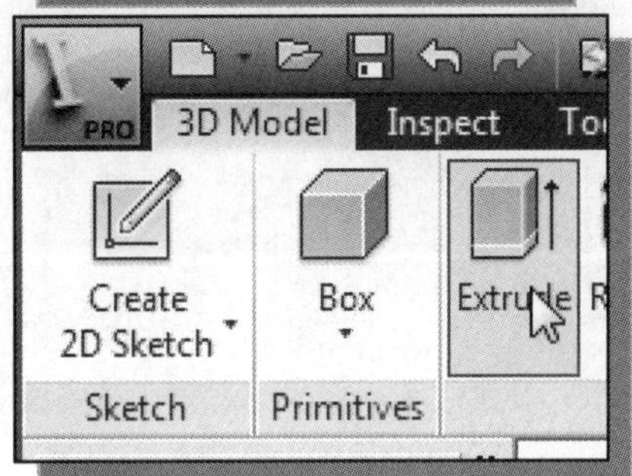

8. In the *Create* toolbar select the **Extrude** command by releasing the left-mouse-button on the icon.

9. In the *Extrude* popup window, the **Profile** button is pressed down; *Autodesk Inventor* expects us to identify the profile to be extruded.

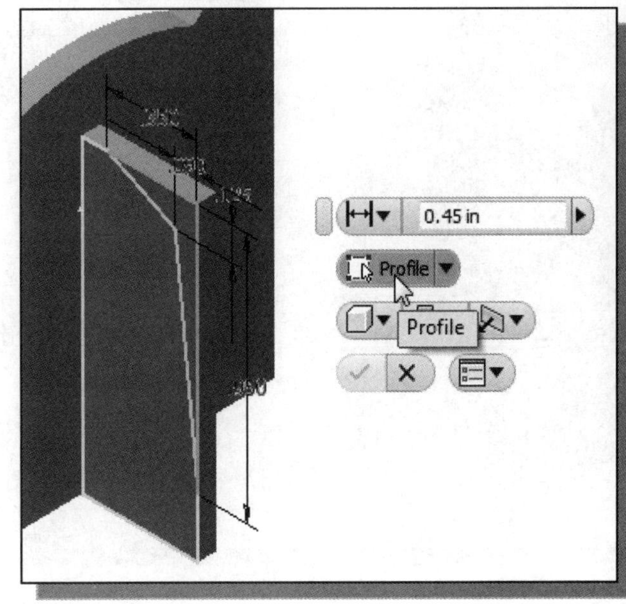

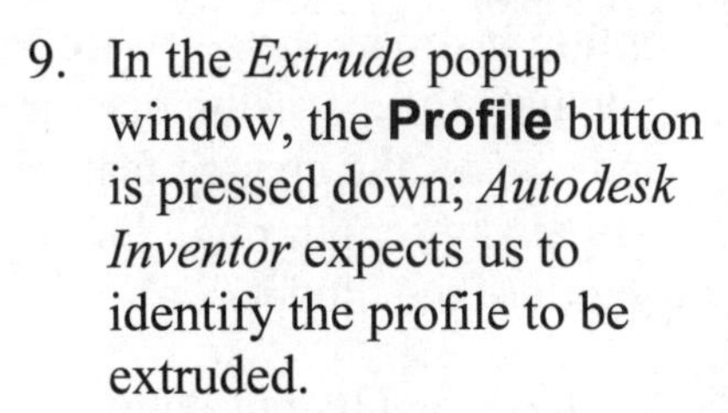

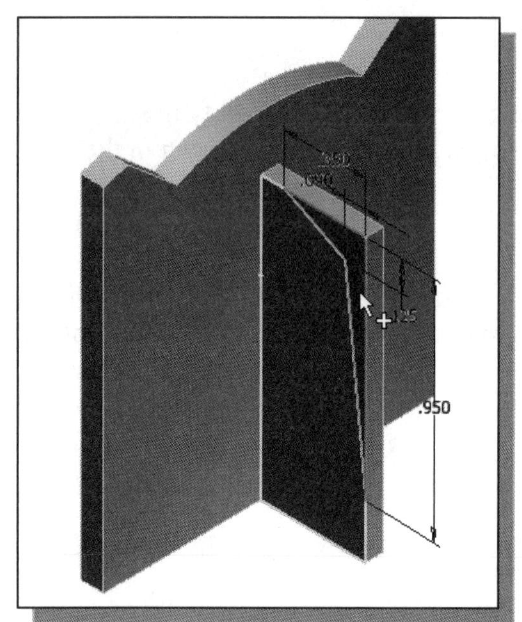

10. Click on the inside of the sketched region as shown.

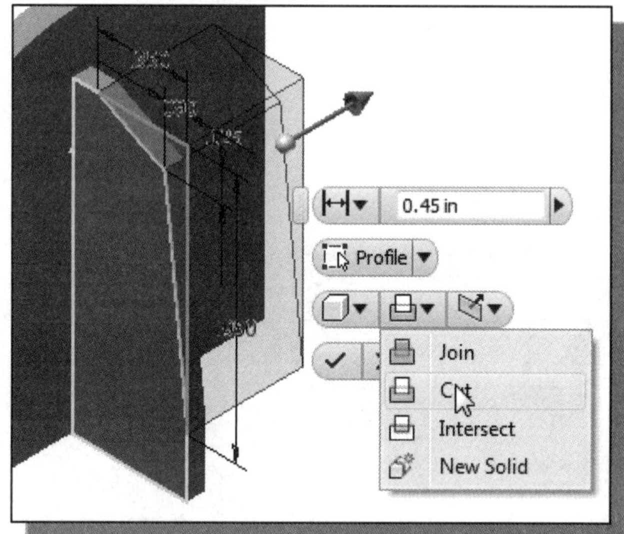

11. Click on the **CUT** icon, as shown, to set the extrusion operation to ***Cut***.

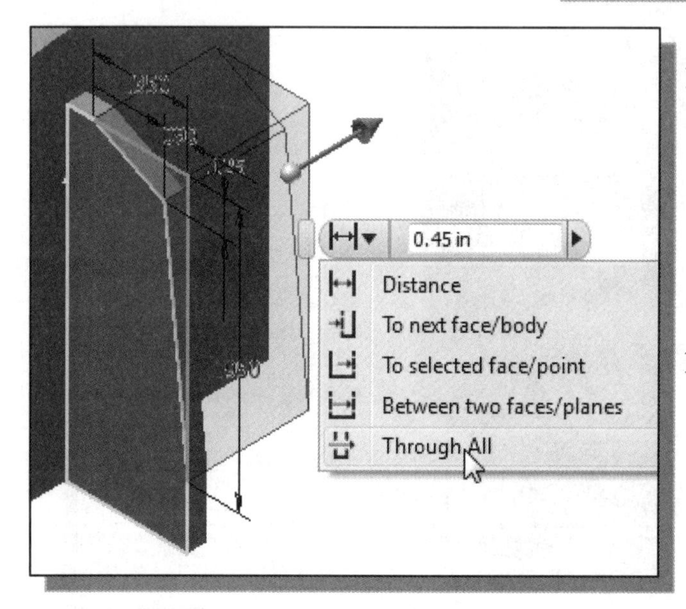

12. Set the *Extents* option to **Through All** as shown. The *All* option instructs the software to calculate the extrusion distance and assures the created feature will always cut through the full length of the model.

13. Click on the **OK** button to proceed with creating the extruded feature.

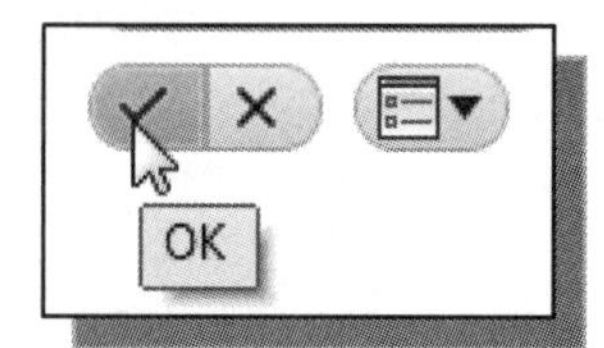

Step 5: Adding Additional Features

- Next, we will create and profile another sketch, a circle, which will be used to create another extrusion feature that will be added to the existing solid object.

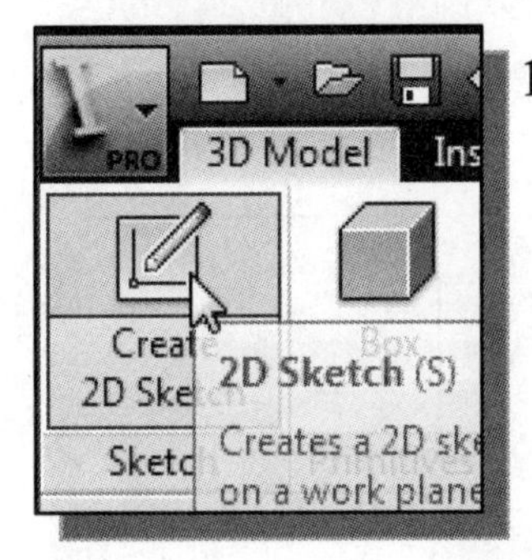

1. In the *Sketch* toolbar select the **Create 2D Sketch** command by left-clicking once on the icon.

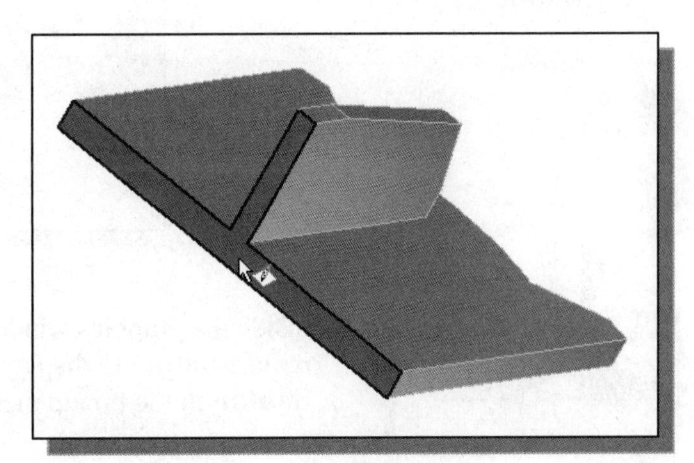

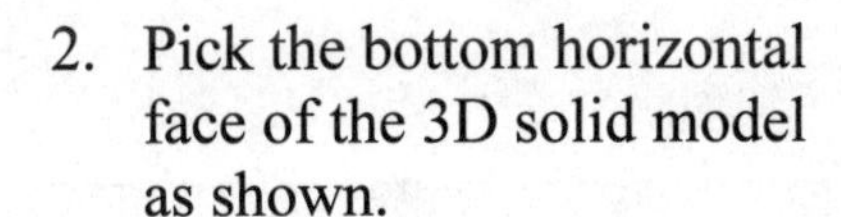

2. Pick the bottom horizontal face of the 3D solid model as shown.

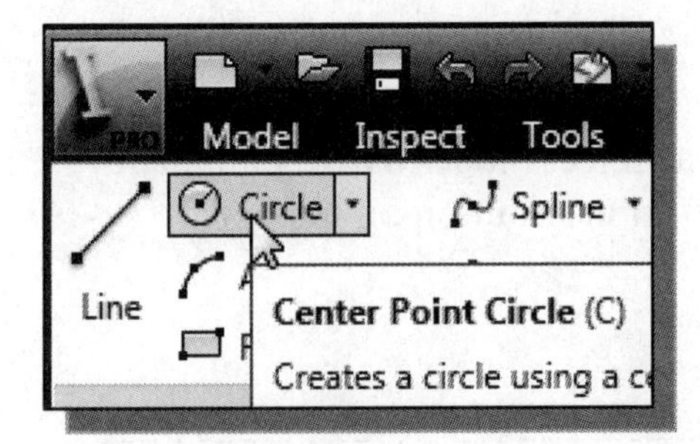

3. Select the **Circle** command by clicking once with the **left-mouse-button** on the icon in the *Draw* toolbar.

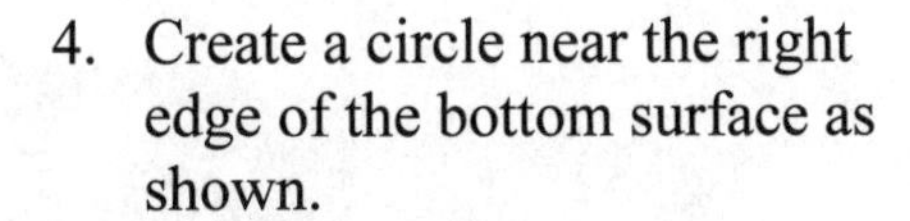

4. Create a circle near the right edge of the bottom surface as shown.

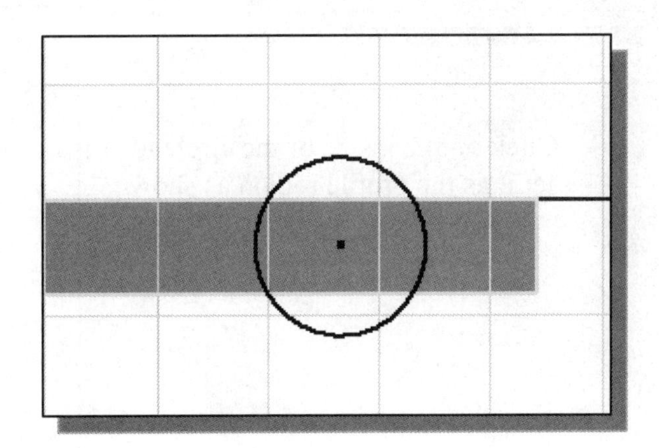

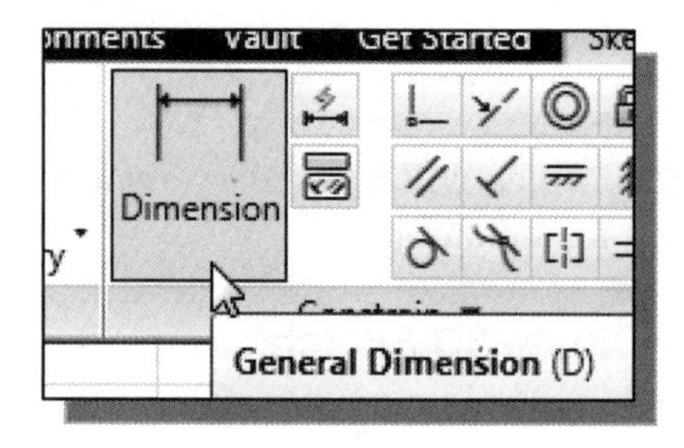

5. Select the **General Dimension** command in the *Constrain* toolbar. The General Dimension command allows us to quickly create and modify dimensions.

6. On your own, create and modify the three dimensions as shown.

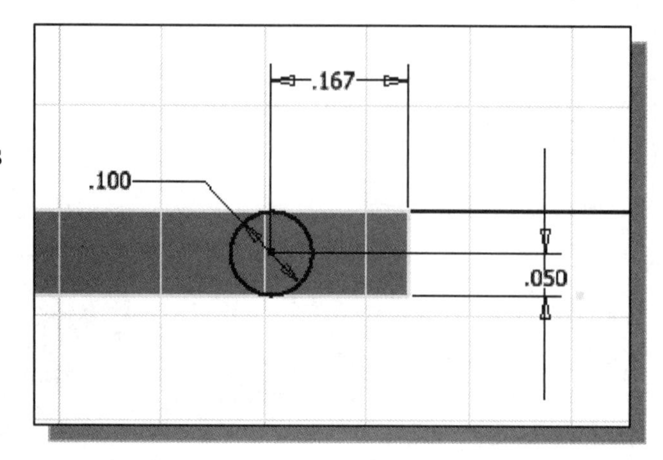

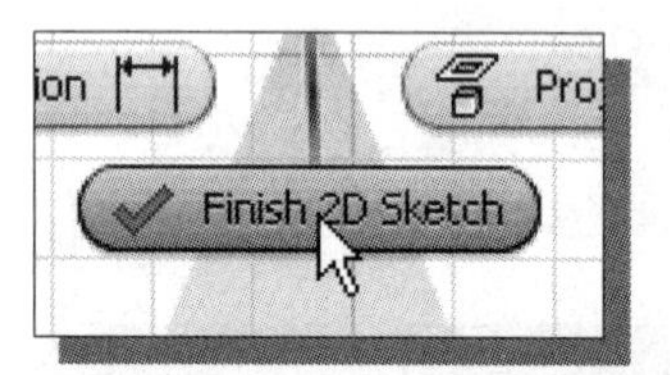

7. Inside the graphics window, click once with the **right-mouse-button** to display the option menu. Select **Finish Sketch** in the popup menu to end the Sketch option.

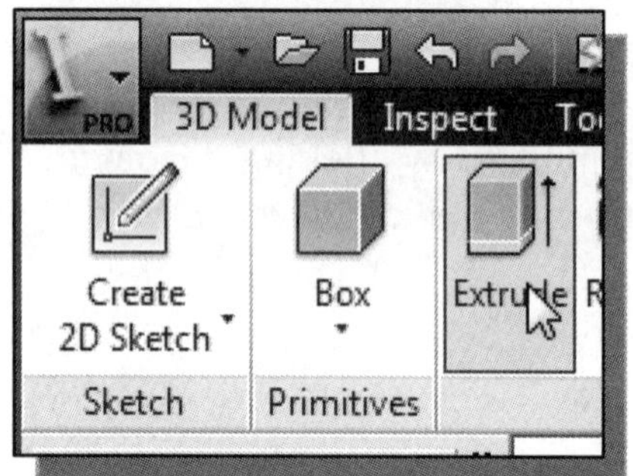

8. In the *Create* toolbar select the **Extrude** command by releasing the left-mouse-button on the icon.

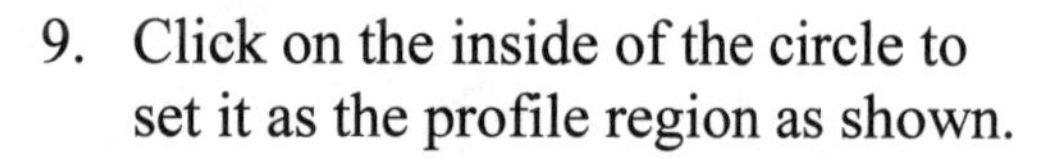

9. Click on the inside of the circle to set it as the profile region as shown.

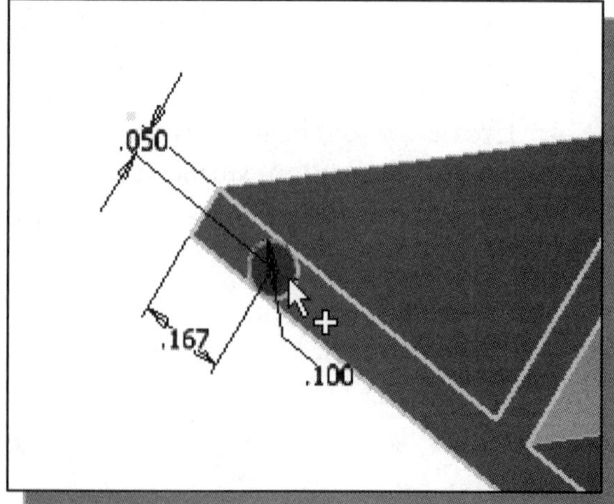

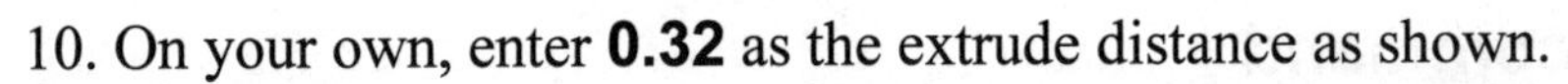

10. On your own, enter **0.32** as the extrude distance as shown.

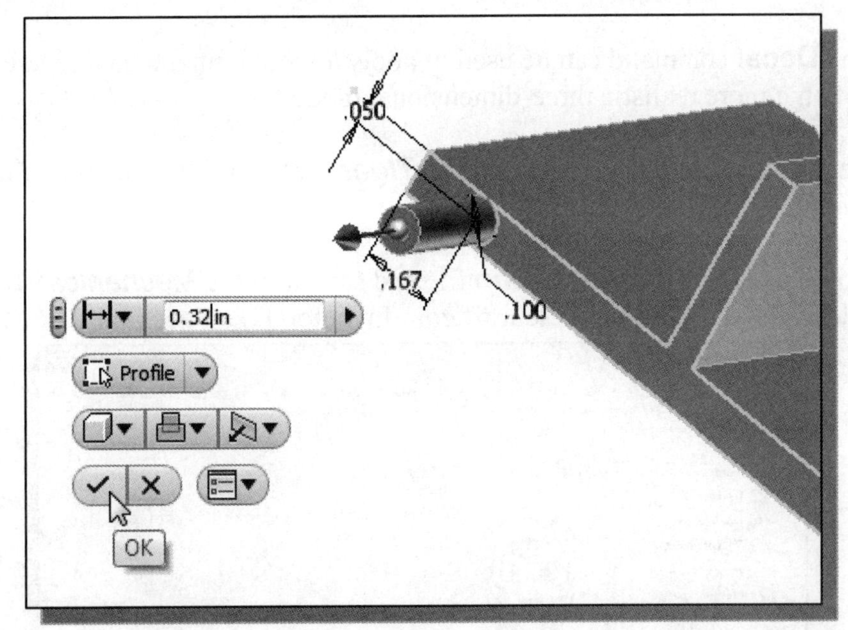

11. Click on the **OK** button to proceed with creating the extruded feature.

12. On your own, repeat the above steps and create another extruded feature on the other side.

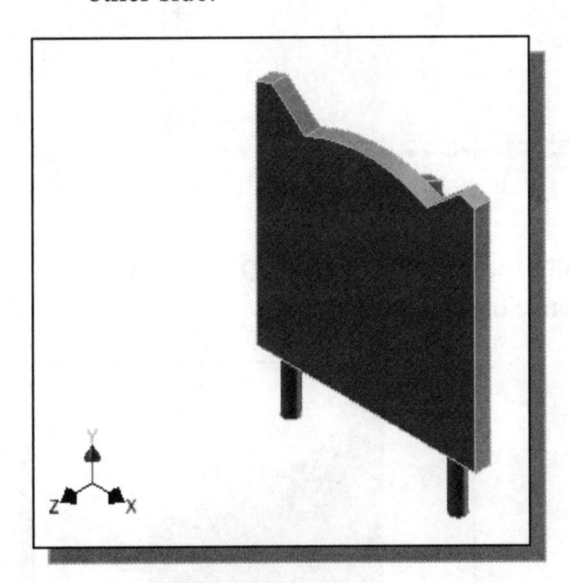

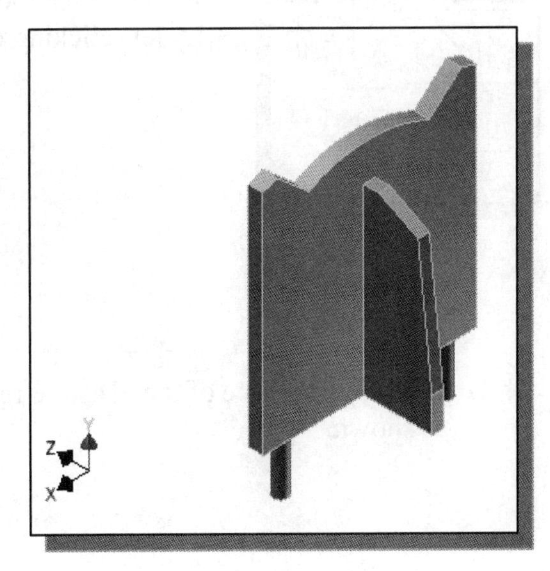

Using the Decal command

❖ The **Decal** command can be used to apply a bitmap image on a flat surface to obtain a more realistic three-dimensional model.

1. On your own, create a ***Mechanical-Tiger*** project folder under the *Parametric Modeling* folder.

2. Download the ***TigerFace.bmp*** file and save it to the *Mechanical-Tiger* project folder. (URL: http://www.schroff.com/Inventor/TigerFace.bmp)

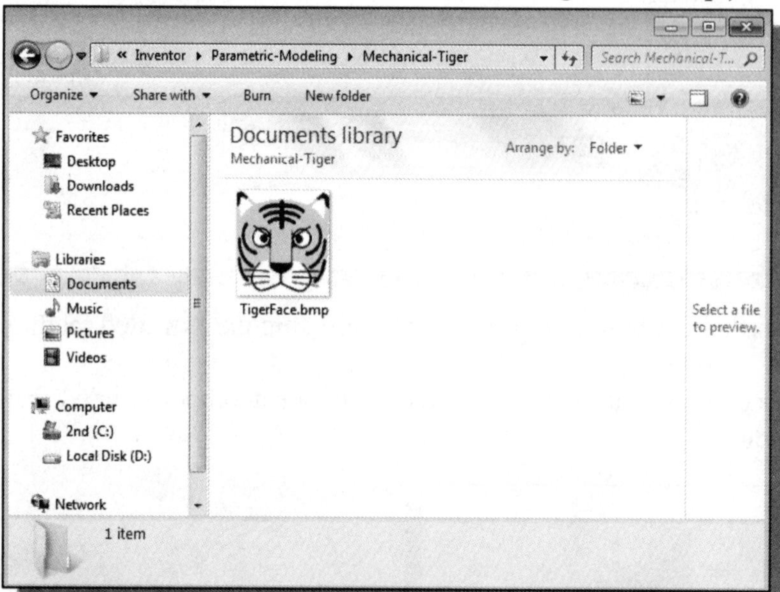

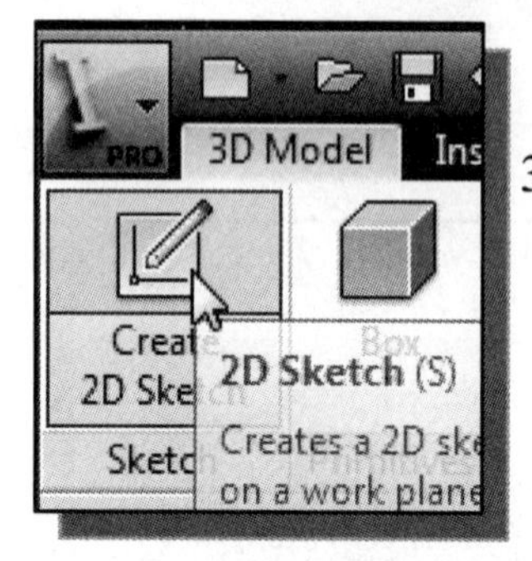

3. In the *Sketch* toolbar select the **Create 2D Sketch** command by left-clicking once on the icon.

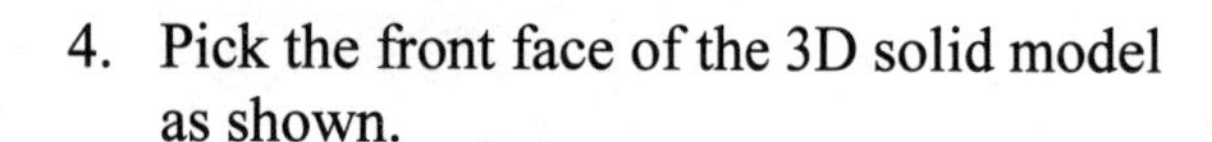

4. Pick the front face of the 3D solid model as shown.

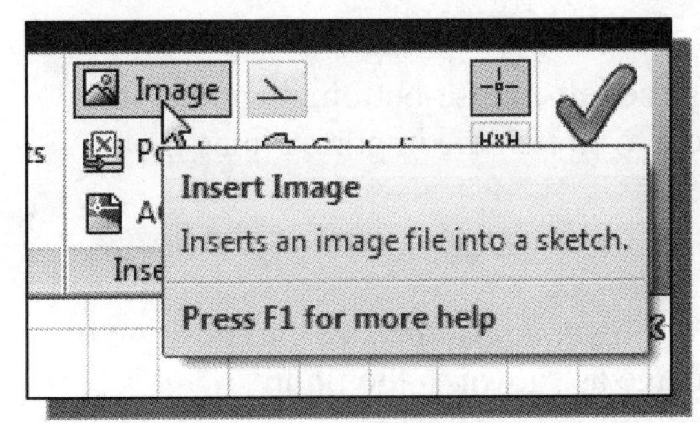

5. Select the **Insert Image** command in the *Insert* toolbar as shown.

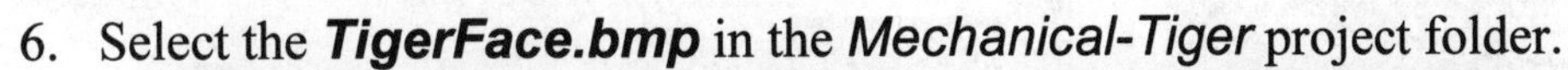

6. Select the ***TigerFace.bmp*** in the *Mechanical-Tiger* project folder.

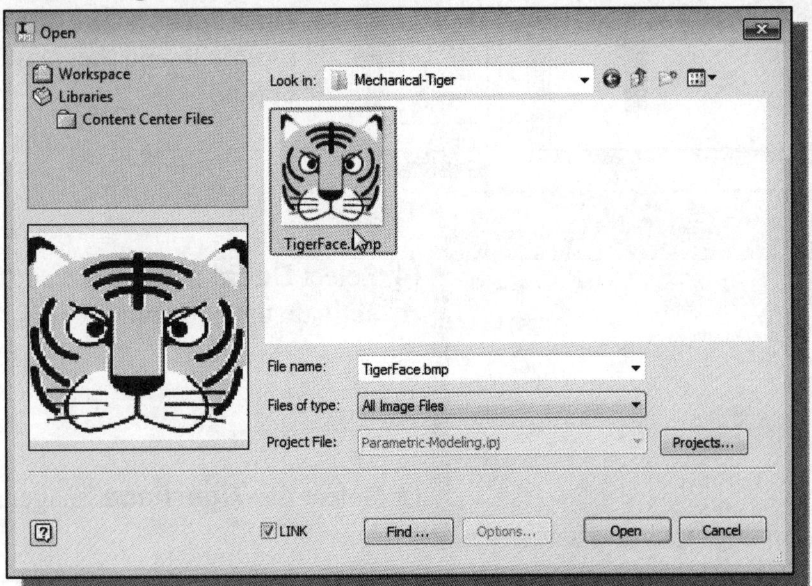

7. Click **Open** to import the image into the current sketch.

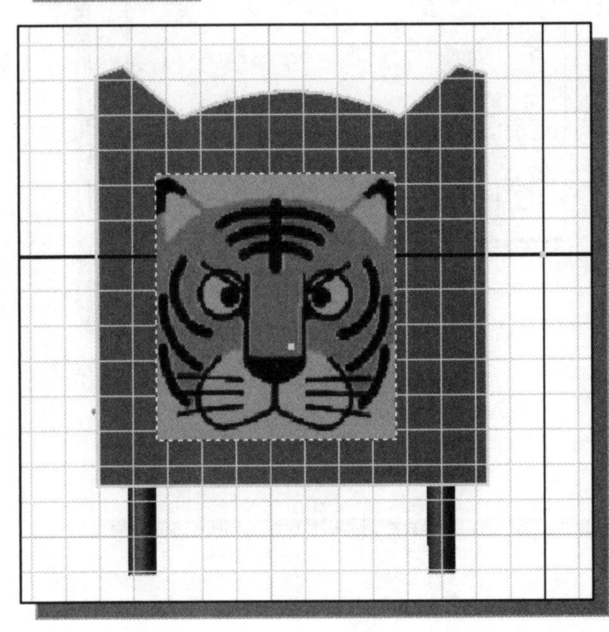

8. Place the image on the **front face** as shown.

9. Click once with the right-mouse-button to bring up the option menu. Select **OK** in the option menu to end the command.

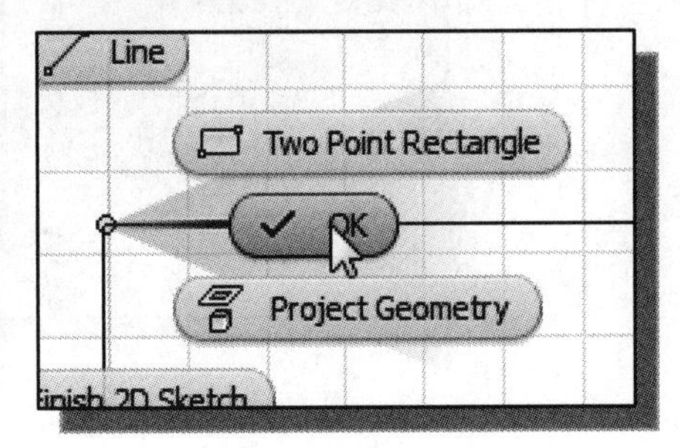

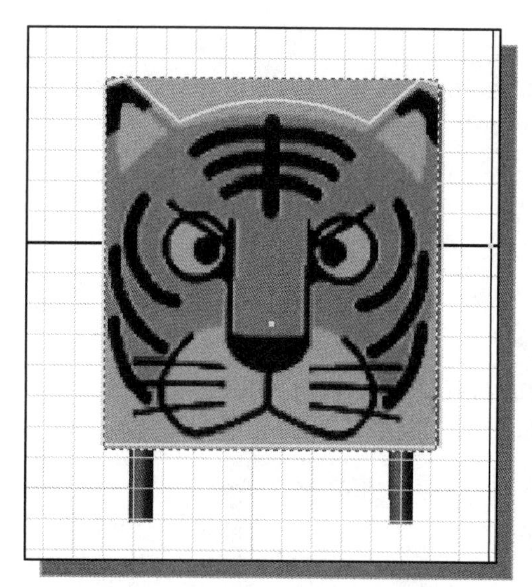

10. Using the left-mouse-button, drag one of the corners to stretch the imported image.

11. Drag on the image to reposition the image.

12. On your own, adjust the size and position of the image as shown in the figure.

13. Click **Finish Sketch** in the *Exit* toolbar to end the **Sketch** option.

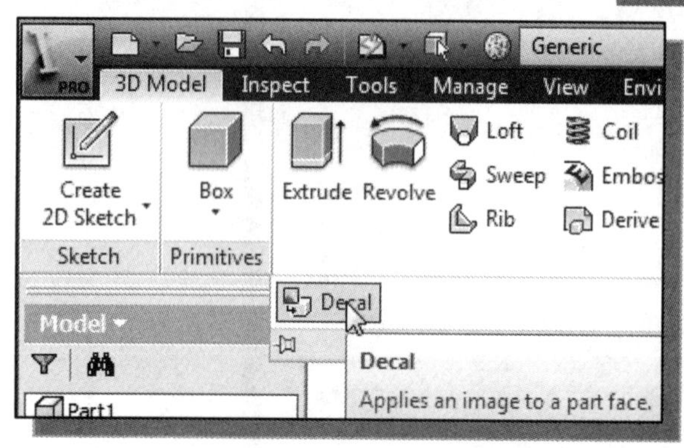

14. Select **Decal** in the *Create* toolbar to activate the command.

15. Select the ***TigerFace*** image as shown.

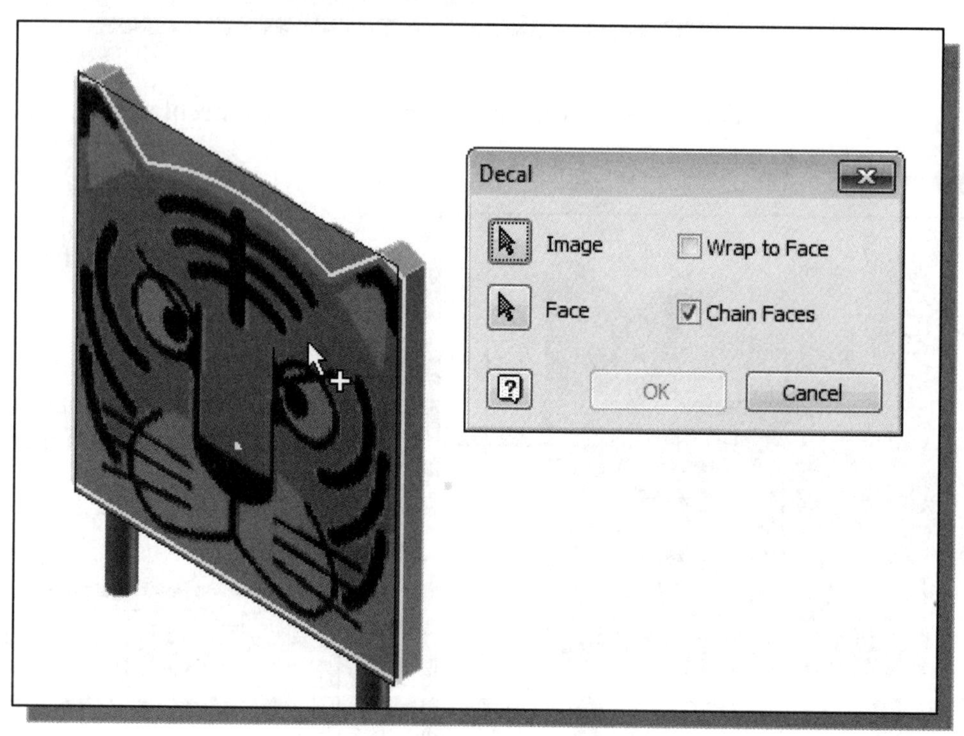

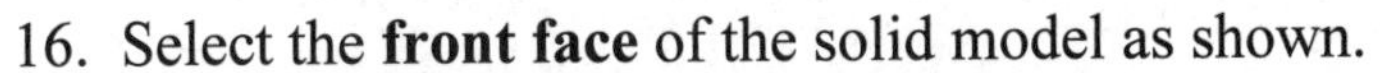

16. Select the **front face** of the solid model as shown.

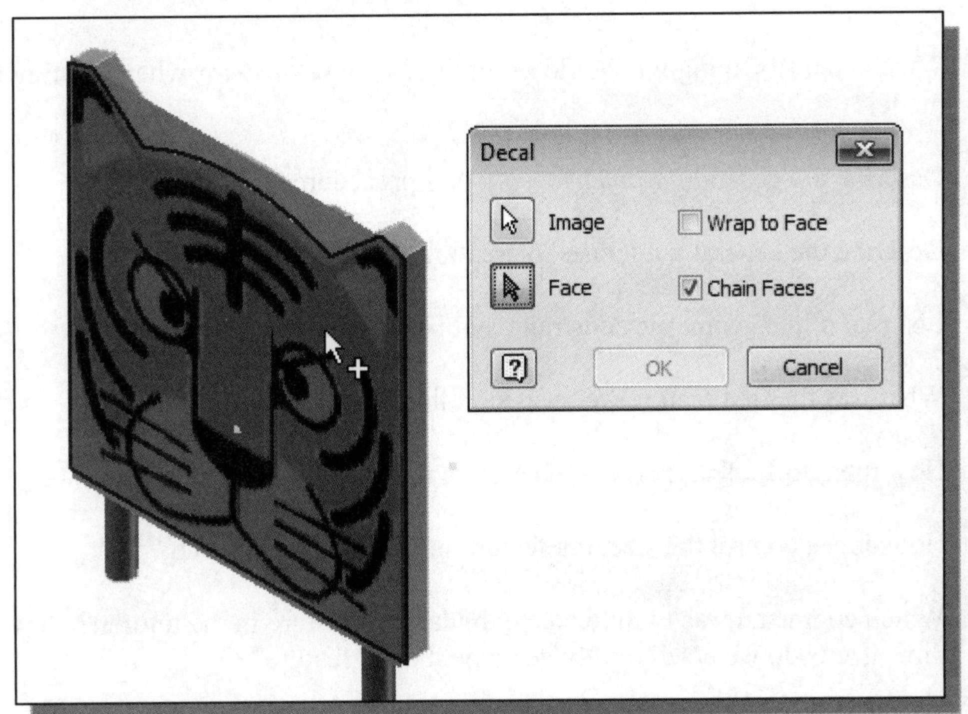

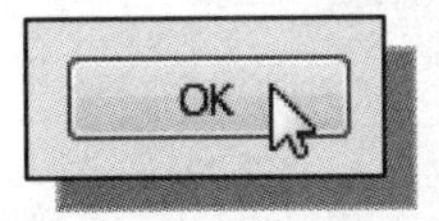

17. Click **OK** to accept the settings and complete the solid model.

Save the Model

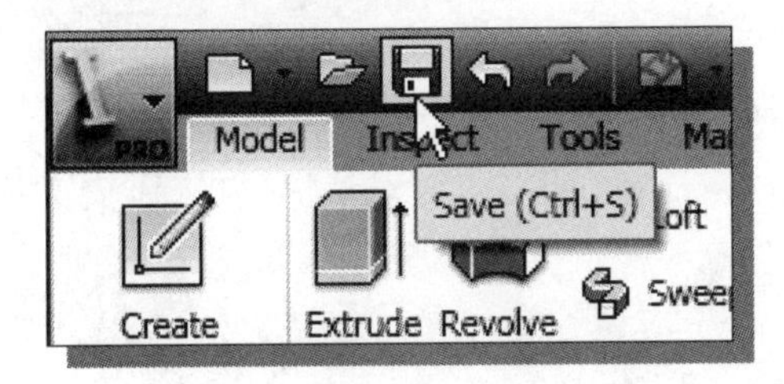

1. Select **Save** in the *Quick Access* toolbar, or you can also use the "**Ctrl-S**" combination (hold down the "Ctrl" key and hit the "S" key once) to save the part.

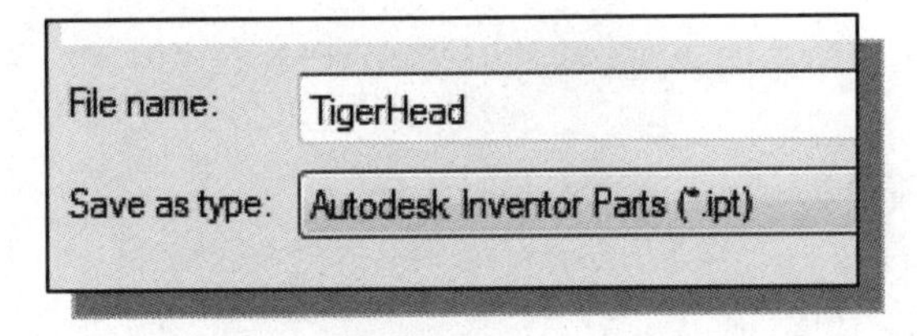

2. In the *File name* editor box, enter **TigerHead** as the file name.

3. Click on the **Save** button to save the file.

❖ You should form a habit of saving your work periodically, just in case something might go wrong while you are working on it. In general, you should save your work at an interval of every 15 to 20 minutes. You should also save before making any major modifications to the model.

Questions:

1. What is the first thing we should set up in *Autodesk Inventor* when creating a new model?

2. Describe the general *parametric modeling* procedure.

3. Describe the general guidelines in creating *rough sketches*.

4. List two of the geometric constraint symbols used by *Autodesk Inventor*.

5. What was the first feature we created in this lesson?

6. How many solid features were created in the tutorial?

7. How do we control the size of a feature in parametric modeling?

8. Which command was used to create the last cut feature in the tutorial? How many dimensions do we need to fully describe the cut feature?

9. List and describe three differences between parametric modeling and traditional 2D Computer Aided Drafting techniques.

Exercises:

1. **Inclined Support** (Thickness: **.5**)

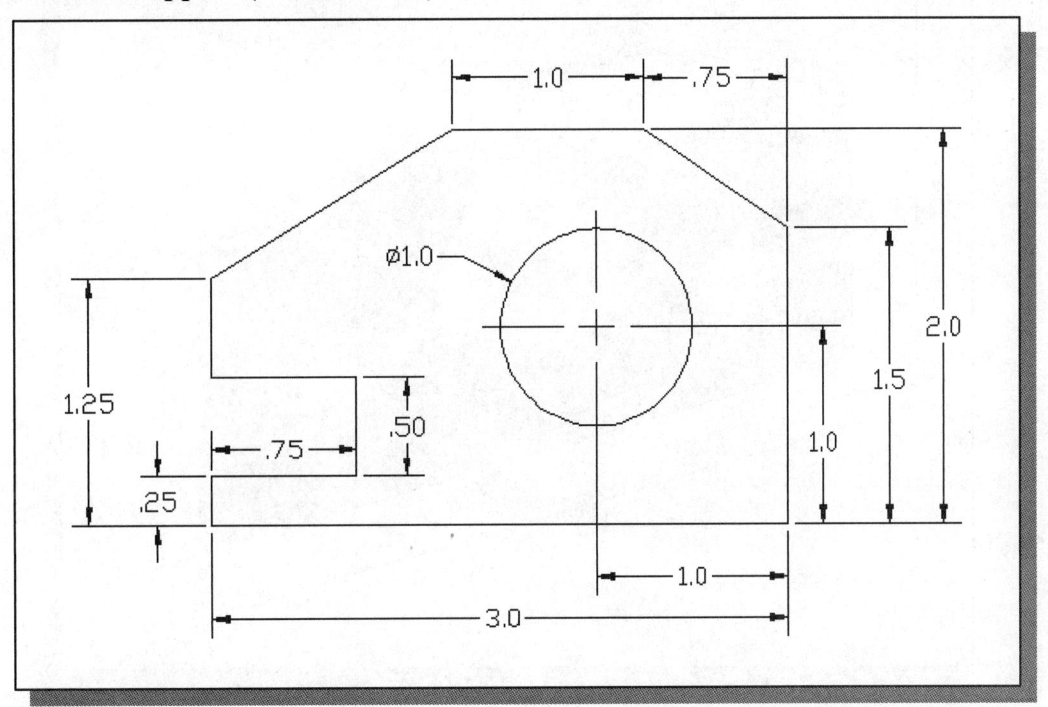

2. **Spacer Plate** (Thickness: **.125**)

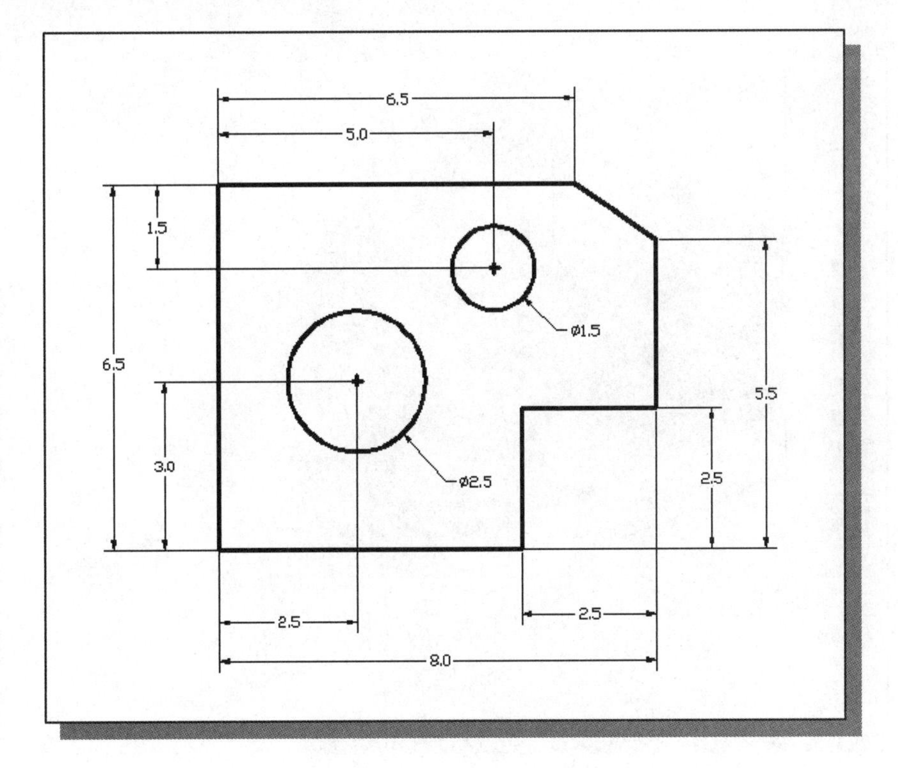

3. **Positioning Stop**

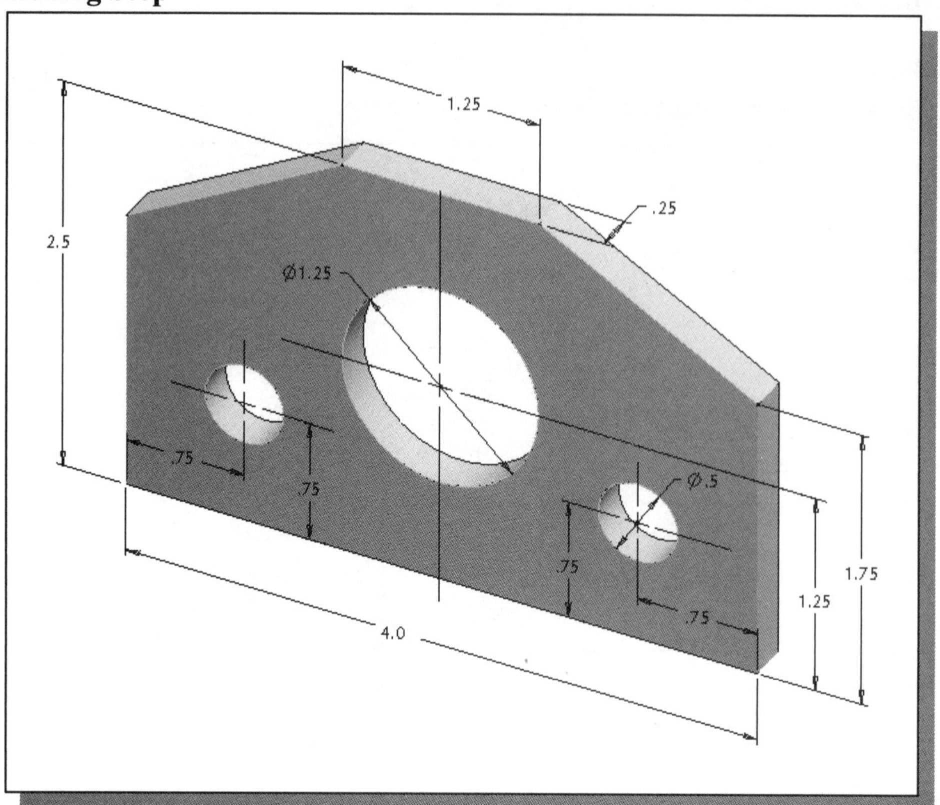

4. **Guide Block**

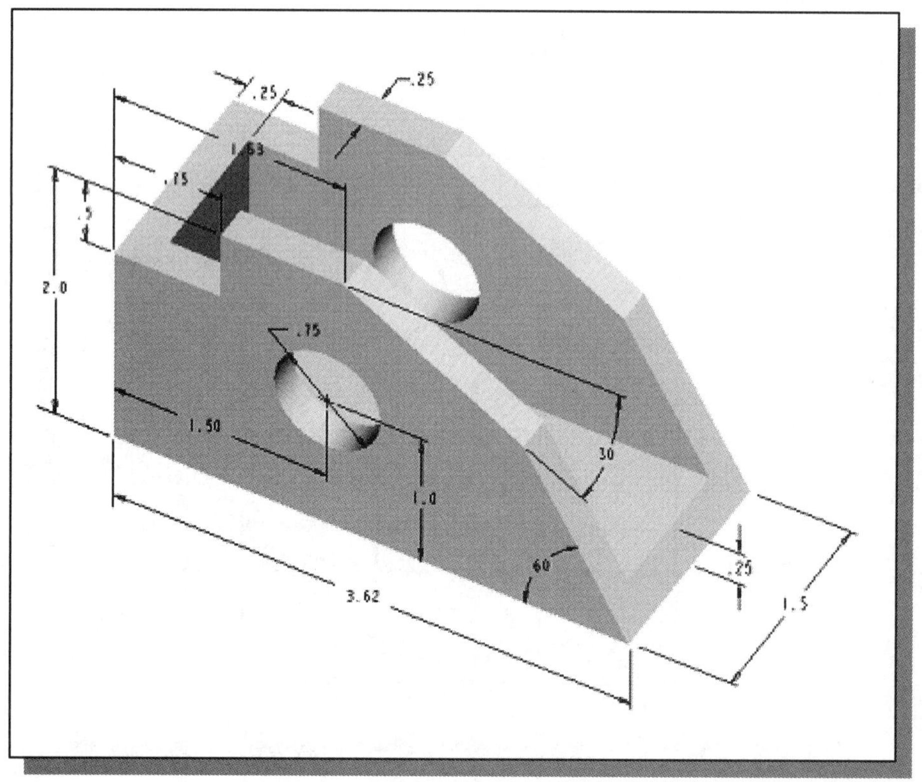

Chapter 3

CSG Concepts and Model History Tree

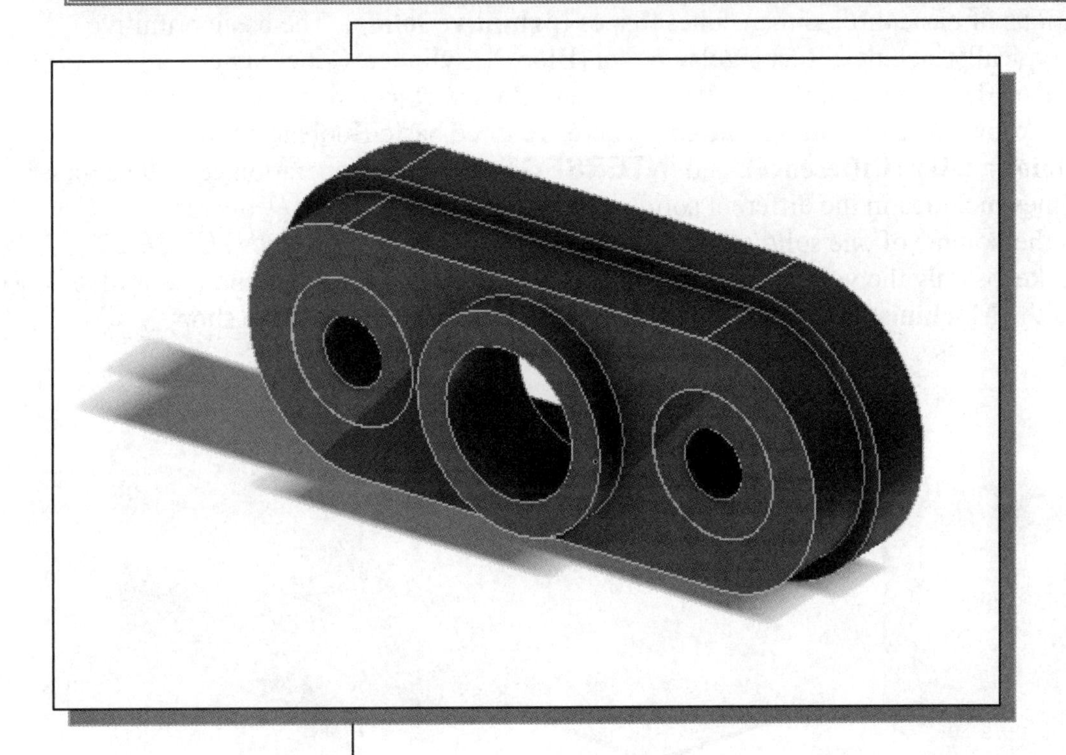

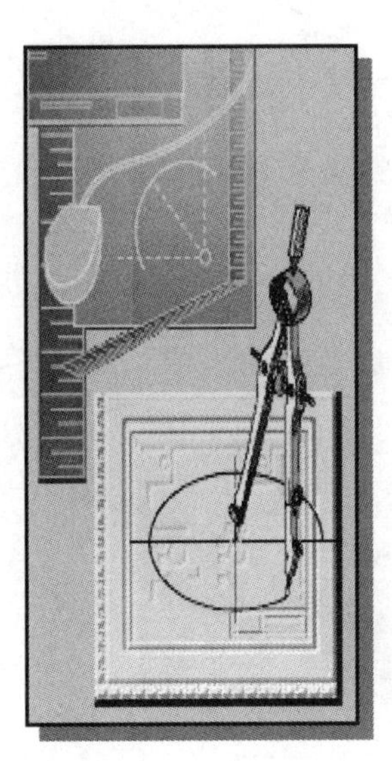

Learning Objectives

- ♦ Understand Feature Interactions
- ♦ Use the Part Browser
- ♦ Modify and Update Feature Dimensions
- ♦ Perform History-Based Part Modifications
- ♦ Change the Names of Created Features
- ♦ Implement Basic Design Changes
- ♦ Calculate the Physical Properties

Introduction

In the 1980s, one of the main advancements in **solid modeling** was the development of the **Constructive Solid Geometry** (CSG) method. CSG describes the solid model as combinations of basic three-dimensional shapes (**primitive solids**). The basic primitive solid set typically includes: Rectangular-prism (Block), Cylinder, Cone, Sphere, and Torus (Tube). Two solid objects can be combined into one object in various ways using operations known as **Boolean operations**. There are three basic Boolean operations: **JOIN (Union)**, **CUT (Difference)**, and **INTERSECT**. The *JOIN* operation combines the two volumes included in the different solids into a single solid. The *CUT* operation subtracts the volume of one solid object from the other solid object. The *INTERSECT* operation keeps only the volume common to both solid objects. The CSG method is also known as the **Machinist's Approach**, as the method is parallel to machine shop practices.

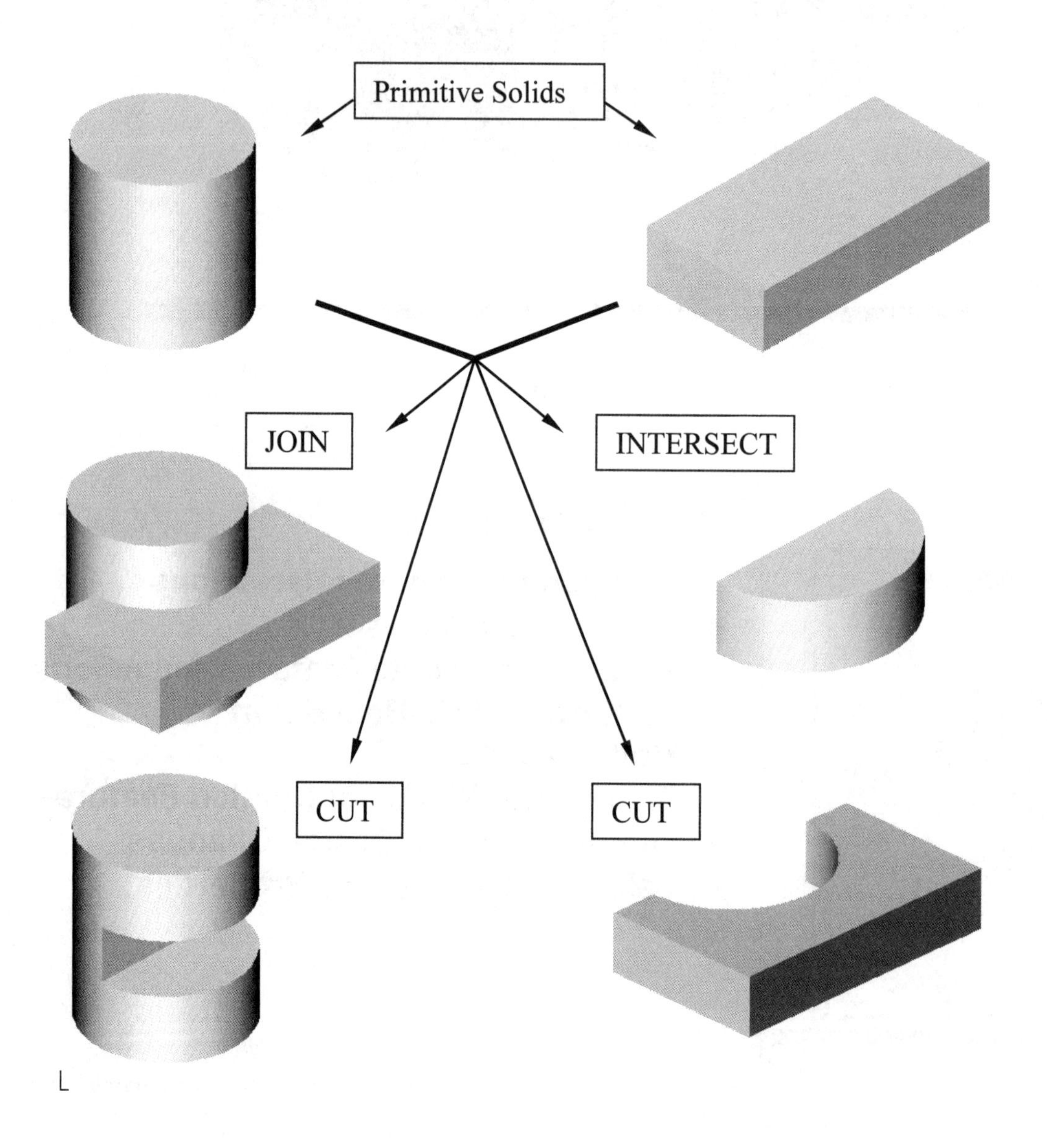

Binary Tree

The CSG is also referred to as the method used to store a solid model in the database. The resulting solid can be easily represented by what is called a **binary tree**. In a binary tree, the terminal branches (leaves) are the various primitives that are linked together to make the final solid object (the root). The binary tree is an effective way to keep track of the *history* of the resulting solid. By keeping track of the history, the solid model can be re-built by re-linking through the binary tree. This provides a convenient way to modify the model. We can make modifications at the appropriate links in the binary tree and re-link the rest of the history tree without building a new model.

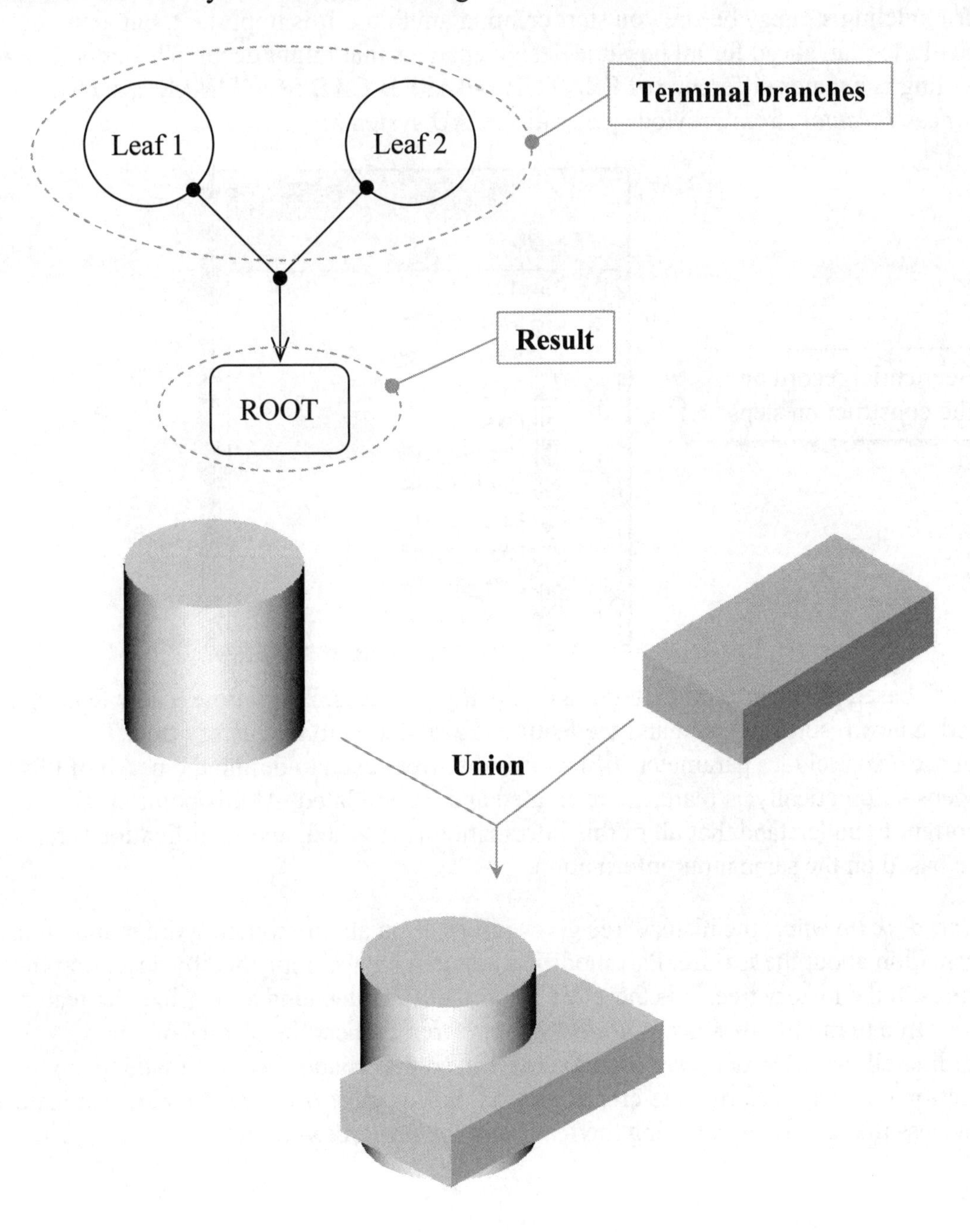

Model History Tree

In *Autodesk Inventor*, the **design intents** are embedded into features in the **history tree.** The structure of the model history tree resembles that of a **CSG binary tree**. A CSG binary tree contains only *Boolean relations*, while the ***Autodesk Inventor* history tree** contains all features, including *Boolean relations*. A history tree is a sequential record of the features used to create the part. This history tree contains the construction steps, plus the rules defining the design intent of each construction operation. In a history tree, each time a new modeling event is created, previously defined features can be used to define information such as size, location, and orientation. It is therefore important to think about your modeling strategy before you start creating anything. It is important, but also difficult, to plan ahead for all possible design changes that might occur. This approach in modeling is a major difference of **FEATURE-BASED CAD SOFTWARE**, such as *Autodesk Inventor*, from previous generation CAD systems.

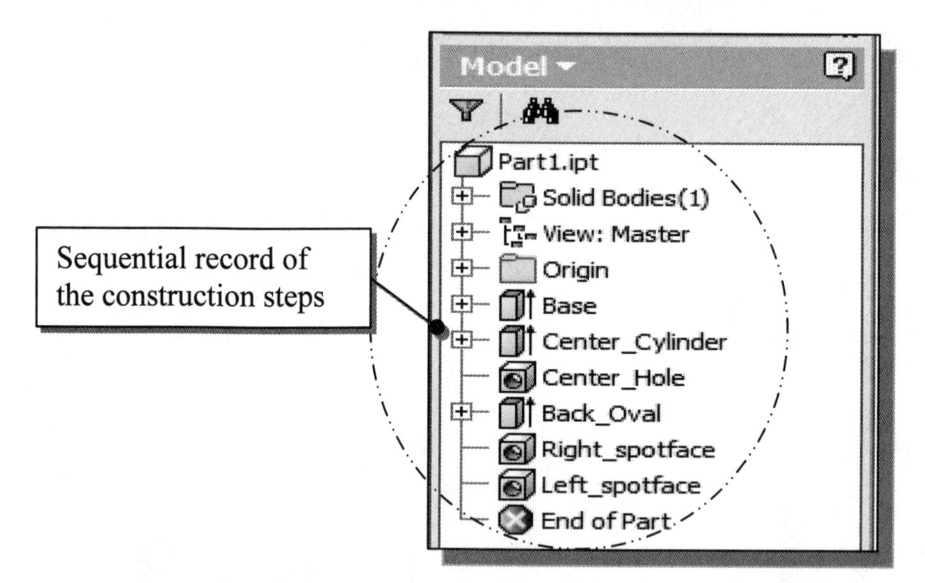

Feature-based parametric modeling is a cumulative process. Every time a new feature is added, a new result is created and the feature is also added to the history tree. The database also includes parameters of features that were used to define them. All of this happens automatically as features are created and manipulated. At this point, it is important to understand that all of this information is retained, and modifications are done based on the same input information.

In *Autodesk Inventor*, the history tree gives information about modeling order and other information about the feature. Part modifications can be accomplished by accessing the features in the history tree. It is therefore important to understand and utilize the feature history tree to modify designs. *Autodesk Inventor* remembers the history of a part, including all the rules that were used to create it, so that changes can be made to any operation that was performed to create the part. In *Autodesk Inventor*, to modify a feature, we access the feature by selecting the feature in the ***browser*** window.

The A6-Knee Part

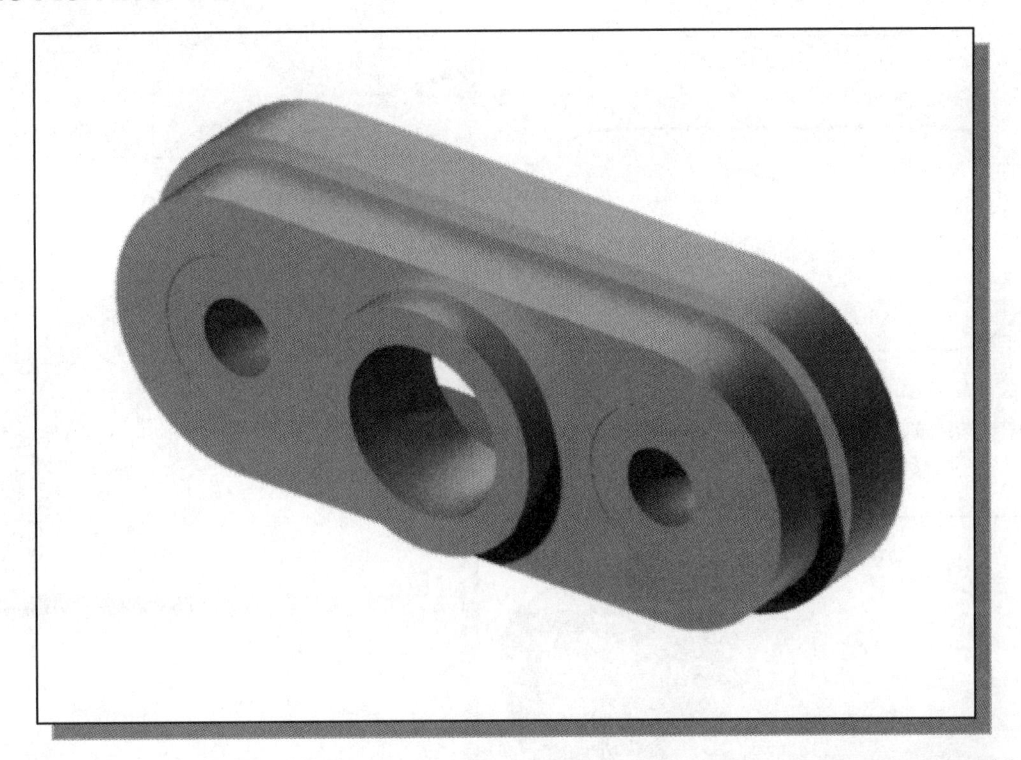

❖ Based on your knowledge of *Autodesk Inventor* so far, how many features would you use to create the design? Which feature would you choose as the **BASE FEATURE**, the first solid feature, of the model? What is your choice in arranging the order of the features? Take a few minutes to consider these questions and do preliminary planning by sketching on a piece of paper. You are also encouraged to create the model on your own prior to following through the tutorial.

Starting *Autodesk Inventor*

1. Select the **Autodesk Inventor** option on the *Start* menu or select the **Autodesk Inventor** icon on the desktop to start *Autodesk Inventor*. The *Autodesk Inventor* main window will appear on the screen.

2. Once the program is loaded into memory, select the **New File** icon with a single click of the left-mouse-button in the *Launch* toolbar.

3. Select the **English** tab and in the *Template* area, then select **Standard(in).ipt**.

4. Pick **Create** in the *New File* dialog box to accept the selected settings.

Modeling Strategy

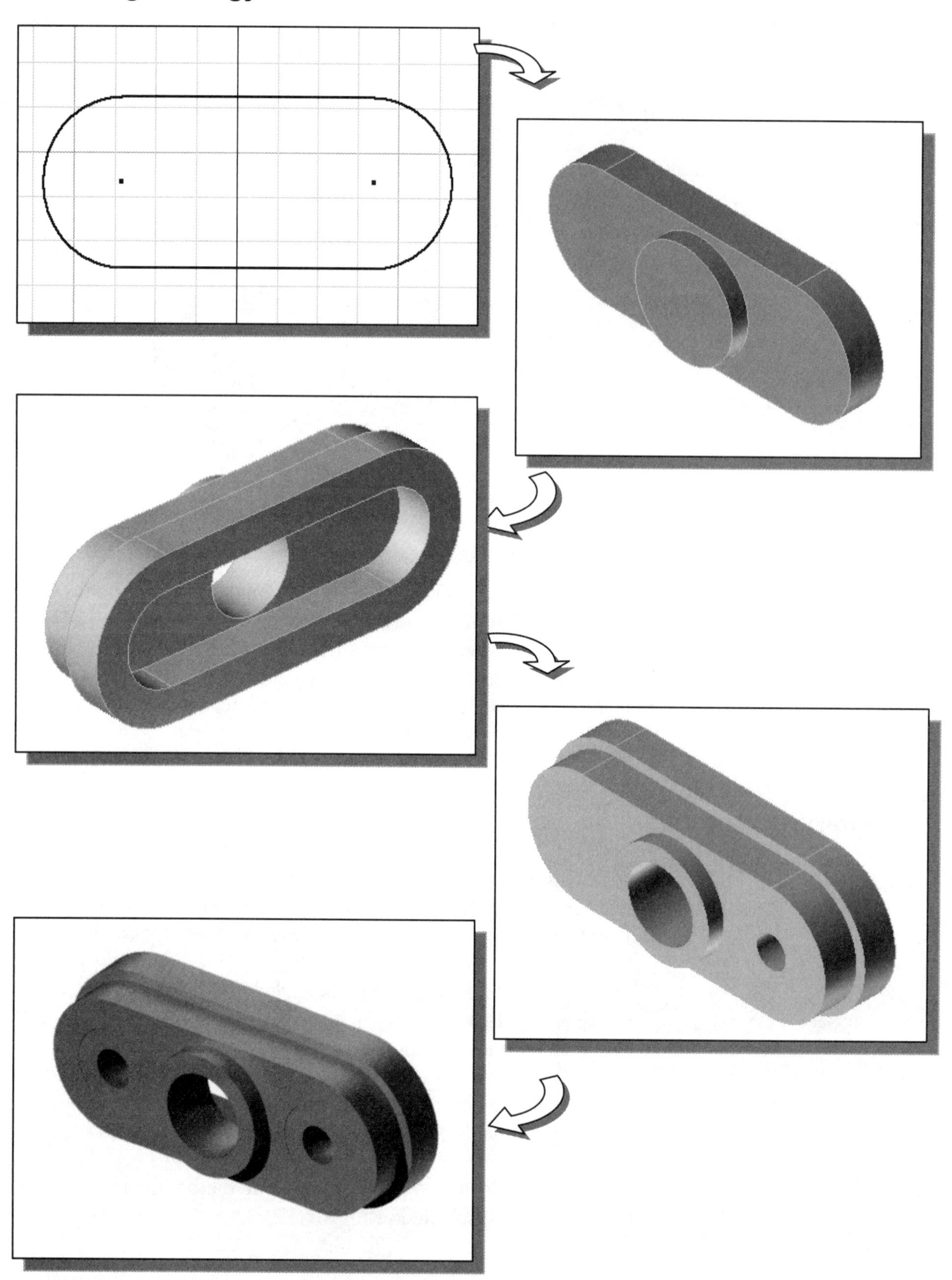

The *Autodesk Inventor* Browser

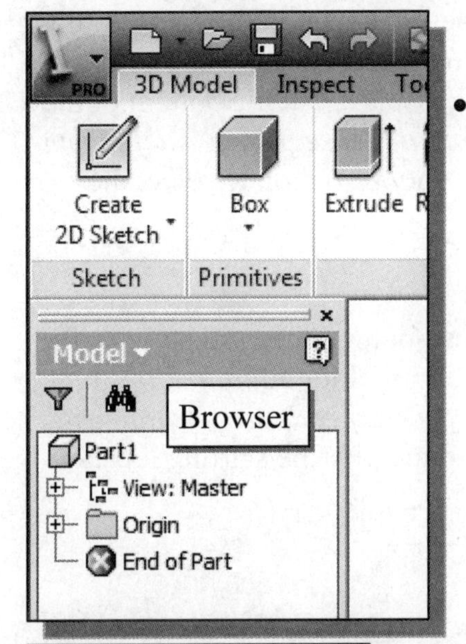

- In the *Autodesk Inventor* screen layout, the ***browser*** is located to the left of the graphics window. *Autodesk Inventor* can be used for part modeling, assembly modeling, part drawings, and assembly presentation. The *browser* window provides a visual structure of the features, constraints, and attributes that are used to create the part, assembly, or scene. The *browser* also provides right-click menu access for tasks associated specifically with the part or feature, and it is the primary focus for executing many of the *Autodesk Inventor* commands.

- The first item displayed in the *browser* is the name of the part, which is also the file name. By default, the name "Part1" is used when we first started *Autodesk Inventor*. The *browser* can also be used to modify parts and assemblies by moving, deleting, or renaming items within the hierarchy. Any changes made in the *browser* directly affect the part or assembly and the results of the modifications are displayed on the screen instantly. The *browser* also reports any problems and conflicts during the modification and updating procedure.

Create the Base Feature

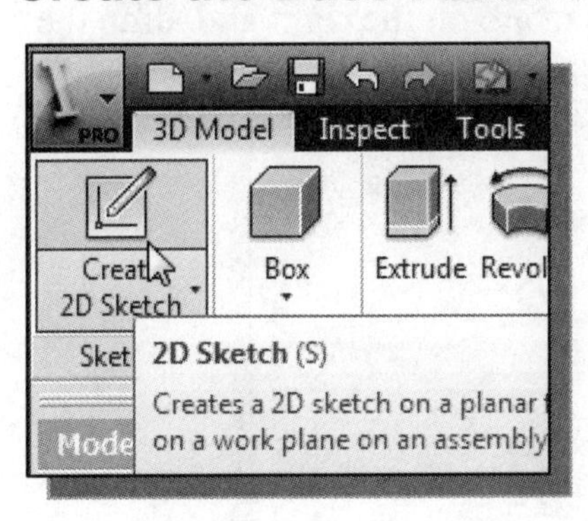

1. Move the graphics cursor to the **Line** icon in the *2D Sketch* toolbar. A *Help-tip box* appears next to the cursor and a brief description of the command is displayed at the bottom of the drawing screen: "*Creates Straight line segments and tangent arcs.*" Click once with the left-mouse-button to select the command.

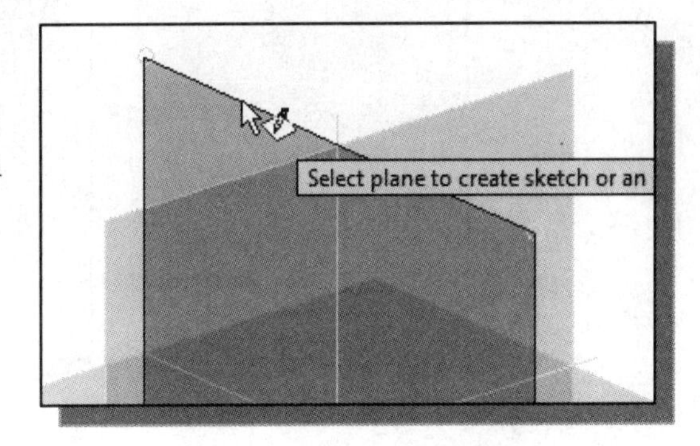

2. Move the cursor over the edge of the *XY Plane* in the graphics area. When the *XY Plane* is highlighted, click once with the **left-mouse-button** to select the *Plane* as the sketch plane for the new sketch.

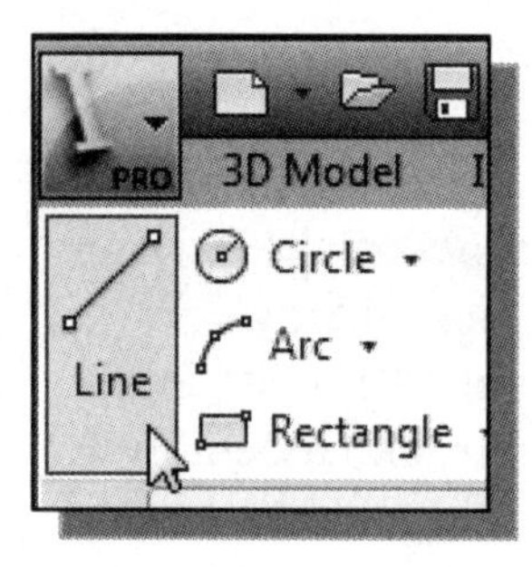

3. Move the graphics cursor to the **Line** icon in the *Draw* toolbar. A *Help tip* box appears next to the cursor and a brief description of the command is displayed at the bottom of the drawing screen: "*Creates Straight line segments and tangent arcs.*" Click once with the left-mouse-button to select the command.

4. Select the icon by clicking once with the left-mouse-button; this will activate the Line command. In the *Status Bar* area, near the bottom of the *Autodesk Inventor* drawing screen, the message "*Specify start point, drag off endpoint for tangent arcs*" is displayed. *Autodesk Inventor* expects us to identify the starting location of a straight line.

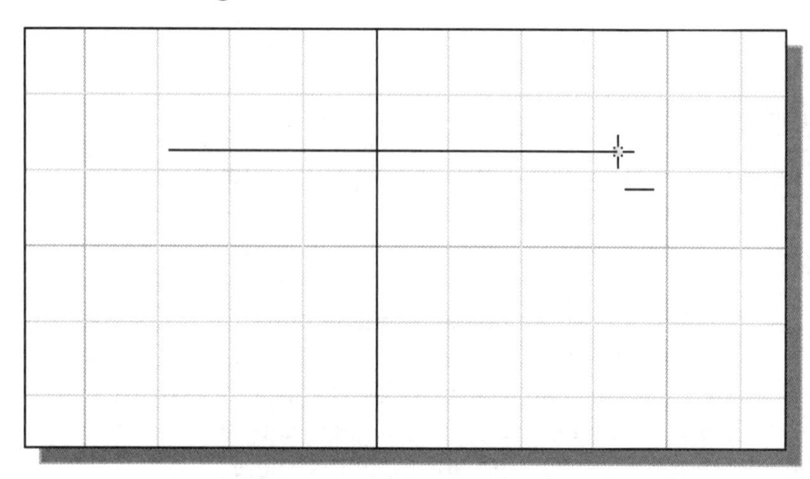

5. On your own, create a **horizontal line** from left to right as shown. Do not exit the Line command yet.

6. Move the cursor on top of the last point of the sketch and **drag** with the **left mouse button** to activate the **Arc** option. Create an arc on the right and align the other end of the arc as shown.

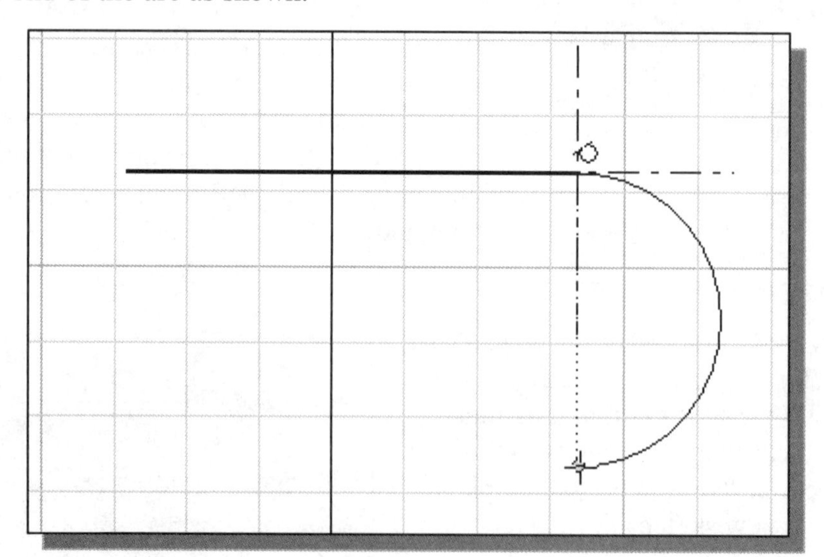

7. Create another horizontal line aligned to the arc as shown.

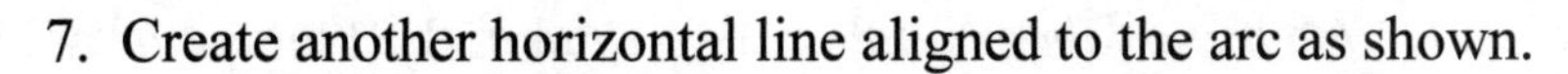

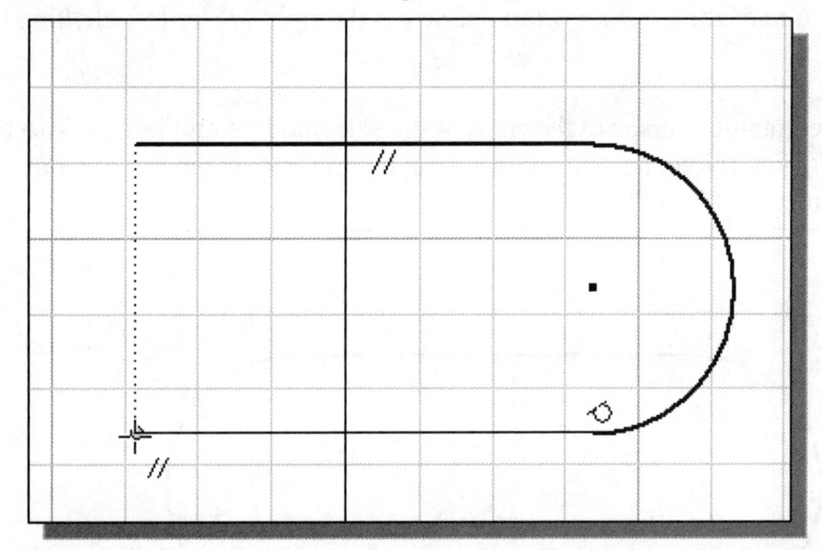

8. Move the cursor on top of the last point of the sketch and drag with the left-mouse-button to activate the **Arc** option. Create another arc on the left to form a closed region as shown.

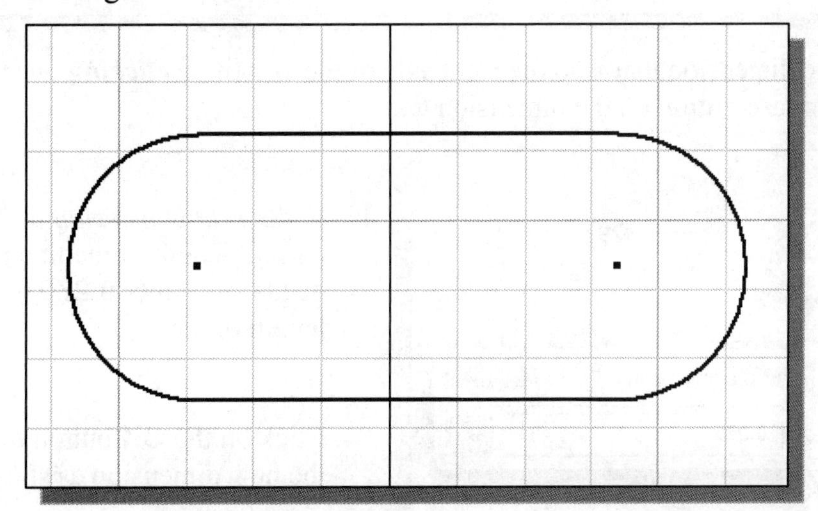

9. Activate the **General Dimension** command by clicking once with the left-mouse-button. The General Dimension command allows us to quickly create and modify dimensions.

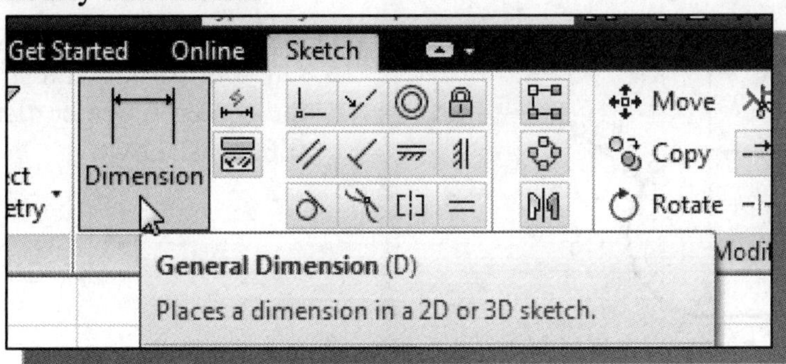

10. The message "*Select Geometry to Dimension*" is displayed in the *Status Bar* area at the bottom of the *Inventor* window. Select the right arc by left-clicking once on the arc.

11. Move the graphics cursor to the right of the selected line and left-click to place the dimension. (Note that the value displayed on your screen might be different than what is shown in the figure below.)

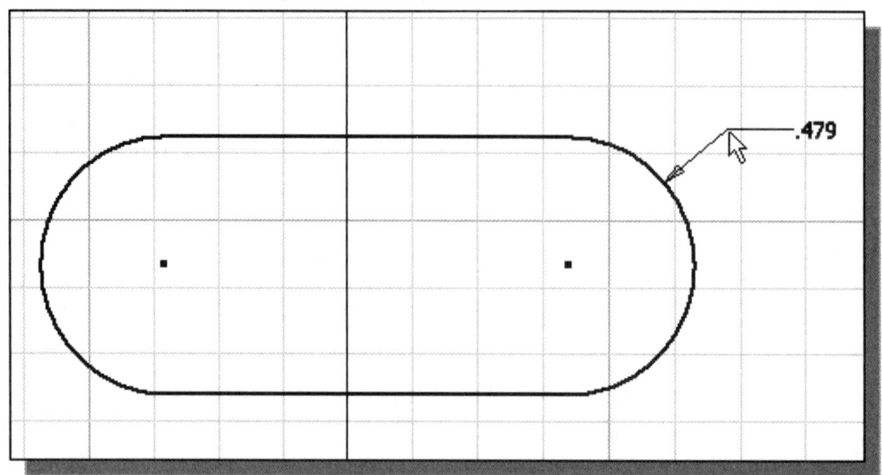

12. Select the dimension that is to the right side of the sketch by ***clicking*** once with the left-mouse-button on the dimension text

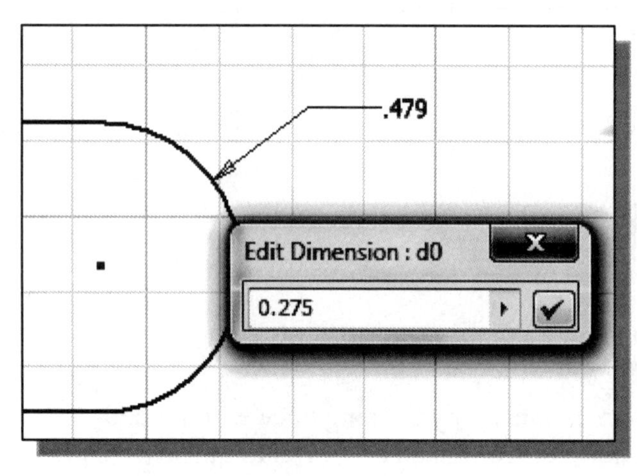

13. In the *Edit Dimension* box, the current length of the line is displayed. Enter **0.275** to set the length of the line.

14. Click on the **OK** button to accept the new dimension as shown.

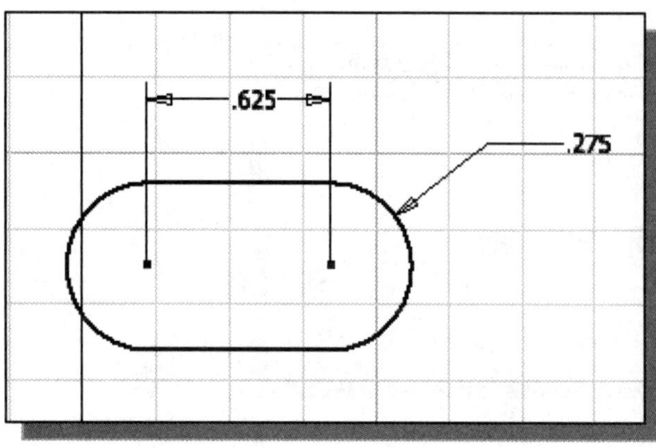

15. On your own, create and modify the center to center distance to **0.625** as shown.

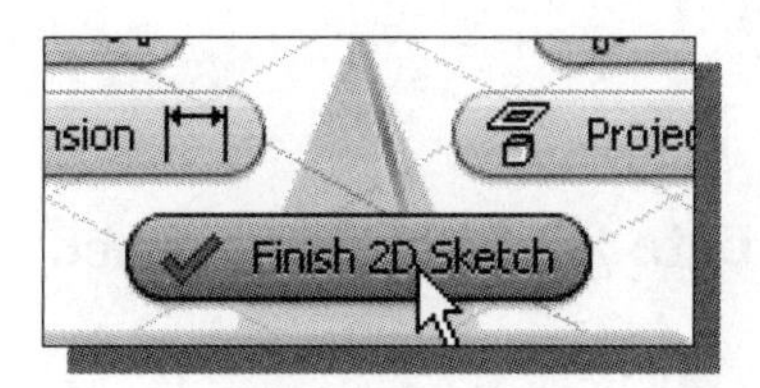

16. Inside the graphics window, click once with the right-mouse-button to display the option menu. Select **Finish 2D Sketch** in the popup menu to end the Sketch option.

17. In the *Create* toolbar (the toolbar that is located to the right side of the *Sketch* toolbar in the *Ribbon*), select the **Extrude** command by releasing the left-mouse-button on the icon.

18. In the *Distance* option box, enter **.081** as the total extrusion distance.

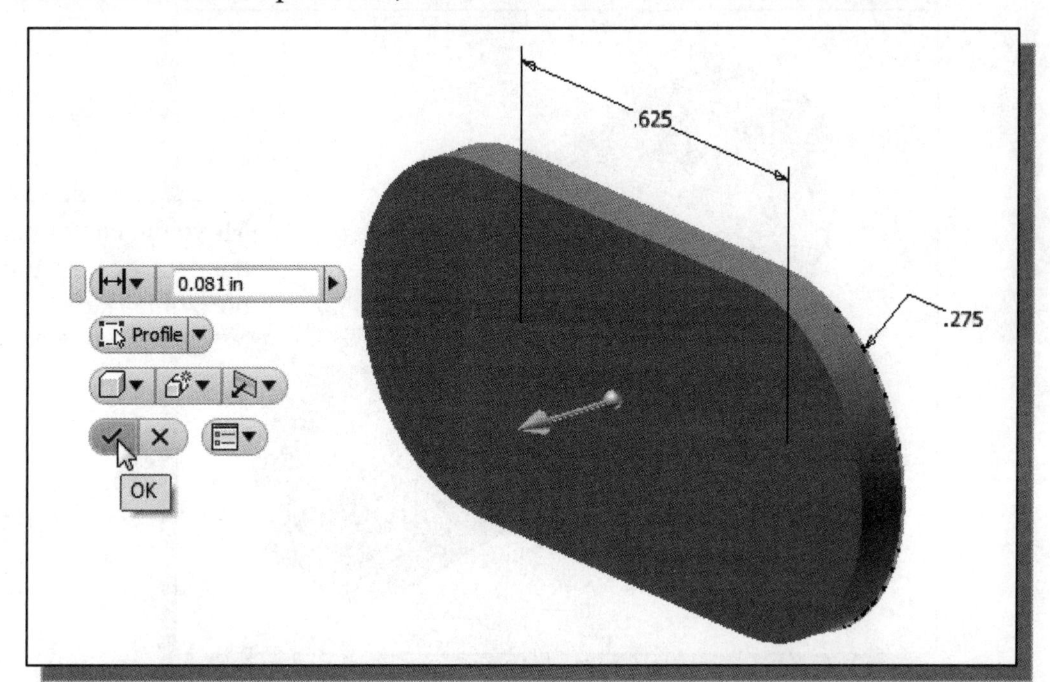

19. Click on the **OK** button to accept the settings and create the base feature.

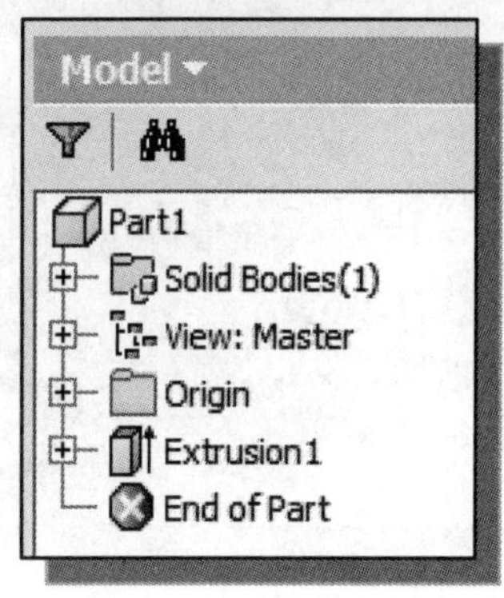

- On your own, use the *Dynamic Viewing* functions to view the 3D model. Also note the extrusion feature is added to the *Model Tree* in the *browser* area.

Add the Second Solid Feature

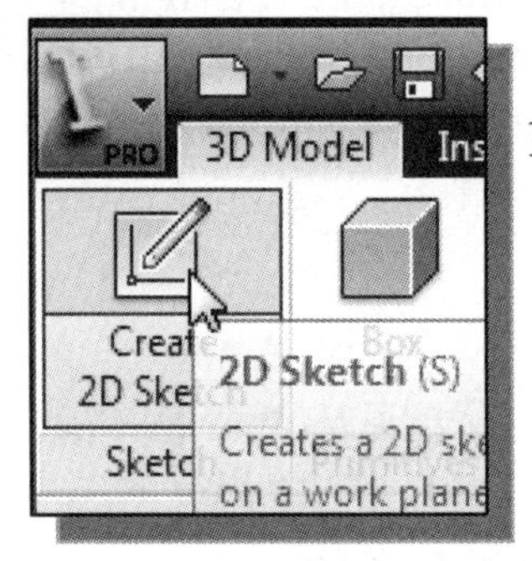

1. In the *Sketch* toolbar select the **Create 2D Sketch** command by left-clicking once on the icon.

2. In the *Status Bar* area, the message: "*Select face, workplane, sketch or sketch geometry.*" is displayed. Move the graphics cursor on the 3D part and notice that *Autodesk Inventor* will automatically highlight feasible planes and surfaces as the cursor is on top of the different surfaces. Move the cursor inside the front face of the 3D object as shown below.

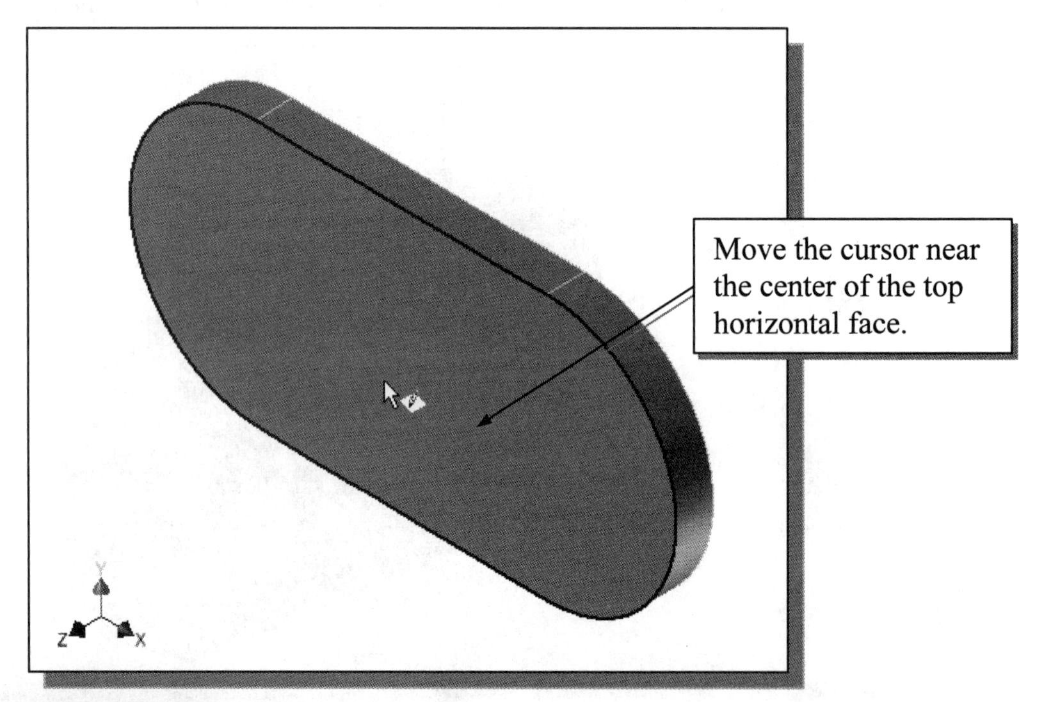

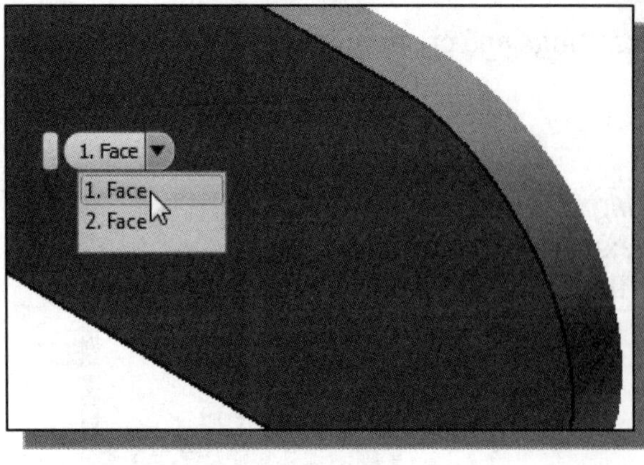

3. Pause the cursor on the surface until a selection list appears, click on the down arrow to examine all possible surface selections.

4. Select the **front face** of the solid model when it is highlighted as shown in the figure.

Create a 2D Sketch

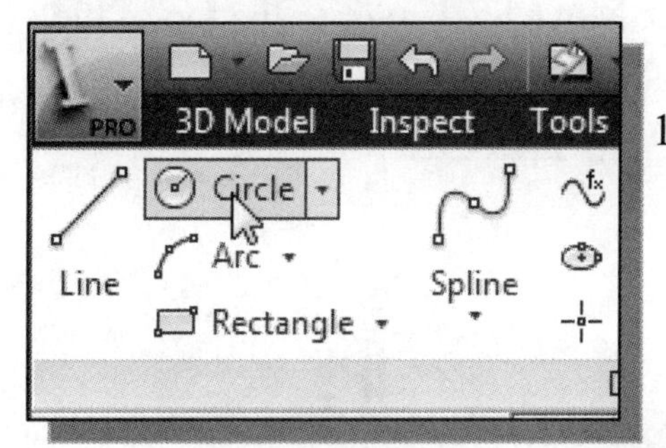

1. Select the **Center point circle** command by clicking once with the left-mouse-button on the icon in the *Draw* toolbar.

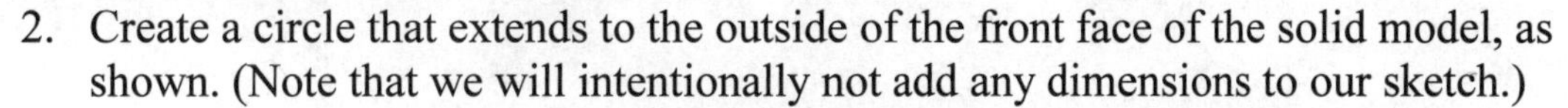

2. Create a circle that extends to the outside of the front face of the solid model, as shown. (Note that we will intentionally not add any dimensions to our sketch.)

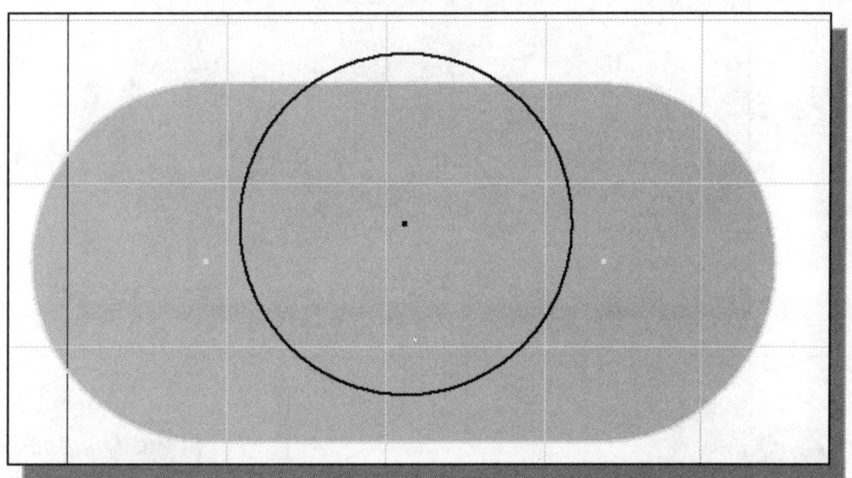

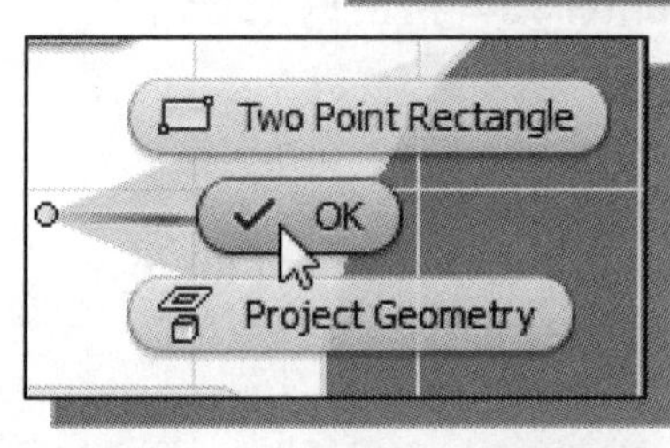

3. Inside the graphics window, click once with the right-mouse-button to display the option menu. Select **OK** in the popup menu to end the Circle command.

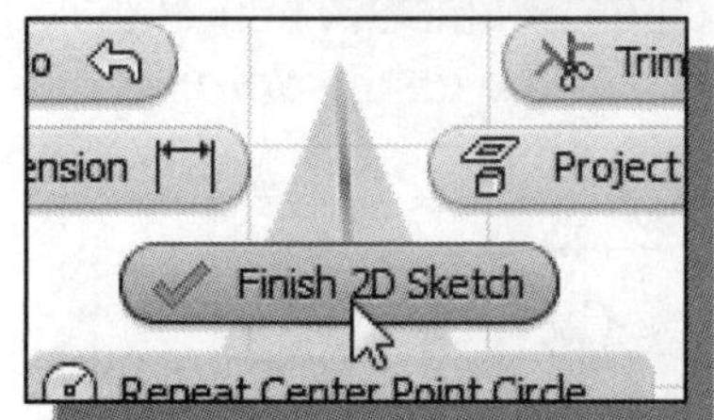

4. Inside the graphics window, click once with the right-mouse-button to display the option menu. Select **Finish Sketch** in the popup menu to end the Sketch option.

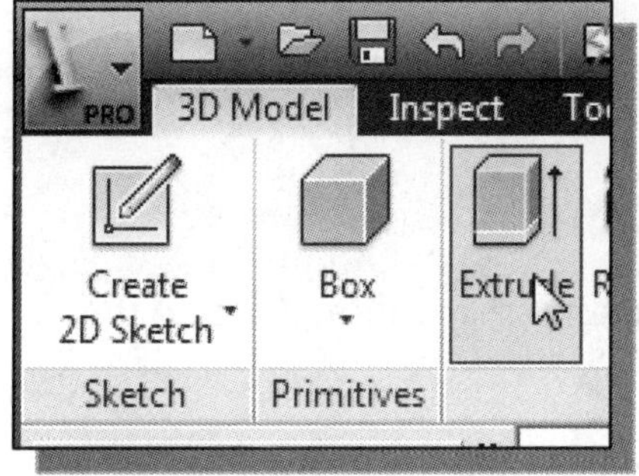

5. In the *Create* toolbar (the toolbar that is located to the right side of the *Sketch* toolbar in the *Ribbon*), select the **Extrude** command by releasing the left-mouse-button on the icon.

6. In the *Extrude* popup window, the **Profile** option is activated; *Autodesk Inventor* expects us to identify the profile to be extruded. Move the cursor to the top of the circle we just created and left-click once to select the region as the profile to be extruded.

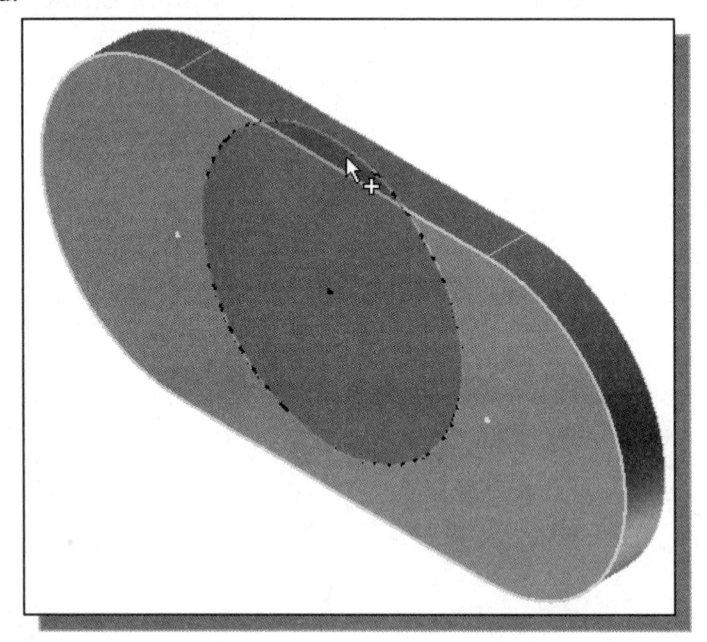

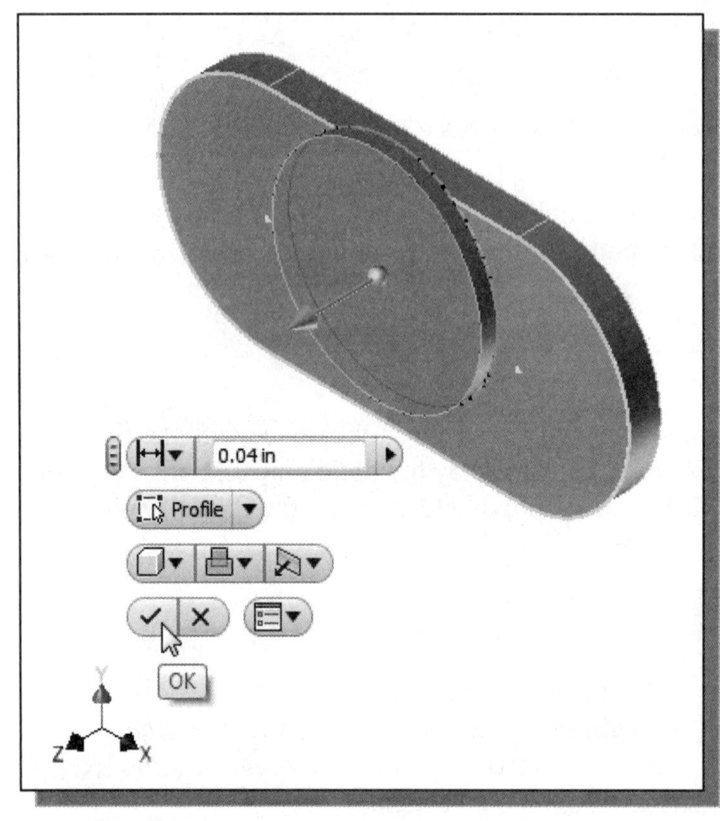

7. In the *Distance* option box, enter **.04** as the extrusion distance.

8. Confirm the extrusion direction is toward the front side as shown.

9. Click on the **OK** button to proceed with the *Join* operation.

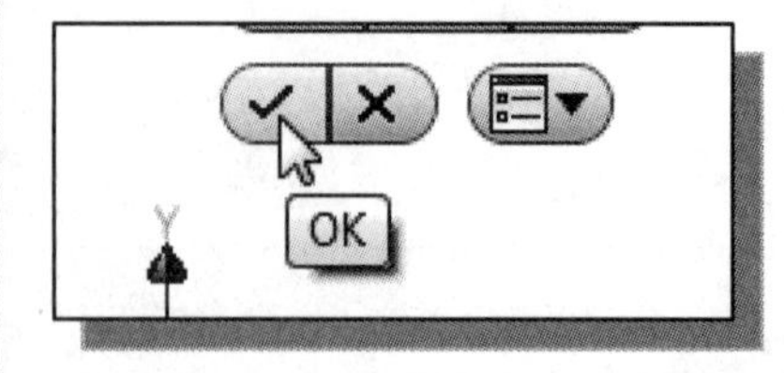

➢ Note that the cylindrical feature is created without adding any dimension to the 2D sketch.

Rename the Part Features

- Currently, our model contains two extruded features. The feature is highlighted in the display area when we select the feature in the *browser* window. Each time a new feature is created, the feature is also displayed in the *Model Tree* window. By default, *Autodesk Inventor* will use generic names for part features. However, when we begin to deal with parts with a large number of features, it will be much easier to identify the features using more meaningful names. Two methods can be used to rename the features: (1) **Clicking** twice on the name of the feature, and (2) using the **Properties** option. In this example, the use of the first method is illustrated.

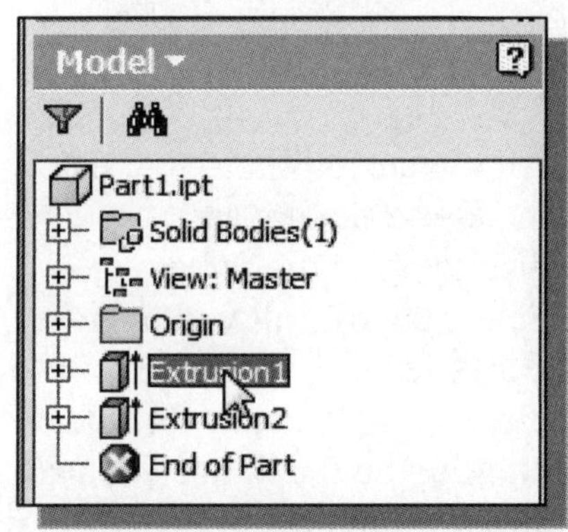

1. Select the first extruded feature in the *model browser* area by left-clicking once on the name of the feature, **Extrusion1**. Notice the selected feature is highlighted in the graphics window.

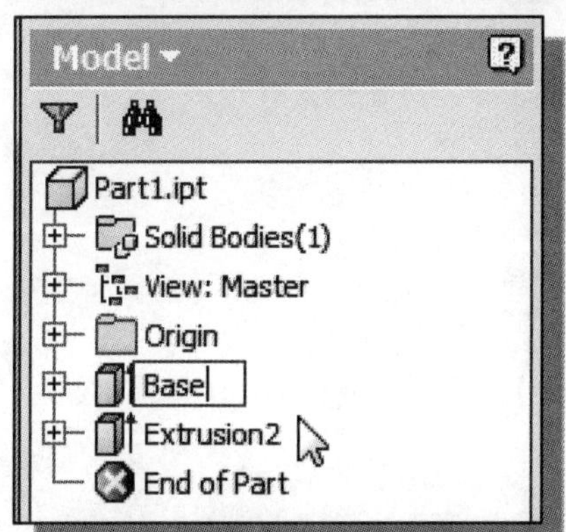

2. Left-mouse-click on the feature name again to enter the *Edit* mode as shown.

3. Enter **Base** as the new name for the first extruded feature.

4. On your own, rename the second extruded feature to **Center_Cylinder**.

Adjusting the Dimensions of the Base Feature

❖ One of the main advantages of parametric modeling is the ease of performing part modifications at any time in the design process. Part modifications can be done through accessing the features in the history tree. *Autodesk Inventor* remembers the history of a part, including all the rules that were used to create it, so that changes can be made to any operation that was performed to create the part.

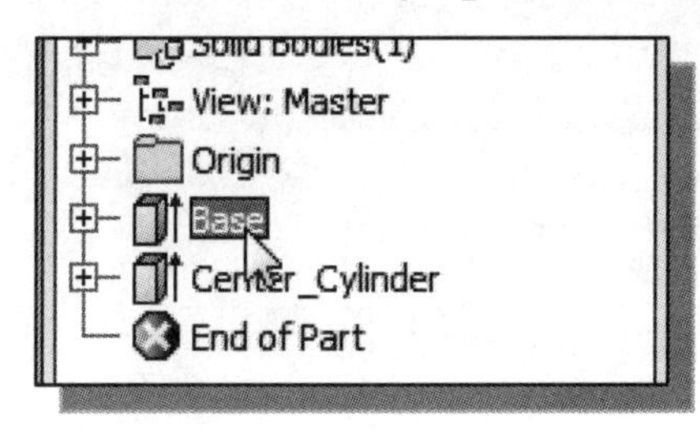

1. Select the first extruded feature, **Base**, in the *browser* area. Notice the selected feature is also highlighted in the graphics window.

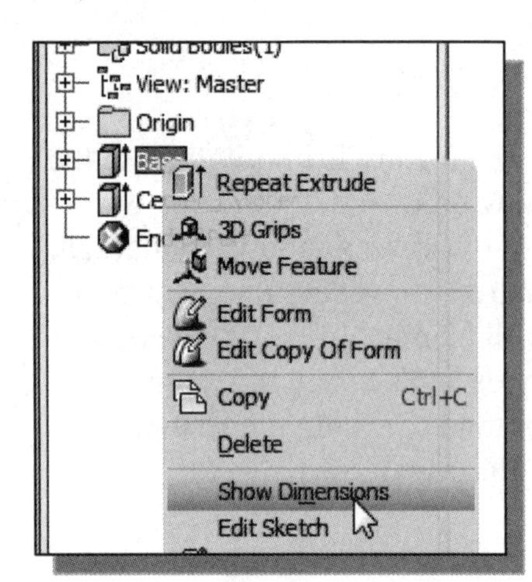

2. Inside the *browser* area, **right-mouse-click** on the first extruded feature to bring up the option menu and select the **Show Dimensions** option in the pop-up menu.

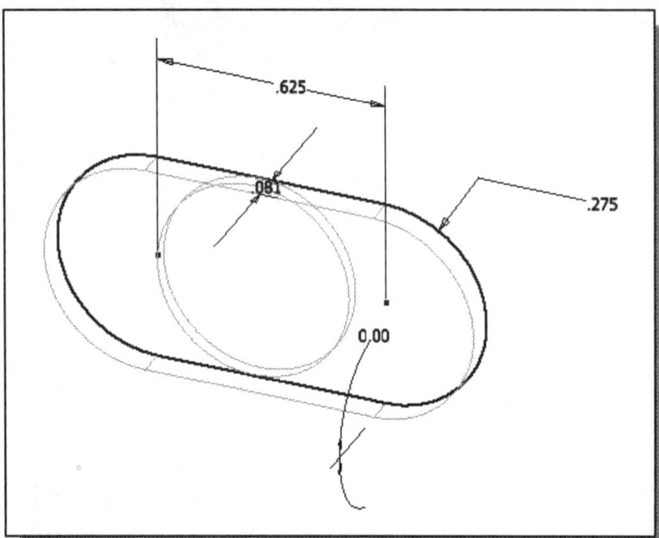

3. All dimensions used to create the **Base** feature are displayed on the screen. Select the overall width of the **Base** feature, the **0.625** dimension value, by **double-clicking** on the dimension text as shown below.

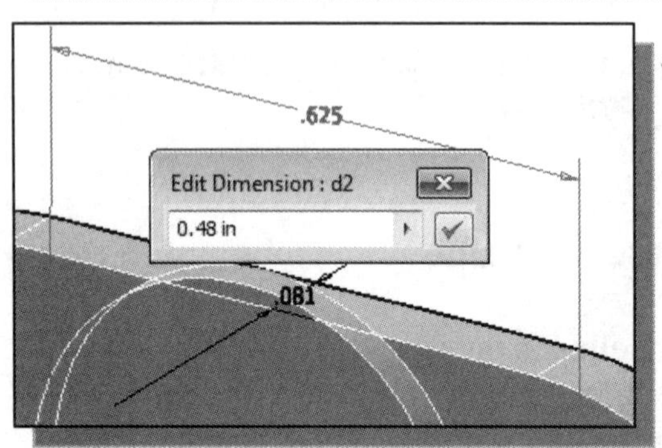

4. Enter **0.48** in the *Edit Dimension* window.

5. On your own, repeat the above steps and modify the radius to half its current value.

6. Enter **0.275/2** in the *Edit Dimension* window. (Note that *Inventor* allows the input of mathematical operations for dimensions.)

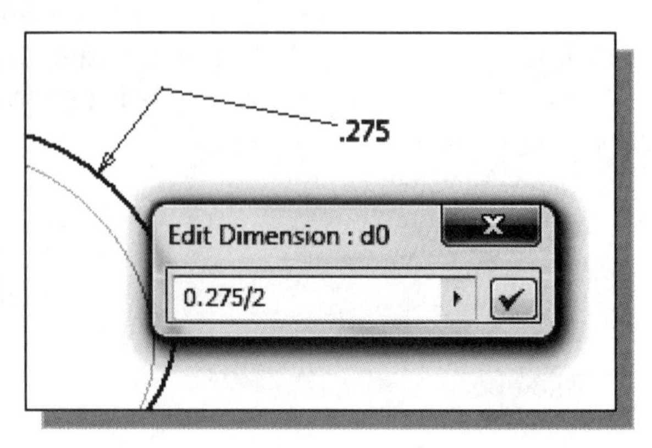

7. The 2D sketch has been updated to reflect the changes, but the solid feature still needs to be updated.

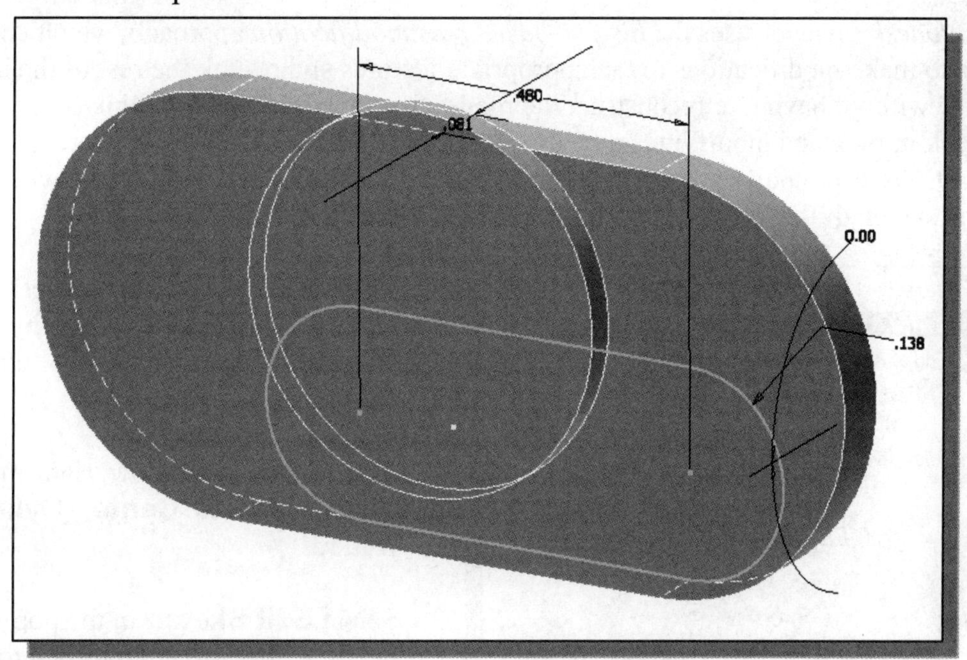

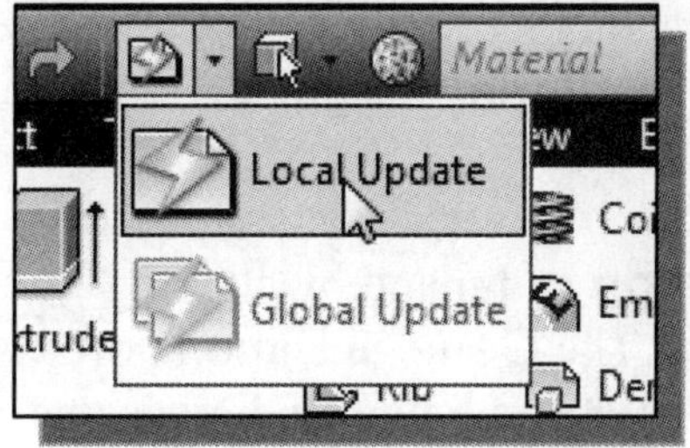

8. Click **Local Update** in the *Quick Access* toolbar.

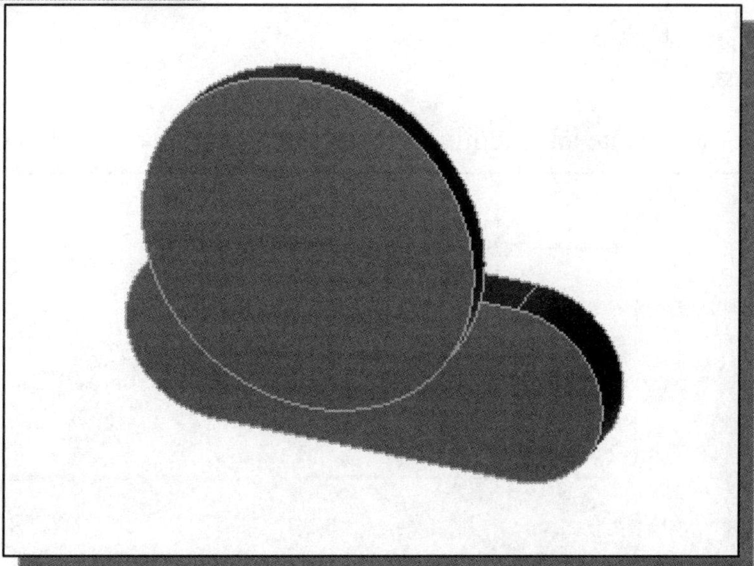

➢ Note that *Autodesk Inventor* updates the model by re-linking all elements used to create the model. Any problems or conflicts that occur will also be displayed during the updating process.

History-Based Part Modifications

- *Autodesk Inventor* uses the *history-based part modification* approach, which enables us to make modifications to the appropriate features and re-link the rest of the history tree without having to reconstruct the model from scratch. We can think of it as going back in time and modifying some aspects of the modeling steps used to create the part. We can modify any feature that we have created. As an example, we will adjust the sketch of the Center_Cylinder feature.

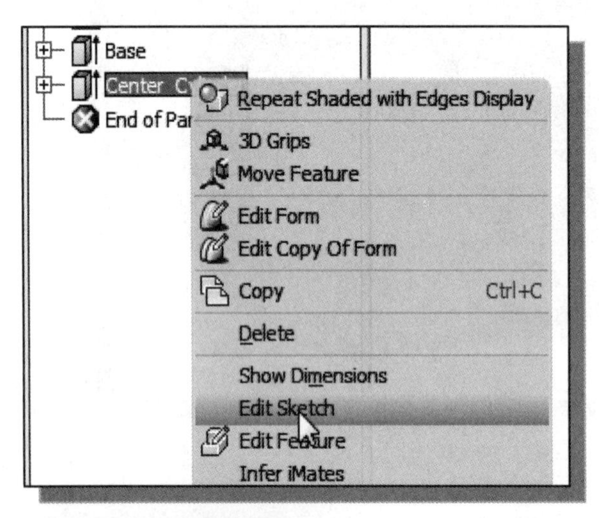

1. In the *browser* window, select the last feature, **Center_Cylinder**, by left-clicking once on the name of the feature.

2. In the *browser* window, right-mouse-click once on the **Center_Cylinder** feature.

3. Select **Edit Sketch** in the pop-up menu. Notice we are returned to the *2D sketch* mode of the **Center_Cylinder** feature.

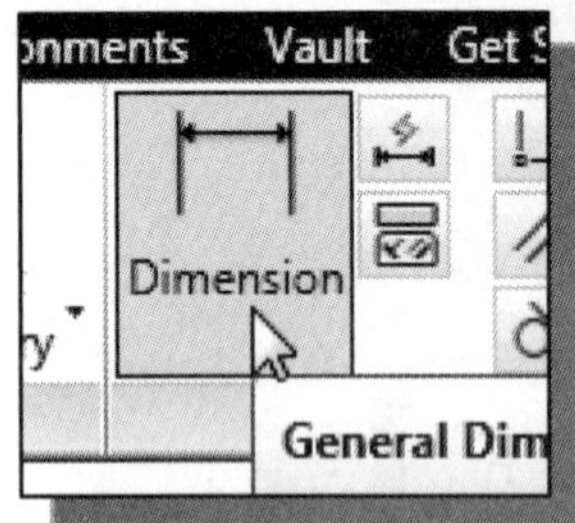

4. Activate the **General Dimension** command by clicking once with the left-mouse-button. The General Dimension command allows us to quickly create and modify dimensions

5. On your own, create and modify the three dimensions as shown in the figure.

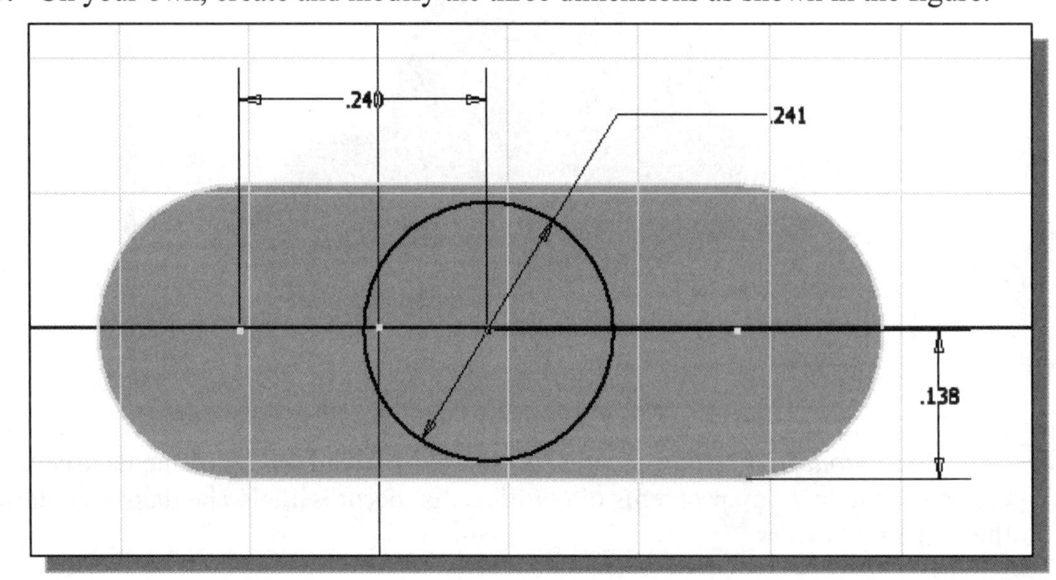

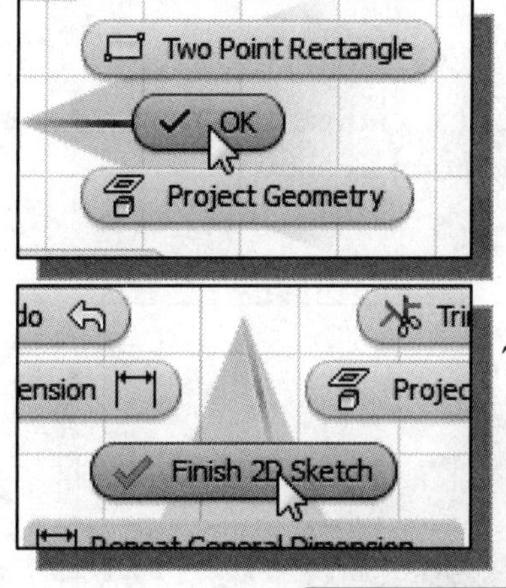

6. Inside the graphics window, click once with the right-mouse-button to display the option menu. Select **OK** in the popup menu to end the **Circle** command.

7. Inside the graphics window, click once with the right-mouse-button to display the option menu. Select **Finish Sketch** in the popup menu to end the **Sketch** option.

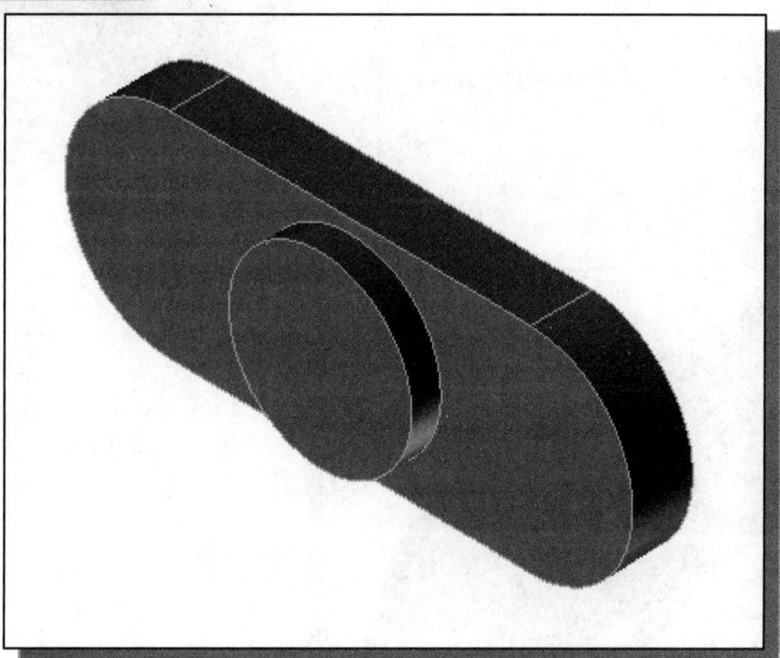

Add a Placed Feature

- In *Autodesk Inventor*, there are two types of geometric features: **placed features** and **sketched features**. The last feature we created is a *sketched feature*, where we created a rough sketch and performed an extrusion operation. We can also create a feature without creating a 2D sketch, which is known as a *placed feature*. A *placed feature* is a feature that does not need a sketch and can be created automatically. In parametric modeling, holes, fillets, chamfers, and shells are all placed features.

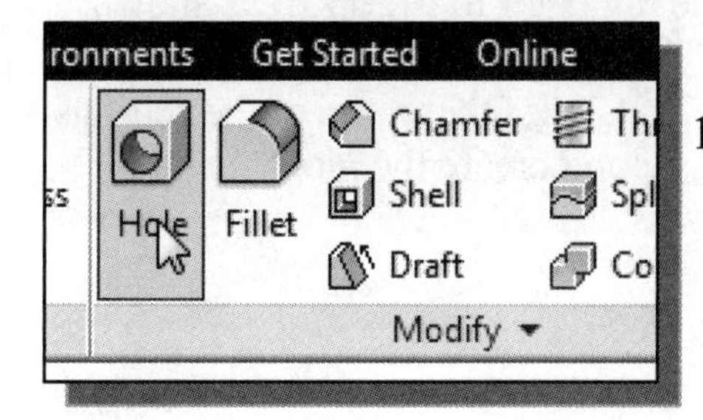

1. In the *Modify* toolbar, select the **Hole** command by clicking the left-mouse-button on the icon.

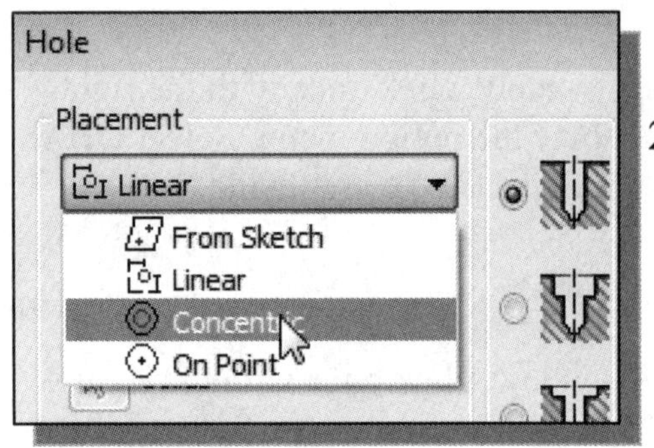

2. In the *Hole* dialog box, choose **Concentric** for the *Placement* option as shown.

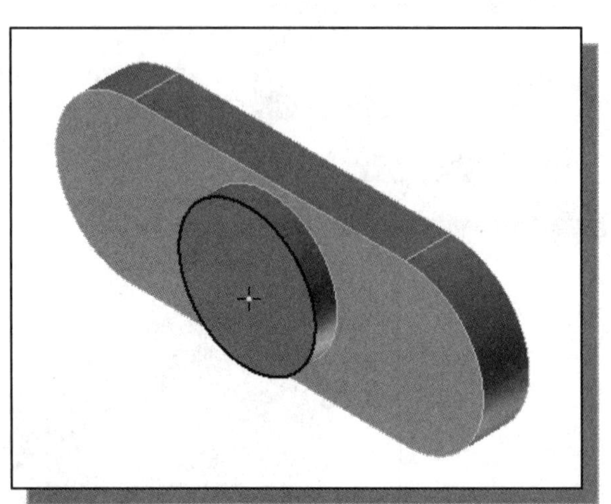

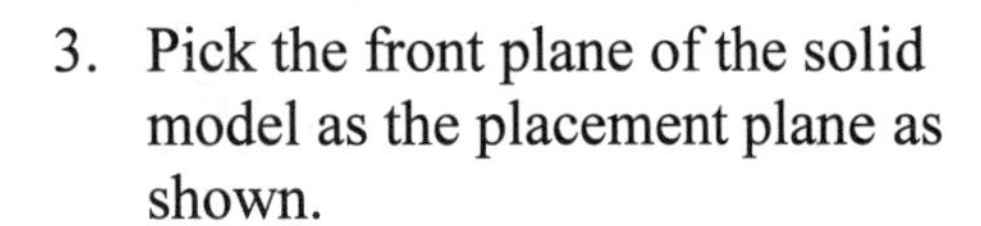

3. Pick the front plane of the solid model as the placement plane as shown.

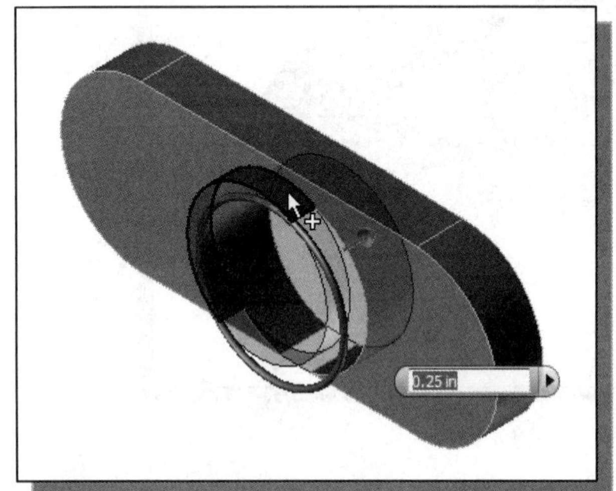

4. Pick the center cylindrical surface to use as the concentric reference.

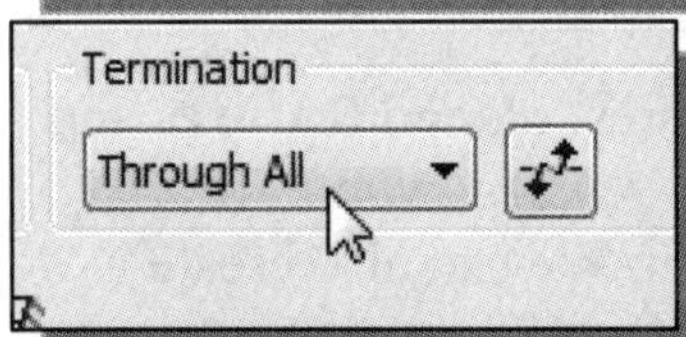

5. Set the *Termination* option to **Through All** as shown.

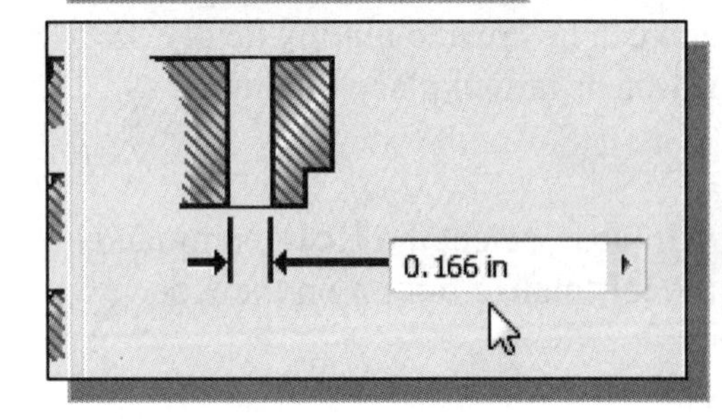

6. Set the hole diameter to **0.166 in** as shown.

7. Click **OK** to accept the settings and create the *Hole* feature.

Using the Offset Command to Create a Feature

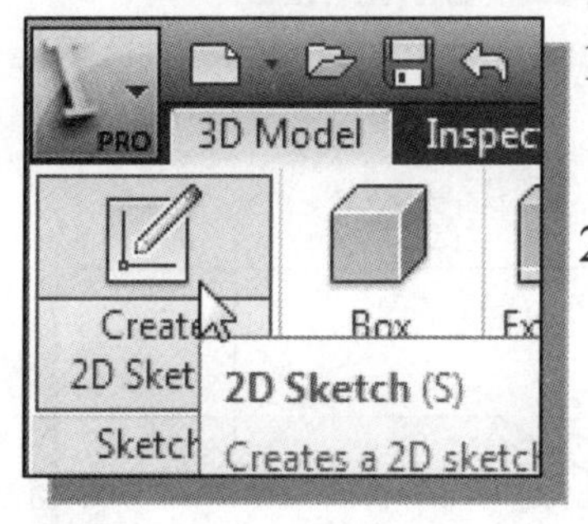

1. In the *Sketch* toolbar select the **Create 2D Sketch** command by left-clicking once on the icon.

2. Pick the **back face** of the solid as shown.

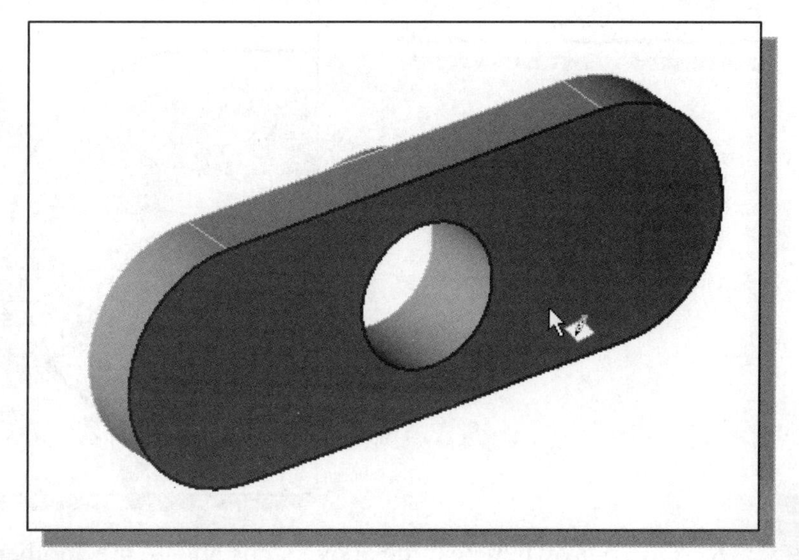

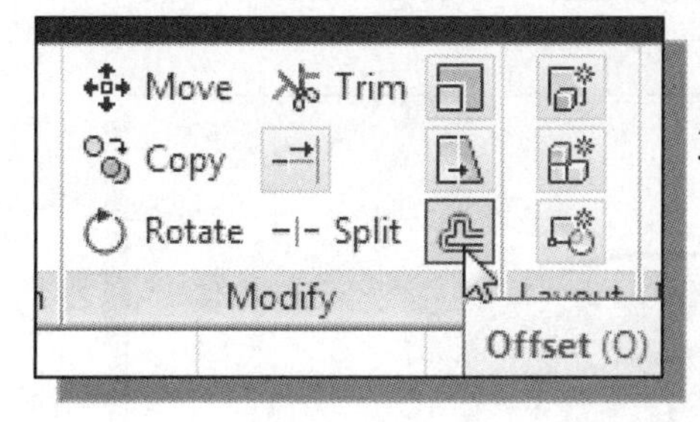

3. Click on the **Offset** icon in the *Modify* panel

4. Select any edge of the **back face** of the 3D model. *Autodesk Inventor* will automatically select all of the connecting geometry to form a closed region.

5. Move the cursor toward the outside of the selected region and notice an offset copy of the outline is displayed.

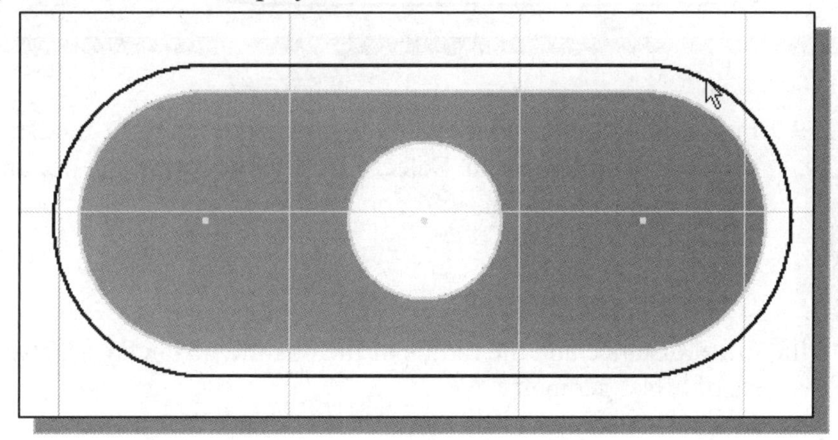

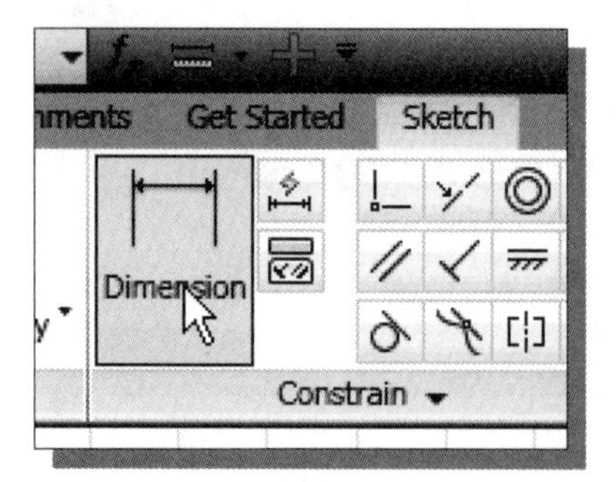

6. On your own, use the **General Dimension** command to create the radius dimension as shown in the figure below.

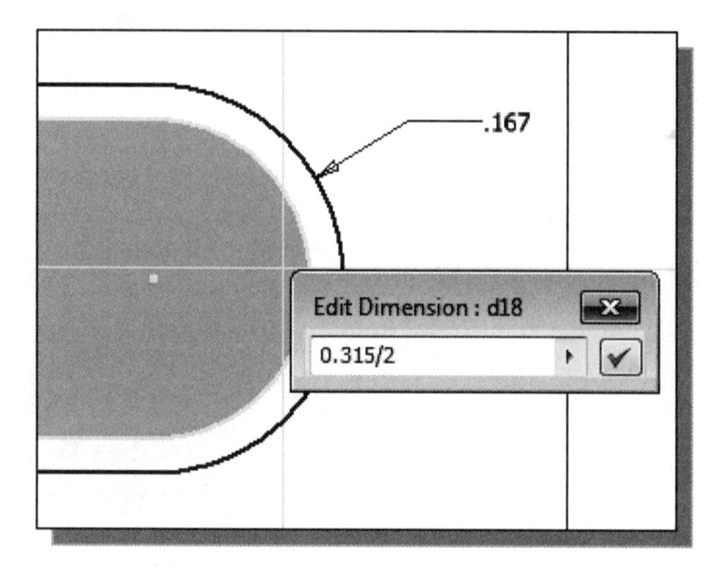

7. Modify the dimension to **0.315/2** as shown in the figure.

8. On your own, repeat the above steps and create another offset geometry; also create the offset dimension as shown.

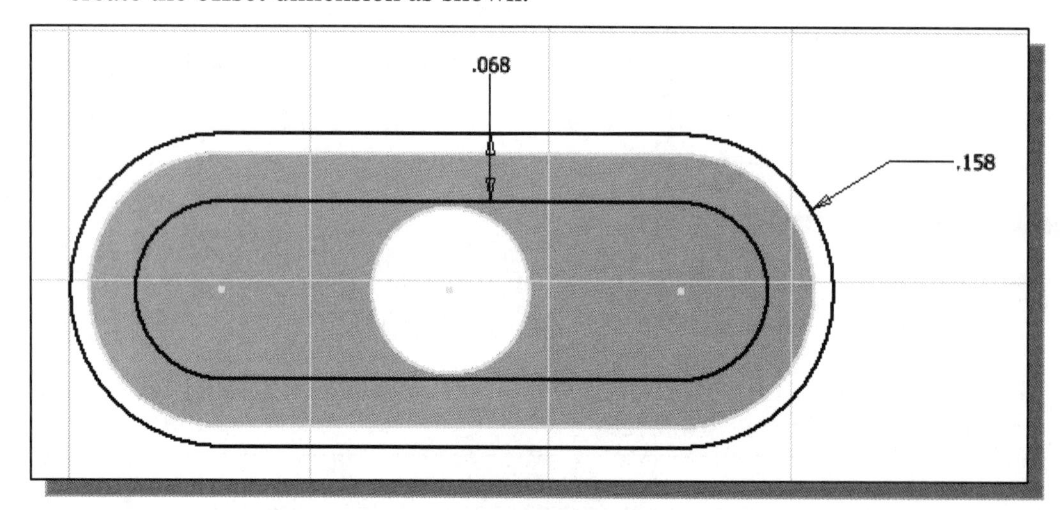

9. Inside the graphics window, click once with the right-mouse-button to display the option menu. Select **Finish Sketch** in the popup menu to end the Sketch option.

➢ Note the offset distance and the radius of the outside geometry are used to control the two sets of offset geometry.

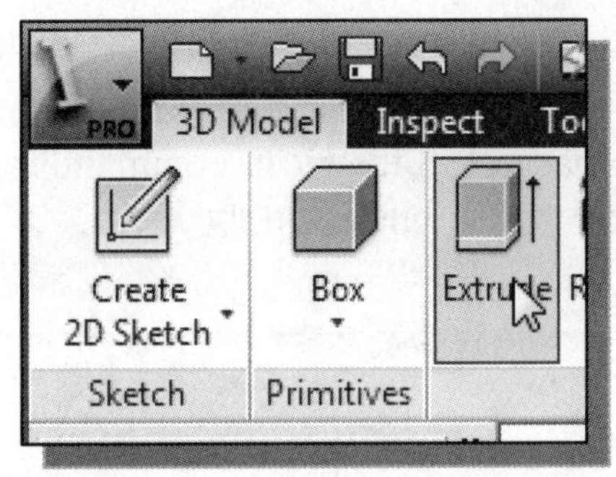

10. In the *Create* toolbar (the toolbar that is located to the right side of the *Sketch* toolbar in the *Ribbon*), select the **Extrude** command by releasing the left-mouse-button on the icon.

11. In the *Extrude* popup window, the **Profile** option is activated down; *Autodesk Inventor* expects us to identify the profile to be extruded. Select the **TWO regions**, located in between the two offset geometry we just created, as the profile to be extruded.

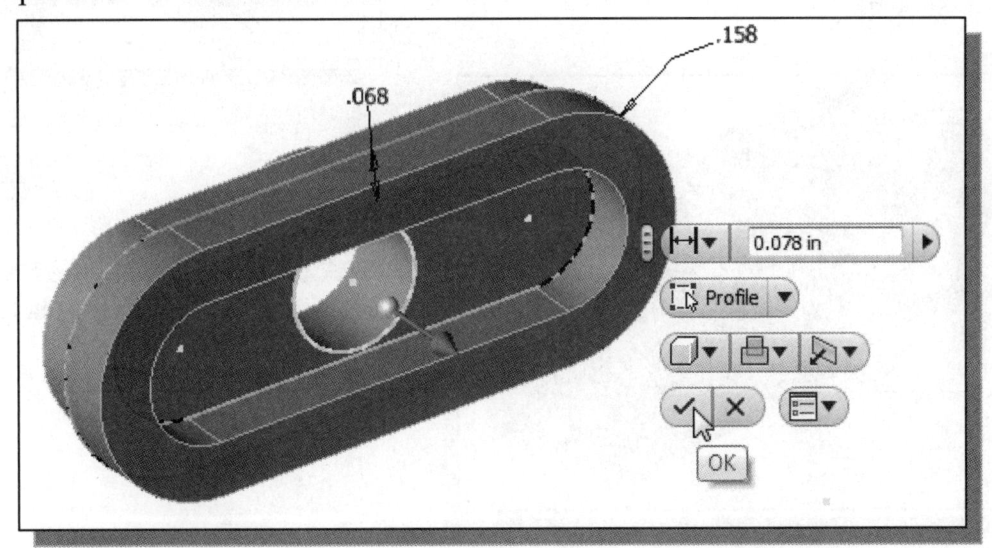

12. In the *Distance* option box, enter **.078** as the extrusion distance. Also confirm the extrusion direction is toward the back of the base feature as shown.

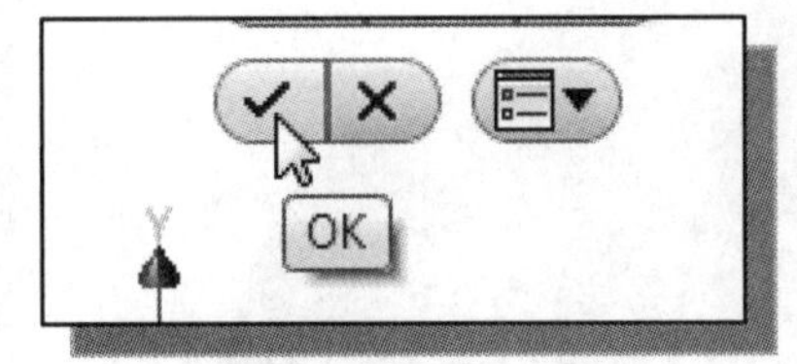

13. Click on the **OK** button to proceed with the *Join* operation.

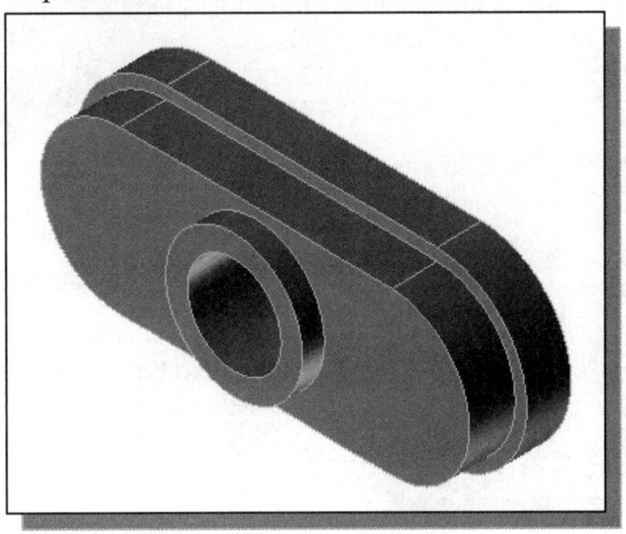

Add Another Hole Feature

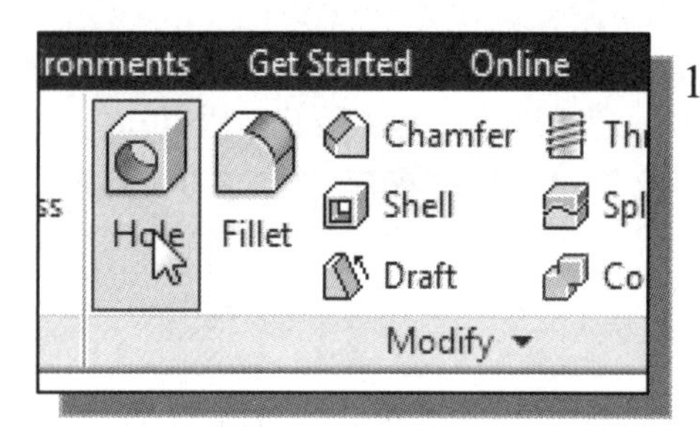

1. In the *Modify* toolbar, select the **Hole** command by clicking the left-mouse-button on the icon.

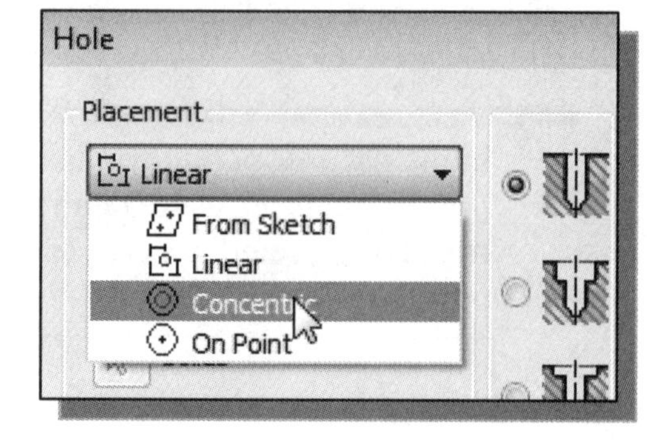

2. In the *Hole* dialog box, choose **Concentric** as the *Placement* option as shown.

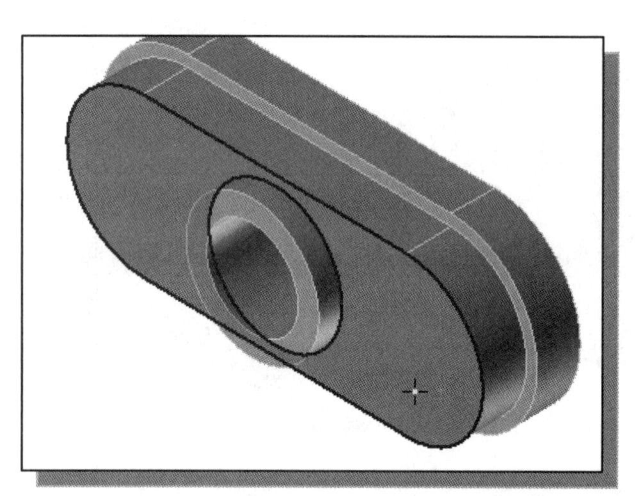

3. Pick the front plane of the **Base** feature as the placement plane as shown.

4. Pick one of the cylindrical surfaces or arcs on the right to use as the concentric reference.

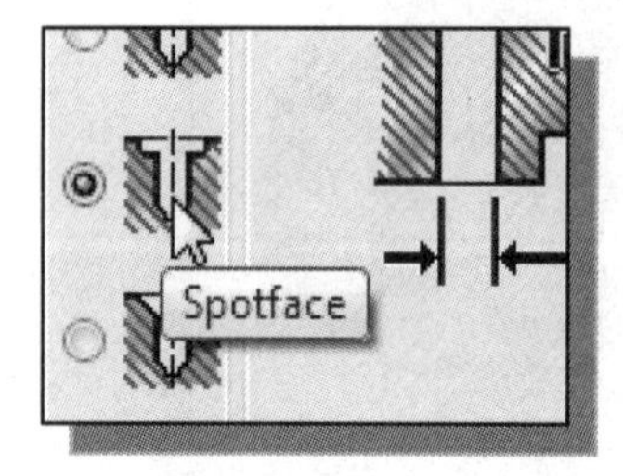

5. Set the hole type to **Spotface** as shown.

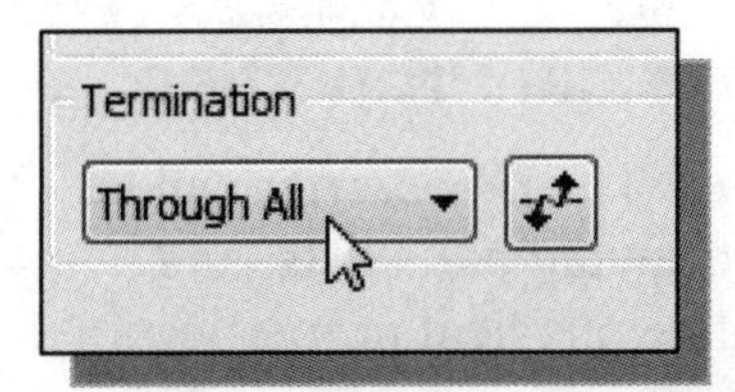

6. Set the *Termination* option to **Through All** as shown.

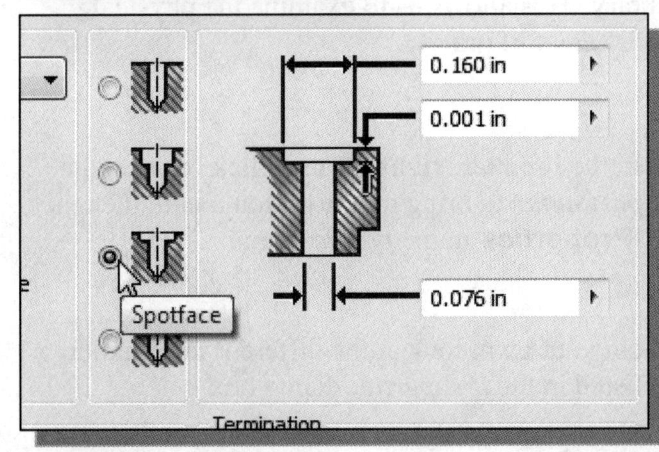

7. Set the hole diameter to **0.076 in** as shown.

8. Set the spotface diameter to **0.160 in** as shown

9. Set the spotface depth to **0.001 in** as shown.

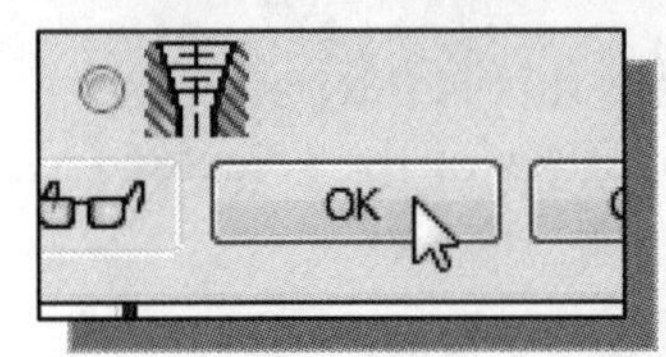

10. Click **OK** to accept the settings and create the *Hole* feature.

11. On your own, repeat the above steps and create another *spotface* feature on the other side as shown.

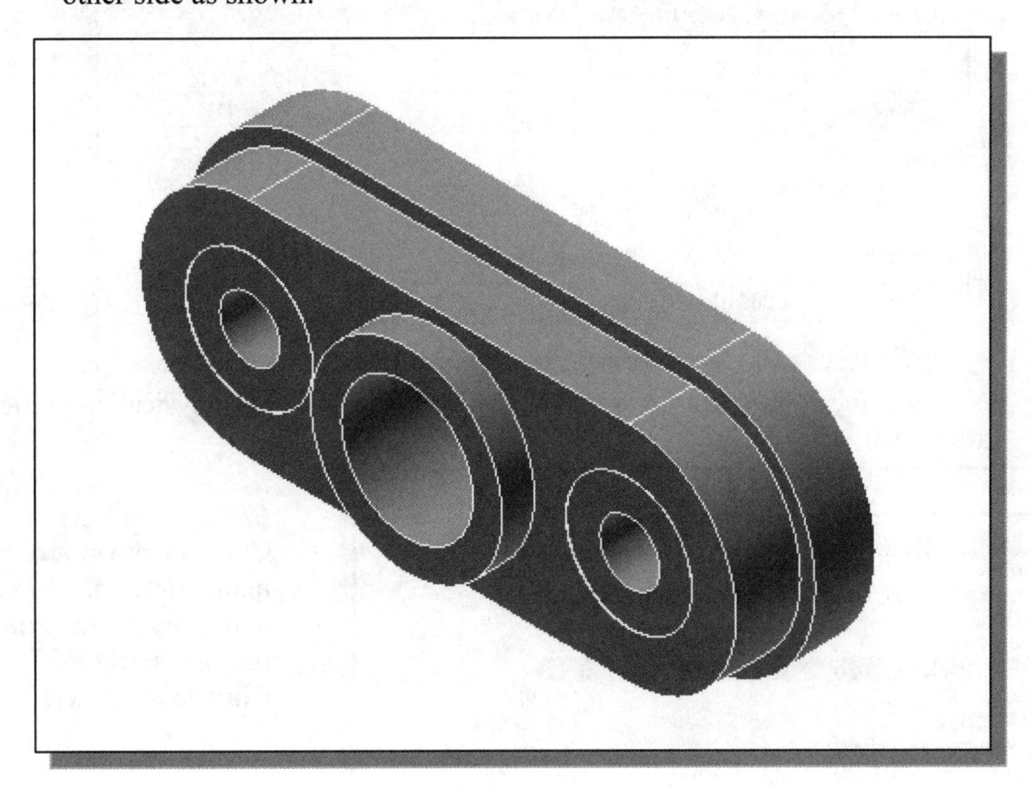

Assigning and Calculating the Associated Physical Properties

❖ *Autodesk Inventor* models have properties called ***iProperties***. The *iProperties* can be used to create reports, and update assembly bills of materials, drawing parts lists, and other information. With *iProperties*, we can also set and calculate physical properties for a part or assembly using the material library. This allows us to examine the physical properties of the model, such as weight or center of gravity.

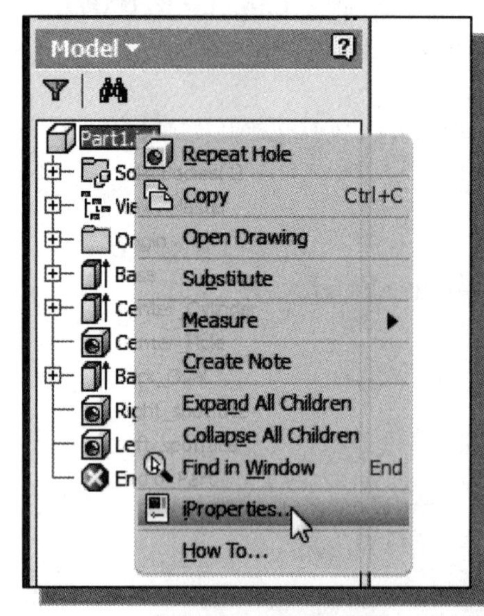

1. In the *browser*, **right-mouse-click** once on the ***part name*** to bring up the option menu; then pick **iProperties** in the *pop-up* menu.

2. On your own, look at the different information listed in the *iProperties* dialog box.

3. Click on the **Physical** tab; this is the page that contains the physical properties of the selected model.

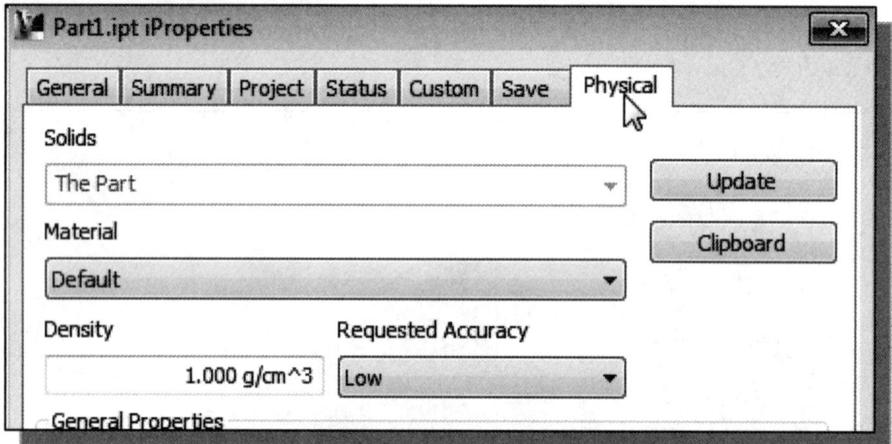

- Note that the *Material* option is not assigned, and none of the physical properties are shown.

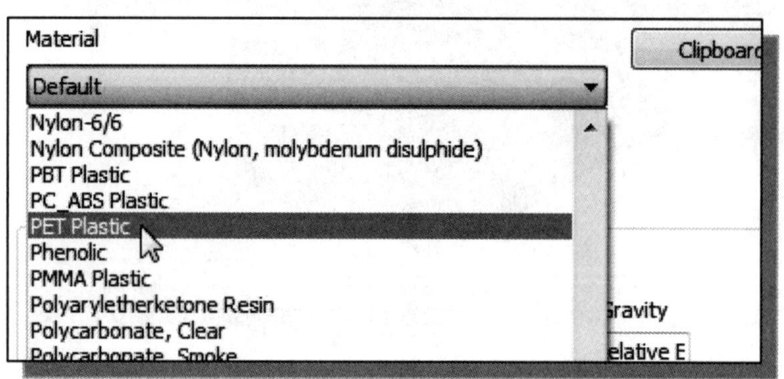

4. Click the down-arrow in the *Material* option to display the material list, and select **PET Plastic** as shown.

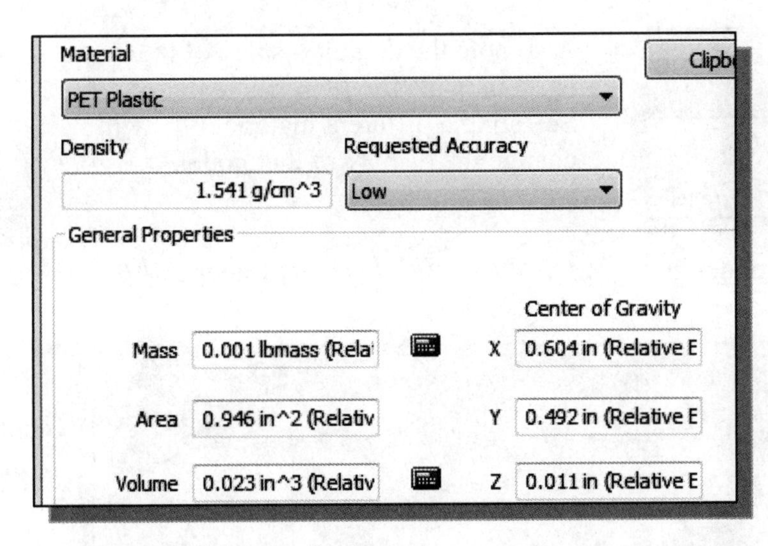

❖ Note the *General Properties* area now has the *Mass*, *Area*, *Volume* and *Center of Gravity* information of the model, based on the density of the selected material.

5. Note the *Mass Moments Inertia* of the design, with respect to the different axes, are also available.

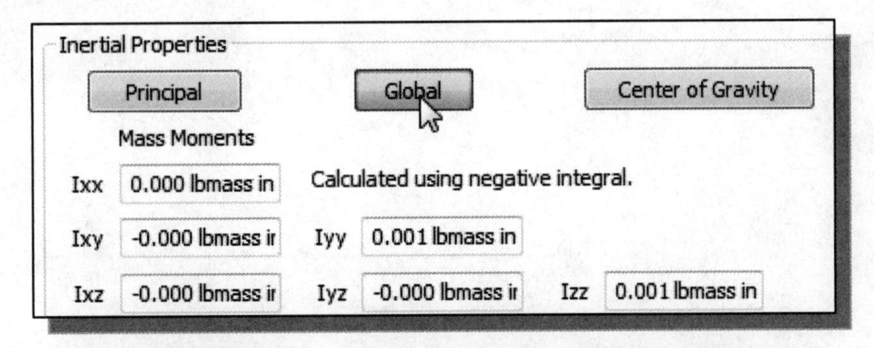

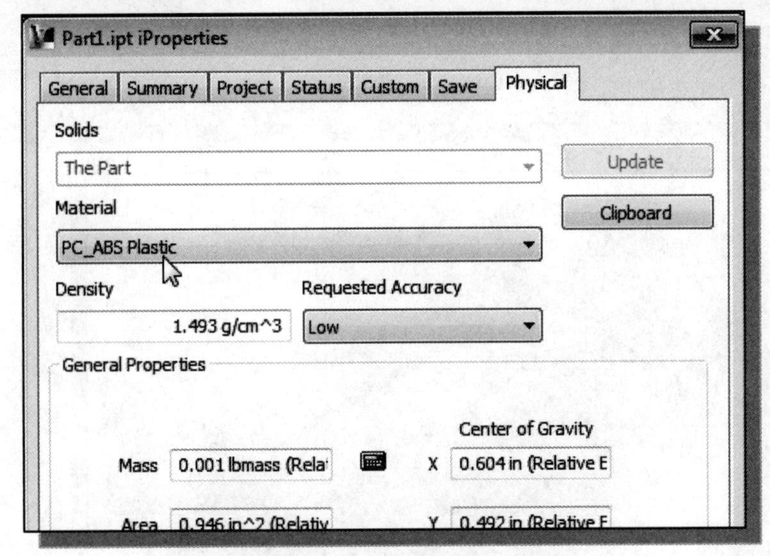

6. On your own, select **PC_ABS Plastic** as the *Material* type and compare the differences in using the different materials.

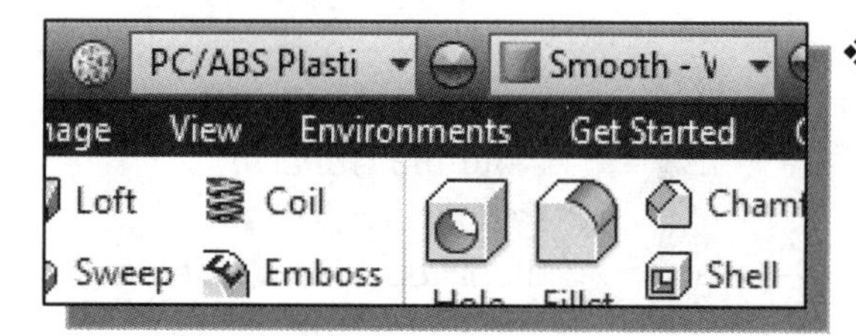

❖ Also note the default display of the model is set to using the *Material* assigned. Selecting a different material type will change the display of the model.

7. On your own, save the model in the *Mechanical-Tiger* project folder as **A6-Knee.ipt**.

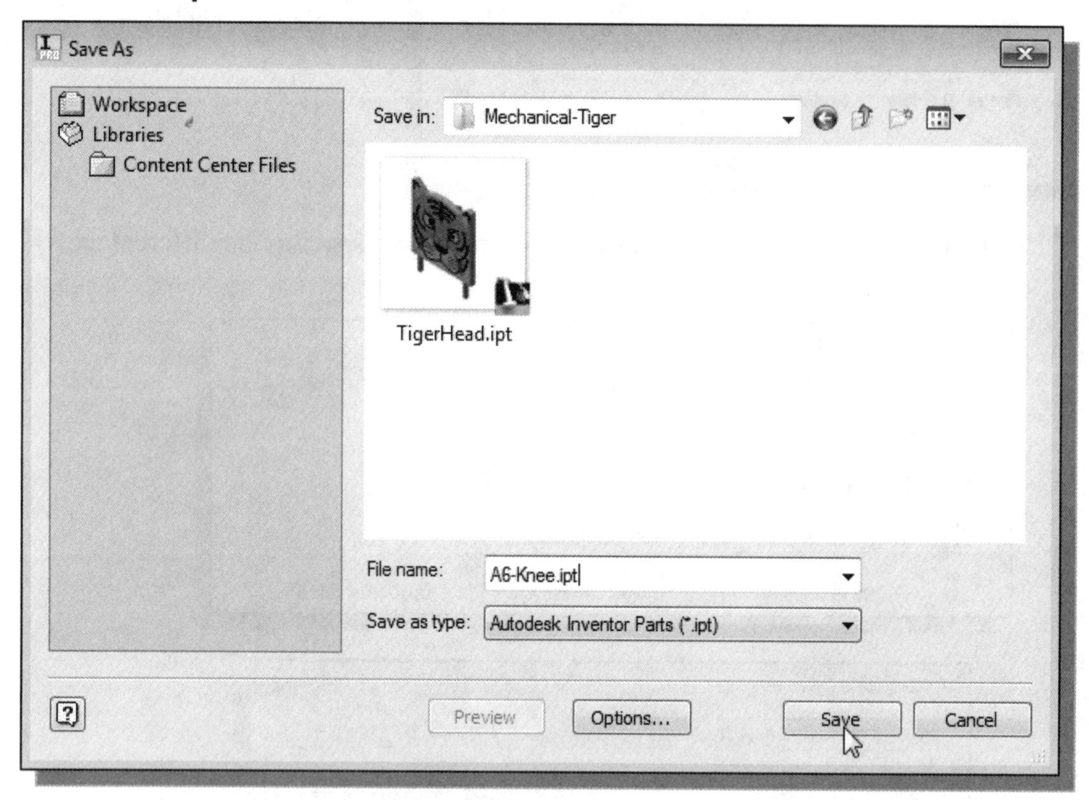

Questions:

1. What are stored in the *Autodesk Inventor History Tree*?

2. When extruding, what is the difference between *Distance* and *Through All*?

3. Describe the *history-based part modification* approach.

4. What determines how a model reacts when other features in the model change?

5. Describe the steps to rename existing features.

6. Describe two methods available in *Autodesk Inventor* to *modify the dimension values* of parametric sketches.

7. Create *History Tree sketches* showing the steps you plan to use to create the two models shown in the next pages:

Ex.3)

Ex.4)

Exercises: (All dimensions are in inches.)

1. **Latch Clip** (thickness: **0.25** inches. Material: **Cast Iron**. Mass and Volume =?)

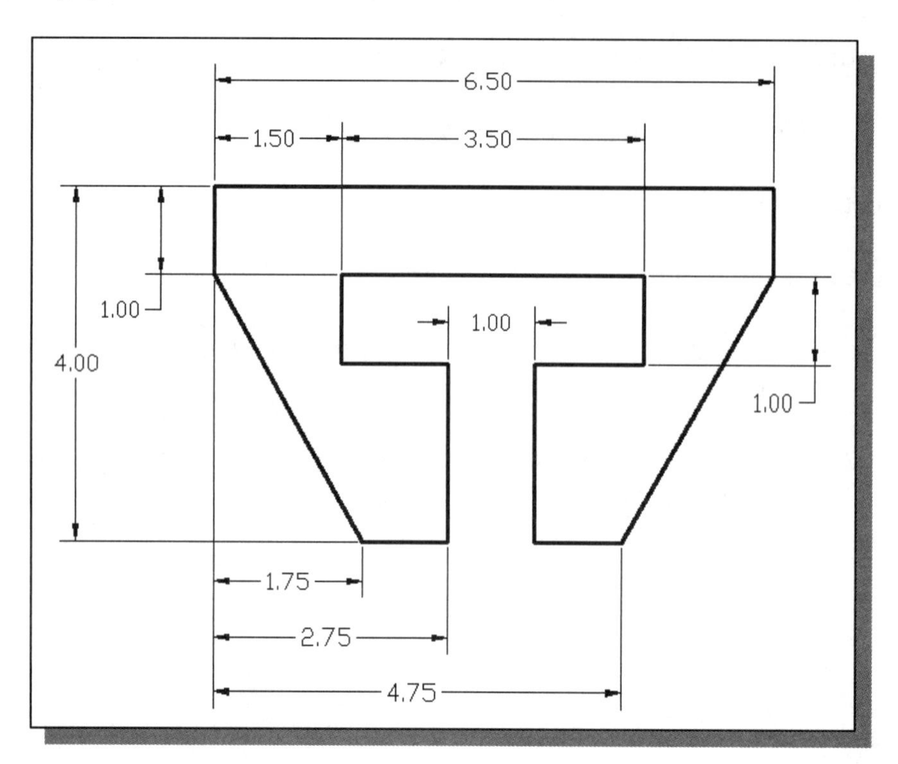

2. **Guide Plate** (thickness: **0.25** inches. Boss height **0.125** inches.)

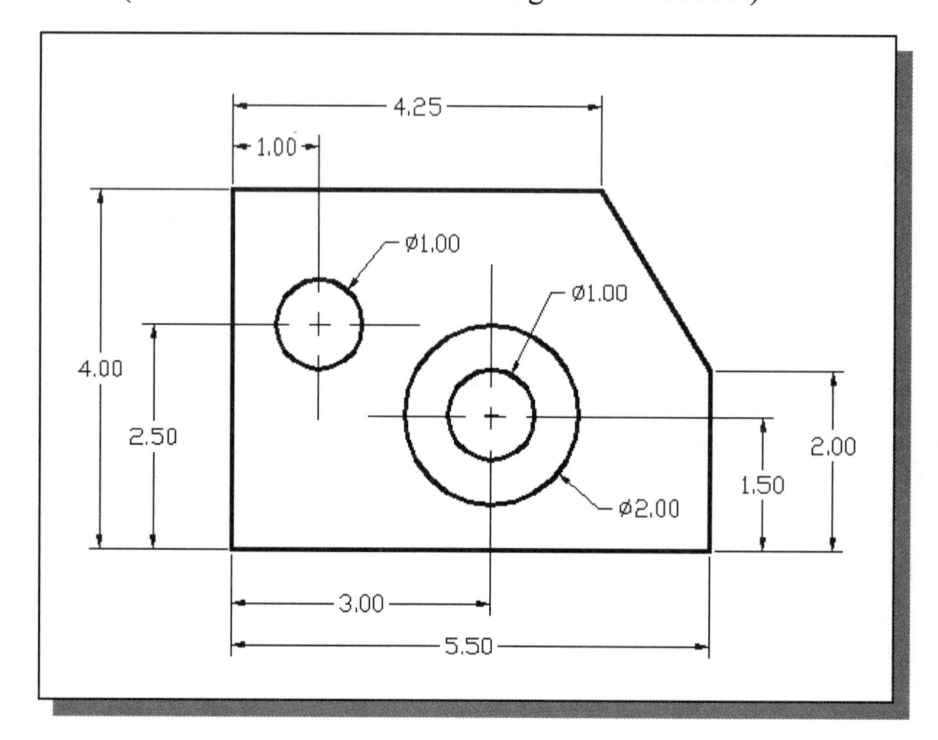

3. **Swivel Yoke** (Material: **Aluminum-6061**. Weight and Volume =?)

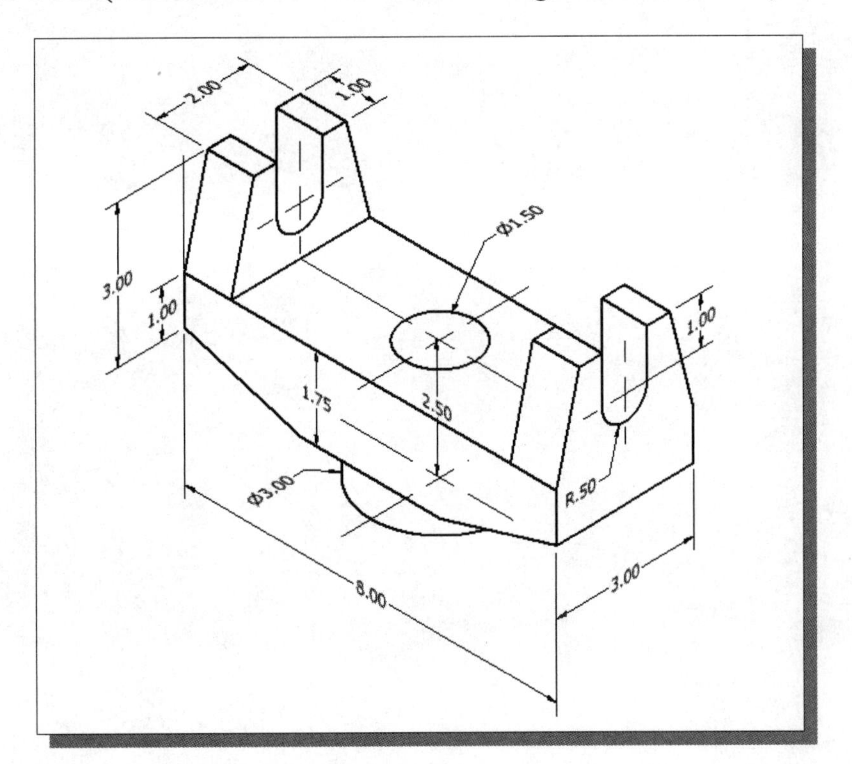

4. **Hanger Jaw**

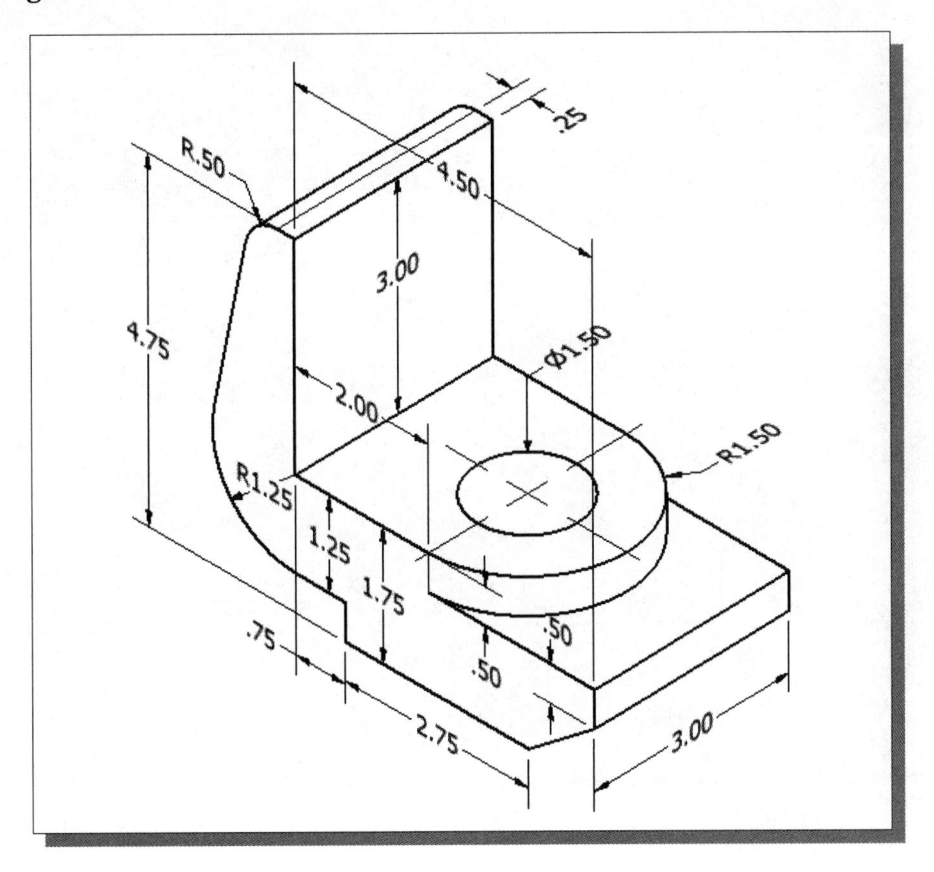

NOTES:

Chapter 4
Parametric Constraints Fundamentals

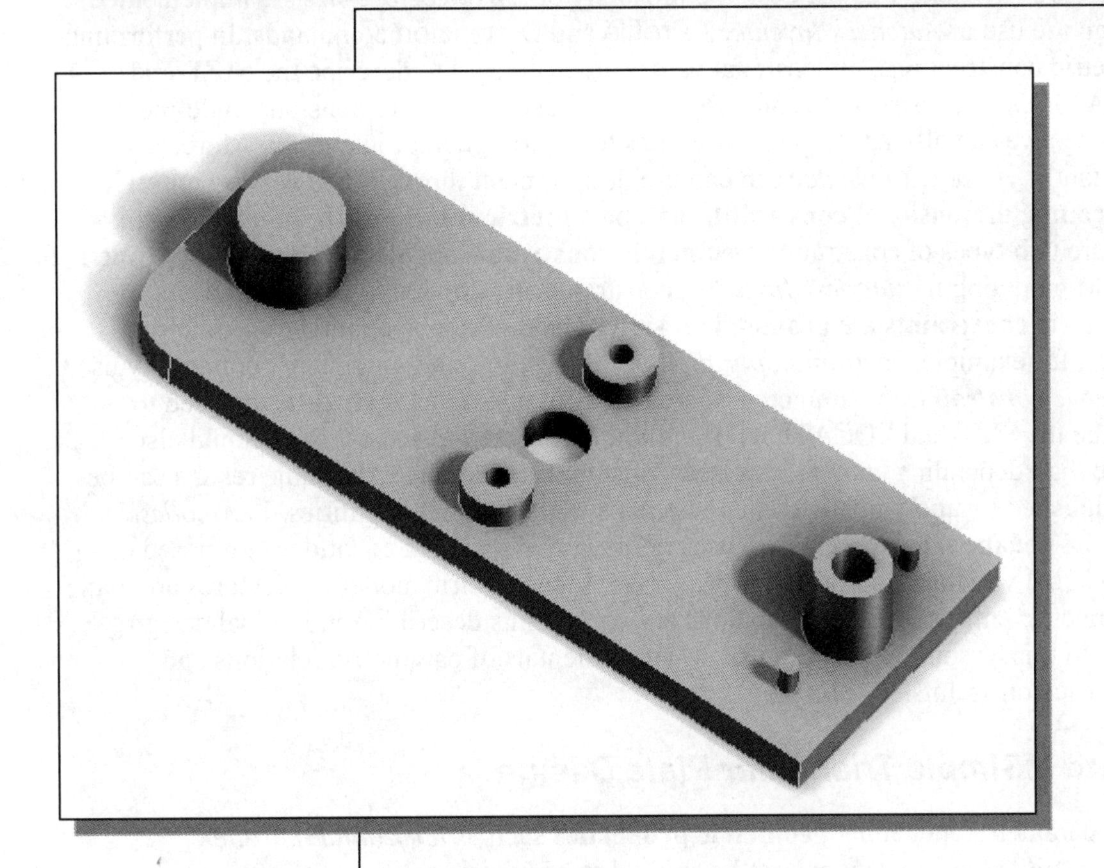

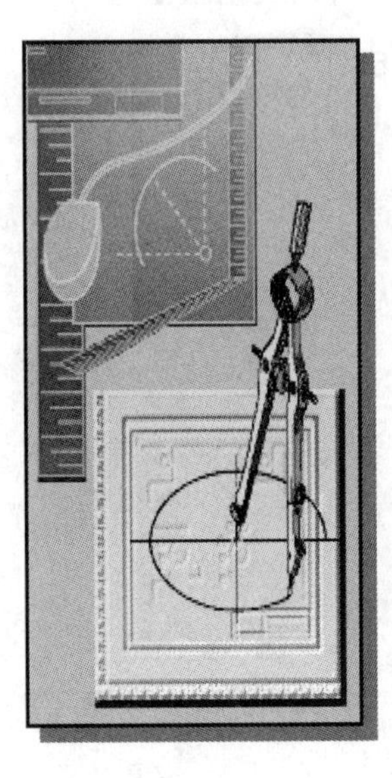

Learning Objectives

- Create Parametric Relations
- Use Dimensional Variables
- Display, Add, and Delete Geometric Constraints
- Understand and Apply Different Geometric Constraints
- Display and Modify Parametric Relations
- Create Fully Constrained Sketches

CONSTRAINTS and RELATIONS

A primary and essential difference between parametric modeling and previous generation computer modeling is that parametric modeling captures the *design intent*. In the previous lessons, we have seen that the design philosophy of "*shape before size*" is implemented through the use of *Autodesk Inventor's* Profile and Dimension commands. In performing geometric constructions, dimensional values are necessary to describe the **SIZE** and **LOCATION** of constructed geometric entities. Besides using dimensions to define the geometry, we can also apply geometric rules to control geometric entities. More importantly, *Autodesk Inventor* can capture design intent through the use of **geometric constraints, dimensional constraints**, and **parametric relations**. In *Autodesk Inventor*, there are two types of constraints: **geometric constraints** and **dimensional constraints**. For part modeling in *Autodesk Inventor*, constraints are applied to *2D sketches*. **Geometric constraints** are **geometric restrictions** that can be applied to geometric entities; for example, *horizontal*, *parallel*, *perpendicular*, and *tangent* are commonly used *geometric constraints* in parametric modeling. **Dimensional constraints** are used to describe the SIZE and LOCATION of individual geometric shapes. One should also realize that, depending upon the way the constraints are applied, the same results can be accomplished by applying different constraints to the geometric entities. In *Autodesk Inventor*, **parametric relations** are user-defined mathematical equations composed of dimensional variables and/or *design variables*. In parametric modeling, features are made of geometric entities with both relations and constraints describing individual design intent. In this lesson, we will discuss the fundamentals of parametric relations and geometric constraints.

Create a *Simple Triangular Plate* Design

➢ In parametric modeling, **geometric properties** such as *horizontal*, *parallel*, *perpendicular*, and *tangent* can be applied to geometric entities automatically or manually. By carefully applying proper **geometric constraints**, very intelligent models can be created. This concept is illustrated by the following example.

Fully Constrained Geometry

In *Autodesk Inventor*, as we create 2D sketches, geometric constraints such as *horizontal* and *parallel* are automatically added to the sketched geometry. In most cases, additional constraints and dimensions are needed to fully describe the sketched geometry beyond the geometric constraints added by the system. Although we can use *Autodesk Inventor* to build partially constrained or totally unconstrained solid models, the models may behave unpredictably as changes are made. In most cases, it is important to consider the design intent and to add proper constraints to geometric entities. In the following sections, a simple triangle is used to illustrate the different tools that are available in *Autodesk Inventor* to create/modify geometric and dimensional constraints.

Starting *Autodesk Inventor*

1. Select the **Autodesk Inventor** option on the *Start* menu or select the **Autodesk Inventor** icon on the desktop to start *Autodesk Inventor*. The *Autodesk Inventor* main window will appear on the screen. Once the program is loaded into memory, the *Startup* dialog box appears at the center of the screen.

2. Select the **New** icon with a single click of the left-mouse-button in the *Launch* toolbar as shown.

3. Select the **English** units set and in the *Part template* area, select **Standard(in).ipt**.

4. Click **Create** in the *New File* dialog box to accept the selected settings to start a new model.

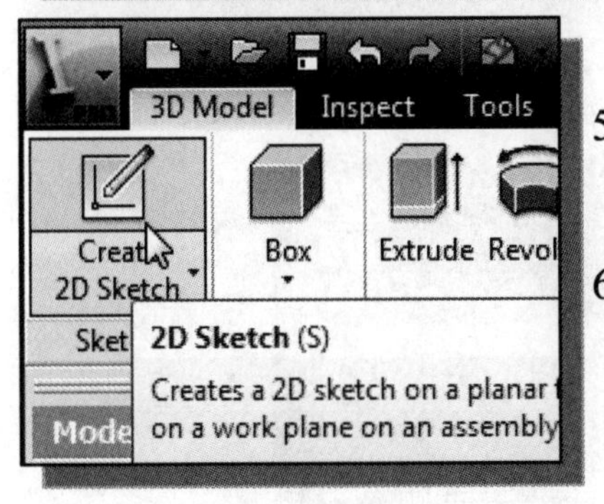

5. Click once with the left-mouse-button to select the **Create 2D Sketch** command.

6. Click once with the **left-mouse-button** to select the *XY Plane* as the sketch plane for the new sketch.

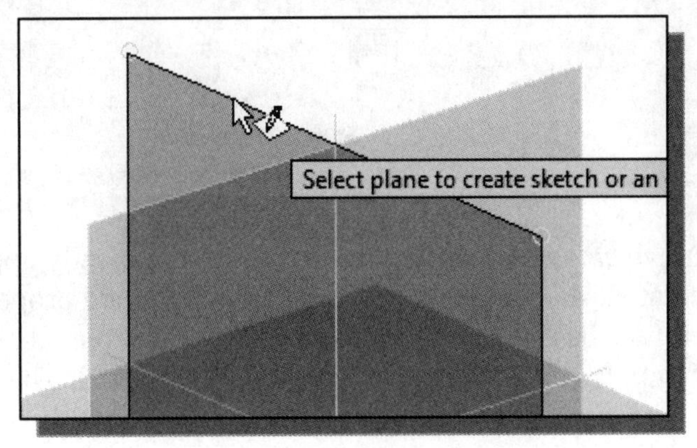

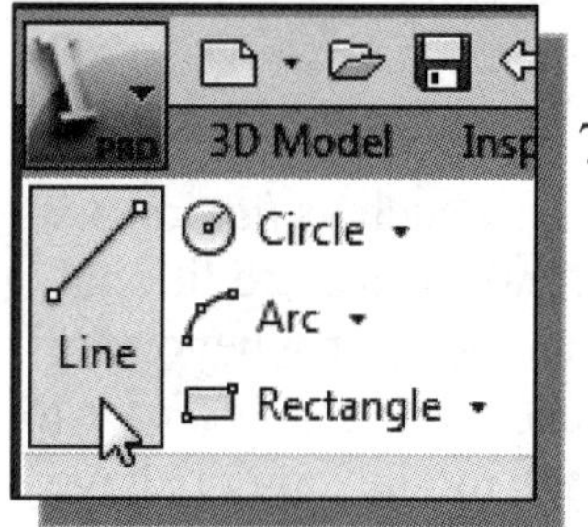

7. Click the **Line** icon in the *Sketch* toolbar to activate the command. A *Help-tip box* appears next to the cursor and a brief description of the command is displayed at the bottom of the drawing screen: "*Creates Straight line segments and tangent arcs.*"

8. Create a triangle of arbitrary size positioned near the lower left corner of the screen as shown below. (Note that the base of the triangle is horizontal.)

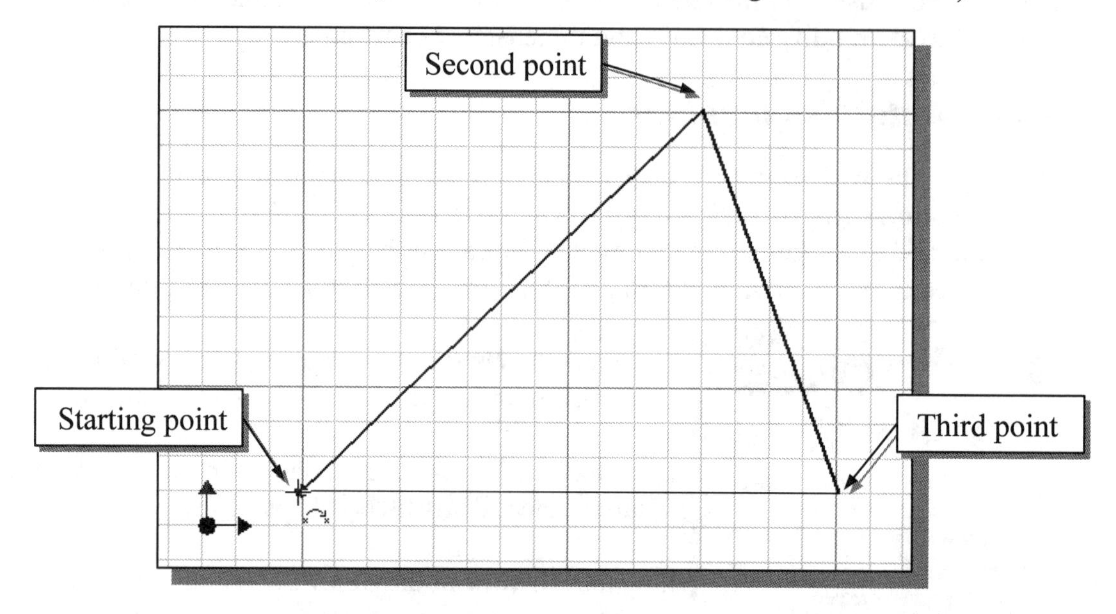

Displaying Existing Constraints

1. Select the **Show Constraints** command in the *Constrain* toolbar. This icon allows us to display constraints that are already applied to the 2D profiles. Left-click once on the icon to activate **Show Constraints**.

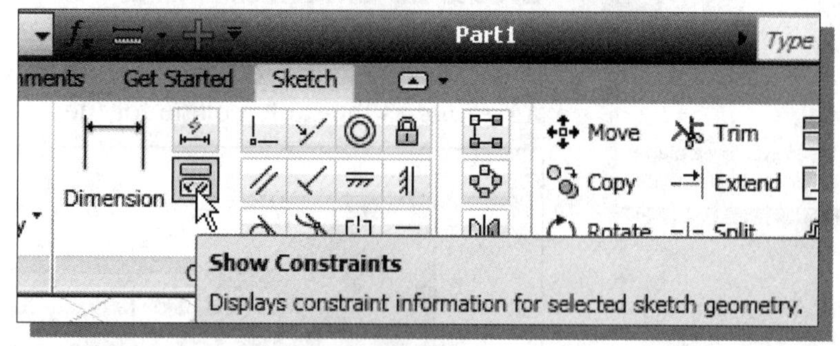

➢ In *parametric modeling*, constraints are typically applied as geometric entities are created. *Autodesk Inventor* will attempt to add proper constraints to the geometric entities based on the way the entities were created. Constraints are displayed as symbols next to the entities as they are created. The current profile consists of three line entities, three straight lines. The horizontal line has a **Horizontal** ***constraint*** applied to it.

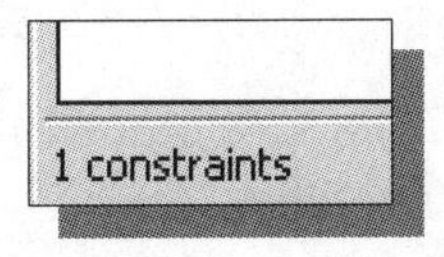

2. Move the cursor on top of the horizontal line and notice the number of constraints applied is displayed in the message area.

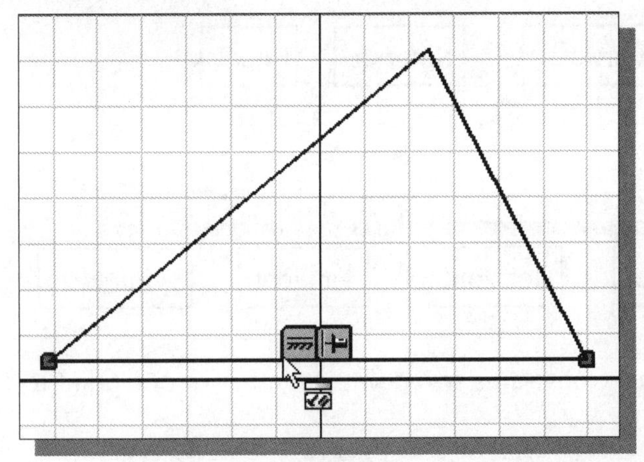

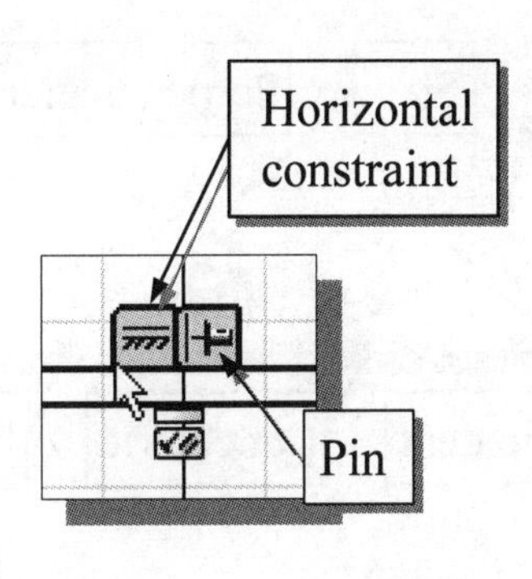

3. Click on the **Pin** icon, the last icon in the *active constraint box*, to lock the display of the *Active Constraints* at its current location.

4. On your own, move the cursor on top of the other two lines and notice no additional constraints exist on the other two entities.

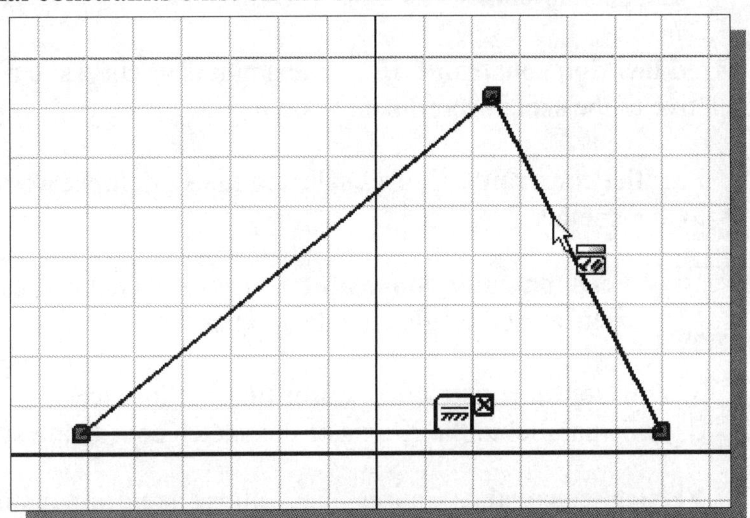

5. Move the cursor on top of the constraints displayed and notice the highlighted endpoint/line indicating the location/entities where the constraints are applied.

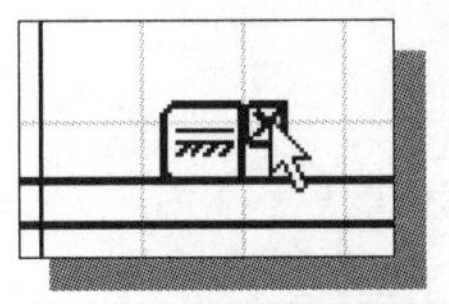

6. Click on the **X** icons, the last icon in the active constraint box, to unlock and turn off the display of the constraints.

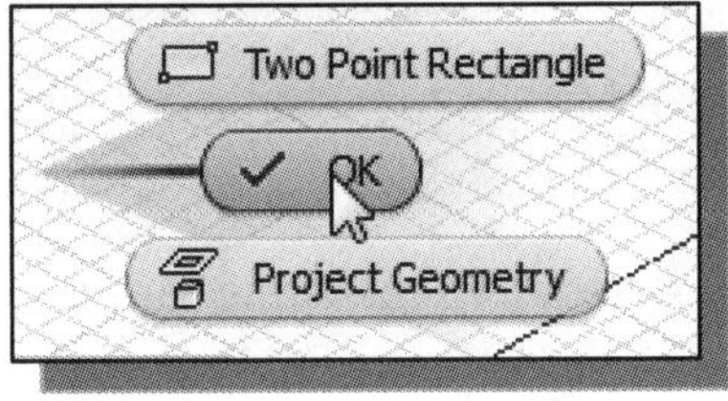

7. Inside the graphics window, right-mouse-click to bring up the option menu and select **OK** to end the Show Constraints command.

Applying Geometric/Dimensional Constraints

- In *Autodesk Inventor*, twelve types of constraints are available for 2D sketches.

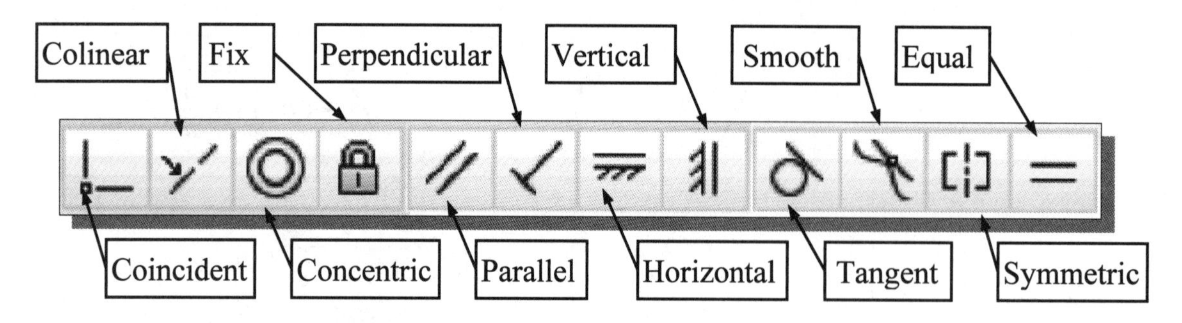

Coincident constraint: Constrains two points together or one point to a curve.

Colinear constraint: Causes two lines or ellipse axes to lie along the same line.

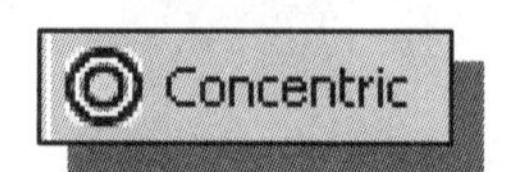

Concentric constraint: Constrains two arcs, circles, or ellipses to the same center point.

Fixed location constraint: Constrains points or curves to a fixed location relative to the sketch coordinate system.

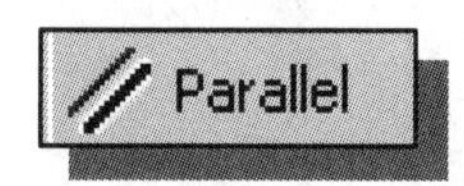

Parallel constraint: Causes selected lines or ellipse axes to lie parallel to one another.

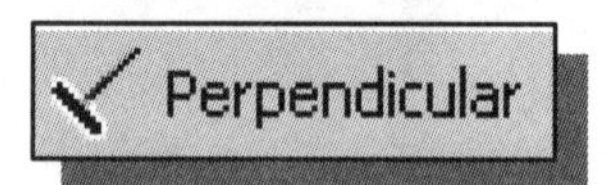

Perpendicular constraint: Causes selected curves or ellipse axes to lie at right angles to one another.

Horizontal constraint: Causes lines, ellipse axes, or pairs of points to lie parallel to the X-axis of the sketch coordinate system.

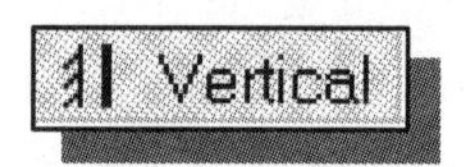

Vertical constraint: Causes lines, ellipse axes, or pairs of points to lie parallel to the Y-axis of the sketch coordinate system.

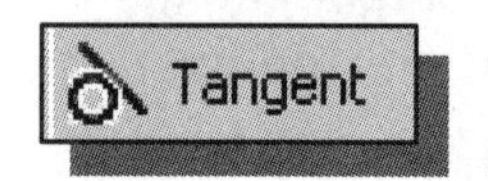

Tangent constraint: Constrains two curves to be tangent to one another.

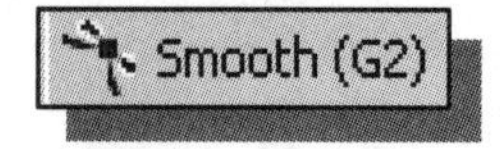

Smooth (G2) constraint: Creates a curvature continuous (G2) condition between a spline and a line, arc, or spline.

Symmetric constraint: Constrains lines or curves to become symmetrically constrained about a selected line.

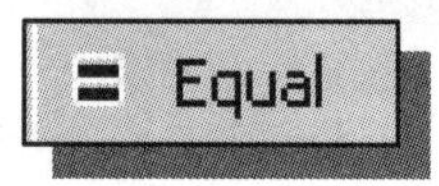

Equal constraint: Selected arcs/circles constrained to the same radius or selected lines to the same length.

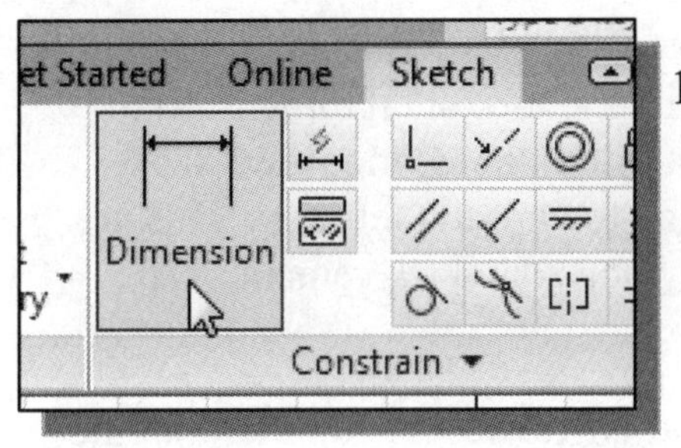

1. Select the **General Dimension** command in the *Sketch* toolbar. The **General Dimension** command allows us to quickly create and modify dimensions. Left-click once on the icon to activate the **General Dimension** command.

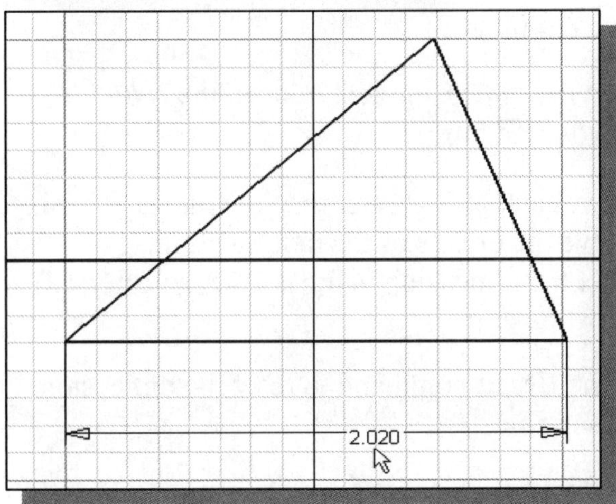

2. On your own, create the dimension as shown in the figure below. (Note that the displayed value might be different on your screen.)

❖ Note *Inventor* identifies the need for additional dimensions at the bottom of the screen.

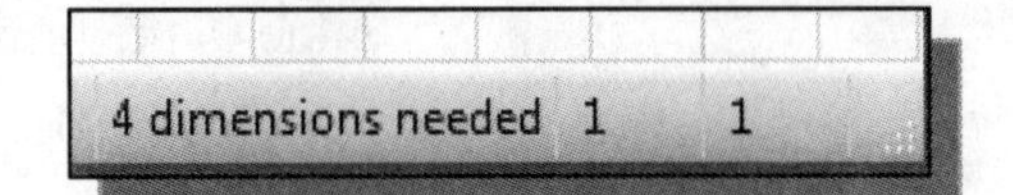

3. Move the cursor on top of the different **Constraint** icons. A *Help-tip box* appears next to the cursor and a brief description of the command is displayed at the bottom of the drawing screen as the cursor is moved over the different icons.

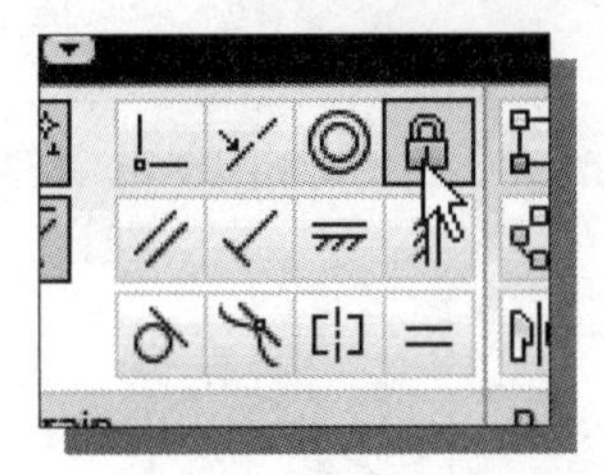

4. Click on the **Fix** constraint icon to activate the command.

5. Pick the **lower right corner** of the triangle to make the corner a fixed point.

6. On your own, use the **Show Constraints** command to confirm the **Fix** constraint is properly applied.

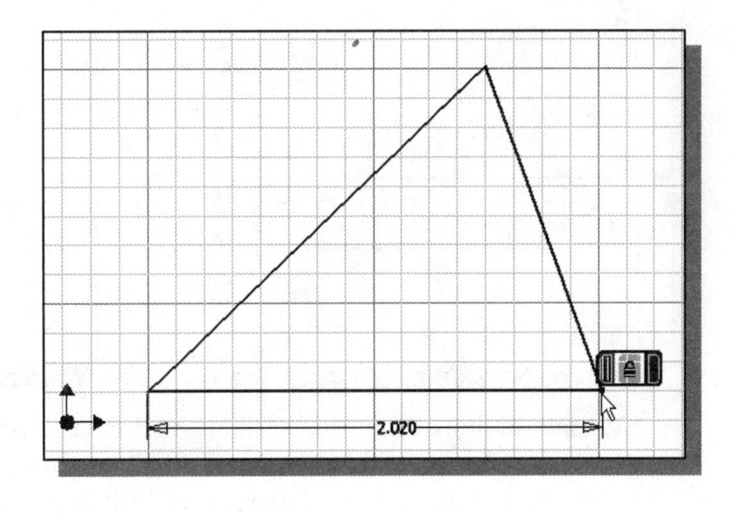

➢ Geometric constraints can be used to control the direction in which changes can occur. For example, in the current design we are adding a horizontal dimension to control the length of the horizontal line. If the length of the line is modified to a greater value, *Autodesk Inventor* will lengthen the line toward the left side. This is due to the fact that the **Fix** constraint will restrict any horizontal movement of the horizontal line toward the right side.

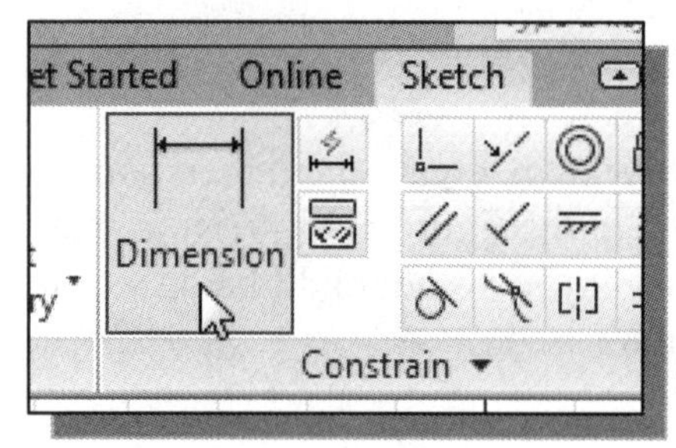

7. Select the **General Dimension** command in the *2D Sketch* toolbar.

8. Click on the dimension text to open the *Edit Dimension* window.

9. Enter a value that is greater than the displayed value to observe the effects of the modification. (For example, the dimension value is 2.02, so enter **3.0** in the text box area.)

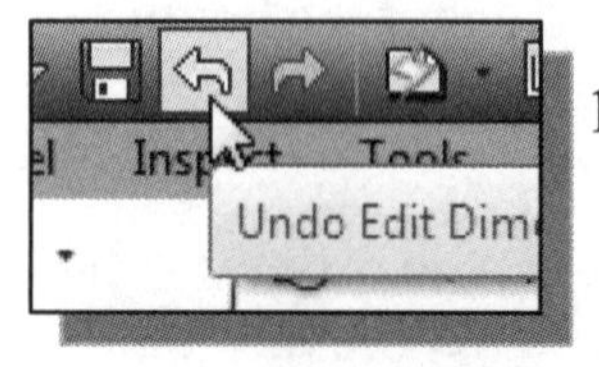

10. On your own, use the **Undo** command to reset the dimension value to the previous value.

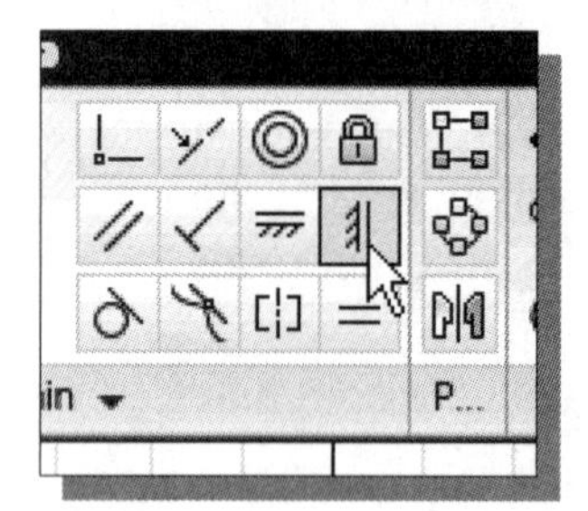

11. Select the **Vertical** constraint icon in the *2D Constraints* toolbar.

12. Pick the inclined line on the right to make the line vertical as shown in the figure below.

13. Hit the [**Esc**] key once to end the **Vertical Constraint** command.

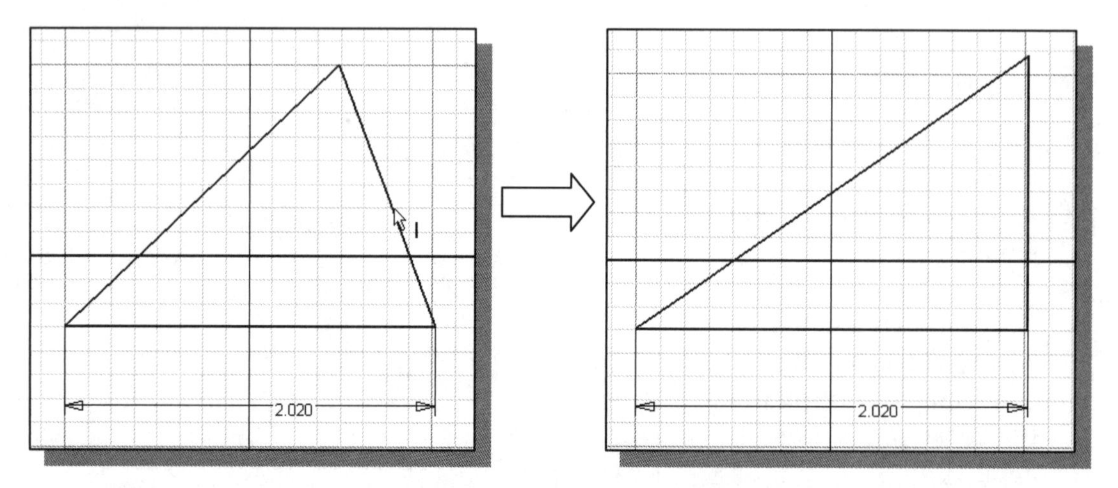

➢ One should think of the constraints and dimensions as defining elements of the geometric entities. How many more constraints or dimensions will be necessary to fully constrain the sketched geometry? Which constraints or dimensions would you use to fully describe the sketched geometry?

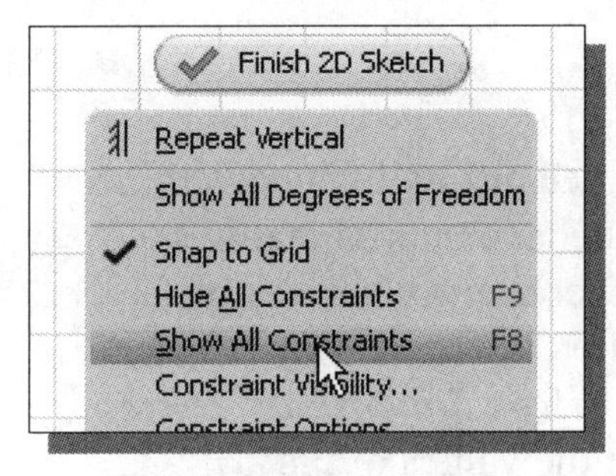

14. Inside the graphics window, click once with the **right-mouse-button** to display the option menu. Select **Show All Constraints** in the popup menu to show all the applied constraints. (Note that function key **F8** can also be used to activate this command.)

15. Move the cursor on top of the top corner of the triangle. (Note the display of the two Coincident constraints at the top corner.)

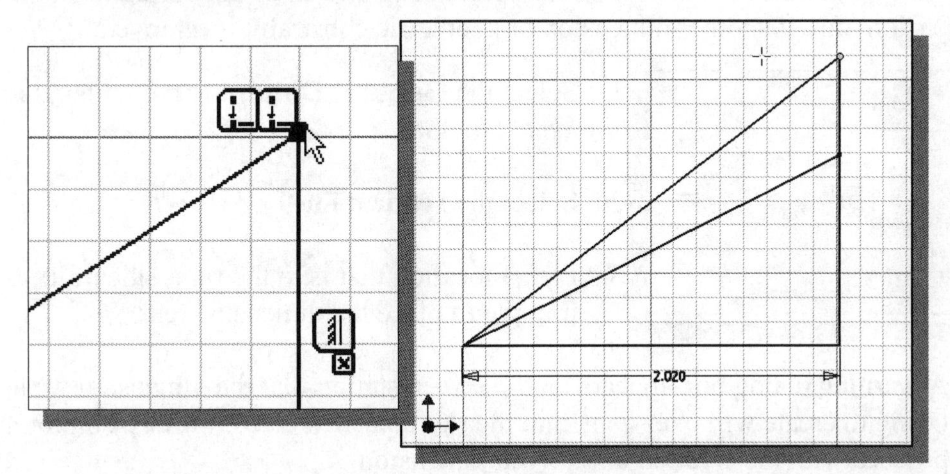

16. Drag the top corner of the triangle and note that the corner can be moved to a new location. Release the mouse button at a new location and notice the corner is adjusted only in an upward or downward direction. Note that the two adjacent lines are automatically adjusted to the new location.

17. On your own, experiment with dragging the other corners to new locations.

- The three constraints that are applied to the geometry provide a full description for the location of the two lower corners of the triangle. The Vertical constraint, along with the Fix constraint at the lower right corner, does not fully describe the location of the top corner of the triangle. We will need to add additional information, such as the length of the vertical line or an angle dimension.

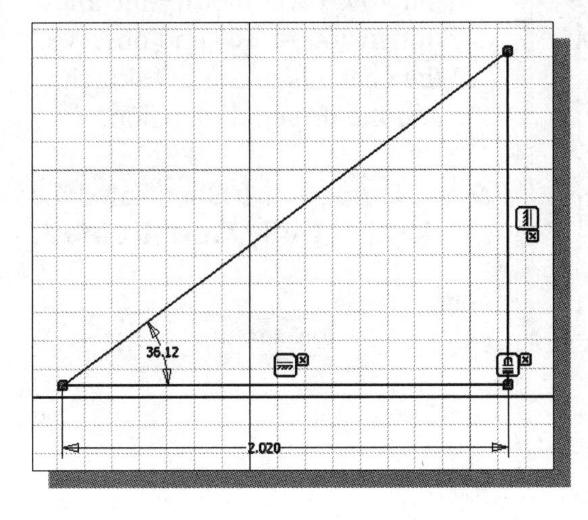

18. On your own, add an angle dimension to the left corner of the triangle.

19. Press the [**Esc**] key once to exit the General Dimension command.

20. On your own, modify the angle to **45°** and then experiment with dragging the top corner of the triangle.

- Note the sketched geometry is fully constrained with the added dimension.

Over-Constraining and Driven Dimensions

- We can use *Autodesk Inventor* to build partially constrained or totally unconstrained solid models. In most cases, these types of models may behave unpredictably as changes are made. However, *Autodesk Inventor* will not let us over-constrain a sketch; additional dimensions can still be added to the sketch, but they are used as references only. These additional dimensions are called ***driven dimensions***. *Driven dimensions* do not constrain the sketch; they only reflect the values of the dimensioned geometry. They are enclosed in parentheses to distinguish them from normal (parametric) dimensions. A *driven dimension* can be converted to a normal dimension only if another dimension or geometric constraint is removed.

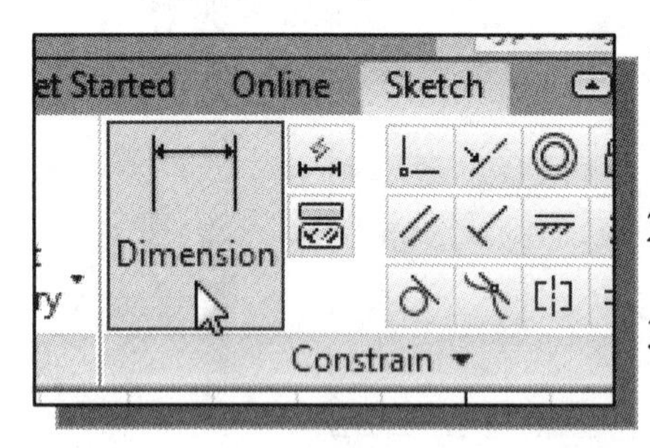

1. Select the **General Dimension** command in the *Sketch* toolbar.

2. Select the **vertical line**.

3. Pick a location that is to the right side of the triangle to place the dimension text.

4. A warning dialog box appears on the screen stating that the dimension we are trying to create will over-constrain the sketch. Click on the **Accept** button to proceed with the creation of a driven dimension.

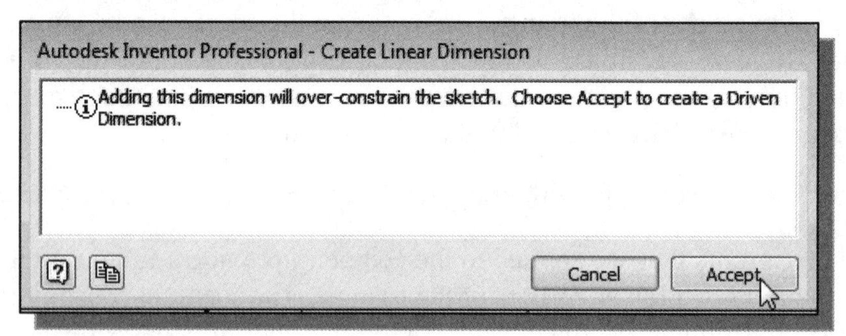

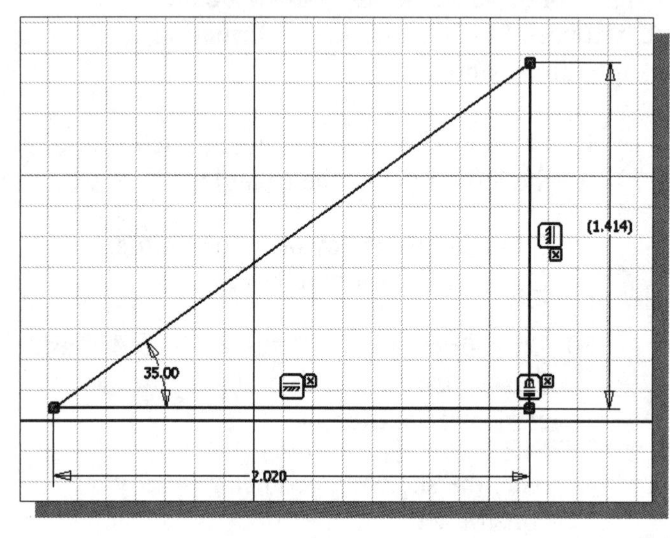

5. On your own, modify the angle dimension to **35°** and observe the changes of the 2D sketch and the driven dimension.

❖ Note *Inventor* has indicated the sketch is **Fully Constrained**.

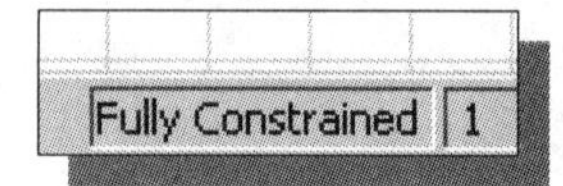

Deleting Existing Constraints

1. On your own, display all the active constraints if they are not already displayed. (Hint: Use the **Show Constraints** command in the *Sketch* toolbar or **Show All Constraints** in the option menu.)

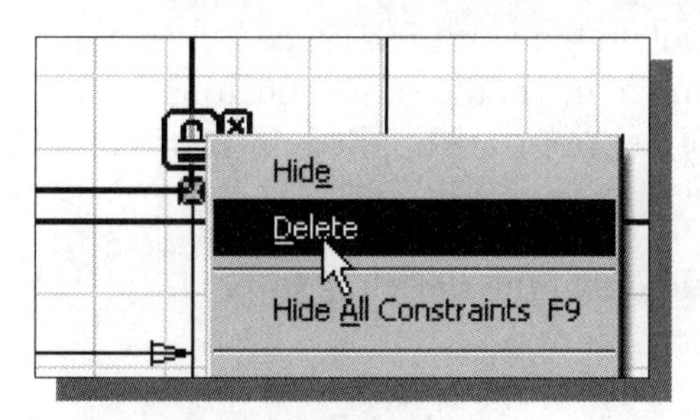

2. Move the cursor on top of the **Fix** constraint icon and **right-mouse-click** once to bring up the option menu.

3. Select **Delete** to remove the Fix constraint that is applied to the lower right corner of the triangle.

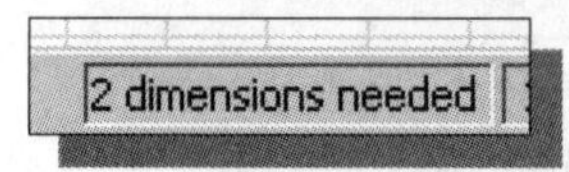

❖ Note the removal of the **Fix** constraint has caused the need for **two additional dimensions.**

4. Drag the top corner of the triangle and note that the entire triangle is free to move in all directions. Drag the corner toward the top right corner of the graphics window as shown in the above figure. Release the mouse button to move the triangle to the new location.

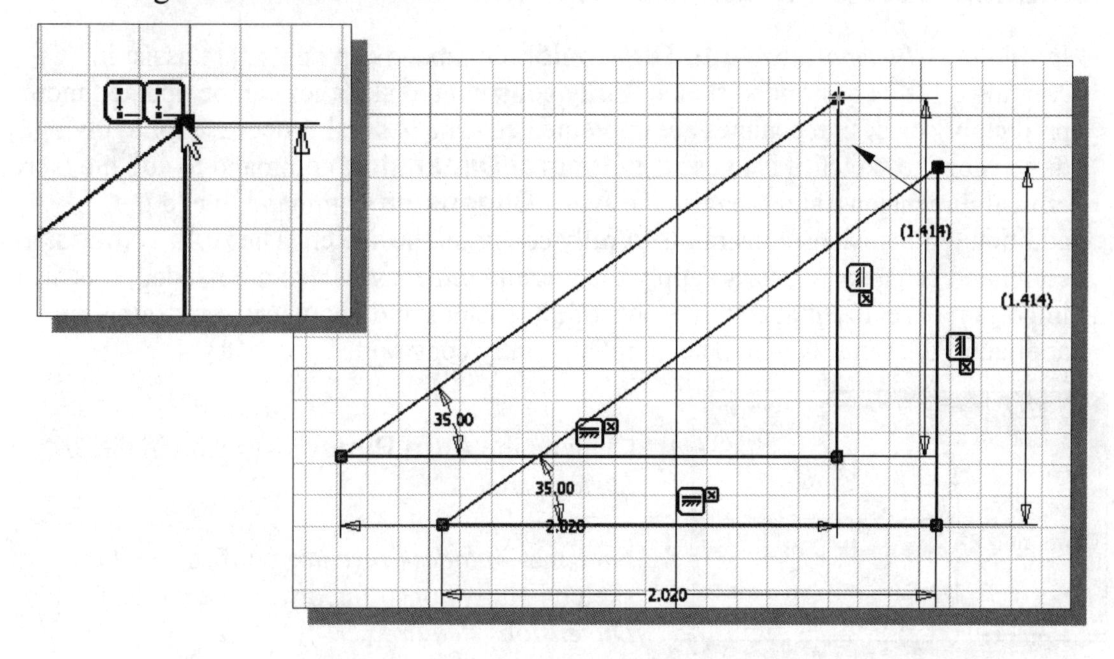

5. On your own, experiment with dragging the other corners and/or the three line segments to new locations on the screen.

❖ **Dimensional constraints** are used to describe the SIZE and LOCATION of individual geometric shapes. **Geometric constraints** are **geometric restrictions** that can be applied to geometric entities. The constraints applied to the triangle are sufficient to maintain its size and shape, but the geometry can be moved around; its location definition isn't complete.

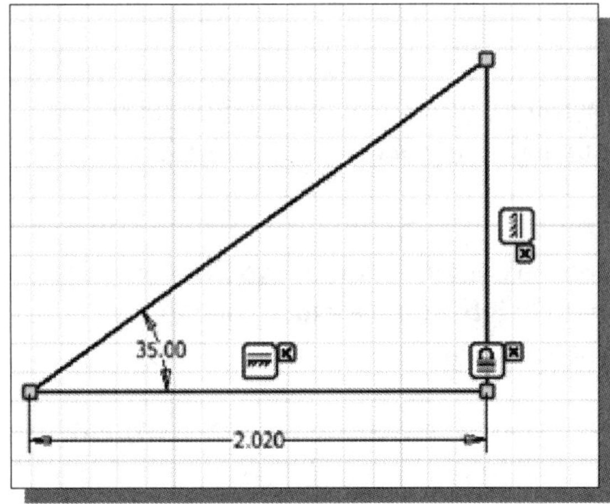

6. On your own, reapply the **Fix** constraint to the lower right corner of the triangle and delete the reference dimension.

7. Delete the extra reference dimension and confirm the same constraints and dimensions are applied on your sketch as shown.

❖ Note that the sketch is fully constrained.

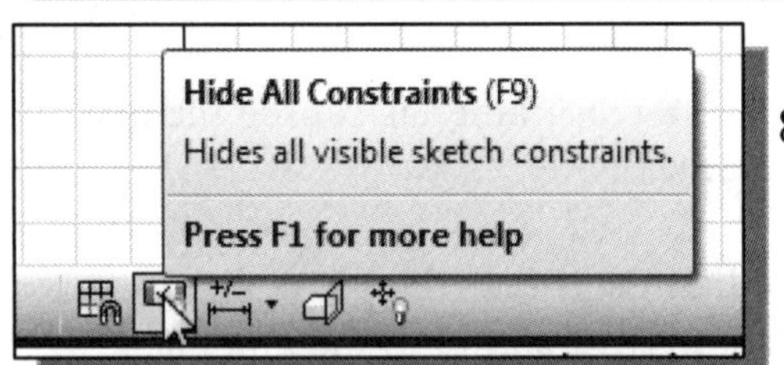

8. Click on the **Display/Hide Constraints** button in the status toolbar to hide all the constraints.

Using the Auto Dimension Command

➢ In *Autodesk Inventor*, the **Auto Dimension** command can be used to assist in creating a fully constrained sketch. **Fully constrained** sketches can be updated more predictably as design changes are implemented. The general procedure for applying dimensions to sketches is to use the **General Dimension** command to add the more critical dimensions, and then use the **Auto Dimension** command to add the additional dimensions/constraints to fully constrain the sketch. The Auto Dimension command can also be used to apply the missing dimensions that are needed. It is also important to realize that different sets of dimensions and geometric constraints can be applied to the same sketch to accomplish a fully constrained geometry.

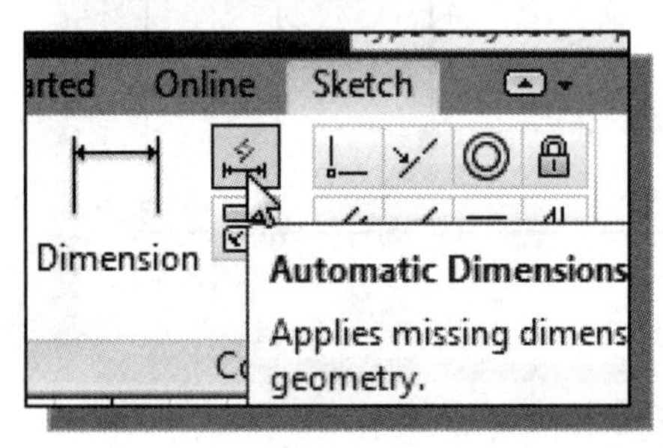

1. Click on the **Auto Dimension** icon in the *2D Sketch* panel.

❖ Note that *Autodesk Inventor* confirms that the sketch is fully constrained with the message "*0 Dimensions Required.*"

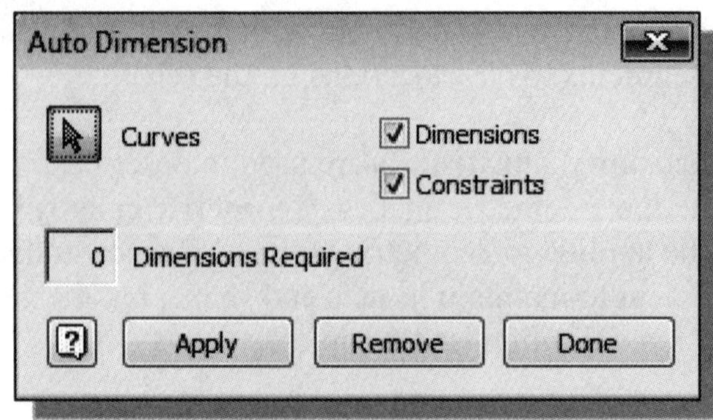

2. Click **Done** to exit the Auto Dimension command.

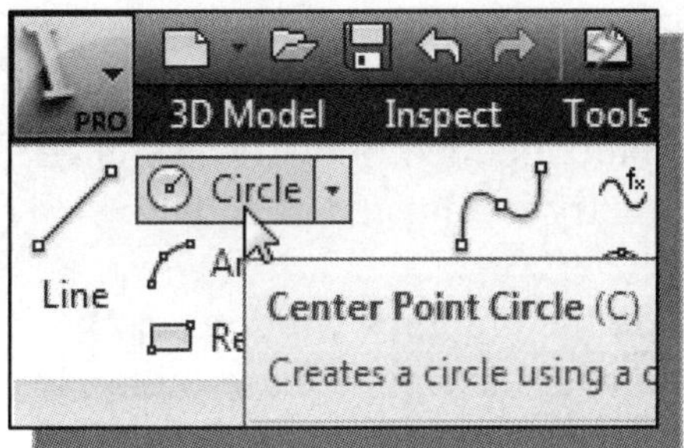

3. Select the **Center point circle** command by clicking once with the left-mouse-button on the icon in the *Sketch* toolbar.

4. On your own, create a circle of arbitrary size inside the triangle as shown below.

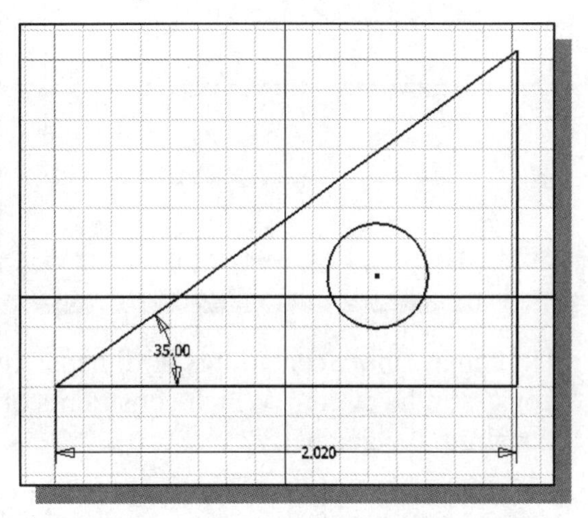

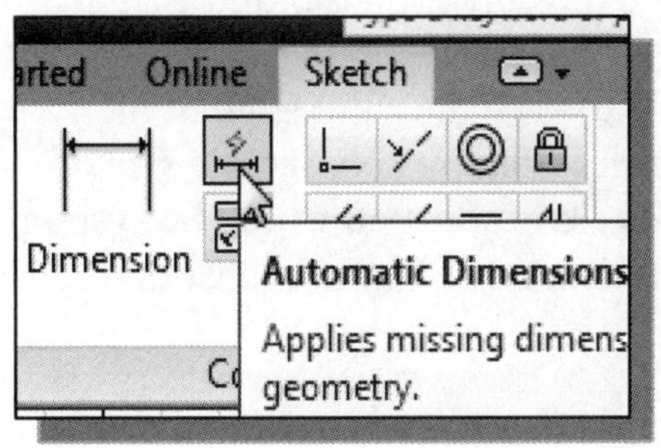

5. Click on the **Auto Dimension** icon in the *2D Sketch* panel.

❖ Note that *Autodesk Inventor* confirms that the sketch is not fully constrained and "*3 Dimensions Required*" to fully constrain the circle. What are the dimensions and/or constraints that can be applied to fully constrain the circle?

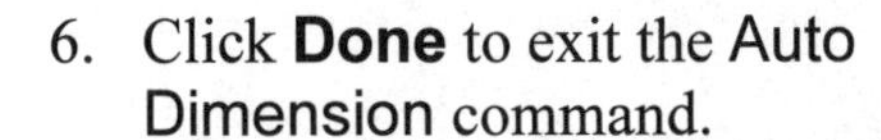

6. Click **Done** to exit the Auto Dimension command.

Auto Dimension

Curves

Dimensions

Constraints

3 Dimensions Required

Apply

Remove

Done

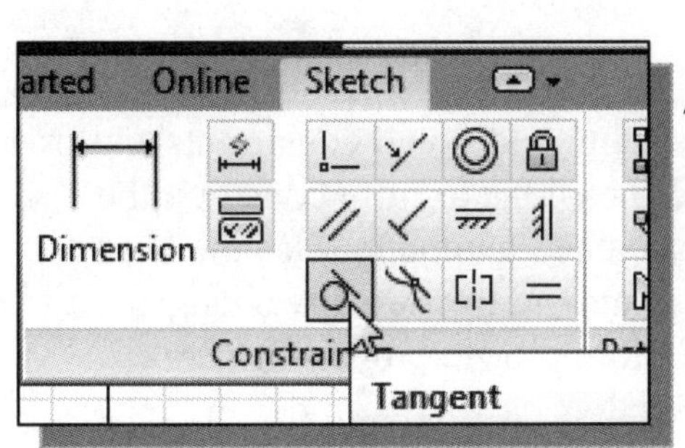

7. Click on the **Tangent** constraint icon in the *Sketch* toolbar.

8. Pick the circle by left-mouse-clicking once on the geometry.

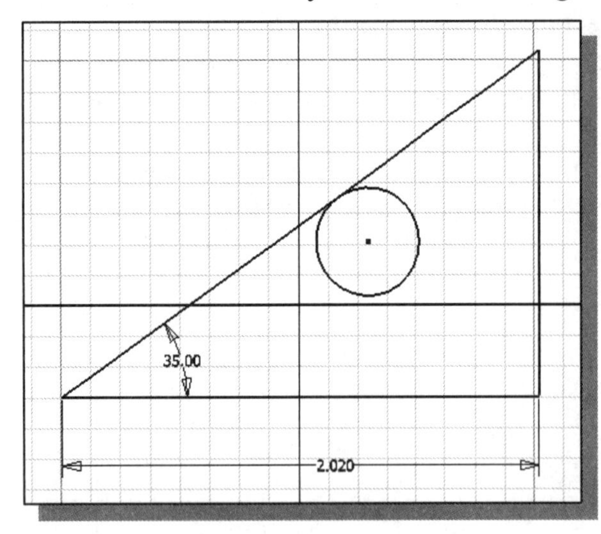

9. Pick the inclined line. The sketched geometry is adjusted as shown.

10. Inside the graphics window, click once with the right-mouse-button to display the option menu. Select **OK** in the popup menu to end the **Tangent** command.

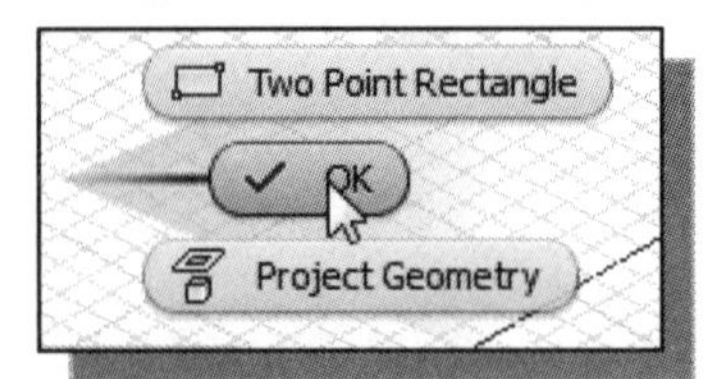

- How many more constraints or dimensions do you think will be necessary to fully constrain the circle? Which constraints or dimensions would you use to fully constrain the geometry?

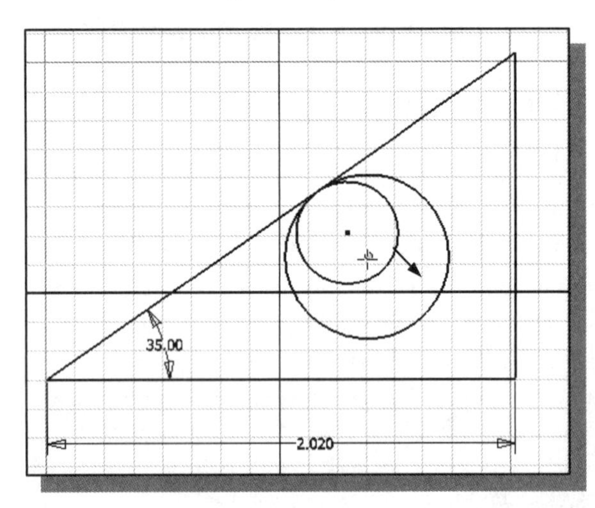

11. Move the cursor on top of the right side of the circle, and then drag the circle toward the right edge of the graphics window. Notice the size of the circle is adjusted while the system maintains the **Tangent** constraint.

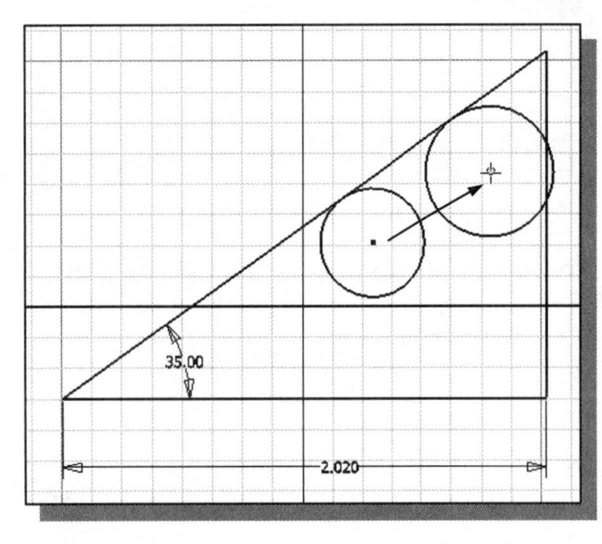

12. Drag the center of the circle toward the upper right direction. Notice the **Tangent** constraint is always maintained by the system.

➢ On your own, experiment with adding additional constraints and/or dimensions to fully constrain the sketched geometry. Use the **Undo** command to undo any changes before proceeding to the next section.

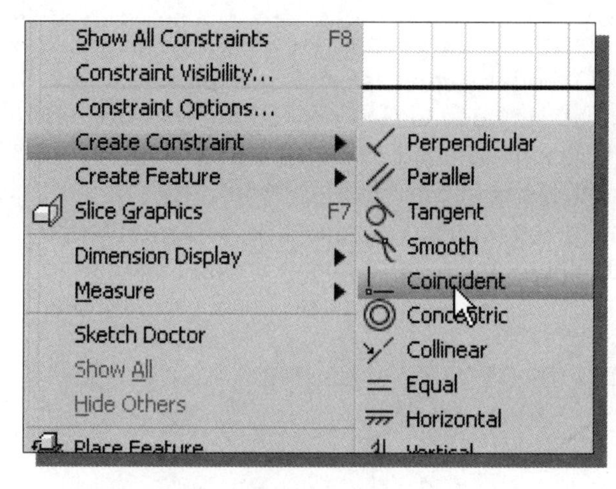

13. Inside the graphics window, click once with the **right-mouse-button** to display the option menu. Select **Create Constraint → Coincident** in the popup menus.

- The option menu is a quick way to access many of the commonly used commands in *Autodesk Inventor*.

14. Pick the vertical line.

15. Pick the center of the circle to align the center of the circle and the vertical line.

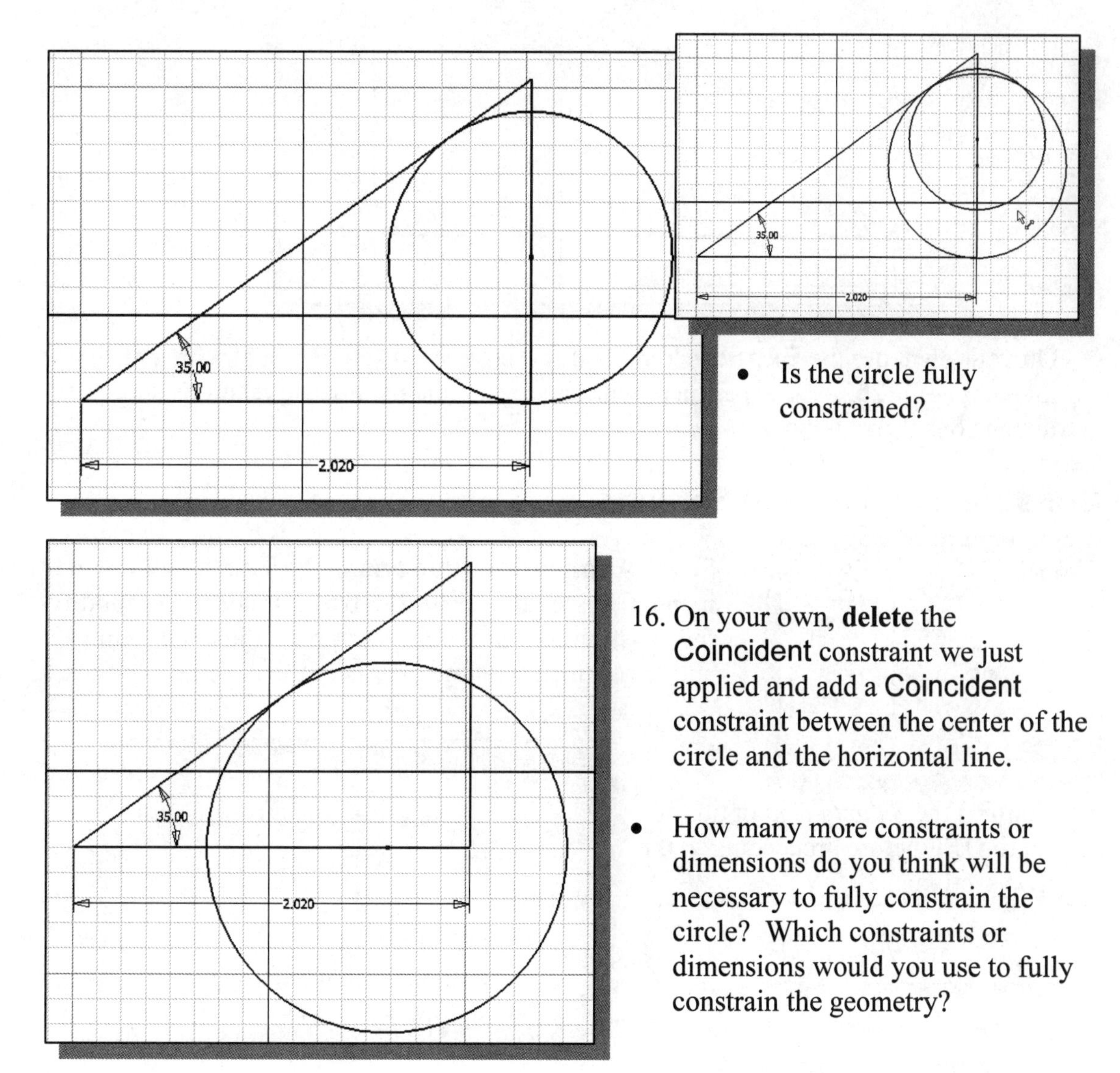

- Is the circle fully constrained?

16. On your own, **delete** the Coincident constraint we just applied and add a Coincident constraint between the center of the circle and the horizontal line.

- How many more constraints or dimensions do you think will be necessary to fully constrain the circle? Which constraints or dimensions would you use to fully constrain the geometry?

❖ The application of different constraints affects the geometry differently. The design intent is maintained in the CAD model's database and thus allows us to create very intelligent CAD models that can be modified/revised fairly easily. On your own, experiment and observe the results of applying different constraints to the triangle. For example: (1) add another **Fix** constraint to the top corner of the triangle; (2) delete the horizontal dimension and add another **Fix** constraint to the left corner of the triangle; and (3) add another **Tangent** constraint and add the size dimension to the circle.

17. On your own, modify the 2D sketch as shown below.

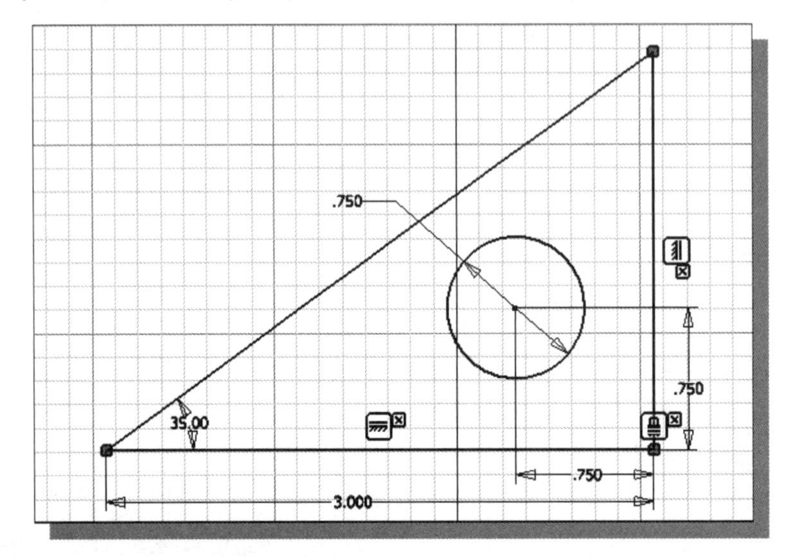

➢ On your own, use the **Extrude** command and create a 3D solid model with a plate thickness of **0.25**. Also experiment with modifying the parametric relations and dimensions through the *part browser*.

Constraint and Sketch Settings

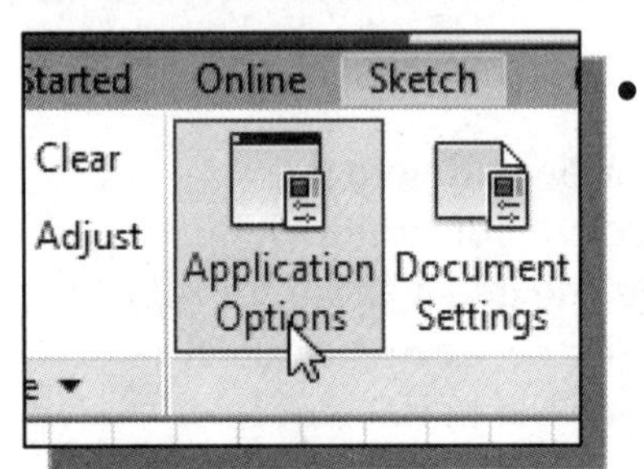

- Select **Application Options** in the **Tools** pull-down menu. Click on the **Sketch** tab to display and/or modify the constraint settings. On your own, adjust the settings and experiment with the effects of the different settings.

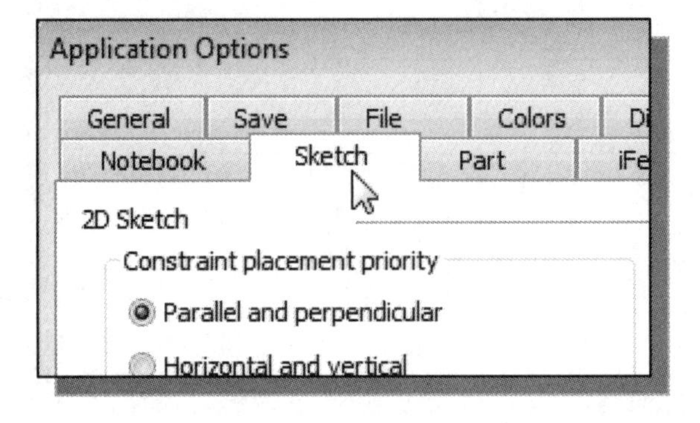

❖ Confirm the *Snap to grid* option is set to **OFF** before proceeding to the next section.

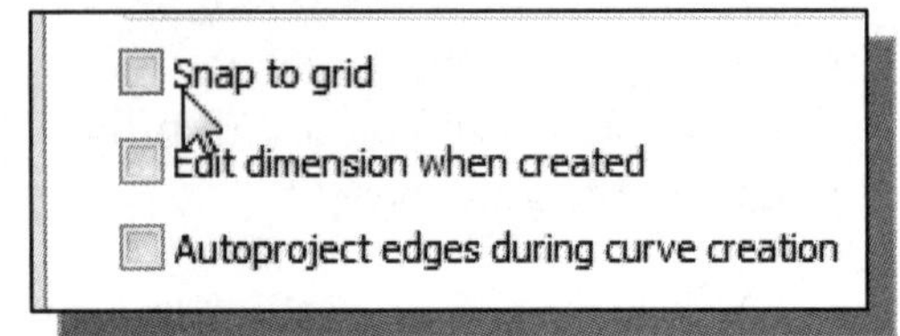

The BORN Technique

In the previous chapter, we have chosen the *first feature* to be an extruded solid object. All subsequent features, therefore, are built by referencing the first feature, or ***base feature***. The *base feature* is the center of all features and is considered to be the key feature of the design. This approach places much emphasis on the selection of the *base feature*. In most cases, this approach is quite adequate and proper in creating the solid models.

A more advanced technique of creating solid models is what is known as the "**Base Orphan Reference Node**" (**BORN**) technique. The basic concept of the BORN technique is to use a *Cartesian coordinate system* as the first feature prior to creating any solid features. With the *Cartesian coordinate system* established, we then have three mutually perpendicular datum planes (namely the *XY*, *YZ*, and *ZX planes*) available to use as sketching planes. The three datum planes can also be used as references for dimensions and geometric constructions. Using this technique, the first node in the history tree is called an "orphan," meaning that it has no history to be replayed. The technique of creating the reference geometry in this "base node" is therefore called the "Base Orphan Reference Node" (BORN) technique.

Autodesk Inventor automatically establishes a set of reference geometry when we start a new part, namely a *Cartesian coordinate system* with three work planes, three work axes, and a work point. All subsequent solid features can then use the coordinate system and/or reference geometry as sketching planes. The *base feature* is still important, but the *base feature* is no longer the ONLY choice for selecting the sketching plane for subsequent solid features. This approach provides us with more options while we are creating parametric solid models. More importantly, this approach provides greater flexibility for part modifications and design changes. This approach is also very useful in creating assembly models.

Sketch Plane Settings

1. Select **Application Options** in the **Tools** menu as shown.

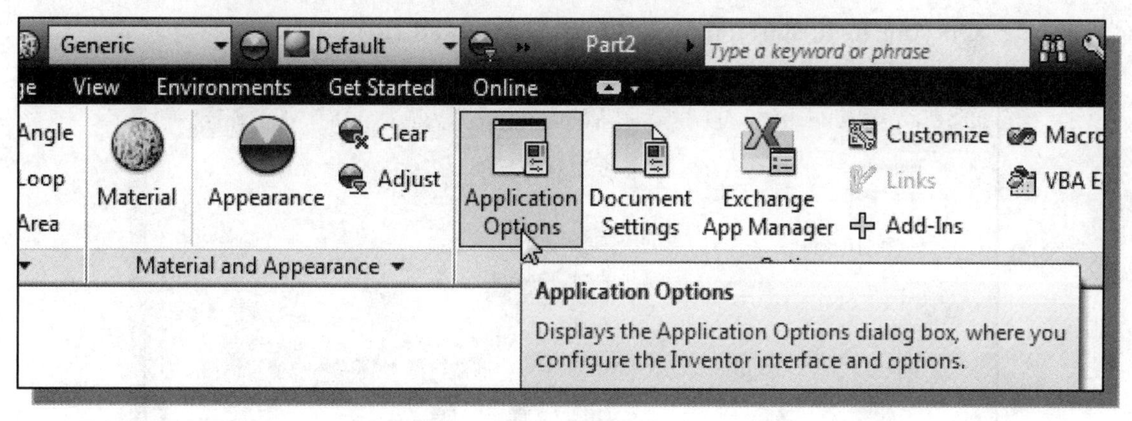

- The **Application Options** menu allows us to set behavioral options, such as *color*, *file locations*, etc.

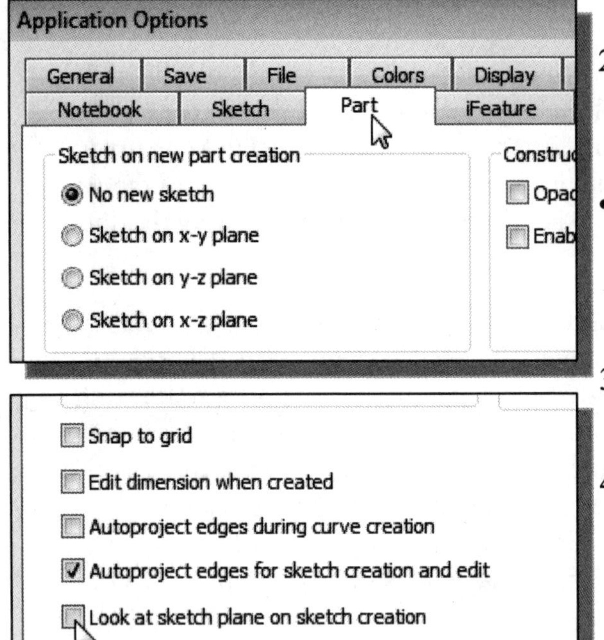

2. Click on the **Part** tab to display and/or modify the default sketch plane settings.

- Note the option to setup the sketch plane to be used during new part creation is available. Confirm the **No new sketch** option is set as shown.

3. Click on the **Sketch** tab to examine/ modify the default sketching settings.

4. Turn ***OFF*** the *Snap to grid* option, if you have not already, and *Look at sketch plane on sketch creation* options.

5. Click on the **OK** button to accept the setting.

- Note that the new settings do not change the current part file; the settings will take effect when a new part file is opened.

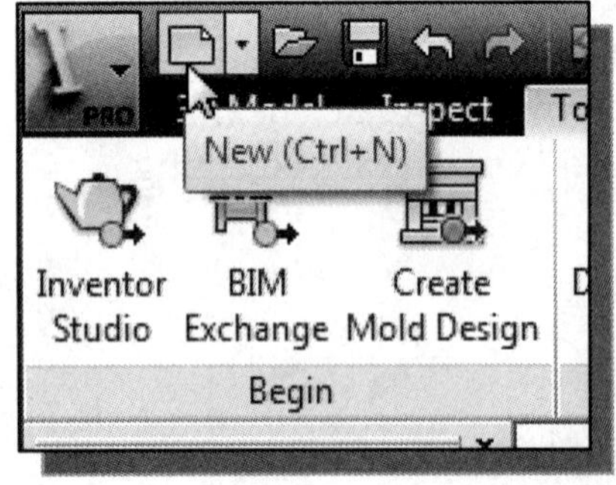

6. Click on the **New** icon in the *Standard* toolbar.

7. On your own, start a new **English – standard (in) part** file.

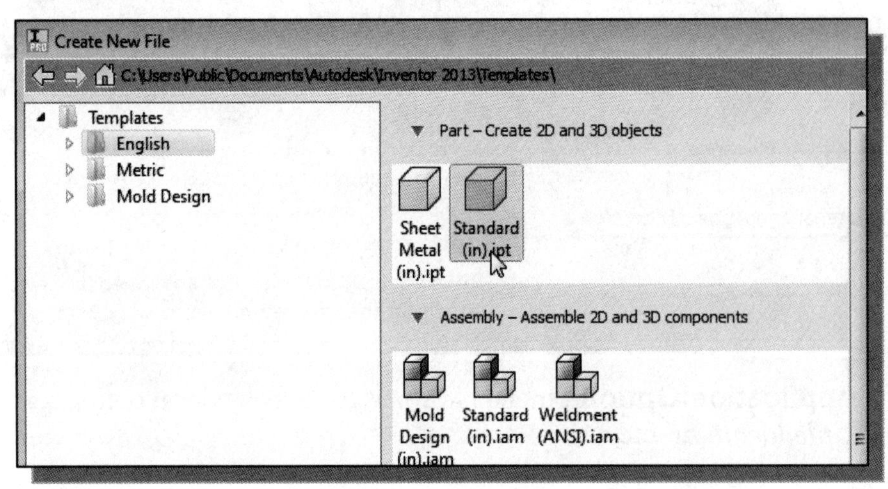

Applying the BORN Technique

1. In the *part browser* window, click on the [+] symbol in front of the **Origin** feature to display more information on the feature.

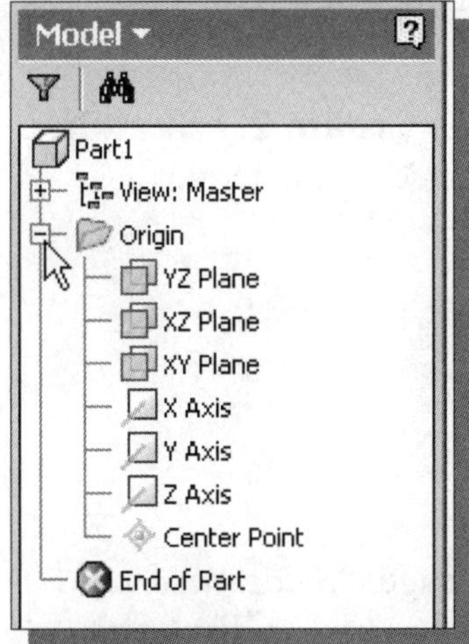

❖ In the *part browser* window, notice a new part name appeared with seven work features established. The seven work features include three *workplanes*, three *work axes*, and a *work point*. By default, the three workplanes and work axes are aligned to the **world coordinate system** and the work point is aligned to the *origin* of the **world coordinate system**.

2. Inside the *browser* window, move the cursor on top of the third work plane, the **XY Plane**. Notice a rectangle, representing the workplane, appears in the graphics window.

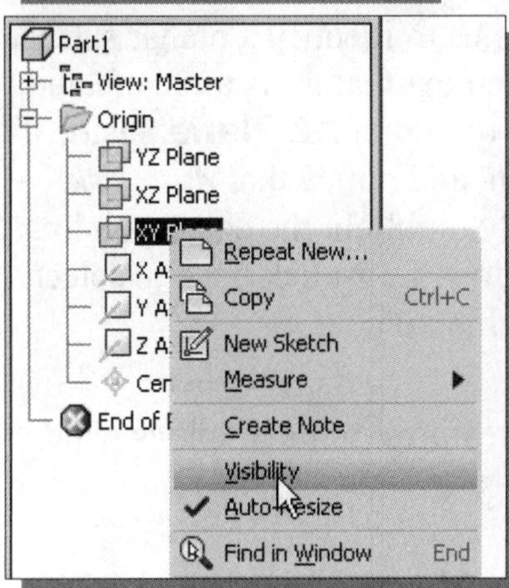

3. Inside the *browser* window, click once with the right-mouse-button on **XY Plane** to display the option menu. Click on **Visibility** to toggle on the display of the plane.

4. On your own, repeat the above steps and toggle ***ON*** the display of all of the *workplanes*, *work axes*, and the *center point* on the screen.

5. On your own, use the *Dynamic Viewing* options (**ViewCube, 3D Orbit, Zoom** and **Pan**) to view the default work features.

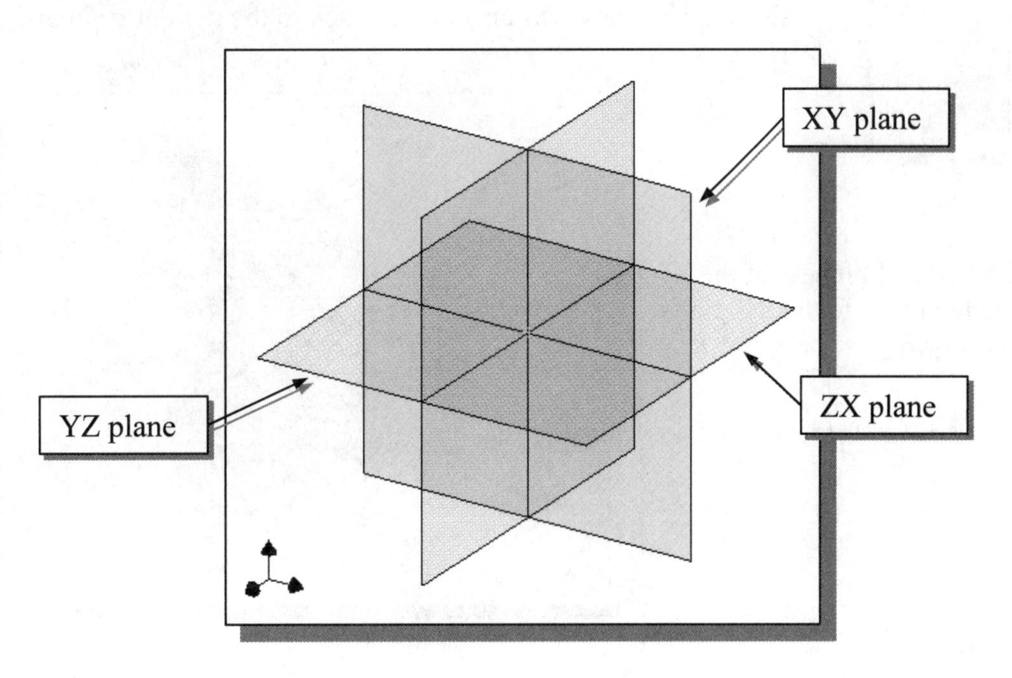

❖ By default, the basic set of work planes is aligned to the world coordinate system; the work planes are the first features of the part. We can now proceed to create solid features referencing the three mutually perpendicular datum planes. Instead of using only the default sketching plane as the starting point, we can now select any of the work planes as the sketching planes for subsequent solid features.

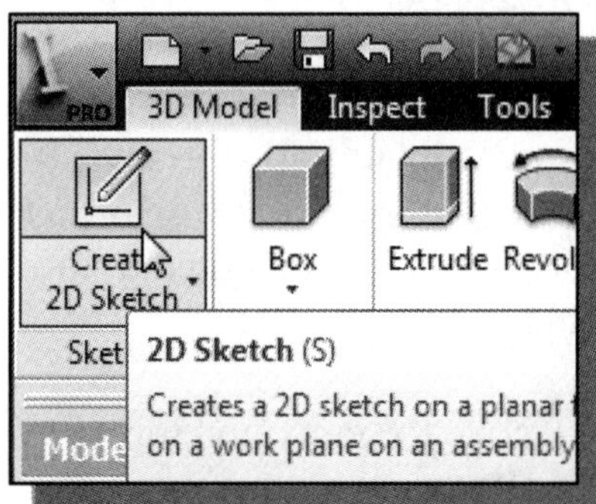

6. In the *Sketch* panel select the **Create 2D Sketch** command by left-clicking once on the icon.

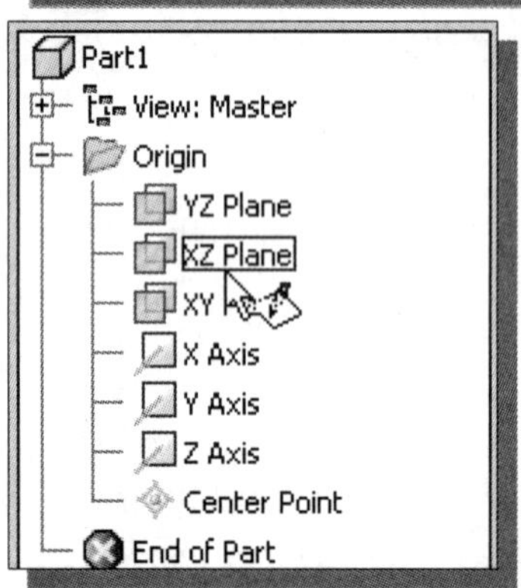

7. In the *Status Bar* area, the message: "*Select face, workplane, sketch or sketch geometry.*" is displayed. *Autodesk Inventor* expects us to identify a planar surface where the 2D sketch of the next feature is to be created. Move the graphics cursor on top of **XZ Plane**, inside the *browser* window as shown, and notice that *Autodesk Inventor* will automatically highlight the corresponding plane in the graphics window. Left-click once to select the **XZ Plane** as the sketching plane.

❖ *Autodesk Inventor* allows us to identify and select features in the graphics window as well as in the *browser* window.

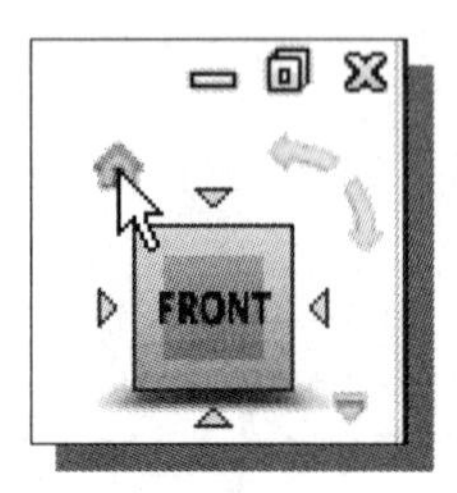

8. Single left-mouse-click to activate the **Home View** option as shown. The view will be adjusted back to the default ***isometric view***.

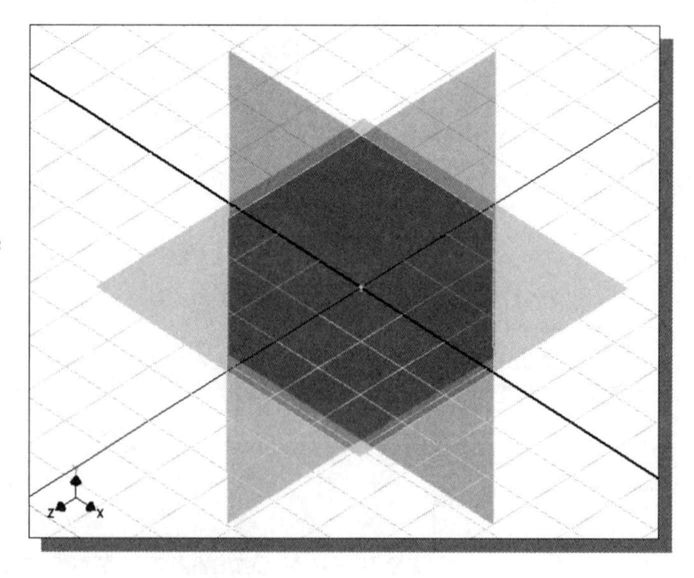

➢ Note the alignment of the sketch plane to the XZ plane as shown.

Create the 2D Sketch for the Base Feature

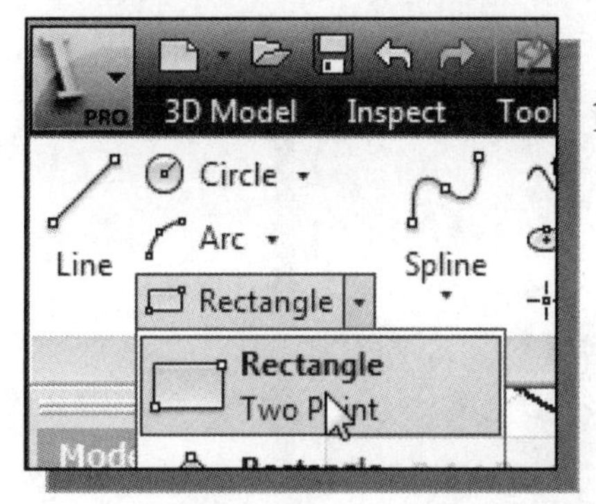

1. Select the **Two point rectangle** command by clicking once with the **left-mouse-button** on the icon in the *Draw* toolbar.

2. Create a rectangle of arbitrary size by selecting two locations on the screen as shown below. To demonstrate the effects of parametric equations, we will intentionally position the rectangle away from the *center point* of the coordinate system.

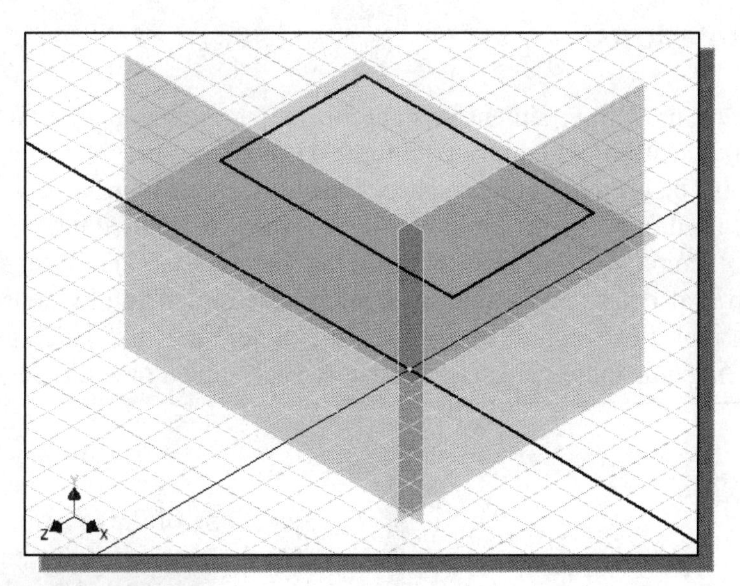

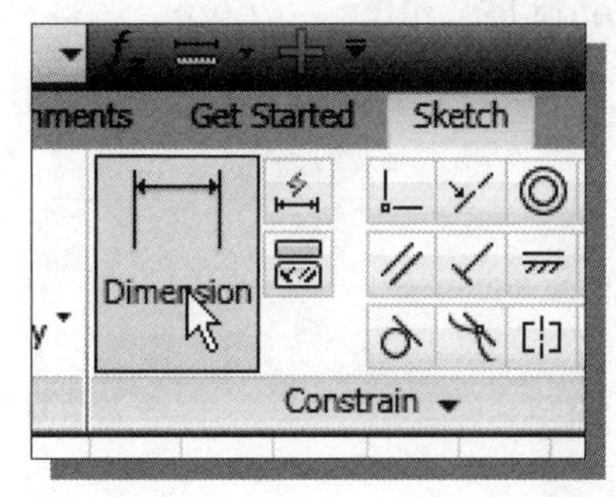

3. On your own, use the **General Dimension** command and create the size dimension of the rectangle as shown in the figure below.

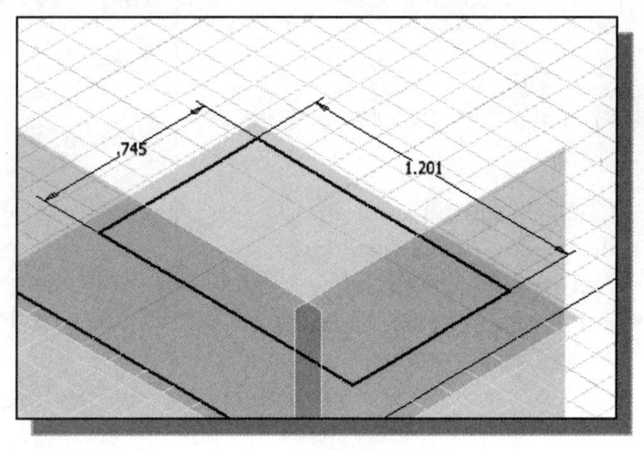

4. Repeat the above steps and also create the location dimensions, measured to the center point as shown. Do not exit the **Dimension** command.

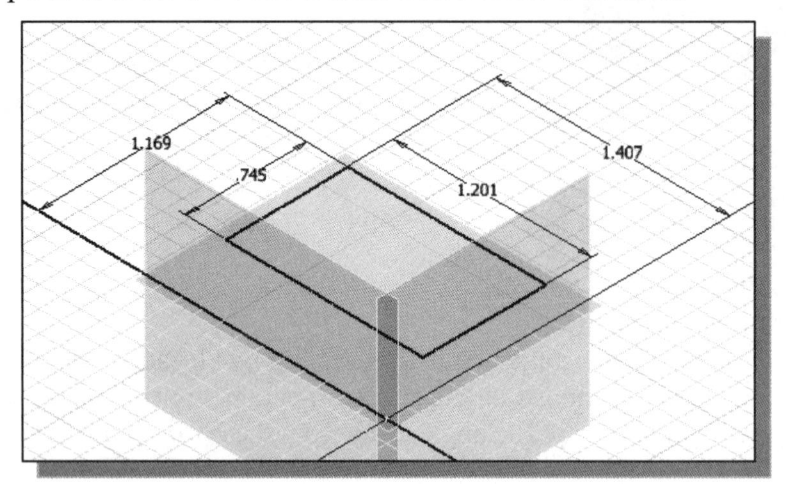

Parametric Relations

- In parametric modeling, dimensions are design parameters that are used to control the size and location of geometric features. Dimensions are more than just values; they can also be used as feature control variables. Initially in *Autodesk Inventor*, values are used to create different geometric entities. The text created by the **Dimension** command also reflects the actual location or size of the entity. Each dimension is also assigned a name that allows the dimension to be used as a control variable. The default format is "dxx," where the "xx" is a number that *Autodesk Inventor* increments automatically each time a new dimension is added.

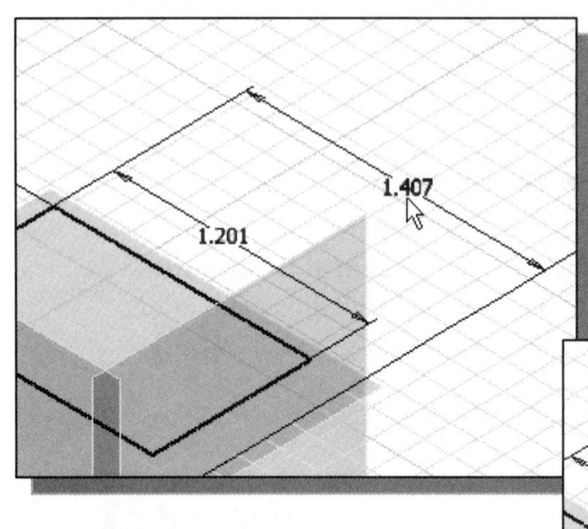

1. Select one of the location dimensions; note the associated variable name is displayed on top of the *Edit box*.

2. Click on the adjacent size dimension of the rectangle. Notice the variable name (**d0**) is automatically entered in the *Edit Dimension* window.

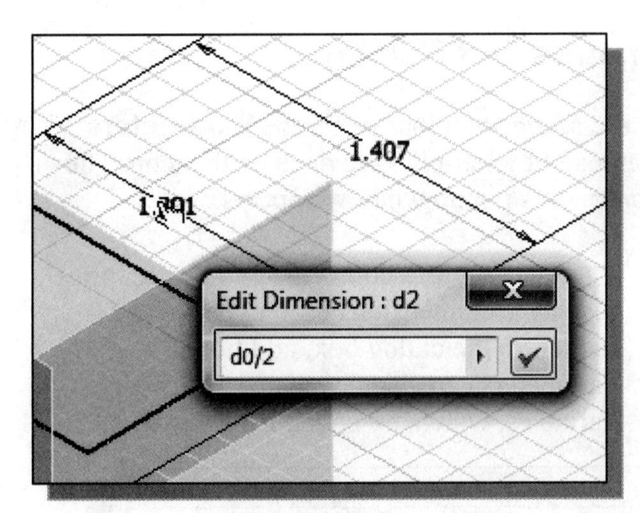

3. In the *Edit Dimension* window, enter **/2** to set the horizontal location dimension of the circle to be one-half of the width of the rectangle.

4. Click on the ***check mark*** button to close the *Edit Dimension* window.

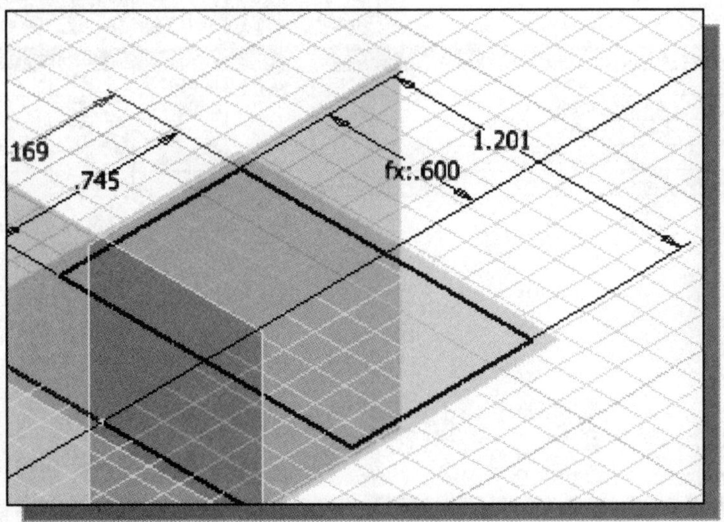

❖ Notice, the derived dimension values are displayed with **fx** in front of the numbers. The parametric relation we entered is used to control the location of the rectangle; the location is based on the size of the rectangle.

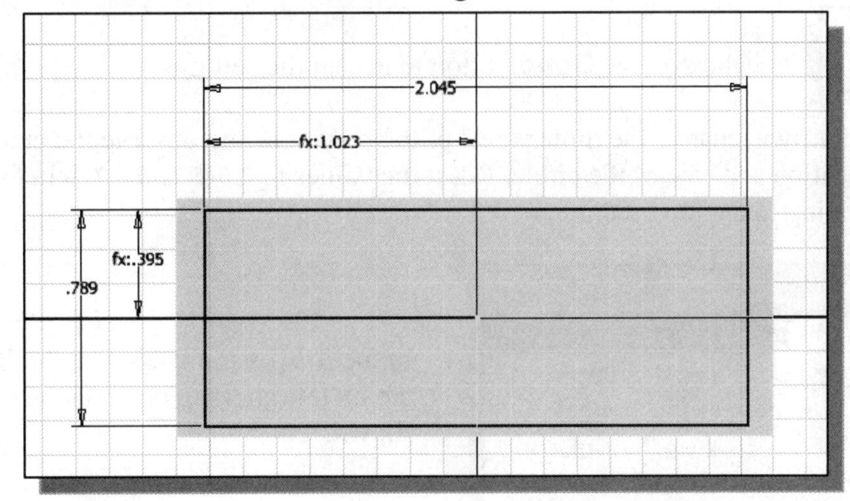

5. Repeat the above steps and create another parametric relation for the other location dimension. Also **modify** the two **size dimensions** as shown.

Viewing the Established Parameters and Relations

1. In the *Manage* toolbar select the **Parameters** command by left-clicking once on the icon. The *Parameters* popup window appears.

   - The **Parameters** command can be used to display all dimensions used to define the model. In the dialog box, additional parameters can also be created as design variables, which are called ***user parameters***.

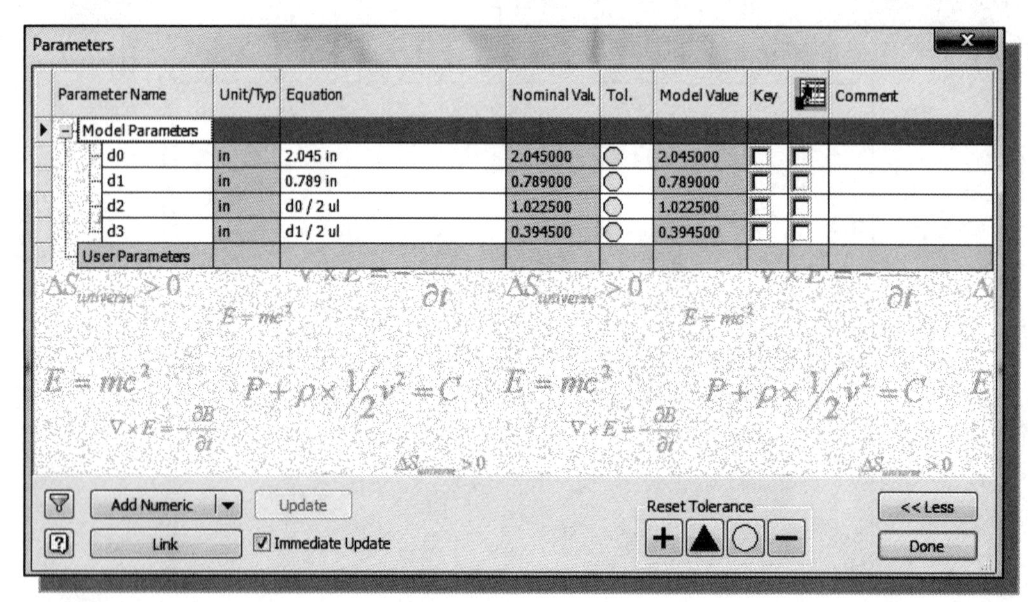

- Note that parametric equations can be entered and/or modified by double-clicking in the *Equation Edit Box*.

2. Click on the **Done** button to accept the settings.

3. On your own, change the dimensions of the rectangle and observe the changes to the location and size of the circle. Reset the values to **2.045 and 0.789** before continuing to the next section.

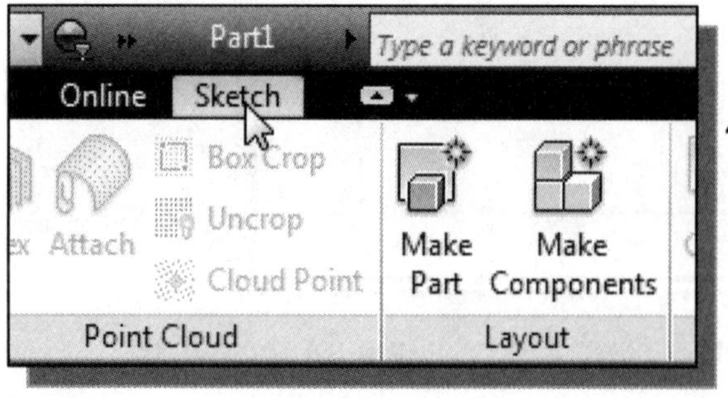

4. In the *Ribbon* toolbar select the **Sketch** tab by left-clicking once on the tab as shown.

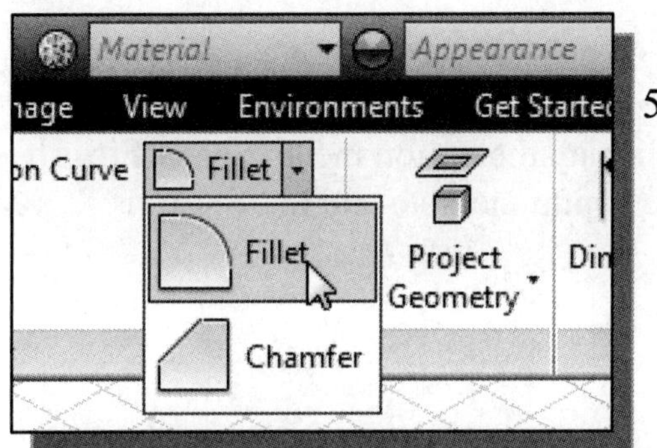

5. Select the **Fillet** command by clicking once with the left-mouse-button on the icon in the *Draw* toolbar.

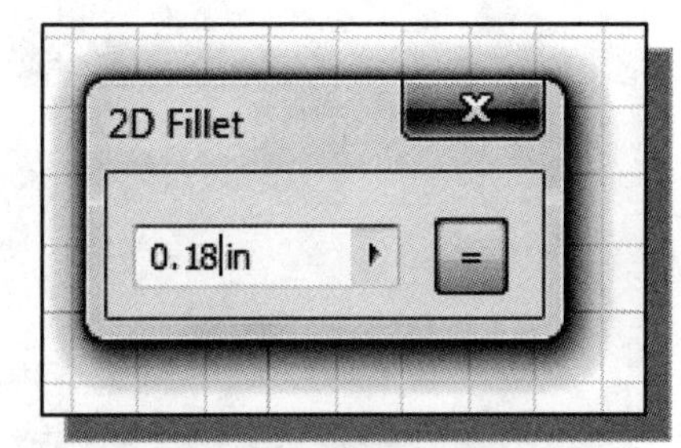

6. Enter **0.18** as the radius value of the fillet.

7. Create the two rounded corners toward the left side of the sketched geometry as shown.

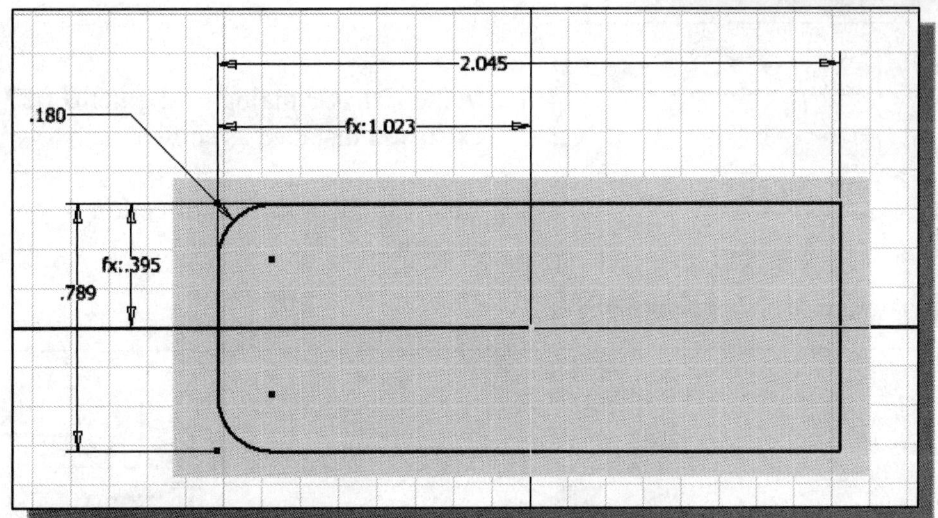

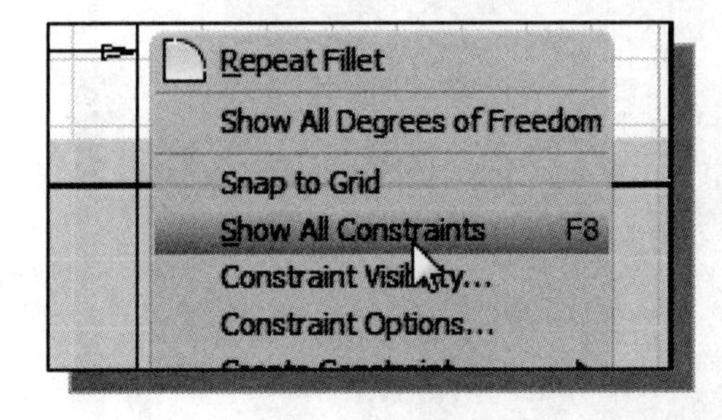

8. Inside the graphics window, click once with the **right-mouse-button** to display the option menu. Select **Show All Constraints** in the popup menu to show all the applied constraints.

- Note only one radius dimension is shown, as the **Equal length constraint** is applied to the two arcs.

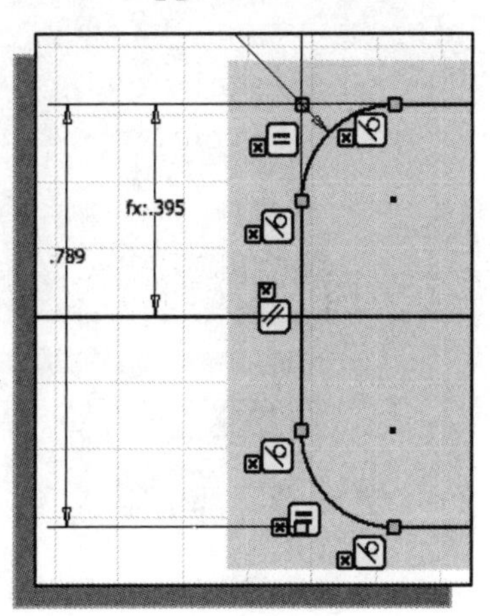

9. Hit the function key [**F9**] once to **turn off** the display of constraints.

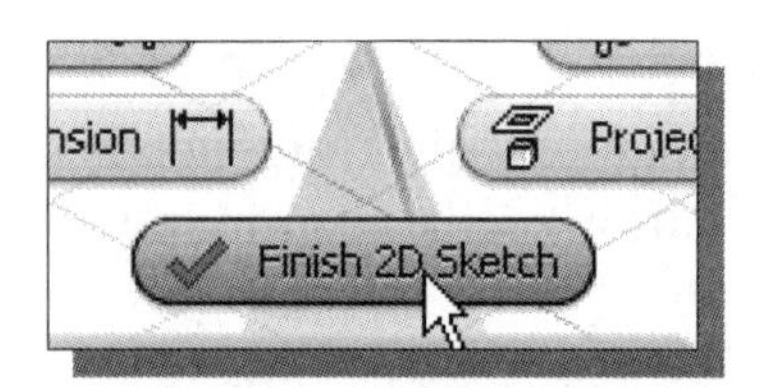

10. Inside the graphics window, click once with the right-mouse-button to display the option menu. Select **Finish 2D Sketch** in the popup menu to end the Sketch option.

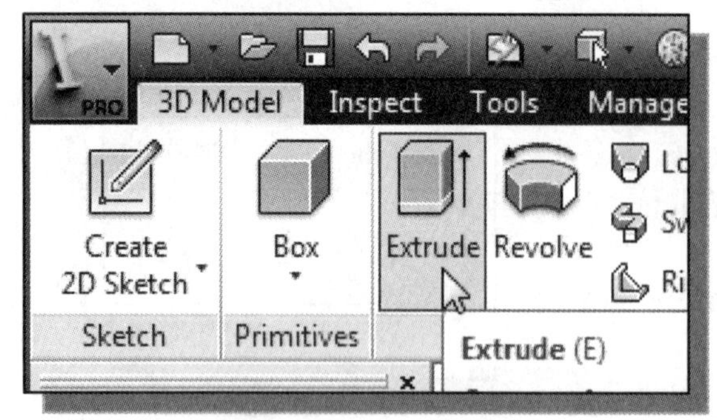

11. In the *Create* toolbar, select the **Extrude** command by left-mouse-clicking once on the icon.

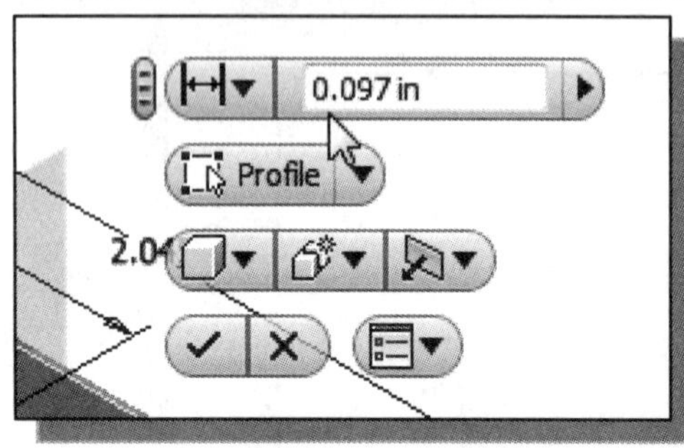

12. In the *Extrude* dialog box, enter **0.097** as the extrusion distance as shown.

13. Confirm the extrusion direction is upward as shown.

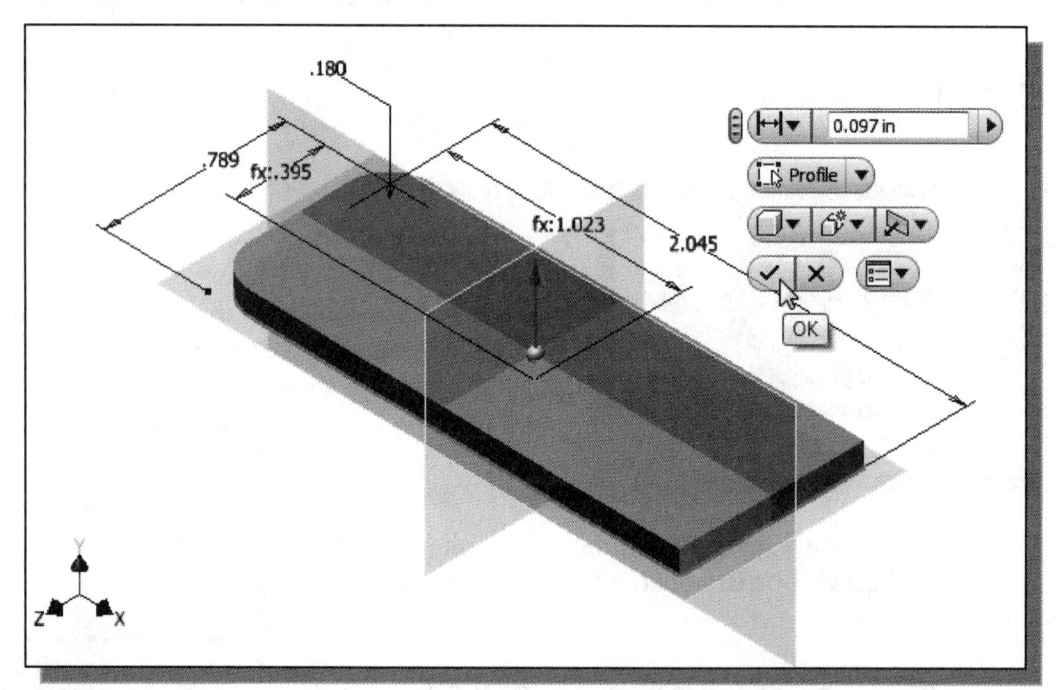

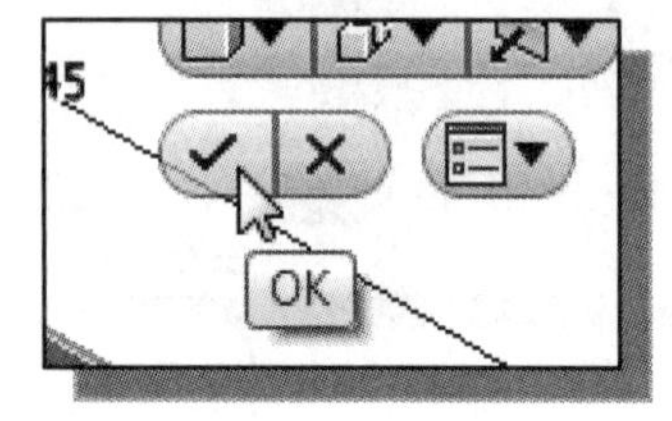

14. Click on the **OK** button to accept the settings and create the solid feature.

Sketches vs. Profiles

❖ In *Autodesk Inventor*, ***profiles*** are closed regions that are defined from ***sketches***. Profiles are used as cross sections to create solid features. For example, **Extrude**, **Revolve**, **Sweep**, **Loft**, and **Coil** operations all require the definition of at least a single profile. The sketches used to define a profile can contain additional geometry since the additional geometry entities are consumed when the feature is created. To create a profile we can create single or multiple closed regions, or we can select existing solid edges to form closed regions. A profile cannot contain self-intersecting geometry; regions selected in a single operation form a single profile. As a general rule, we should dimension and constrain profiles to prevent them from unpredictable size and shape changes. *Autodesk Inventor* does allow us to create under-constrained or non-constrained profiles; the dimensions and/or constraints can be added and edited later.

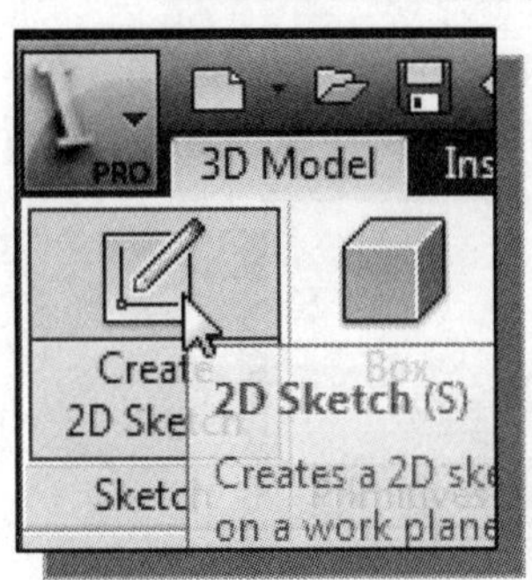

1. In the *Sketch* toolbar select the **Create 2D Sketch** command by left-clicking once on the icon.

2. Select the **top plane**, by clicking once with the left-mouse-button, as the sketching plane.

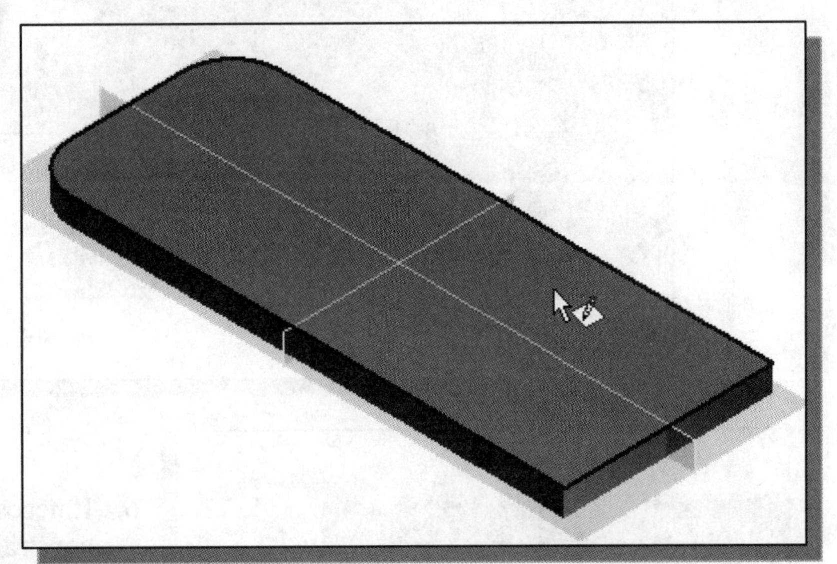

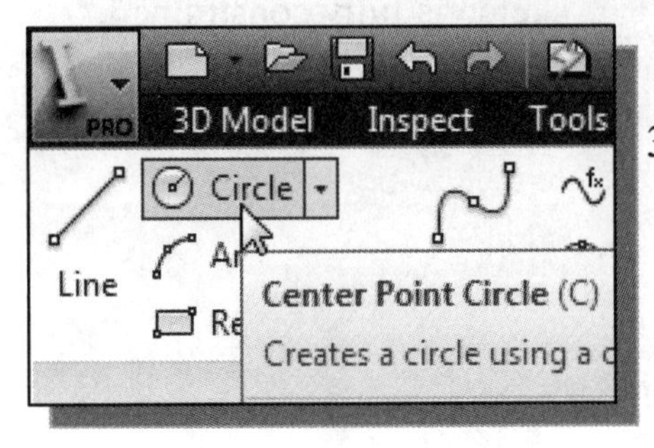

3. Select the **Center point circle** command by clicking once with the left-mouse-button on the icon in the *Draw* panel.

4. On your own, create four circles of arbitrary sizes as shown.

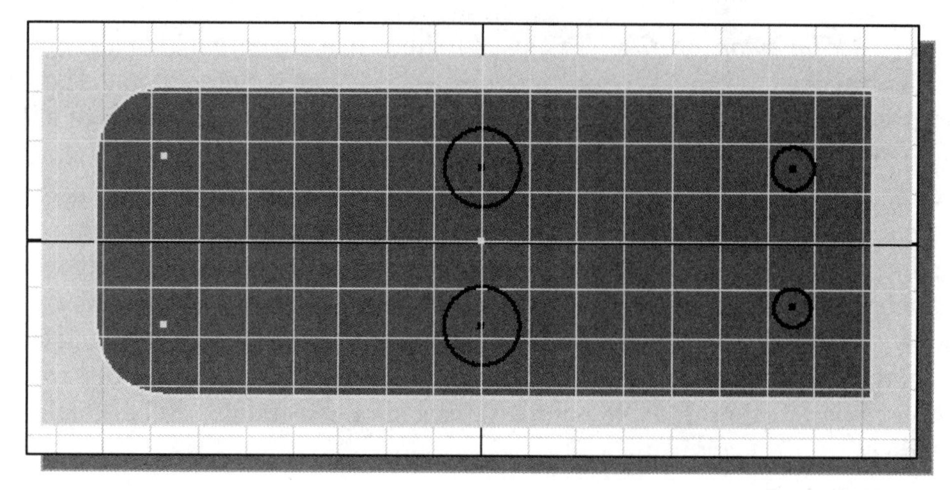

5. Also create the dimensions to define the sizes and location of the circles as shown.

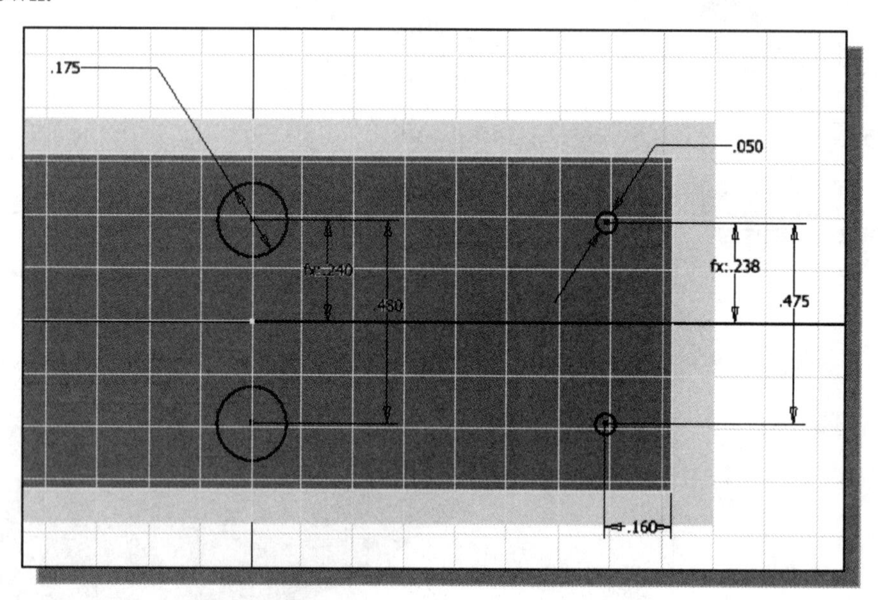

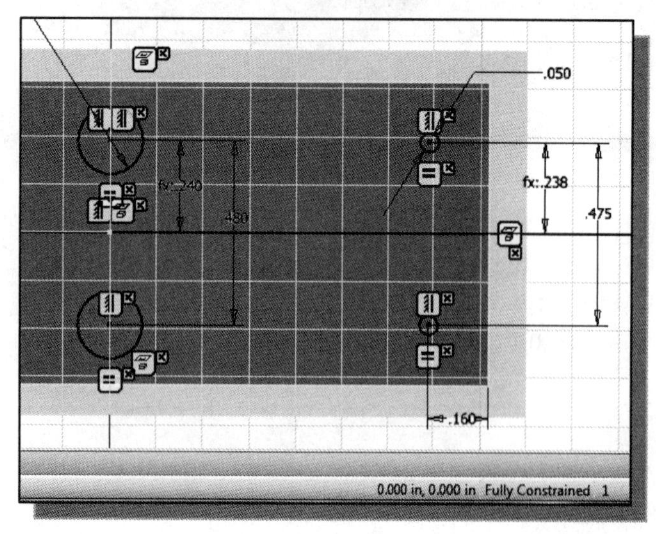

6. If necessary, apply additional constraints to assure the sketch is fully constrained.

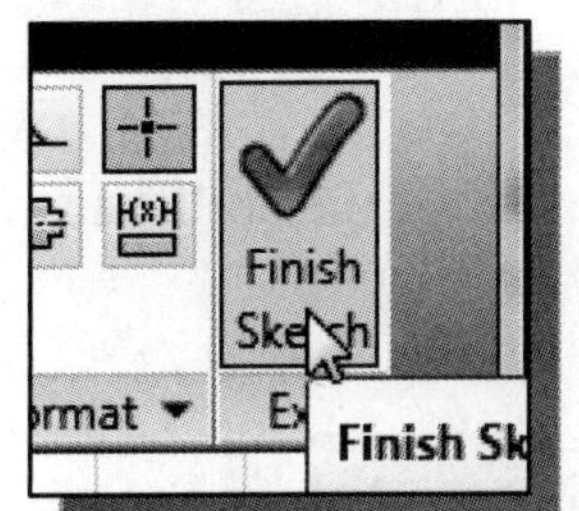

7. Select **Finish Sketch** in the *Ribbon* toolbar to exit the 2D Sketch module.

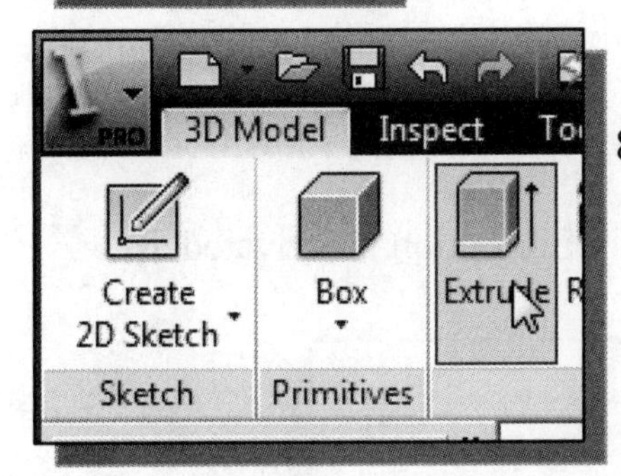

8. In the *Create* toolbar, select the **Extrude** command by left-mouse-clicking once on the icon.

9. Select two of the circular regions, by clicking on the inside of the circles, as shown. (Use the dynamic Zoom option to aid the selection.)

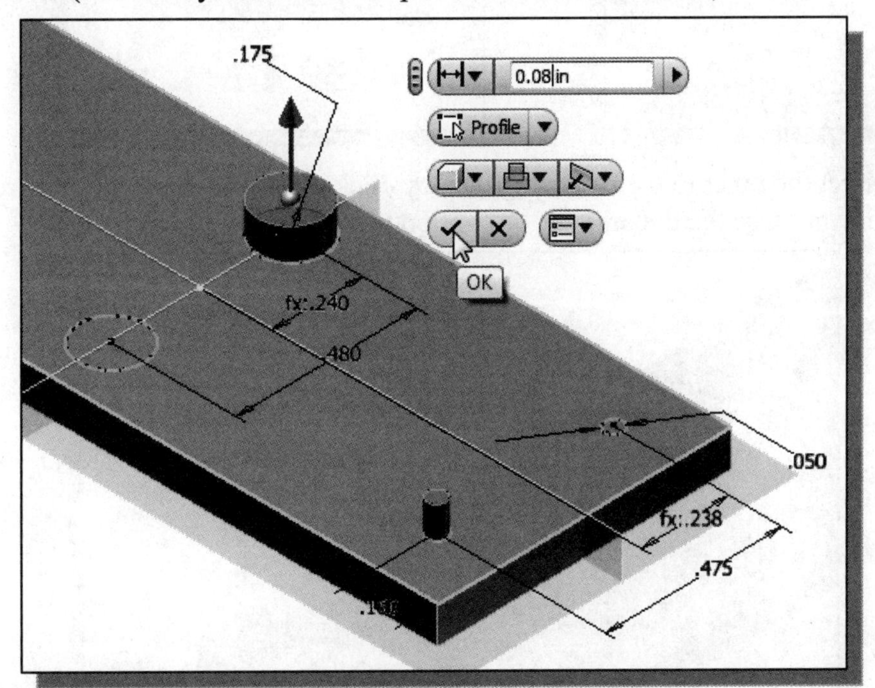

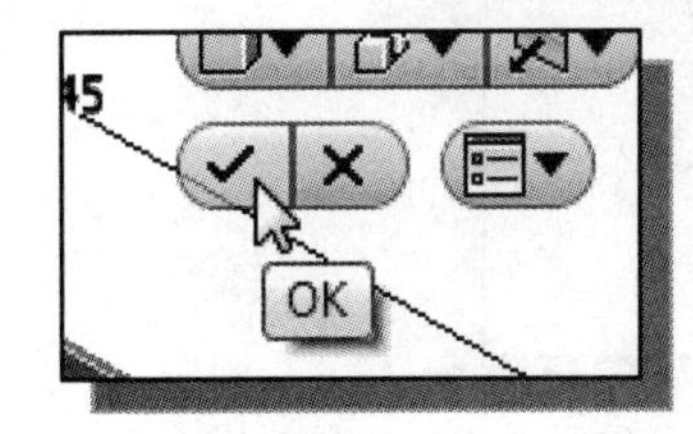

10. Set the extrusion distance to **0.08**, and click on the **OK** button to accept the settings and create the solid feature.

- Note that only the selected regions were used to define the profile, which was then used to create the solid feature. The additional geometric entities still exist in the 2D sketch.

Modify the Profile

- ***Profiles*** can also be adjusted by using the **Edit Feature** option. With the ability to edit the profiles, parametric modeling provides us the very powerful option of embedding design alternatives within the same model.

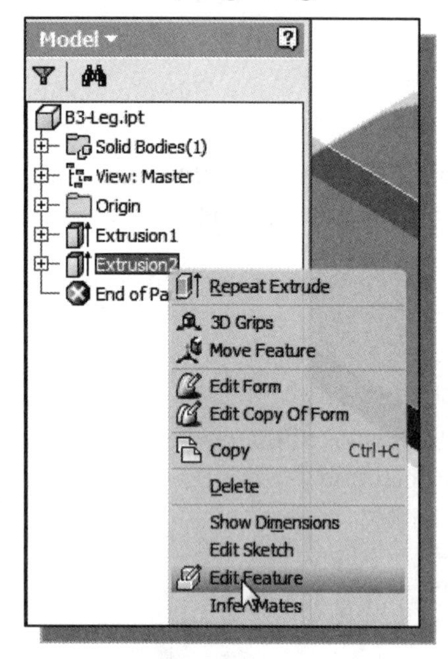

1. In the *Model Tree* area, right-mouse-click on the **Extrusion2** feature to bring up the option menu.

2. Select the **Edit Feature** option as shown.

3. Note the **Select Profile** option is activated as shown.

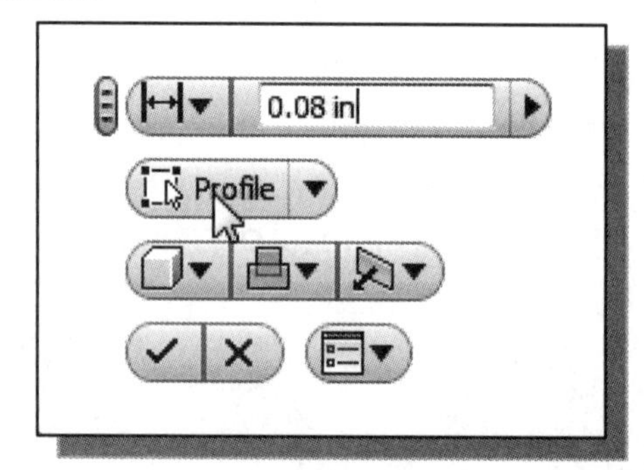

4. Select the other two circular regions, by clicking on the inside of the circles, as shown. (Use the dynamic Zoom option to aid the selection.)

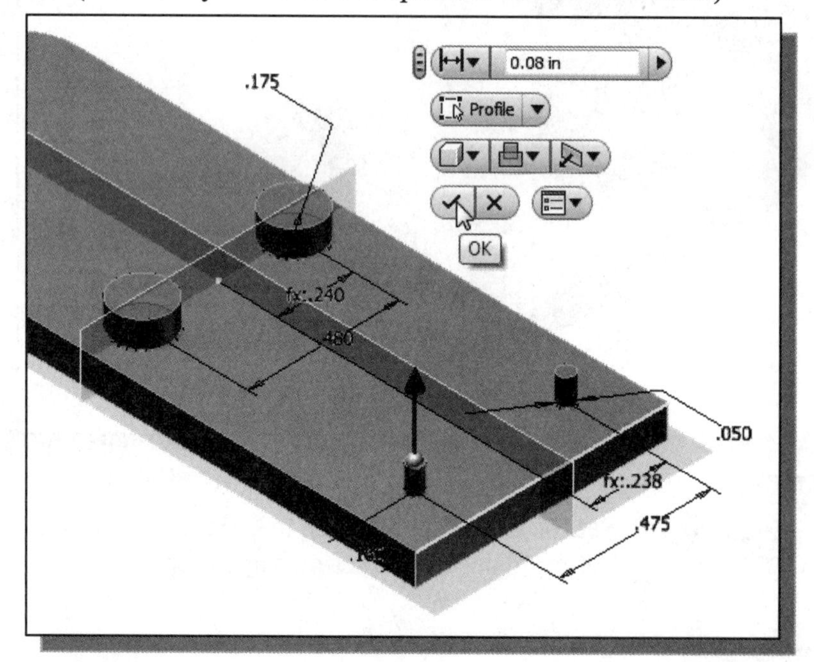

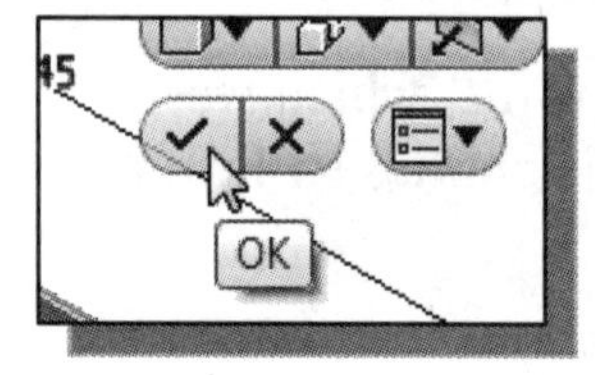

5. Click on the **OK** button to proceed with modifying the solid feature, which now includes the additional geometry.

Extrusion with the Taper Angle option

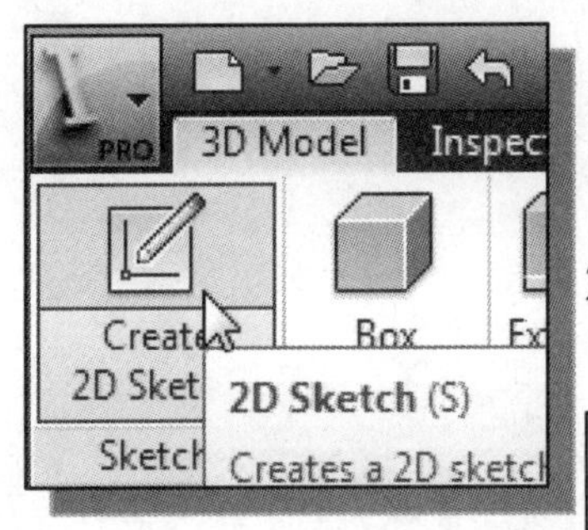

1. In the *Sketch* toolbar select the **Create 2D Sketch** command by left-clicking once on the icon.

2. Select the **top plane of the base feature**, by clicking once with the left-mouse-button, as the sketching plane.

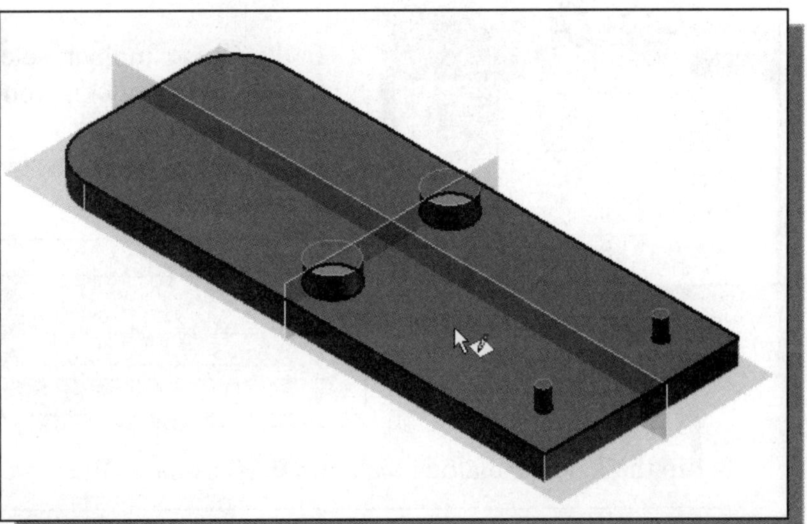

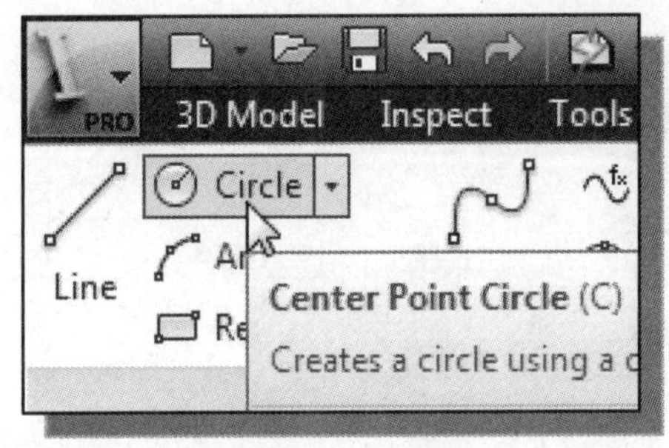

3. Select the **Center point circle** command by clicking once with the left-mouse-button on the icon in the *Draw* panel.

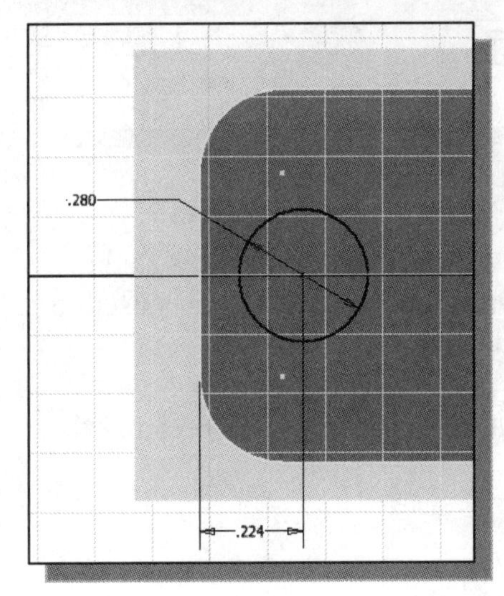

4. On your own, create a circle with the associated dimensions (diameter **0.28**) as shown.

5. Select **Finish Sketch** in the *Ribbon* toolbar to exit the 2D Sketch module.

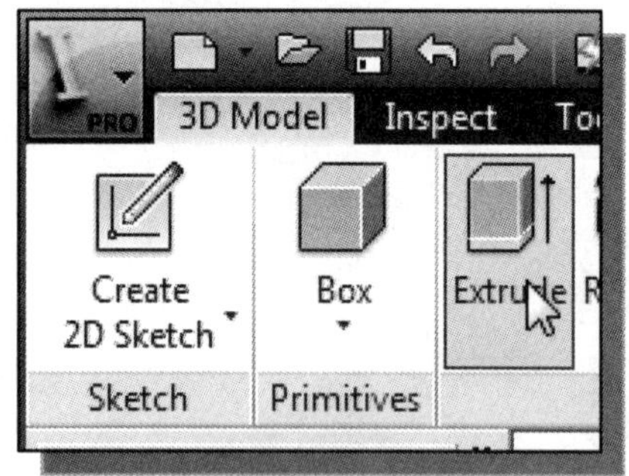

6. In the *Create* toolbar, select the **Extrude** command by left-mouse-clicking once on the icon.

7. Click on the **down arrow** icon to expand the *Extrude* dialog box.

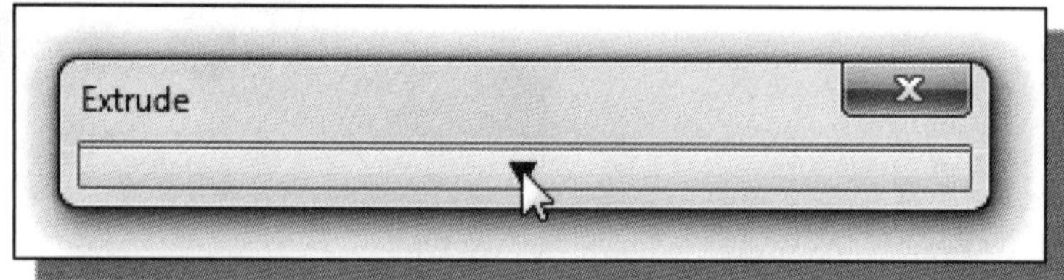

8. In the *Extrude* dialog box, enter **0.18** as the extrusion distance as shown.

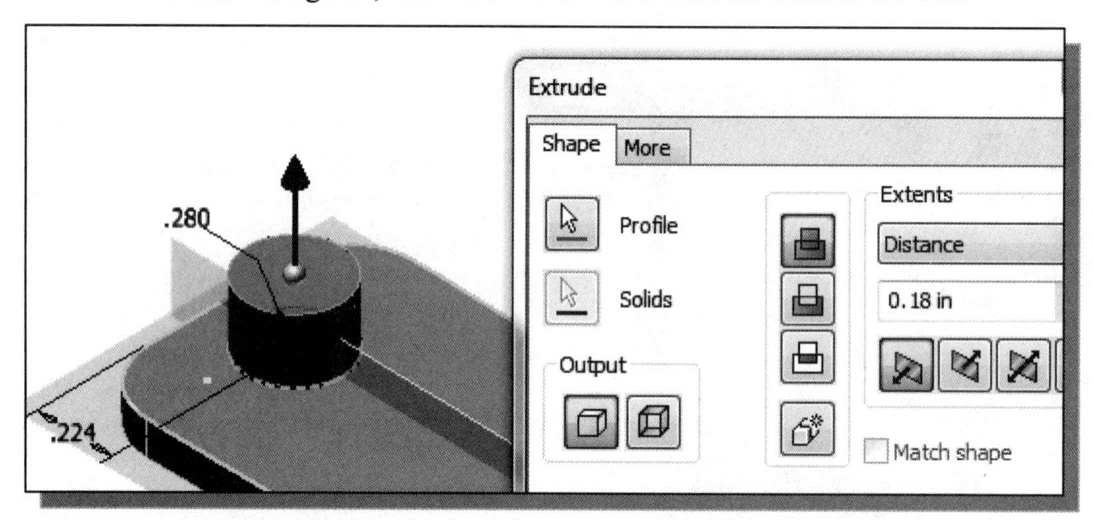

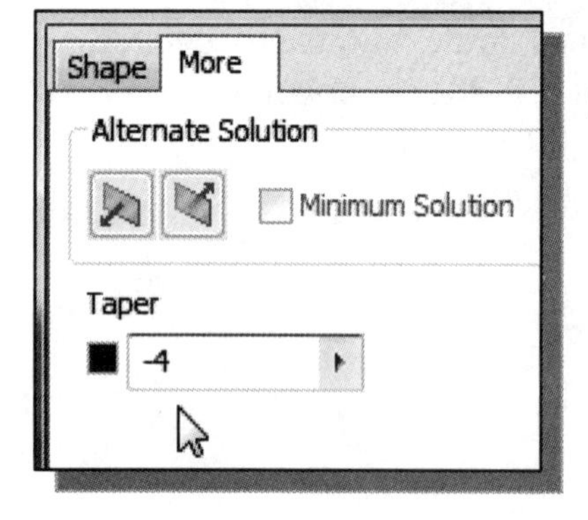

9. Click on the **More** tab and enter **-4** degrees as the taper angle as shown. (The negative value is used to set the direction of the taper.)

10. Click on the **OK** button to proceed with creating the solid feature.

11. On your own, repeat the above steps and create another tapered solid feature as shown. (**Diameter 0.22, extrusion distance 0.217, Taper angle 4 degrees.**)

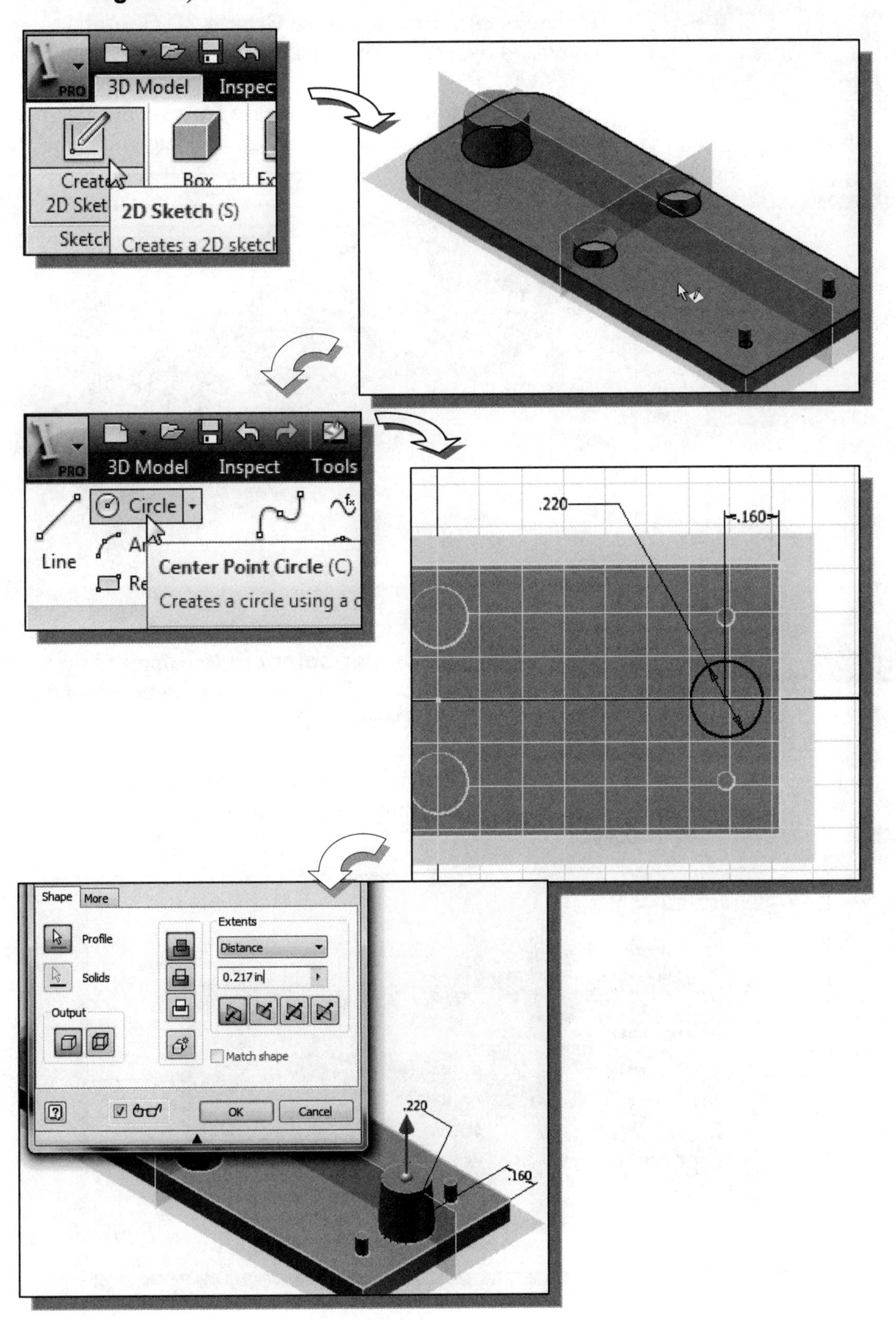

A Profile Containing Multiple Closed Regions

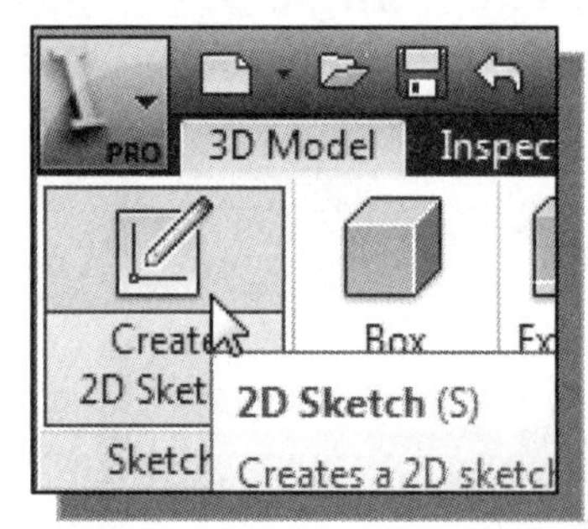

1. In the *Sketch* toolbar select the **Create 2D Sketch** command by left-clicking once on the icon.

2. Select the **bottom plane of the base feature**, by clicking once with the left-mouse-button, as the sketching plane.

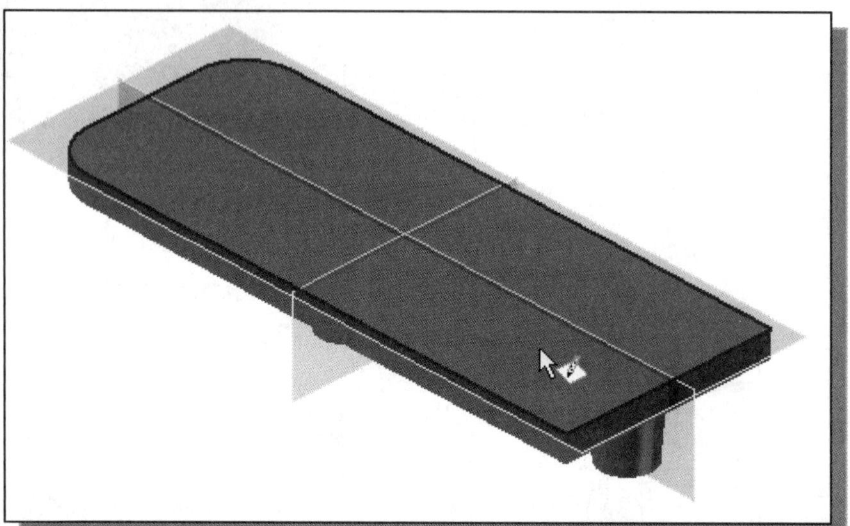

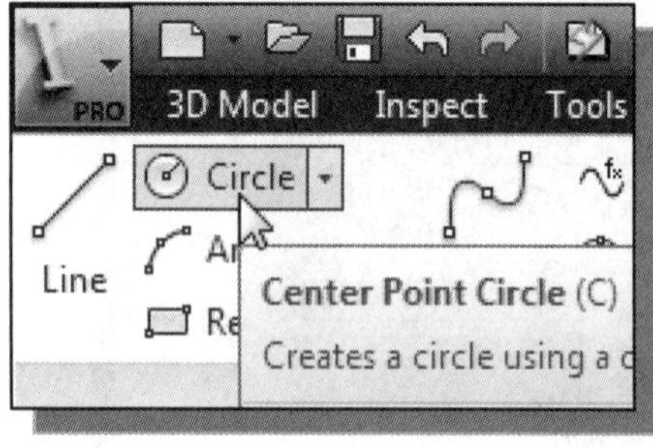

3. Select the **Center point circle** command by clicking once with the left-mouse-button on the icon in the *Draw* panel.

4. On your own, create the **five circles** and also the associated dimensions as shown.

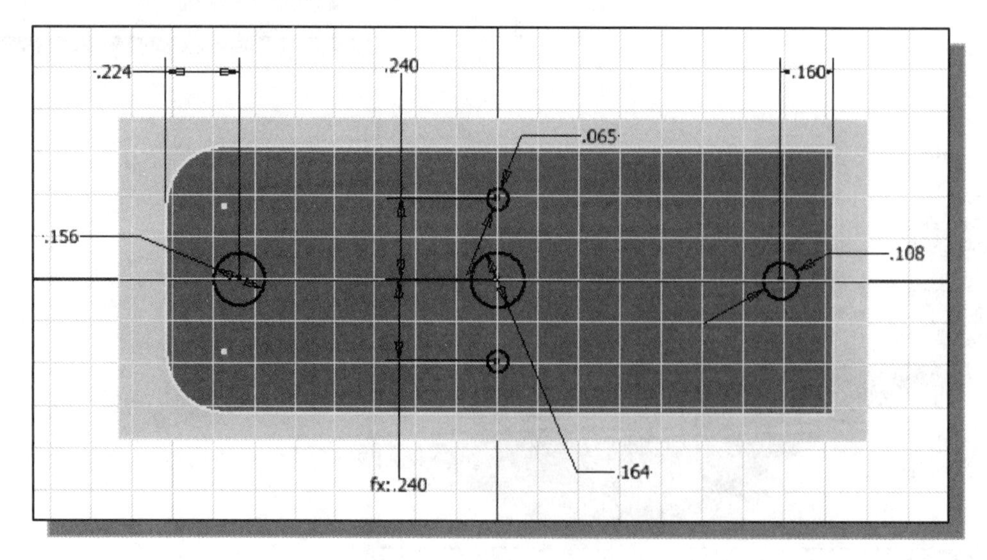

5. Select **Finish Sketch** in the *Ribbon* toolbar to exit the **2D Sketch** module.

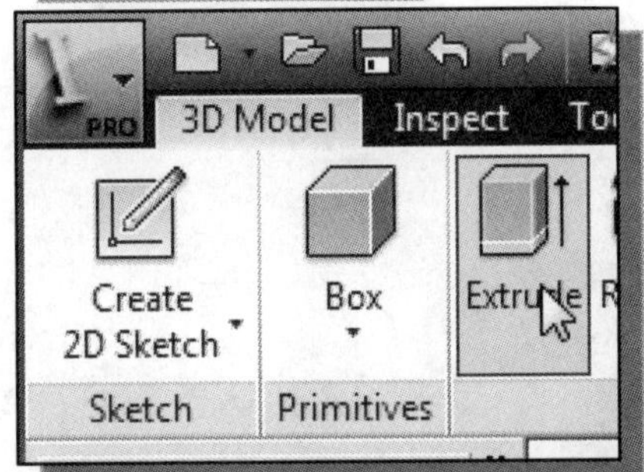

6. In the *Create* toolbar, select the **Extrude** command by left-mouse-clicking once on the icon.

7. Select the five closed regions we just created by clicking on the inside of the circles.

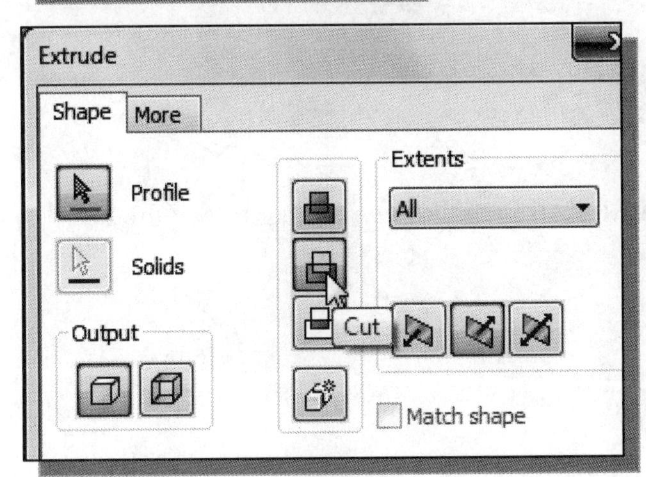

8. In the *Extrude* dialog box, set the *Extents* option to **All** as shown.

9. Set the *Extrude* option to **Cut** as shown.

10. Click on the **OK** button to proceed with creating the solid feature.

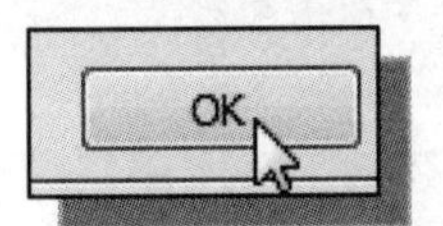

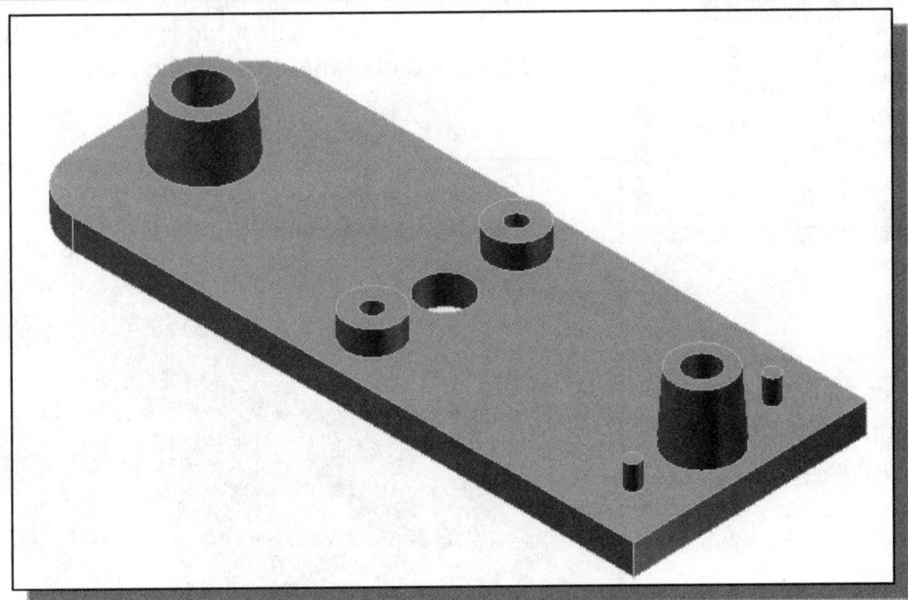

Add a Feature Using Existing Geometry

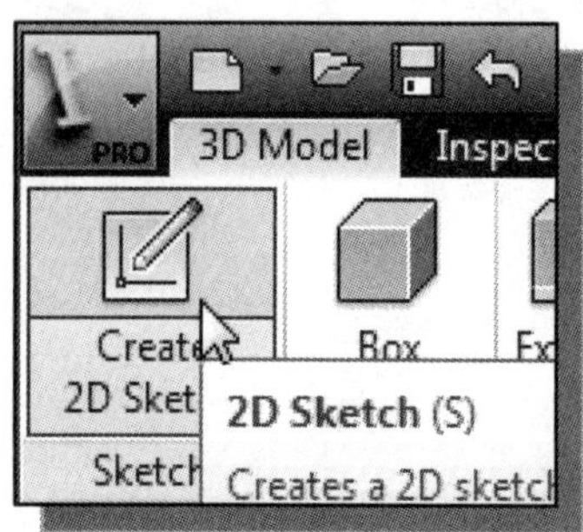

1. In the *Sketch* toolbar select the **Create 2D Sketch** command by left-clicking once on the icon.

2. Select the **top plane of the left extrude feature**, by clicking once with the left-mouse-button, as the sketching plane.

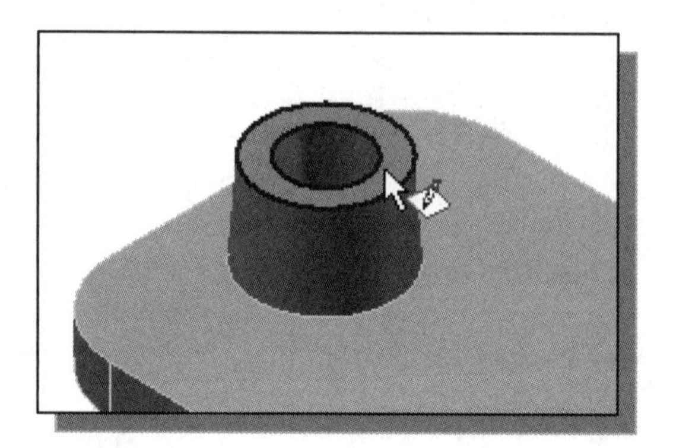

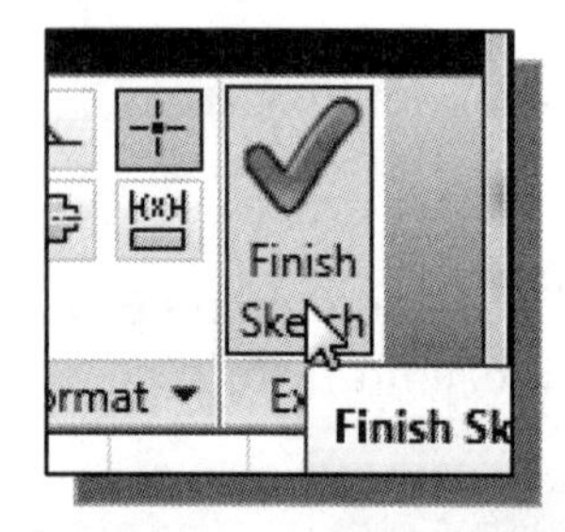

3. Select **Finish Sketch** in the *Ribbon* toolbar to exit the 2D Sketch module.

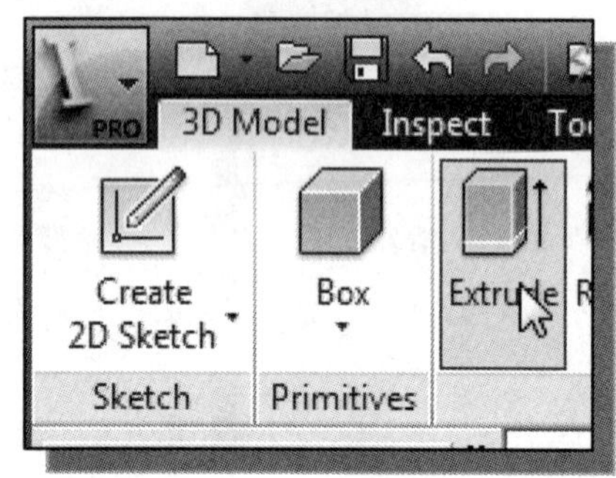

4. In the *Create* toolbar, select the **Extrude** command by left-mouse-clicking once on the icon.

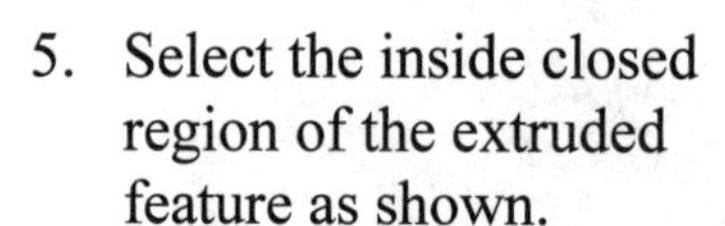

5. Select the inside closed region of the extruded feature as shown.

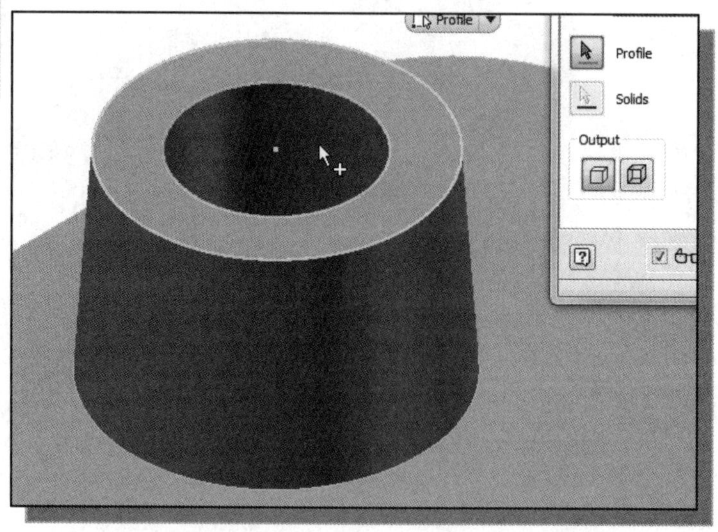

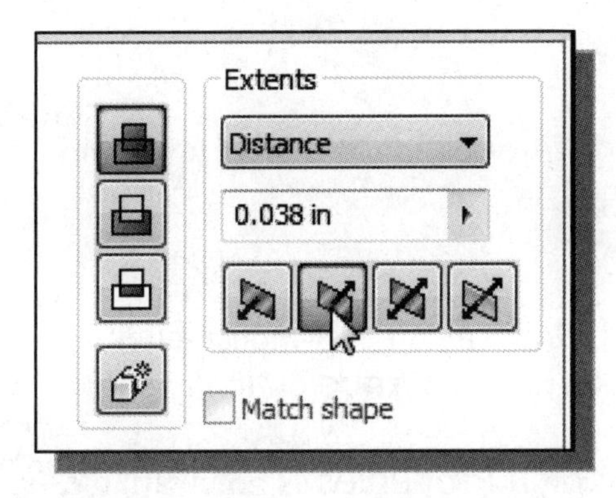

6. In the *Extrude* dialog box, set the extrusion distance to **0.038** as shown.

7. Flip the extrude direction by clicking on the second icon as shown.

8. Click on the **OK** button to proceed with creating the solid feature.

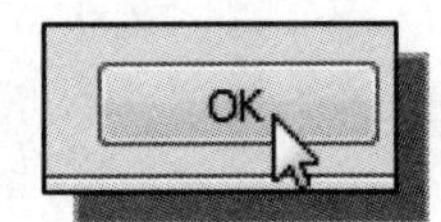

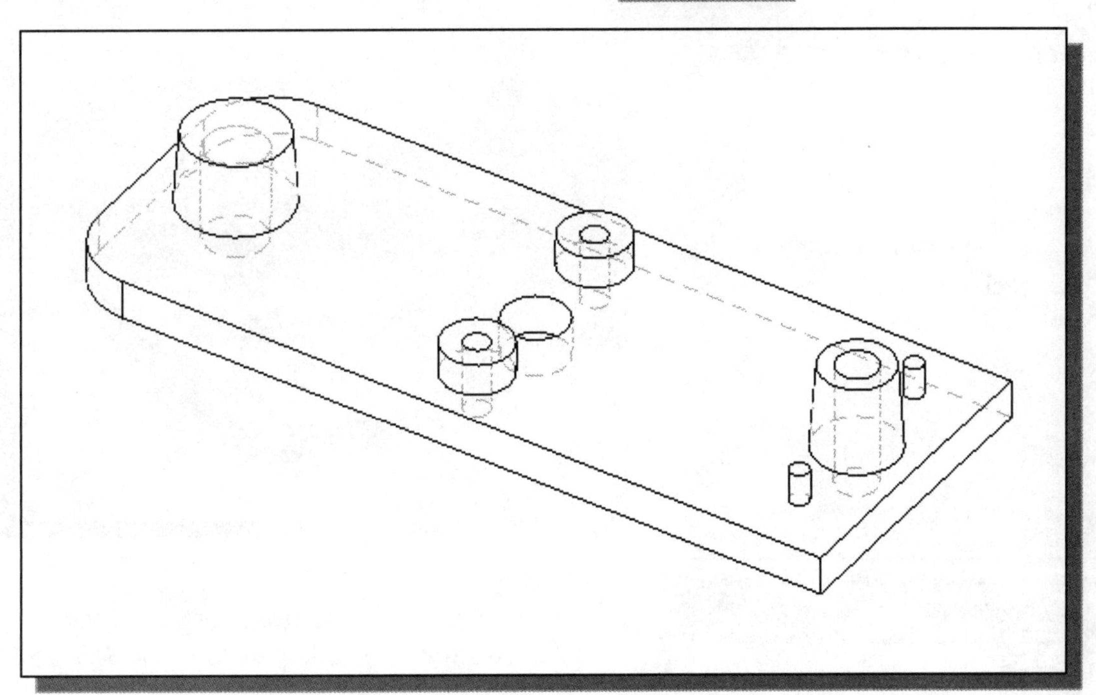

Save the Model File

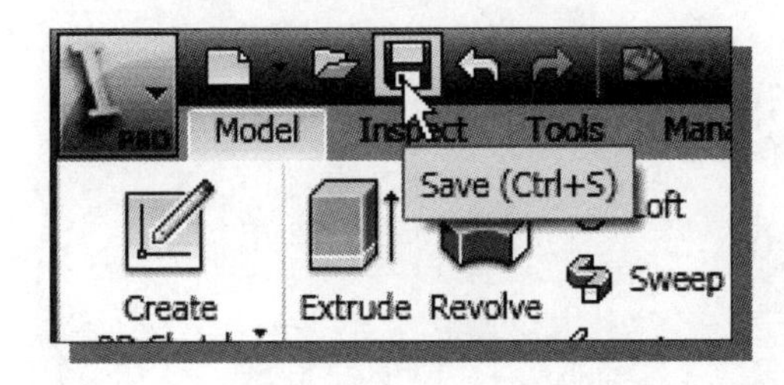

1. Select **Save** in the *Standard* toolbar. We can also use the "**Ctrl-S**" combination (press down the "**Ctrl**" key and hit the "**S**" key once) to save the part.

2. In the popup window, save the model in the *Mechanical-Tiger* project folder using **B3-Leg** as the name of the file.

3. Click on the **SAVE** button to save the file.

Using the Measure Tools

- Besides using the measure tools to get geometric information at the 2D level, the measure tools can also be used on 3D models.

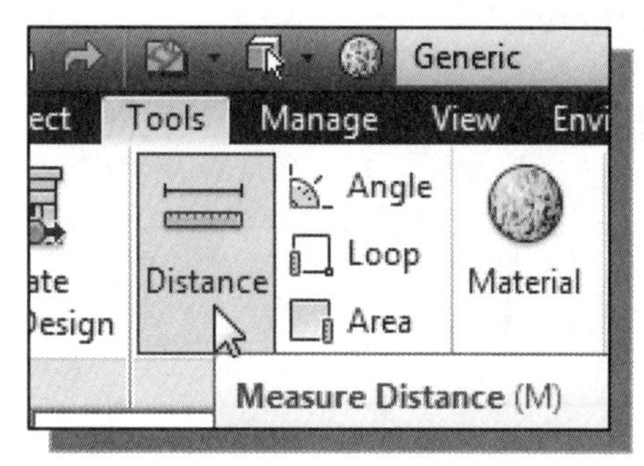

1. In the *Tools* pull-down menu, left-mouse-click once on the **Measure Distance** option as shown.

   - Note that Region Properties is not available at the 3D level.

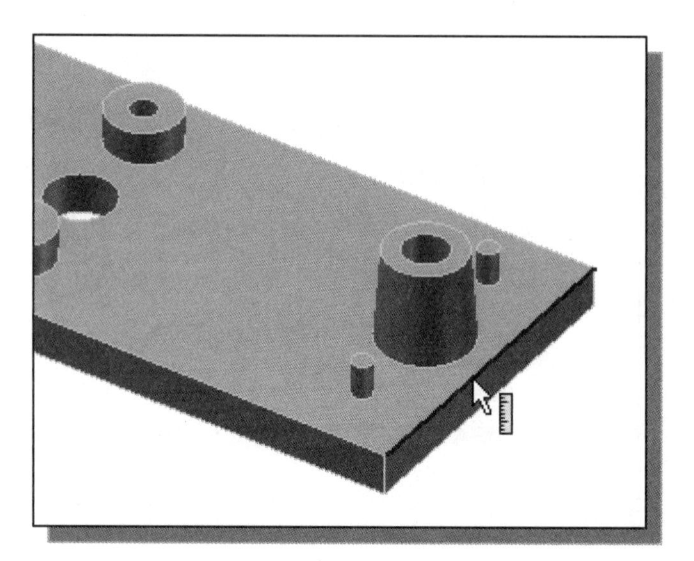

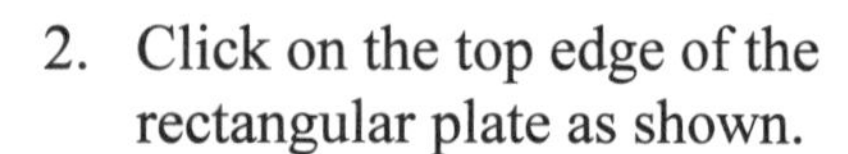

2. Click on the top edge of the rectangular plate as shown.

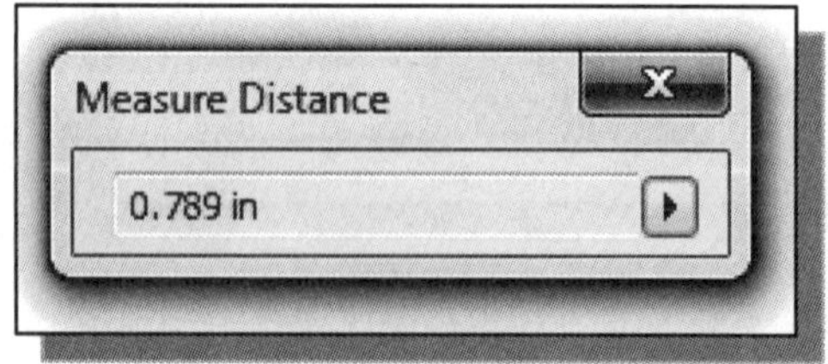

3. The associated length measurement of the selected geometry is displayed in the *Measure Distance* dialog box as shown.

4. Click on the **triangular icon** in the *Measure Distance* dialog box to display the available options.

5. Choose [**Dual Unit**] → [**Millimeter**] to also display the measurement in mm.

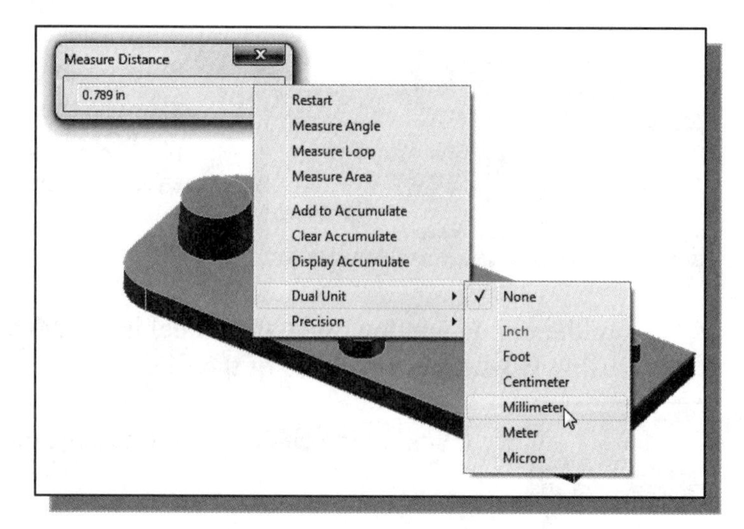

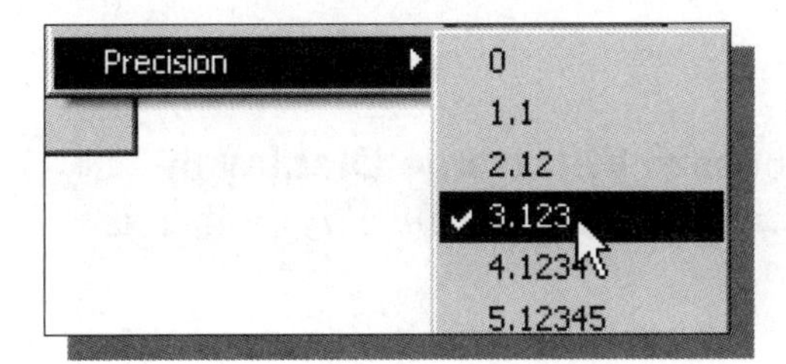

6. On your own, set the *Precision* to show 3 decimal places as shown.

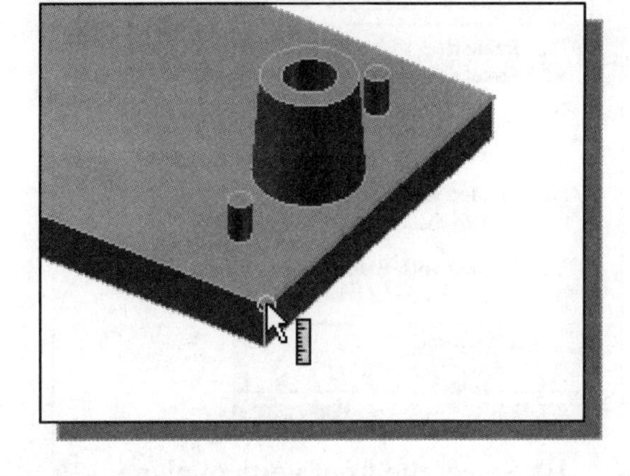

7. Click the **right corner** of the 3D model as shown.

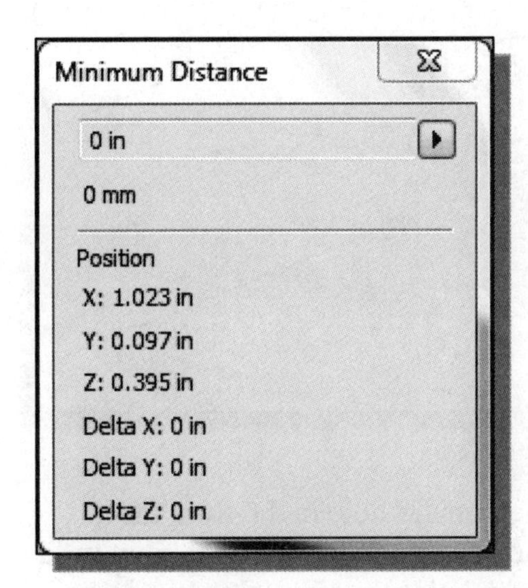

❖ Notice the information regarding the selected point is displayed in the *Measure Distance* box. The absolute position of the point is displayed. (Note the displayed numbers may be different on your screen.)

8. Select the center point of the left circular feature as the second location for the **Measure Length** command. The distance in between the two selected objects is calculated.

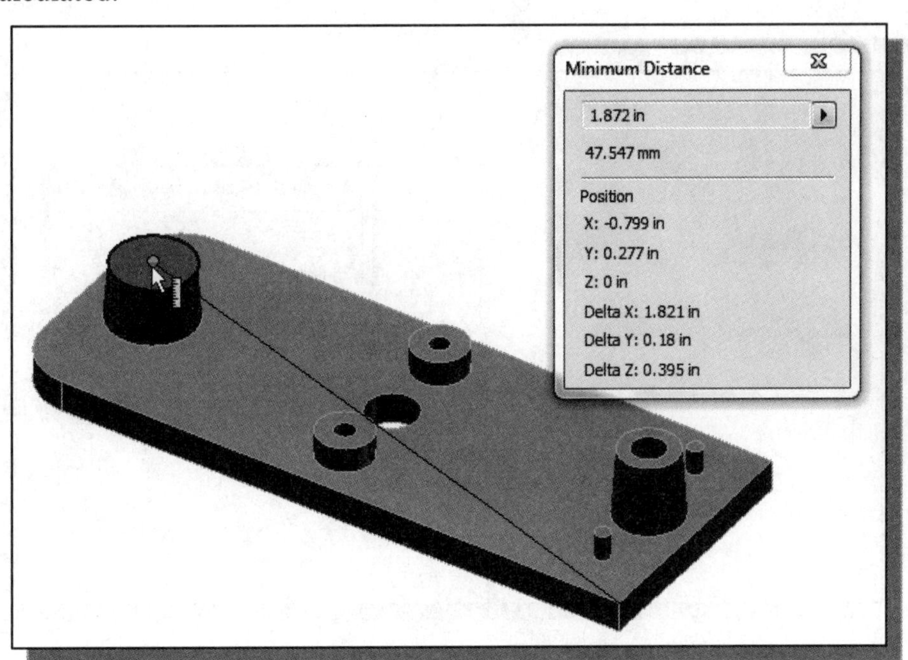

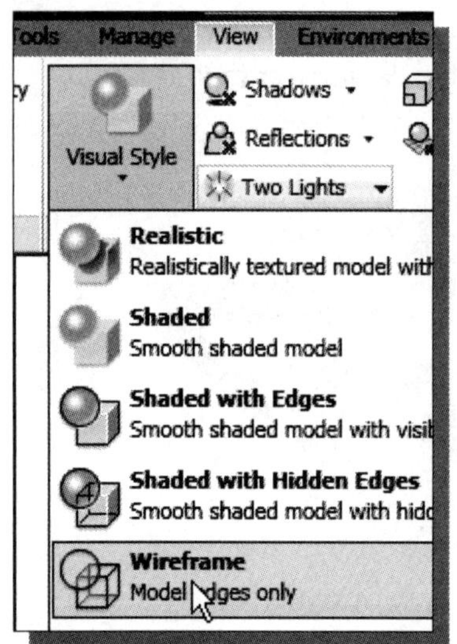

9. Set the display option to **Wireframe Display** by clicking the associated icon in the *Display* toolbar as shown.

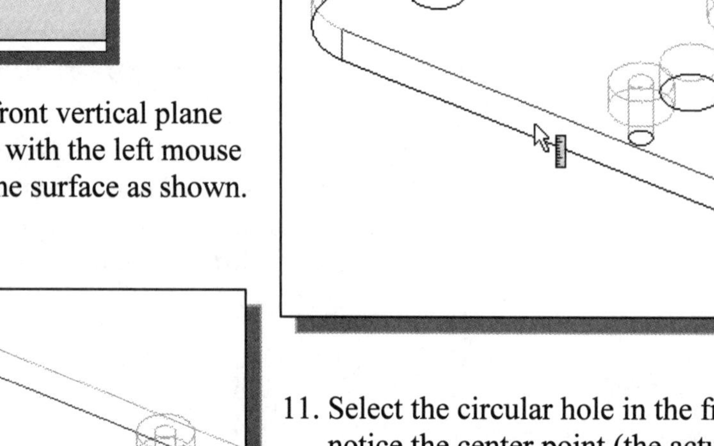

10. Select the front vertical plane by clicking with the left mouse button on the surface as shown.

11. Select the circular hole in the front face; notice the center point (the actual point used for the calculation) is highlighted as shown.

❖ Different types of entities can be selected and the measurements are calculated accordingly.

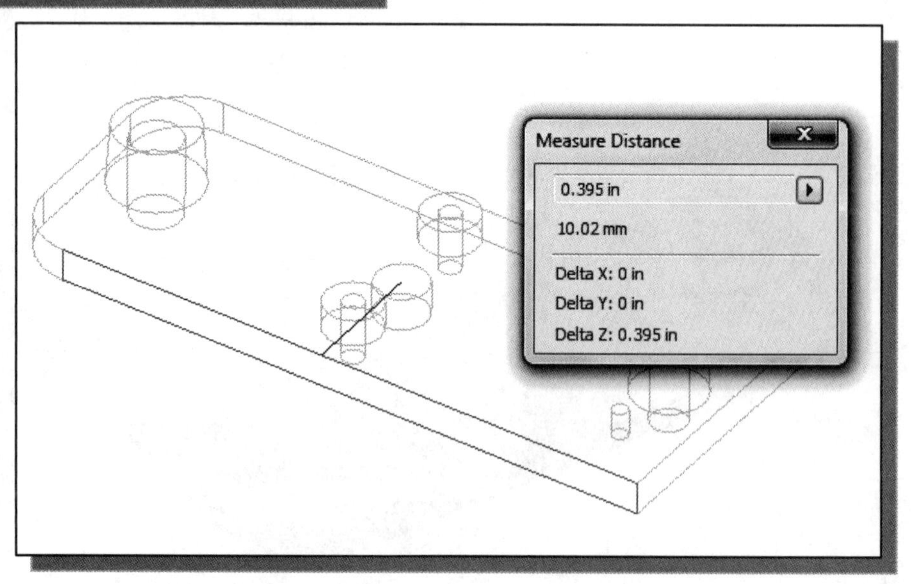

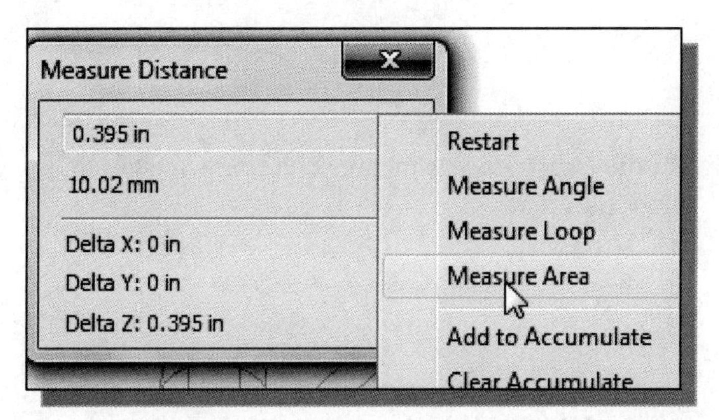

12. In the Measure Distance window, switch to the **Measure Area** option as shown.

13. Click on the front face of the plate model to display the area of the selected surface.

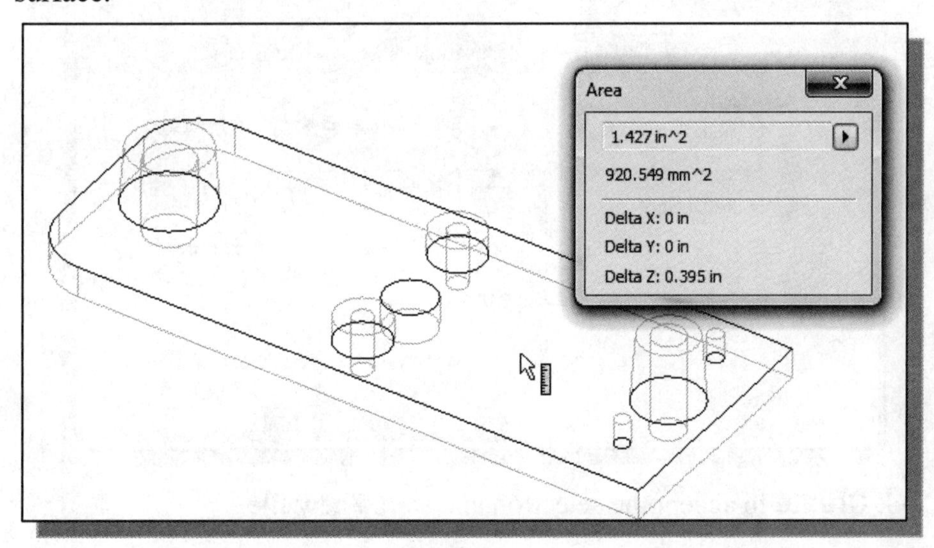

14. Note that we can also select cylindrical surfaces; select the cylindrical surface as shown.

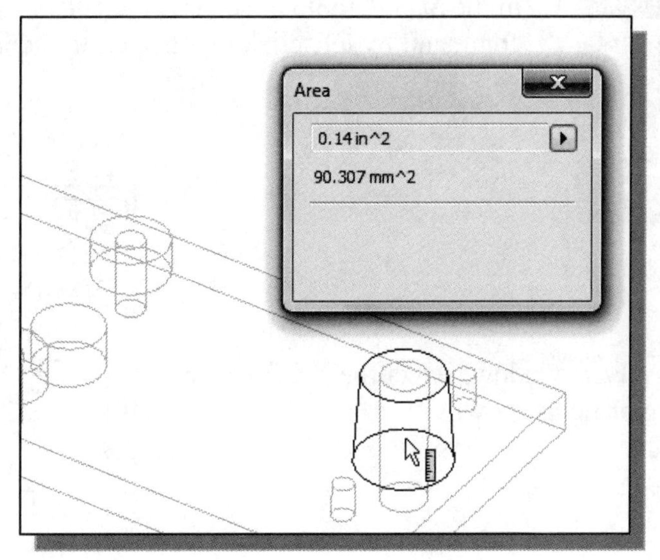

15. On your own, experiment with the different **measure tools** available.

The *Boot* Part

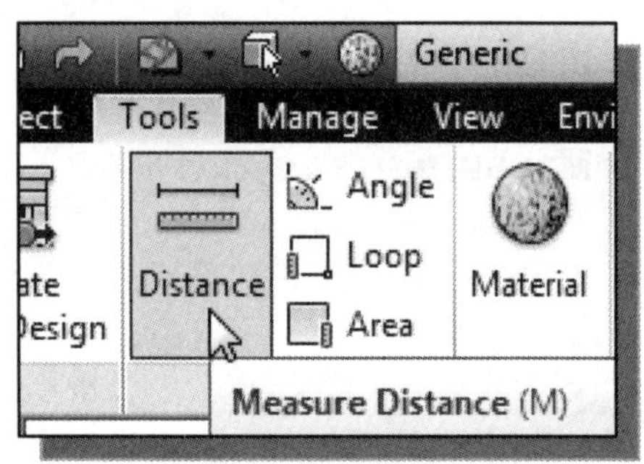

1. In the *Quick Access* menu, select *New* to start a new file.

2. In the *New File* dialog box, select the **Metric Standard (mm).ipt** template by left-mouse-clicking once on the icon.

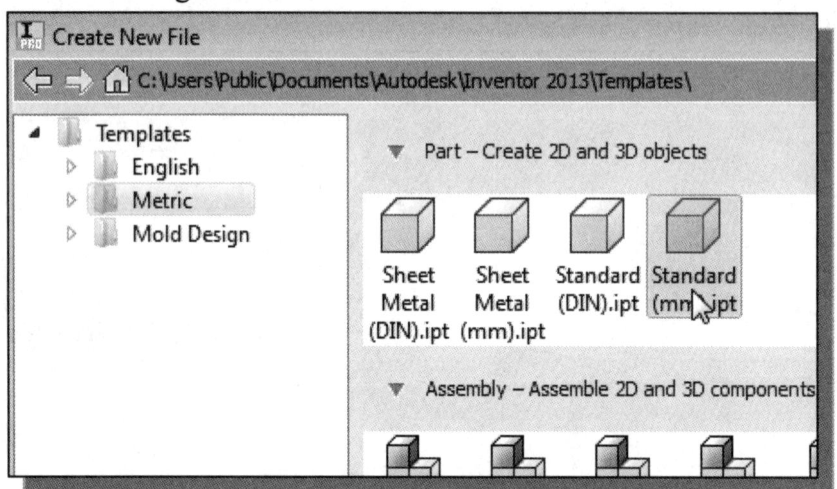

3. Click **Create** to accept the selection and start a new file.

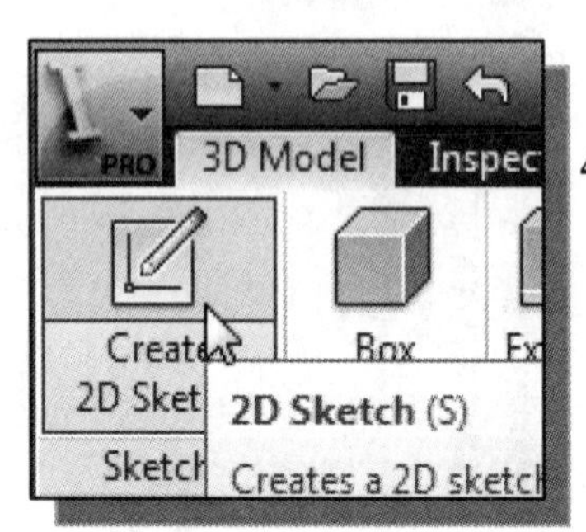

4. In the *Sketch* toolbar, select the **Create 2D Sketch** command by left-clicking once on the icon.

5. Inside the *browser* window, select the XZ Plane as the sketching plane as shown.

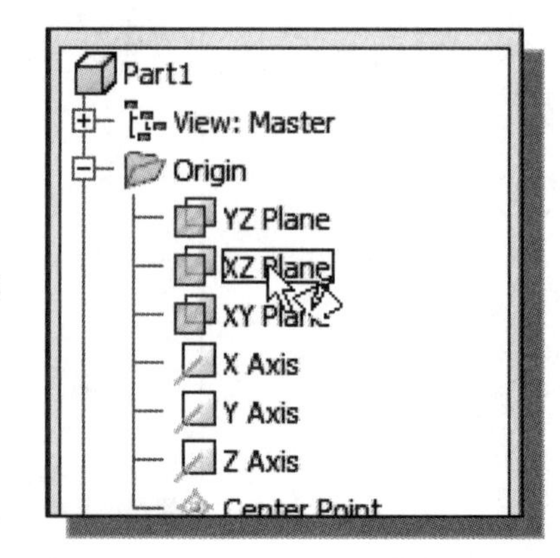

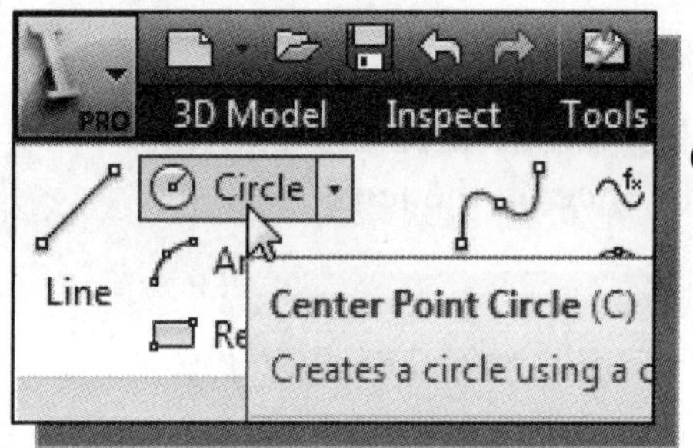

6. Select the **Center point circle** command by clicking once with the left-mouse-button on the icon in the *Draw* toolbar.

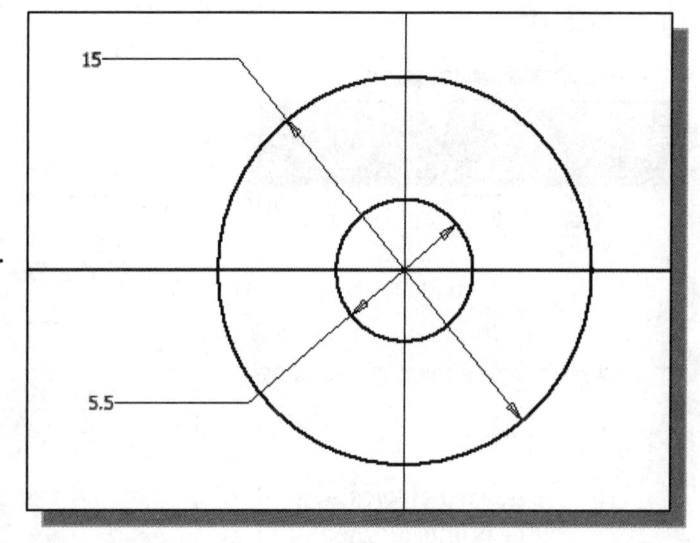

7. Create **two circles**, with the center points aligned to the *Center Point* of the coordinate system as shown.

8. Select **Finish Sketch** in the *Ribbon* toolbar to exit the **Sketch** mode.

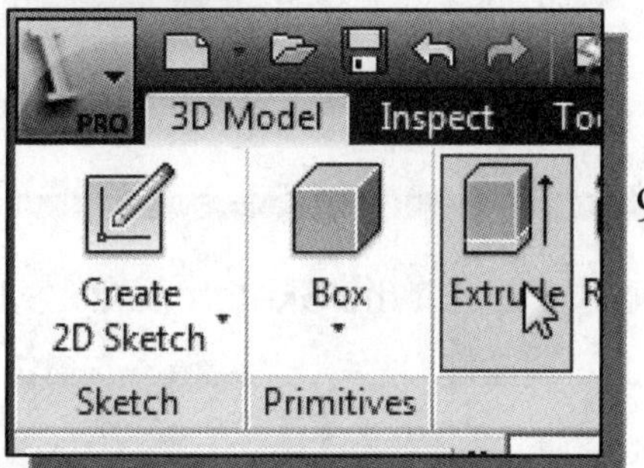

9. In the *Create* toolbar, select the **Extrude** command by left-mouse-clicking once on the icon.

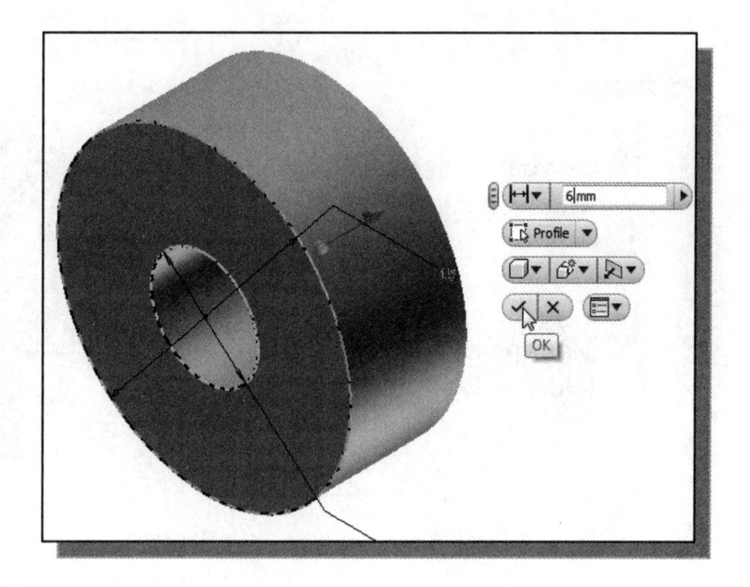

10. Select the region in between the two circles as shown.

11. Set the extrusion distance to **6 mm** as shown.

12. Click **OK** to create the feature.

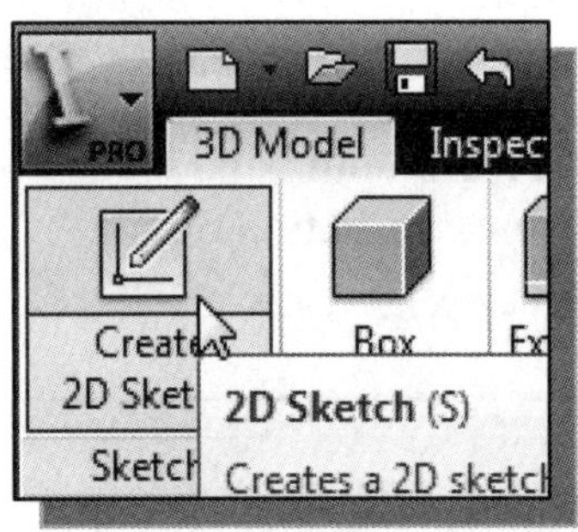

13. In the *Sketch* toolbar, select the **Create 2D Sketch** command by left-clicking once on the icon.

14. Select the front face of the solid model to align the sketching plane.

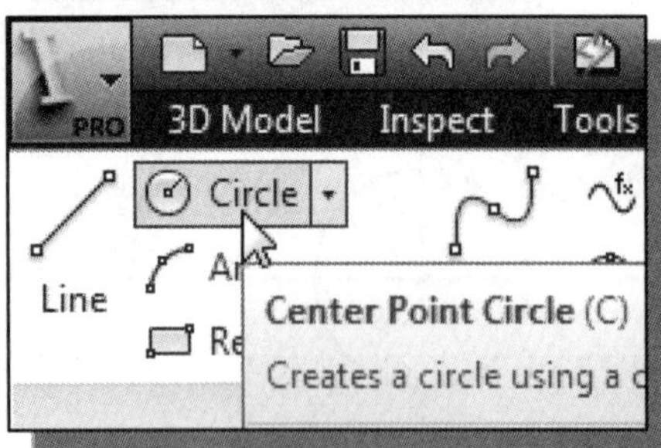

15. Select the **Center point circle** command by clicking once with the left-mouse-button on the icon in the *Sketch* toolbar.

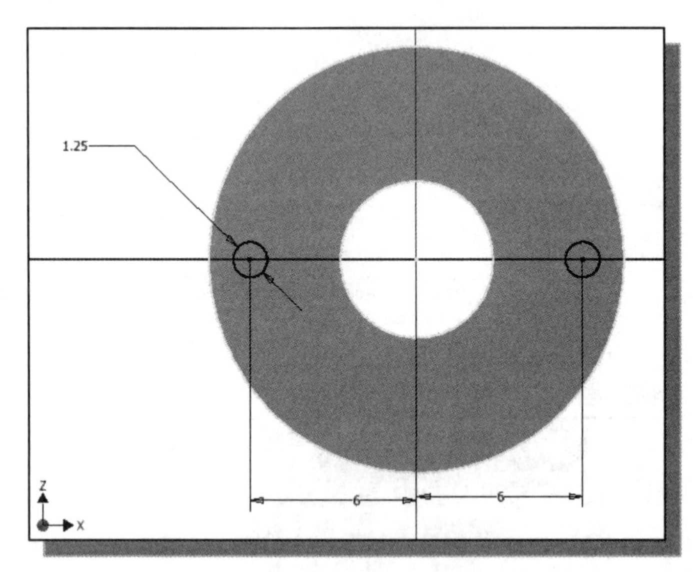

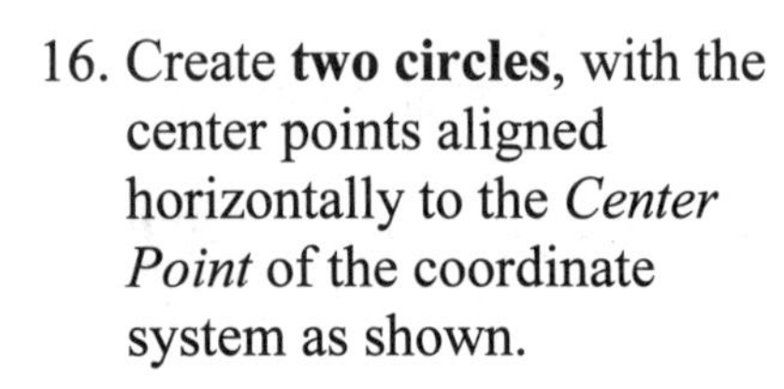

16. Create **two circles**, with the center points aligned horizontally to the *Center Point* of the coordinate system as shown.

17. Select **Finish Sketch** in the *Ribbon* toolbar to exit the Sketch mode.

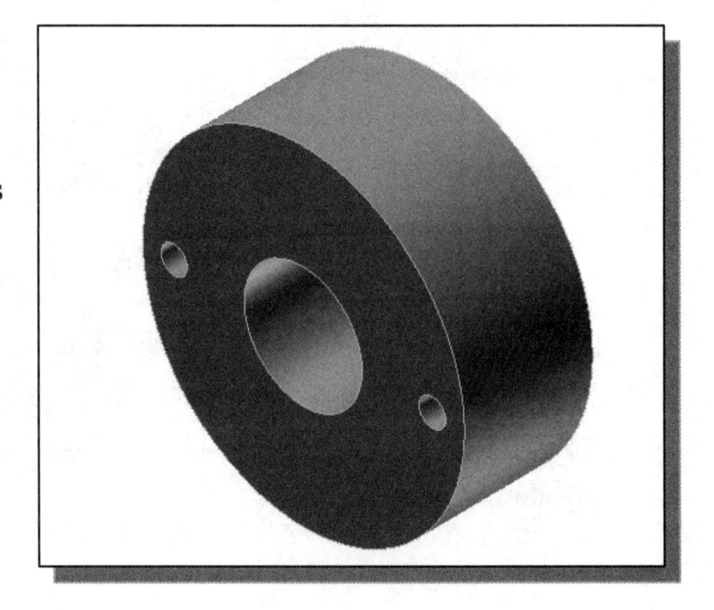

18. On your own, create a cut feature that is **2 mm** deep as shown.

19. Save the model in the *Mechanical-Tiger* project folder as **Boot.ipt**.

Questions:

1. What is the difference between *dimensional* constraints and *geometric* constraints?

2. How can we confirm that a sketch is fully constrained?

3. How do we distinguish between derived dimensions and regular dimensions on the screen?

4. Describe the procedure to **Display/Edit** user-defined equations.

5. List and describe three different geometric constraints available in *Autodesk Inventor*.

6. Does *Autodesk Inventor* allow us to build partially constrained or totally unconstrained solid models? What are the advantages and disadvantages of building these types of models?

7. How do we display and examine the existing constraints that are applied to the sketched entities?

8. Describe the advantages of using parametric equations.

9. Can we delete an applied constraint? How?

10. Create the following 2D sketch and measure the associated area and perimeter.

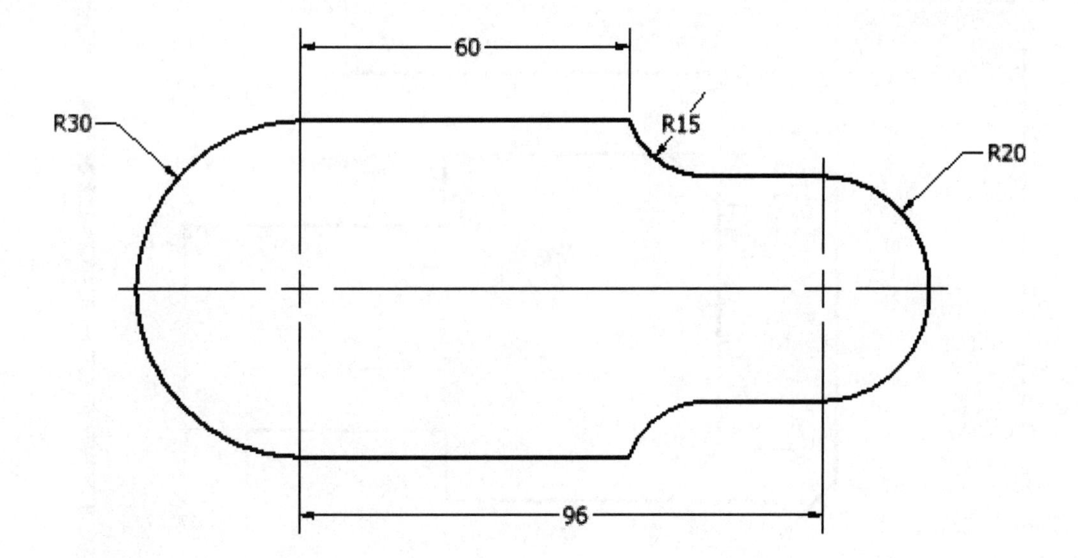

11. Describe the purpose and usage of the Auto Dimension command.

Exercises:

(Create and establish three parametric relations for each of the following designs.)

1. **Swivel Base** (Dimensions are in millimeters. Base thickness:**10 mm.** Boss: **5 mm.**)

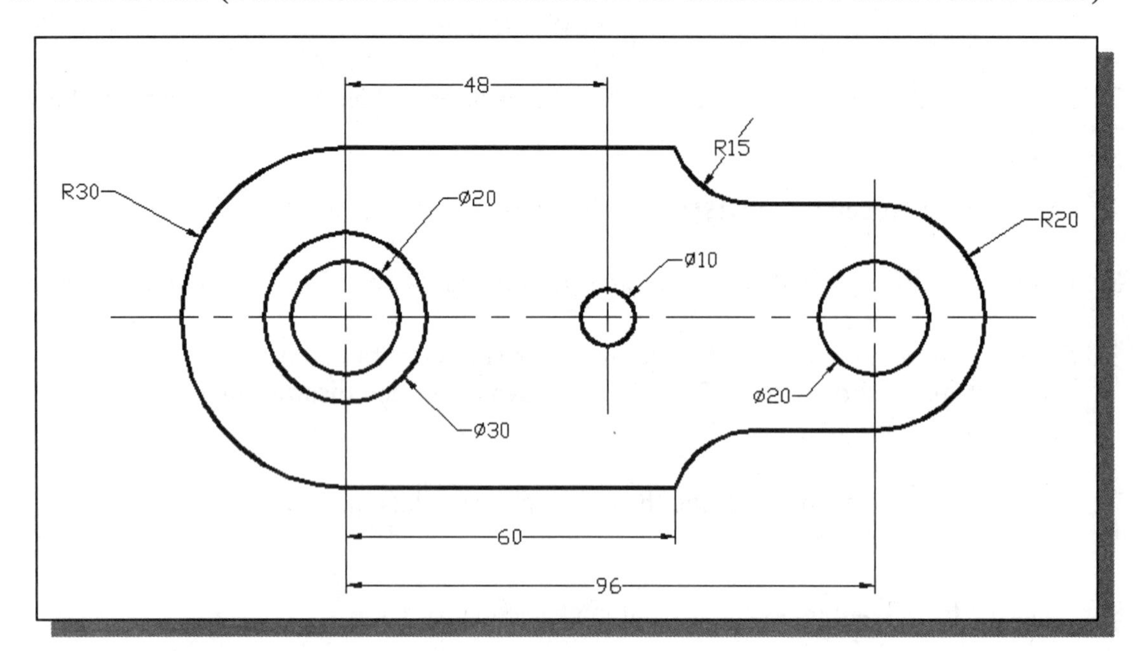

2. **C-Clip** (Dimensions are in inches. Plate thickness: **0.25 inches.**)

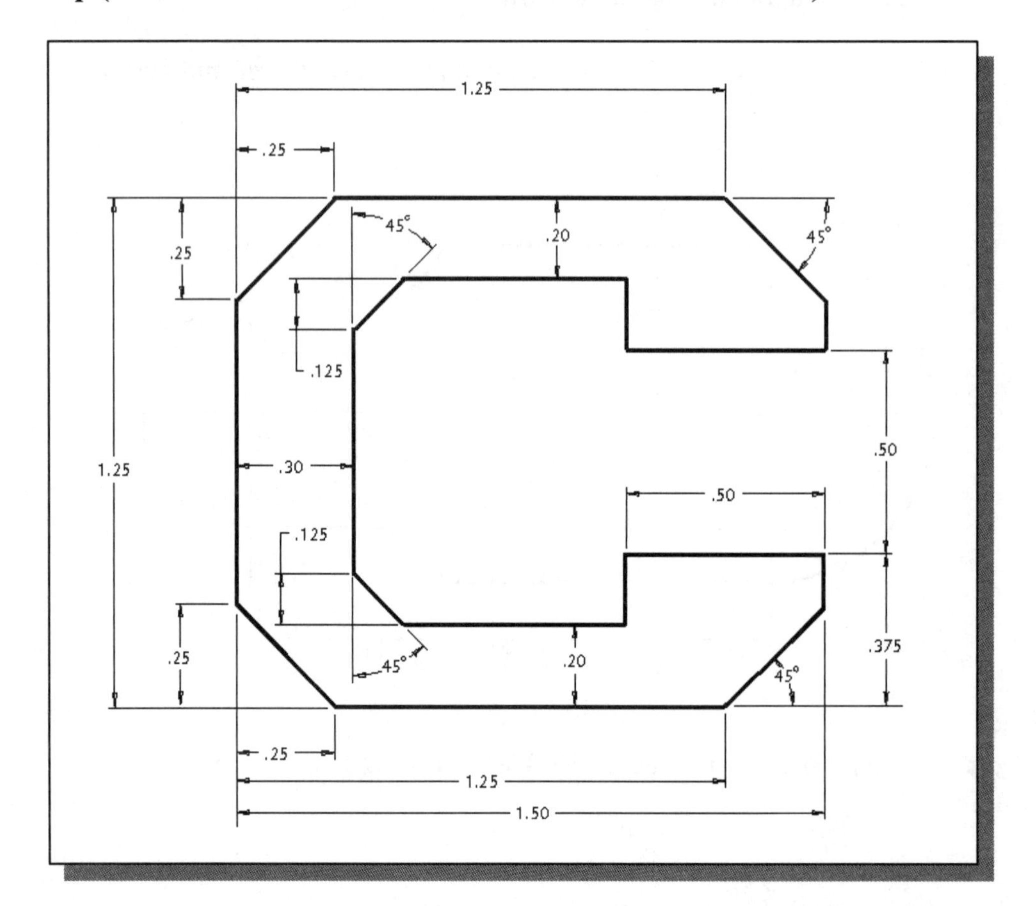

3. **Wedge Block** (Dimensions are in inches)

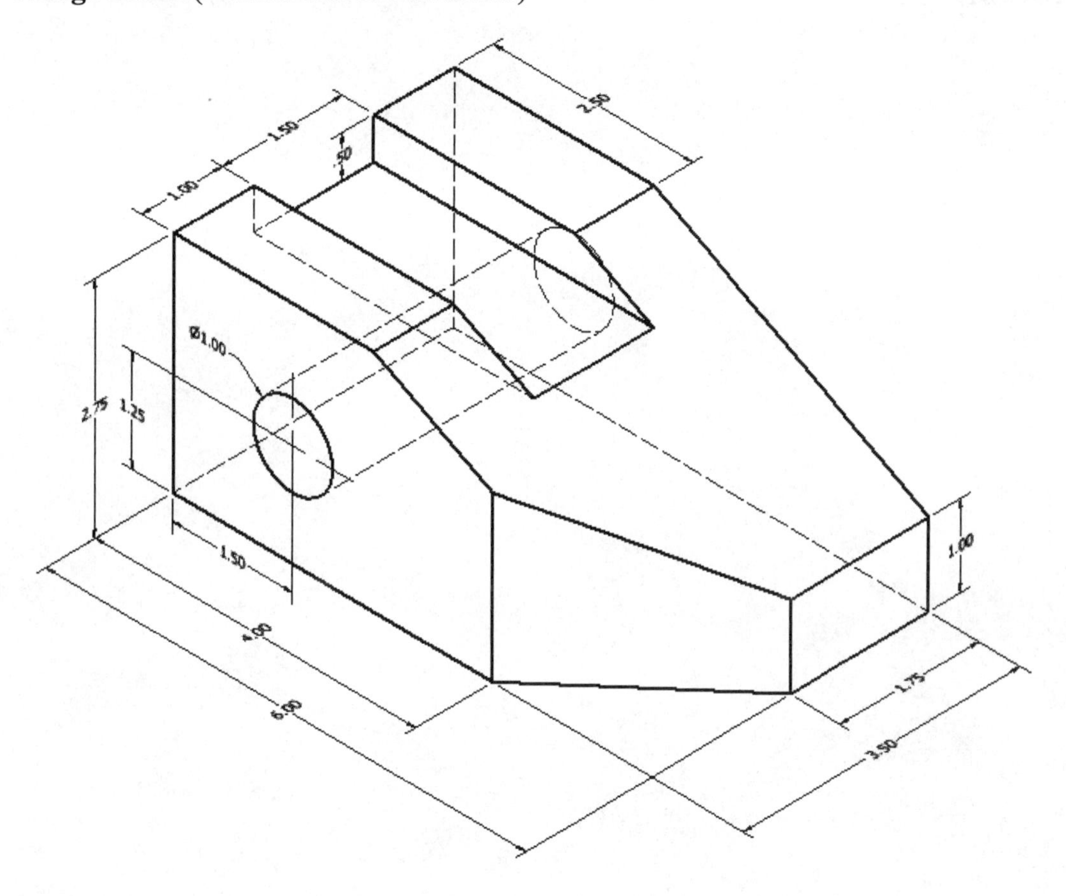

4. **Hinge Guide** (Dimensions are in inches.)

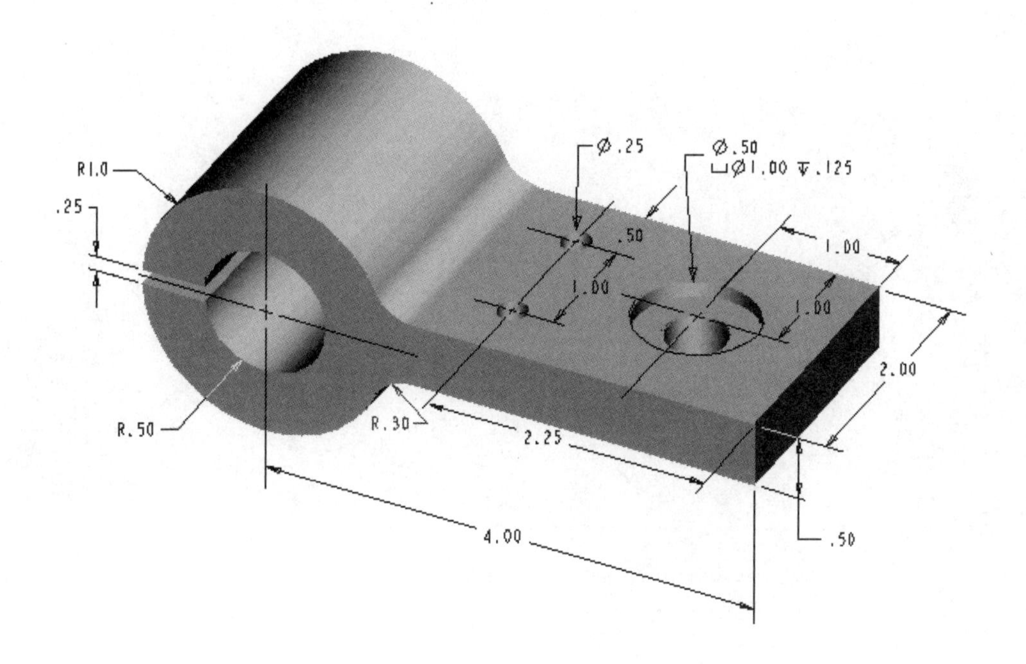

Notes:

Chapter 5
Pictorials and Sketching

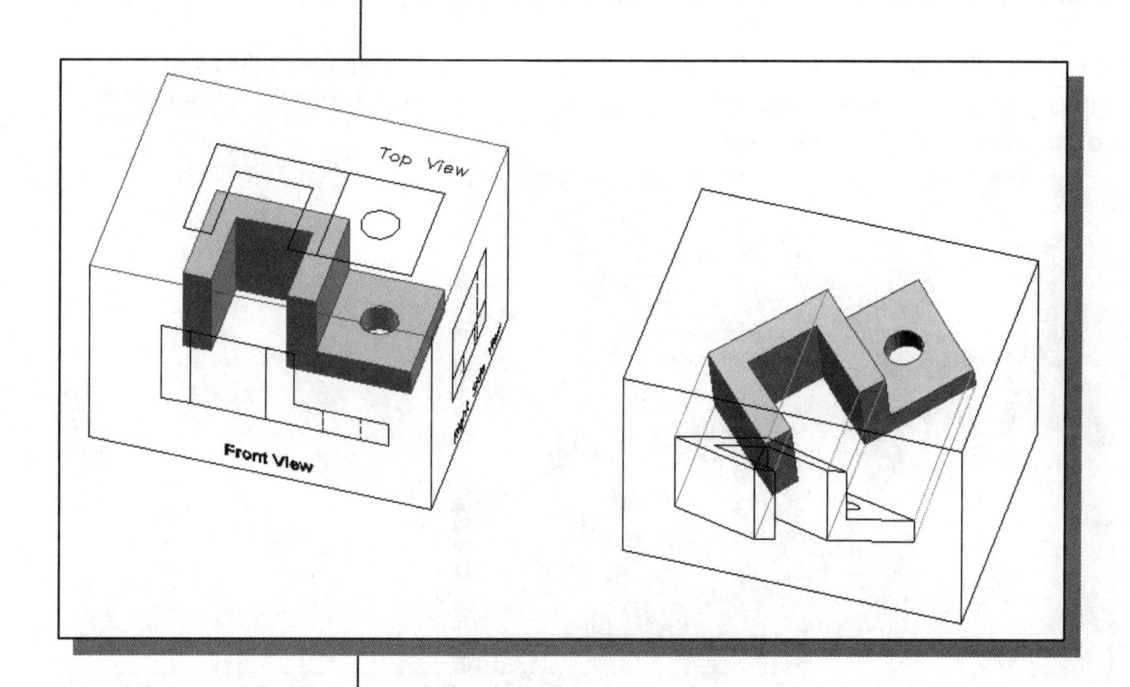

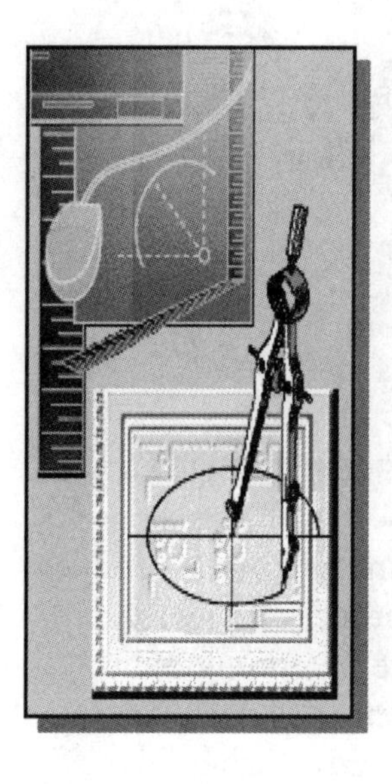

Learning Objectives

- ♦ **Understand the Importance of Freehand Sketching**
- ♦ **Understand the Terminology Used in Pictorial Drawings**
- ♦ **Understand the Basics of the Following Projection Methods: Axnonometric, Oblique and Perspective**
- ♦ **Be Able to Create Freehand 3D Pictorials**

Engineering Drawings, Pictorials and Sketching

One of the best ways to communicate one's ideas is through the use of a picture or a drawing. This is especially true for engineers and designers. Without the ability to communicate well, engineers and designers will not be able to function in a team environment and therefore will have only limited value in the profession.

For many centuries, artists and engineers used drawings to express their ideas and inventions. The two figures below are drawings by da Vinci (1453-1528) illustrating some of his engineering inventions.

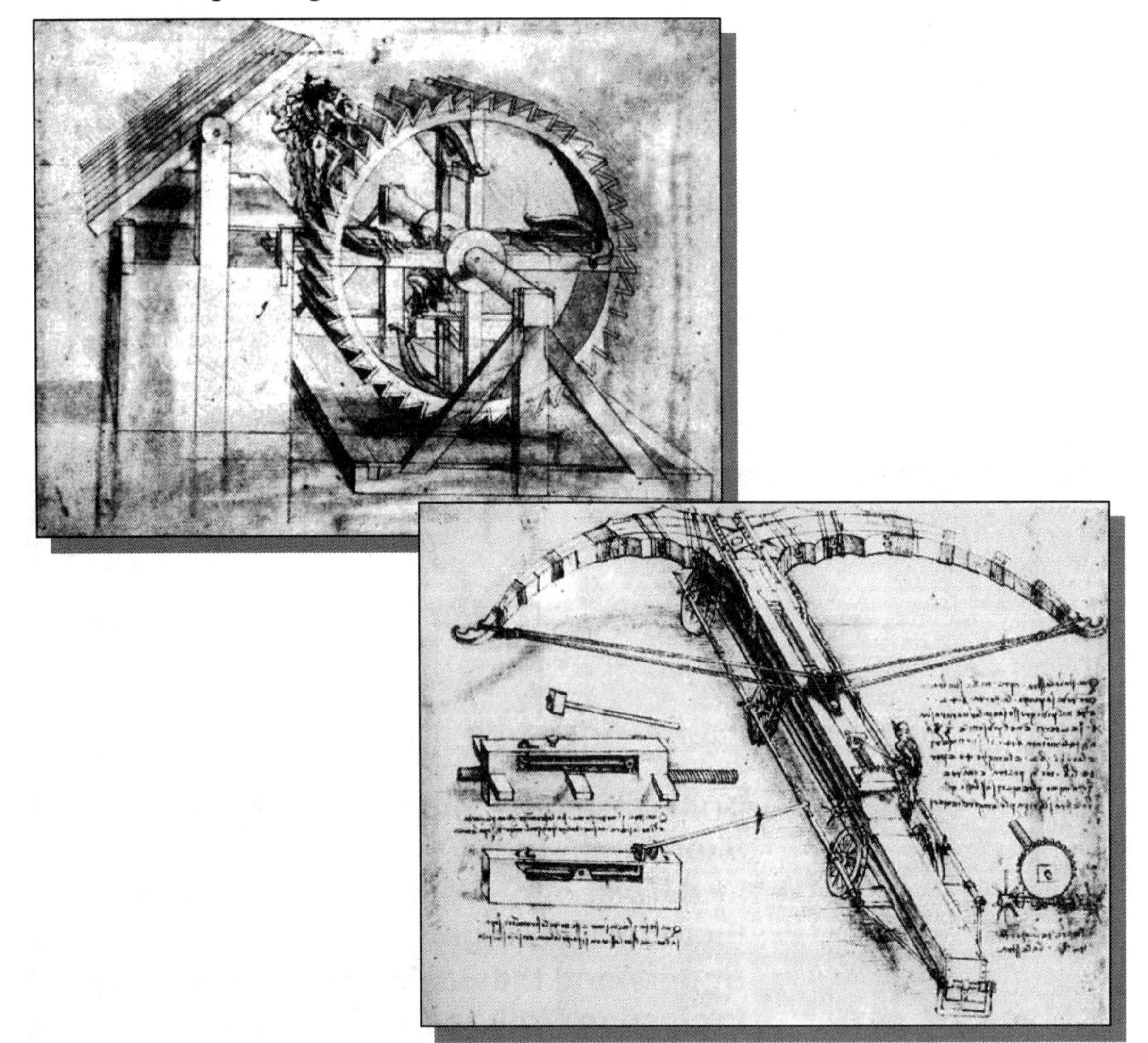

Engineering design is a process to create and transform ideas and concepts into a product definition that meets the desired objective. The engineering design process typically involves three stages: (1) Ideation/conceptual design stage: this is the beginning of an engineering design process, where basic ideas and concepts take shape. (2) Design development stage: the basic ideas are elaborated and further developed. During this stage, prototypes and testing are commonly used to ensure the developed design meets the desired objective. (3) Refine and finalize design stage: This stage of the design process is the last stage of the design process, where the finer details of the design are further refined. Detailed information of the finalized design is documented to assure the design is ready for production.

Two types of drawings are generally associated with the three stages of the engineering process: (1) Freehand Sketches and (2) Detailed Engineering Drawings.

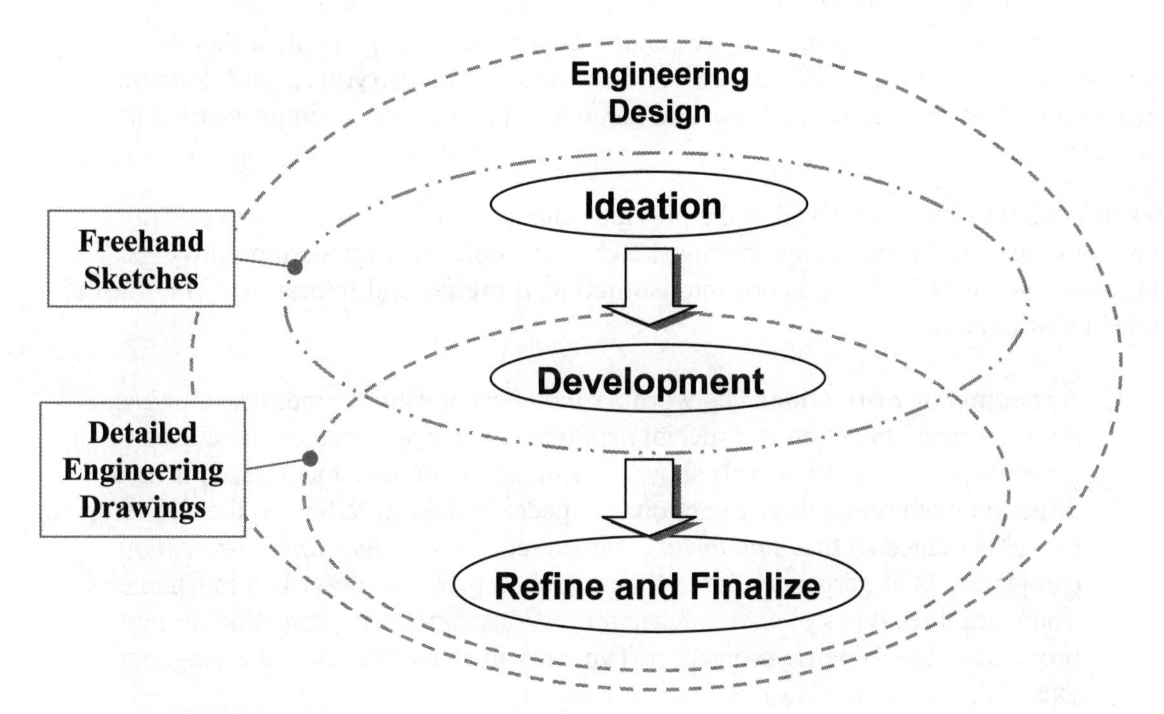

Freehand sketches are generally used in the beginning stages of a design process: (1) to quickly record the designer's ideas and help in formulating different possibilities, (2) to communicate the designer's basic ideas with others, and (3) to develop and elaborate further the designer's ideas/concepts.

During the initial design stage, an engineer will generally picture the ideas in his/her head as three-dimensional images. The ability to think visually, specifically three-dimensional visualization, is one of the most essential skills for an engineer/designer. And freehand sketching is considered as one of the most powerful methods to help develop visualization skills.

Detailed engineering drawings are generally created during the second and third stages of a design process. The detailed engineering drawings are used to help refine and finalize the design and also to document the finalized design for production. Engineering drawings typically require the use of drawing instruments, from compasses to computers, to bring precision to the drawings.

Freehand Sketches and **Detailed Engineering Drawings** are essential communication tools for engineers. By using the established conventions, such as perspective and isometric drawings, engineers/designers are able to quickly convey their design ideas to others.

The ability to sketch ideas is absolutely essential to engineers. The ability to sketch is helpful, not just to communicate with others, but also to work out details in ideas and to identify any potential problems. Freehand sketching requires only simple tools, a pencil and a piece of paper, and can be accomplished almost anywhere and anytime. Creating freehand sketches does not require any artistic ability. Detailed engineering drawing is employed only for those ideas deserving a permanent record.

Freehand sketches and engineering drawings are generally composed of similar information, but there is a tradeoff between the time required to generate a sketch/ drawing versus the level of design detail and accuracy. In industry, freehand sketching is used to quickly document rough ideas and identify general needs for improvement in a team environment.

Besides the 2D views, described in the previous chapter, there are three main divisions commonly used in freehand engineering sketches and detailed engineering drawings: (1) **Axonometric**, with its divisions into **isometric, dimetric** and **trimetric**; (2) **Oblique**; and (3) **Perspective**.

1. **Axonometric projection**: The word *Axonometric* means "to measure along axes." Axonometric projection is a special *orthographic projection* technique used to generate ***pictorials***. **Pictorials** show a 2D image of an object as viewed from a direction that reveals three directions of space. In the figure below, the *Adjuster* model is rotated so that a *pictorial* is generated using *orthographic projection* (projection lines perpendicular to the projection plane) as described in Chapter 4. There are three types of axonometric projections: isometric projection, dimetric projection, and trimetric projection. Typically in an axonometric drawing, one axis is drawn vertically.

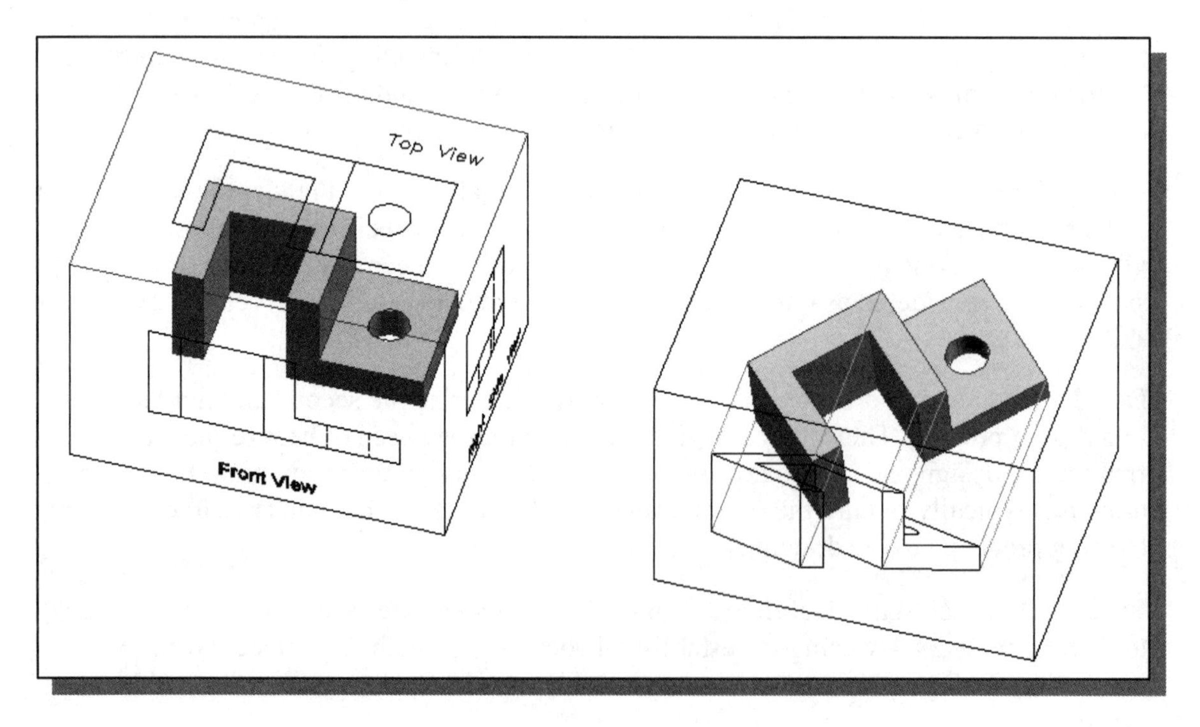

In **isometric projections**, the direction of viewing is such that the three axes of space appear equally foreshortened, and therefore the angles between the axes are equal. In **dimetric projections**, the directions of viewing are such that two of the three axes of space appear equally foreshortened. In **trimetric projections**, the direction of viewing is such that the three axes of space appear unequally foreshortened.

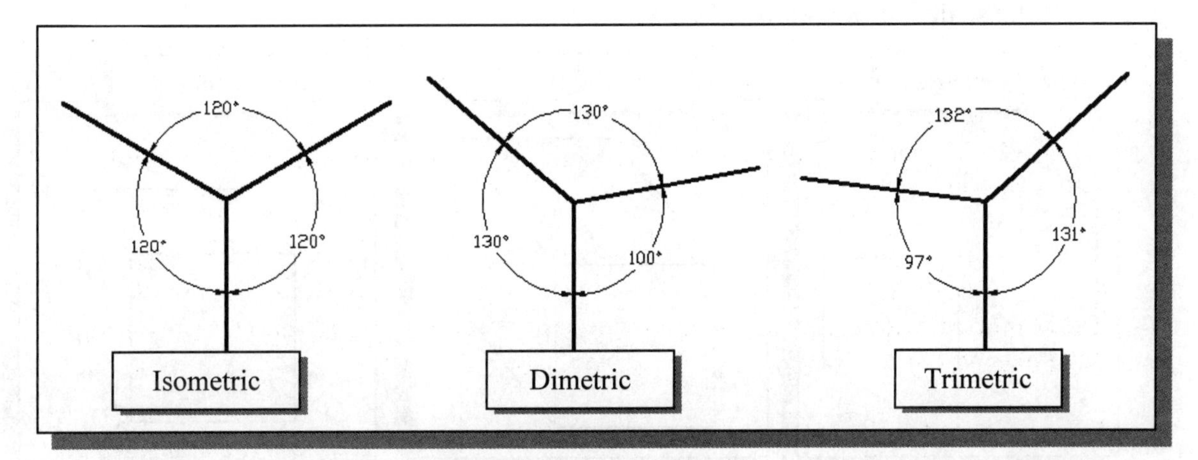

Isometric projection is perhaps the most widely used for pictorials in engineering graphics, mainly because isometric views are the most convenient to draw. Note that the different projection options described here are not particularly critical in freehand sketching as the emphasis is generally placed on the proportions of the design, not the precision measurements. The general procedure for constructing isometric views is illustrated in the following sections.

2. **Oblique Projection** represents a simple technique of keeping the front face of an object parallel to the projection plane and still reveals three directions of space. An **orthographic projection** is a parallel projection in which the projection lines are perpendicular to the plane of projection. An **oblique projection** is one in which the projection lines are other than perpendicular to the plane of projection.

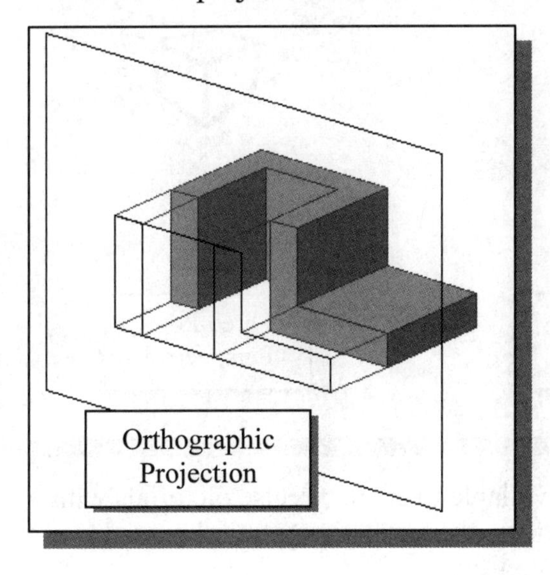

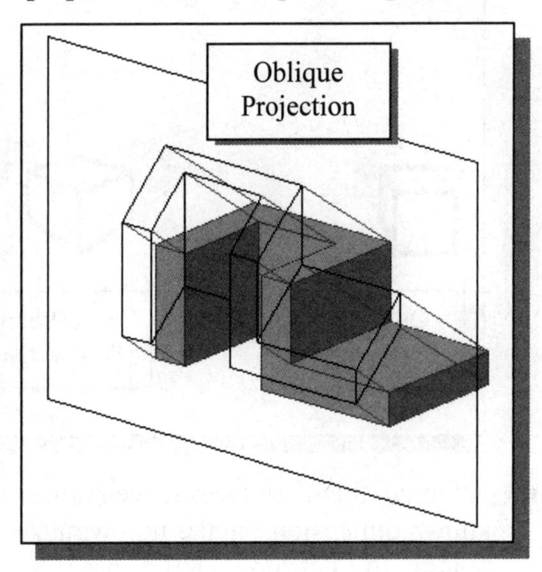

In an oblique drawing, geometry that is parallel to the frontal plane of projection is drawn true size and shape. This is the main advantage of the oblique drawing over the axonometric drawings. The three axes of the oblique sketch are drawn horizontally, vertically, and the 3rd axis can be at any convenient angle (typically between 30 and 60 degrees). The proportional scale along the 3rd axis is typically a scale anywhere between ½ and 1. If the scale is ½, then it is a ***Cabinet*** oblique. If the scale is 1, then it is a ***Cavalier*** oblique.

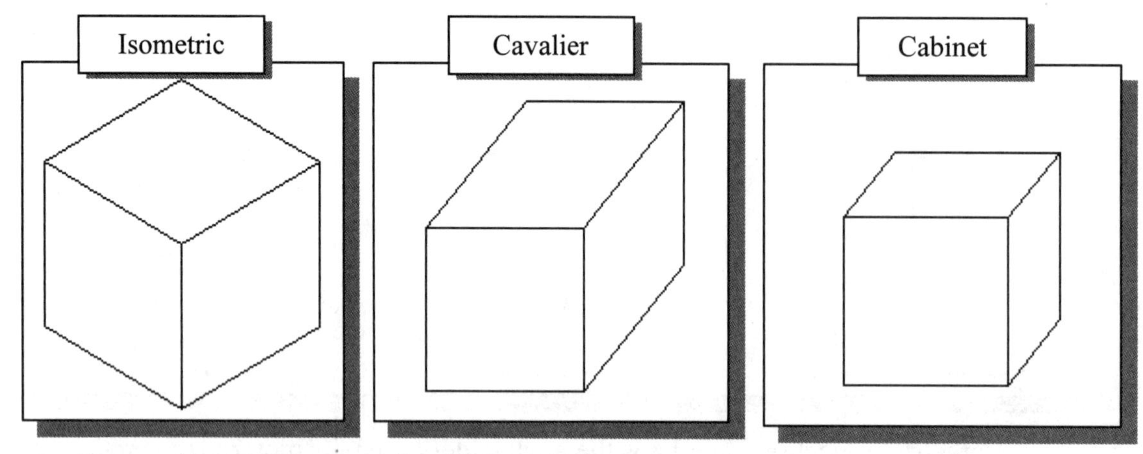

3. **Perspective Projection** adds realism to the three-dimensional pictorial representation. A perspective drawing represents an object as it appears to an observer; objects that are closer to the observer will appear larger to the observer. The key to the perspective projection is that parallel edges converge to a single point, known as the **vanishing point**. If there is just one vanishing point, then it is called a one-point perspective. If two sets of parallel edge lines converge to their respective vanishing points, then it is called a two-point perspective. There is also the case of a three-point perspective in which all three sets of parallel lines converge to their respective vanishing points.

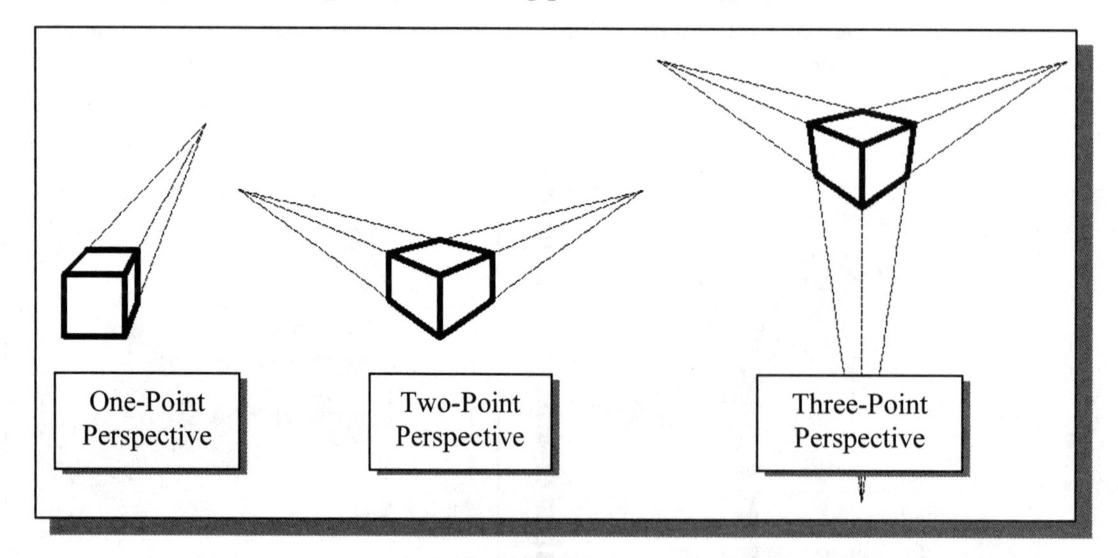

- Although there are specific techniques available to create precise pictorials with known dimensions, in the following sections, the basic concepts and procedures relating to freehand sketching are illustrated.

Isometric Sketching

Isometric drawings are generally done with one axis aligned to the vertical direction. A **regular isometric** is when the viewpoint is looking down on the top of the object, and a **reversed isometric** is when the viewpoint is looking up on the bottom of the object.

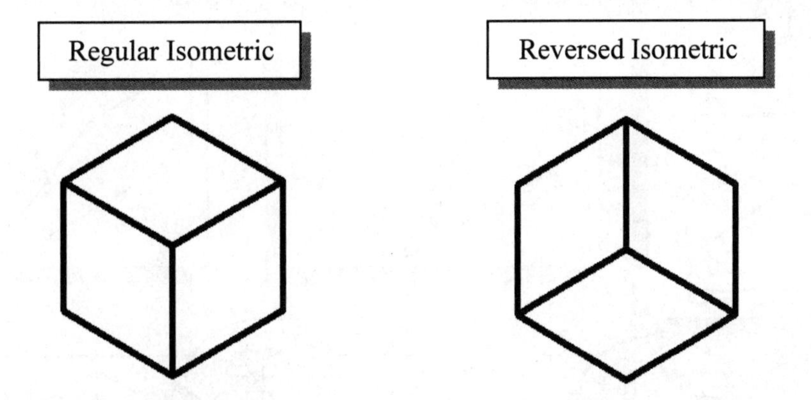

Two commonly used approaches in creating isometric sketches are: (1) the **enclosing box** method and (2) the **adjacent surface** method. The enclosing box method begins with the construction of an isometric box showing the overall size of the object. The visible portions of the individual 2D-views are then constructed on the corresponding sides of the box. Adjustments of the locations of surfaces are then made, by moving the edges, to complete the isometric sketch.

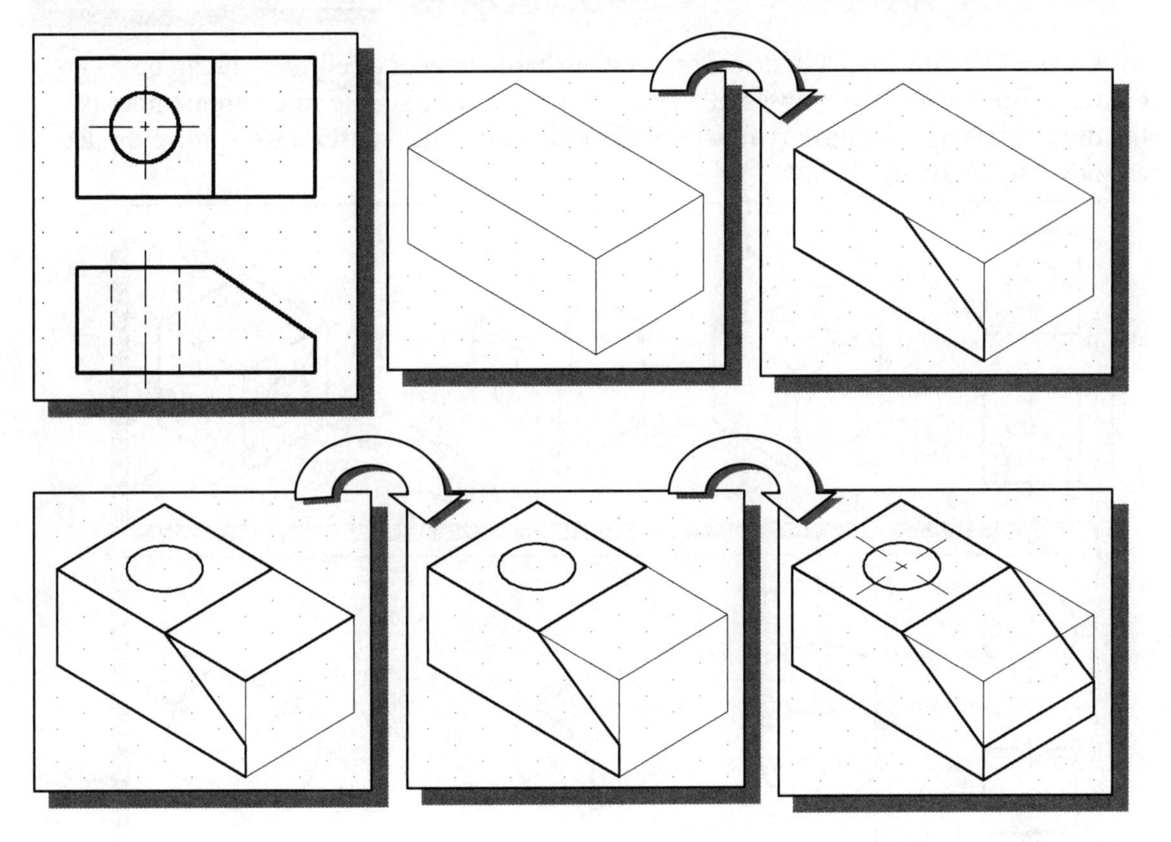

The adjacent surface method begins with one side of the isometric drawing, again with the visible portion of the corresponding 2D-view. The isometric sketch is completed by identifying and adding the adjacent surfaces.

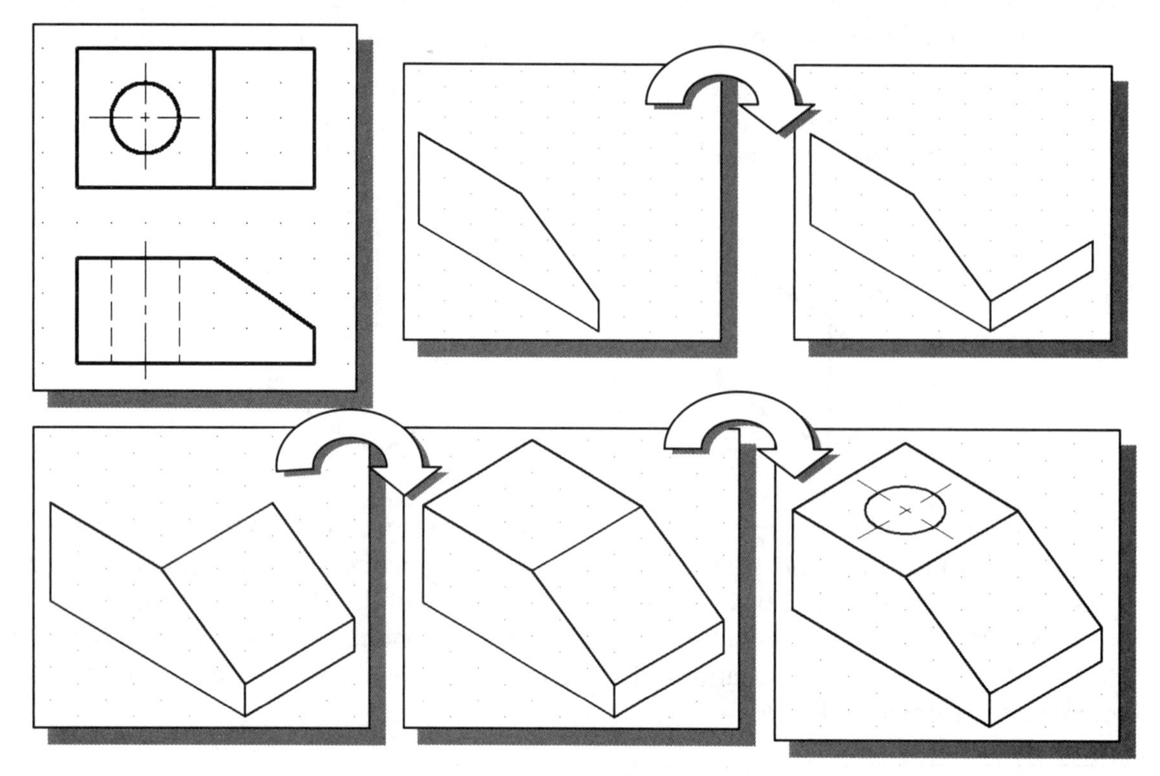

In an isometric drawing, cylindrical or circular shapes appear as ellipses. It can be confusing in drawing the ellipses in an isometric view; one simple rule to remember is the **major axis** of the ellipse is always **perpendicular** to the **center axis** of the cylinder as shown in the figures below.

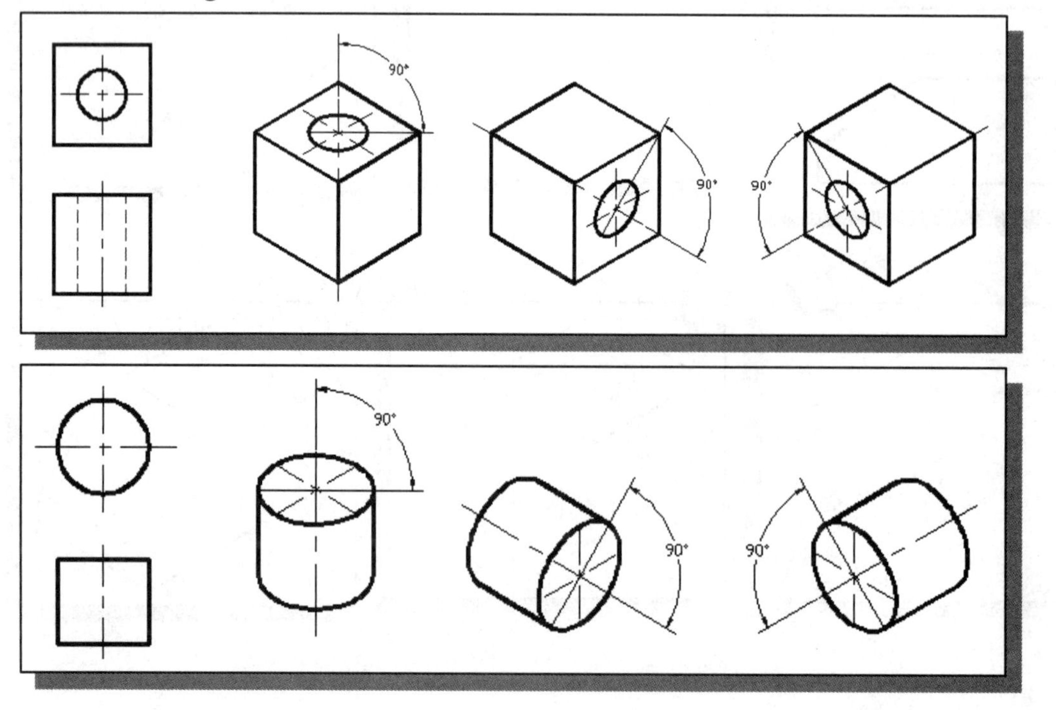

Isometric Sketching Exercises:

Given the Orthographic Top view and Front view, create the isometric view.

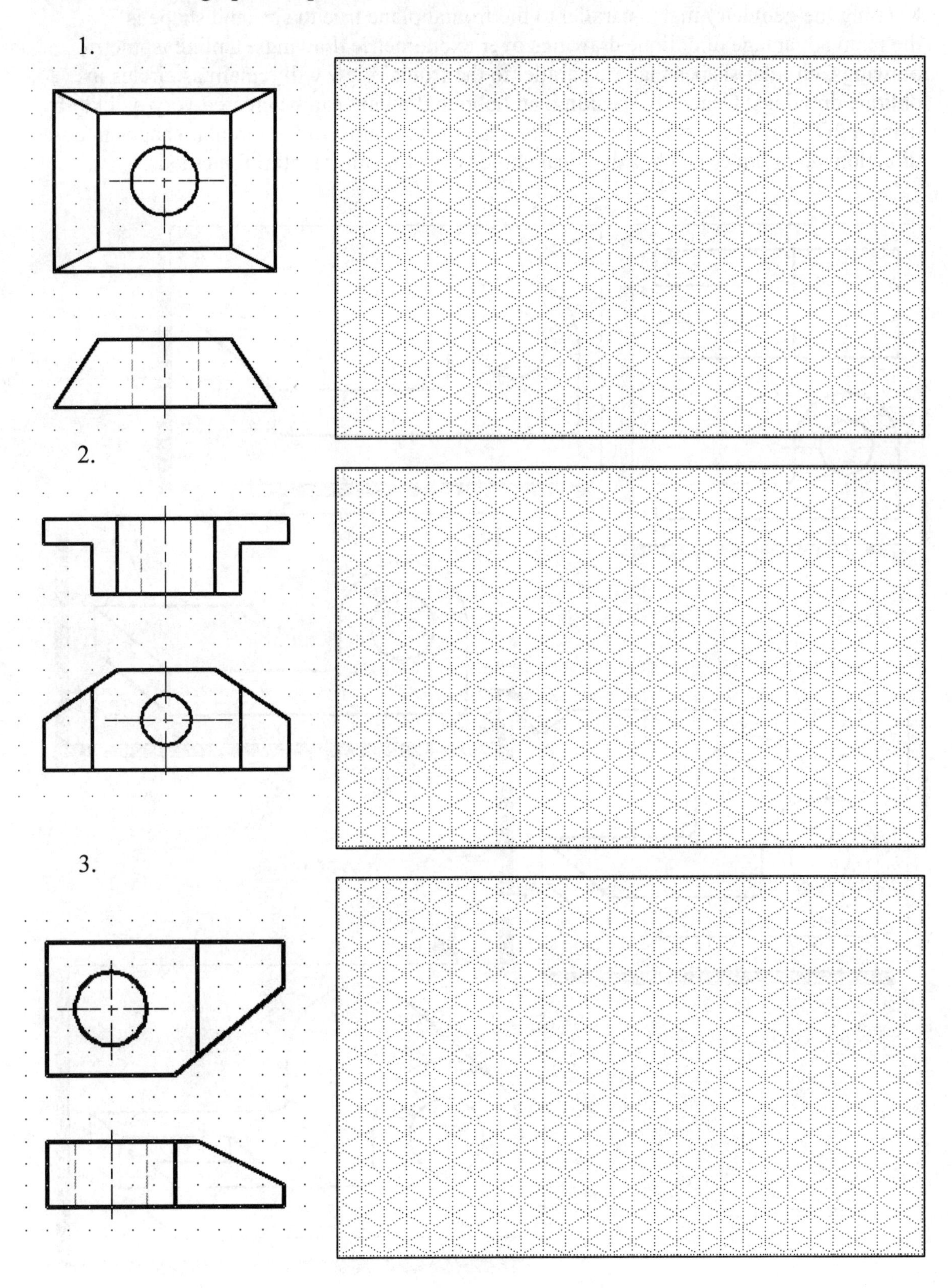

Oblique Sketching

Keeping the geometry that is parallel to the frontal plane true to size and shape is the main advantage of oblique drawings over axonometric drawings. Unlike isometric drawings, circular shapes that are parallel to the frontal view will remain as circles in oblique drawings. Generally speaking, an oblique drawing can be created very quickly by using a 2D view as the starting point. For designs with most of the circular shapes in one direction, an oblique sketch is the ideal choice over the other pictorial methods.

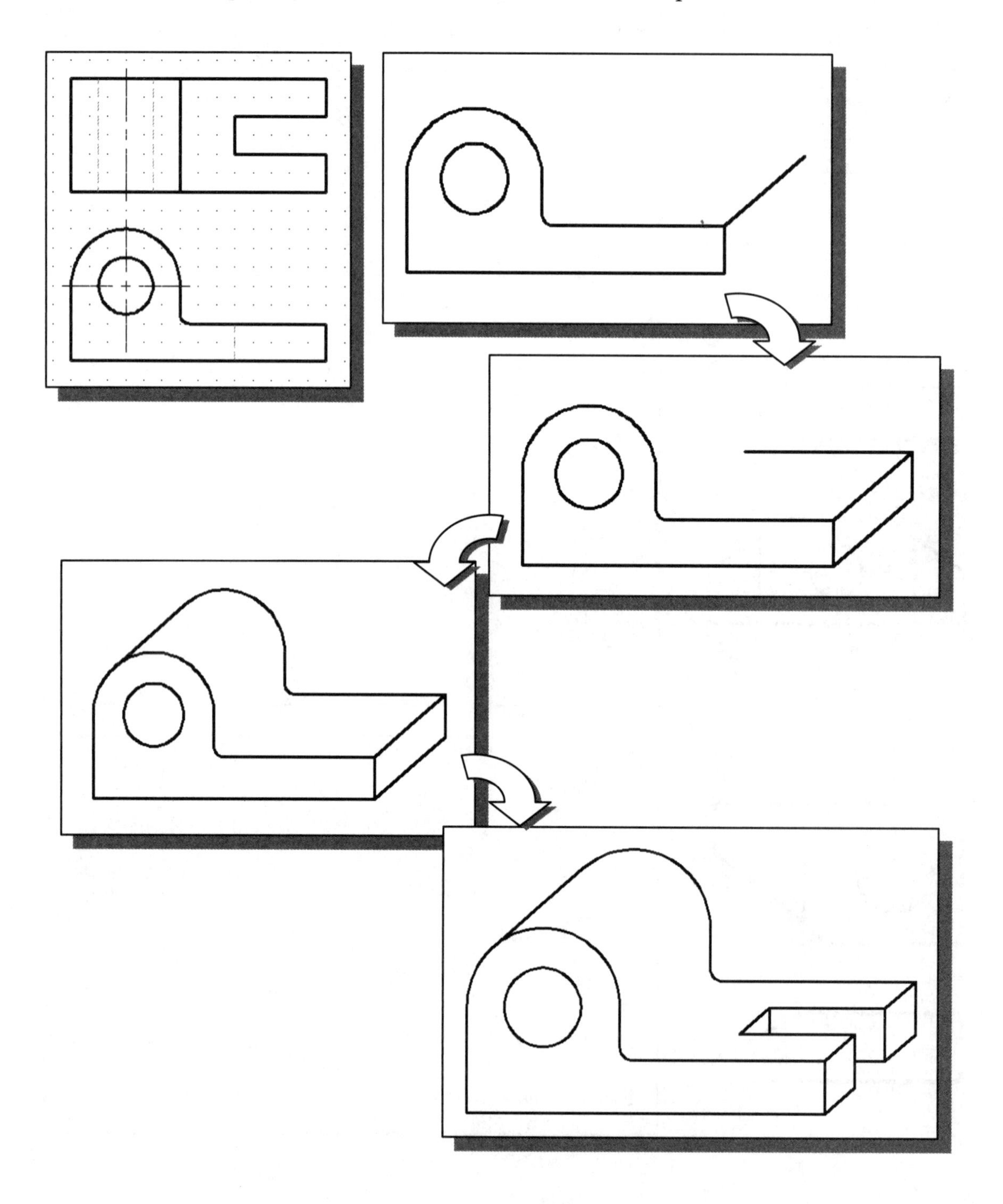

Oblique Sketching Exercises:

Given the Orthographic Top view and Front view, create the oblique view.

1.

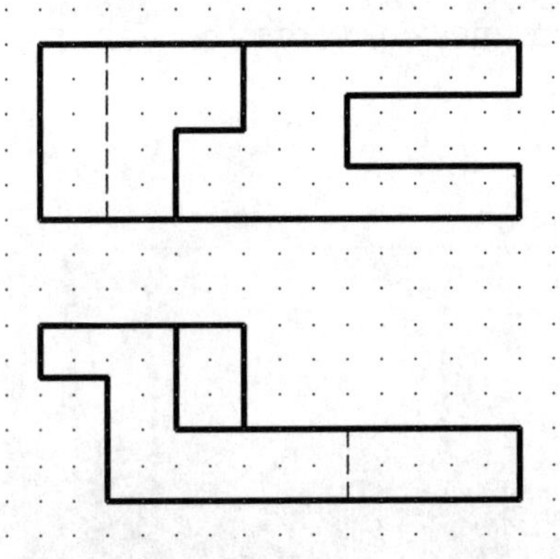

2.

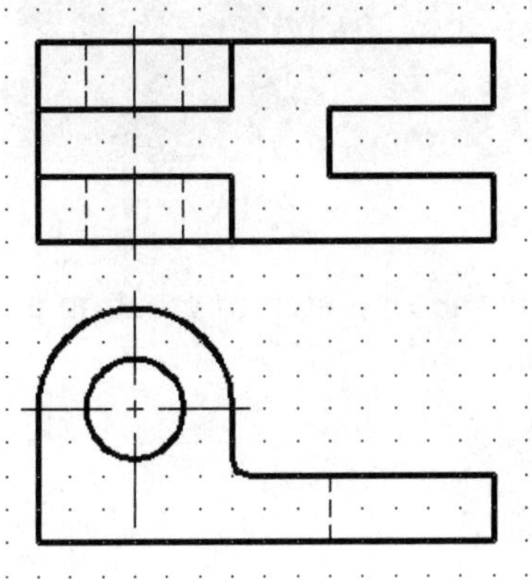

3.

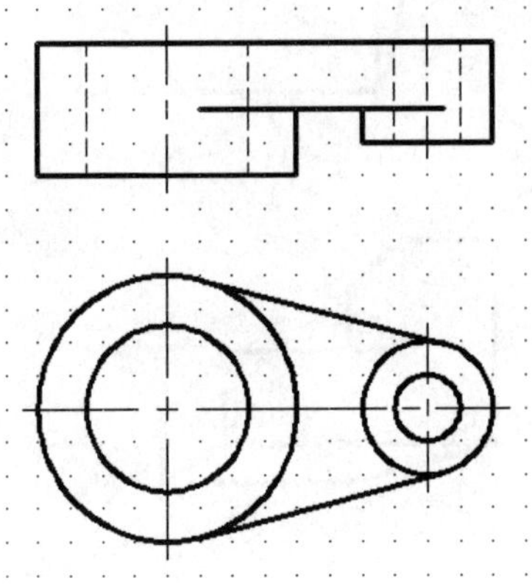

Perspective Sketching

A perspective drawing represents an object as it appears to an observer; objects that are closer to the observer will appear larger to the observer. The key to perspective projection is that parallel edges converge to a single point, known as the **vanishing point**. The vanishing point represents the position where projection lines converge.

The selection of the locations of the vanishing points, which is the first step in creating a perspective sketch, will affect the look of the resulting images.

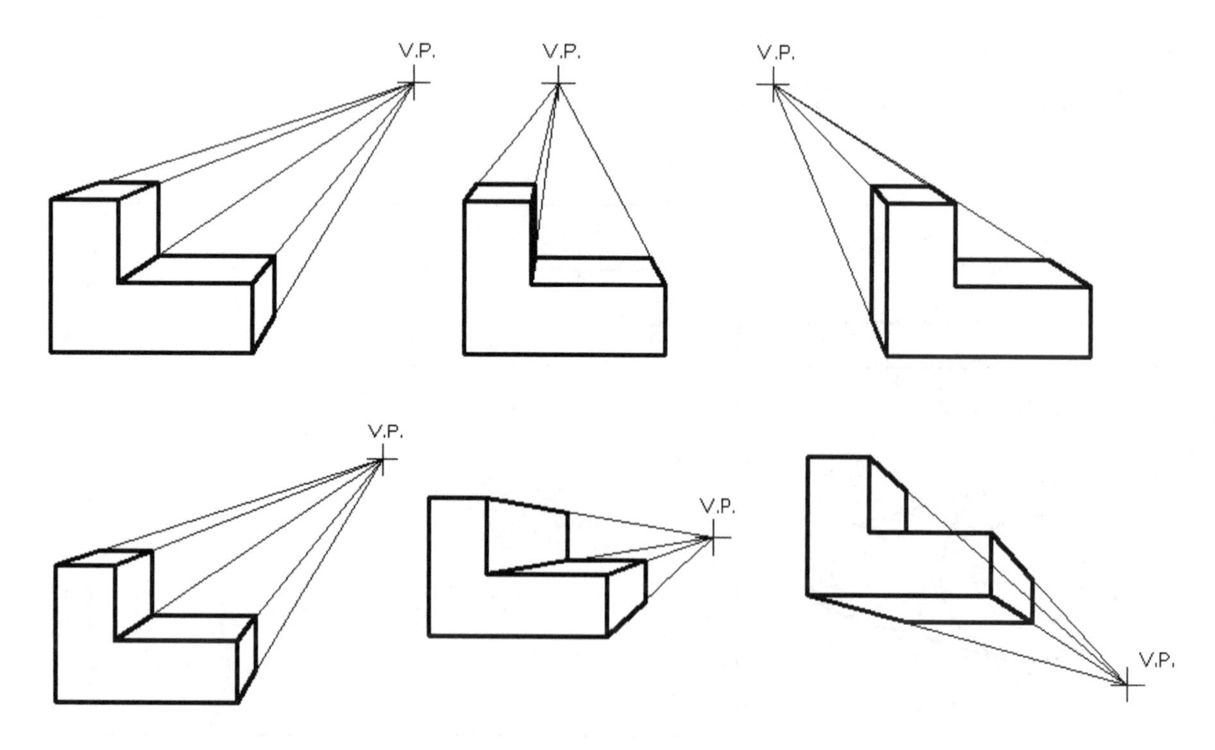

Autodesk Inventor Orthographic vs. Perspective

1. Orthographic View

2. Perspective View

One-point Perspective

One-point perspective is commonly used because of its simplicity. The first step in creating a one-point perspective is to sketch the front face of the object just as in oblique sketching, followed by selecting the position for the vanishing point. For *mechanical* designs, the vanishing point is usually placed above and to the right of the picture. The use of construction lines can be helpful in locating the edges of the object and to complete the sketch.

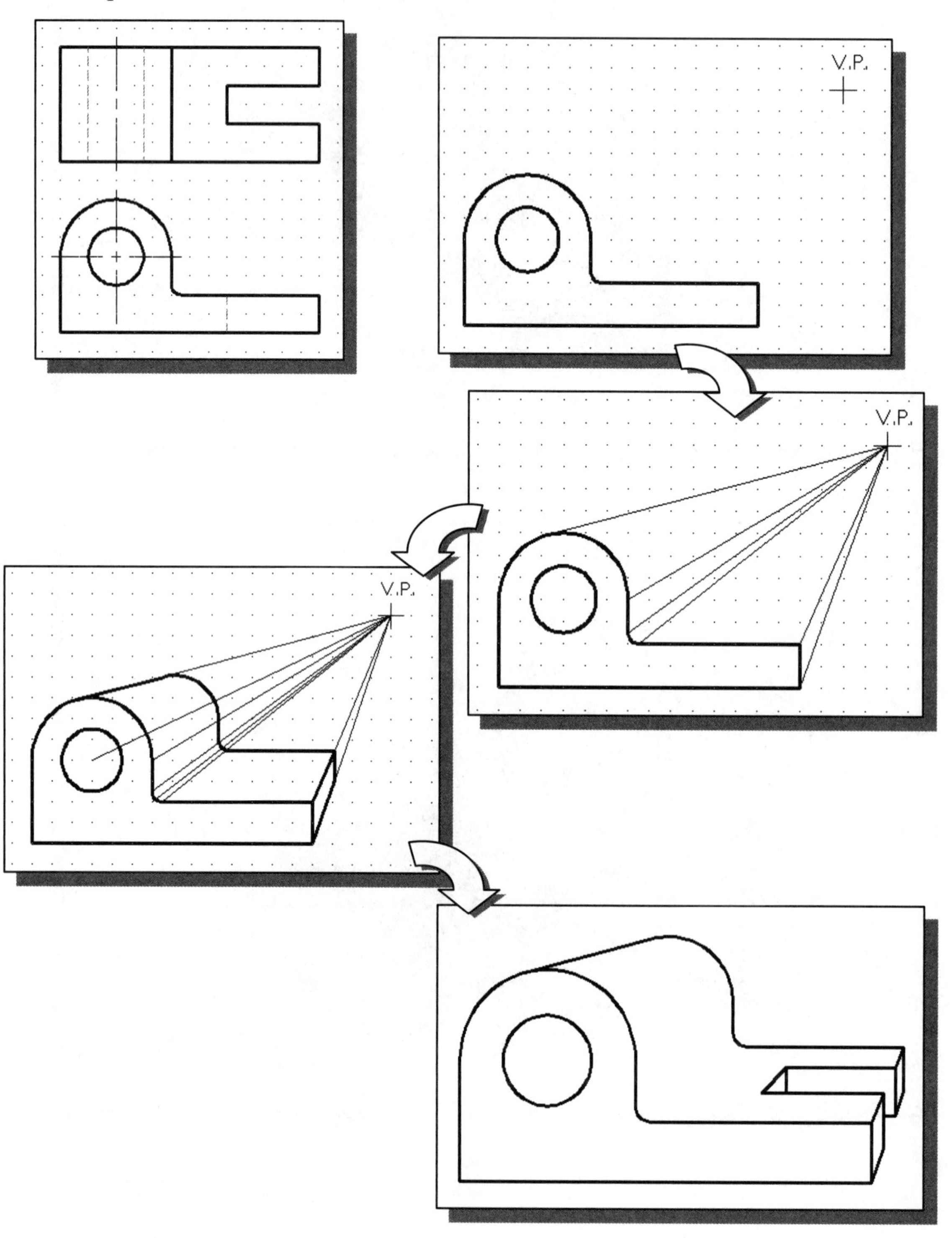

Two-point Perspective

Two-point perspective is perhaps the most popular of all perspective methods. The use of the two vanishing points creates very true-to-life images. The first step in creating a two-point perspective is to select the locations for the two vanishing points, followed by sketching an enclosing box to show the outline of the object. The use of construction lines can be very helpful in locating the edges of the object and to complete the sketch.

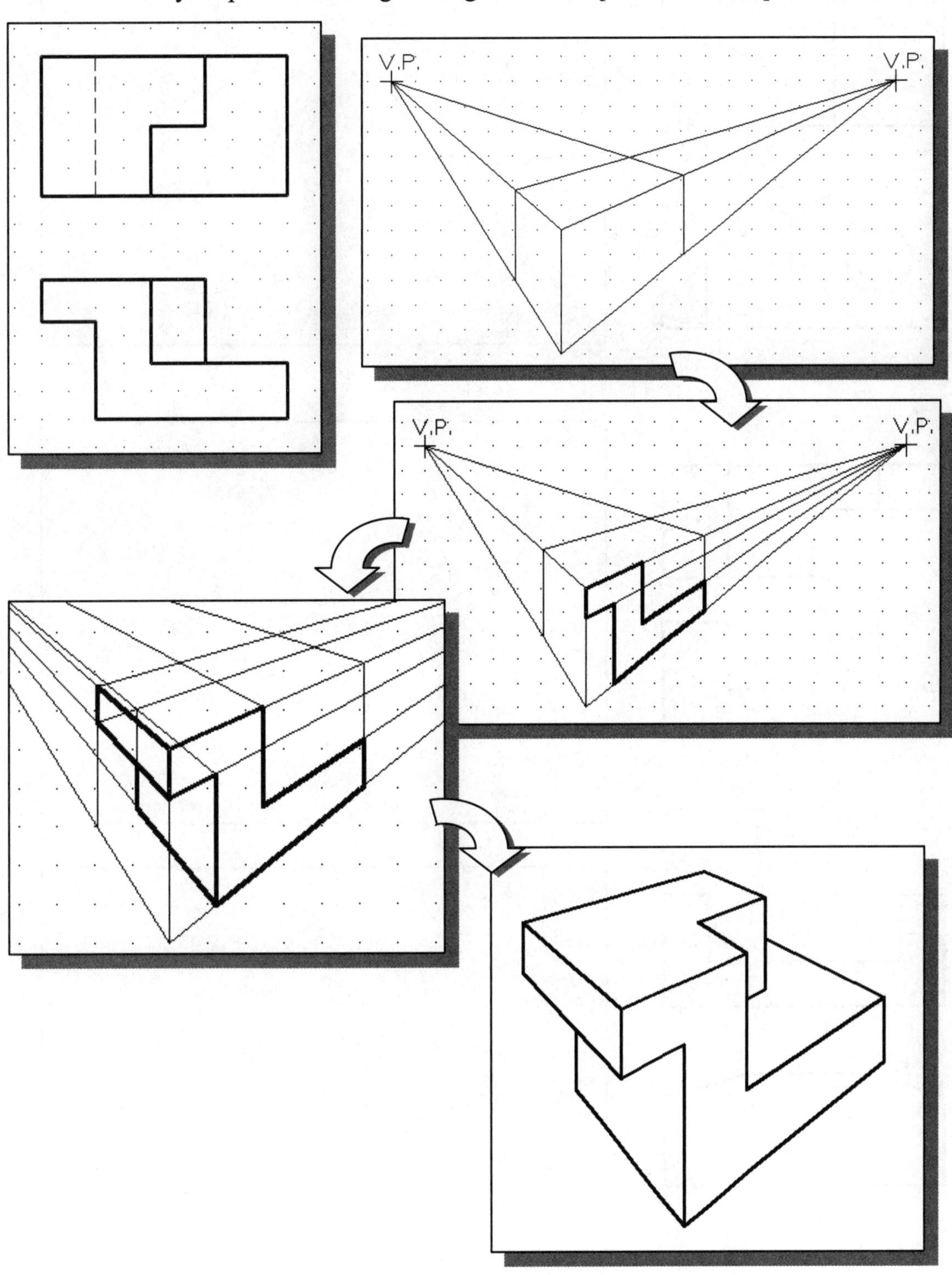

Perspective Sketching Exercises:

Given the Orthographic Top view and Front view, create one-point or two-point perspective views.

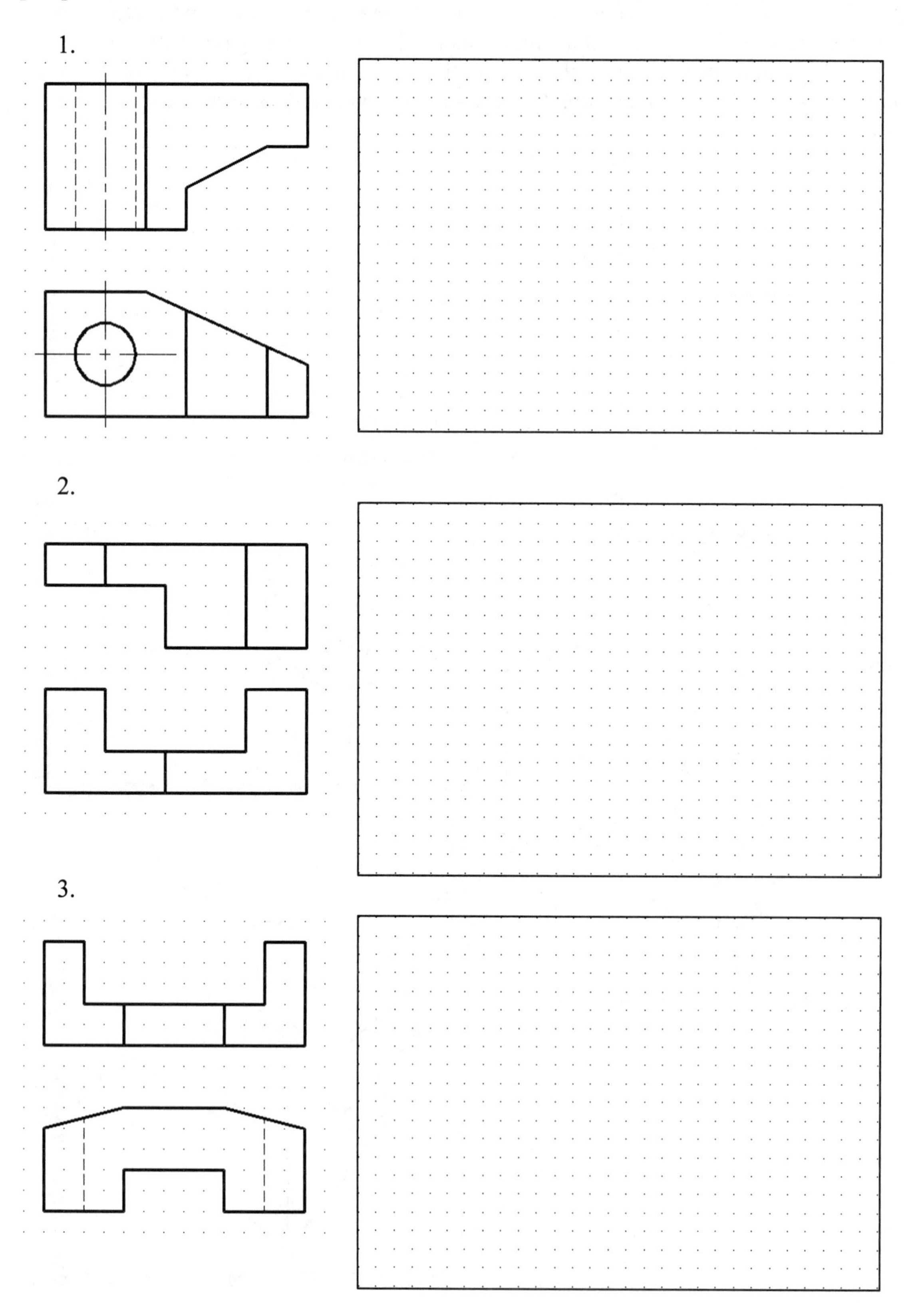

Questions:

1. What are the three types of Axonometric projection?

2. Describe the differences between an Isometric drawing and a Trimetric drawing.

3. What is the main advantage of Oblique projection over Isometric projection?

4. Describe the differences between a one-point perspective and a two-point perspective.

5. Which pictorial methods maintain true size and shape of geometry on the frontal plane?

6. What is a vanishing point in a perspective drawing?

7. What is a Cabinet Oblique?

8. What is the angle between the three axes in an Isometric drawing?

9. In an Axonometric drawing, are the projection lines perpendicular to the projection plane?

10. A cylindrical feature, in a frontal plane, will remain a circle in which pictorial methods?

11. In an Oblique drawing, are the projection lines perpendicular to the projection plane?

12. Create freehand pictorial sketches of:
    - Your desk
    - Your computer
    - One corner of your room
    - The tallest building in your area

Exercises:

1. Given the 2D views, construct the associated isometric views.

2. Given the 2D views, construct the associated two-point perspective views.

3. Given the 2D views, construct the associated oblique views.

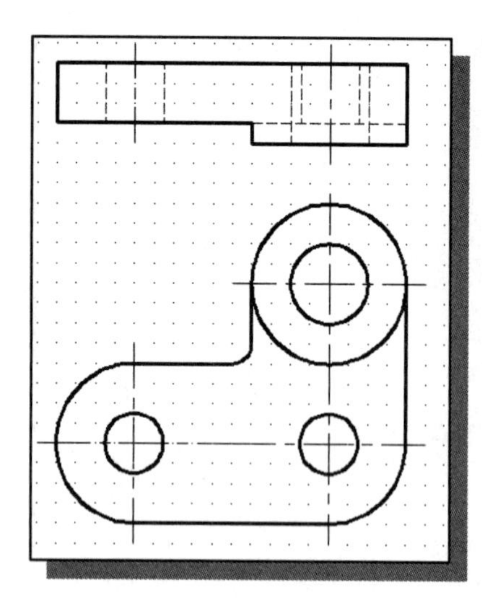

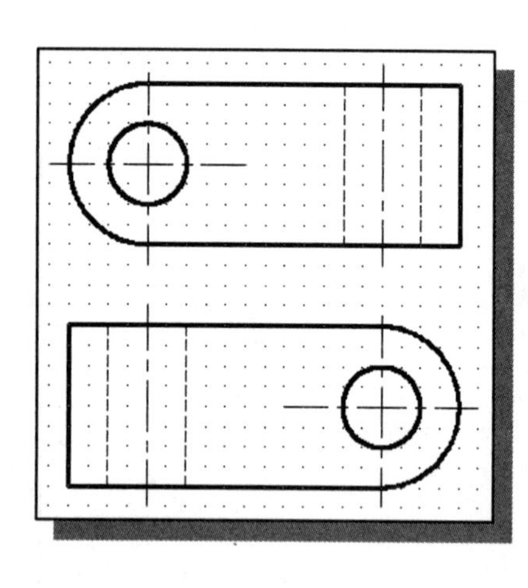

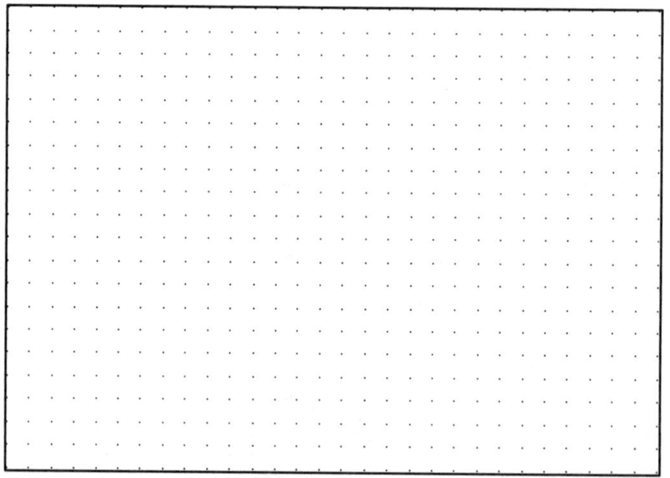

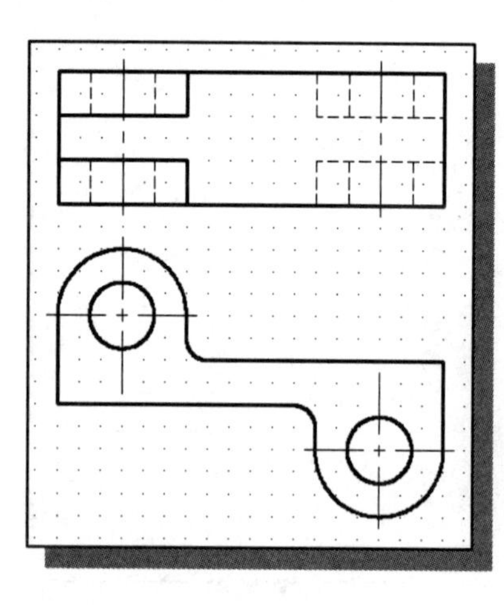

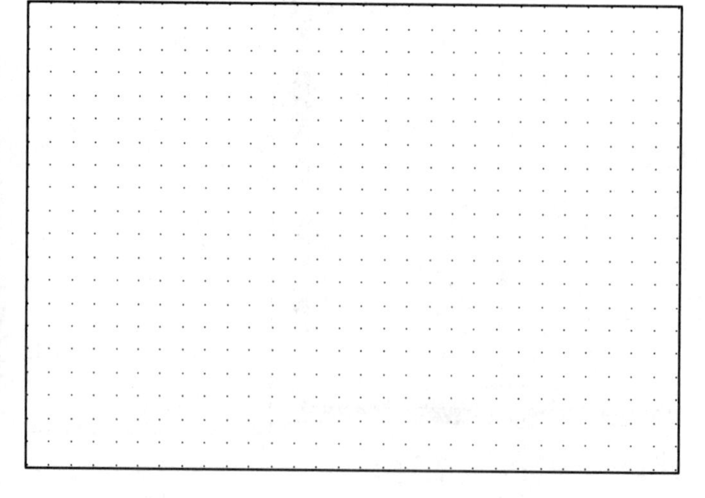

4. Given the 2D views, construct the associated one-point perspective views.

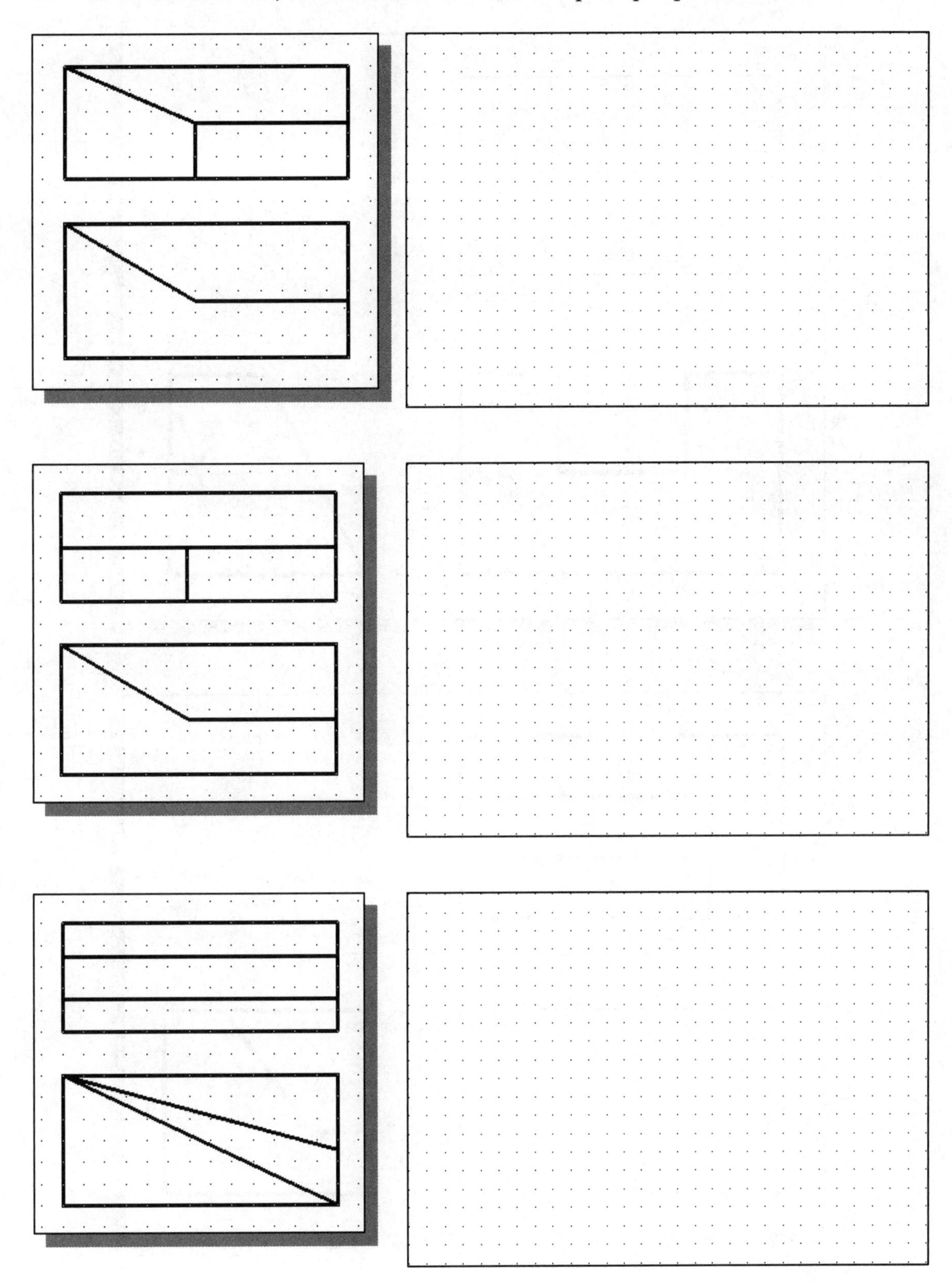

5. Complete the missing views. (Create a pictorial sketch as an aid in reading the views.)

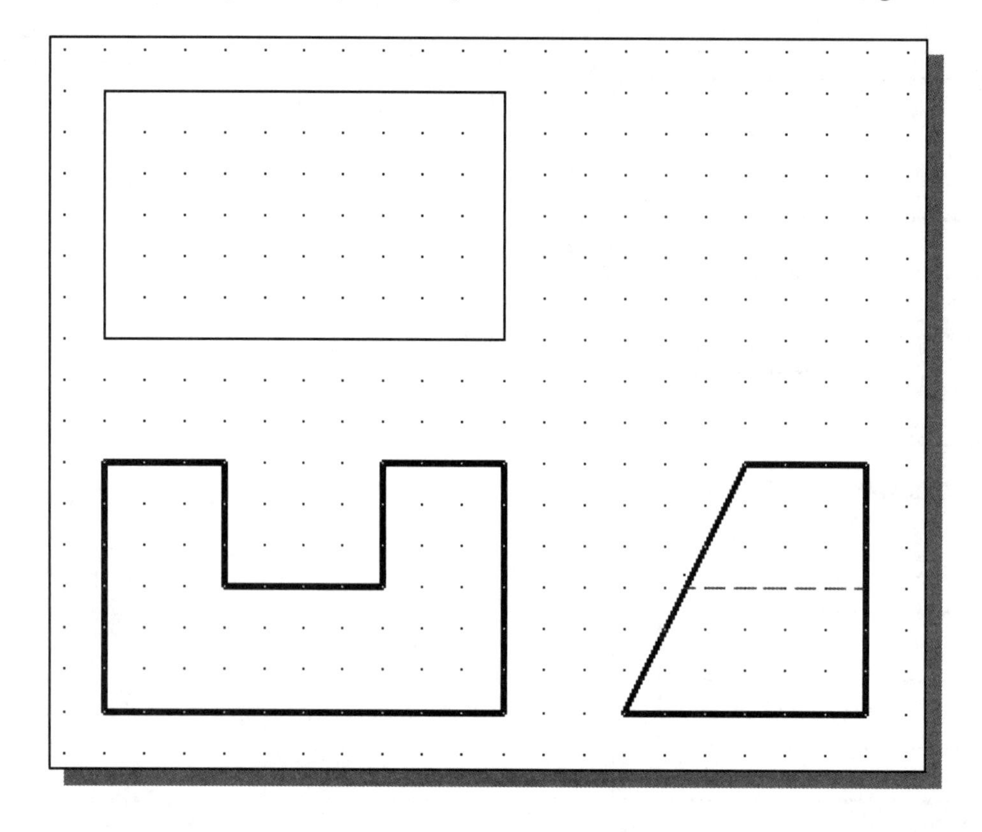

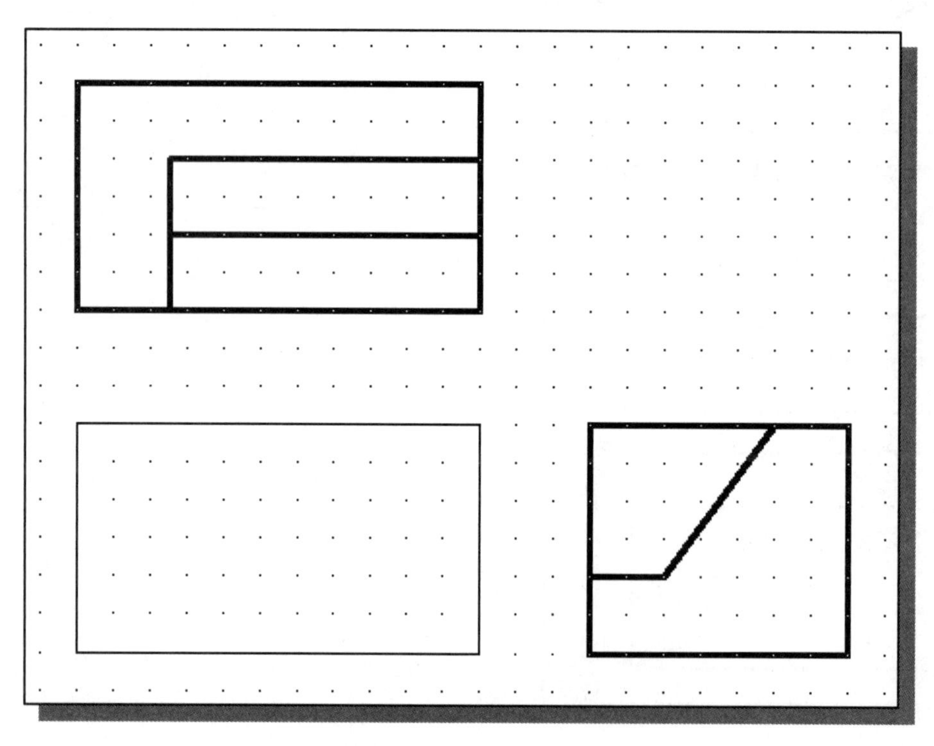

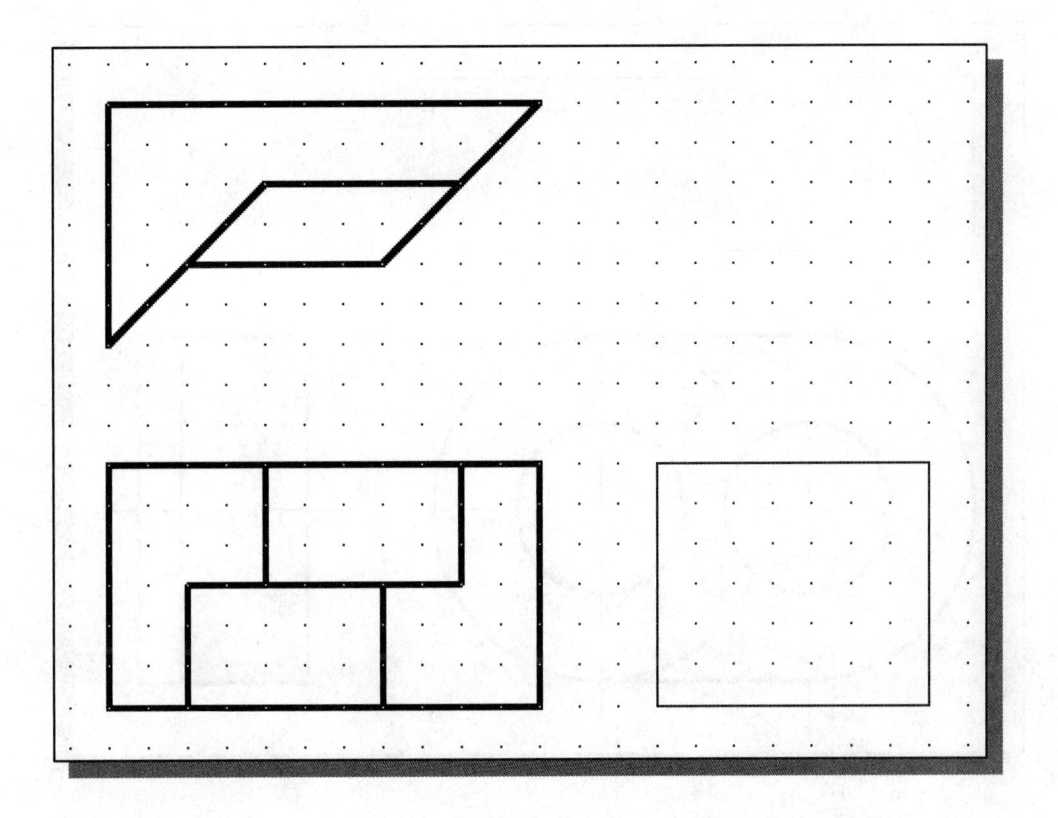

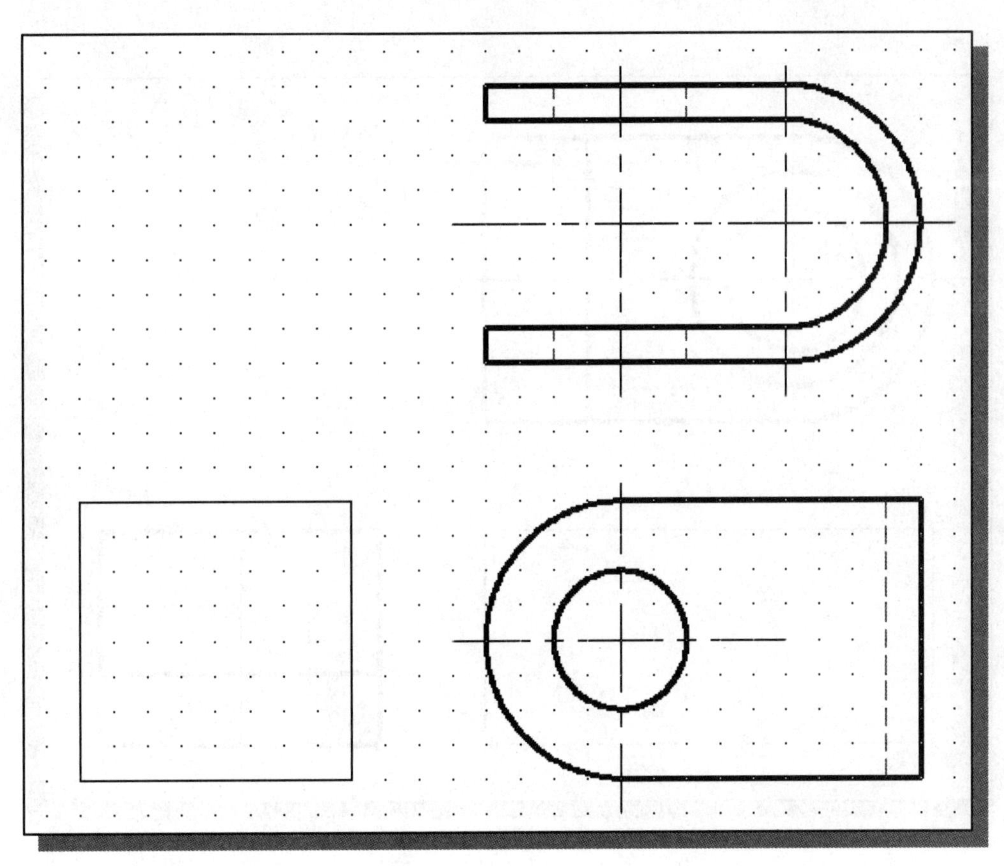

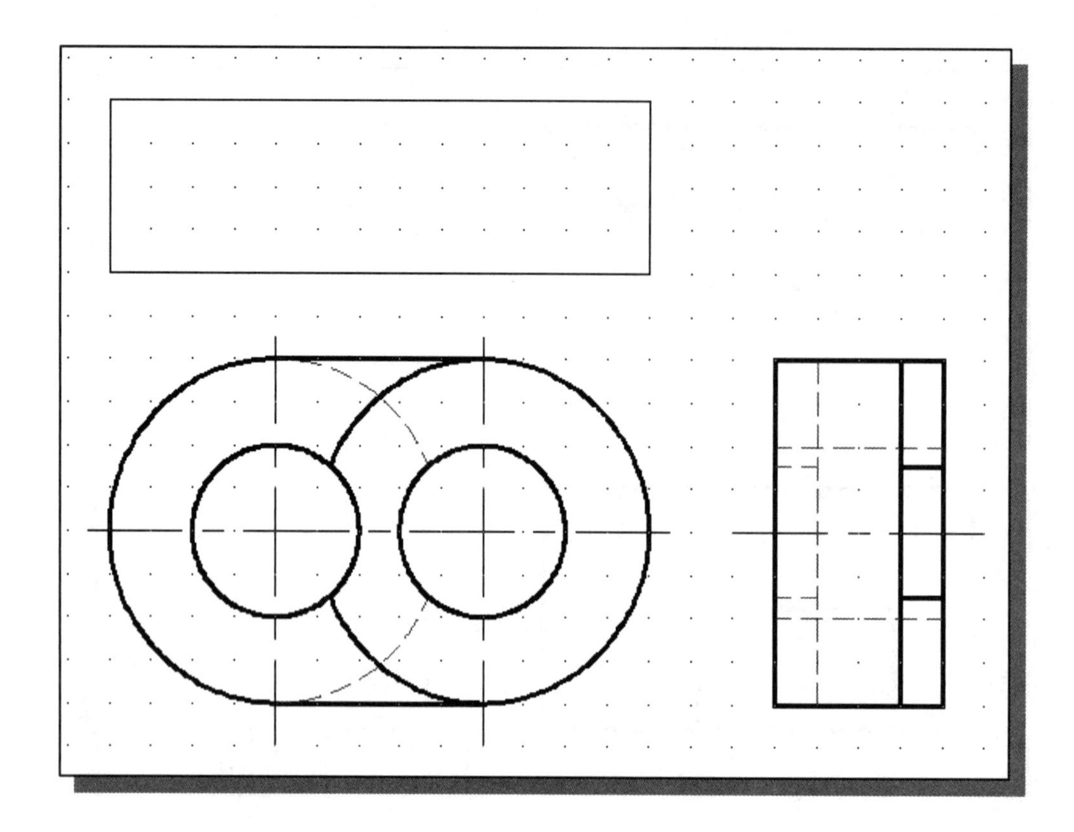

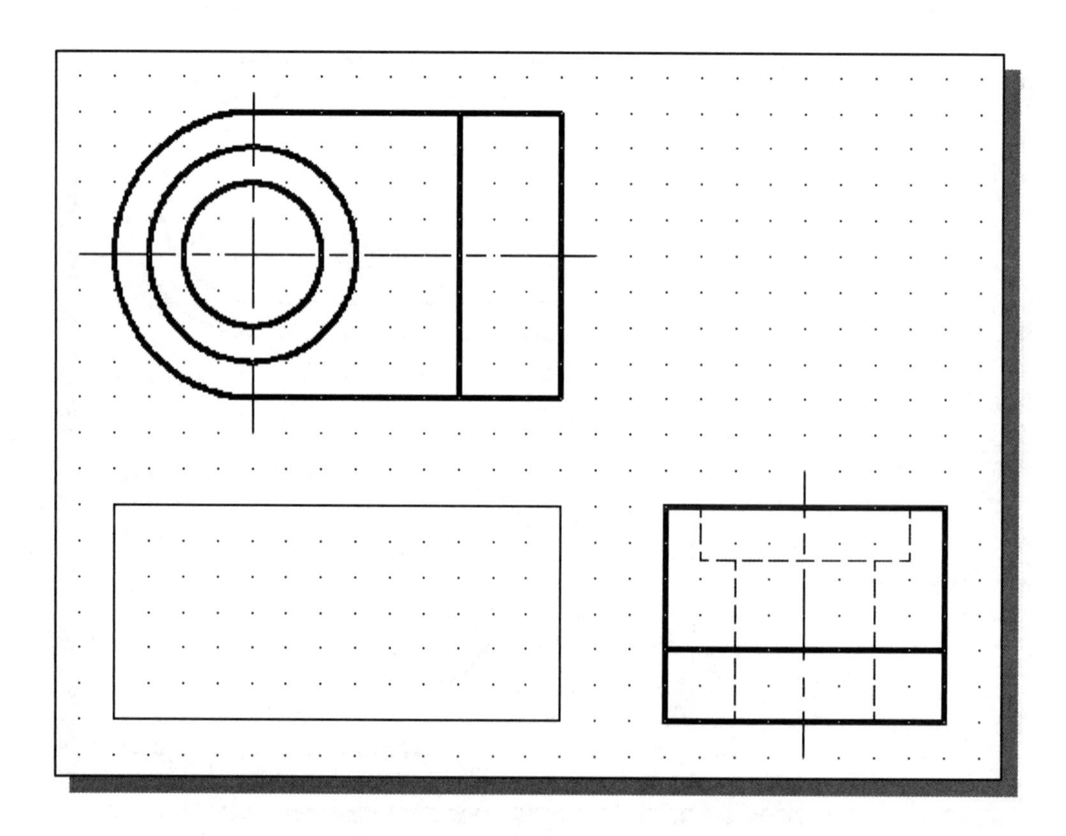

Chapter 6
Symmetrical Features and Part Drawings

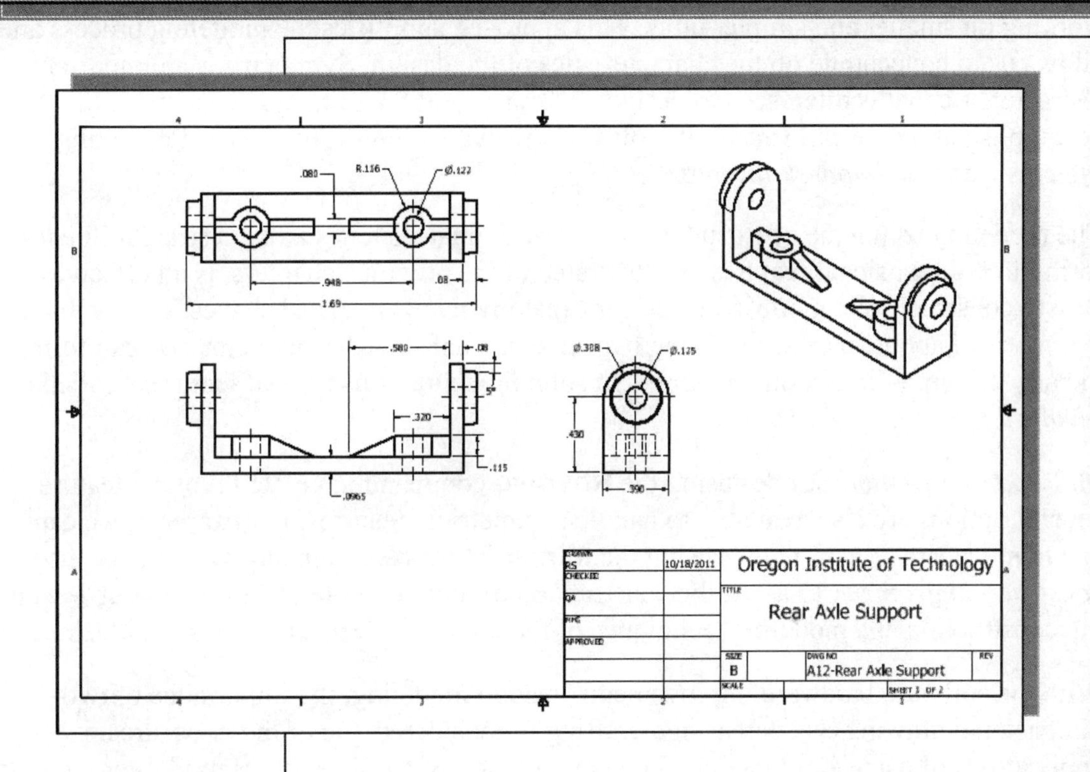

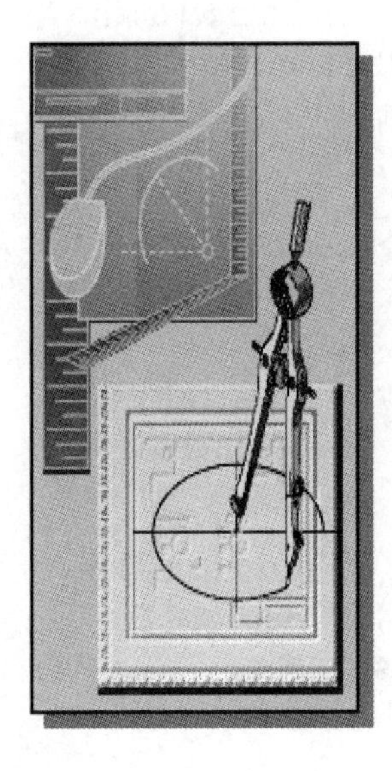

Learning Objectives

- ♦ **Create Drawing Layouts from Solid Models**
- ♦ **Understand Associative Functionality**
- ♦ **Use the Default Borders and Title Block in the Layout Mode**
- ♦ **Arrange and Manage 2D Views in Drawing Mode**
- ♦ **Display and Hide Feature Dimensions**
- ♦ **Create Reference Dimensions**
- ♦ **Create 3D Annotations in Isometric Views**

Drawings from Parts and Associative Functionality

In parametric modeling, it is important to identify and determine the features that exist in the design. *Feature-based parametric modeling* enables us to build complex designs by working on smaller and simpler units. This approach simplifies the modeling process and allows us to concentrate on the characteristics of the design. Symmetry is an important characteristic that is often seen in designs. Symmetrical features can be easily accomplished by the assortment of tools that are available in feature-based modeling systems, such as *Autodesk Inventor*.

The modeling technique of extruding two-dimensional sketches along a straight line to form three-dimensional features, as illustrated in the previous chapters, is an effective way to construct solid models. For designs that involve cylindrical shapes, shapes that are symmetrical about an axis, revolving two-dimensional sketches about an axis can form the needed three-dimensional features. In solid modeling, this type of feature is called a ***revolved feature***.

In *Autodesk Inventor*, besides using the **Revolve** command to create revolved features, several options are also available to handle symmetrical features. For example, we can create mirror images of models using the **Mirror Feature** command. We can also use *construction geometry* to assist the construction of more complex features. In this lesson, the construction and modeling techniques of these more advanced options are illustrated.

With the software/hardware improvements in solid modeling, the importance of two-dimensional drawings is decreasing. Drafting is considered one of the downstream applications of using solid models. In many production facilities, solid models are used to generate machine tool paths for *computer numerical control* (CNC) machines. Solid models are also used in *rapid prototyping* to create 3D physical models out of plastic resins, powdered metal, etc. Ideally, the solid model database should be used directly to generate the final product. However, the majority of applications in most production facilities still require the use of two-dimensional drawings. Using the solid model as the starting point for a design, solid modeling tools can easily create all the necessary two-dimensional views. In this sense, solid modeling tools are making the process of creating two-dimensional drawings more efficient and effective.

Autodesk Inventor provides associative functionality in the different *Autodesk Inventor* modes. This functionality allows us to change the design at any level, and the system reflects it at all levels automatically. For example, a solid model can be modified in the *Part Modeling Mode* and the system automatically reflects that change in the *Drawing Mode*. And we can also modify a feature dimension in the *Drawing Mode*, and the system automatically updates the solid model in all modes.

In this lesson, the general procedure of creating multi-view drawings is illustrated. The *A12- Rear Axle Support* part of the Mechanical Tiger design is first created, and the associated 2D drawing is then generated from the solid model. The associative functionality between the model and drawing views is also examined.

The *A12- Rear Axle Support Design*

Starting *Autodesk Inventor*

1. Select the **Autodesk Inventor** option on the *Start* menu or select the **Autodesk Inventor** icon on the desktop to start *Autodesk Inventor*. The *Autodesk Inventor* main window will appear on the screen. Once the program is loaded into memory, the *Startup* dialog box appears at the center of the screen.

2. Select the **New** icon with a single click of the left-mouse-button in the *Launch* toolbar as shown.

3. Select the **English** units set and in the *Template* area select **Standard(in).ipt**.

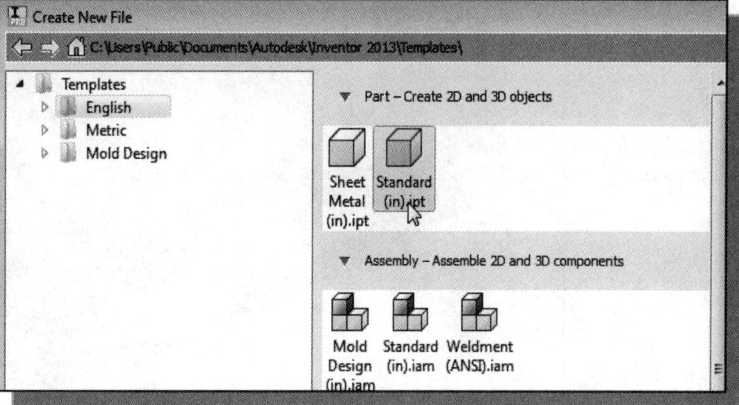

4. Click **Create** in the *New File* dialog box to accept the selected settings.

Modeling Strategy

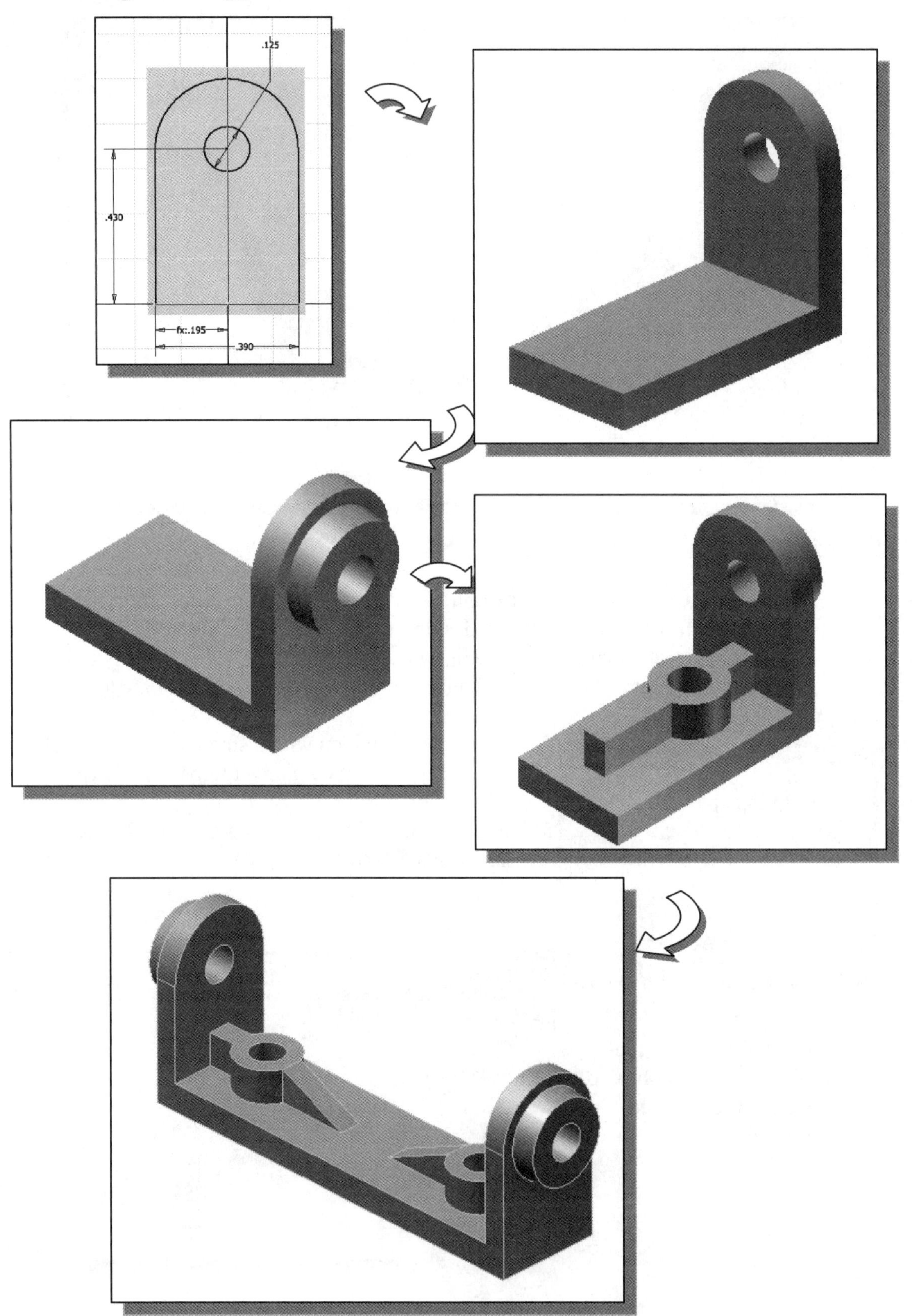

Set Up the Display of the Sketch Plane

1. In the *part browser* window, click on the [+] symbol in front of the **Origin** feature to display more information on the feature.

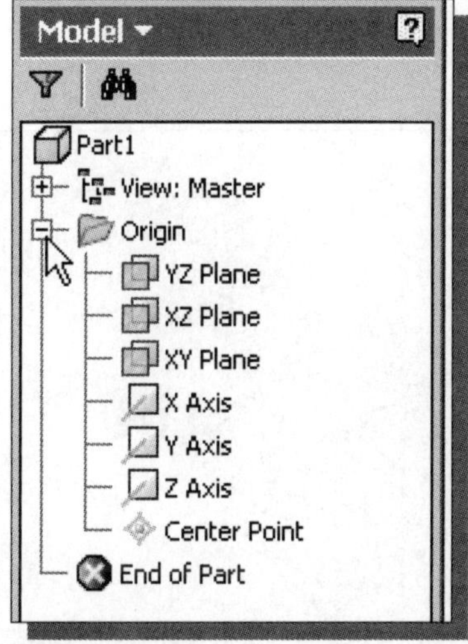

❖ In the *browser* window, notice a new part name appeared with seven work features established. The seven work features include three *workplanes*, three *work axes* and a *work point*. By default, the three workplanes and work axes are aligned to the **world coordinate system** and the work point is aligned to the *origin* of the **world coordinate system**.

2. Inside the *browser* window, move the cursor on top of the third work plane, **XY Plane**. Notice a rectangle, representing the workplane, appears in the graphics window.

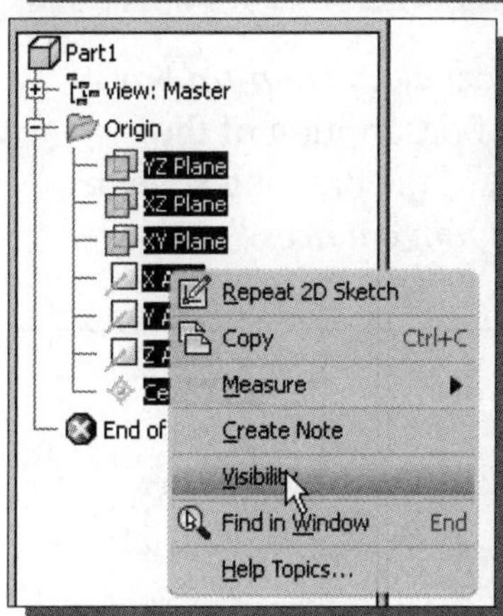

3. Inside the *browser* window, select all of the work features by holding down the **[Shift]** key and clicking with the left-mouse-button.

4. Click the right-mouse-button on any of the work features to display the option menu. Click on **Visibility** to toggle ***ON*** the display of the plane.

5. On your own, use the *Dynamic Viewing* options (ViewCube, 3D Orbit, Zoom and Pan) to view the default work features.

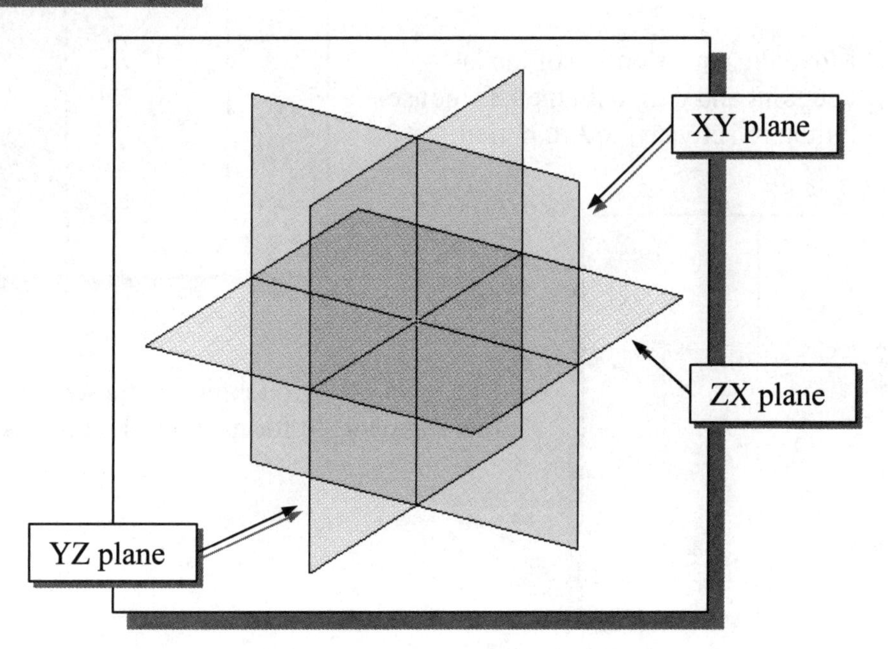

Creating the Base Feature

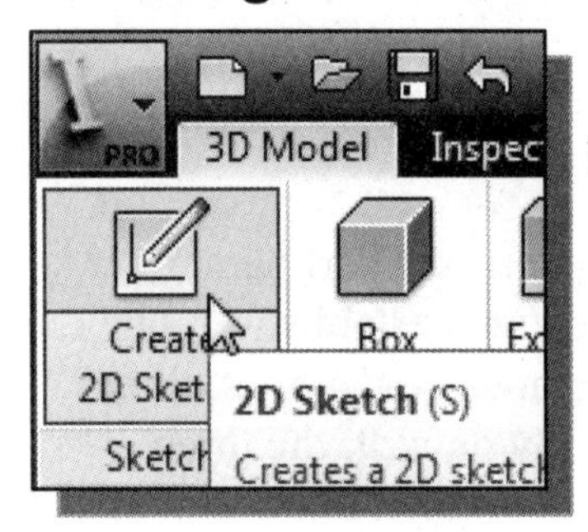

1. In the *Sketch* toolbar select the **Create 2D Sketch** command by left-clicking once on the icon.

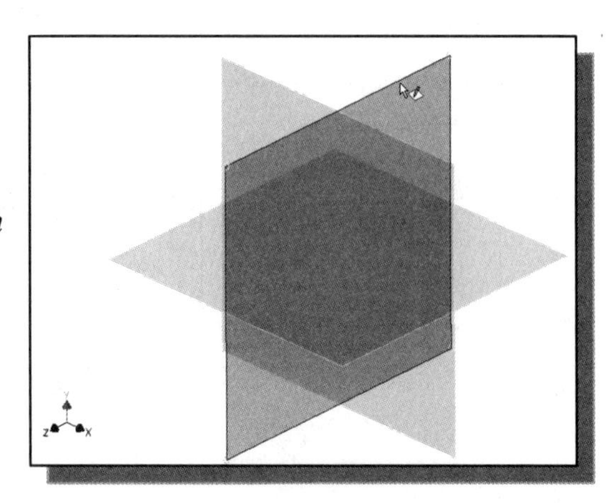

2. In the *Status Bar* area, the message: "*Select face, workplane, sketch or sketch geometry.*" is displayed. Select the **YZ Plane** by clicking on any edge of the plane inside the graphics window, as shown.

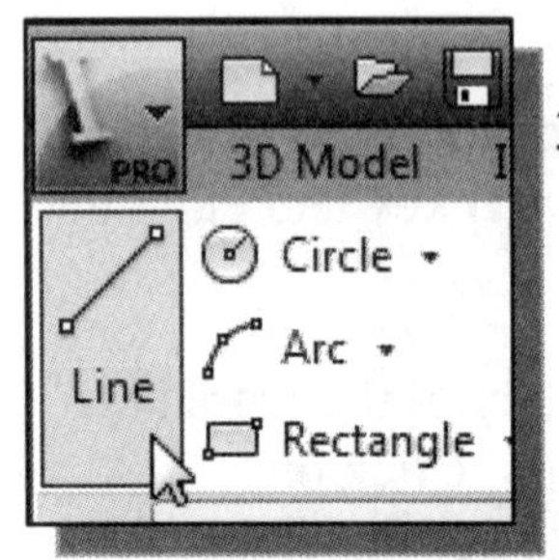

3. Select the **Line** option in the *Draw* panel. A *Help-tip box* appears next to the cursor and a brief description of the command is displayed at the bottom of the drawing screen: "*Creates Straight line segments and tangent arcs.*"

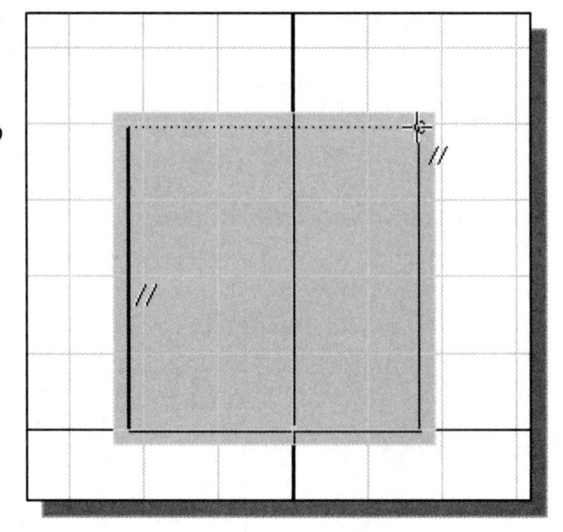

4. Create three line segments, lines are either vertical or horizontal, with the two endpoints aligned as shown. Do not exit the Line command.

5. Move the cursor on top of the last endpoint and drag with the left-mouse-button to activate the Arc option.

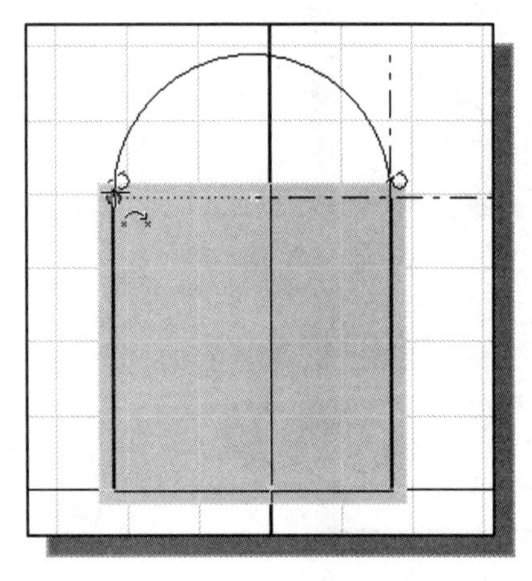

6. Create an arc by clicking on the starting points of the line segments to form a closed region as shown.

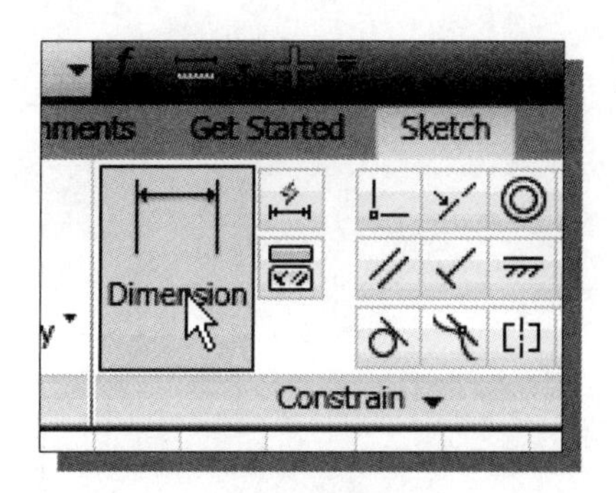

7. On your own, use the **General Dimension** command and create the size and location dimensions of the sketch as shown in the figure below.

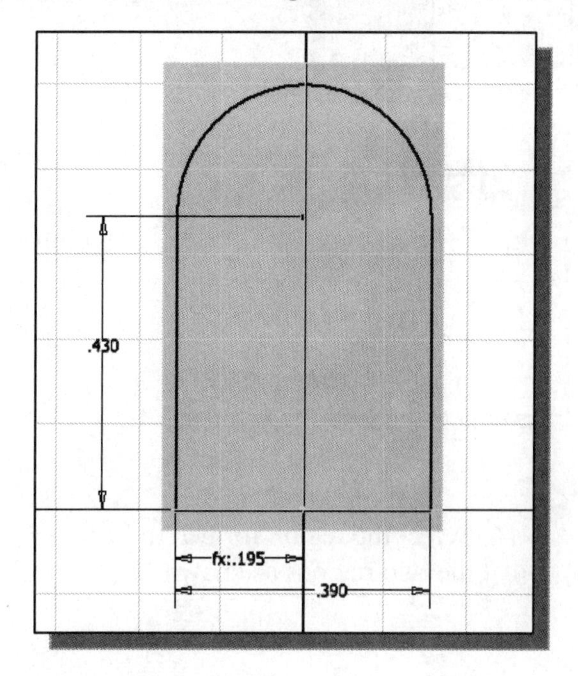

8. On your own, set up a relation to position the sketch aligned to the center point as shown.

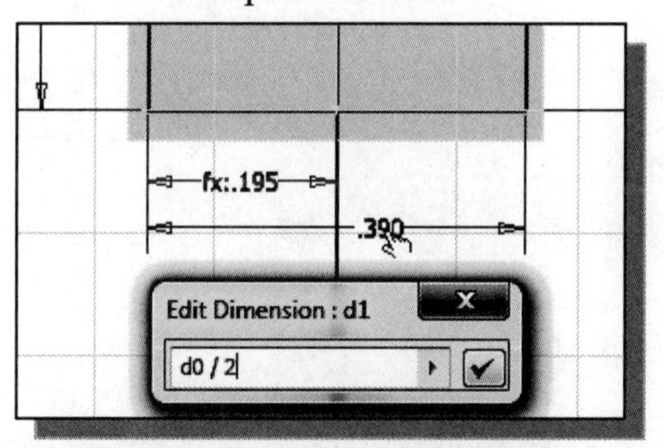

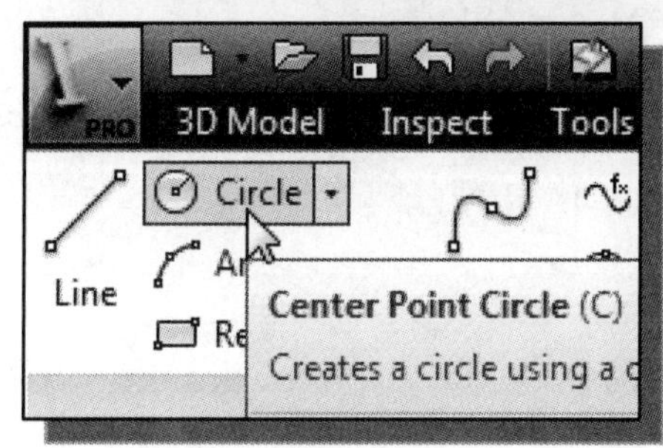

9. Select the **Center point circle** command by clicking once with the left-mouse-button on the icon in the *Draw* panel.

10. On your own, create a circle aligned to the center of the upper arc as shown.

11. Create and modify the diameter dimension to **0.125** as shown.

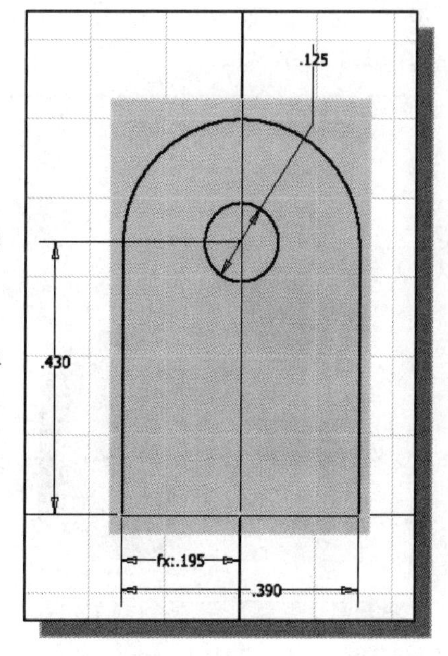

12. Select **Finish Sketch** in the *Ribbon* toolbar to exit the 2D Sketch module.

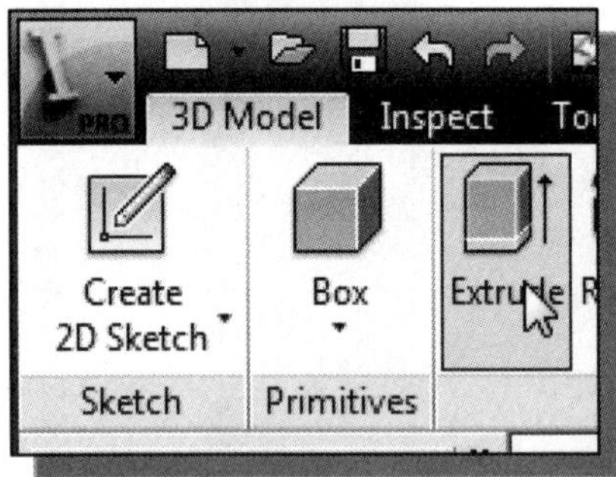

13. In the *Create* toolbar, select the **Extrude** command by left-mouse-clicking once on the icon.

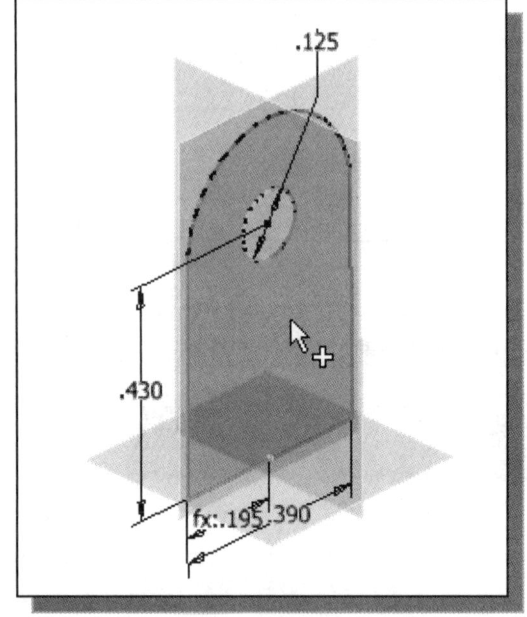

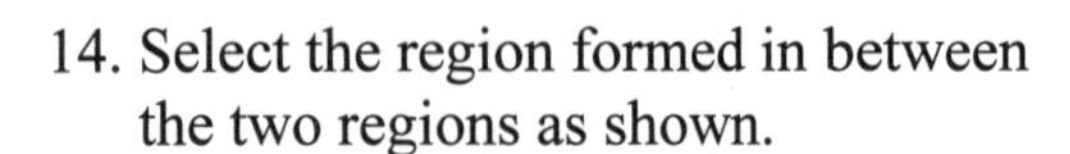

14. Select the region formed in between the two regions as shown.

15. Set the extrusion distance to **0.764**, and set the extrusion direction toward the right as shown.

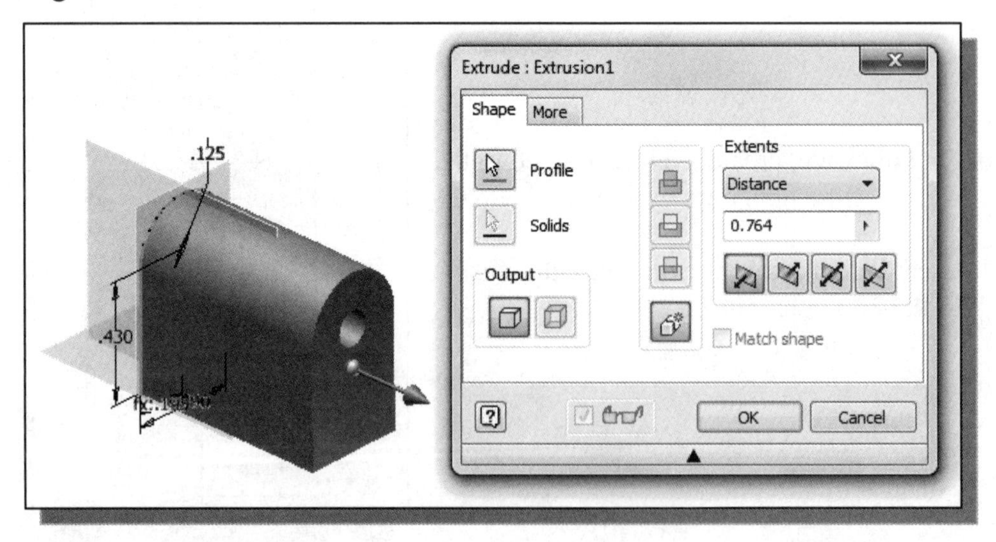

16. Click on the **OK** button to accept the settings and create the solid feature.

Creating a Symmetrical Cut Feature

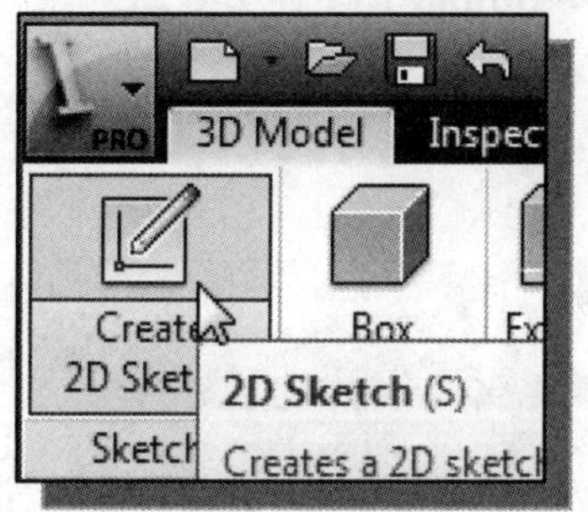

1. In the *Sketch* toolbar select the **Create 2D Sketch** command by left-clicking once on the icon.

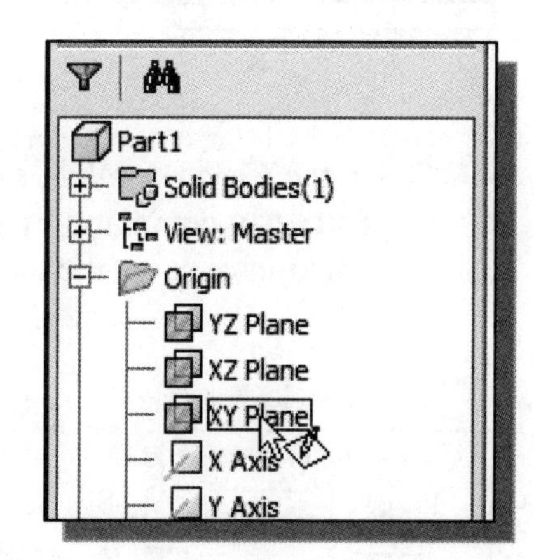

2. In the *Status Bar* area, the message: "*Select face, workplane, sketch or sketch geometry.*" is displayed. Select the **XY Plane** by clicking on the *Model Tree* as shown.

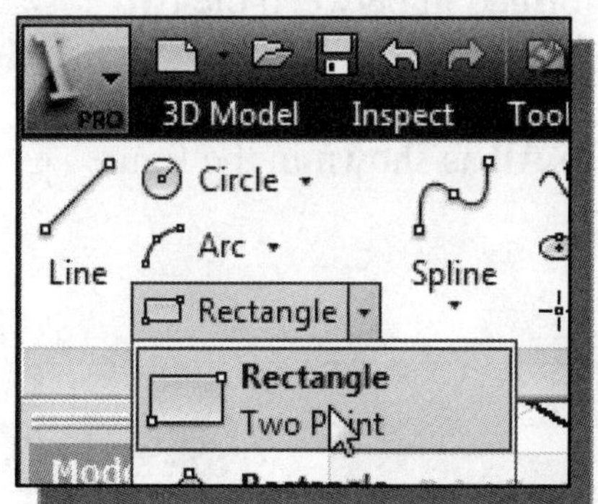

3. Select the **Two point Rectangle** command by clicking once with the **left-mouse-button** on the icon in the *Draw* toolbar.

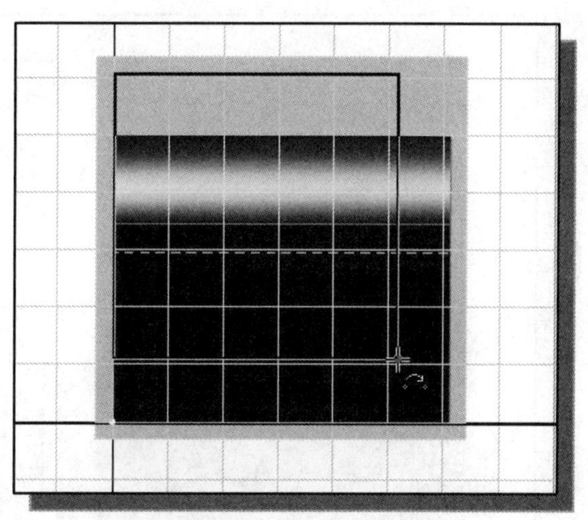

4. Create a rectangle with one edge aligned to the vertical Y axis as shown.

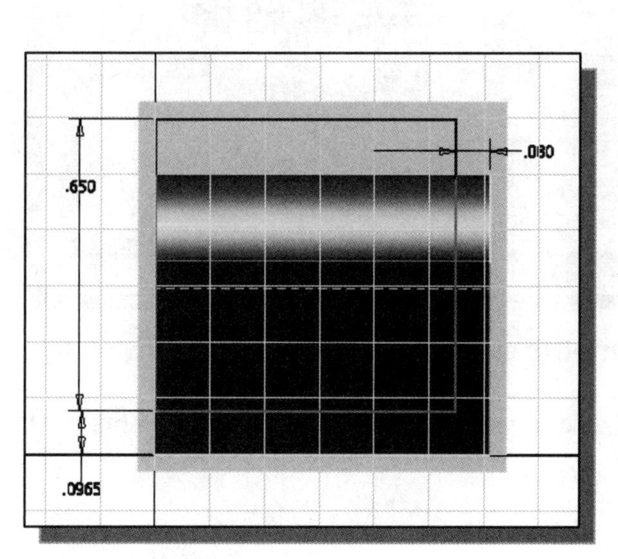

5. Create and modify the three associated dimensions, **0.65**, **0.0965**, and **0.08**, as shown.

6. Select **Finish Sketch** in the *Ribbon* toolbar to exit the 2D Sketch module.

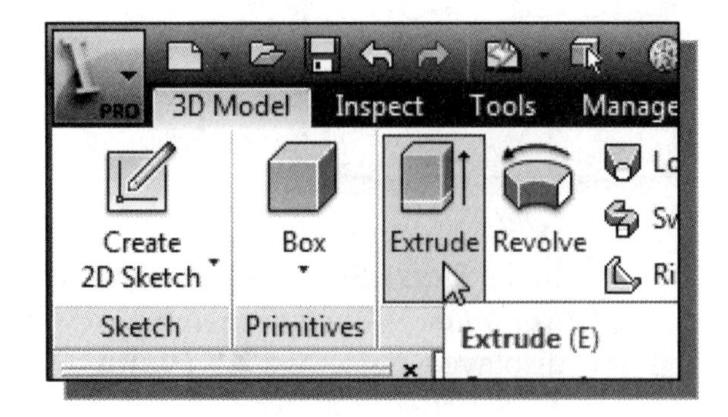

7. In the *Create* toolbar, select the **Extrude** command by left-mouse-clicking once on the icon.

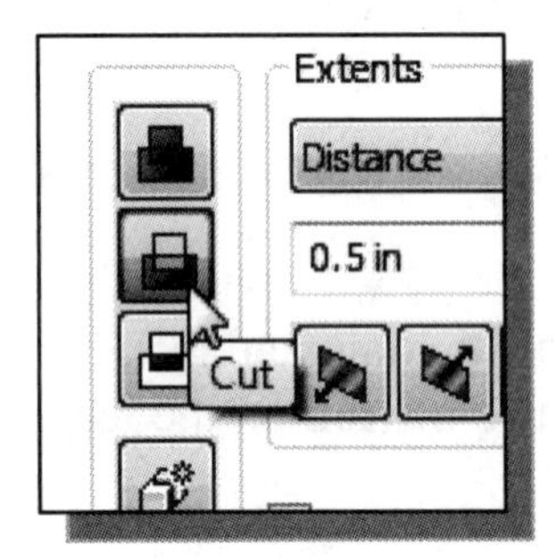

8. Set the extrude option to **Cut** formed in between the two regions as shown.

9. Also set the extrusion extents to **All** as shown in the figure below.

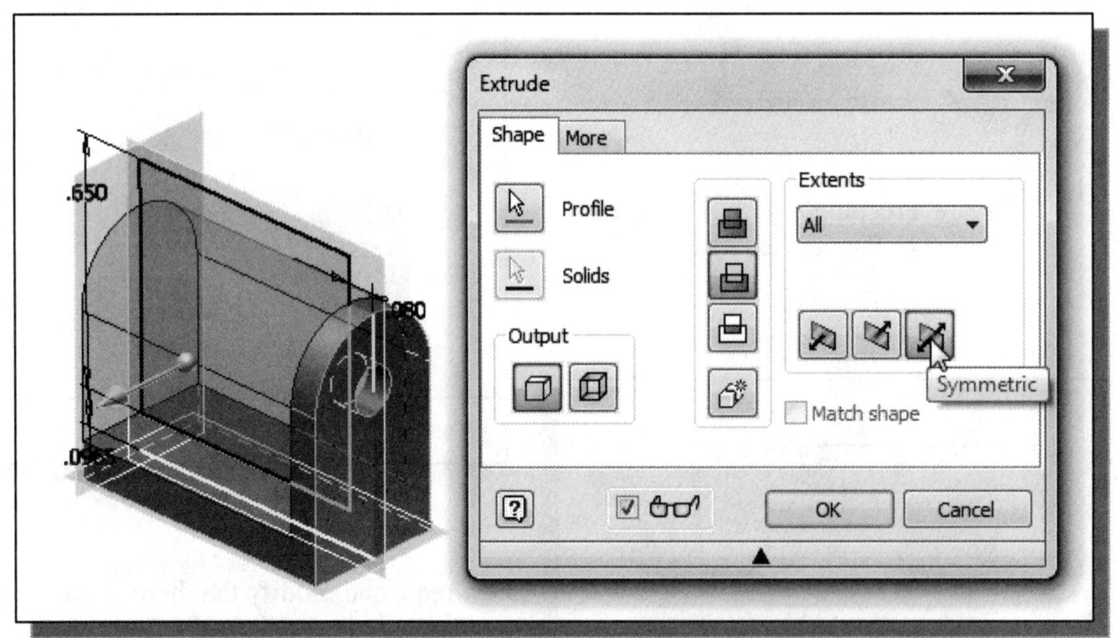

10. Set the extrusion direction to **Symmetric** as shown.

11. Click on the **OK** button to accept the settings and create the cut feature in both directions.

Using the Projected Geometry Option

❖ **Projected geometry** is another type of *reference geometry*. The **Project Geometry** tool can be used to project geometry from previously defined sketches or features onto the sketch plane. The position of the projected geometry is fixed to the feature from which it was projected. We can use the **Project Geometry** tool to project geometry from a sketch or feature onto the active sketch plane.

Typical uses of projected geometry include:

- Project a silhouette of a 3D feature onto the sketch plane for use in a 2D profile.
- Project the default center point onto the sketch plane to constrain a sketch to the origin of the coordinate system.
- Project a sketch from a feature onto the sketch plane so that the projected sketch can be used to constrain a new sketch.

Creating a Revolved Feature

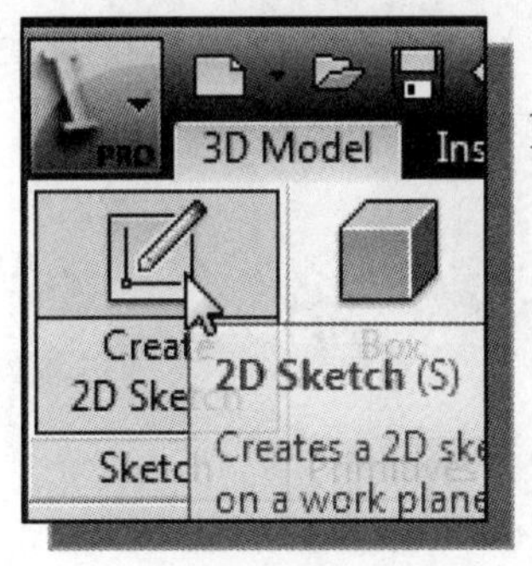

1. In the *Sketch* toolbar select the **Create 2D Sketch** command by left-clicking once on the icon.

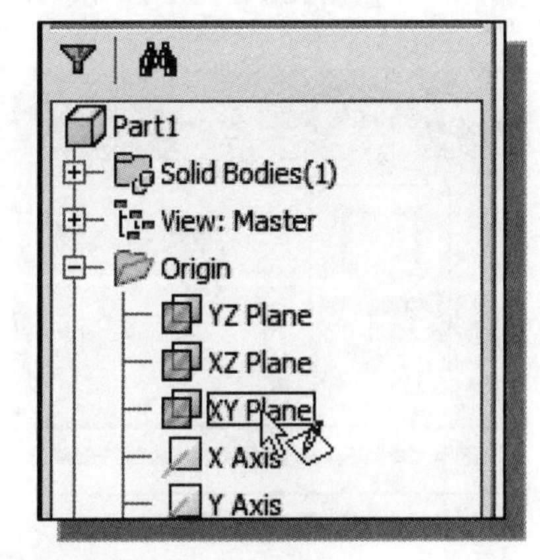

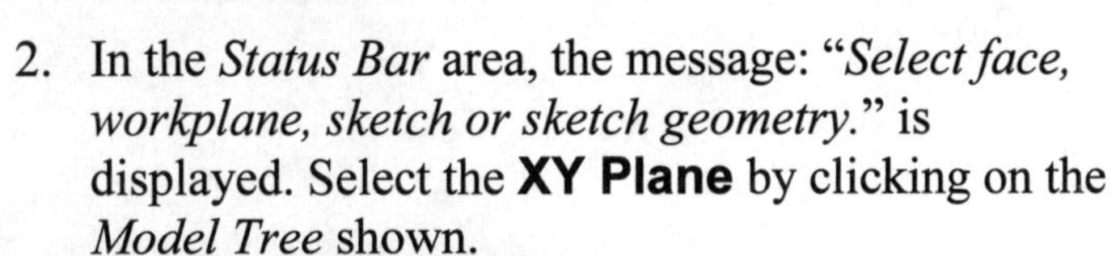

2. In the *Status Bar* area, the message: "*Select face, workplane, sketch or sketch geometry.*" is displayed. Select the **XY Plane** by clicking on the *Model Tree* shown.

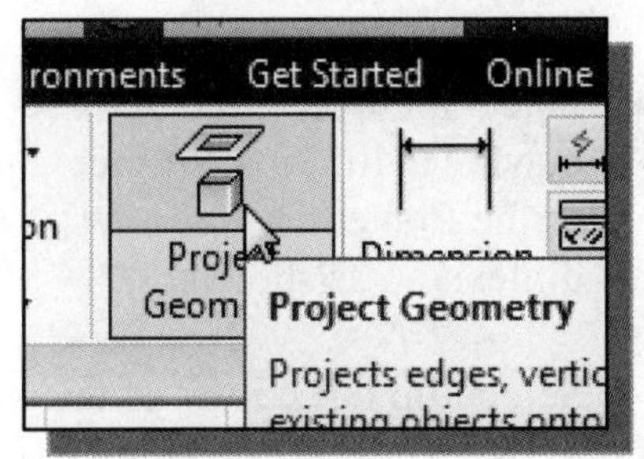

3. Select the **Project Geometry** command in the *Draw* panel.

4. Select the **top edge** of the cylindrical surface to project. (Hint: Use the *Dynamic Zoom* function to aid the selection.)

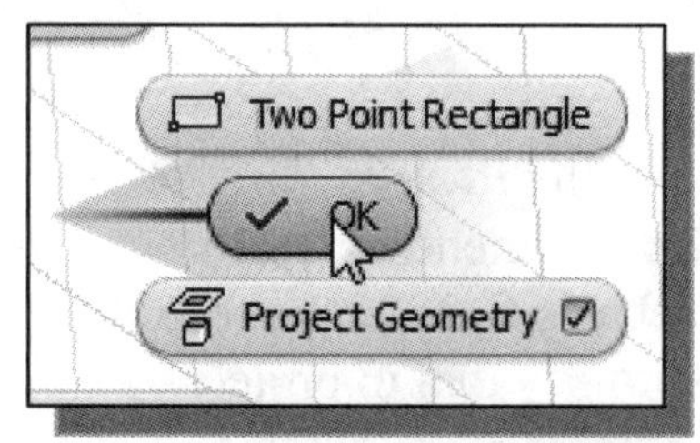

5. Inside the graphics window, right-click once to bring up the option menu and select **OK** to end the command.

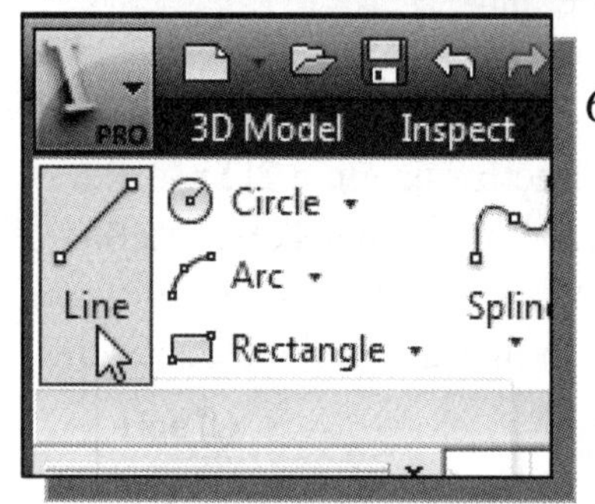

6. Select the **Line** option in the *Draw* panel.

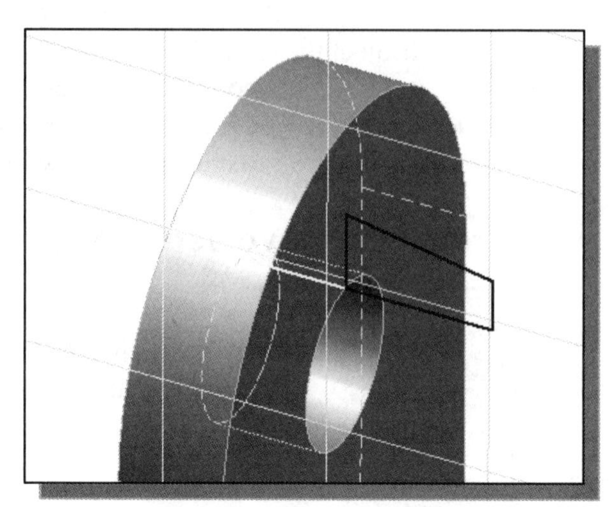

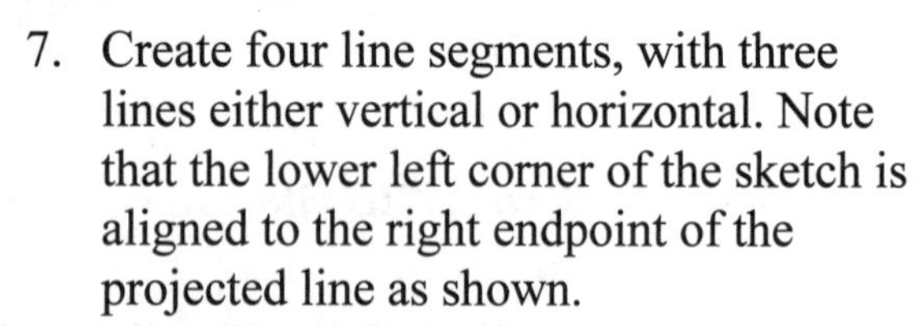

7. Create four line segments, with three lines either vertical or horizontal. Note that the lower left corner of the sketch is aligned to the right endpoint of the projected line as shown.

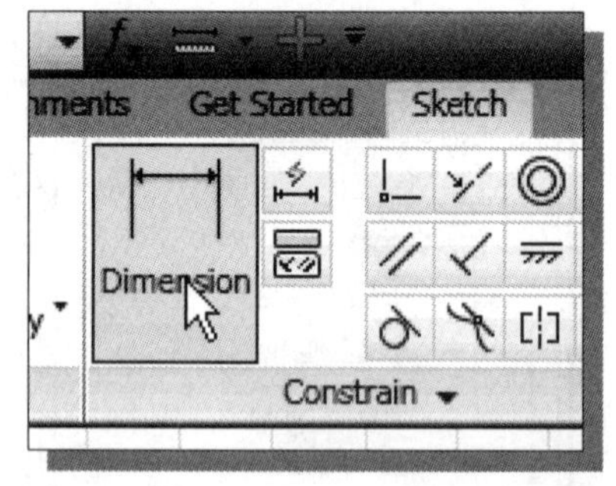

8. Select the **General Dimension** command as shown.

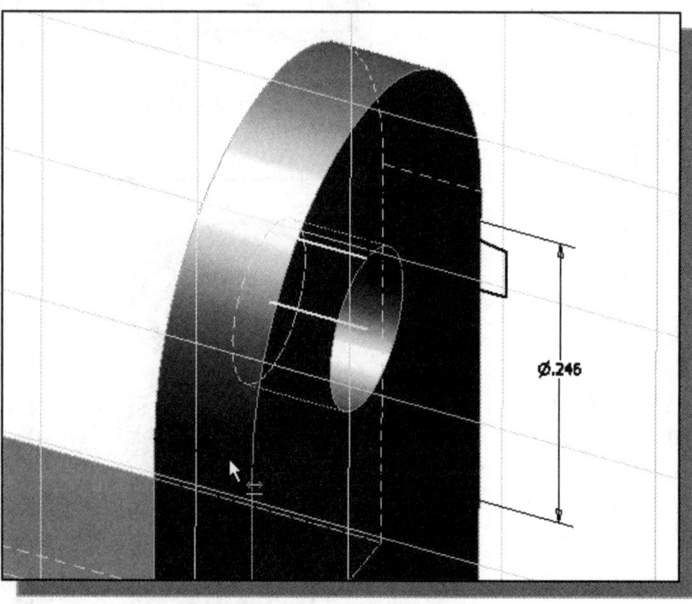

9. Select the front edge of outside cylindrical surface as the first entity to dimension as shown.

10. Select top left corner of the sketch as the second entity to dimension.

11. Move the cursor to the right side of the sketch. Do not click on any location yet.

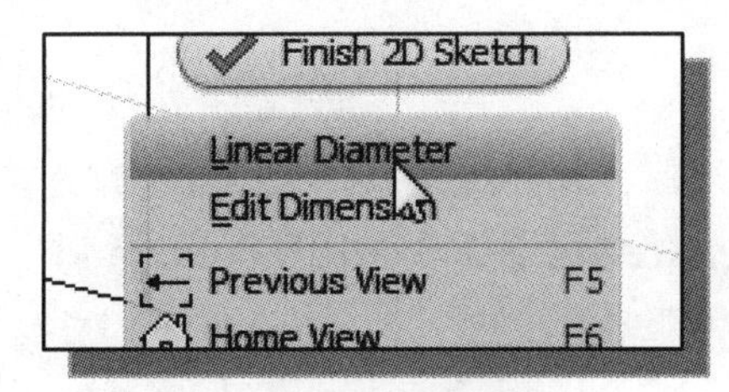

12. Inside the graphics window, right-click once to bring up the option menu and select **Linear Diameter** as shown.

13. Place the diameter dimension to the right side of the sketch.

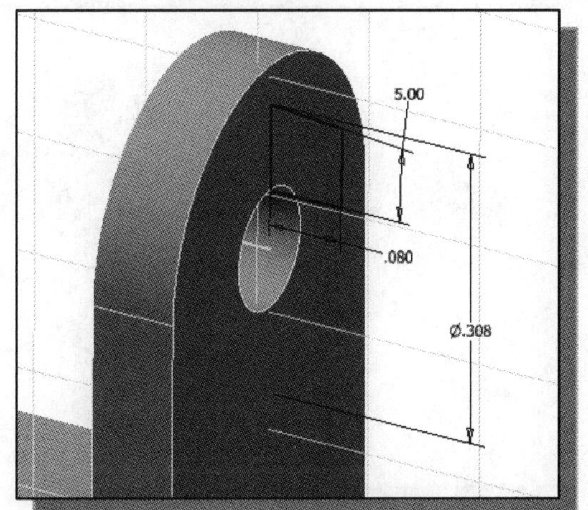

14. On your own, create and modify the three dimensions with an angle dimension of 5 degrees, as shown.

15. Select **Finish Sketch** in the *Ribbon* toolbar to exit the **2D Sketch** module.

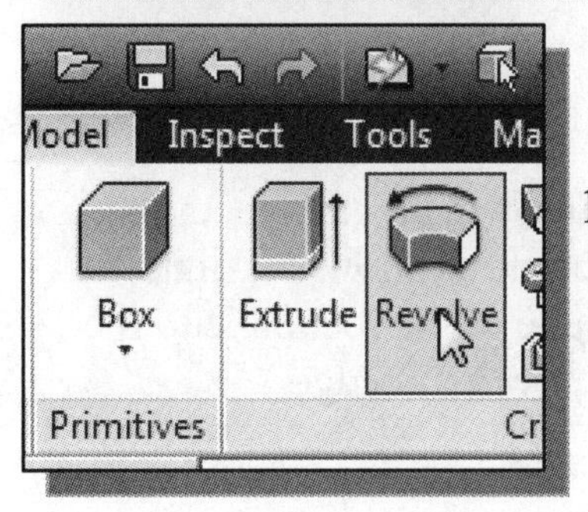

16. In the *Create* toolbar, select the **Revolve** command by left-mouse-clicking once on the icon.

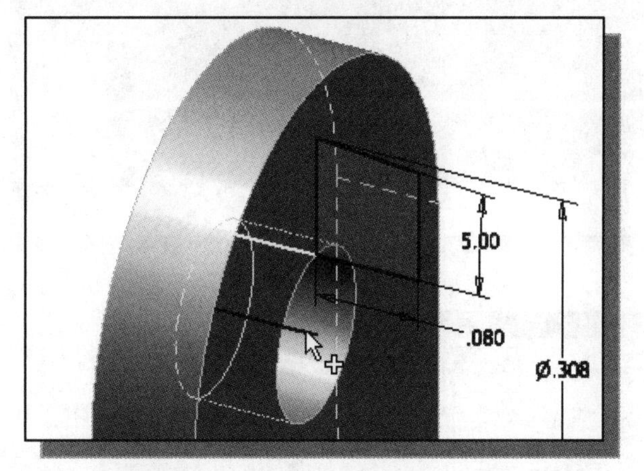

17. Select the **center axis** of the cylinder as the axis of rotation for the revolved feature.

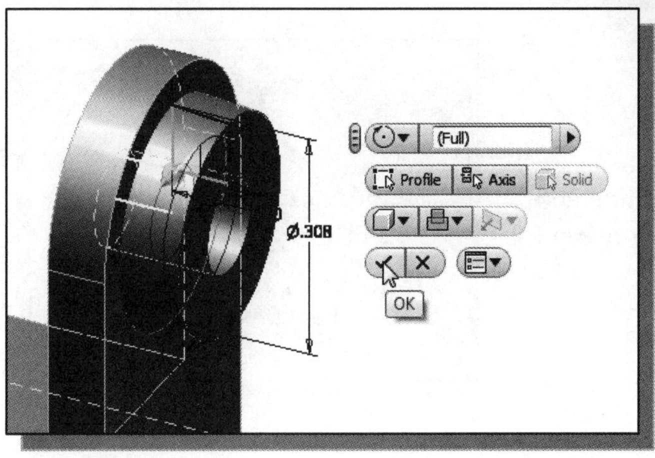

18. Confirm the revolve angle option is set to **(Full)** as shown.

19. Click **OK** to accept the settings and create the revolved feature.

Creating another Extrude Feature

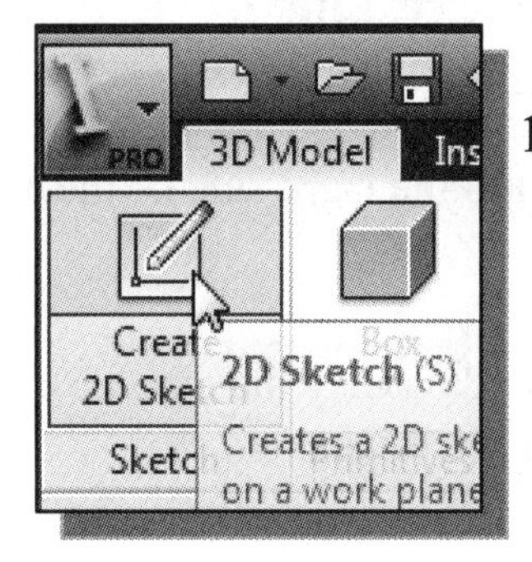

1. In the *Sketch* toolbar select the **Create 2D Sketch** command by left-clicking once on the icon.

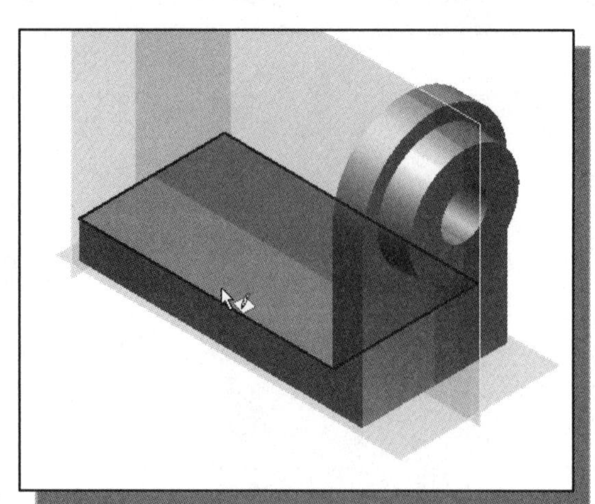

2. In the *Status Bar* area, the message: "*Select face, workplane, sketch or sketch geometry.*" is displayed. Select the **top plane** of the flat base as shown.

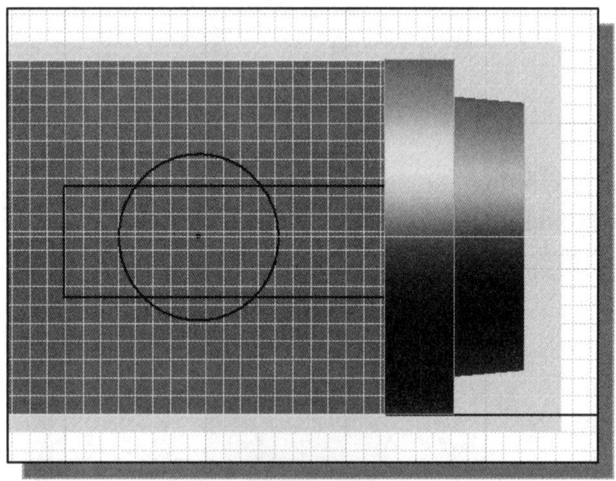

3. On your own, create a rectangle and a circle as shown. (Align the right edge of the rectangle to the adjacent vertical edge.)

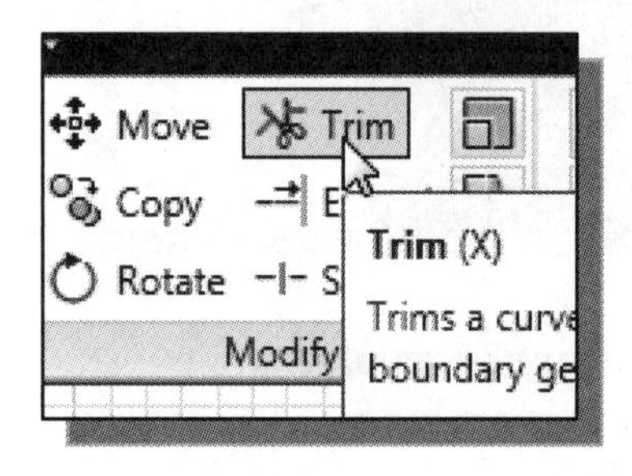

4. In the *Modify* toolbar, select the **Trim** command by left-mouse-clicking once on the icon.

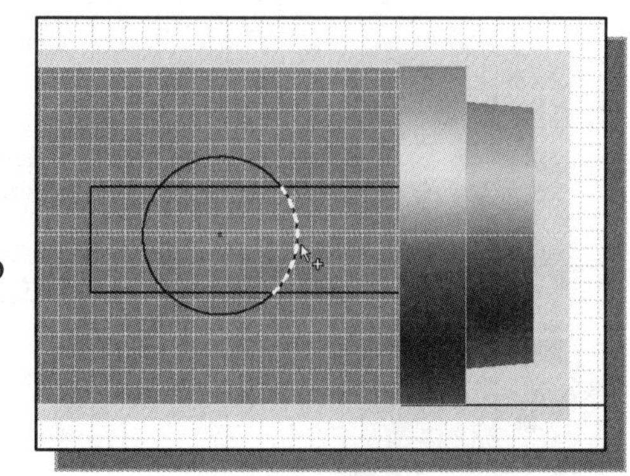

5. Move the cursor on top of the circle and notice *Inventor* will display the segment to be removed.

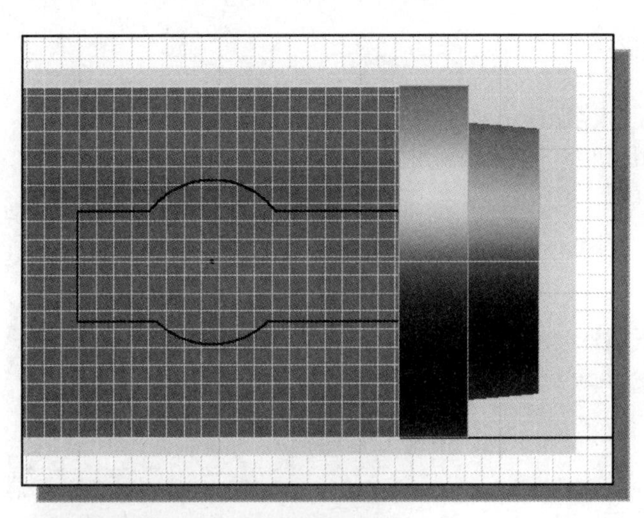

6. On your own, trim the geometry so that the sketch contains two horizontal lines and an arc as shown.

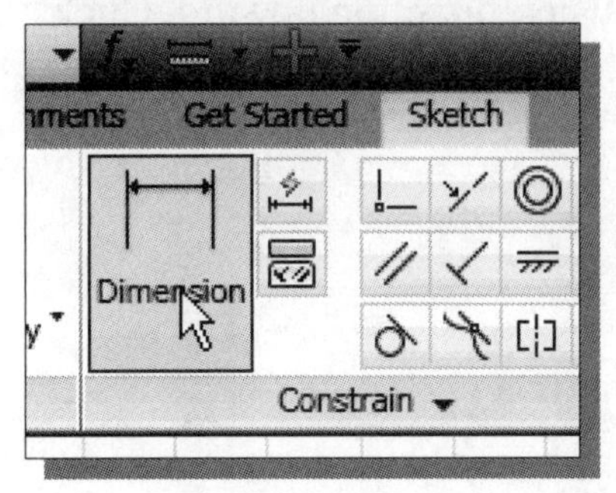

7. On your own, use the **General Dimension** command and create the size and location dimensions of the sketch as shown in the figure below.

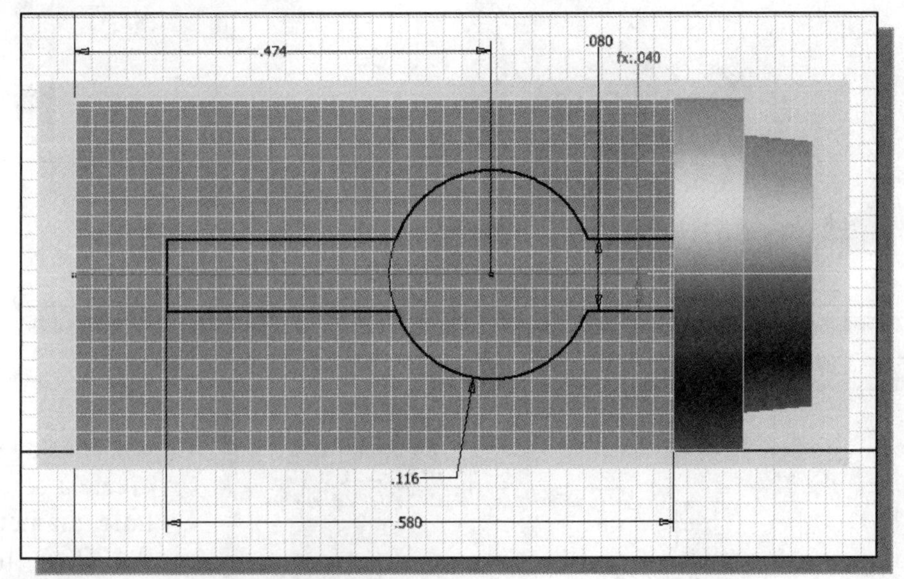

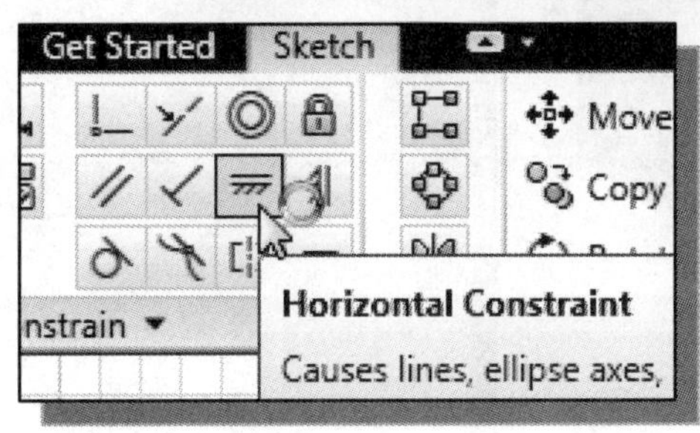

8. Apply a **Horizontal** constraint to align the **arc center** to the **center point**.

9. On your own, also apply Equal Radii constraints to the two arcs and collinear constraints to the horizontal lines.

10. Select **Finish Sketch** in the *Ribbon* toolbar to exit the **2D Sketch** module.

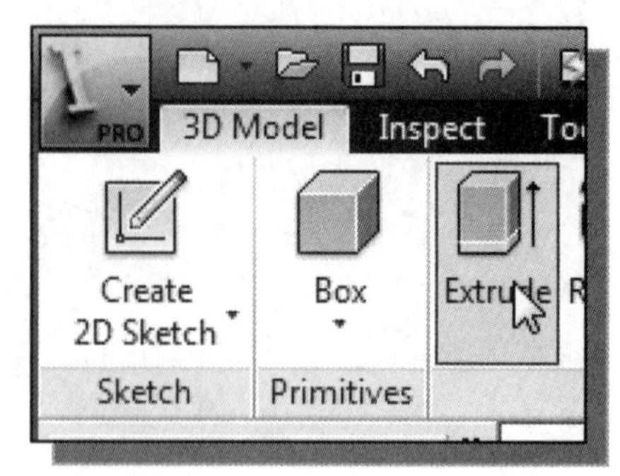

11. In the *Create* toolbar, select the **Extrude** command by left-mouse-clicking once on the icon.

12. Set the extrusion distance to **0.115**, and set the extrusion direction upward. Click on the **OK** button to accept the settings and create the solid feature.

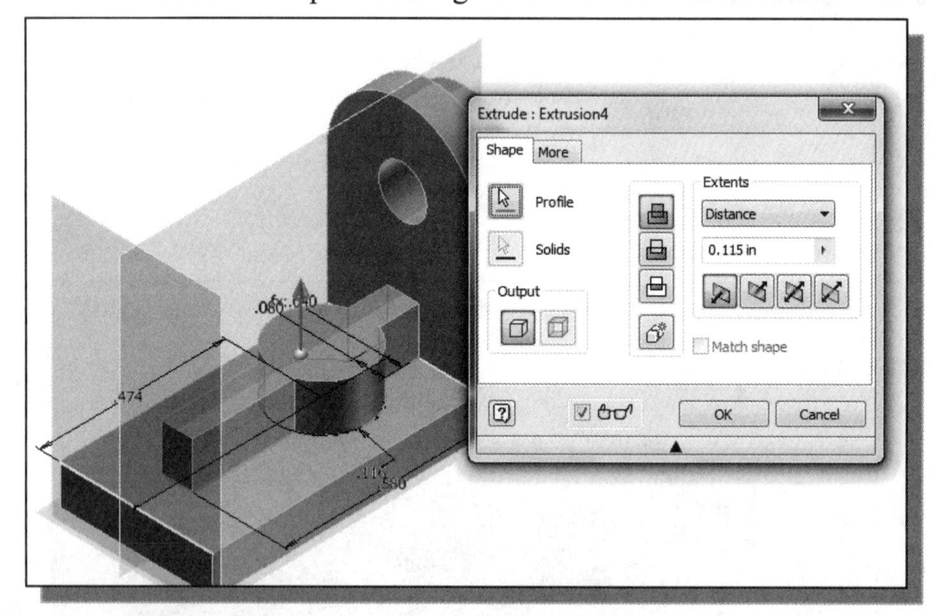

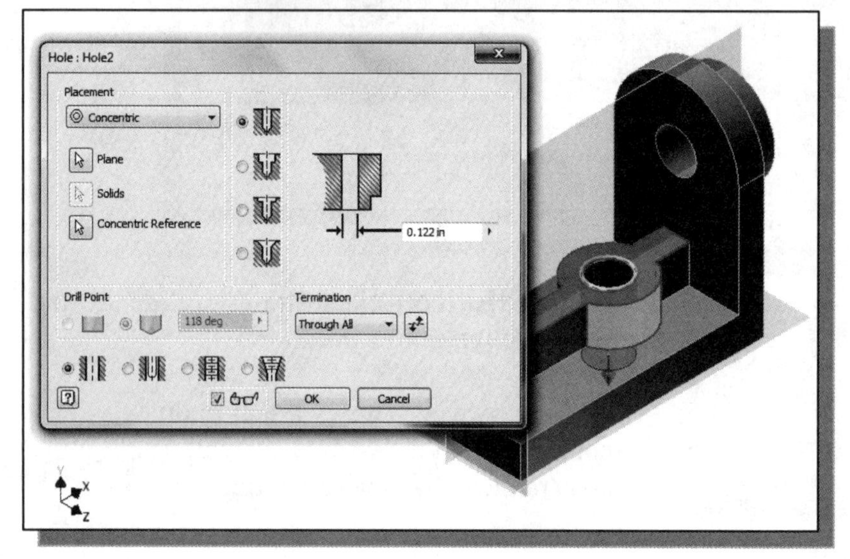

13. On your own, create a concentric hole, diameter **0.122**, as shown.

Create a Cut Feature

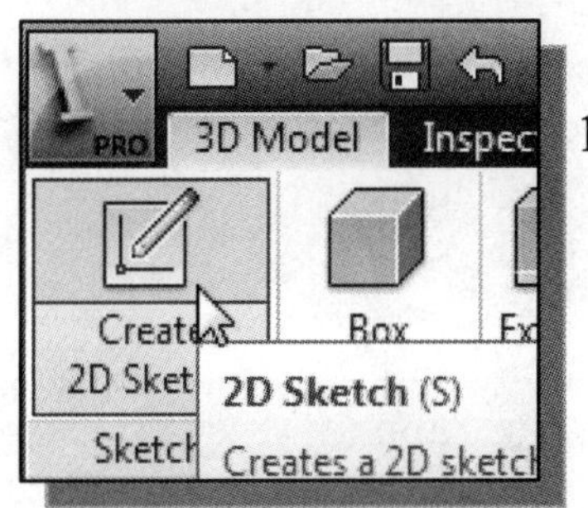

1. In the *Sketch* toolbar select the **Create 2D Sketch** command by left-clicking once on the icon.

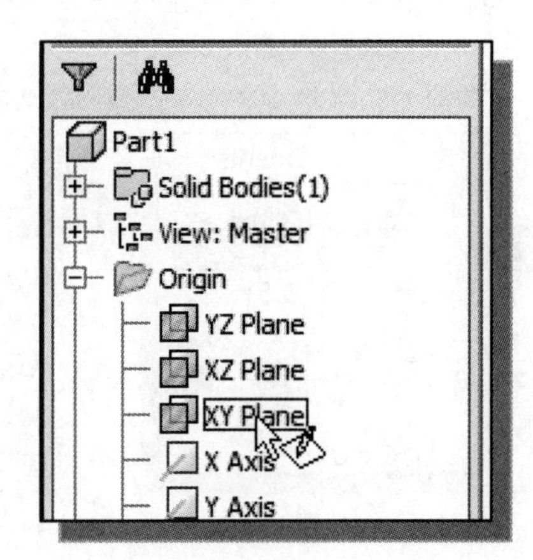

2. In the *Status Bar* area, the message: "*Select face, workplane, sketch or sketch geometry.*" is displayed. Select the **XY Plane** by clicking on the *Model Tree* as shown.

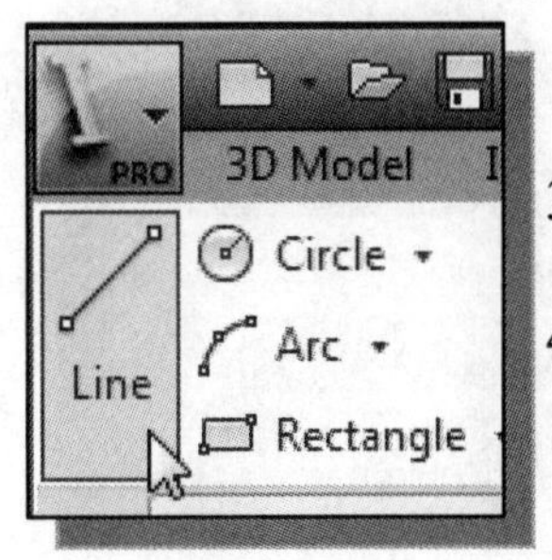

3. Select the **Line** option in the *Draw* panel.

4. On your own, create a triangle aligned to the left of the last extruded feature, as shown in the figure below.

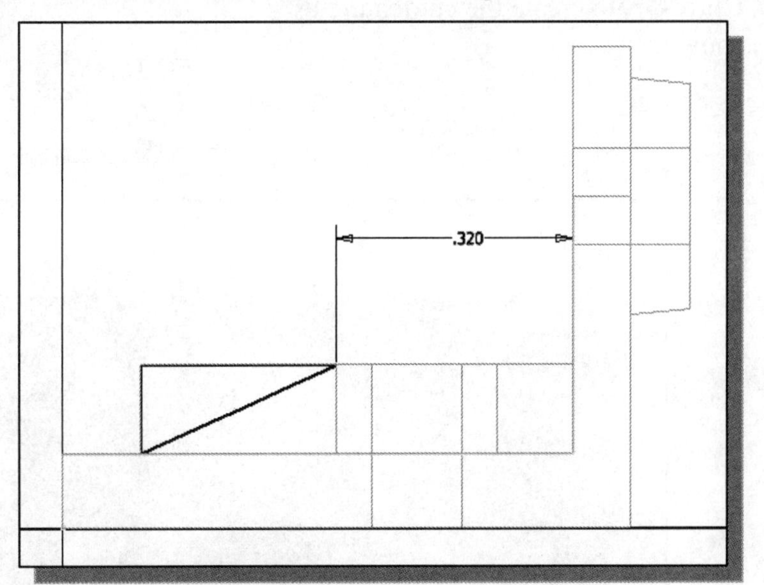

5. Select **Finish Sketch** in the *Ribbon* toolbar to exit the 2D Sketch module.

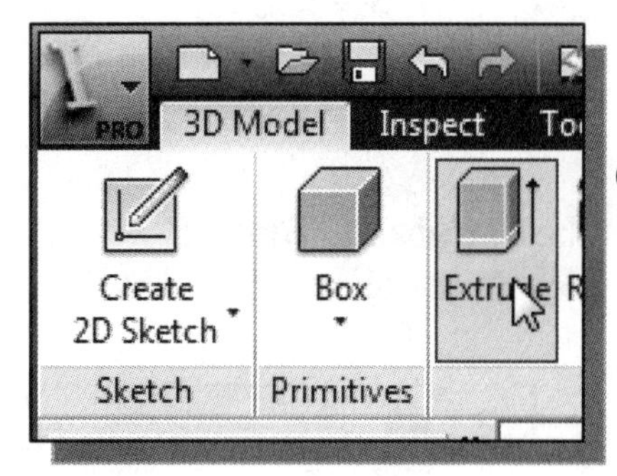

6. In the *Create* toolbar, select the **Extrude** command by left-mouse-clicking once on the icon.

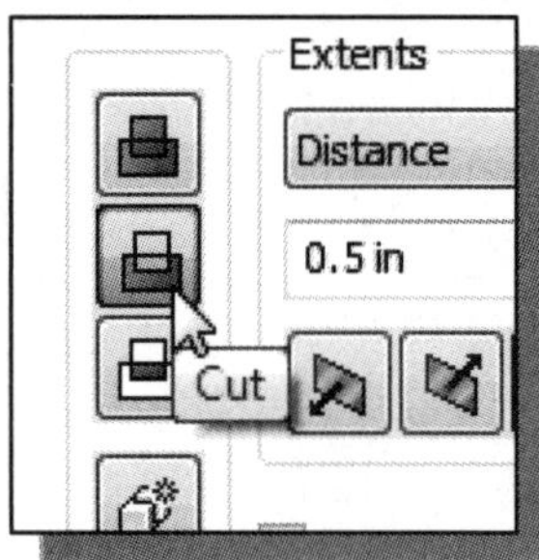

7. Set the extrude option to **Cut** formed in between the two regions as shown.

8. Also set the extrusion extents to **All** as shown in the figure below.

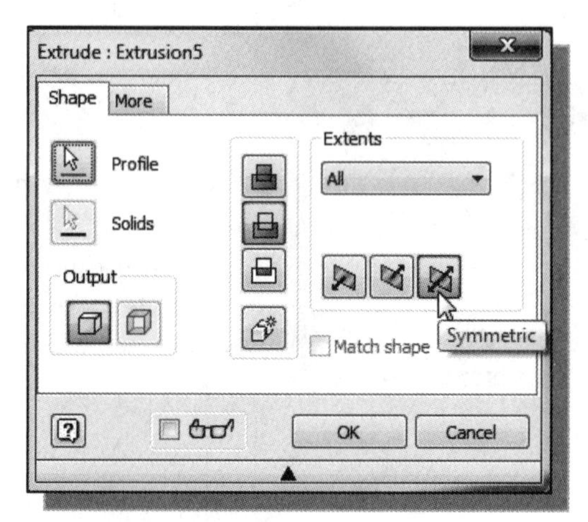

9. Select the **Symmetric** option for the extrude direction as shown.

10. Click **OK** to create the cut feature as shown.

Create a Mirrored Feature

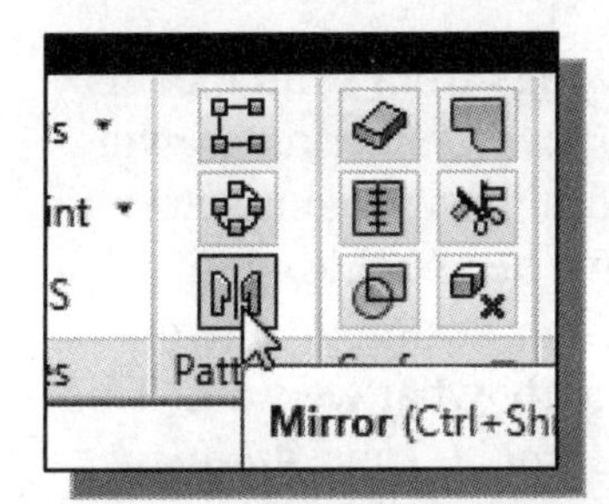

1. In the *Pattern* toolbar select the **Mirror** command by left-clicking once on the icon.

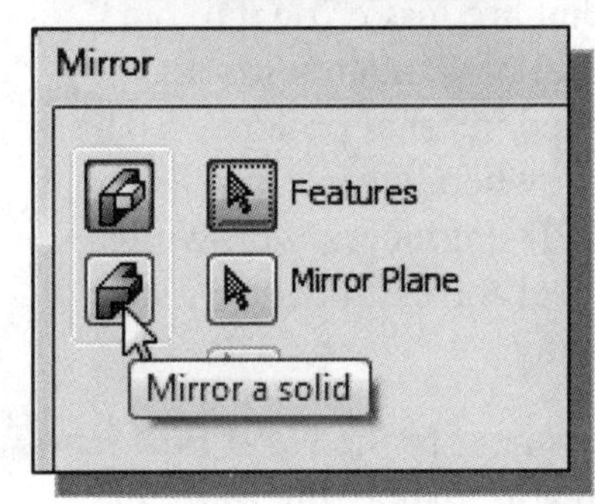

2. Click on the **Mirror a solid** option in the *Mirror* dialog box as shown. Note the entire solid is now preselected indicating the solid will be mirrored.

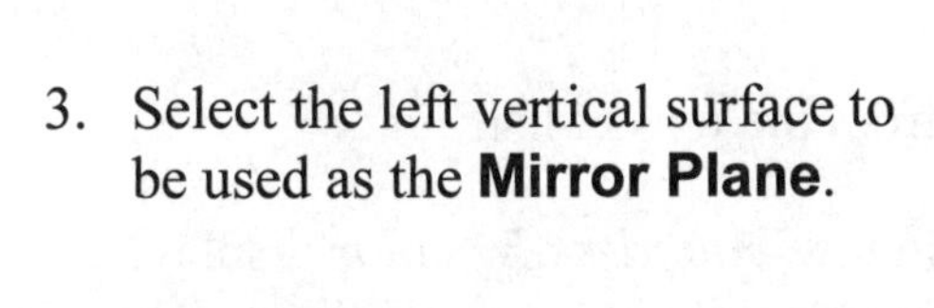

3. Select the left vertical surface to be used as the **Mirror Plane**.

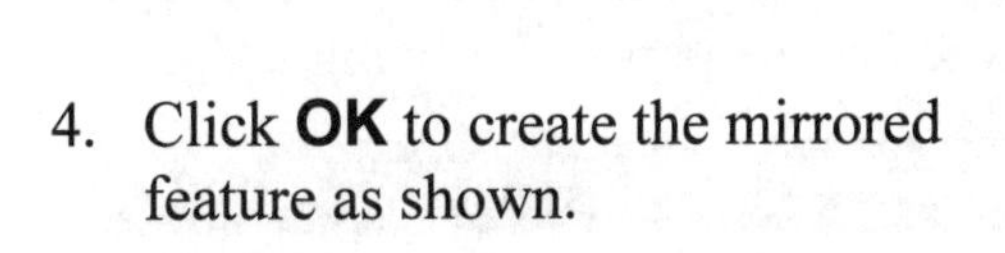

4. Click **OK** to create the mirrored feature as shown.

5. Select **Save** in the *Standard* toolbar.

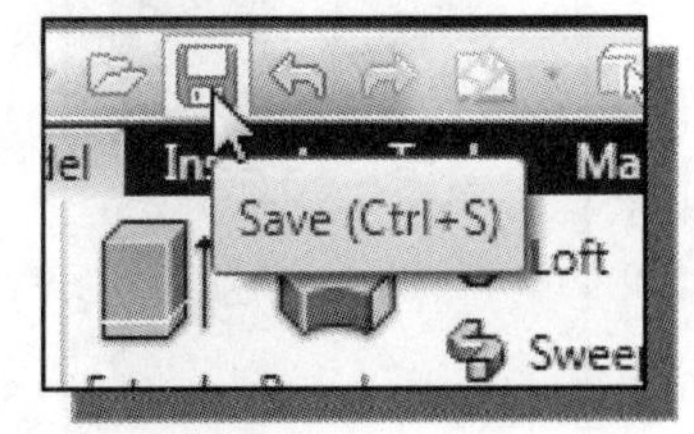

6. In the popup window, enter ***A12- Rear Axle Support*** as the name of the file.

7. Click on the **SAVE** button and save the file.

Drawing Mode – 2D Paper Space

➢ *Autodesk Inventor* allows us to generate 2D engineering drawings from solid models so that we can plot the drawings to any exact scale on paper. An engineering drawing is a tool that can be used to communicate engineering ideas/designs to manufacturing, purchasing, service, and other departments. Until now we have been working in ***model space*** to create our design in ***full size***. We can arrange our design on a two-dimensional sheet of paper so that the plotted hardcopy is exactly what we want. This two-dimensional sheet of paper is known as ***paper space*** in *AutoCAD* and *Autodesk Inventor*. We can place borders and title blocks, objects that are less critical to our design, on *paper space*. In general, each company uses a set of standards for drawing content, based on the type of product and also on established internal processes. The appearance of an engineering drawing varies depending on when, where, and for what purpose it is produced. In *Autodesk Inventor*, creation of 2D engineering drawings from solid models consists of four basic steps: drawing sheet formatting, creating/positioning views, annotations, and printing/plotting.

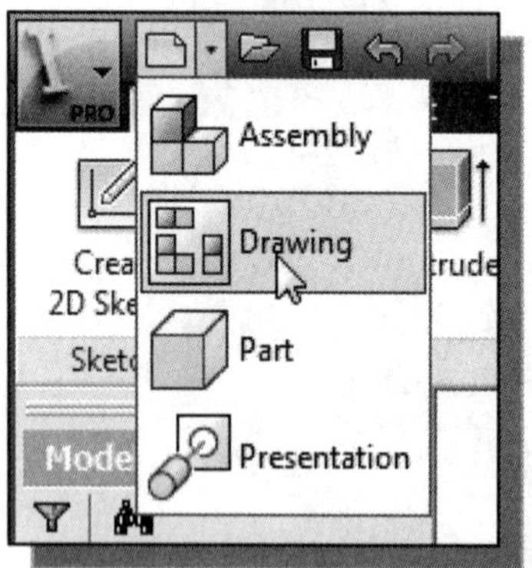

1. Click on the **drop-down arrow** next to the **New File** icon in the *Quick Access* toolbar area to display the available New File options.

2. Select **Drawing** from the option list.

❖ In the graphics window, *Autodesk Inventor* displays a default drawing sheet that includes a title block. The drawing sheet is placed on the 2D paper space, and the title block also indicates the paper size being used.

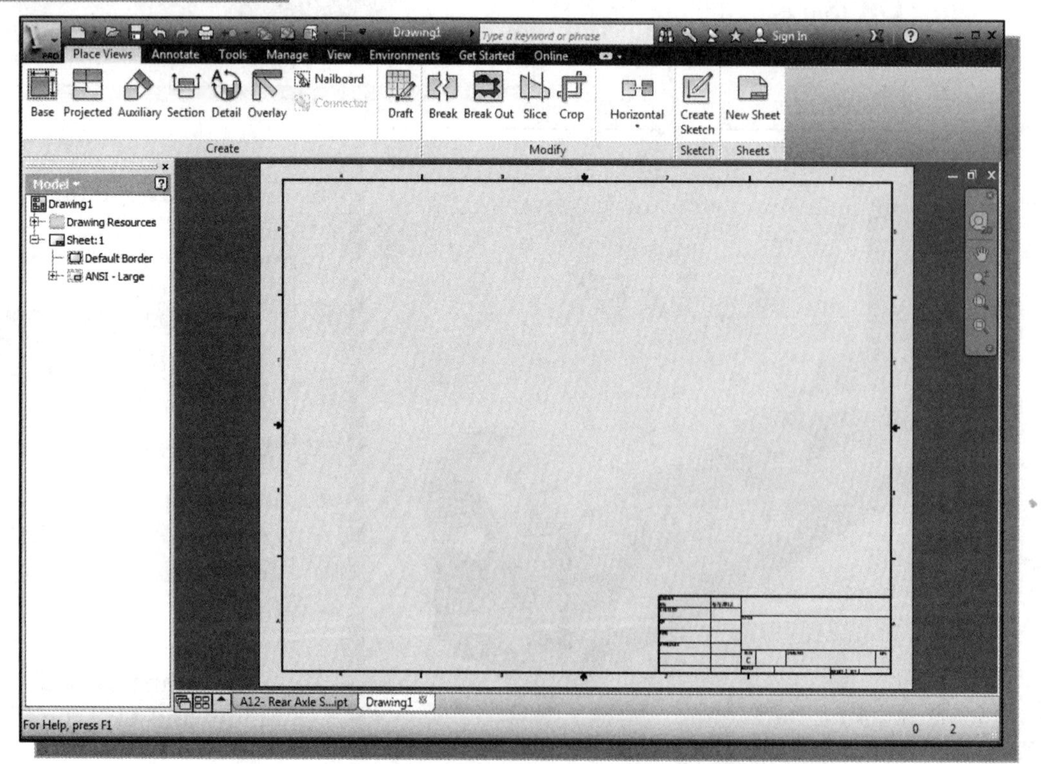

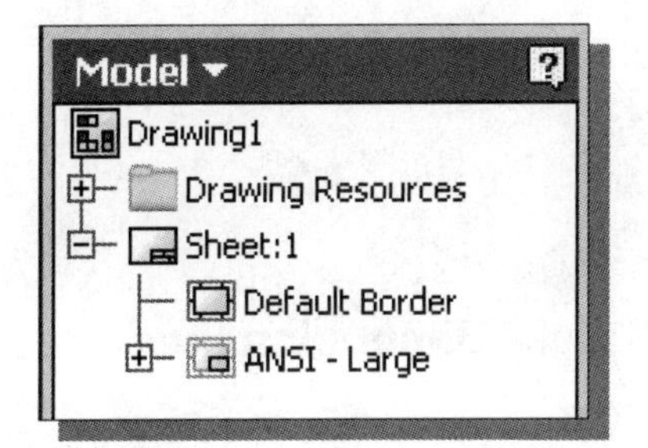

➢ In the *browser* area, the **Drawing1** icon is displayed at the top, which indicates that we have switched to **Drawing** mode. **Sheet1** is the current drawing sheet that is displayed in the graphics window.

Drawing Sheet Format

1. Choose the **Manage** tab in the *Ribbon* toolbar.

2. Click **Styles Editor** in the *Styles and Standards* toolbar as shown.

3. Click on the [+] sign in front of **Standard** to display the current active standard. Note that there can only be one **active standard** for each drawing.

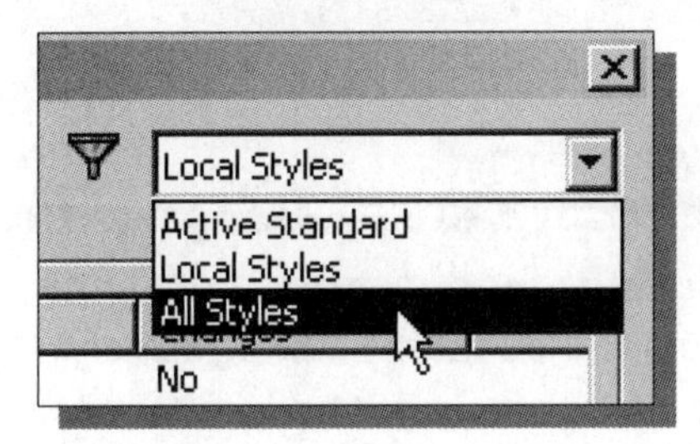

4. Set the *Filter Styles Setting* to display **All Styles**. Note that besides the default ANSI drafting standard, other standards, such as ISO, GB, BSI, DIN, and JIS, are also available.

5. Set the *Filter Styles Setting* to display **Active Standard** and note only the active standard is displayed.

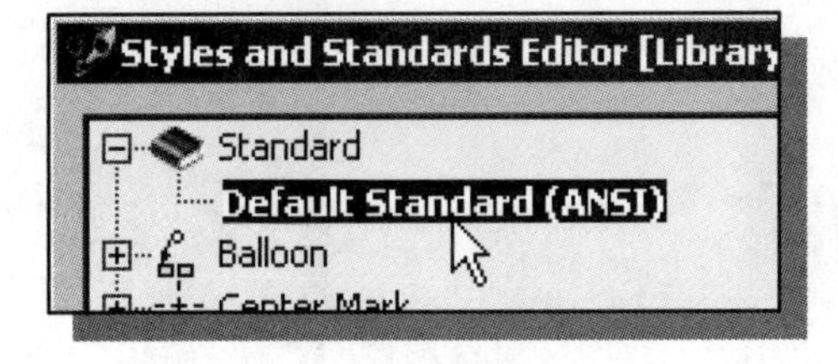

6. Click **Default Standard (ANSI)** to toggle the display of detailed settings for the current standard.

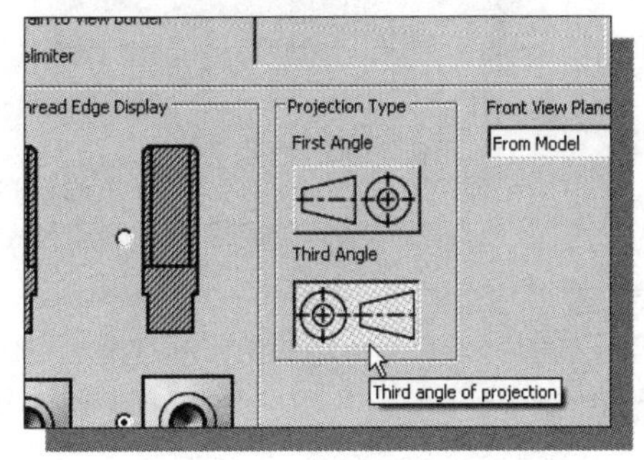

7. In the *View Preferences* page, confirm that the *Projection Type* is set to **Third Angle of projection**.

❖ Notice the different settings available in the *General* option window, such as the *Units* setting and the *Line Weight* setting.

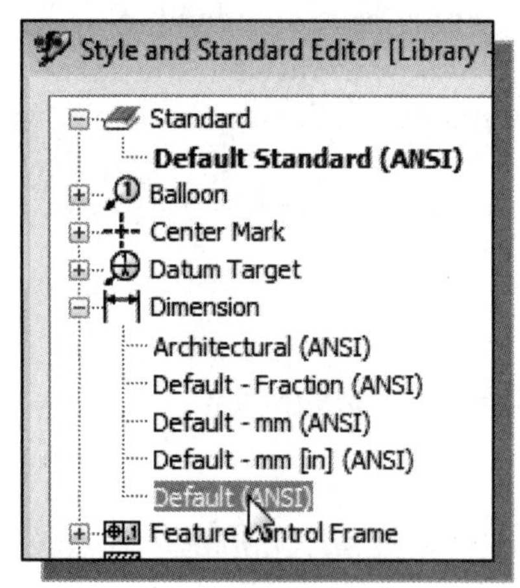

8. Choose **Default (ANSI)** in the **Dimension** list as shown.

➢ Note that the default *Dimension Style* in *Inventor* is based on the ANSI Y14.5-1994 standard.

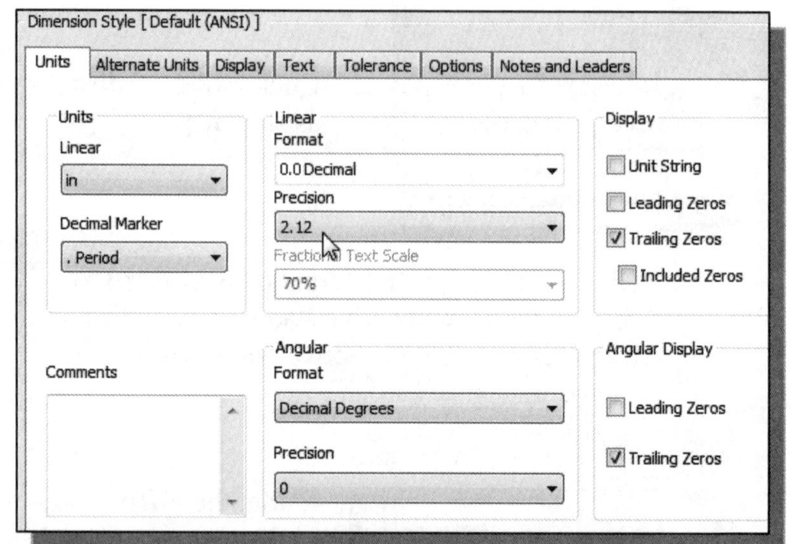

9. The **Units** tab contains the settings for linear/angular units. Note that the *Linear* units are set to Decimal and the *Precision* for linear dimensions is set to two digits after the decimal point.

10. Click on the **Text** tab to display and examine the settings for dimension text. Note that the default *Dimension Style*, Default (ANSI), cannot be modified. However, new *Dimension Styles* can be created and modified.

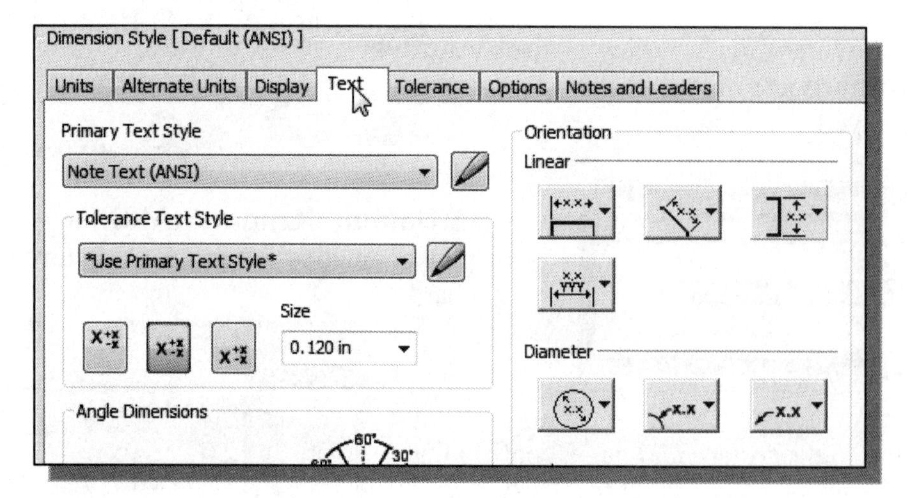

11. On your own, display and examine the other *Settings* available, click on the **Done** button to exit the *Styles and Standards Editor* option.

Using the Pre-defined Drawing Sheet Formats

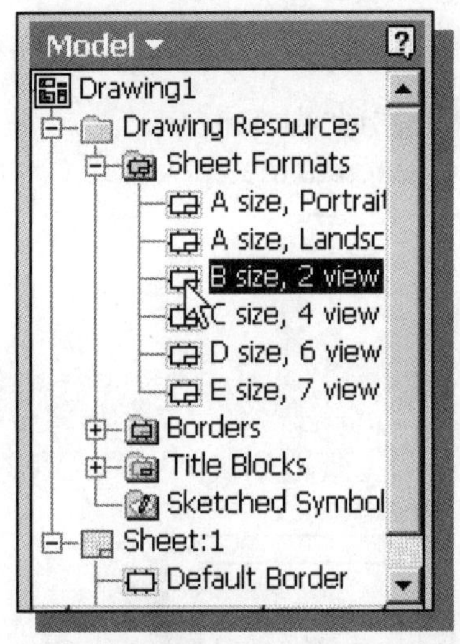

1. Inside the *Drawing Browser* window, click on the **[+]** symbol in front of **Drawing Resources** to display the available options.

2. Click on the **[+]** symbol in front of **Sheet Formats** to display the available pre-defined sheet formats.

❖ Notice several pre-defined *sheet formats*, each with a different view configuration, are available in the *browser* window.

3. Inside the *browser* window, **double-click** on the **B size, 2 view** sheet format.

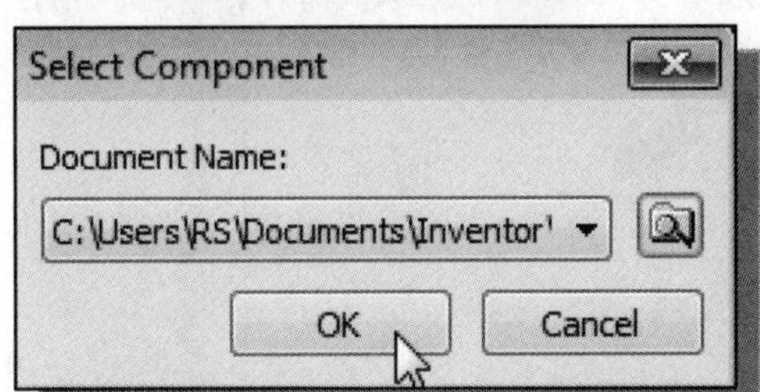

4. Click on the **OK** button to accept the default part file and generate the 2D views.

➢ The ***A12-Rear Axle Support*** model is the only model opened. All of the 2D drawings will be generated from this opened model file.

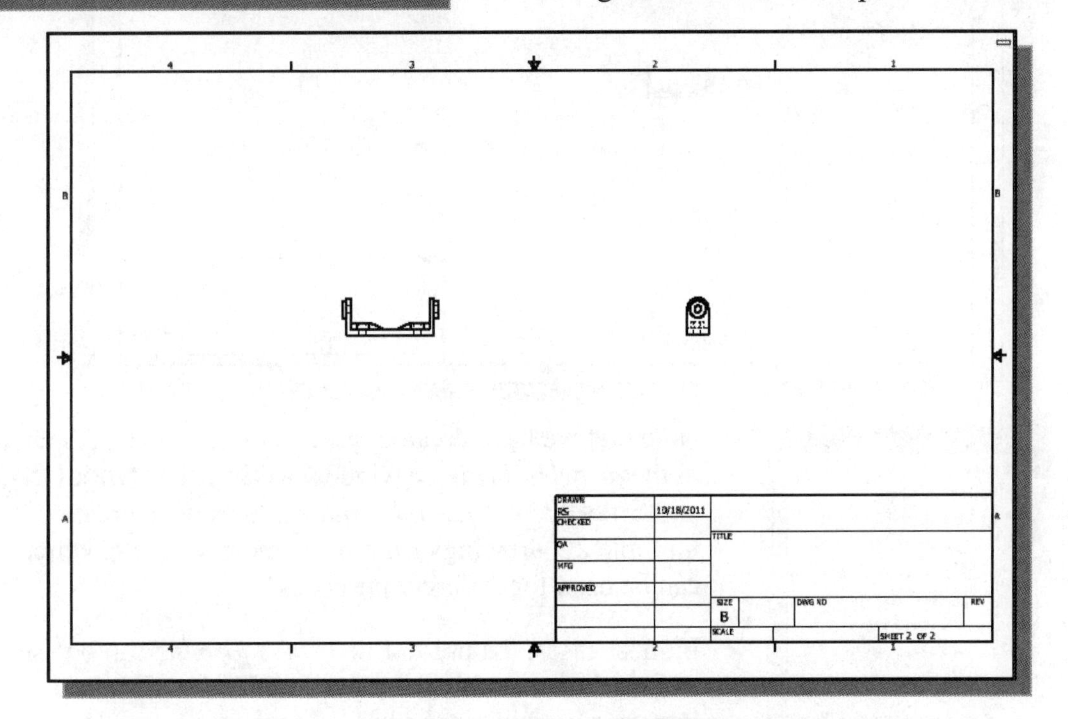

❖ We have created a B-size drawing of the ***A12-Rear Axle Support*** model. *Autodesk Inventor* automatically generates and positions the front view and side view of the model inside the title block.

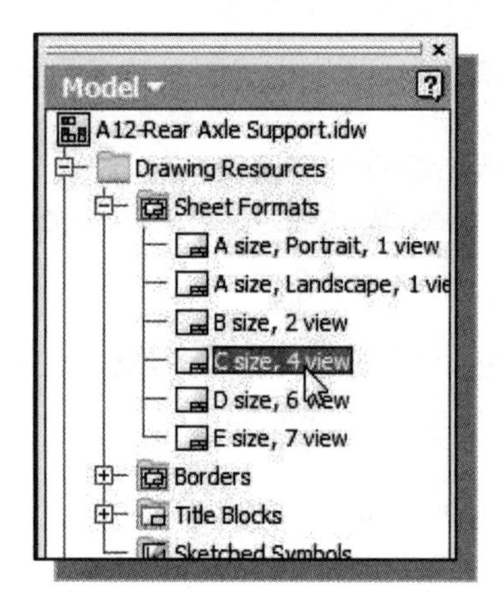

5. Inside the *browser* window, double-click on the **C size, 4 view** sheet format.

6. Click on the **OK** button to use the default part file, ***A12-Rear Axle Support***, to generate the 2D views.

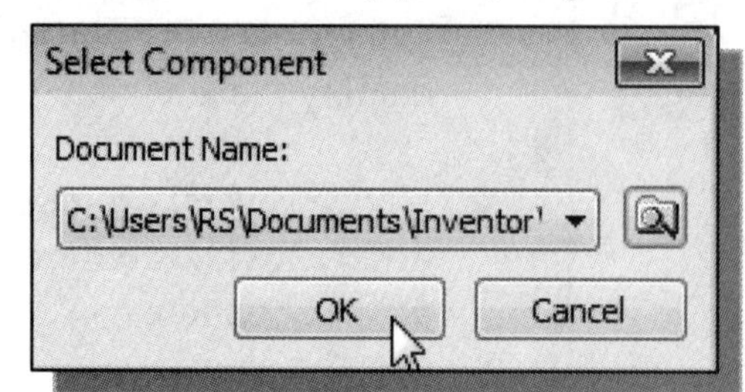

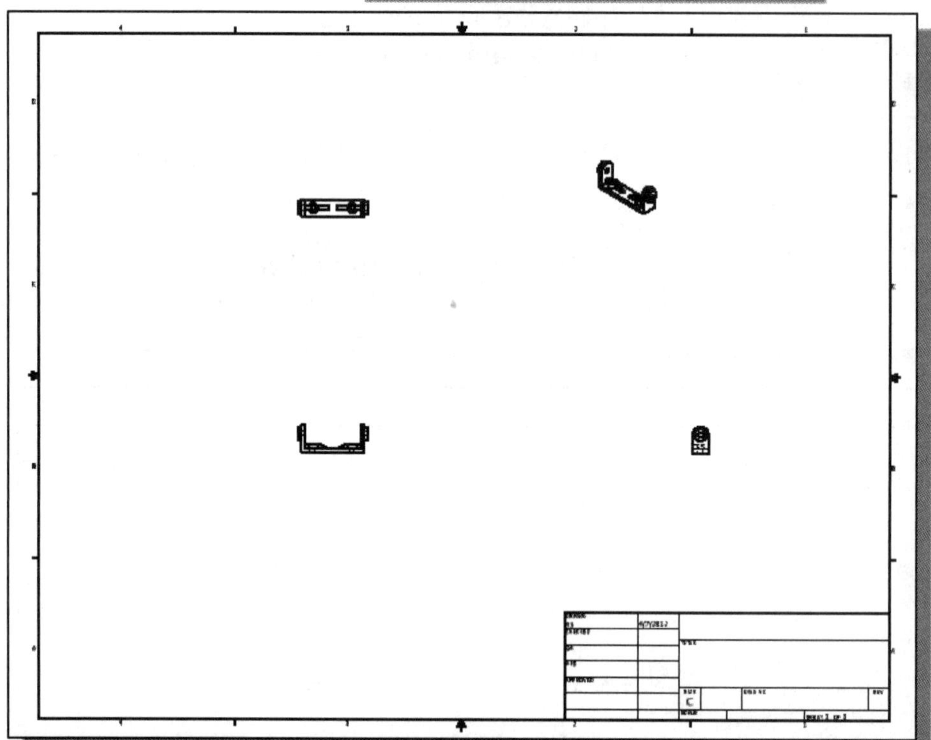

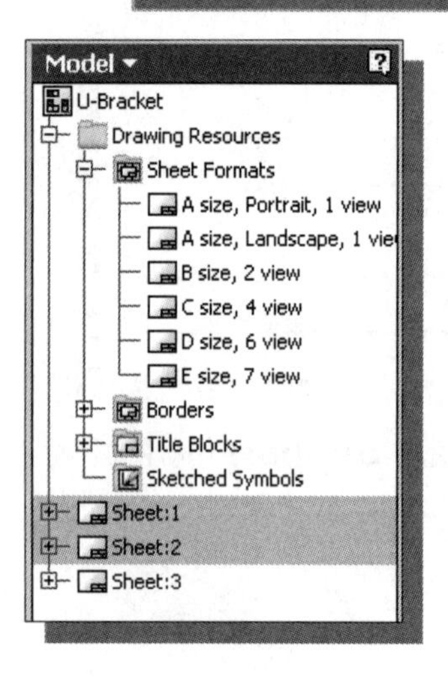

❖ Note that we have created three drawing sheets, displayed in the *drawing browser* window as Sheet:1, Sheet:2, and Sheet:3. *Autodesk Inventor* allows us to create multiple 2D drawings from the same model file, which can be used for different purposes.

➢ In most cases, the pre-defined *sheet formats* can be used to quickly set up the views needed. However, it is also important to understand the concepts and principles involved in setting up the views. In the next sections, the procedures to set up drawing sheets and different types of views are illustrated.

Delete, Activate, and Edit a Drawing Sheet

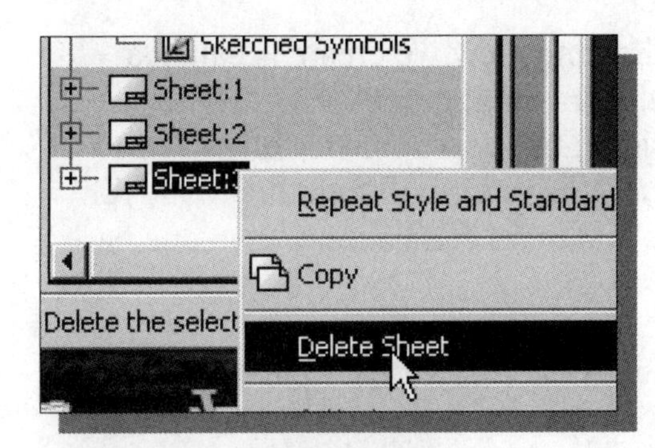

1. Inside the *drawing browser* window, right-mouse-click on **Sheet:3** to display the option menu.

2. Select **Delete Sheet** in the option menu to remove the Sheet:3 drawing.

3. In the *warning window*, click on the **OK** button to proceed with deleting the drawing.

❖ Note that **Sheet:3** is removed and **Sheet:2** now becomes the active drawing sheet.

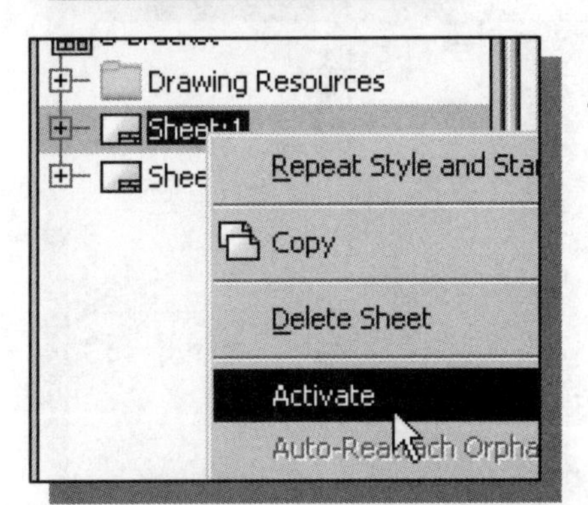

4. Inside the *drawing browser* window, right-mouse-click on **Sheet:1** to display the option menu.

5. Select **Activate** in the option menu to set the Sheet:1 drawing as the active drawing sheet.

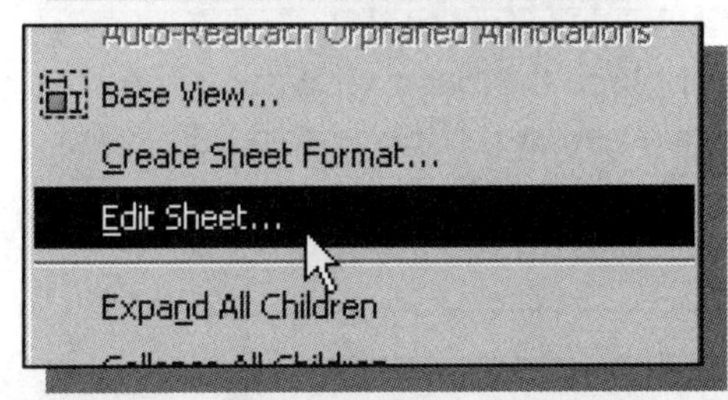

6. Inside the *drawing browser* window, right-mouse-click on **Sheet:1** to display the option menu.

7. Select **Edit Sheet** in the option menu to display the settings for the Sheet:1 drawing.

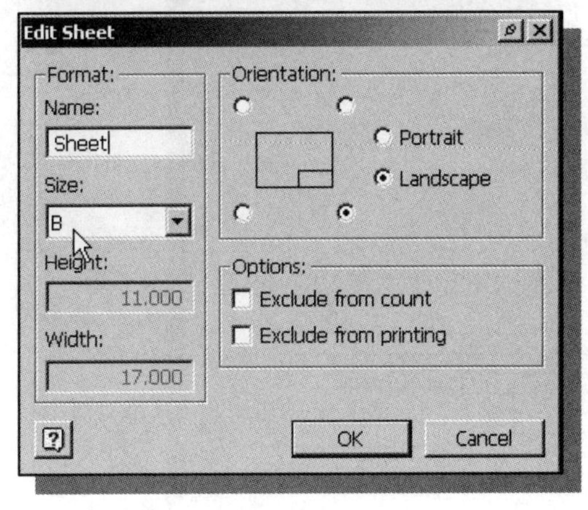

8. Set the sheet *Size* to **B** and click on the **OK** button to exit the *Edit Sheet* dialog box.

Add a Base View

❖ In *Autodesk Inventor* Drawing mode, the first drawing view we create is called a **base view**. A *base view* is the primary view in the drawing; other views can be derived from this view. When creating a base view, *Autodesk Inventor* allows us to specify the view to be shown. By default, *Autodesk Inventor* will treat the *world XY plane* as the front view of the solid model. Note that there can be more than one base view in a drawing.

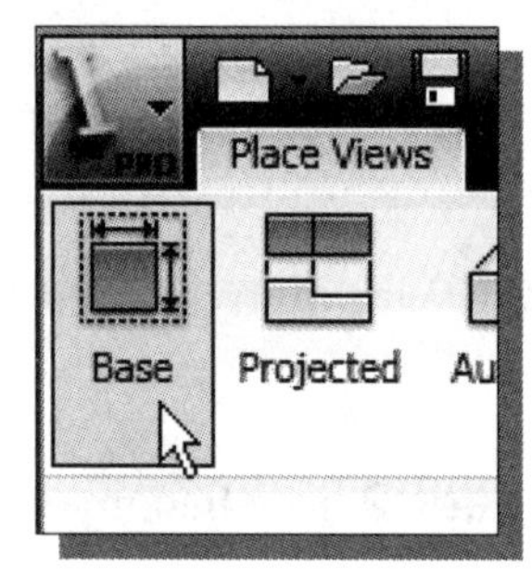

1. Click on the **Base View** in the *Place Views* tab to create a base view.

2. In the *Drawing View* dialog box, confirm that the settings are set to **Front View** and **Hidden Line** as shown in the figure below. (**DO NOT** click on the **OK** button at this point.)

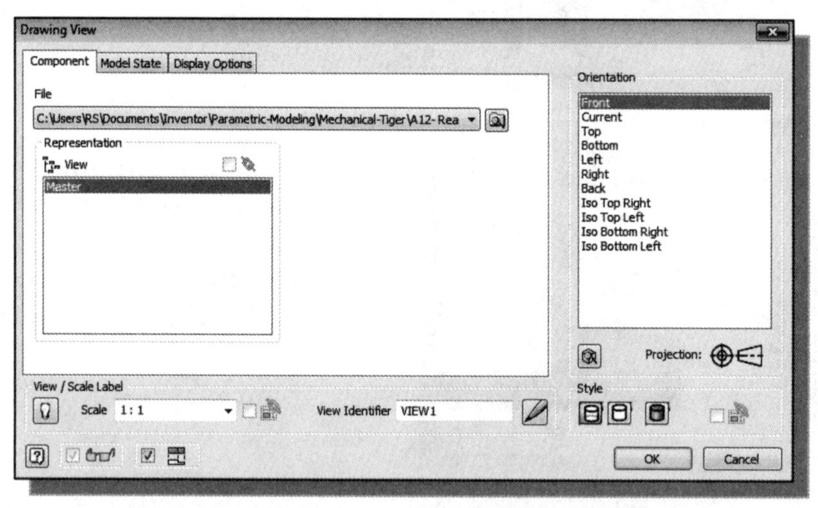

3. Move the cursor inside the graphics window and place the **base view** near the lower left corner of the graphics window as shown below. (If necessary, drag the *Create View* dialog box to another location on the screen.) Note that once the *base* view is placed, the *Drawing View* dialog box is closed automatically.

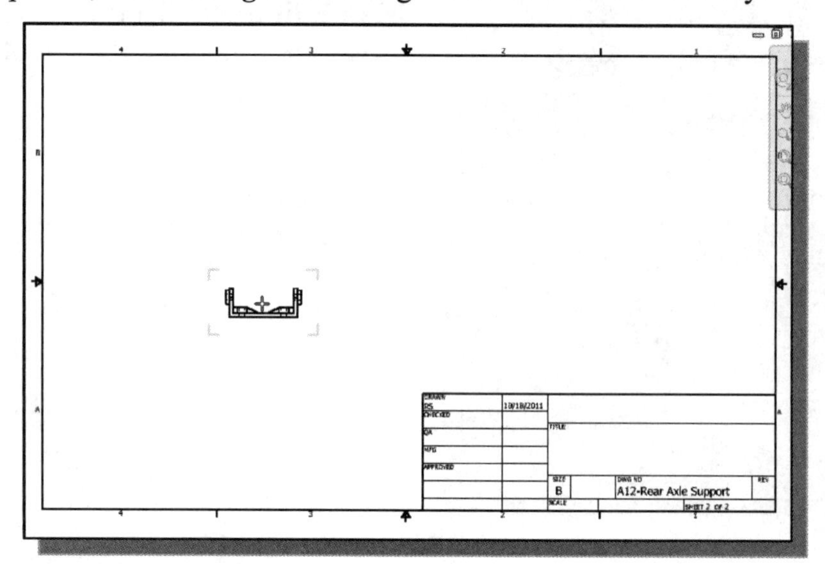

Create Projected Views

❖ In *Autodesk Inventor* Drawing mode, **projected views** can be created with a first-angle or third-angle projection, depending on the drafting standard used for the drawing. We must have a base view before a projected view can be created. Projected views can be orthographic projections or isometric projections. Orthographic projections are aligned to the base view and inherit the base view's scale and display settings. Isometric projections are not aligned to the base view.

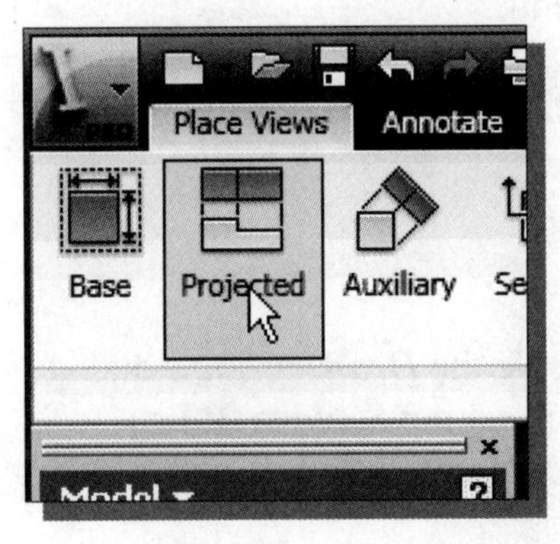

1. Notice the **Projected View** button in the *Drawing Views* panel is activated; this command allows us to create a projected view.

2. Move the cursor above the base view and select a location to position the projected side view of the model.

3. Move the cursor toward the upper right corner of the title block and select a location to position the isometric view of the model as shown below.

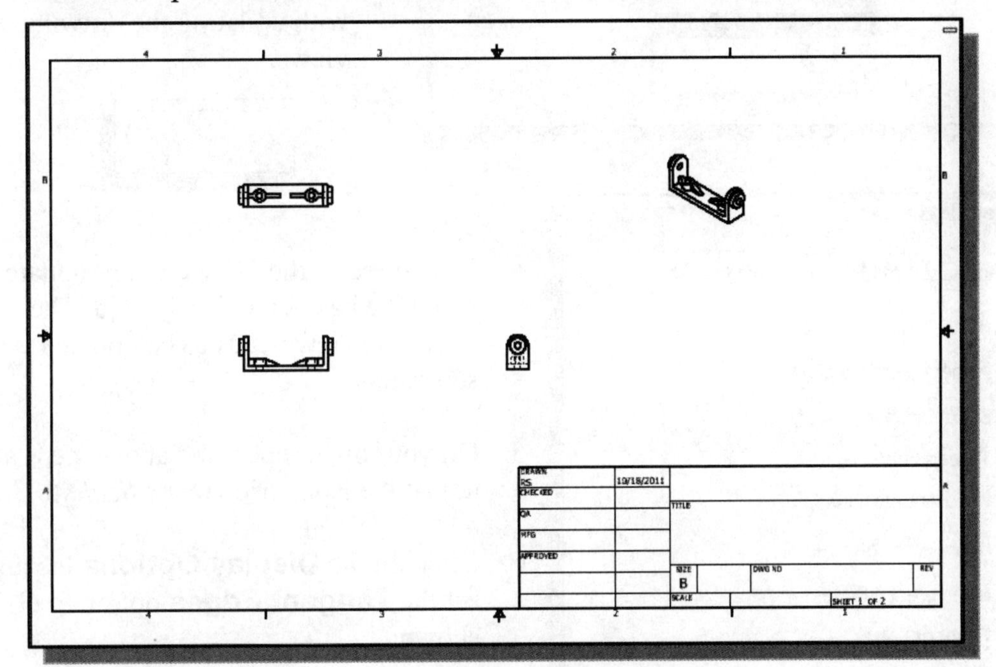

4. Inside the graphics window, right-mouse-click once to bring up the option menu.

5. Select **Create** to proceed with creating the two projected views.

Adjust the View Scale

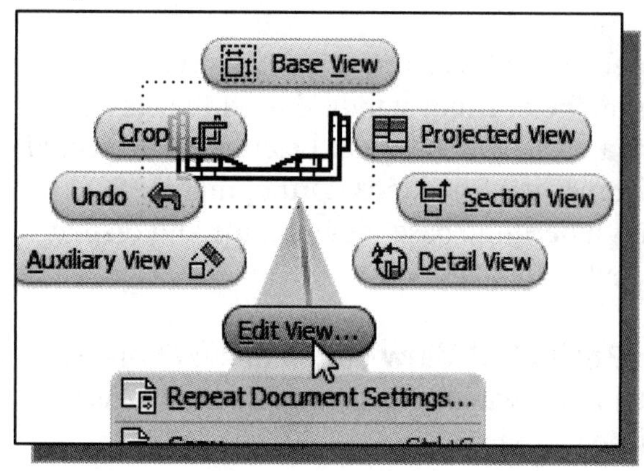

1. Move the cursor on top of the base view and watch for the box around the entire view indicating the view is selectable as shown in the figure. Right-mouse-click once to bring up the option menu.

2. Select **Edit View** in the option menu.

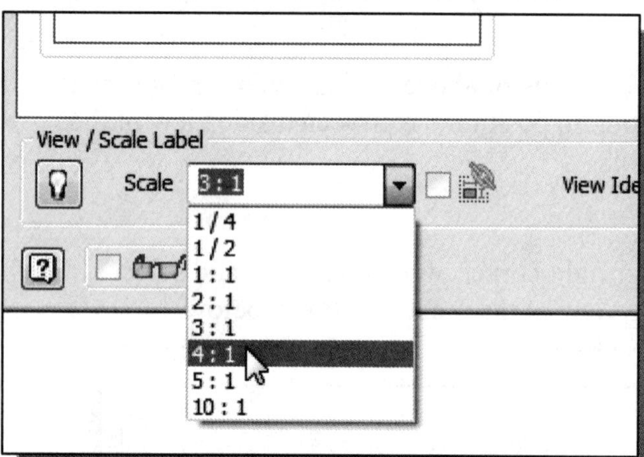

3. Inside the *Drawing View* dialog box, set the *Scale* to **3:1** as shown in the figure.

4. Click on the **OK** button to accept the settings and proceed with updating the drawing views.

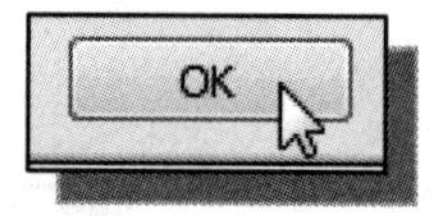

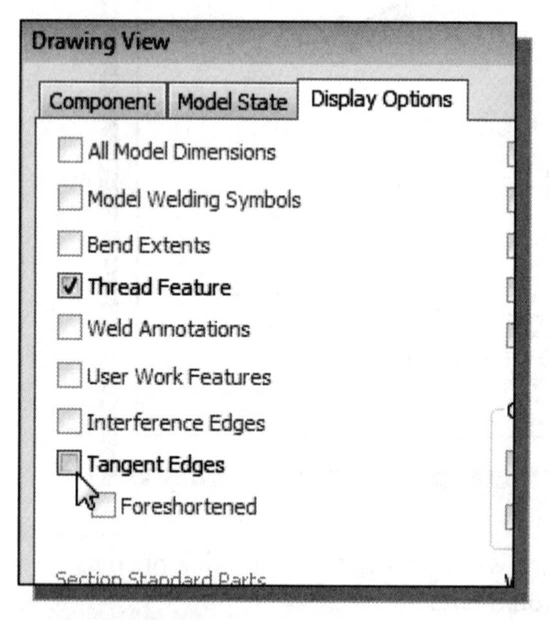

- Note three of the 2D views are updated when the base view is adjusted. The isometric view needs to be updated separately.

5. On your own, repeat the above steps and adjust the isometric view's *Scale* to **3:1**.

6. Click on the **Display Options** tab and set the **Tangent Edges** option to ***OFF*** as shown.

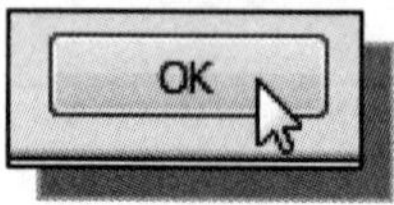

7. Click on the **OK** button to accept the settings and proceed with updating the drawing views.

Reposition Views

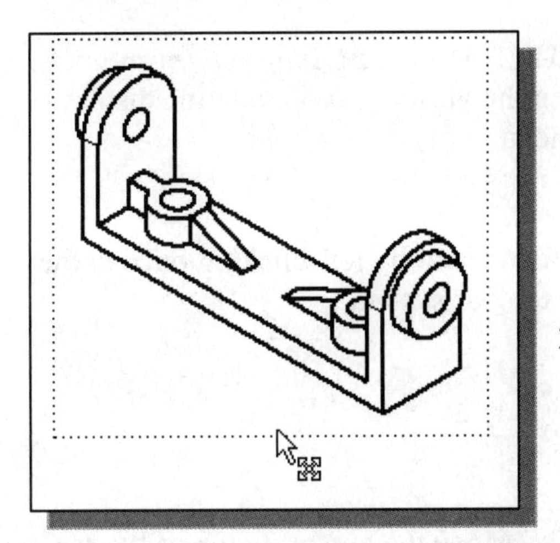

1. Move the cursor on top of the isometric view and watch for the **four-arrow Move symbol** as the cursor is near the border indicating the view can be dragged to a new location as shown in the figure.

2. Press and hold down the left-mouse-button and reposition the view to a new location.

3. On your own, reposition the views we have created so far. Note that the top view can be repositioned only in the vertical direction. The top view remains aligned to the base view, the front view.

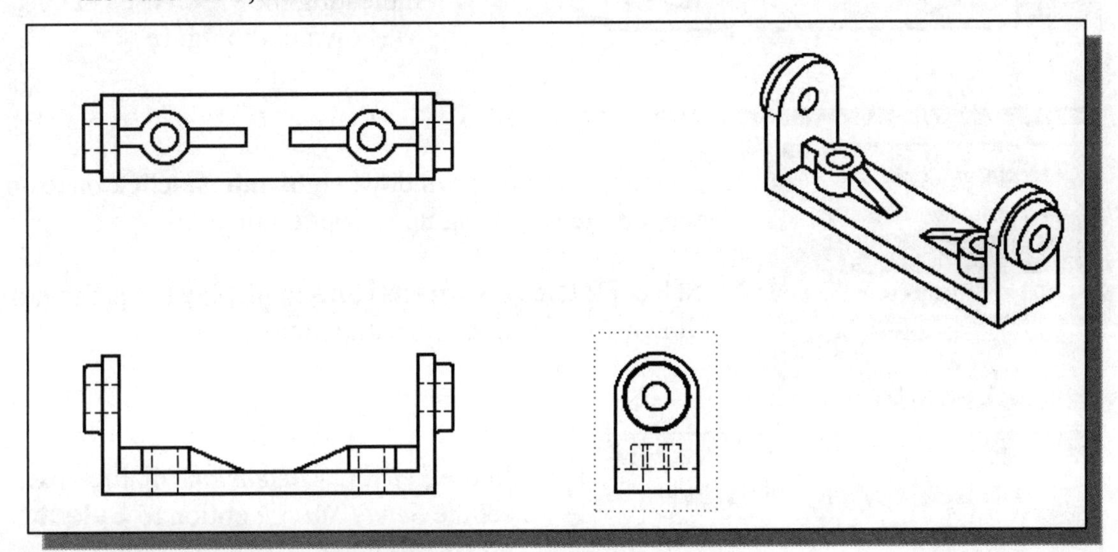

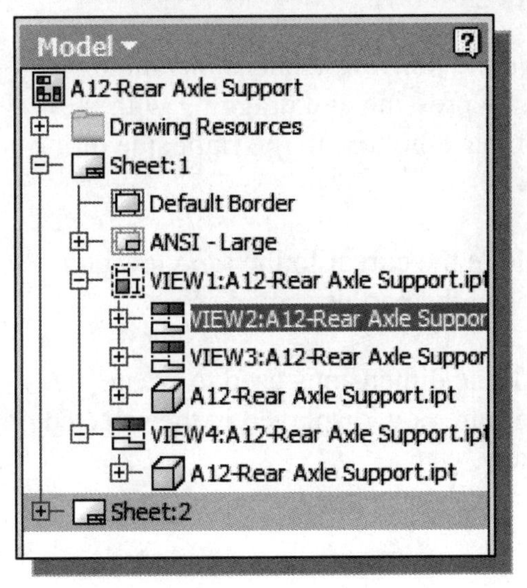

➤ Note that in the *drawing browser* area, a hierarchy of the created views is displayed under Sheet:1. The base view, View1, is listed as the first view created, with View2 linked to it. The top view, View2, is projected from the base view, View1. The implied parent/child relationship is maintained by the system. Drawing views are associated with the model and the drawing sheets. As we create views from the base view, they are nested beneath the base view in the *browser*.

Display Feature Dimensions

- By default, feature dimensions are not displayed in 2D views in *Autodesk Inventor*. We can change the default settings while creating the views or switch on the display of the parametric dimensions using the option menu.

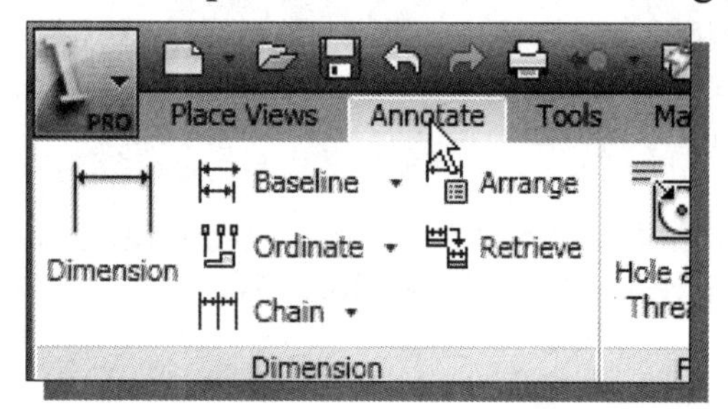

1. Select **Annotate** by left-clicking once in the *Ribbon* toolbar system.

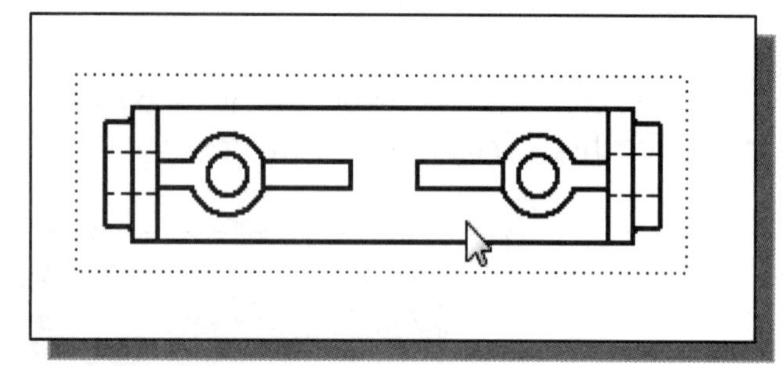

2. Move the cursor on top of the ***top*** view of the model and watch for the box around the entire view indicating the view is selectable, as shown in the figure.

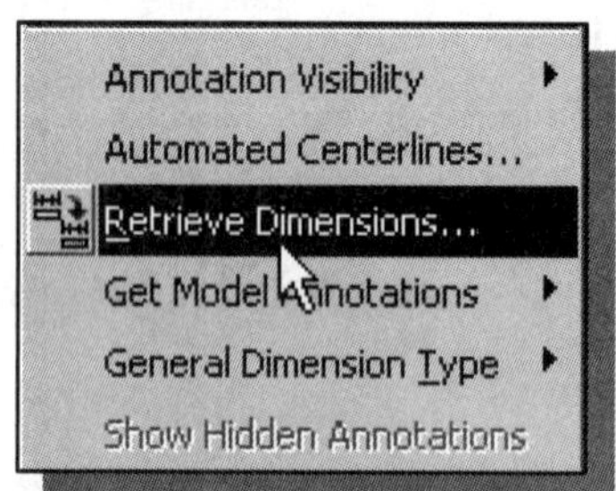

3. Inside the graphics window, **right-mouse-click** once on the top view to bring up the option menu.

4. Select **Retrieve Dimensions** to display the parametric dimensions used to create the model.

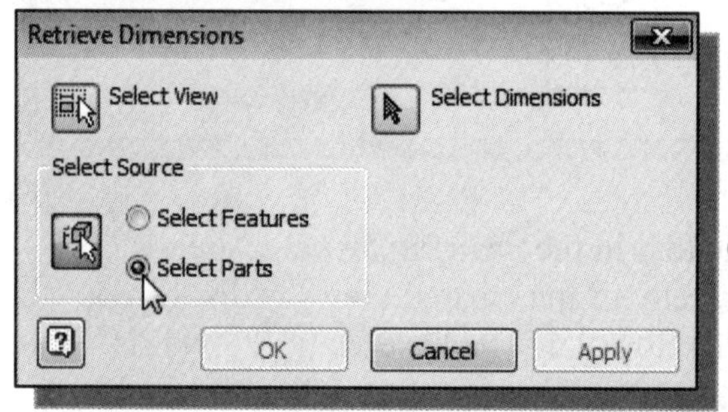

5. In the *Retrieve Dimensions* dialog box, set the *Select Source* option to **Select Parts** as shown.

6. Move the *Retrieve Dimensions* dialog box, by pressing and dragging with the left-mouse-button, to the right side of the view.

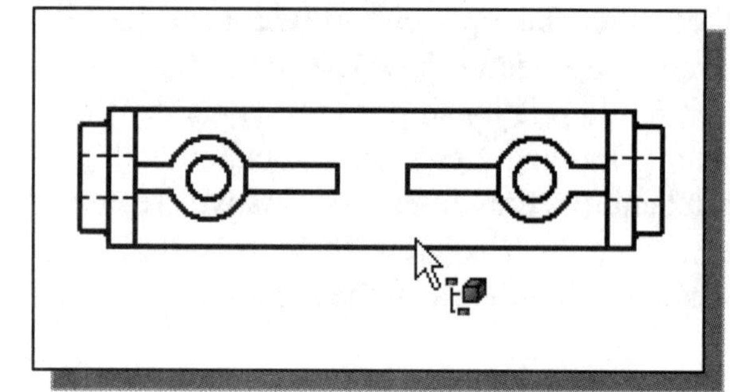

7. Move the cursor to the ***top*** view and select the part as shown.

➢ All the dimensions used to create the part are now displayed in the selected view.

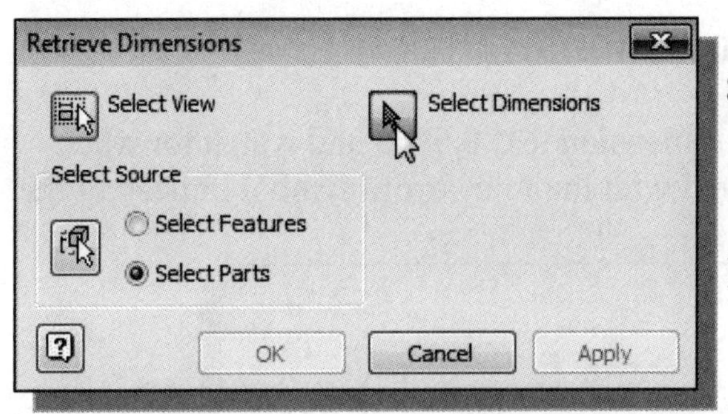

8. In the *Retrieve Dimensions* dialog box, switch on the **Select dimensions** option as shown.

➢ The system now expects us to select the dimensions to retrieve.

9. On your own, select the needed dimensions to retrieve by left-clicking once on the dimensions as shown. (Note that only selected dimensions are retrieved.)

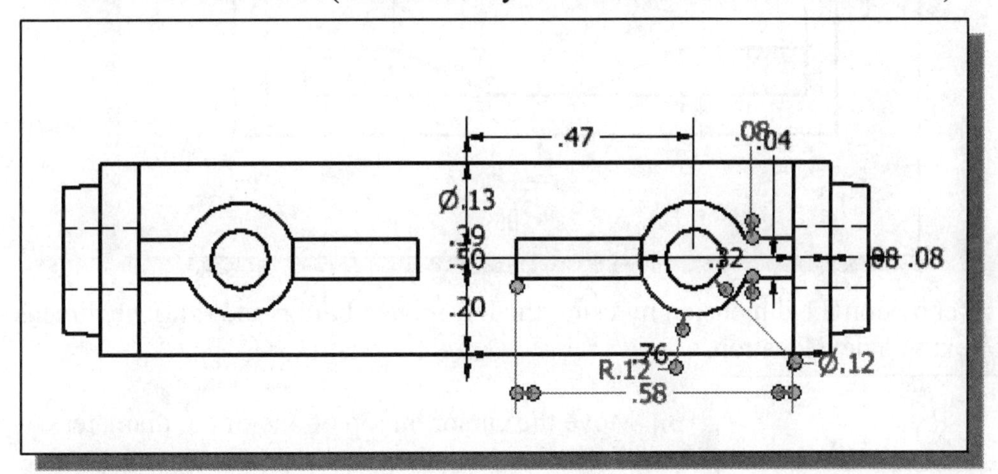

10. Click on the **Apply** button to proceed with retrieving the selected dimensions.

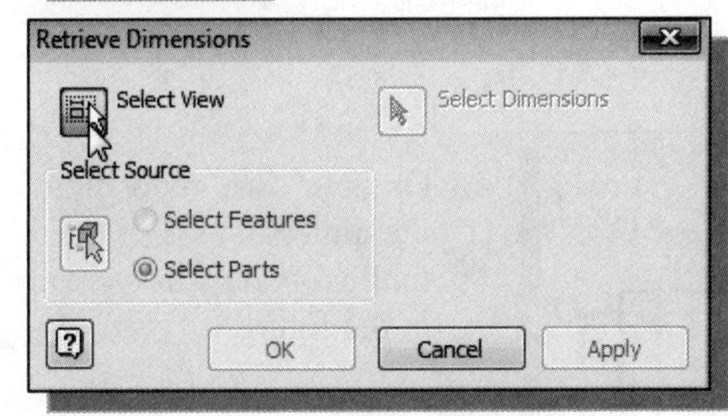

11. In the *Retrieve Dimensions* dialog box, the **Select View** option is switched *ON* as shown.

12. Select the ***front*** view.

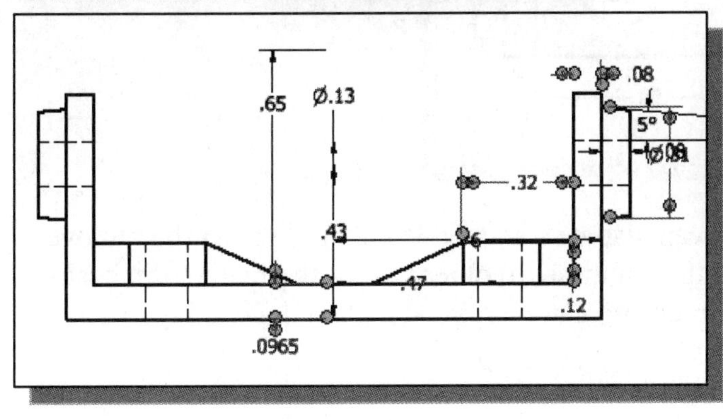

13. On your own, retrieve some of the dimensions as shown by using the **Select Dimensions** option.

14. Click on the **OK** button to end the Retrieve Dimensions command.

Reposition and Hide Feature Dimensions

1. Move the cursor on top of the height dimension text **0.0965** and watch for when the dimension text becomes highlighted with the four-arrow symbol indicating the dimension is selectable.

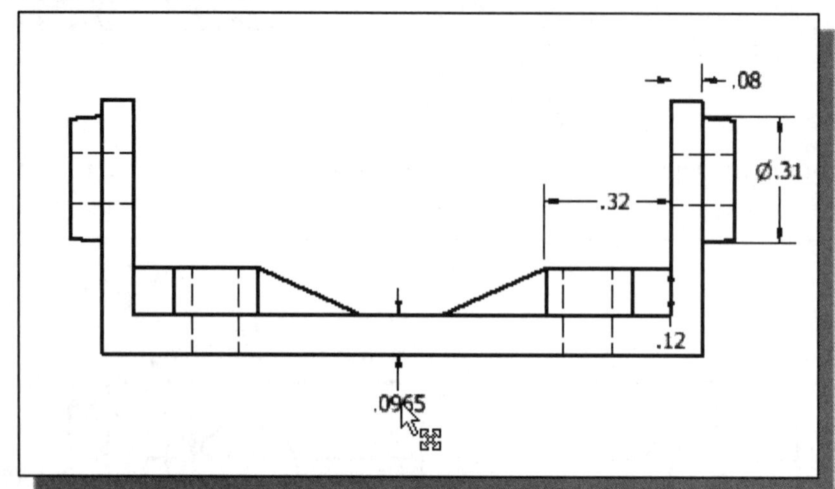

2. Reposition the dimension by using the left-mouse-button and drag the dimension text to a new location.

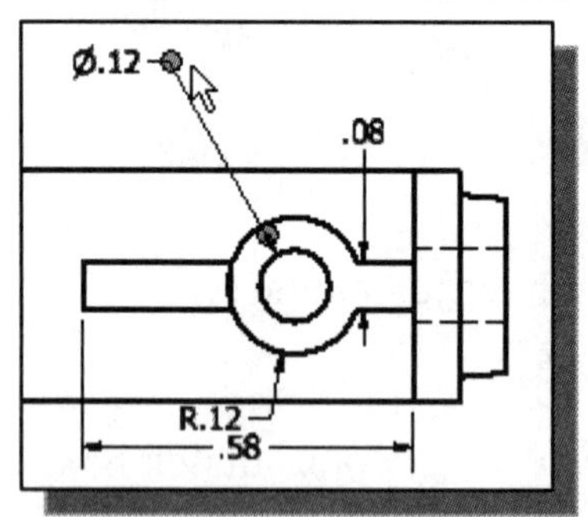

3. Move the cursor on top of one of the diameter dimensions and drag the grip point (green dot) associated with the dimension to reposition the dimension. Note that we can also drag on the dimension text, which only repositions the text.

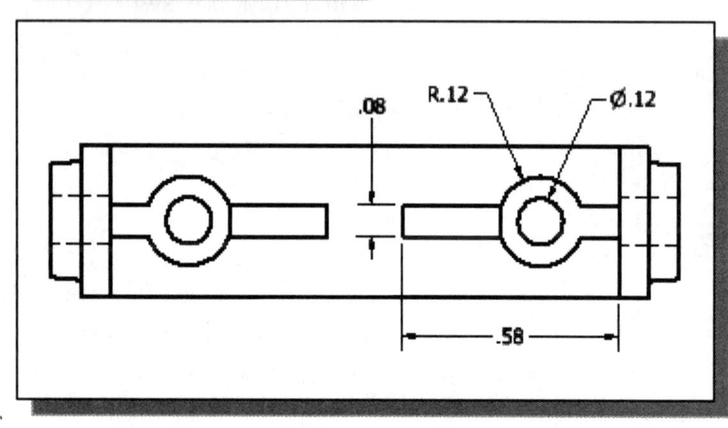

4. On your own, reposition the dimensions displayed in the ***top*** view as shown in the figure.

➢ Note that a feature dimension can also be removed from the display; the removed dimension will still remain in the database. In other words, the feature dimension is turned *OFF*. Any feature dimension can be removed from the display just as easily as it can be displayed.

Add Additional Dimensions – Reference Dimensions

- Besides displaying the **feature dimensions**, dimensions used to create the features, we can also add additional **reference dimensions** in the drawing. *Feature dimensions* are used to control the geometry, whereas *reference dimensions* are controlled by the existing geometry. In the drawing layout, therefore, we can ***add*** or ***delete*** *reference dimensions* but we can only ***hide*** the *feature dimensions.* One should try to use as many *feature dimensions* as possible and add *reference dimensions* only if necessary. It is also more effective to use *feature dimensions* in the drawing layout since they are created when the model was built. Note that additional Drawing mode entities, such as lines and arcs, can be added to drawing views. Before Drawing mode entities can be used in a reference dimension, they must be associated to a *drawing view.*

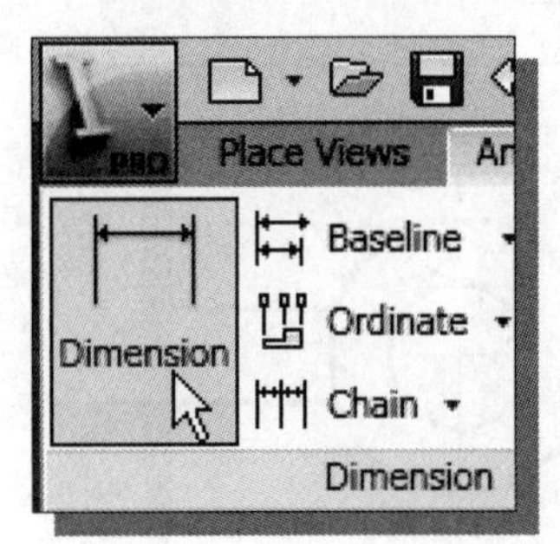

1. Click on the **General Dimension** button.

➤ Note the **General Dimension** command is similar to the **Smart Dimensioning** command in 3D Modeling mode.

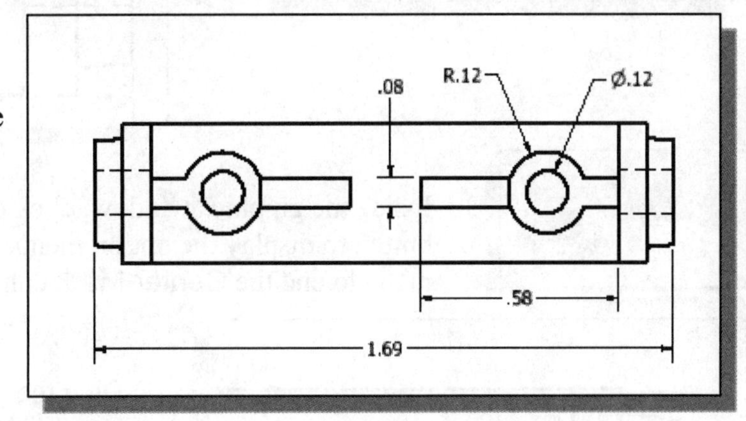

2. In the prompt area, the message "*Select first object*:" is displayed. Create the overall width in the top view.

3. On your own, create and position the center to center dimension in the top view as shown. (Hint: Select the two circles in the top view.)

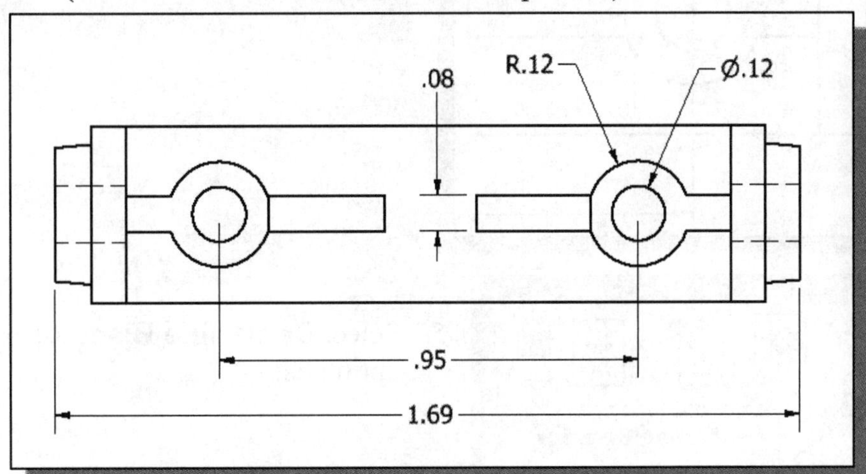

Add Center Marks and Center Lines

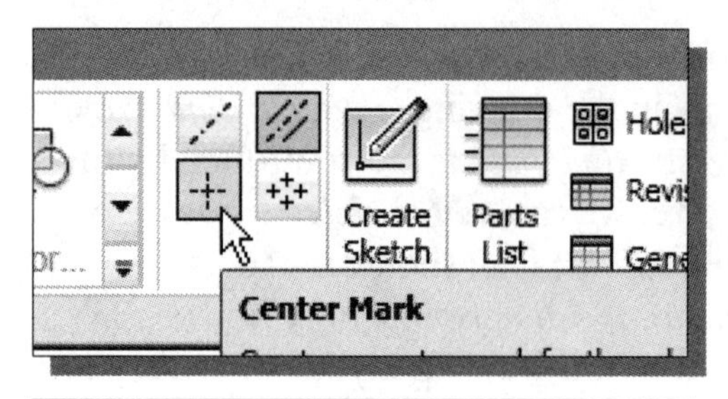

1. Click on the **Center Mark** button in the *Drawing Annotation* window.

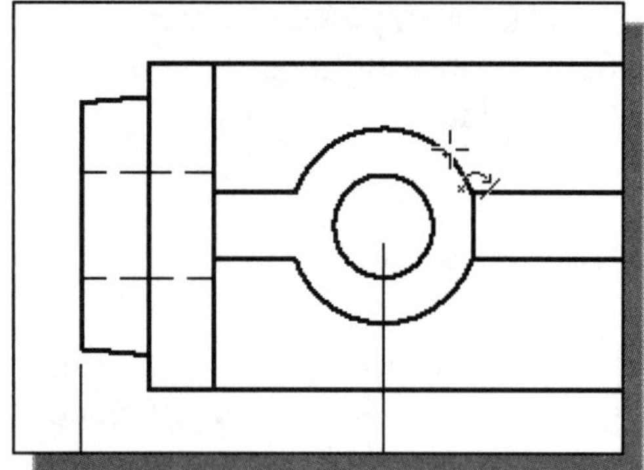

2. Click on one of the arcs in the top view to add the center mark.

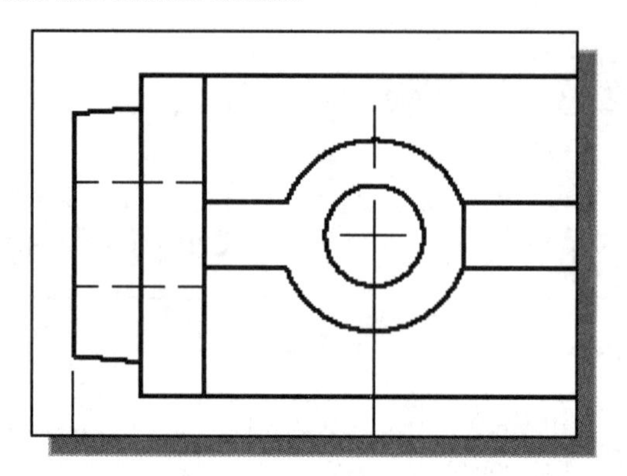

3. Inside the graphics window, click once with the right-mouse-button to display the option menu. Select **OK** in the popup menu to end the Center Mark command.

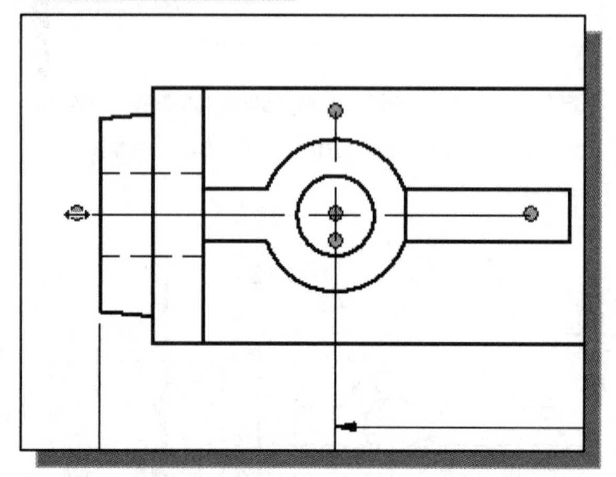

4. Drag the grip points to adjust the length of the center mark as shown.

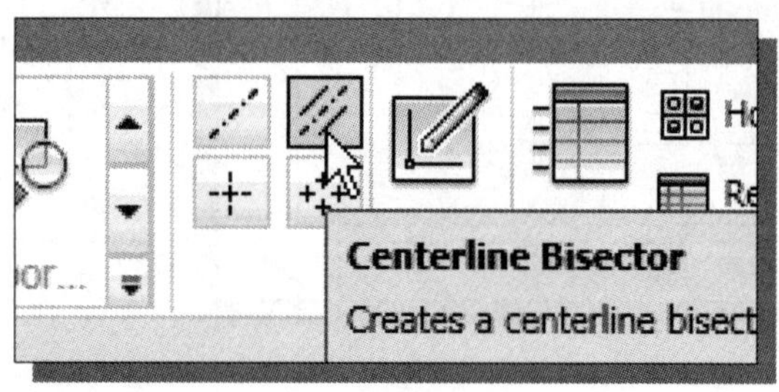

5. Select **Centerline Bisector** from the option list.

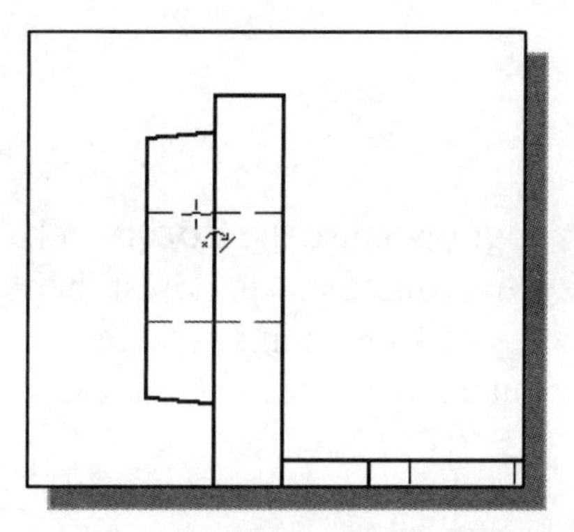

6. Click on the two hidden edges of one of the *revolved* features of the front view as shown in the figure.

7. On your own, repeat the above step and create the other centerlines on the front view as shown.

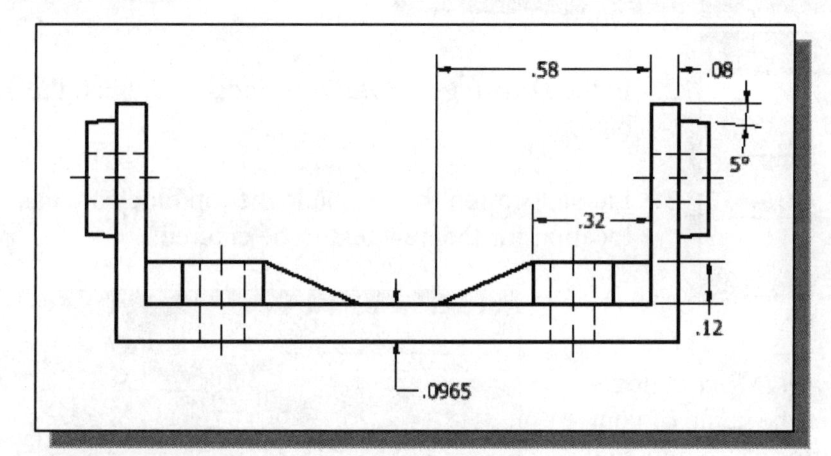

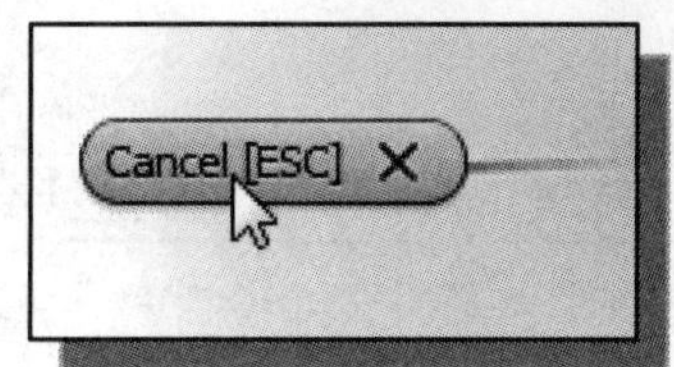

8. Inside the graphics window, click once with the right-mouse-button to display the option menu. Select **Cancel [ESC]** in the popup menu to end the **Centerline Bisector** command.

9. On your own, repeat the above steps and create additional centerlines as shown in the figure below.

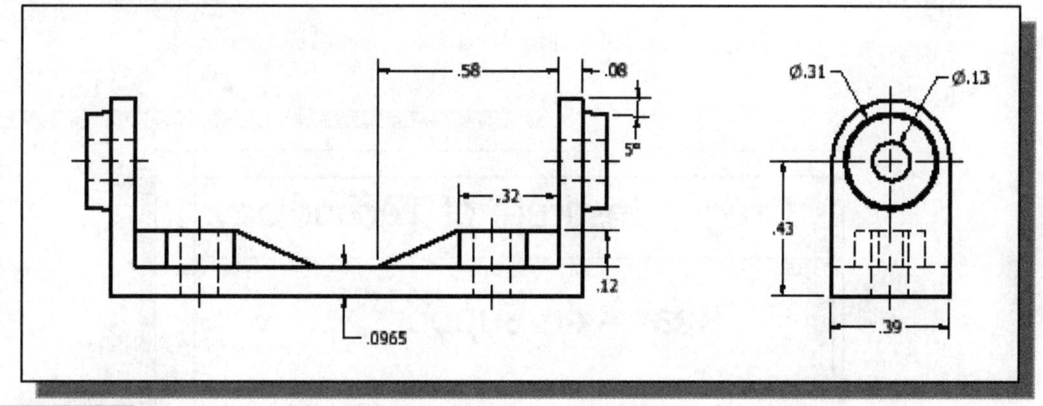

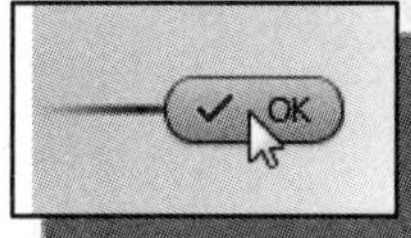

10. Inside the graphics window, click once with the right-mouse-button to display the option menu. Select **OK** in the popup menu to end the **Centerline** command.

Complete the Drawing Sheet

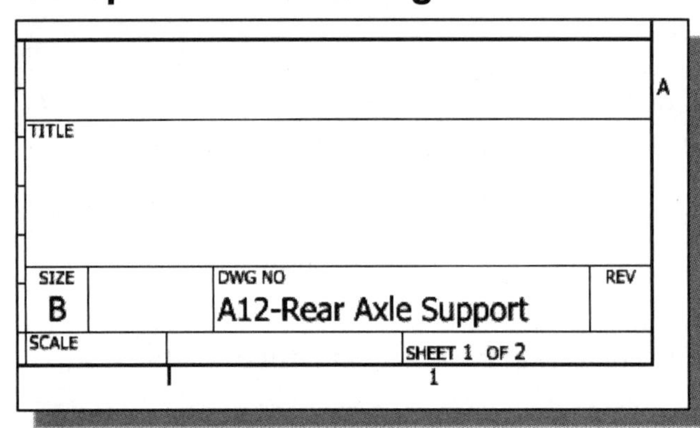

1. On your own, use the **Zoom** and the **Pan** commands to adjust the display as shown; this is so that we can complete the title block.

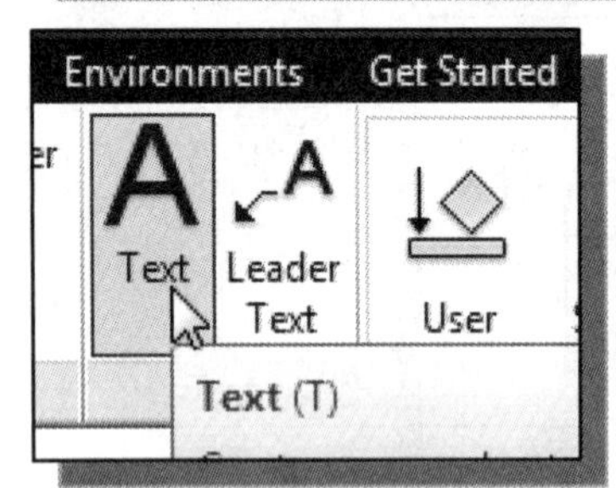

2. In the *Drawing Annotation* window, click on the **Text** button.

3. Pick a location that is inside the top block area as the location for the new text to be entered.

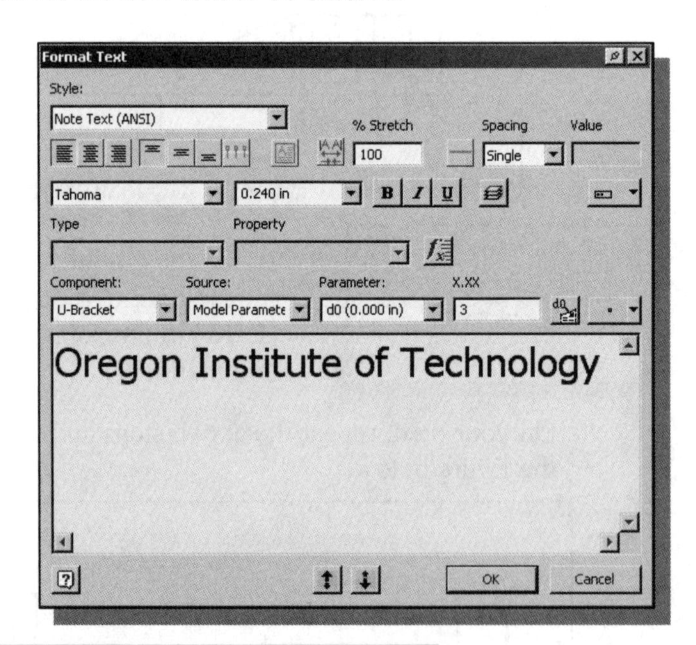

4. In the *Format Text* dialog box, enter the name of your organization. Also note the different settings available.

5. Click **OK** to proceed.

6. On your own, repeat the above steps and complete the title block.

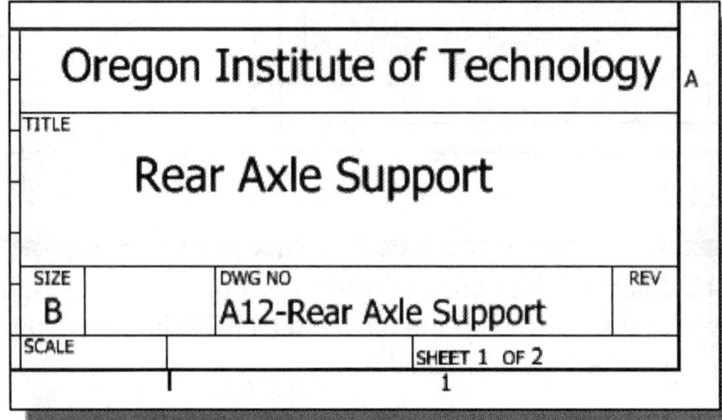

Associative Functionality – Modify Feature Dimensions

- *Autodesk Inventor's associative functionality* allows us to change the design at any level, and the system reflects the changes at all levels automatically.

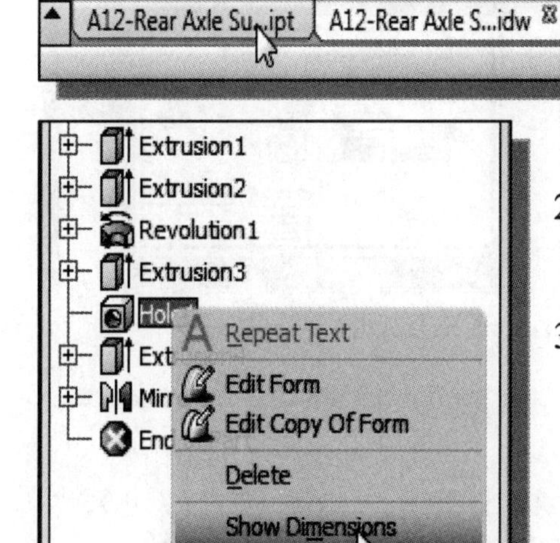

1. Click on the ***A12-Rear axle support*** part window or tab to switch to the Part Modeling mode.

2. In the *browser* window, right-click once on **Hole1** to bring up the option menu.

3. Select **Show Dimensions** in the popup option menu.

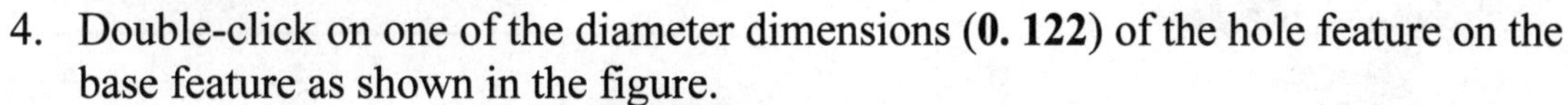

4. Double-click on one of the diameter dimensions (**0. 122**) of the hole feature on the base feature as shown in the figure.

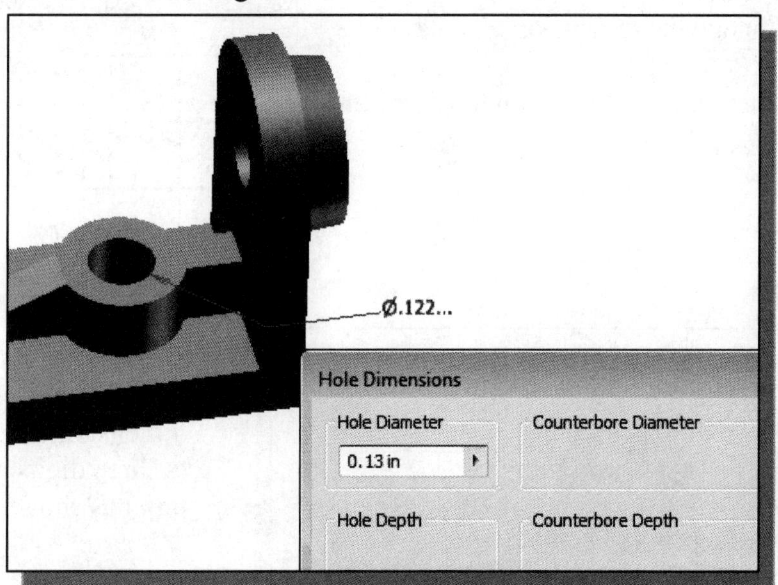

5. In the *Hole Dimensions* dialog box, enter **0.13** as the new diameter dimension.

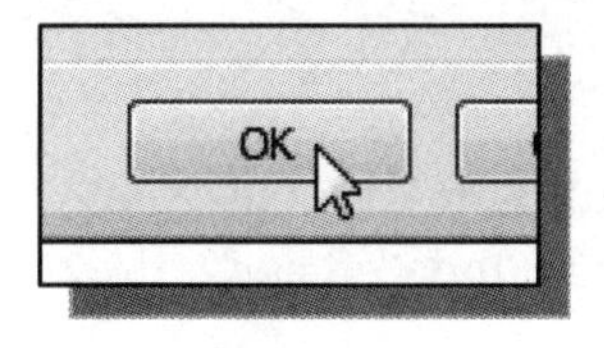

6. Click on the **OK** button to accept the new setting.

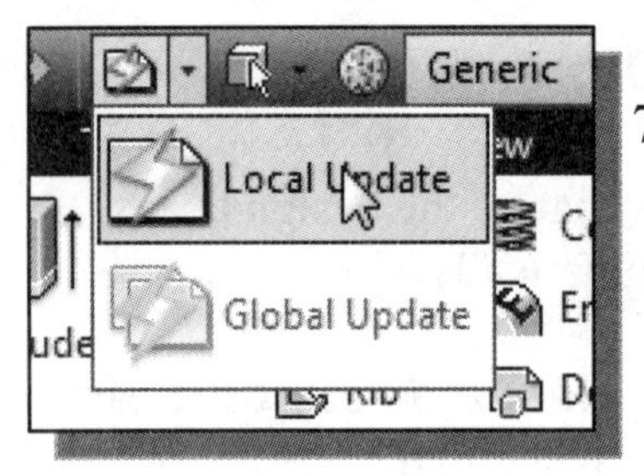

7. Click on the **Update** button in the *Standard* toolbar area to proceed with updating the solid model.

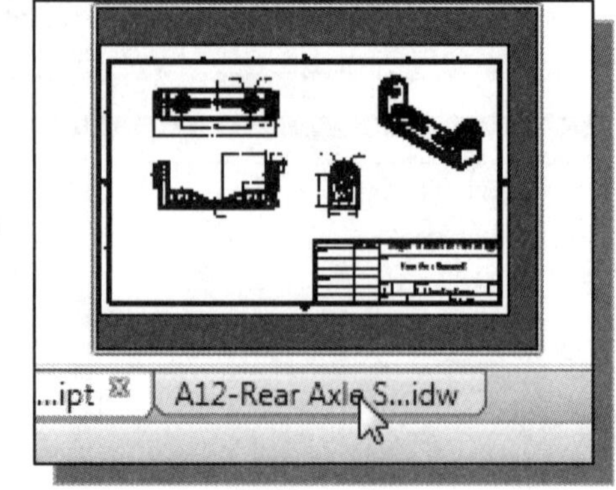

8. Click on the ***A12-Rear Axle Support*** drawing graphics window or tab to switch to the *Paper Space*.

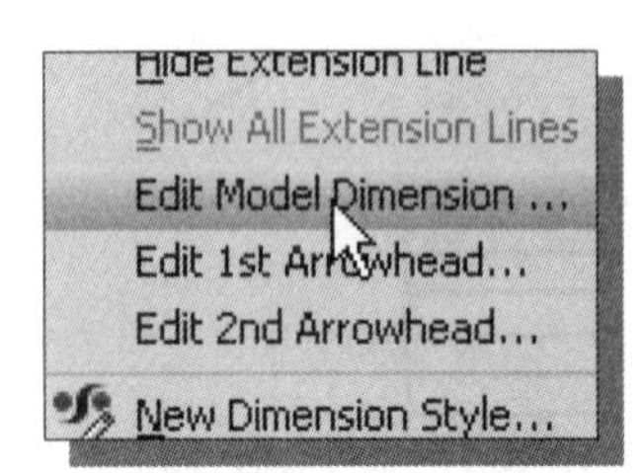

9. Inside the graphics window, right-click once on the **Ø0.13** dimension in the *top* view to bring up the option menu.

10. Select **Edit Model Dimension** in the popup menu.

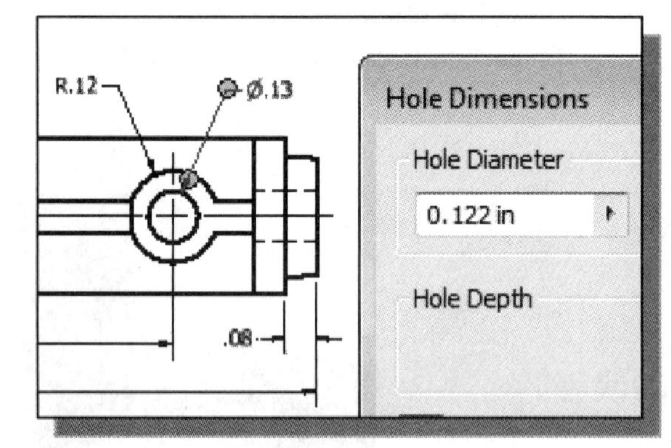

11. Change the dimension to **1.22**.

12. Click on the **OK** button to accept the setting.

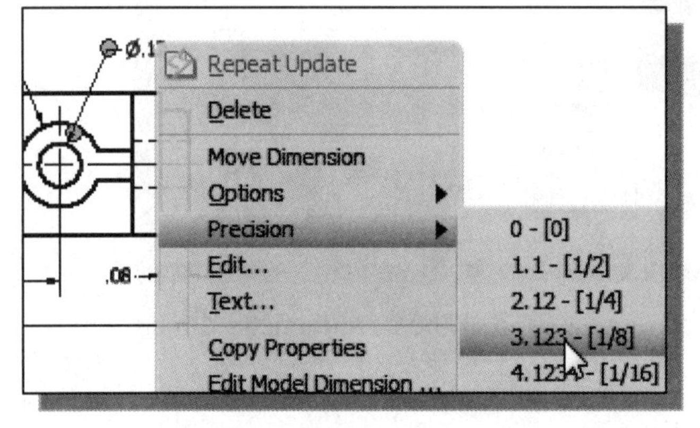

13. Right-mouse-click once on the dimension and set the precision to three digits after the decimal point as shown.

❖ Note the geometry of the hole feature is updated in all views automatically. On your own, switch to the Part Modeling mode and confirm the design is updated as well.

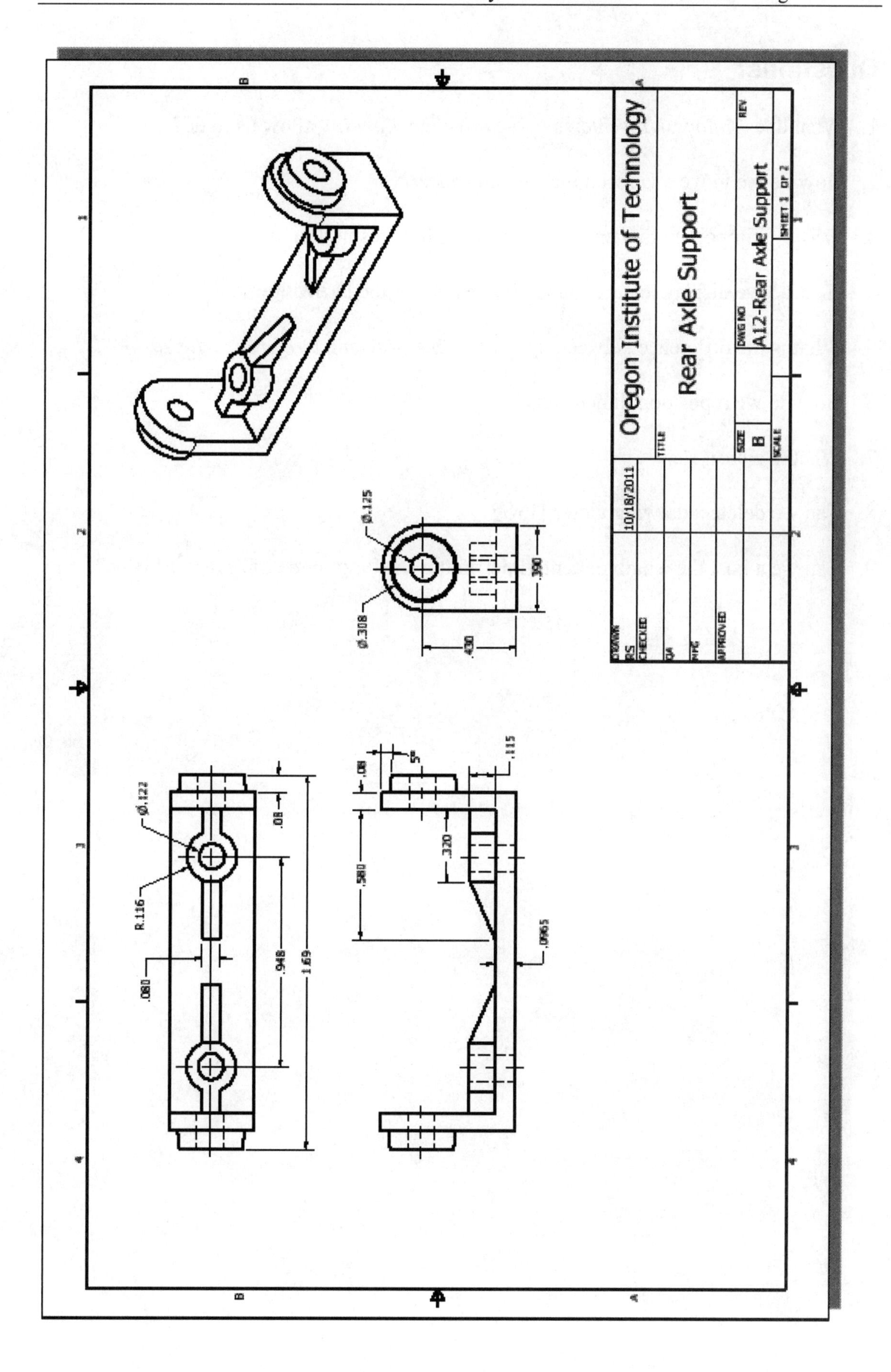
Oregon Institute of Technology
TITLE
Rear Axle Support
DRAWN
RS
10/18/2011
CHECKED
QA
MFG
APPROVED
SIZE
B
DWG NO
A12-Rear Axle Support
REV
SCALE
SHEET 1 OF 2
Ø.125
Ø.308
.390
.430
Ø.122
R.116
.080
.08
.948
1.69
.08
.115
.580
.320
.0965

Questions:

1. What does *Autodesk Inventor*'s *associative functionality* allow us to do?

2. How do we move a view on the *drawing sheet*?

3. Why is it important to identify symmetrical features in designs?

4. How do we display feature/model dimensions in the Drawing mode?

5. What is the difference between a *feature dimension* and a *reference dimension*?

6. How do we reposition dimensions?

7. What is a *base view*?

8. Can we delete a drawing view? How?

9. Can we adjust the length of centerlines in the Drafting mode of *Inventor*? How?

Exercises:

1. **Slide Mount** (Dimensions are in inches.)

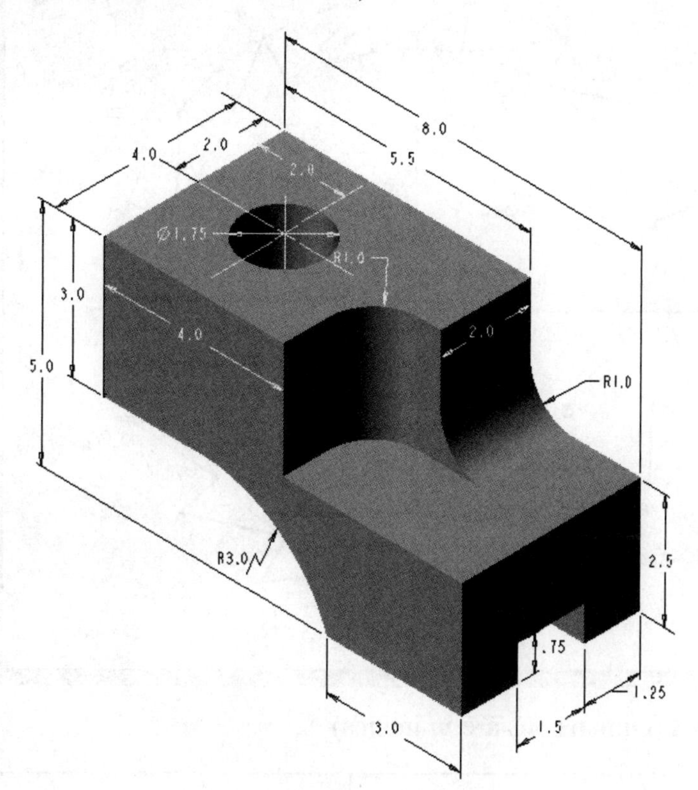

2. **Corner Stop** (Dimensions are in inches.)

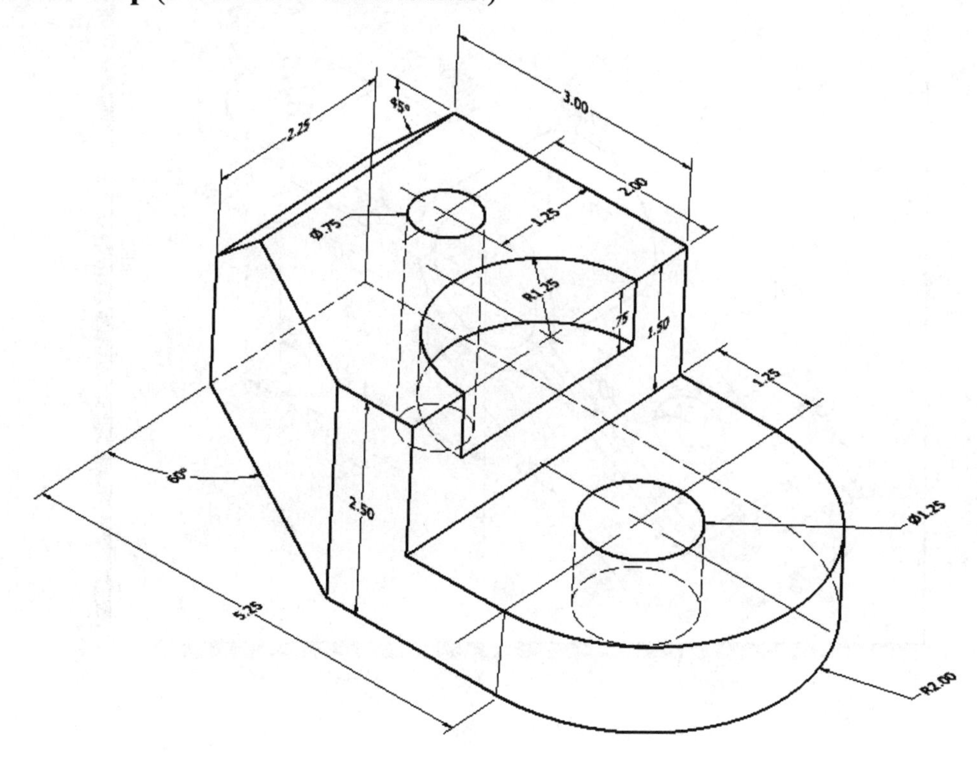

3. **Ratchet Plate** (thickness: **0.125 inch**) Hint: use the Circular Pattern command.

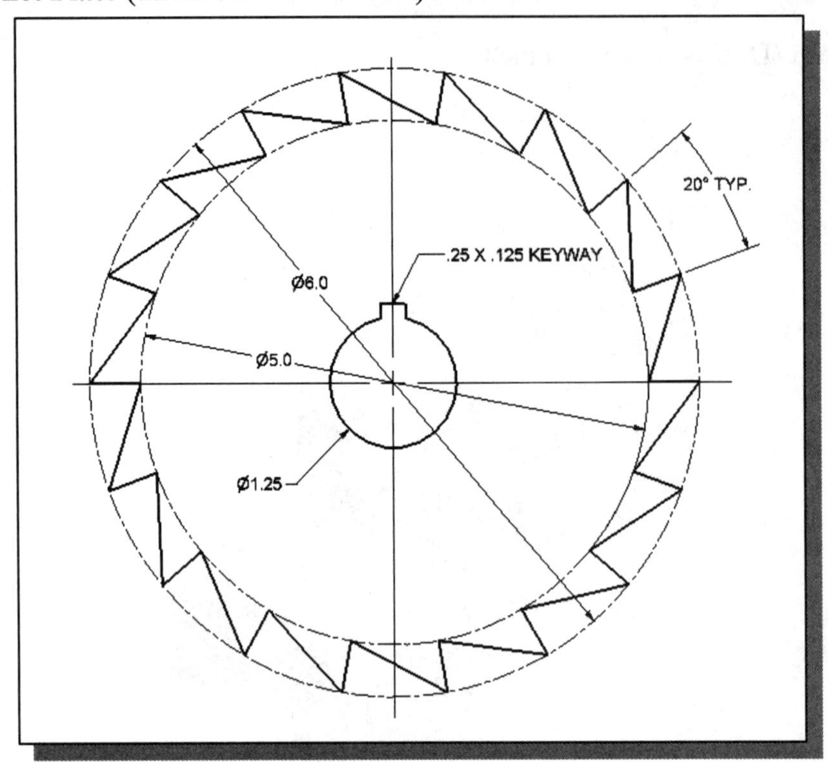

4. **Angle Support** (Dimensions are in inches)

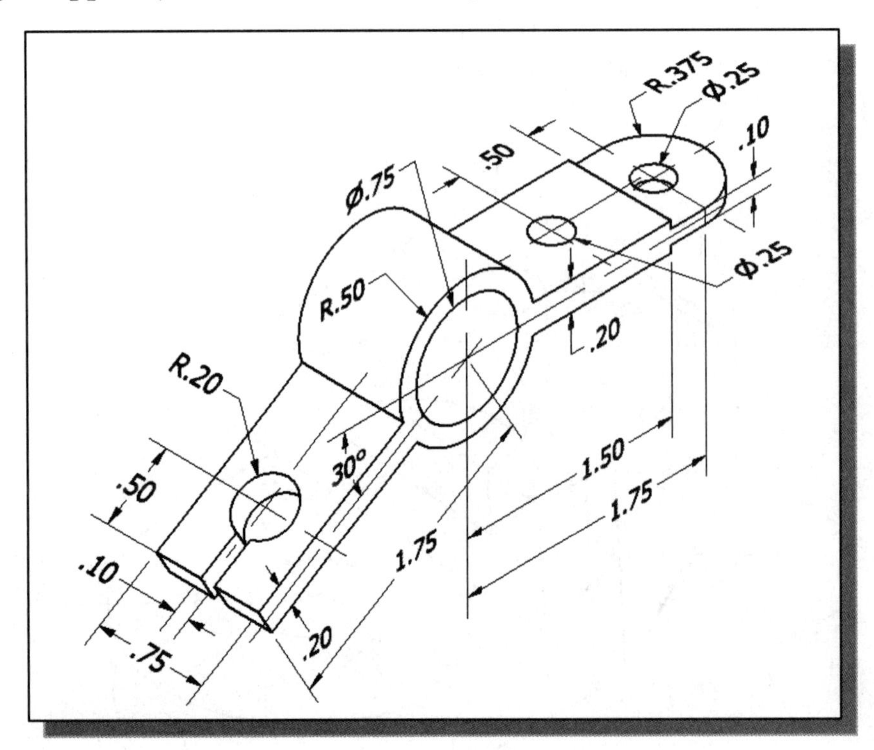

Chapter 7
Datum Features in Designs

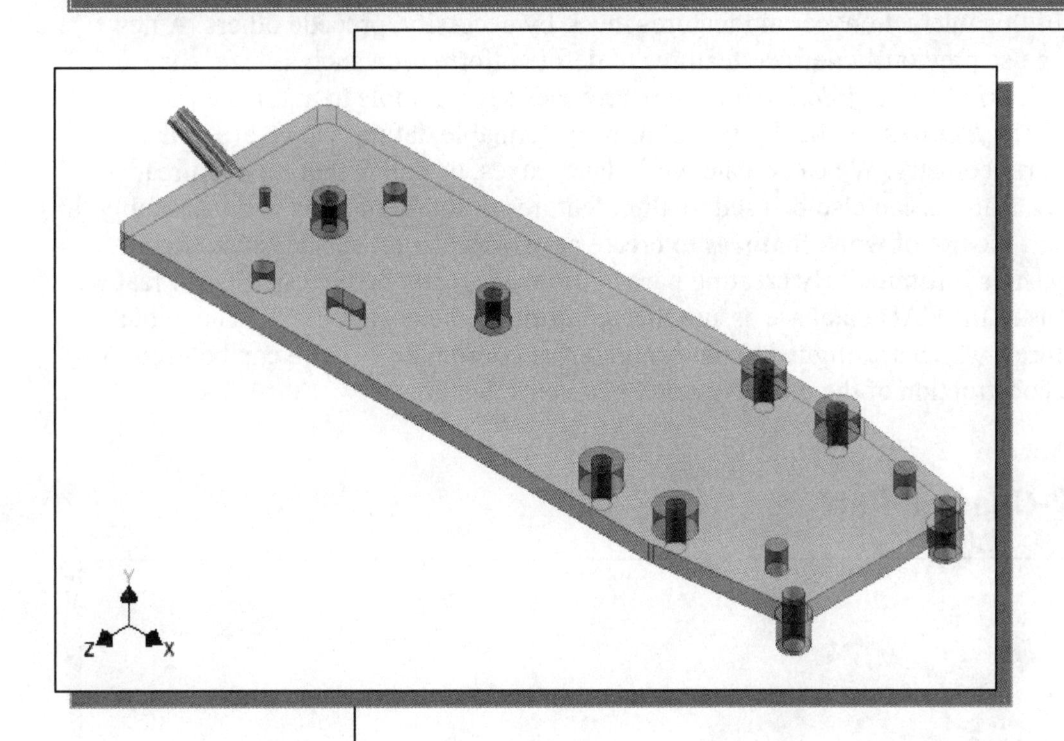

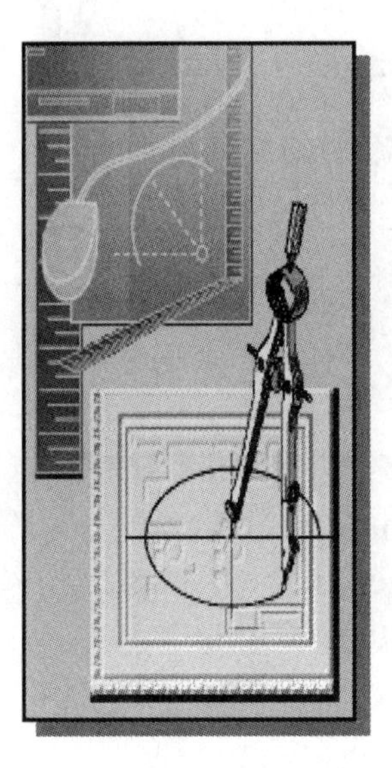

Learning Objectives

- Understand the Concepts and the Use of Work Features
- Using the Different Options to Create Work Features
- Creating Revolved Features
- Creating Tapered Features
- Adjust the Color of the Models

Work Features

Feature-based parametric modeling is a cumulative process. The relationships that we define between features determine how a feature reacts when other features are changed. Because of this interaction, certain features must, by necessity, precede others. A new feature can use previously defined features to define information such as size, shape, location and orientation. *Autodesk Inventor* provides several tools to automate this process. ***Work features*** can be thought of as user-definable datum, which are updated with the part geometry. We can create workplanes, axes, or points that do not already exist. Work features can also be used to align features or to orient parts in an assembly. In this chapter, the use of **work features** to create new workplanes, surfaces that do not already exist, is illustrated. By creating parametric work features, the established feature interactions in the CAD database assure the capturing of the design intent. The default work features, which are aligned to the origin of the coordinate system, can be used to assist the construction of the more complex geometric features.

The *B2-Chassis* Part

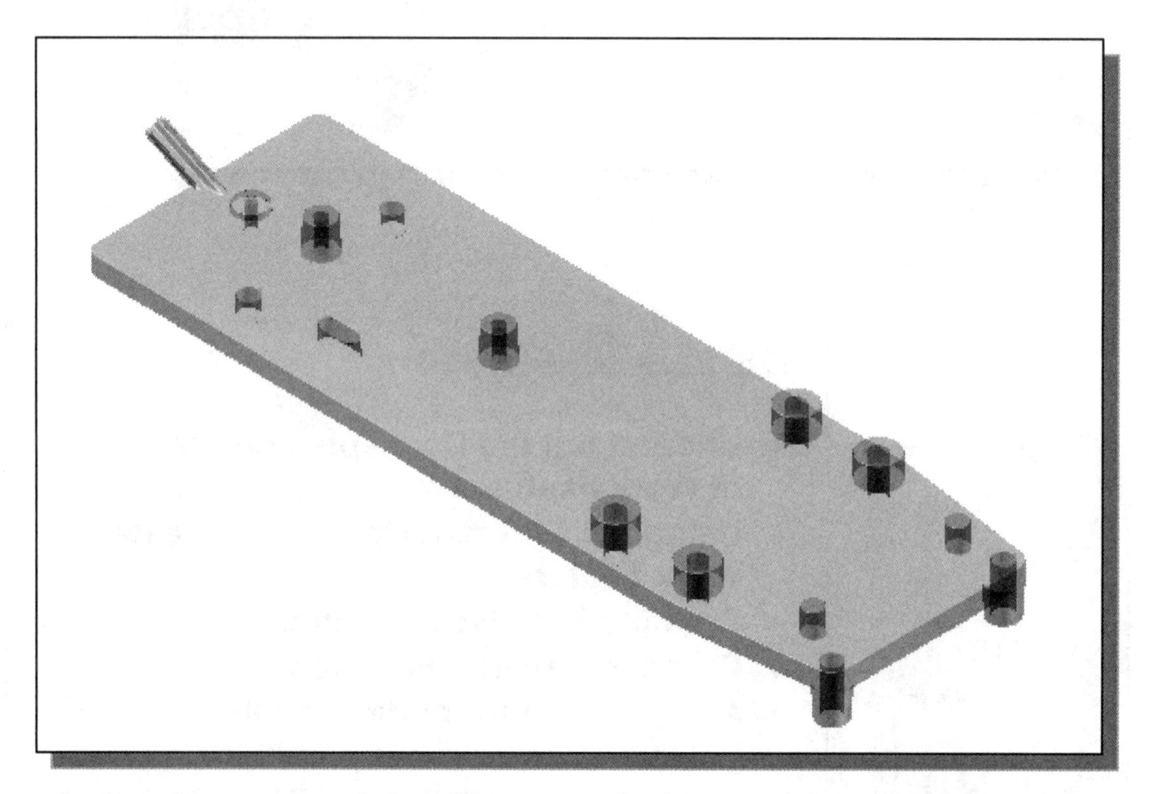

- Based on your knowledge of *Inventor* so far, how many features would you use to create the model? What are the more difficult features involved in the design? Which feature would you choose as the **base feature**? What is your choice for arranging the order of the features?

Modeling Strategy

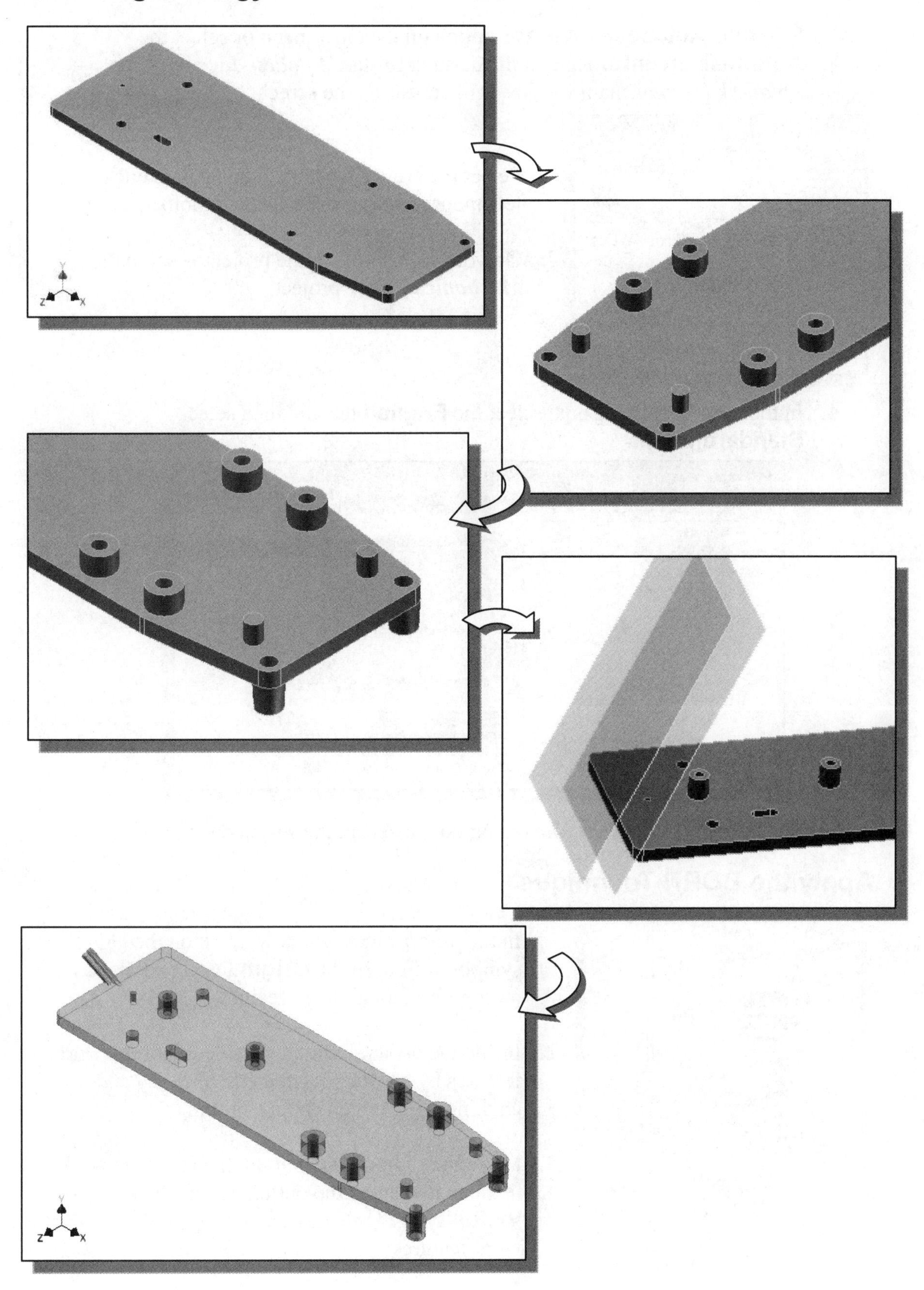

Starting *Autodesk Inventor*

1. Select the **Autodesk Inventor** option on the *Start* menu or select the **Autodesk Inventor** icon on the desktop to start *Autodesk Inventor*. The *Autodesk Inventor* main window will appear on the screen.

2. Select the **New File** icon with a single click of the left-mouse-button in the *Launch* toolbar.

3. On your own, confirm the project is set to the ***Mechanical-Tiger*** project.

4. In the *New File* dialog box, select the **English** tab and then select **Standard(in).ipt**.

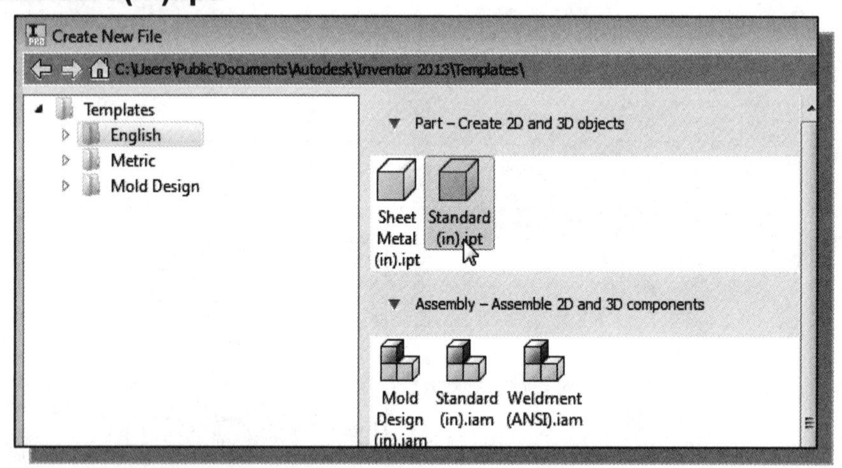

5. Pick **Create** in the *New File* dialog box to accept the selected settings.

Apply the BORN Technique

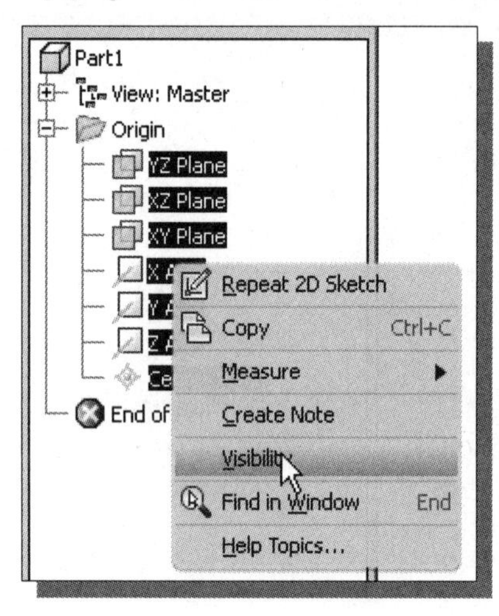

1. In the *part browser* window, click on the [**+**] symbol in front of the **Origin** feature to display more information on the feature.

2. Inside the *browser* window, select all of the work features by holding down the **[Shift]** key and clicking with the left-mouse-button.

3. Move the right-mouse-button on any of the work features to display the option menu. Click on **Visibility** to toggle *ON* the display of the selected work features.

4. On your own, use the dynamic viewing options (**3D Rotate, Zoom** and **Pan**) to view the work features established.

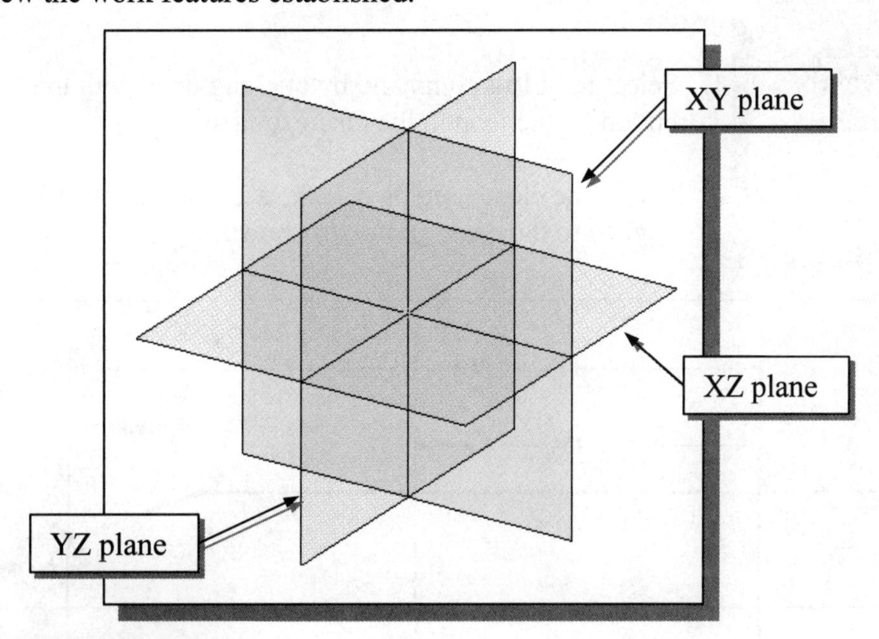

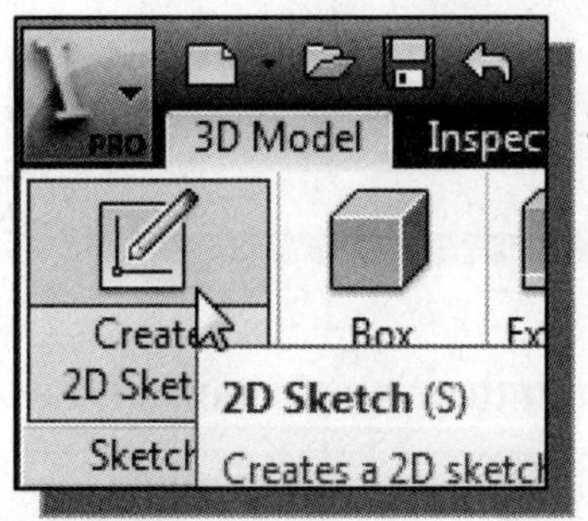

5. In the *Sketch* toolbar select the **Create 2D Sketch** command by left-clicking once on the icon.

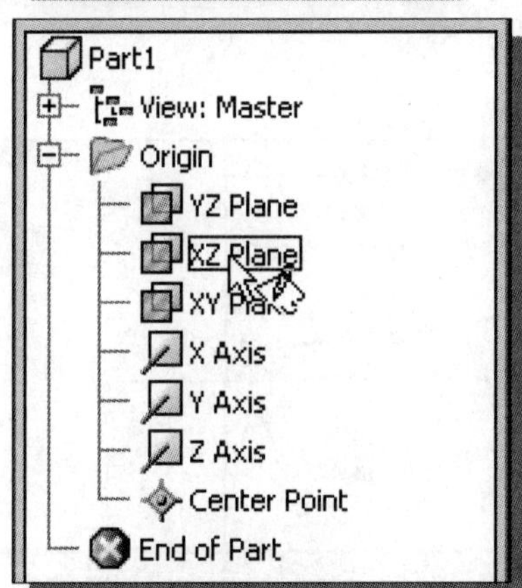

6. In the *Status Bar* area, the message: "*Select face, workplane, sketch or sketch geometry*" is displayed. *Autodesk Inventor* expects us to identify a planar surface where the 2D sketch of the next feature is to be created. Move the graphics cursor on top of **XZ Plane**, inside the *browser* window as shown, and notice that *Autodesk Inventor* will automatically highlight the corresponding plane in the graphics window. Left-click once to select the XZ Plane as the sketching plane.

7. Single left-mouse-click to activate the **Home View** option as shown. The view will be adjusted back to the default ***isometric view***.

Create the Base Feature

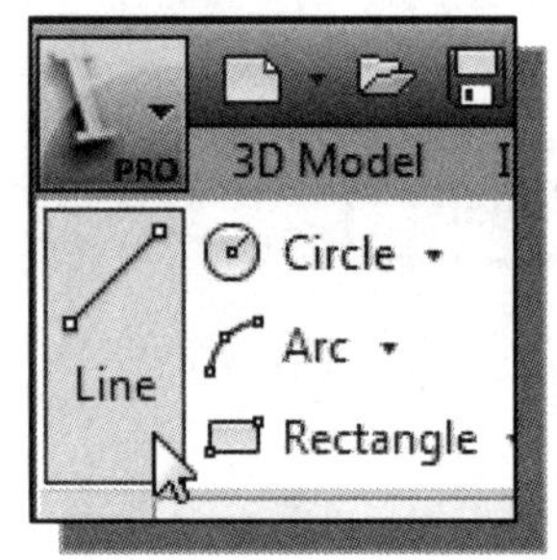

1. Select the **Line** command by clicking once with the left-mouse-button on the icon in the *Draw* toolbar.

2. Create the closed region sketch, with the center of the sketch aligned to the *center point*, as shown.

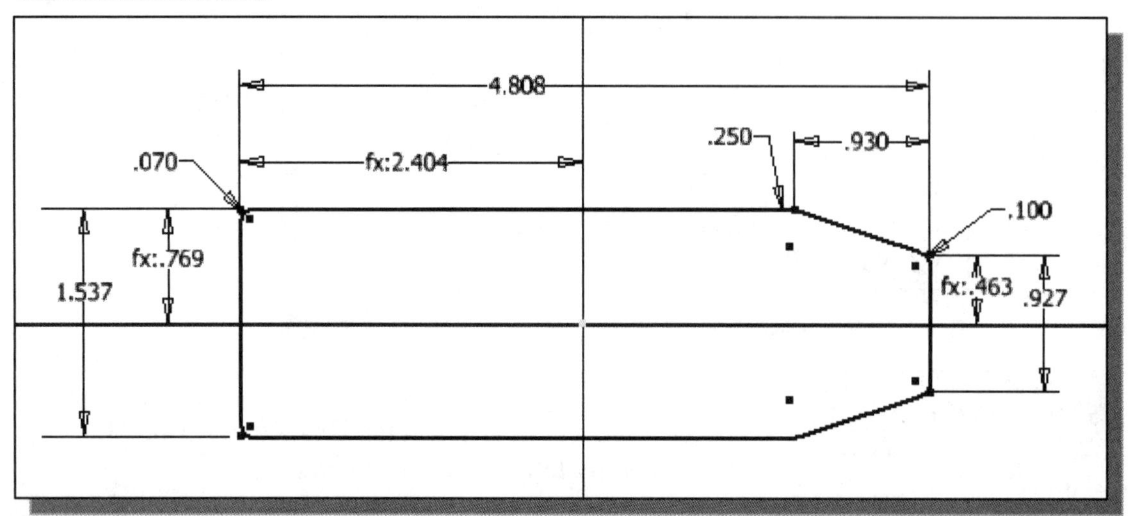

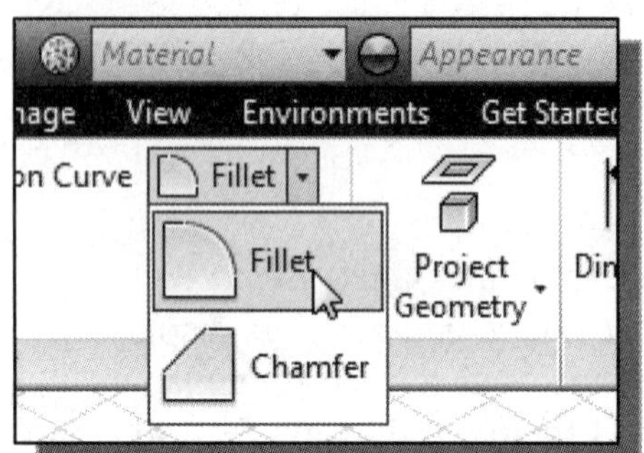

3. Click on the **Fillet** icon in the *Draw* panel.

4. On your own, create the 2D Rounds & Fillets as shown in the figure below.

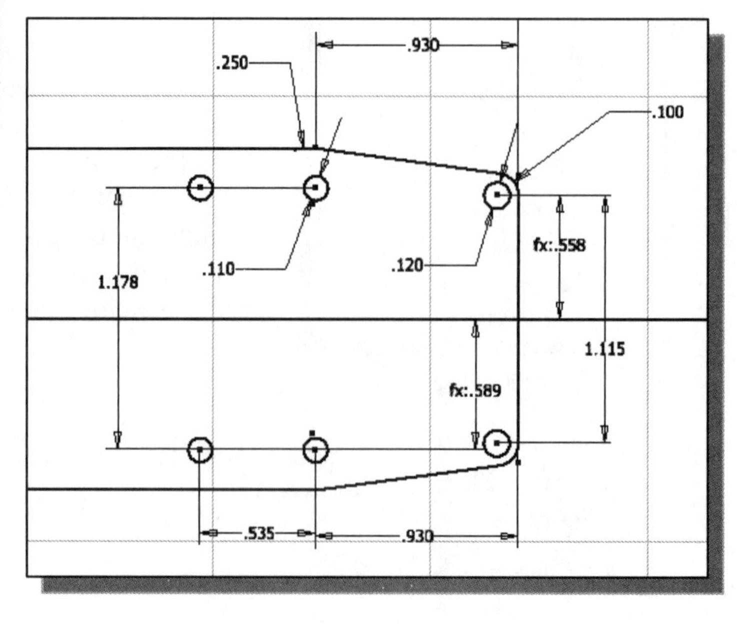

5. On your own, create six circles; the four circles on the left centers are aligned both horizontally and vertically as shown. Also create and modify the associated dimensions as shown in the figure.

6. On your own, create the four closed regions on the left as shown.

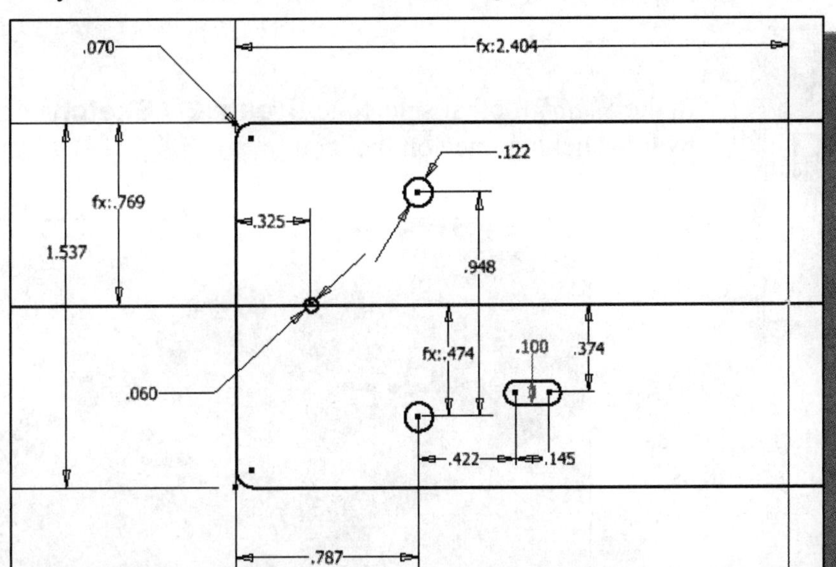

7. Inside the graphics window, click once with the **right-mouse-button** and select **Finish 2D Sketch** in the popup menu to end the Sketch option.

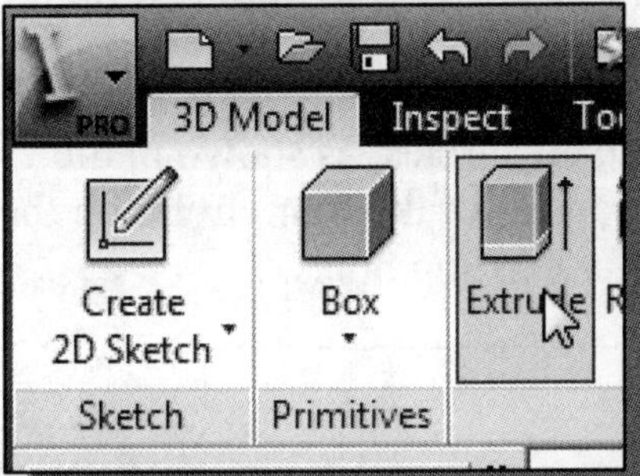

8. In the *Create* toolbar, select the **Extrude** command by releasing the left-mouse-button on the icon.

9. Select the inside region of the 2D sketch to create a profile as shown.

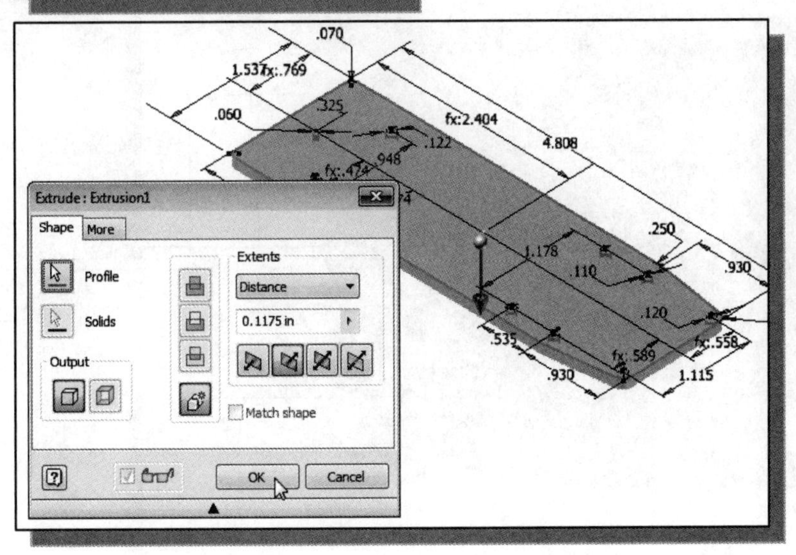

10. In the *Extrude* popup window, enter **0.1175** as the extrusion distance.

11. Set the extrusion direction to downward as shown.

12. Click on the **OK** button to proceed with creating the feature.

Create the Second Extruded Feature

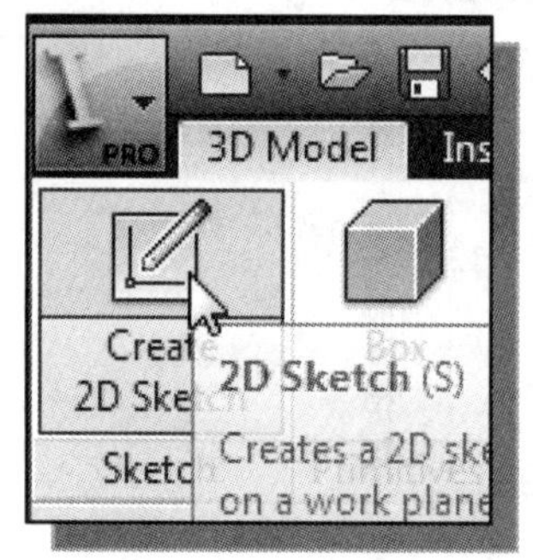

1. In the *Sketch* toolbar select the **Create 2D Sketch** command by left-clicking once on the icon.

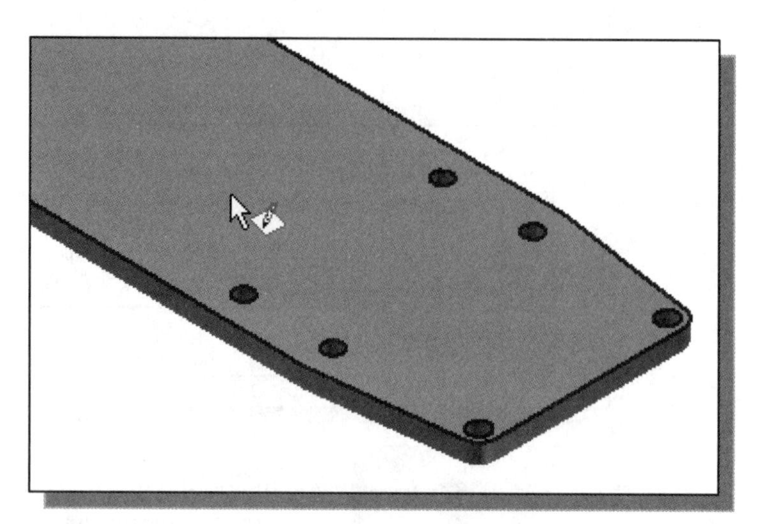

2. Select the top surface as the sketching plane as shown.

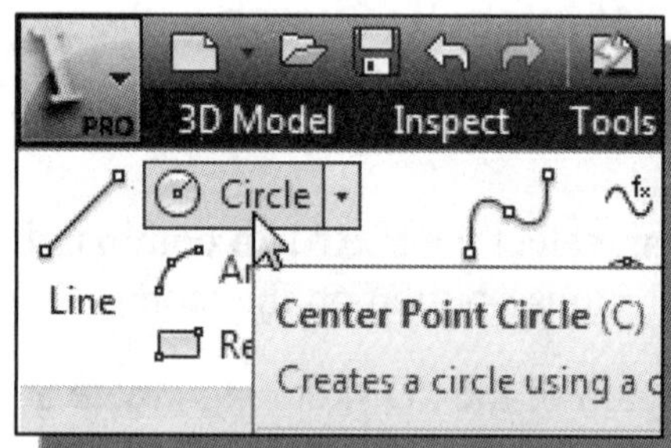

3. Select the **Center point circle** command by clicking once with the left-mouse-button on the icon in the *Draw* toolbar.

4. On your own, create six circles, as shown in the figure below. The centers of the four circles on the left are aligned to the existing holes.

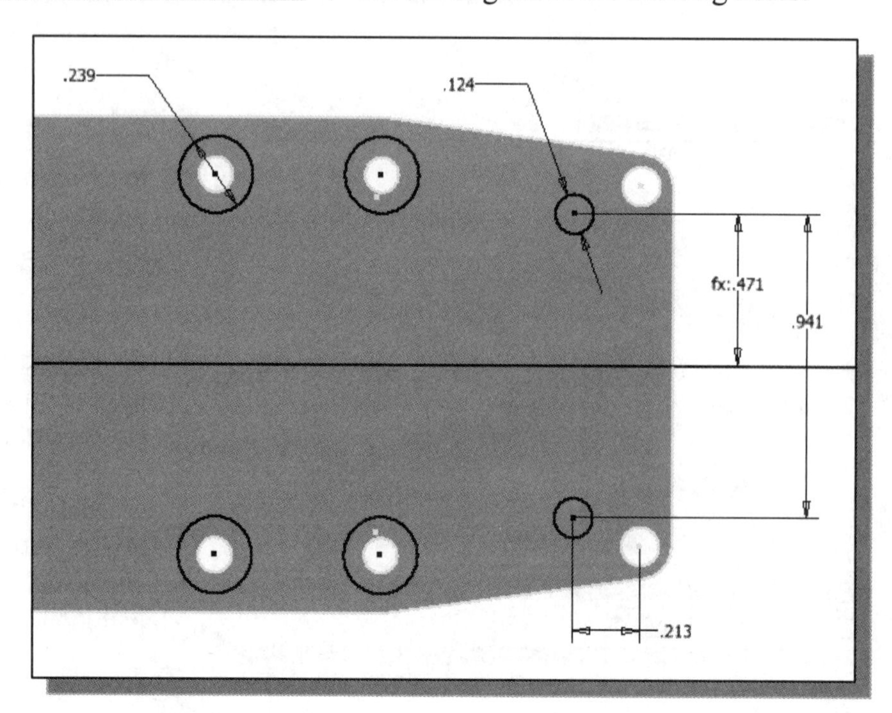

5. Select **Finish Sketch** in the *Ribbon* toolbar to exit the **2D Sketch** module.

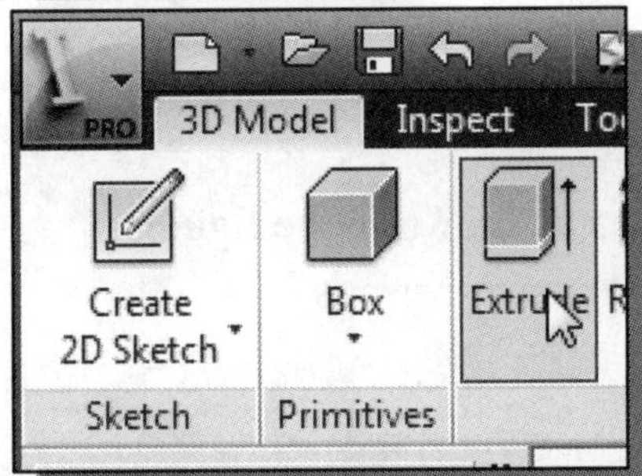

6. In the *Create* toolbar, select the **Extrude** command by releasing the left-mouse-button on the icon.

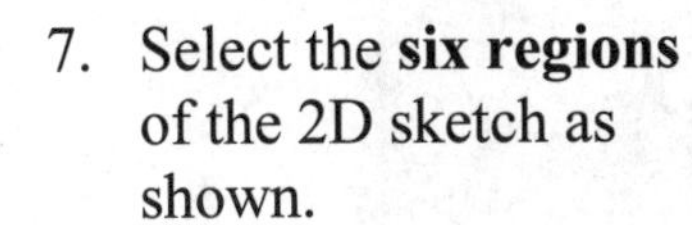

7. Select the **six regions** of the 2D sketch as shown.

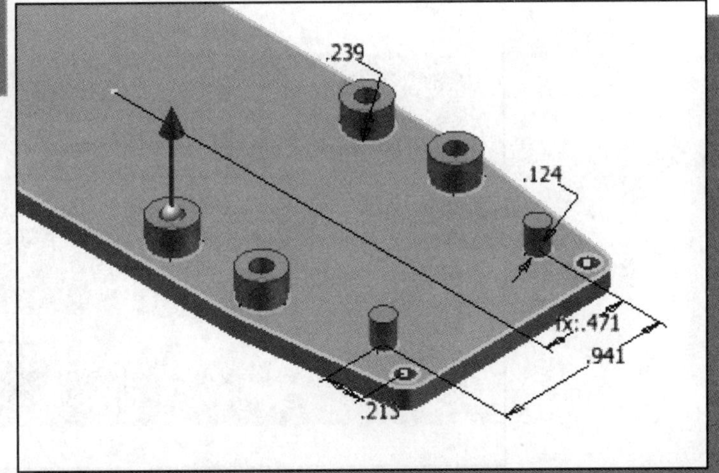

8. In the *Extrude* popup window, enter **0.1485** as the extrusion distance.

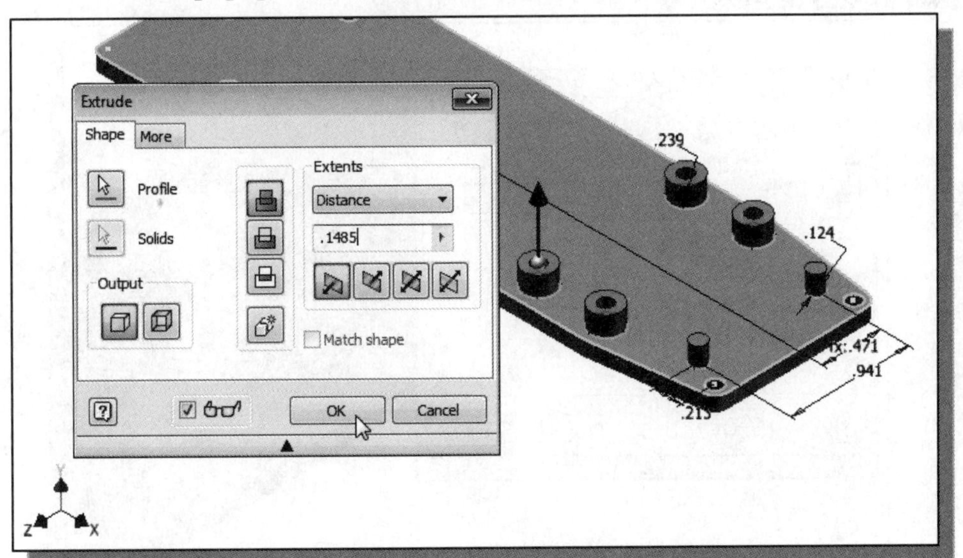

9. Confirm the extrusion direction is set to **up** as shown.

10. Click on the **OK** button to proceed with creating the features.

Create a Tapered Extruded Feature

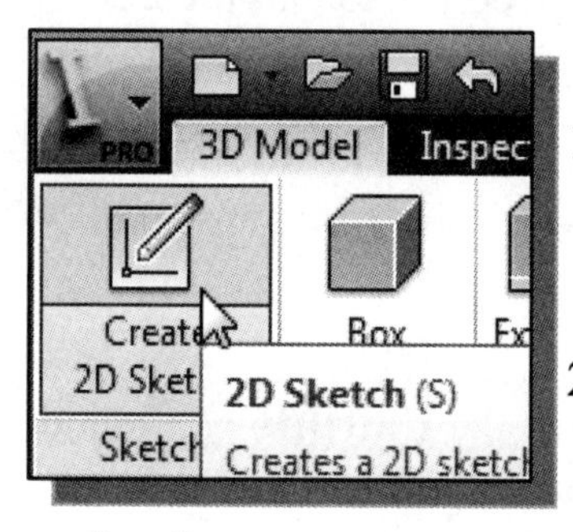

1. In the *Sketch* toolbar select the **Create 2D Sketch** command by left-clicking once on the icon.

2. Select the top surface as the sketching plane.

3. On your own, create four circles with the dimensions as shown in the figure.

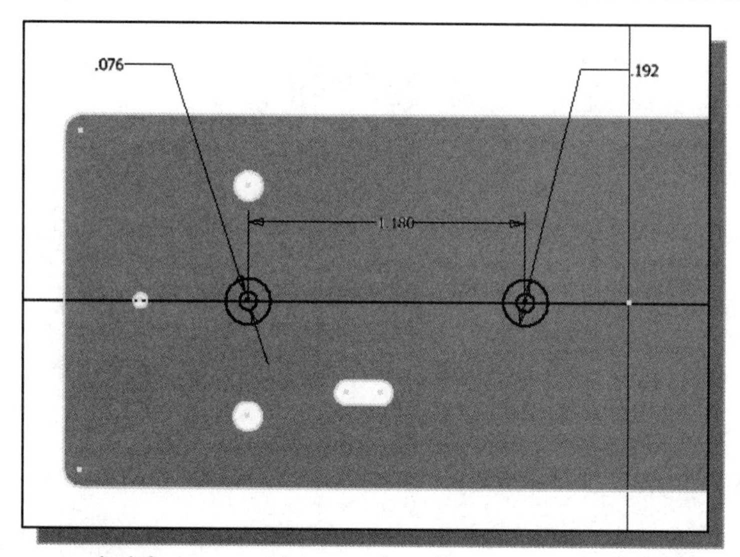

4. Create an extruded feature, with extrusion distance set to **0.1955** and taper angle set to **-2.5** degrees.

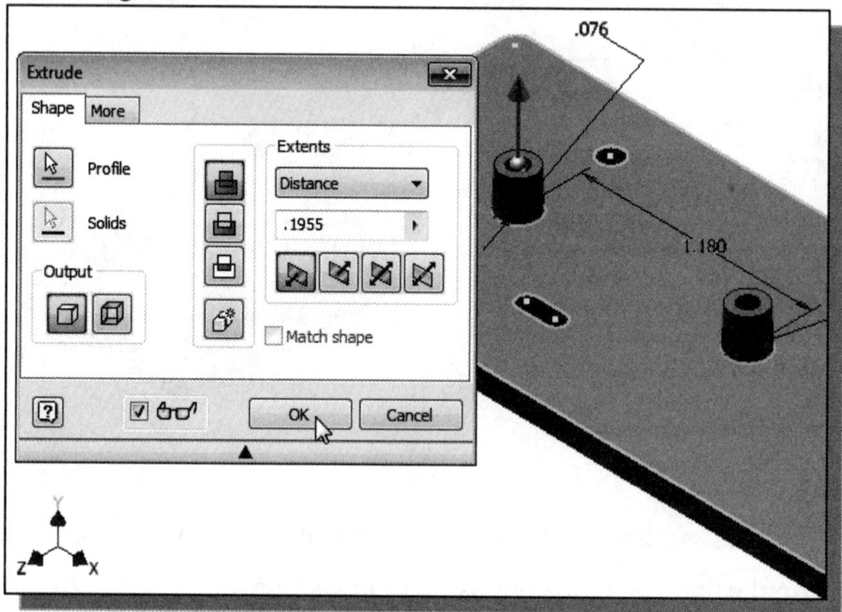

5. Click on the **OK** button to proceed with creating the feature.

Create an Offset Work Plane

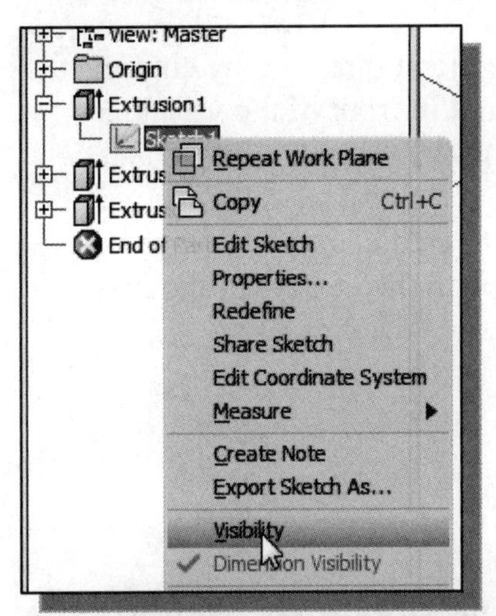

1. Show the sub-list of the first extruded feature in the *model browser* area by left-clicking once on the [**+**] symbol in front of the name of the feature, **Extrusion1**.

2. In the *model browser* area, right-click once on the 2D sketch to display the option menu.

3. Select **Visibility** in the option list. Notice the 2D sketch with all the associated dimensions is displayed in the graphics window.

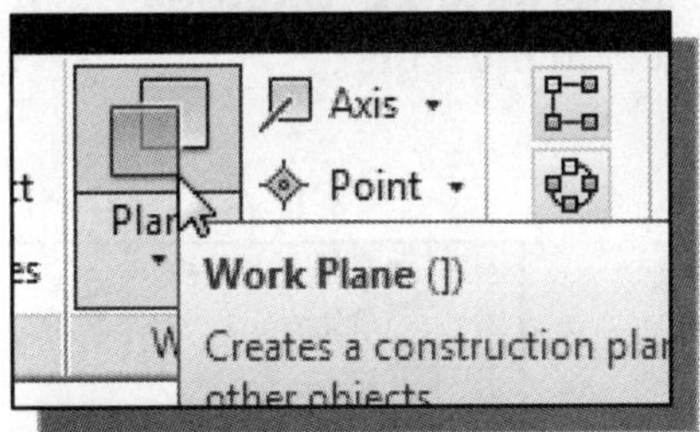

4. In the *Create* toolbar, select the **Work Plane** command by left-clicking the icon.

- Note that clicking on the triangle symbol can also be used to choose the specific creation option.

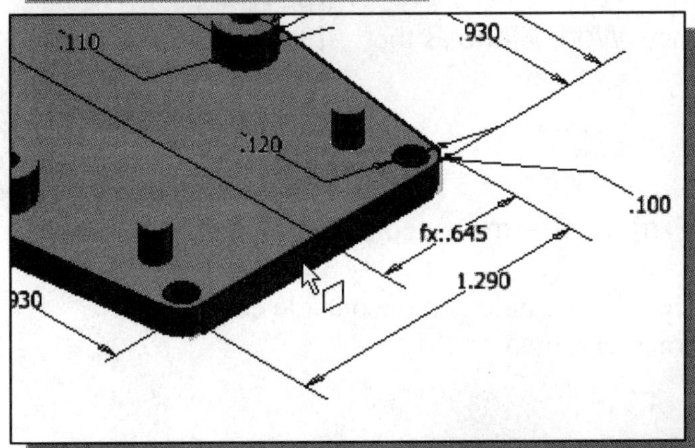

5. Inside the graphics window, select the **right vertical plane** of the base feature as the reference of the new work plane.

6. Drag with the left-mouse-button to enter the **Offset Plane** option.

7. Click inside the edit box and choose the **radius** of the round on the right as shown.

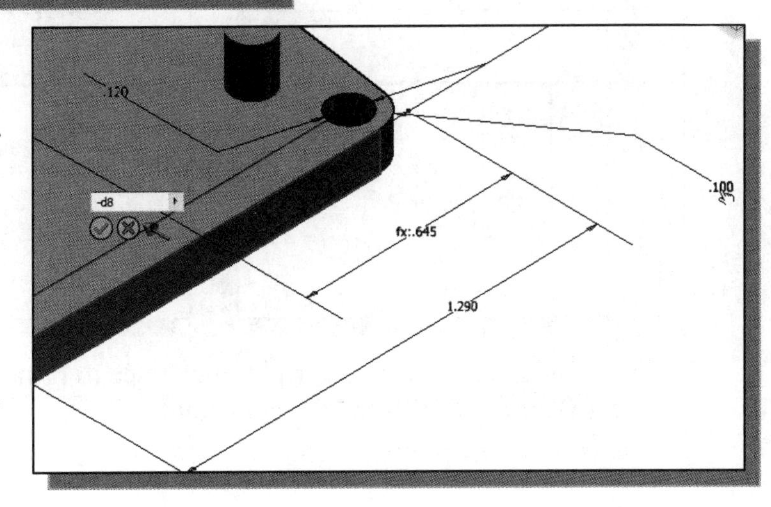

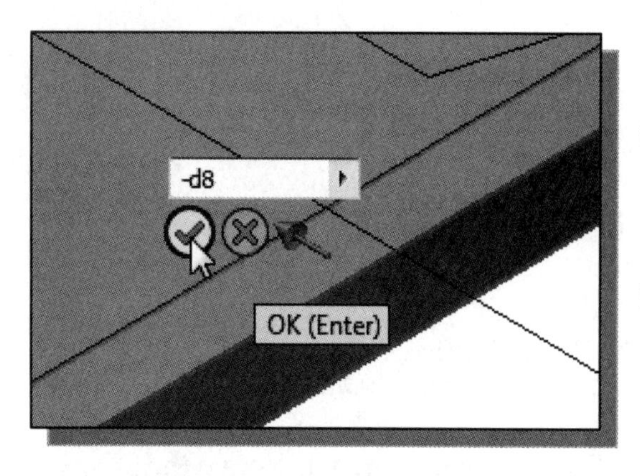

8. Switch the offset direction by entering a negative sign in front of the variable name as shown.

9. Click **OK** to create an offset plane that passes through the centers of the two small holes.

Create a Revolved Feature

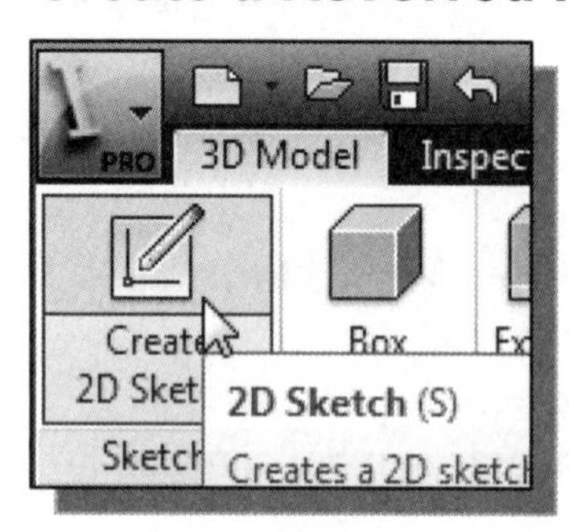

1. In the *Sketch* toolbar, select the **Create 2D Sketch** command by left-clicking once on the icon.

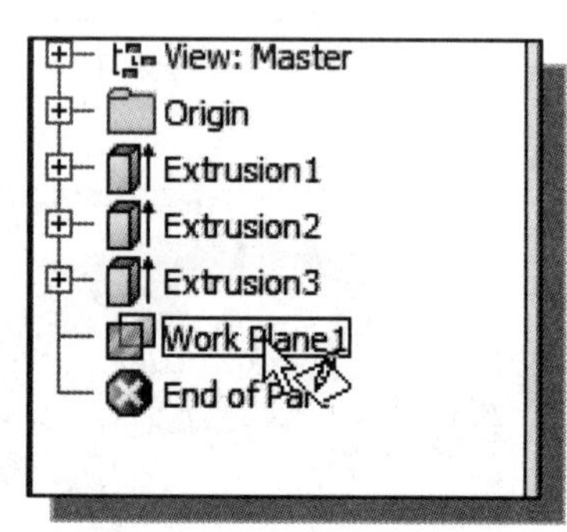

2. In the *browser* area, select the new Work Plane as the sketching plane.

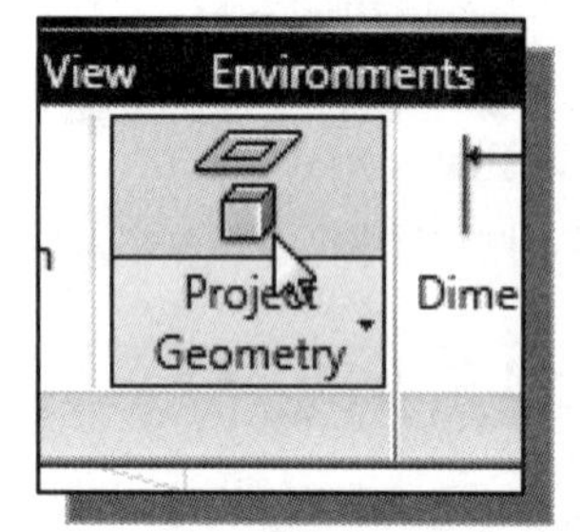

3. Select the **Project Geometry** command in the *Draw* panel.

4. Select one edge of the hole and the outside edge of the rounded corner as shown.

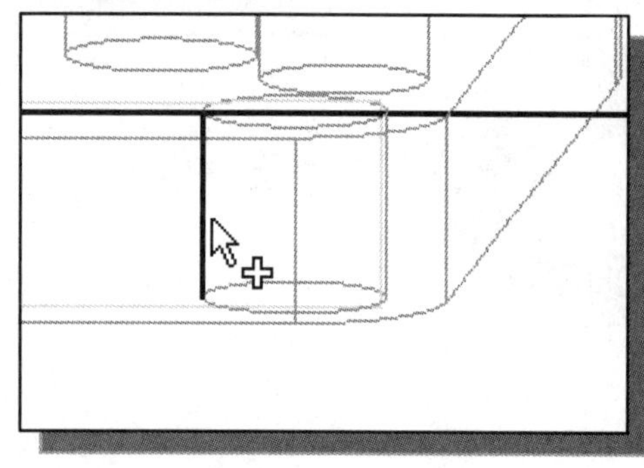

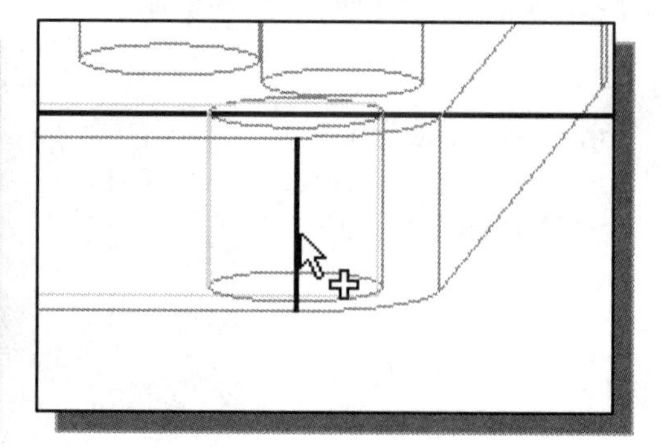

5. Inside the graphics window, right-click once to bring up the option menu and select **Done [Esc]** to end the command.

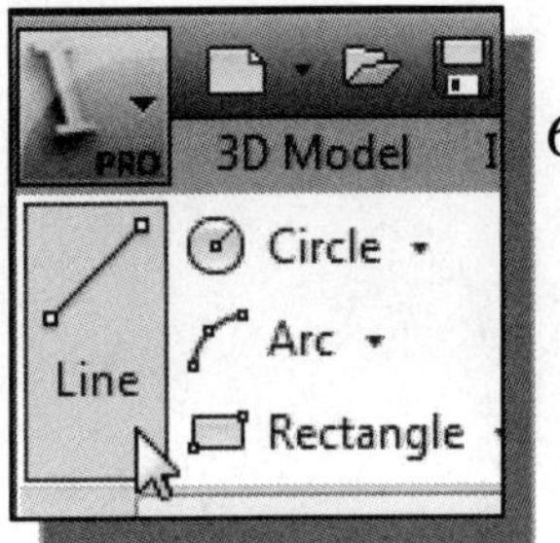

6. Select the **Line** command by clicking once with the left-mouse-button on the icon in the *Draw* toolbar.

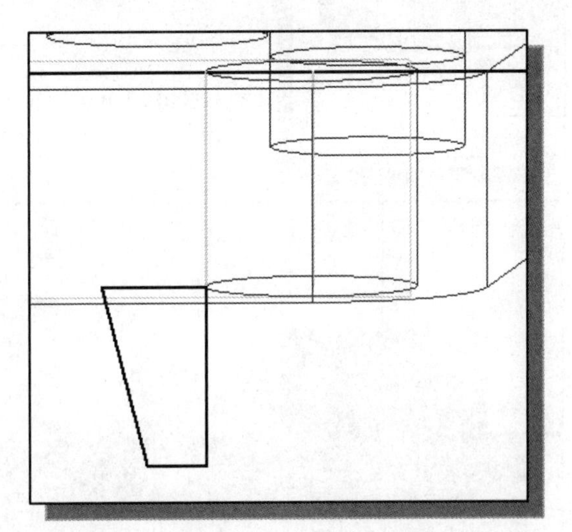

7. Create the closed region sketch, with the upper right corner aligned to the projected edge, as shown.

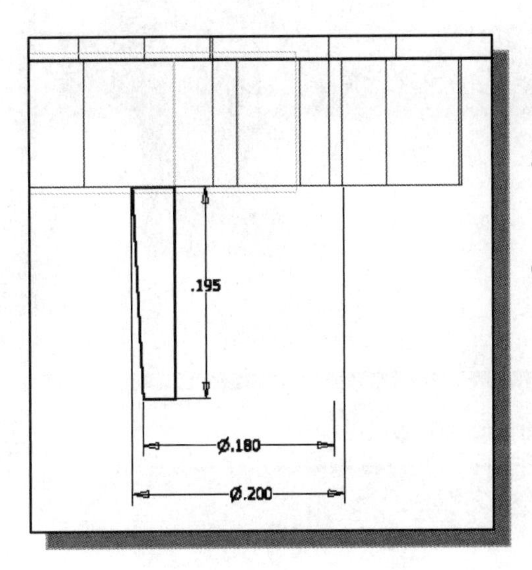

8. On your own, create the three dimensions as shown in the figure.

9. In the *Create* toolbar, select the **Revolve** command by left-mouse-clicking once on the icon.

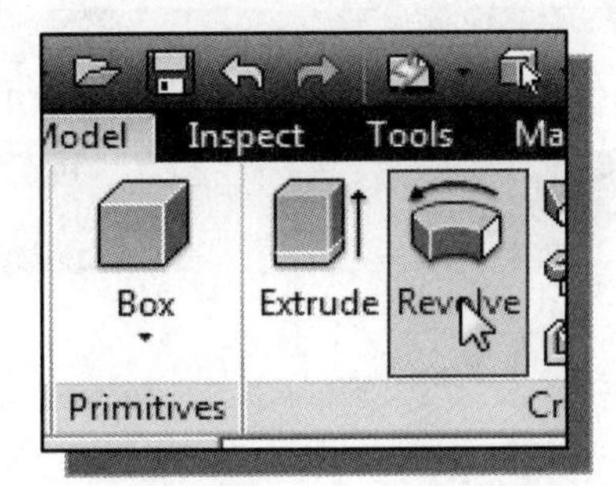

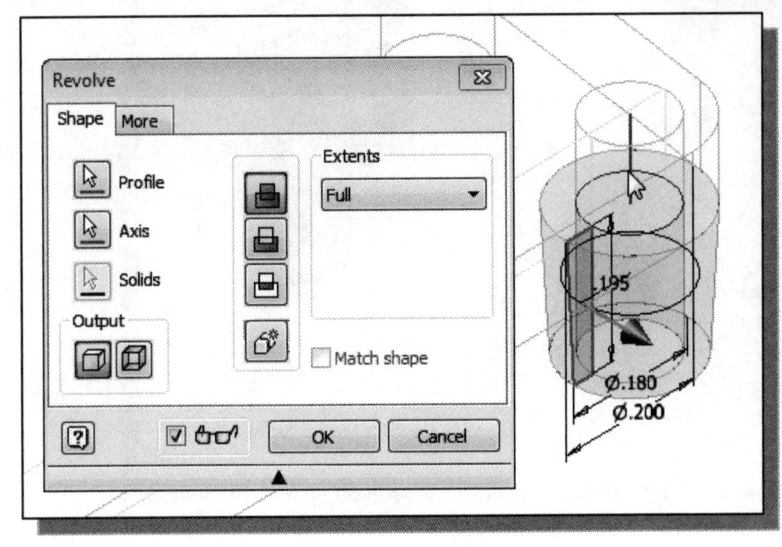

10. Select the center axis of the cylinder as the axis of rotation for the revolved feature

11. Confirm the revolve angle option is set to **Full** as shown.

12. Click **OK** to accept the settings and create the revolved feature.

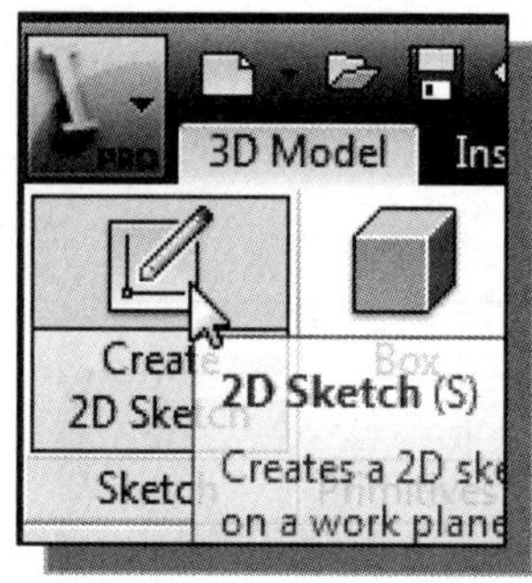

13. In the *Sketch* toolbar select the **Create 2D Sketch** command by left-clicking once on the icon.

14. Create the closed region sketch, with the upper left corner aligned to the projected edge, as shown.

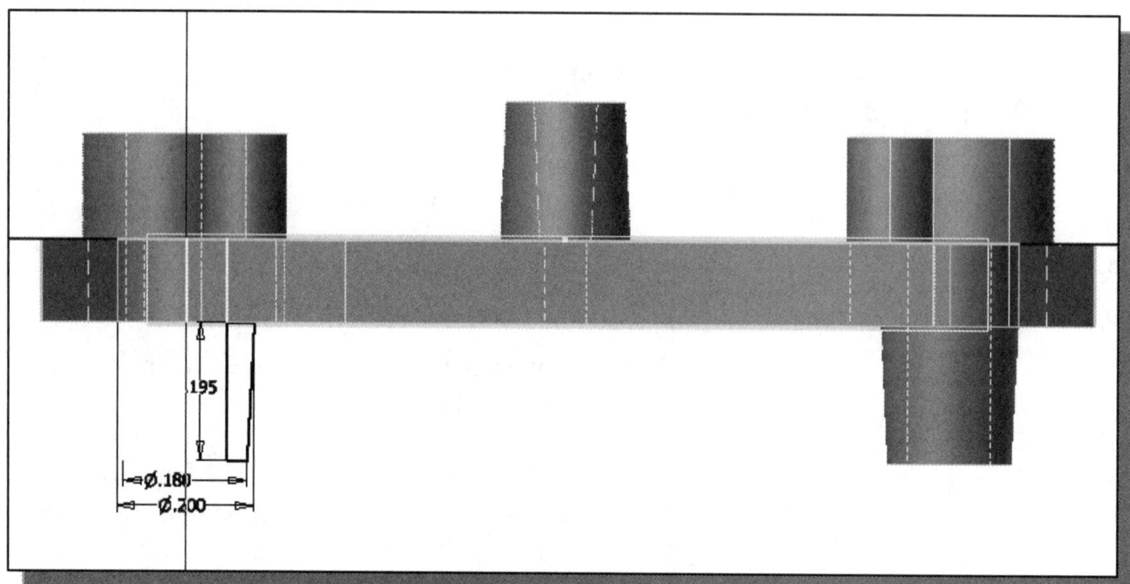

15. On your own, create another revolved feature as shown.

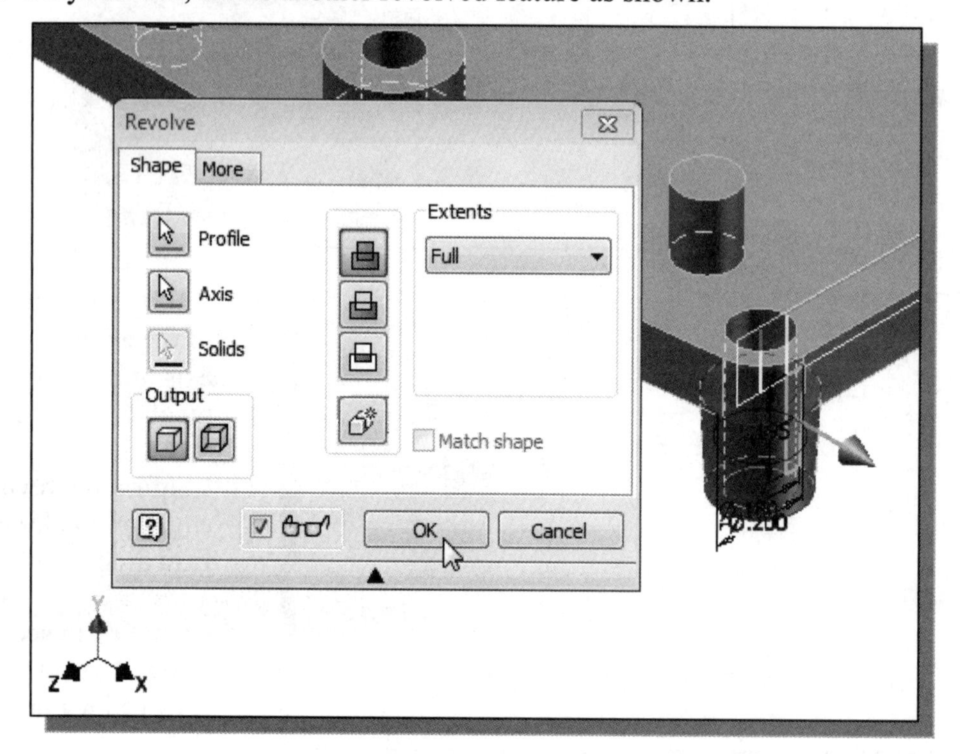

Create an Angled Work Plane

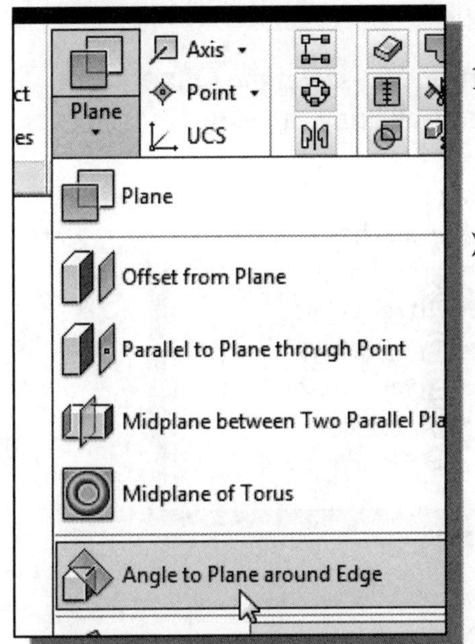

1. In the Work Plane option list, select the **Angle to Plane around Edge** command by left-clicking the icon.

➢ *Autodesk Inventor* expects us to select a line and a plane to be used as references to create the new work plane.

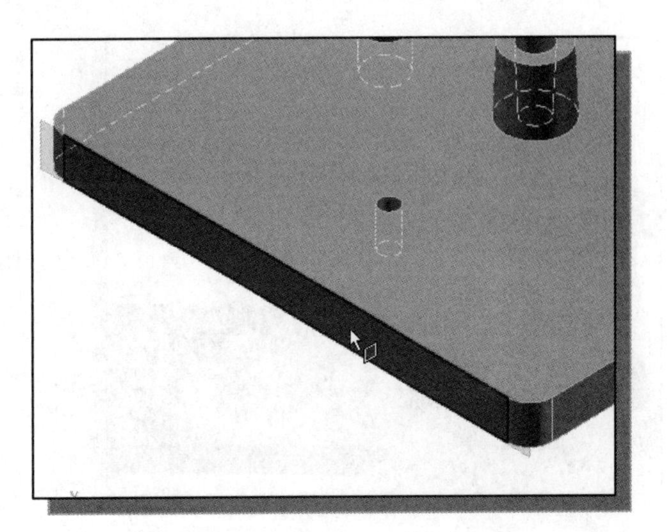

2. Select the **left vertical plane** as the first reference of the new work plane.

3. Select the **top edge** as the second reference of the new work plane.

4. In the *Angle* popup window, enter **25** as the rotation angle for the new work plane.

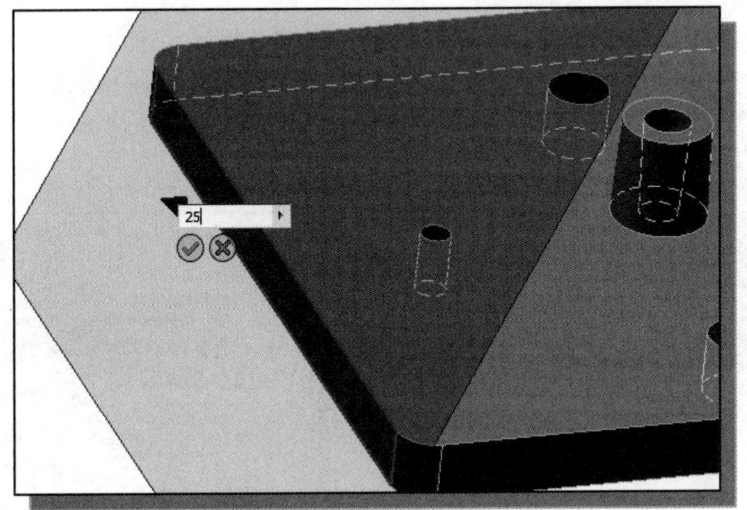

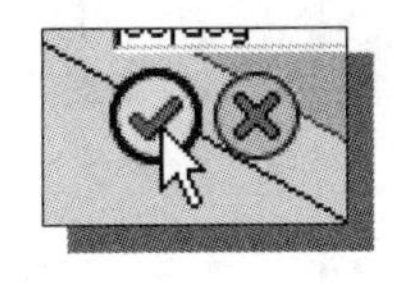

5. Click on the **check mark** button to accept the setting.

Create another Offset Work Plane

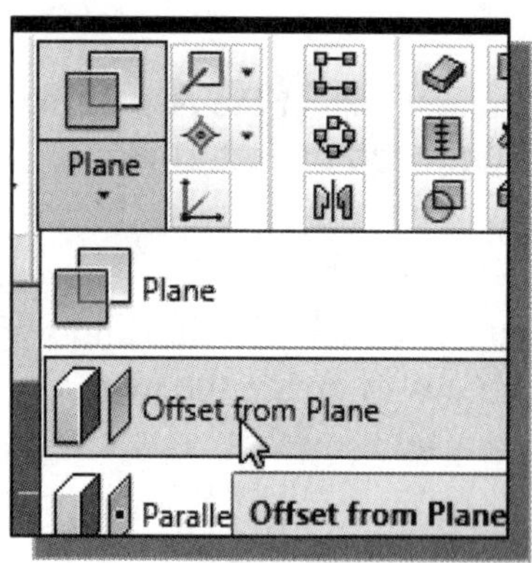

1. In the Work Plane option list, select the **Offset from Plane** command by left-clicking the icon.

2. Pick the new work plane by clicking the work plane name inside the *browser* as shown.

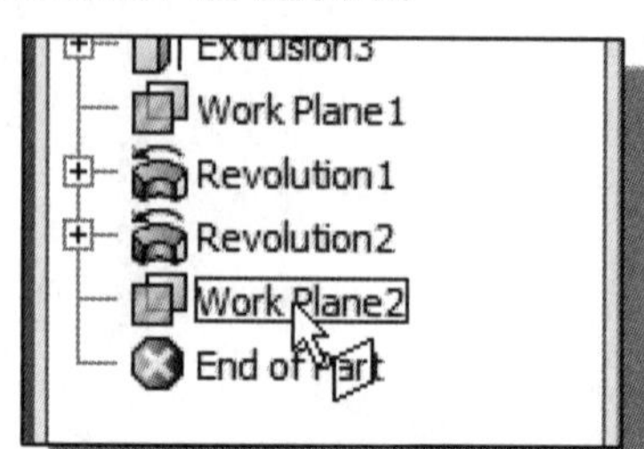

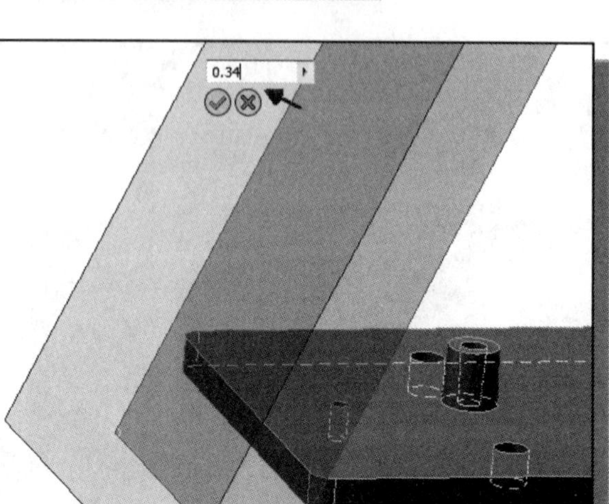

3. In the *Offset* edit box, enter **0.34** as the offset distance as shown.

4. Click on the check mark button to accept the setting.

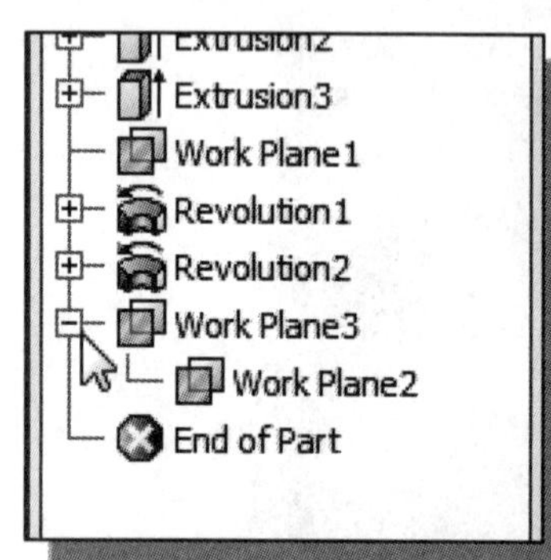

➤ Since Work Plane2 was used to define Work Plane3, the two items are grouped together in the model history tree.

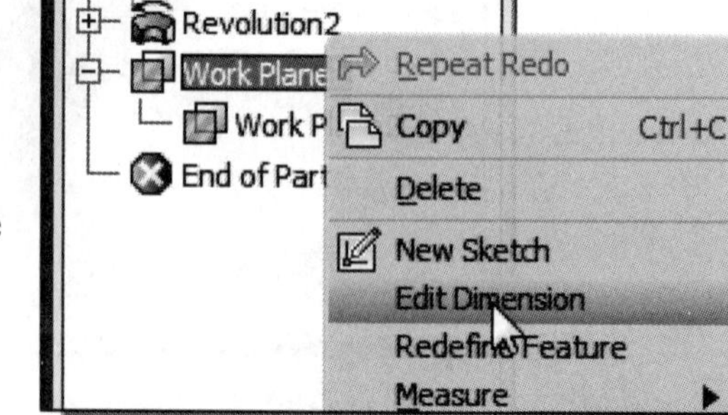

➤ Note that the offset distance can also be adjusted through the Model *Tree* window.

Create a 2D Sketch on Work Plane3

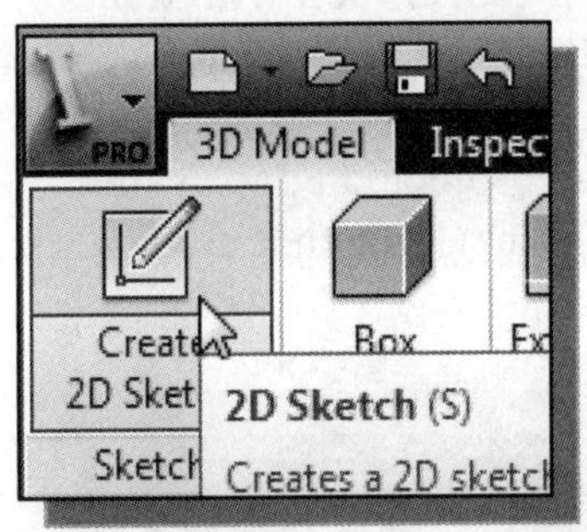

1. In the *Sketch* toolbar select the **Create 2D Sketch** command by left-clicking once on the icon.

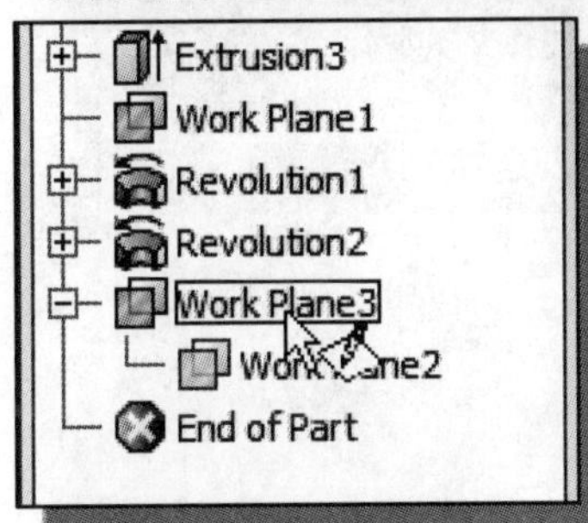

2. In the *Status Bar* area, the message: "*Select face, work plane, sketch or sketch geometry*" is displayed. Pick **Work Plane3** by clicking the work plane name inside the *browser* as shown below.

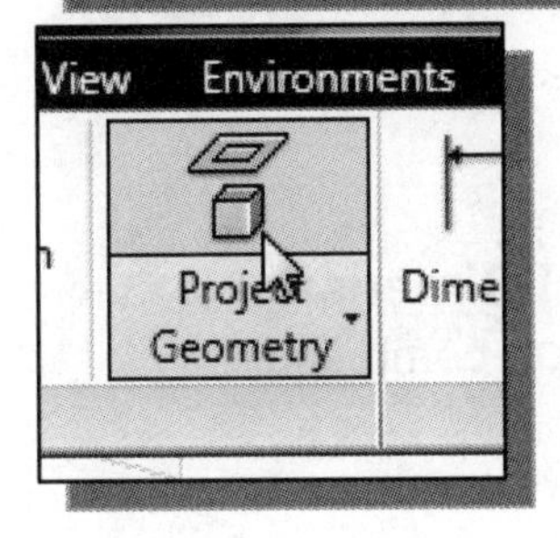

3. Select the **Project Geometry** command in the *Draw* panel.
4. Select the **XY Plane** and the **top edge** of the base feature as shown.

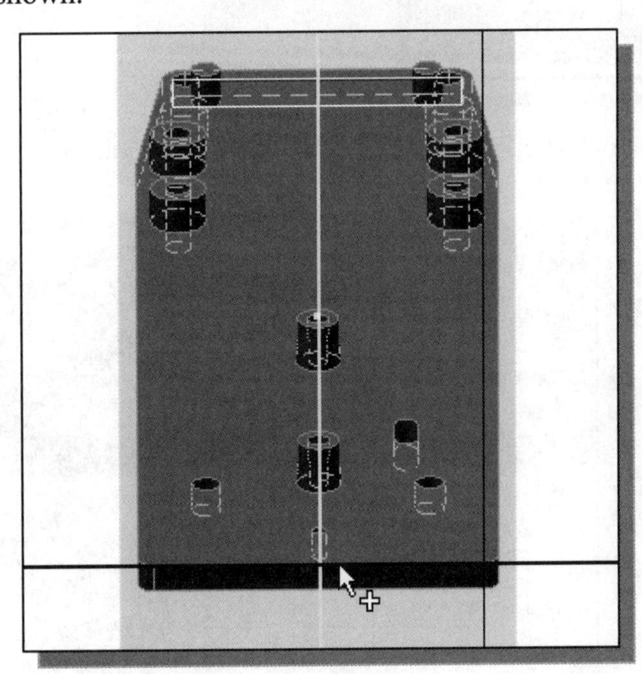

5. Inside the graphics window, right-click once to bring up the option menu and select **OK** to end the command

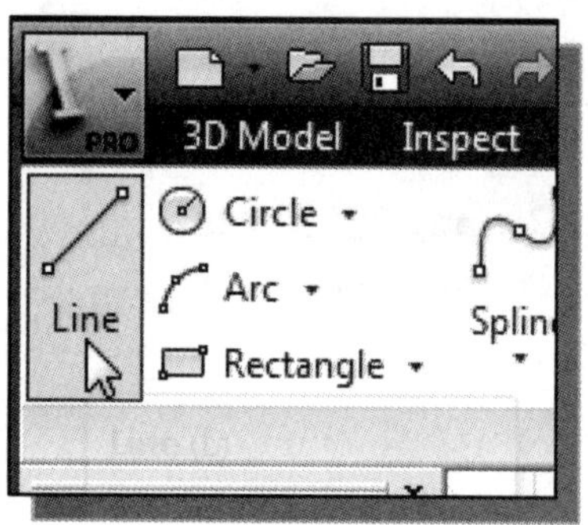

6. In the *Draw* toolbar, click on the **Line** icon with the left-mouse-button to activate the Line command.

7. Create a rough sketch using the projected edge as the bottom line as shown in the figure. (Note that all edges are either horizontal or vertical.)

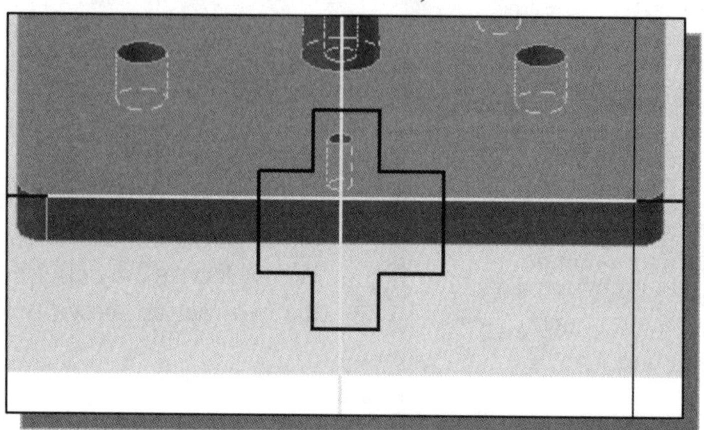

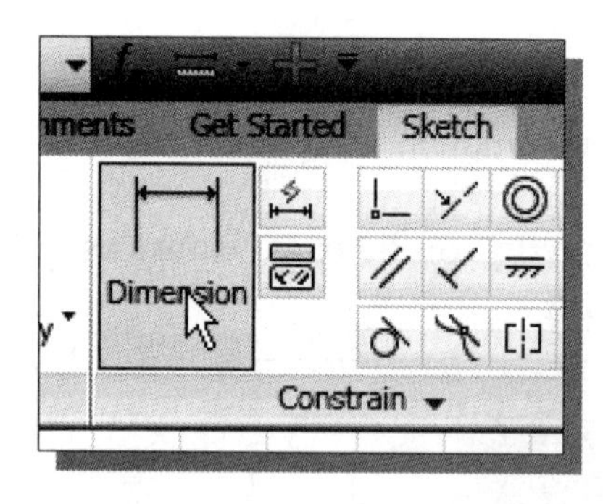

8. Left-click once on the **General Dimension** icon to activate the General Dimension command.

9. On your own, create and modify the dimensions as shown; note that the sketch is symmetrical vertically and horizontally.

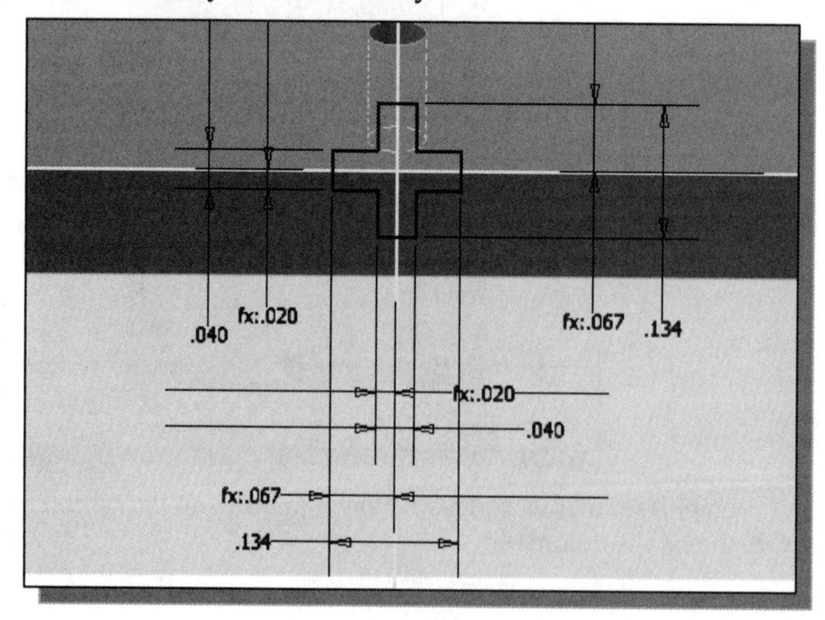

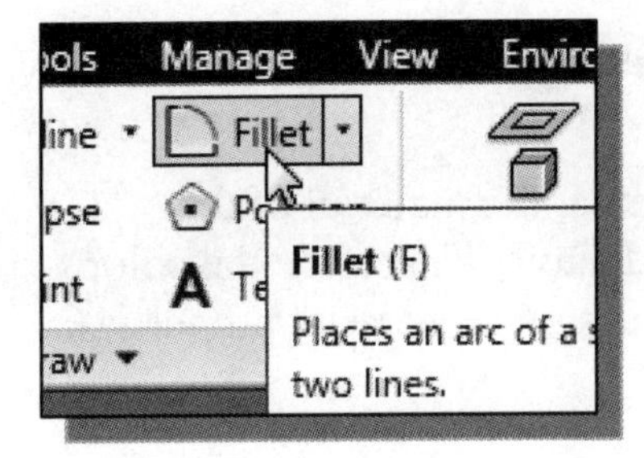

10. Select the **Fillet** option in the *Draw* panel as shown.

11. On your own, add the rounded corners, radius **0.008** to the sketch as shown.

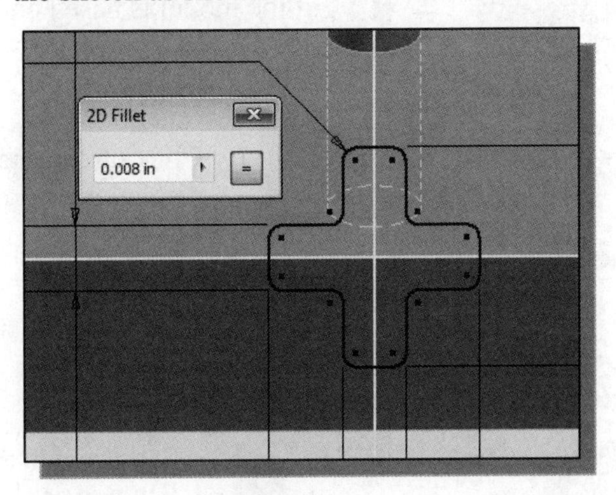

12. Select **Finish Sketch** in the *Ribbon* toolbar to exit the **Sketch** mode.

Complete the Solid Feature

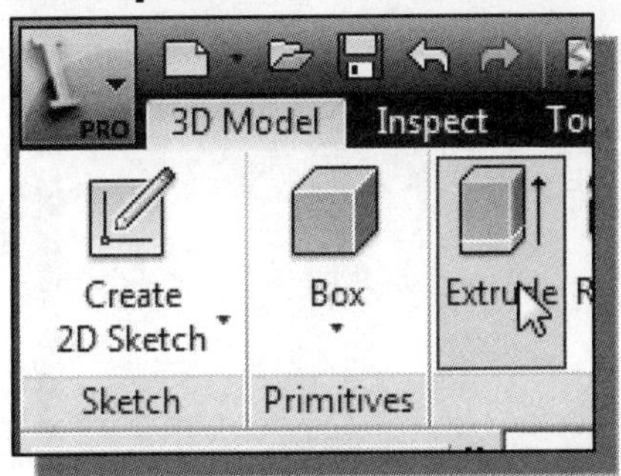

1. In the *Create* toolbar, select the **Extrude** command by releasing the left-mouse-button on the icon.

2. Select the inside region of the 2D sketch to create a profile as shown.

3. Set the extrude extents to **To Next** and create the feature as shown.

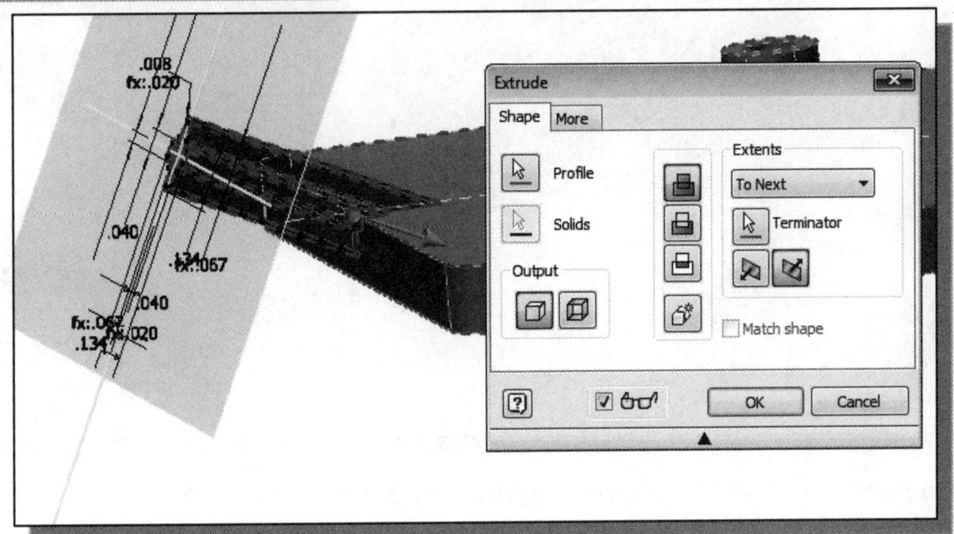

Quick Change of the Appearance of the Solid Model

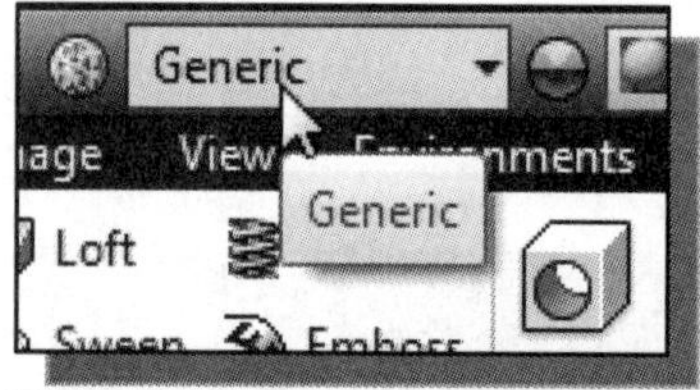

❖ In the *Quick Access* toolbar area, the material of the model is set to **Generic** by default. Therefore the color of the model is based on the assigned material properties.

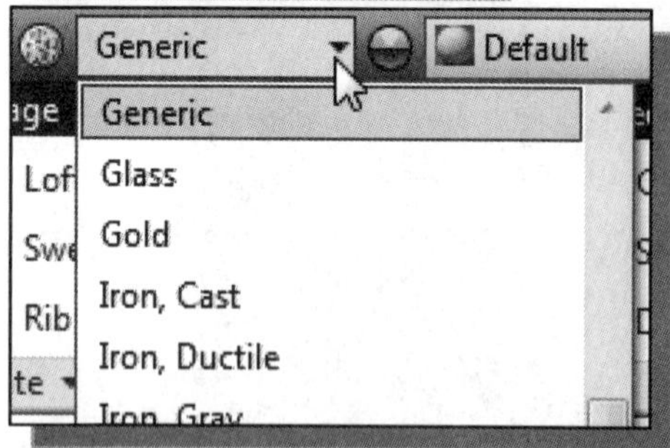

1. Click on the triangle next to the **Material** box to display the list of material library available in *Inventor*.

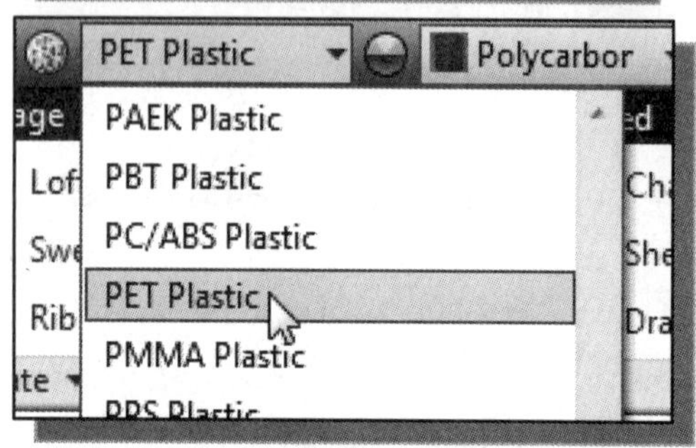

2. Scroll down the color list and choose **PET Plastic** as shown.

➢ The **PET Plastic** material adjusts the model to the clear transparent look as shown.

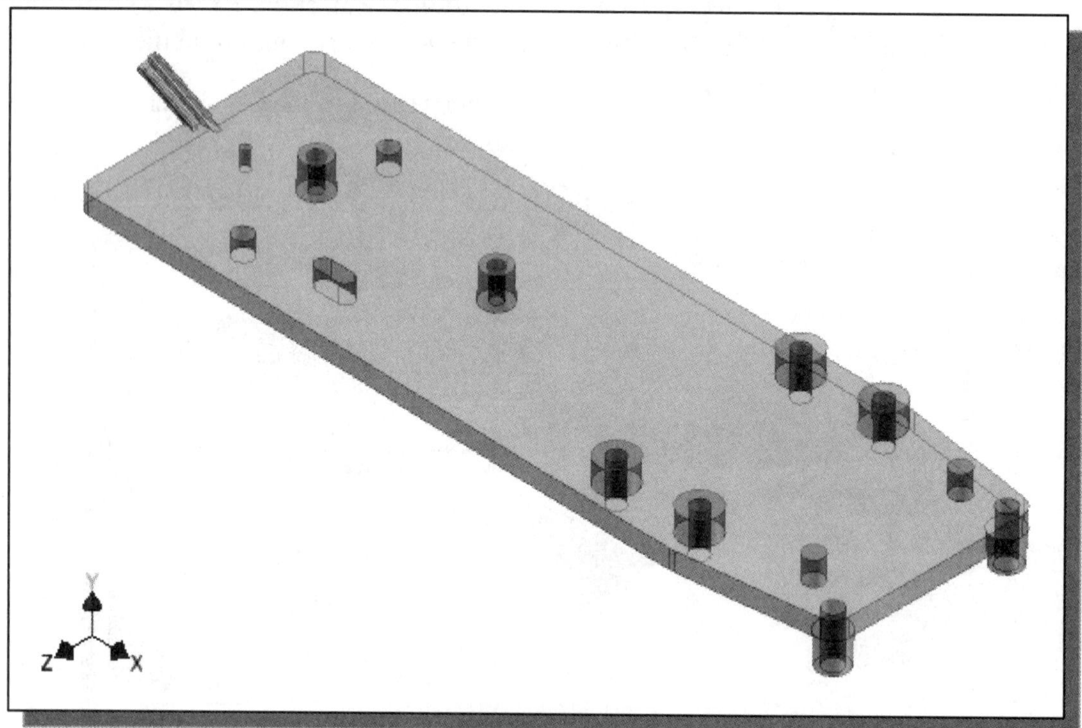

3. On your own, save the model using filename **B2-Chassis.ipt**.

The *Crank Right* Part

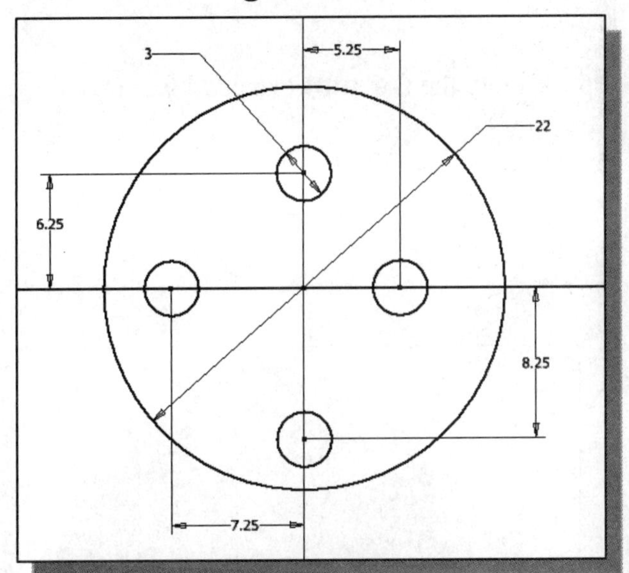

1. Start a new **Metric (mm)** model.

2. Create the base extrude feature, **2mm**, starting on one of the datum planes.

3. Create five diameter **5.0 mm** circles, coinciding with the five existing center points, and extrude **8 mm** using the **Symmetric** option.

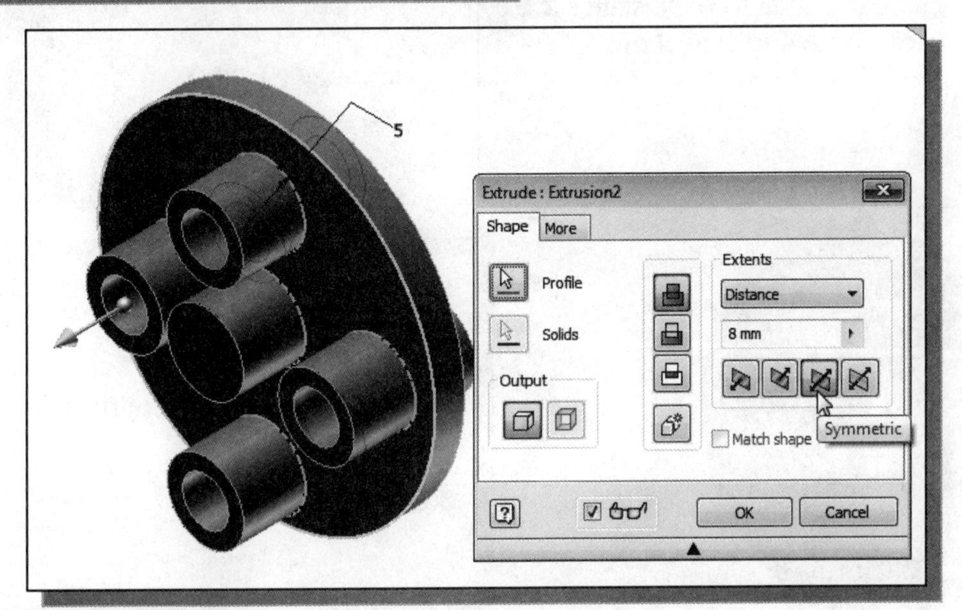

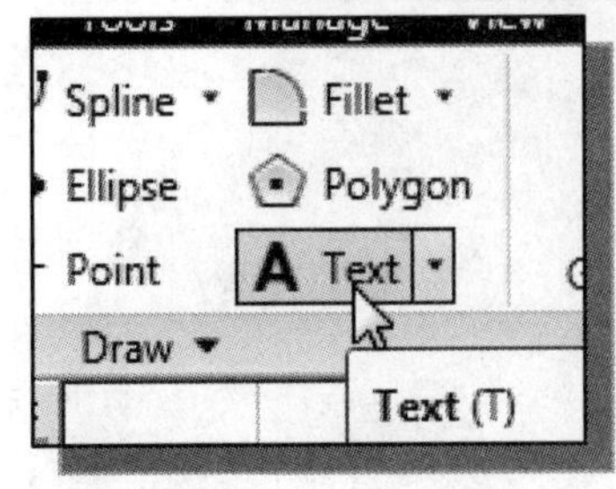

4. On the center cylinder, create a new sketch and use the **Text** command and create the letter **R**.

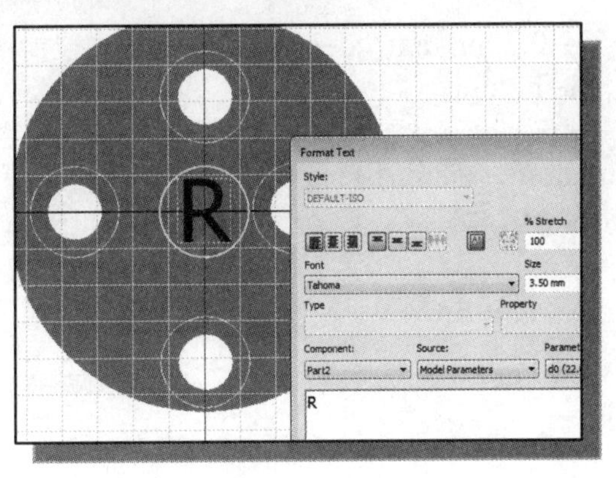

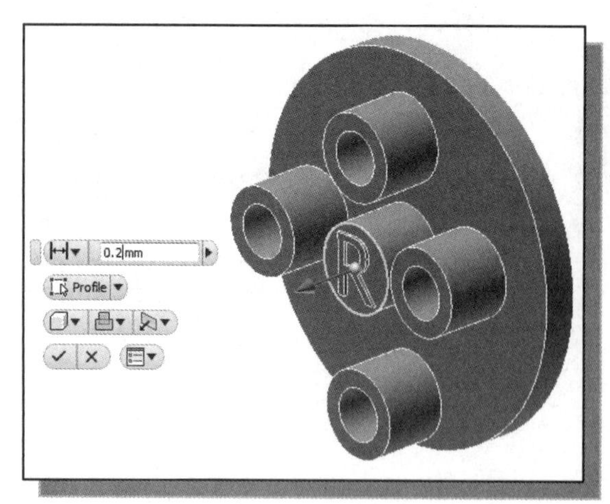

5. Create the **0.2 mm** extruded feature as shown.

6. On the back side, create a hexagon cut feature, flat to flat distance **2.5 mm**, depth **4 mm** as shown.

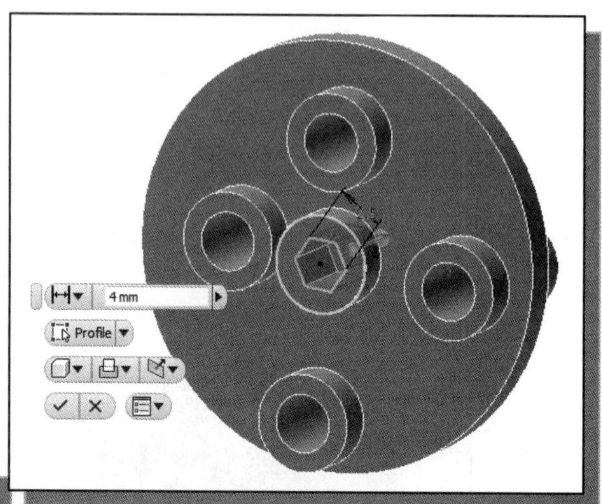

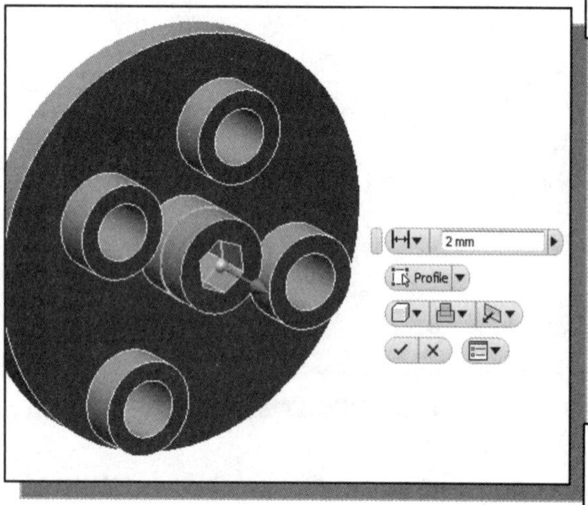

7. Extend the center cylindrical feature by **2 mm**.

8. Create the 0.2 mm extruded labels next to the holes.

9. Save the model using filename **A9-Crank-Right.ipt**.

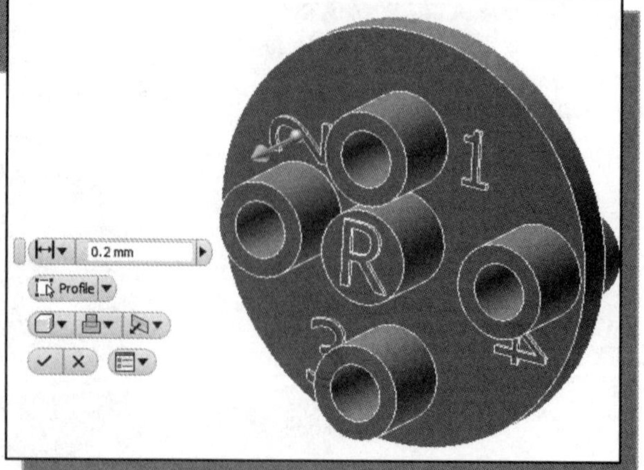

The *A10-Crank Left* Part

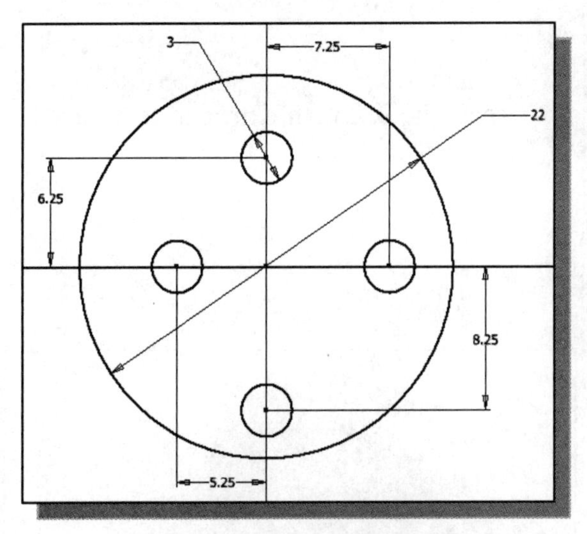

1. Start a new **Metric (mm)** model.

2. Create the base extruded **2 mm** feature, starting on one of the datum planes.

3. Create five diameter **5.0 mm** circles, coinciding with the five existing center points, and extrude **8 mm** using the **Symmetric** option.

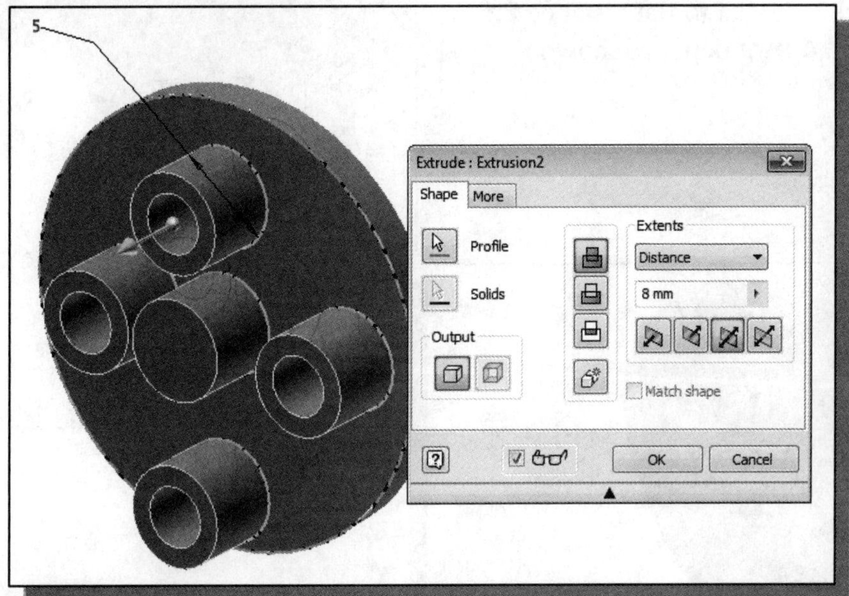

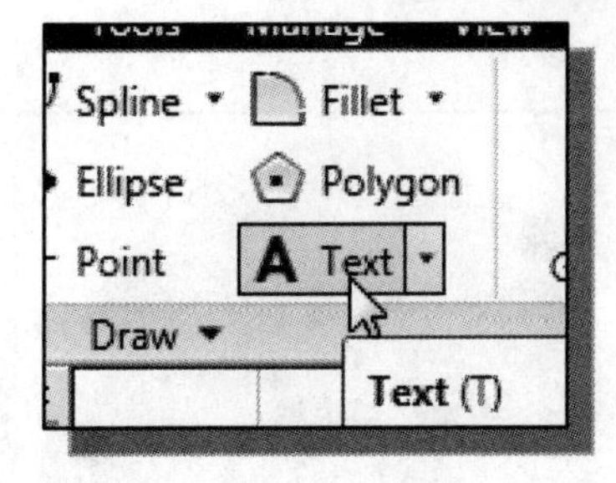

4. On the center cylinder, create a new sketch and use the **Text** command to create the letter **L**.

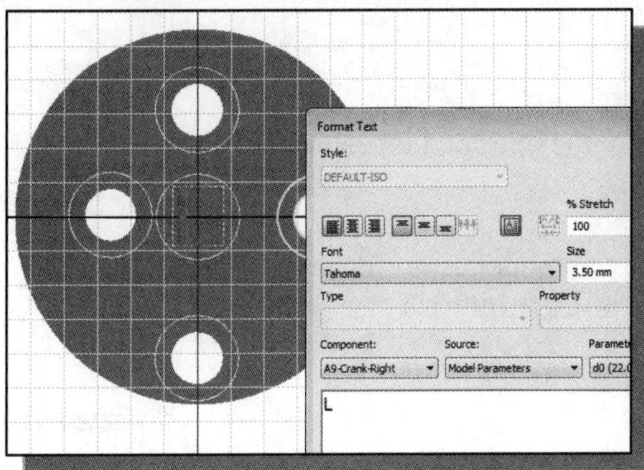

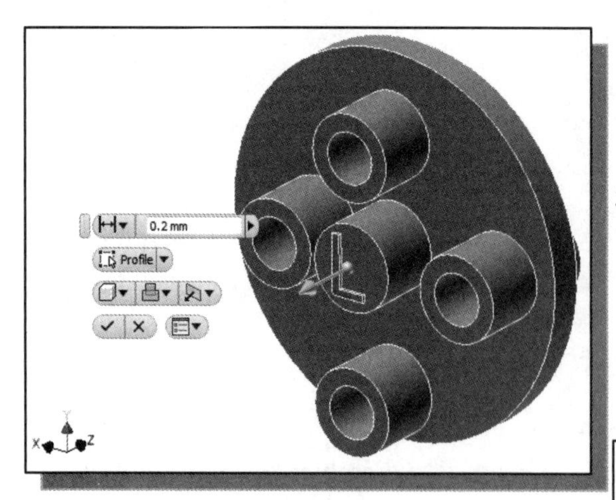

5. Create the **0.2 mm** extruded feature as shown.

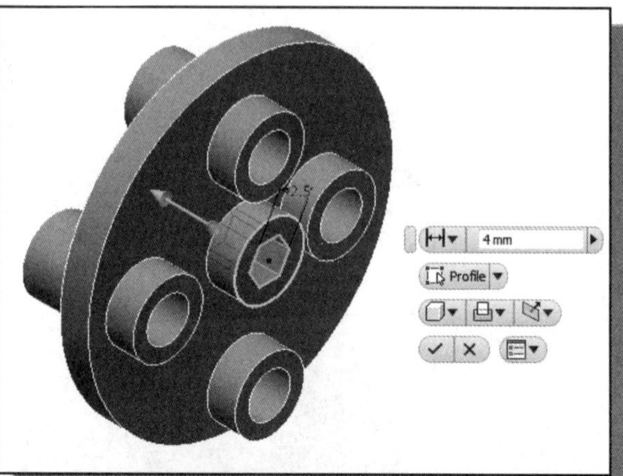

6. Create a hexagon cut feature on the back side, flat to flat distance **2.5 mm**, **4 mm** depth as shown.

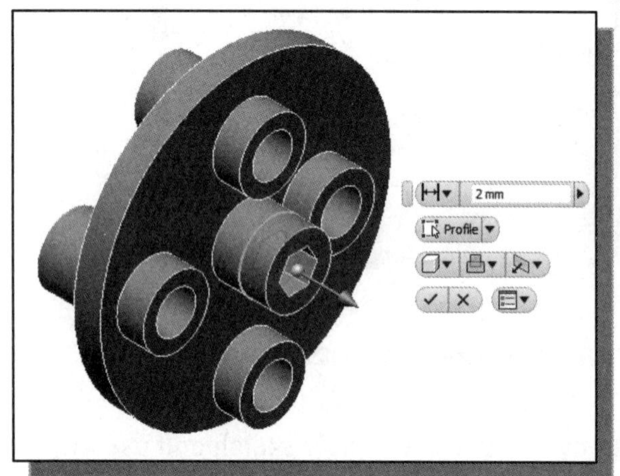

7. Extend the center cylindrical feature by **2 mm**.

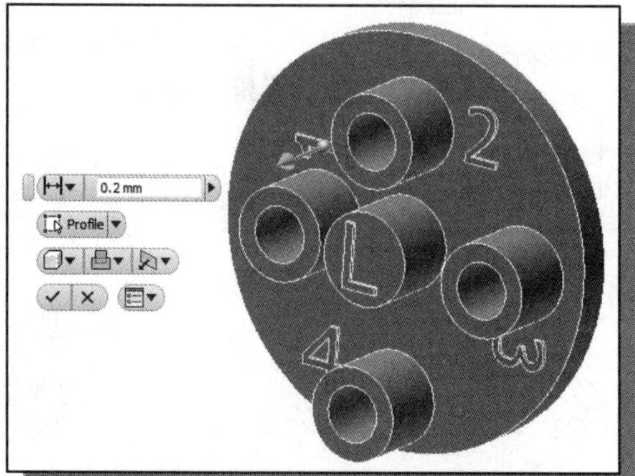

8. Create the extruded labels next to the holes.

9. Save the model using filename **A10-Crank-Left.ipt**.

The *Motor*

1. Start a new **Metric (mm)** model.

2. Create the following 2D sketch on one of the datum planes.

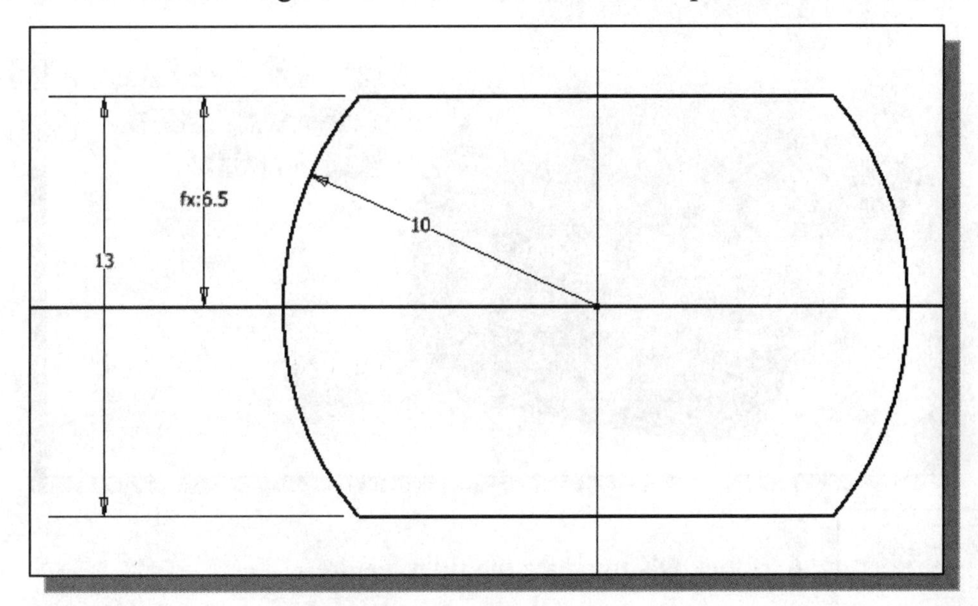

3. Create the base extruded feature, depth **25 mm**, as shown.

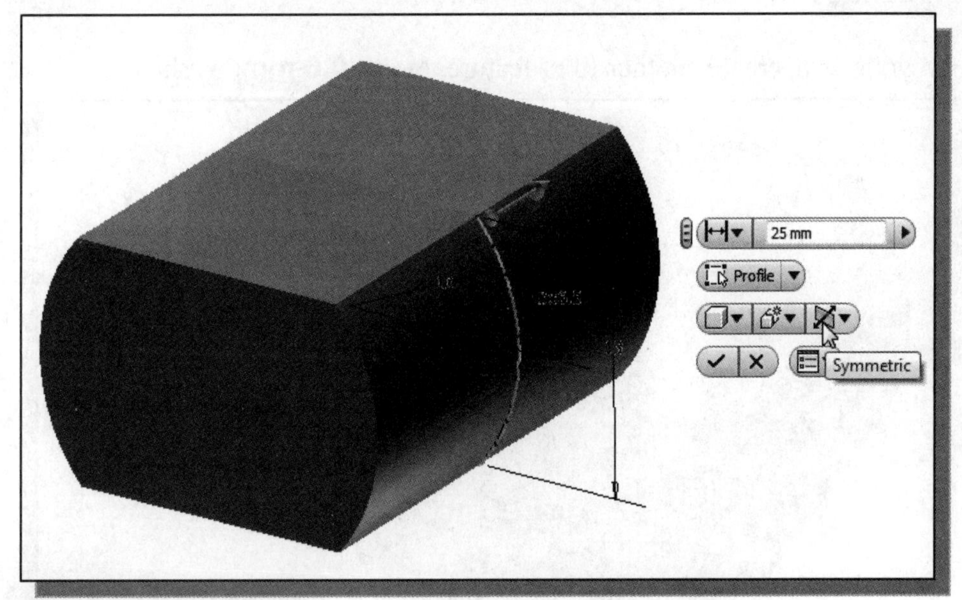

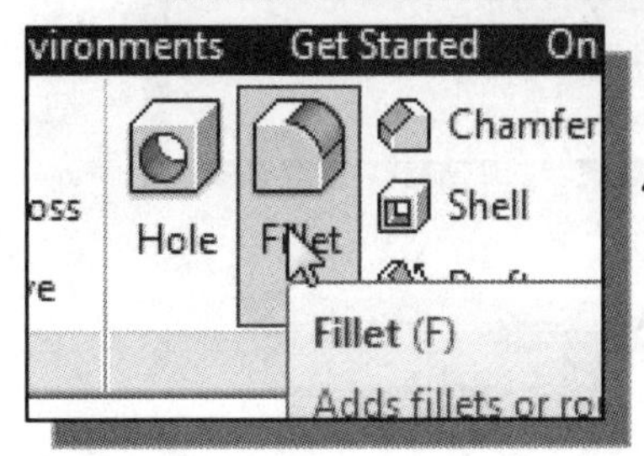

4. In the *Modify* toolbar, select the **Fillet** command by clicking the left-mouse-button on the icon.

5. Enter **2.5 mm** as the radius value, and select the **four arcs** to set the fillets as shown.

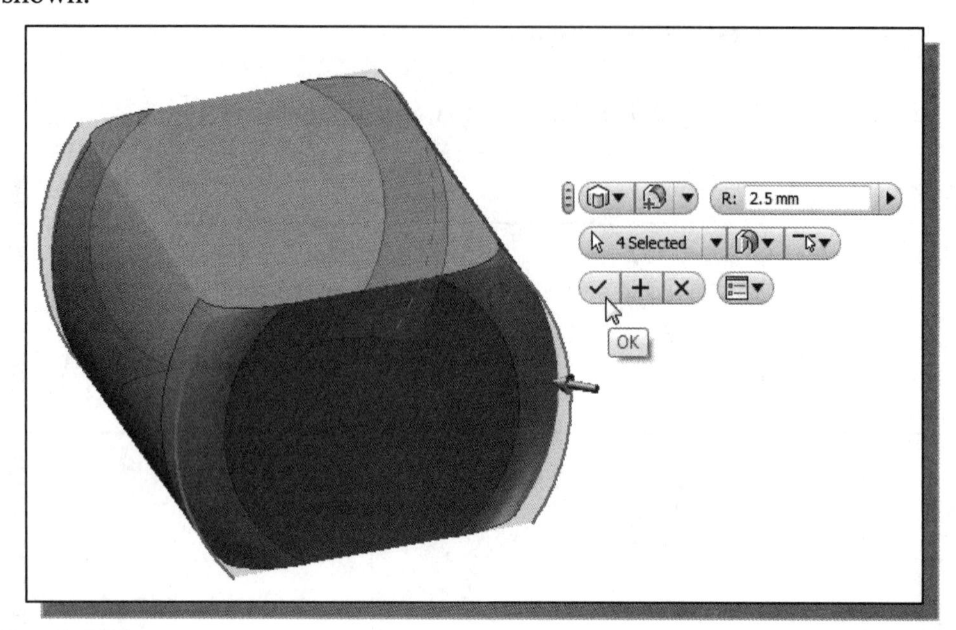

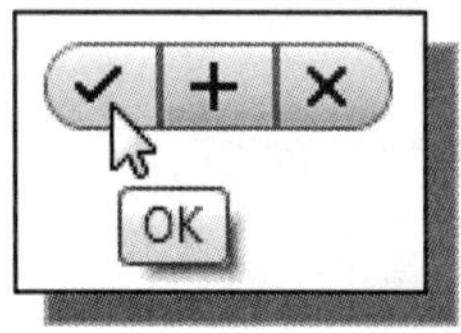

6. Click **OK** to create the fillet feature.

7. On your own, create another fillet feature, radius **0.6 mm**, as shown.

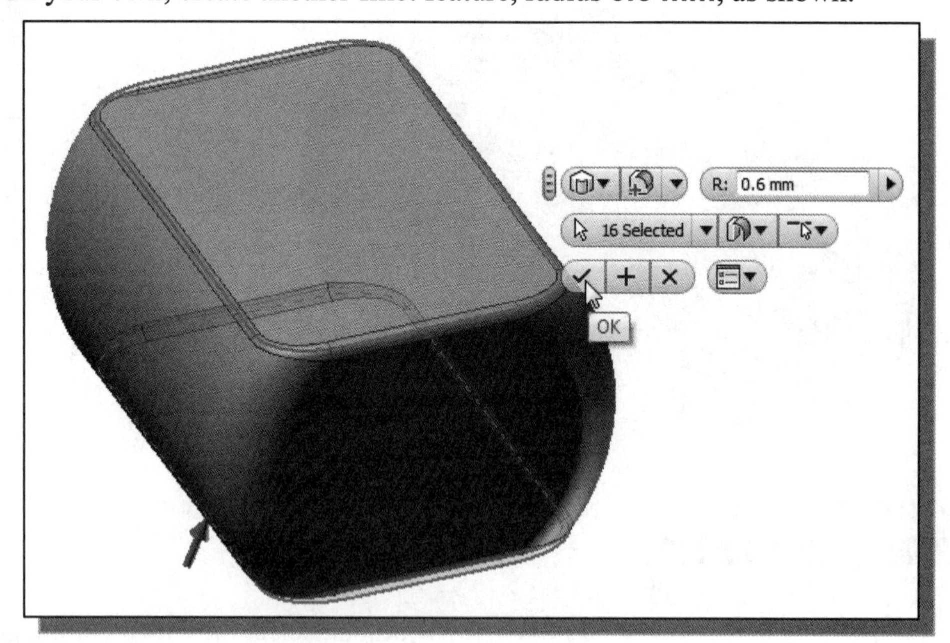

8. Click **OK** to create the fillet feature.

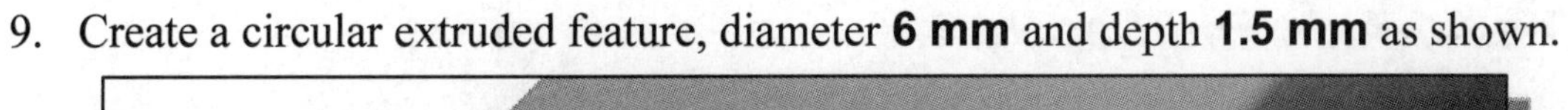

9. Create a circular extruded feature, diameter **6 mm** and depth **1.5 mm** as shown.

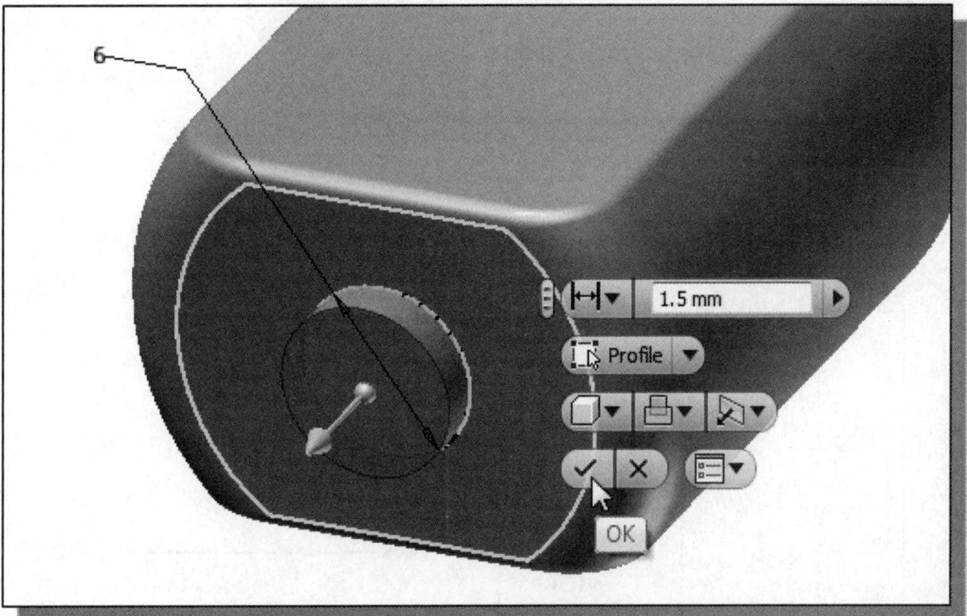

10. On your own, create the shaft portion of the **Motor**; the diameter is **2 mm** and the depth is **7.5 mm** as shown.

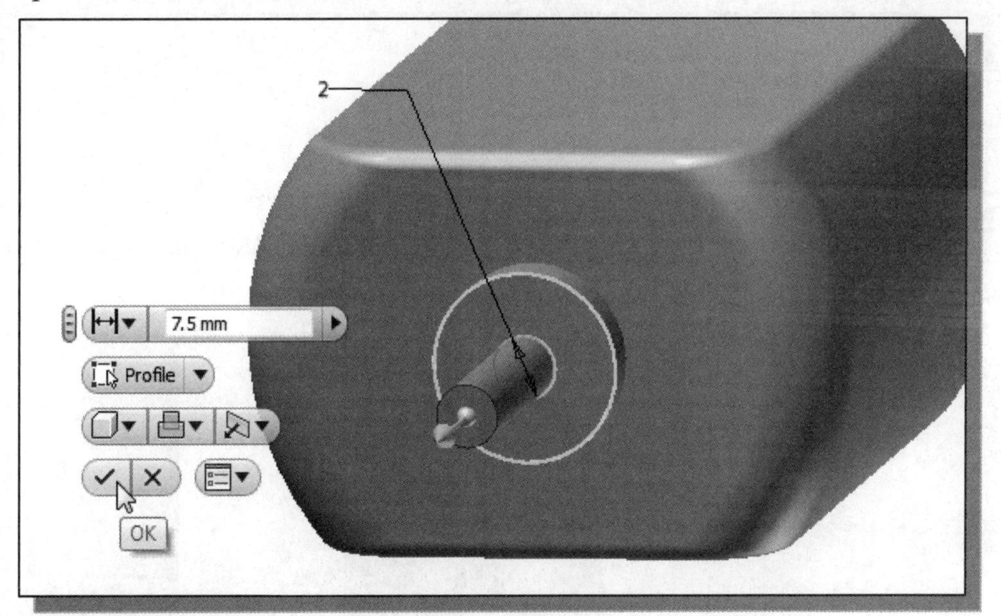

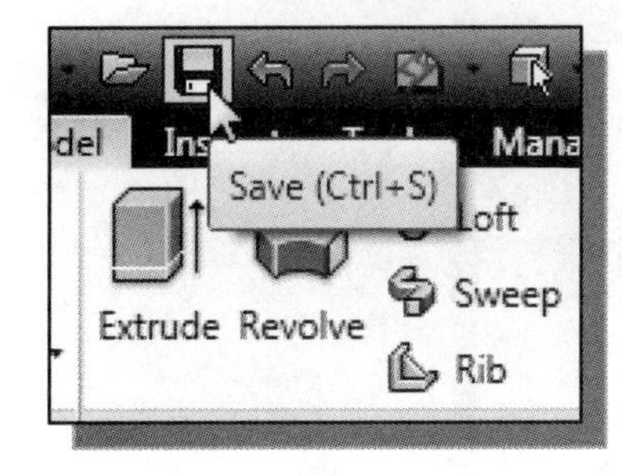

11. In the *Quick Access* menu, select *Save* to save the current model to disk.

12. On your own, save the model using filename **Motor.ipt.**

The *A1-Axle End Cap* Part

1. Start a new **Metric (mm)** model.

2. Create a **revolved** feature, starting on one of the datum planes as shown.

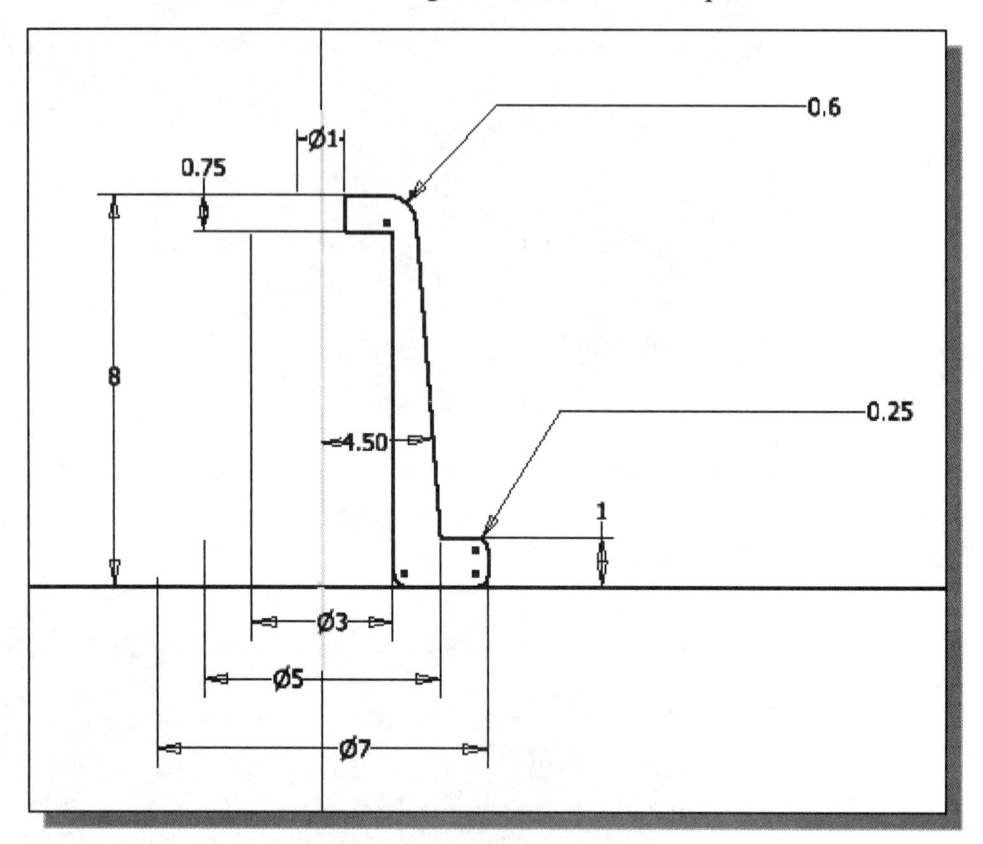

3. Save the model using filename **A1-Axle-EndCap.ipt**.

The *Hex Shaft Collar* Part

1. Start a new **Metric (mm)** model.

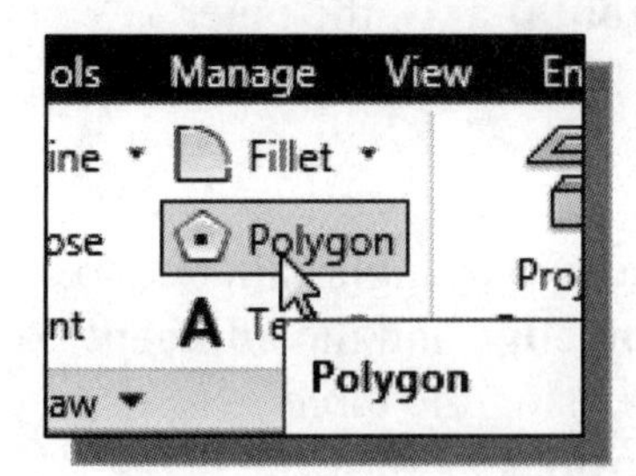

2. Create a new **2D Sketch** on one of the datum planes.

3. Use the **Polygon** command and create two six-sided polygons, with the centers aligned to the origin, as shown.

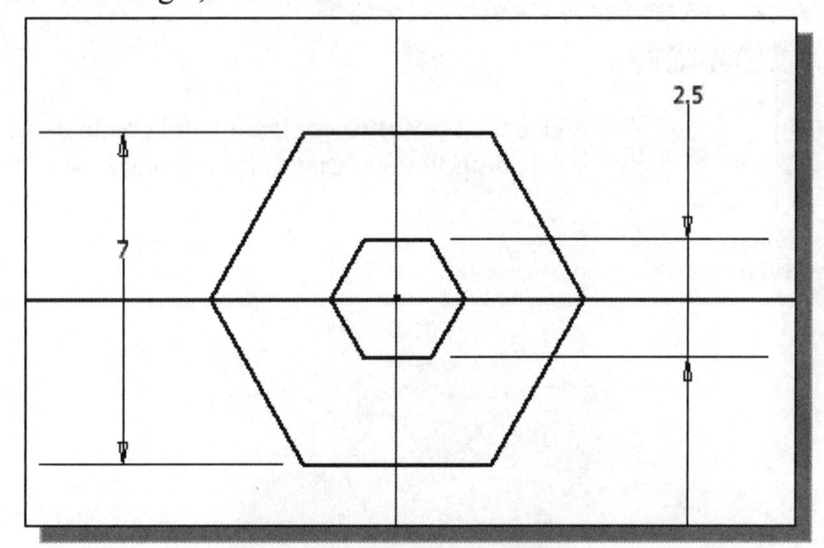

4. On your own, extrude **8 mm** using the **Symmetric** option.

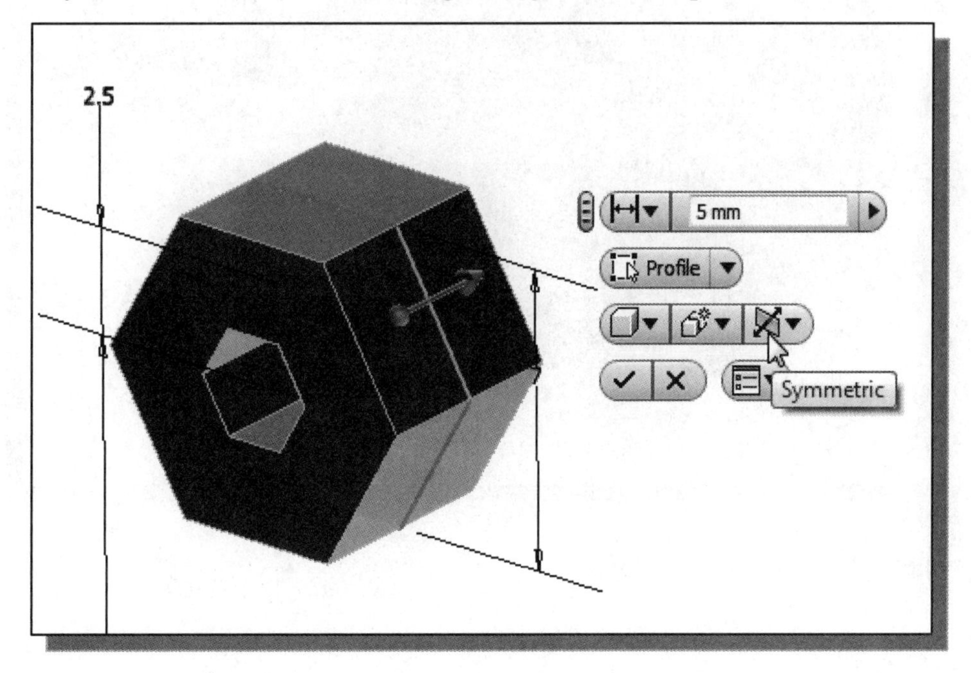

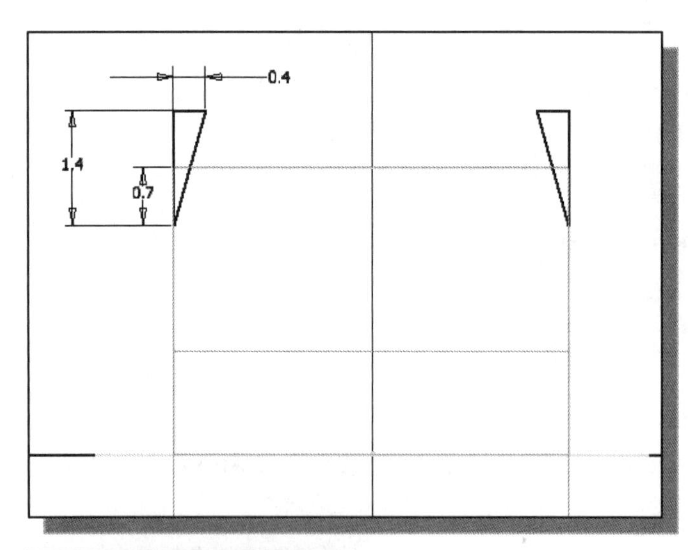

5. Use the **Project Geometry** command to **project the horizontal axis**, the outer edges of the solid model.

6. Create two triangles aligned to the **top edge** and the **adjacent edges** using the three dimensions as shown.

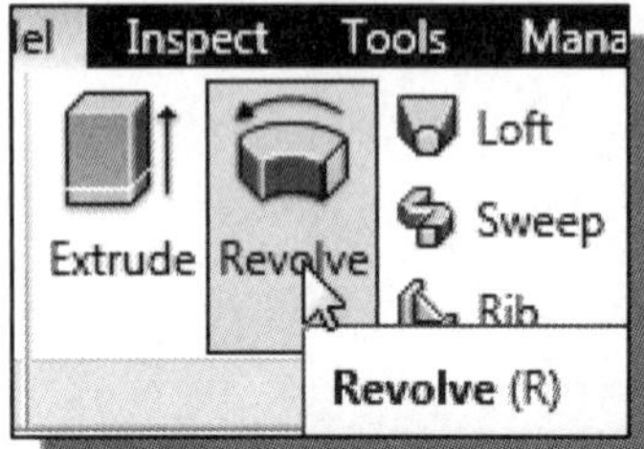

7. Use the **Revolve** command and create a cut feature by selecting the two triangular regions as shown.

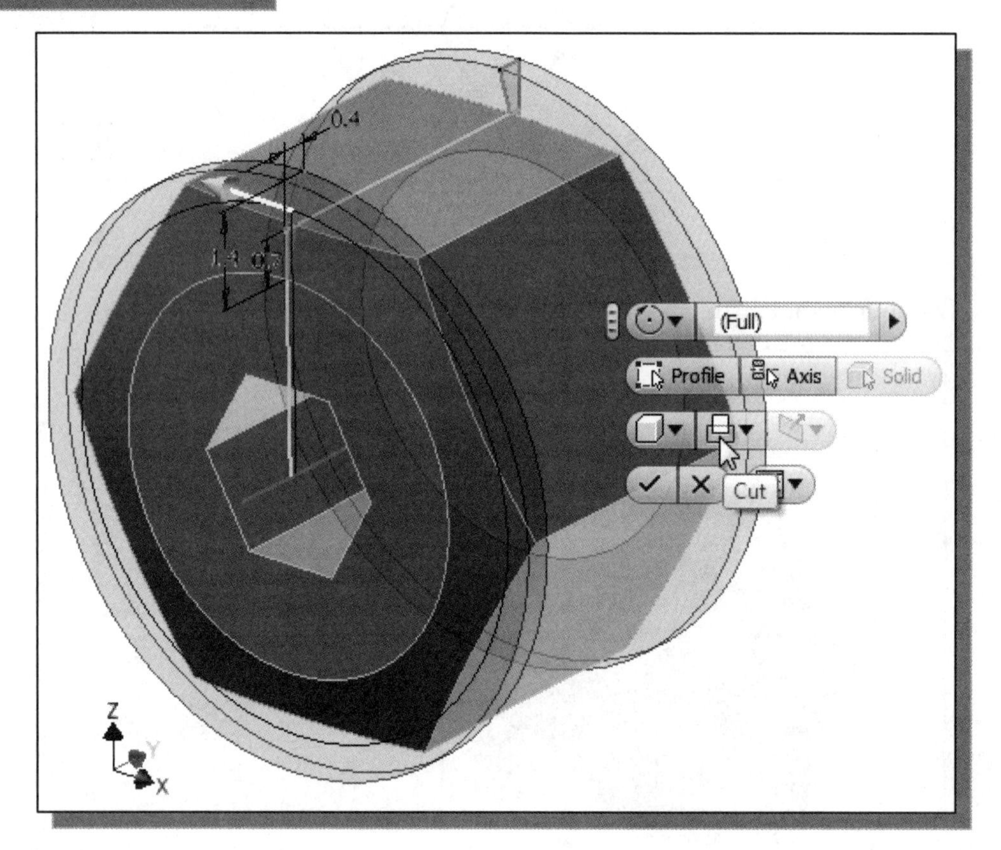

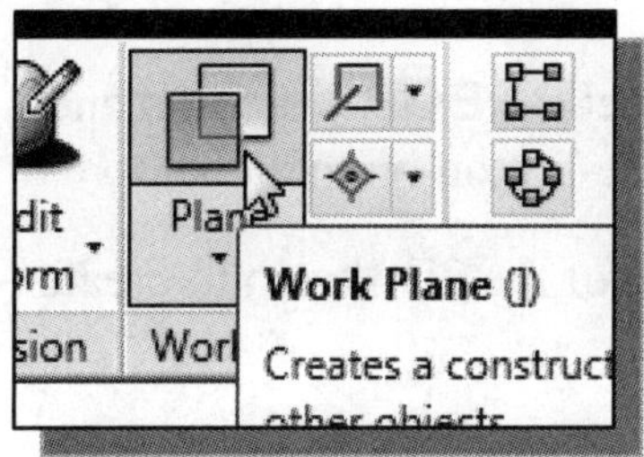

8. Use the **Work Plane** command and create an offset work plane feature that is **17.86 mm** away from the front face of the base feature.

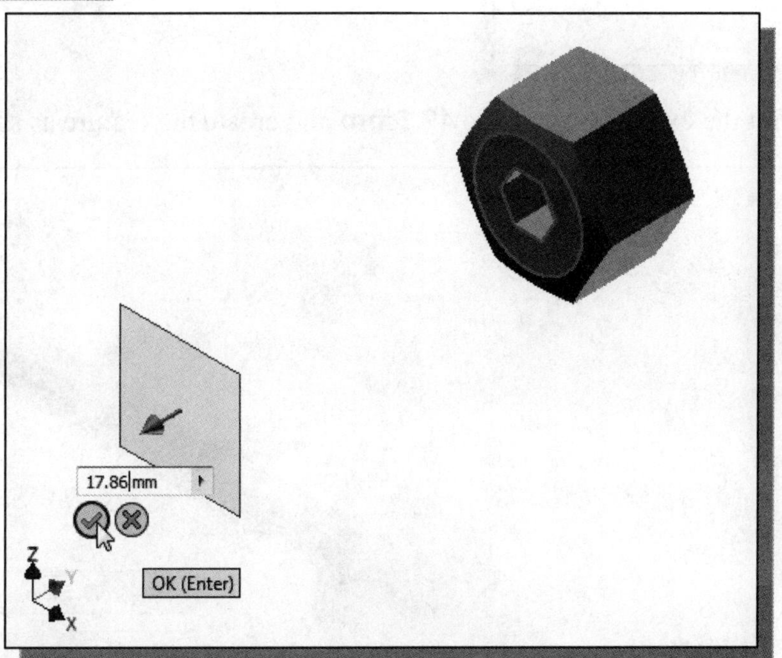

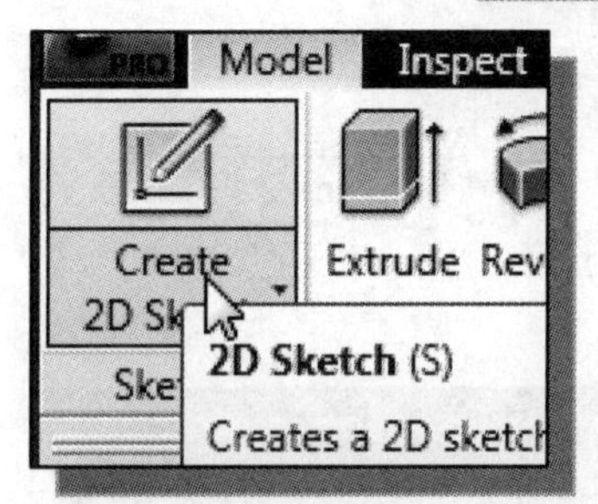

9. Create a new 2D sketch aligned to the new work plane.

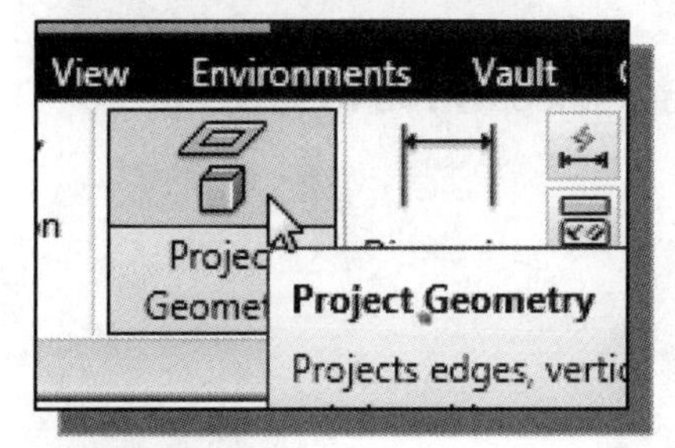

10. Use the **Project Geometry** command and **project the six edges of the inside hexagon** onto the sketch plane.

11. Select **Finish Sketch** in the *Ribbon* toolbar to exit the **Sketch** mode.

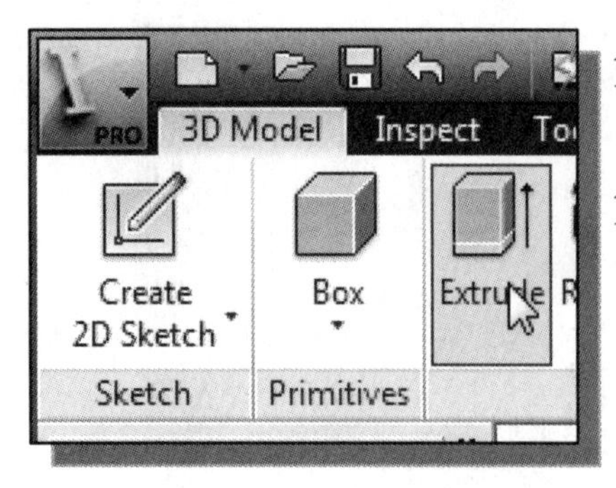

12. In the *Create* toolbar, select the **Extrude** command by releasing the left-mouse-button on the icon.

13. Select the hexagon region of the 2D sketch to create a profile as shown.

14. Set the extrude distance to **49.5mm** and create the feature as shown.

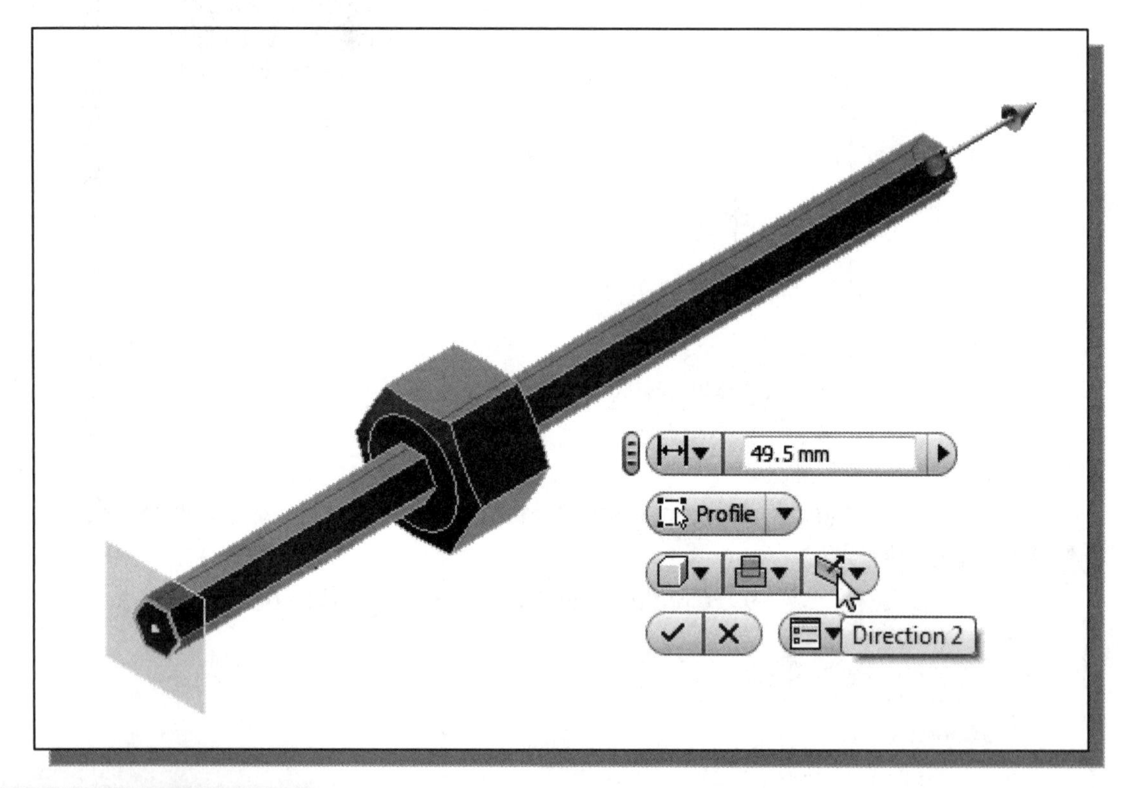

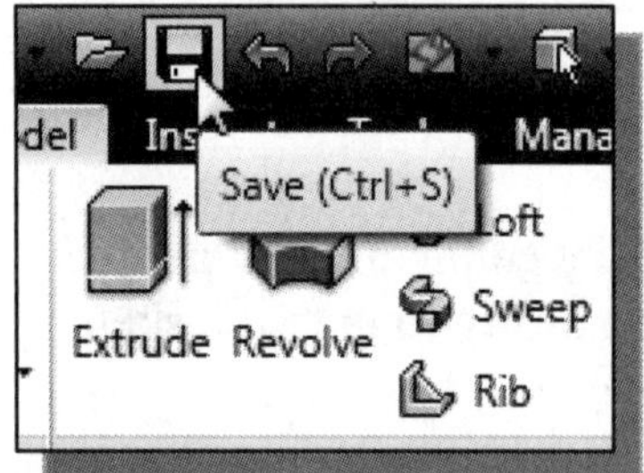

15. In the *Quick Access* menu, select **Save** to save the current model to disk.

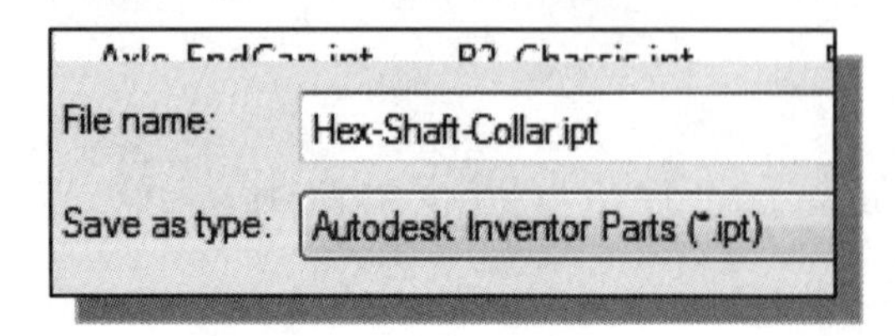

16. On your own, save the model using the filename **Hex-Shaft-Collar.ipt**.

The *A8-Rod Pin* Part

1. Start a new **Metric (mm)** model.

2. Create a revolved feature, starting on one of the datum planes.

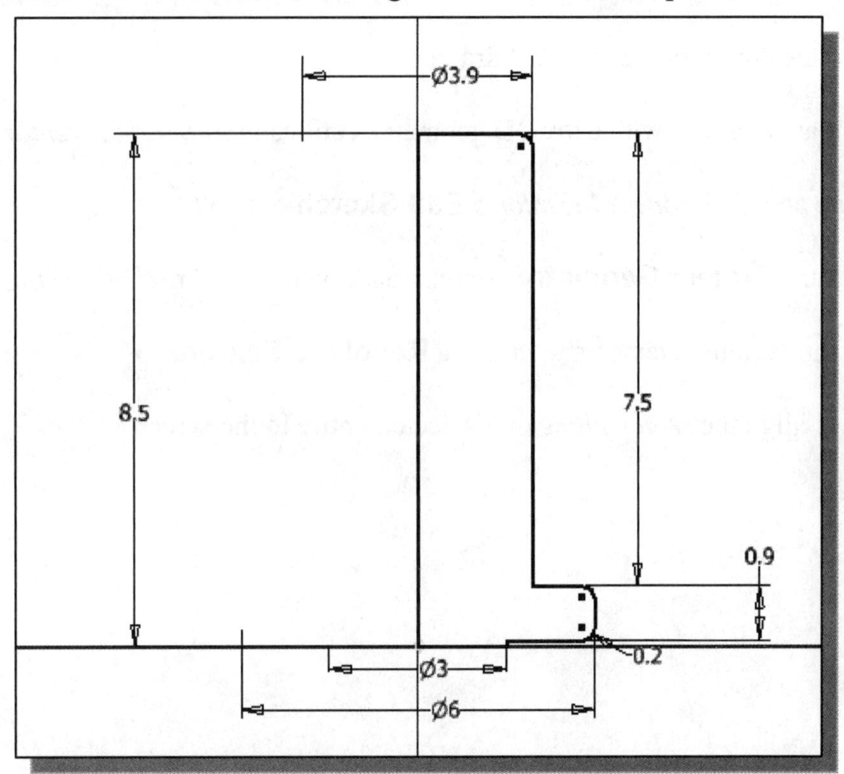

3. Save the model using the filename **A8-RodPin.ipt**.

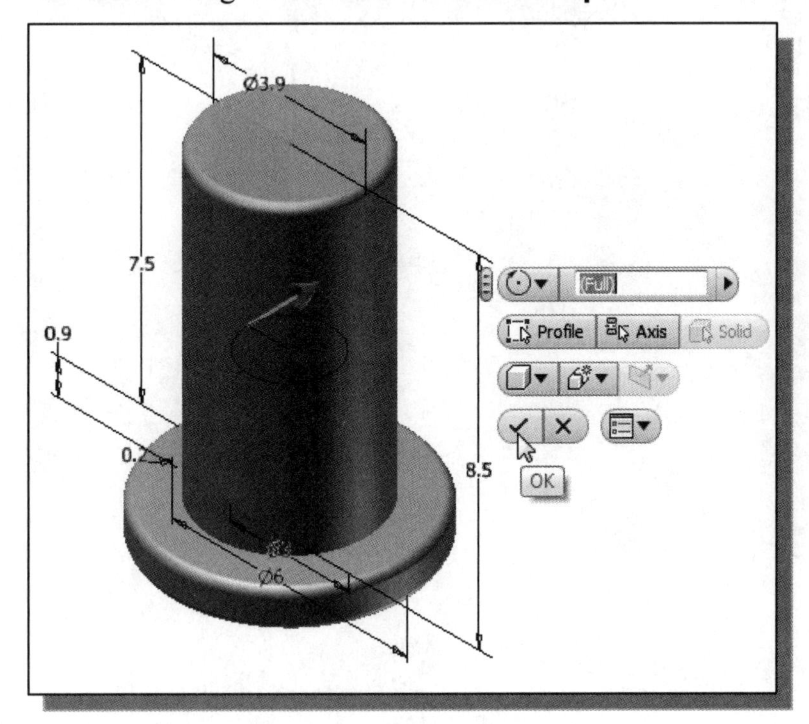

Questions:

1. What are the different types of work features available in *Autodesk Inventor*?

2. Why are work features important in parametric modeling?

3. Can we create auxiliary views in 2D drawings?

4. Can we create a profile with extra 2D geometry entities in *Autodesk Inventor*?

5. How do we access *Autodesk Inventor's* **Edit Sketch** option?

6. What does the **Project Geometry** command allow us to do in *2D Drawing* mode?

7. What are the required elements to create a **Revolved Feature**?

8. How do we align the *sketch plane* of a selected entity to the screen?

Exercises:

1. **Rod Slide** (Dimensions are in inches.)

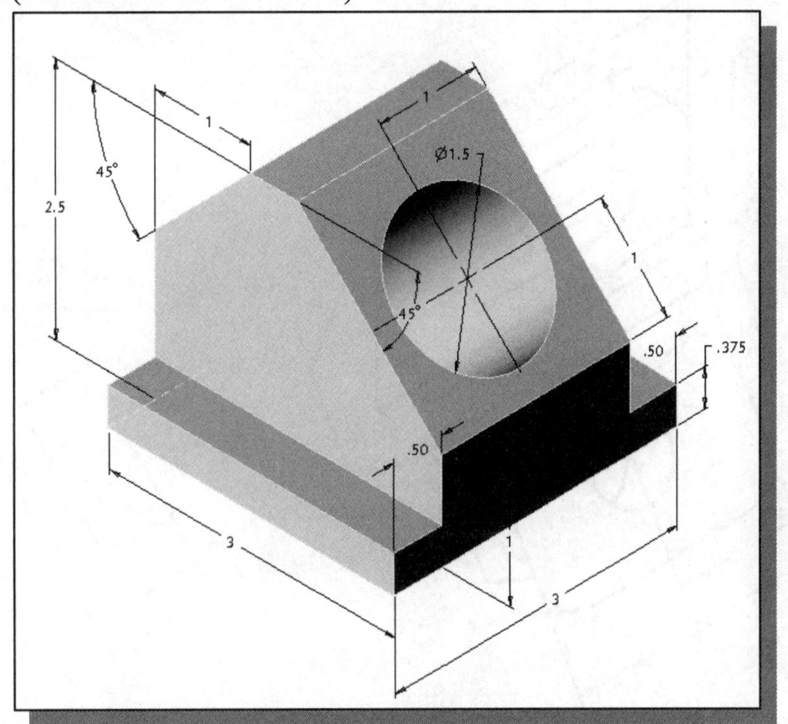

2. **Angle Support** (Dimensions are in millimeters.)

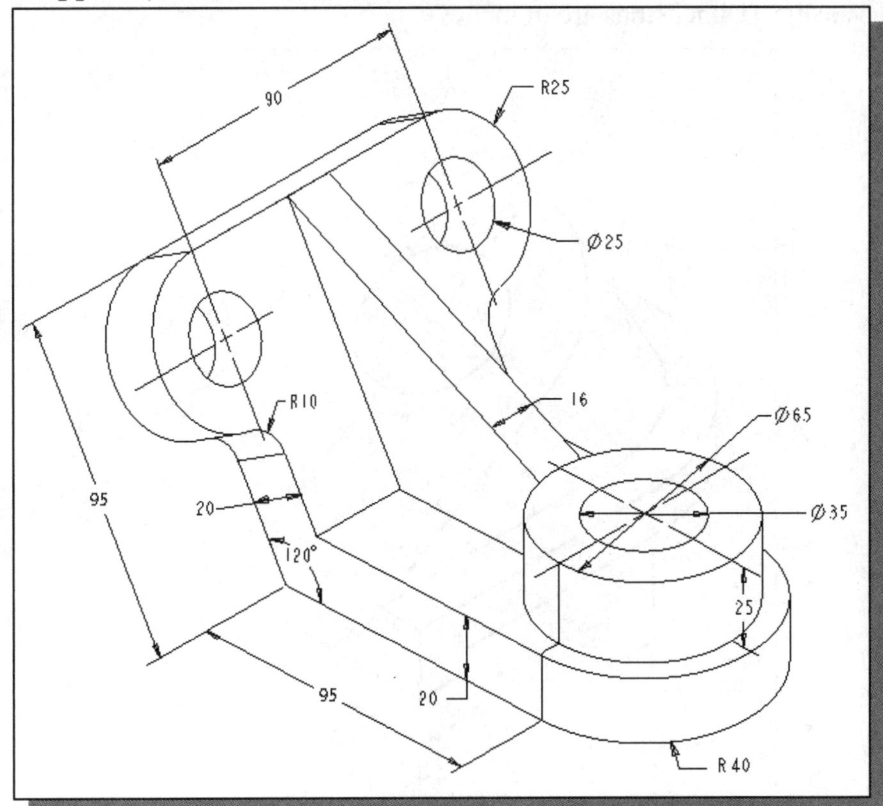

3. **Anchor Base** (Dimensions are in inches)

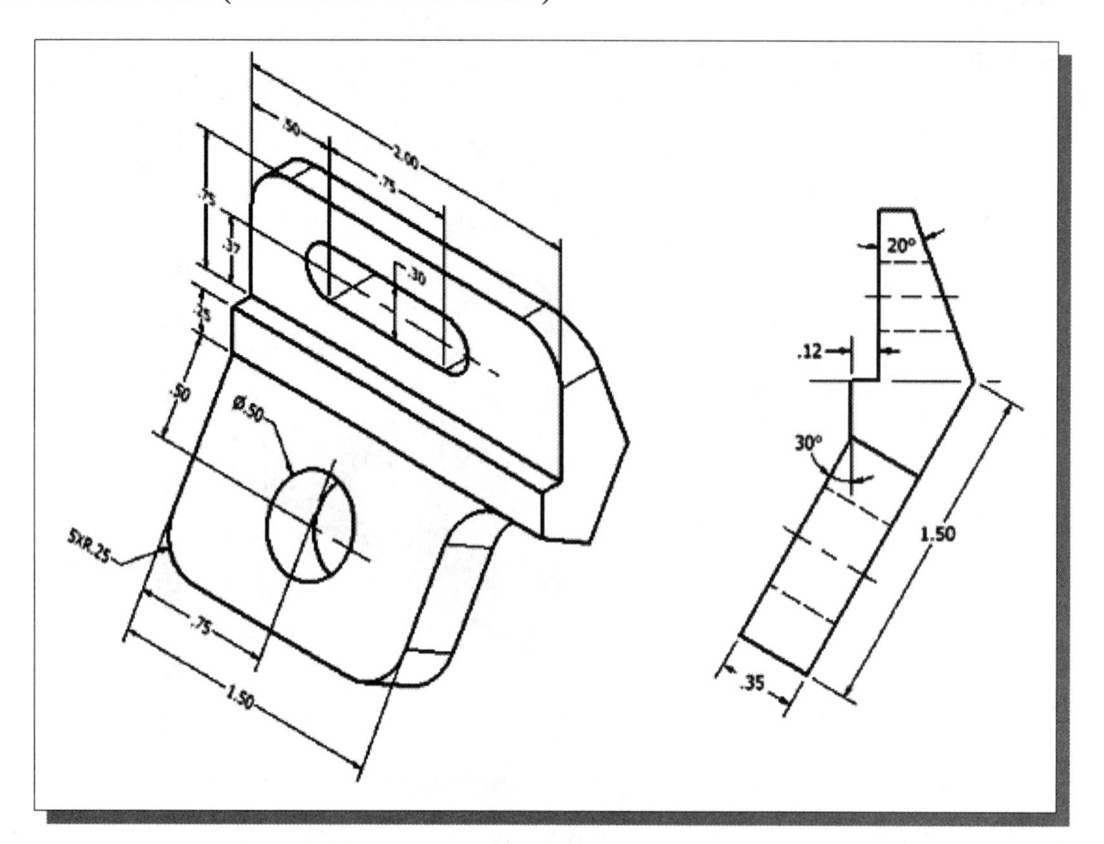

4. **Bevel Washer** (Dimensions are in inches)

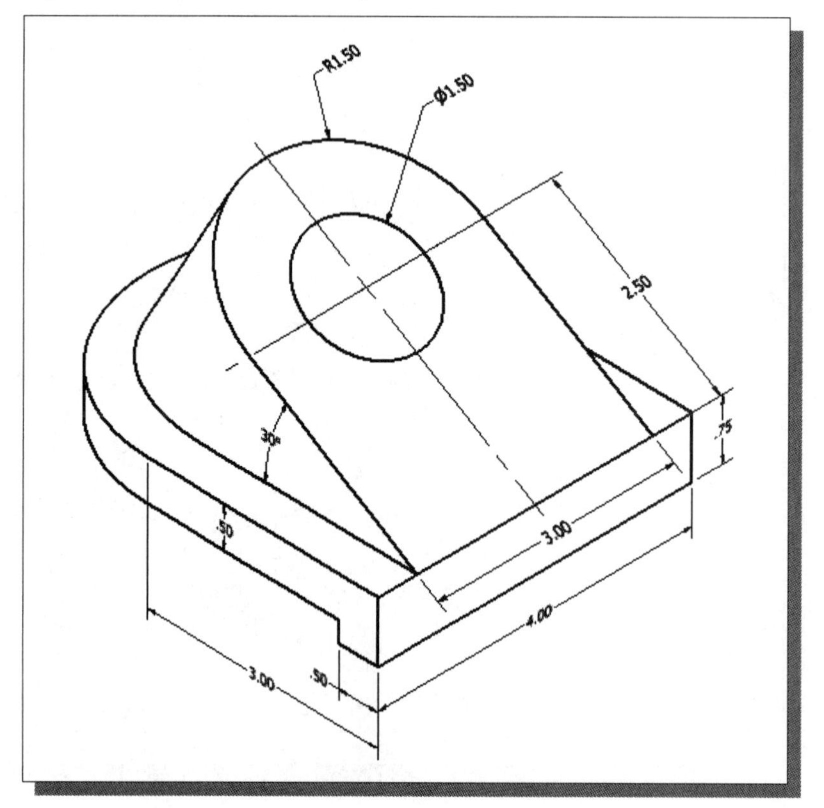

Chapter 8
Gear Generator and Content Center

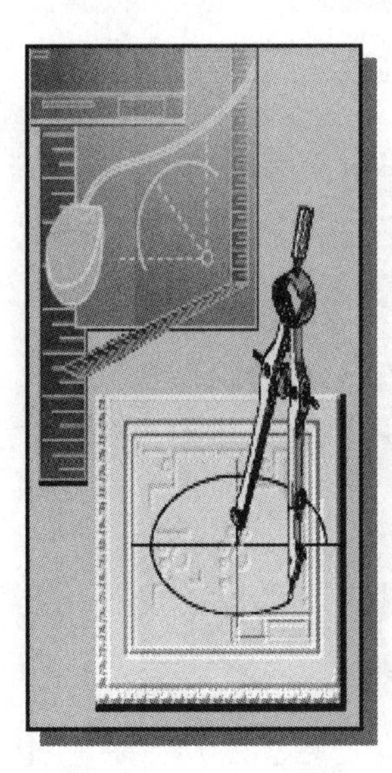

Learning Objectives

- ♦ Understand the Gear Nomenclatures
- ♦ Use the Autodesk Inventor Design Accelerator
- ♦ Use the Spur Gear Component Generator
- ♦ Export and Reuse the Tooth Profile
- ♦ Use the Autodesk Content Center Libraries

Introduction to Gears

- A **gear** is a rotating machine part having cut *teeth*, which *mesh* with another toothed part in order to transmit mechanical motion and power. Gears are very versatile machine elements; they range in size and use from tiny gears in watches to large driving gears in punch presses. Geared devices can be used to change the speed, torque, and direction of a power source. The most common situation is for a gear to mesh with another gear having parallel shaft axes; however a gear can also mesh with another toothed part in a way to producing motion in a nonparallel axis direction. The more commonly used types of gears include the **Spur Gear**, the **Bevel Gear**, the **Helical Gear**, the **Worm Gear** and the **Rack and Pinion**.

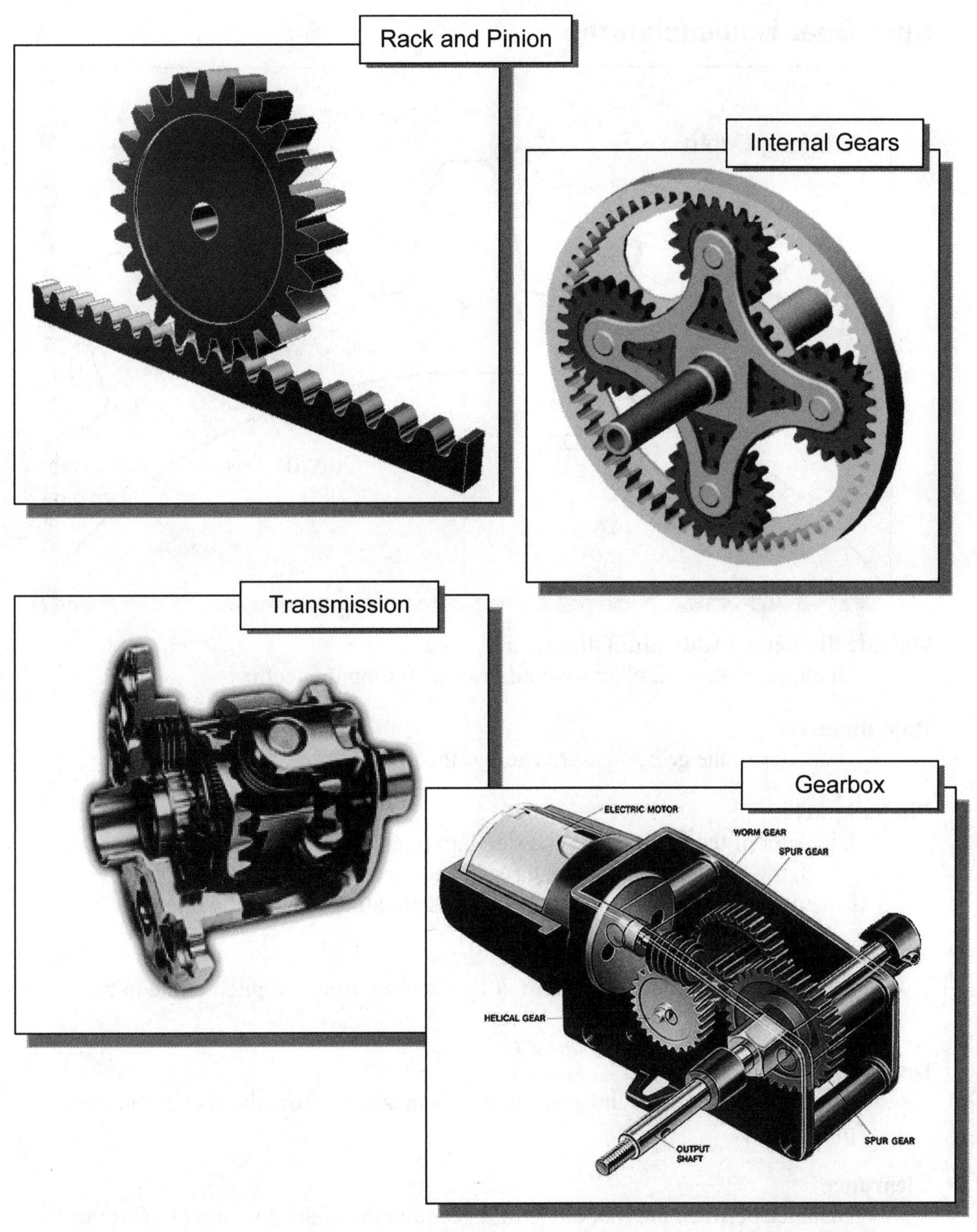

Two or more gears working in tandem are called a ***gear train*** and can produce a mechanical advantage through a designed gear ratio. For gears to function properly, they must fulfill the following basic requirements: (1) Transmit motion smoothly and efficiently; (2) Maintain fixed angular relationships between members; (3) Must be interchangeable with other gears having the same tooth size.

Spur Gear Nomenclatures

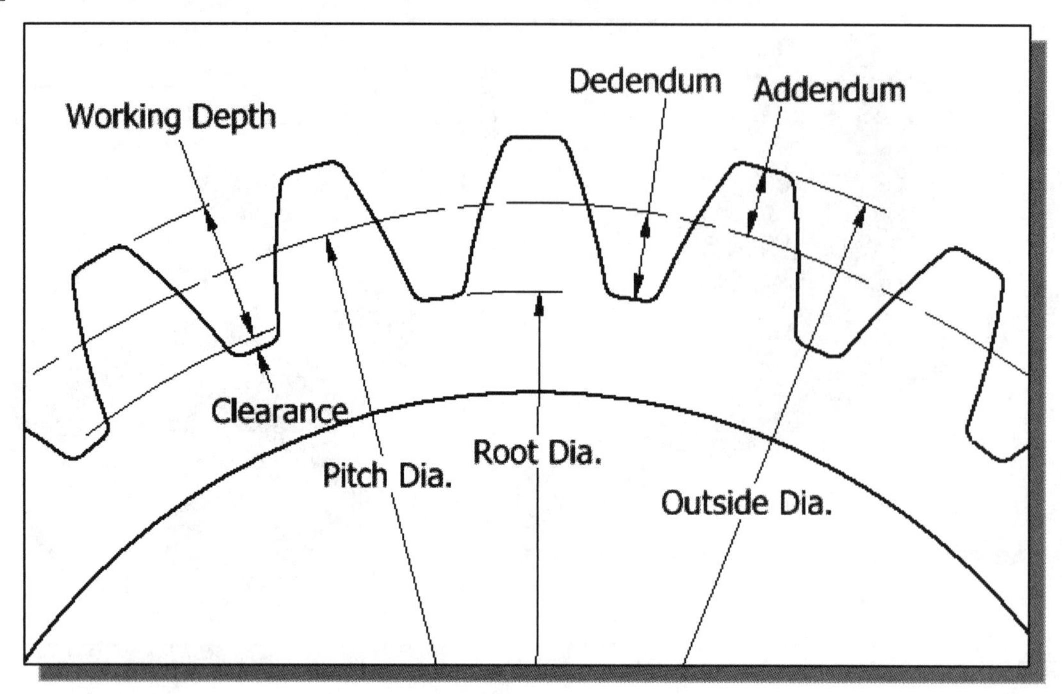

Outside diameter (Addendum diameter)
Diameter of gear blank, measured prior to cutting the teeth.

Root diameter
Diameter of the gear, measured across the bottom of the teeth.

Pitch diameter
Diameter of the pitch circle; the standard pitch diameter is a basic dimension at which the thread tooth and the thread space are equal. The circular tooth thickness, pressure angle and helix angles are all defined at the pitch circle.

Addendum
The top portion of the gear tooth; it is measured from the pitch circle to the outermost point of the tooth.

Dedendum
The bottom portion of the gear tooth; it is measured from the root circle to the pitch circle.

Clearance
Distance between the root circle of a gear and the addendum circle of its mate.

Working depth
Depth of engagement of two gears, that is, the sum of their operating addendums.

Whole depth
The distance from the top of the tooth to the root; it is equal to addendum plus dedendum or to working depth plus clearance.

Circular Pitch

Circular pitch is the distance measured along the pitch circle, from a point on one tooth to the corresponding point on the adjacent tooth.

Circular Pitch = Pitch diameter / No. of teeth

Pitch, Diametral Pitch

Diametral pitch of a gear is an expression of tooth size. It is the ***number of teeth per inch of pitch diameter***. Thus, a gear with a 2-in. pitch diameter and 40 teeth is a 20-pitch gear. Any two gears having the same pitch will operate together, provided they are based on the same gear system.

Diametral Pitch = No. of teeth / Pitch diameter = 40/2 = 20-Pitch

Module

A scaling factor used in **metric** gears with units in millimeters whose effect is to enlarge the gear tooth size as the module increases and reduce the size as the module decreases.

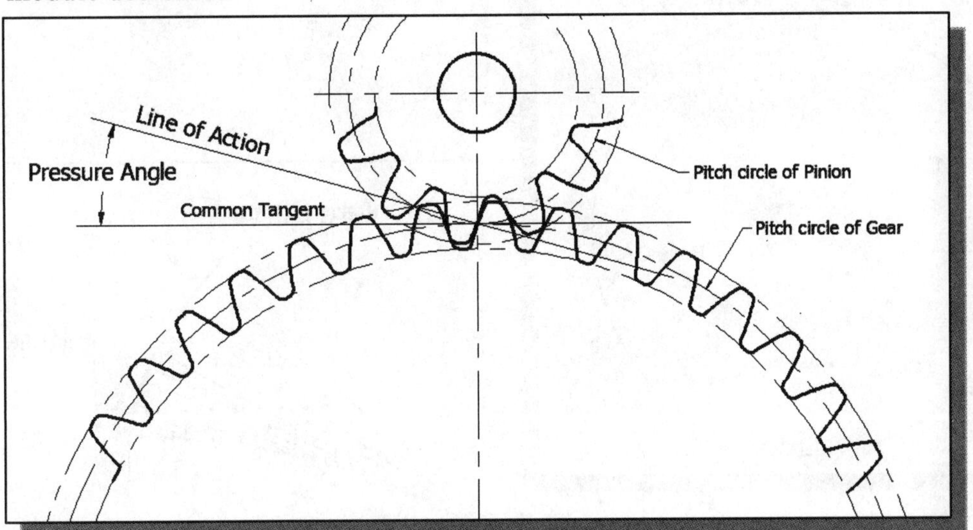

Gear: The larger of two interacting gears or a gear on its own.

Pinion: The smaller of two interacting gears.

Pitch point: Point where the line of action crosses a line joining the two gear axes.

Line of action, pressure line

Line of action is the line along which the force between two meshing gear teeth is directed. In general, the line of action changes from moment to moment during the period of engagement of a pair of teeth. For involute gears, however, the tooth-to-tooth force is always directed along the same line.

Pressure angle

The angle formed between the direction of the teeth exerting force on each other, and the common tangent line of the two pitch circles of the two gears. For involute gears, the teeth always exert force along the line of action, which, for involute gears, is a straight line; and thus, for involute gears, the pressure angle is constant.

Basic Involute Tooth Profile

- For gears to transmit uniform motion from one gear to another, the tooth profiles must have the proper form. There are a number of mathematical curves that can fulfill the requirements of the tooth contact, but the most commonly employed is the ***Involute curve***. The *Involute curve* forms the basis upon which several forms of American Standard gear teeth are based.

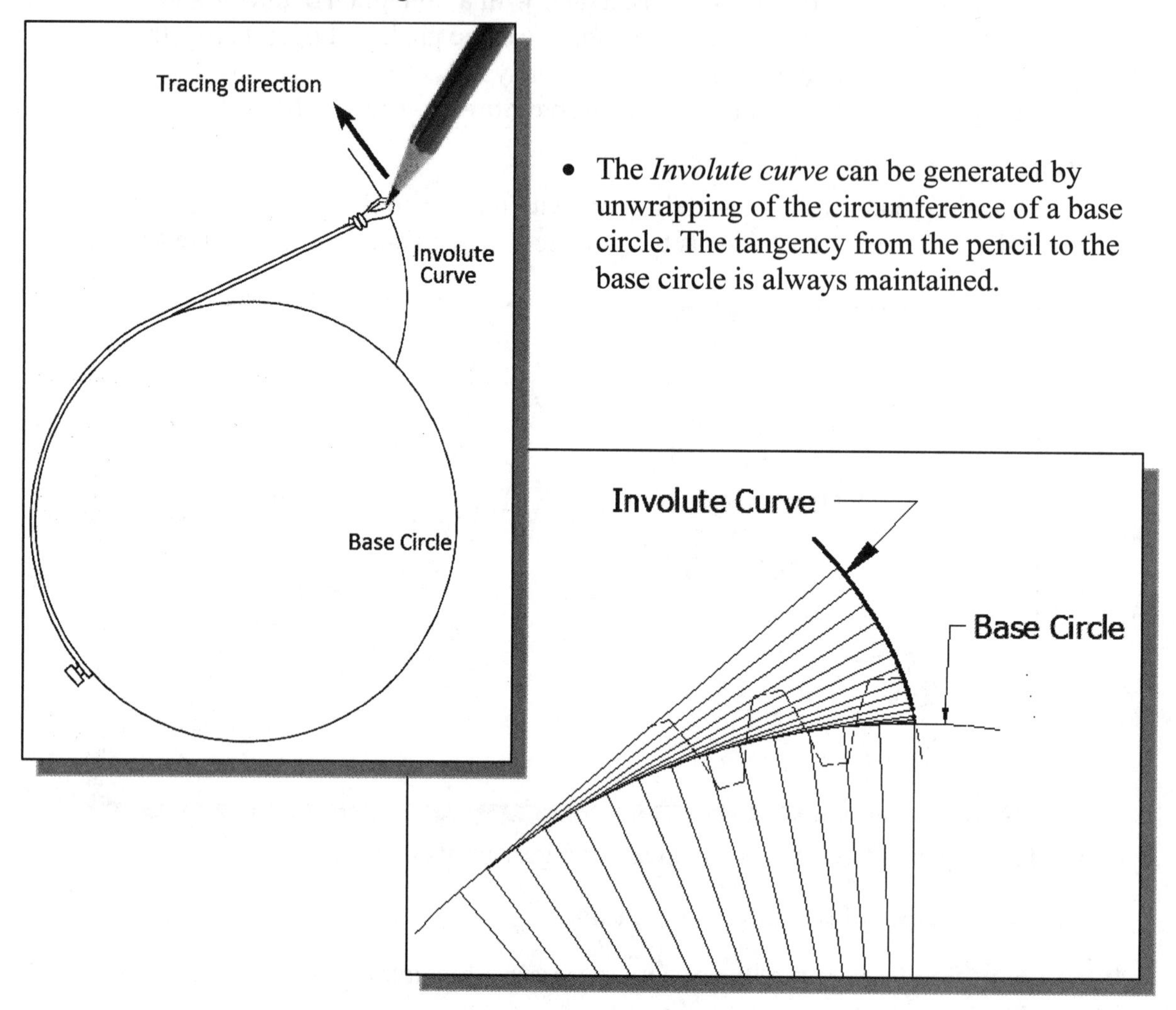

- The *Involute curve* can be generated by unwrapping of the circumference of a base circle. The tangency from the pencil to the base circle is always maintained.

Note that several mathematic equations are available to generate an *Involute curve.*

The basic set of **Parametric Cartesian equations** of an ***Involute curve*** is:

$$x = a(\cos(t) + t\sin(t))$$
$$y = a(\sin(t) - t\cos(t))$$

This set of equations use the radius a of the circle and the angular parameter t to generate the *Involute curve.*

Gear Ratio

A ***gear train*** is two or more gears working in tandem for the purpose of transmitting motion from one axis to another. A gear train is called an **ordinary gear train** if all the rotating shafts are mounted on a common stationary frame. A **simple ordinary train** is seen to be one in which there is only one gear for each axis. A **compound ordinary train** is seen to be one in which two or more gears may rotate about a single axis.

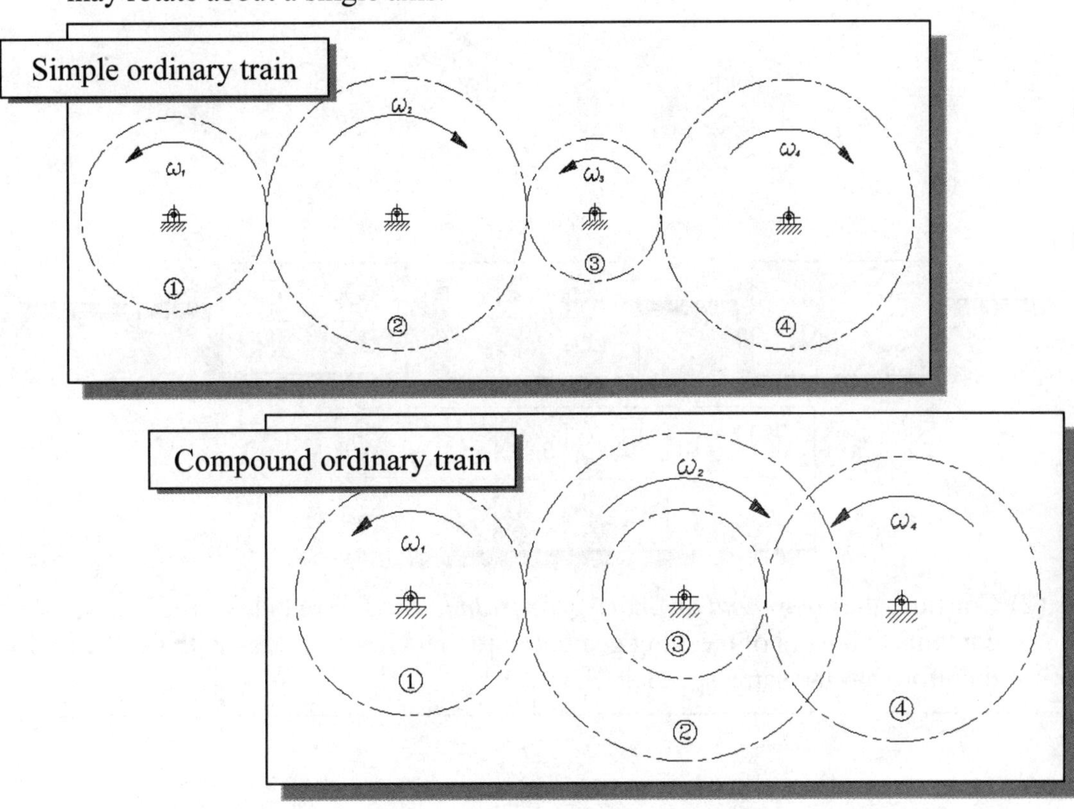

The **gear ratio** of a *gear train* is the ratio of the angular velocity of the input gear to the angular velocity of the output gear, also known as the *speed ratio* of the gear train. The **speed ratio** of a pair of gears is the inverse proportion of the diameters of their pitch circle. The gear ratio can also be computed directly from the numbers of teeth of the various gears that engage to form the gear train. For two meshing gears, the gear ratio is calculated as:

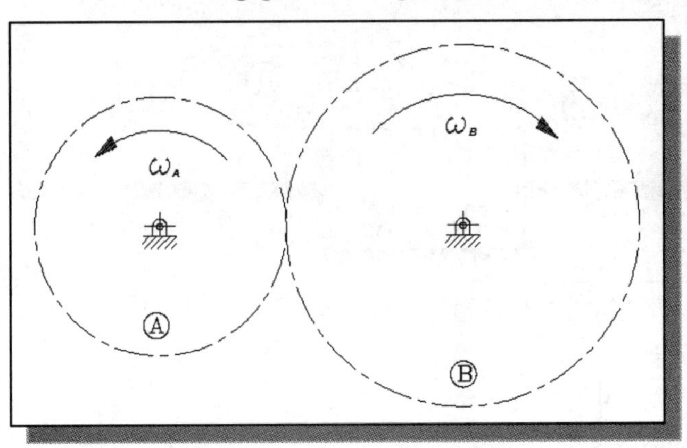

$$\text{Gear Ratio} = \frac{\omega_A}{\omega_B} = \frac{N_B}{N_A}$$

(1) Consider the *simple ordinary gear train* in the figure below; the gear train contains three pairs of meshing gears. (Note that Gear 2 and Gear 3 are *idler gears*. An intermediate gear which does not drive a shaft to perform any work is called an *idler gear.*)

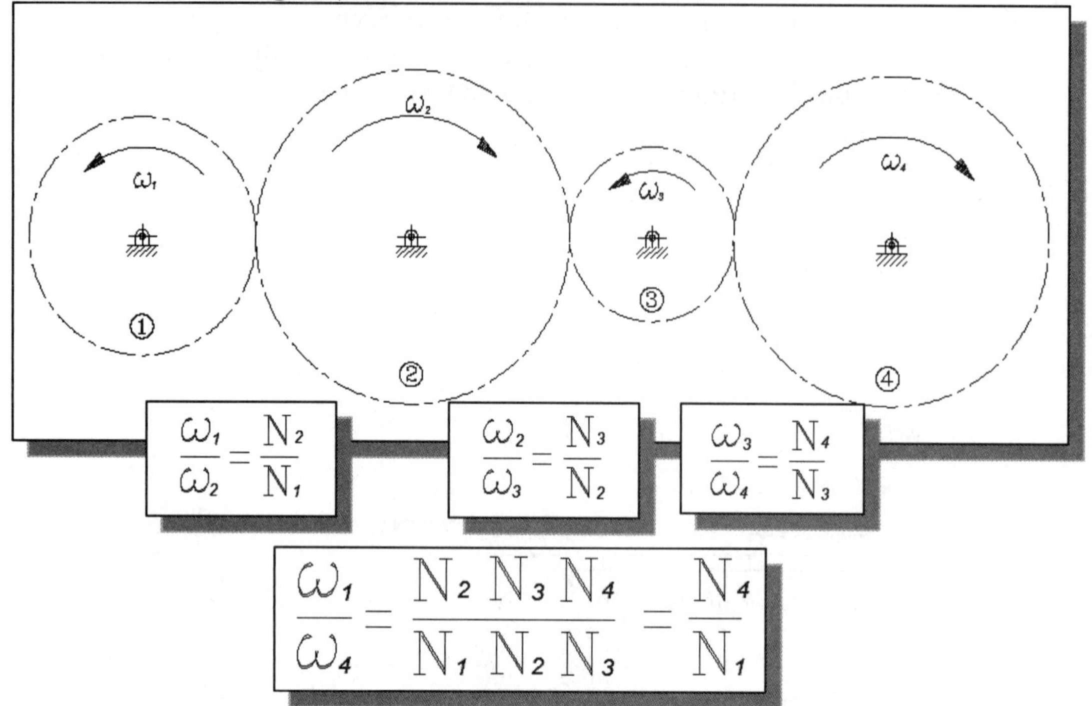

(2) Consider the *compound ordinary gear train* in the figure below; the gear train contains two pairs of meshing gears. (Note that Gear 3 moves with Gear 2 and therefore has the same speed as Gear 2.)

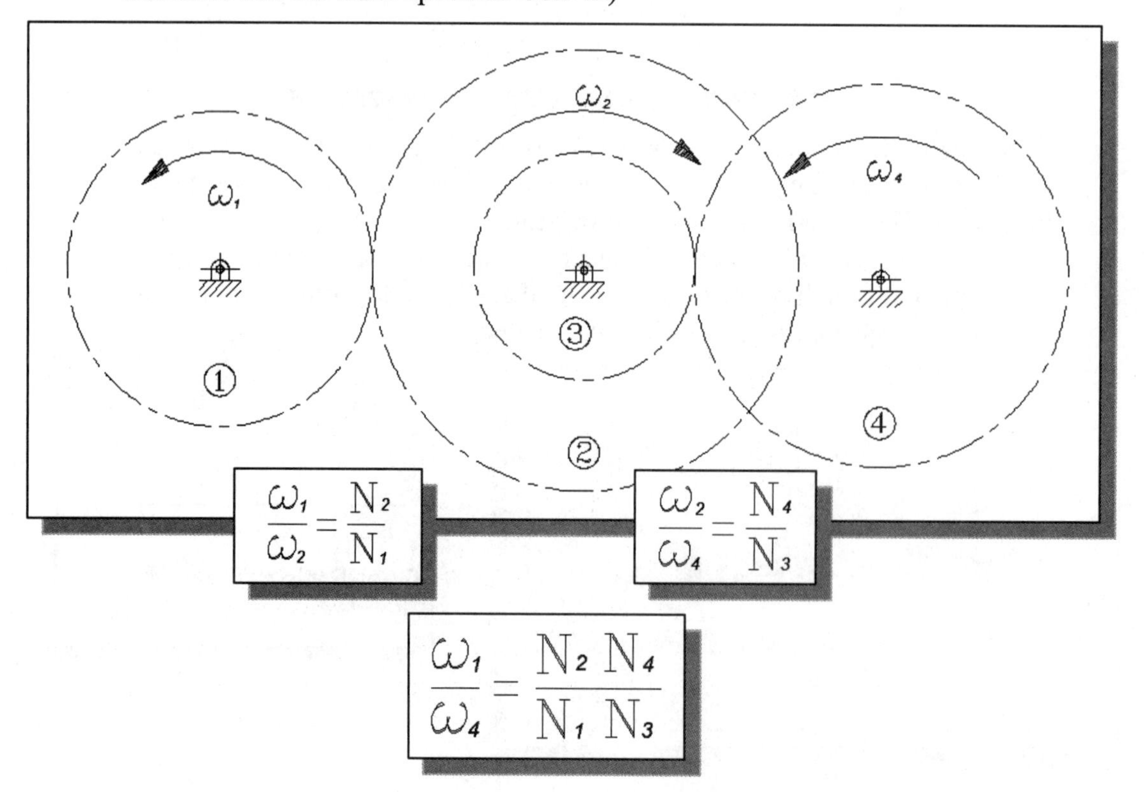

Starting *Autodesk Inventor*

1. Select the **Autodesk Inventor** option on the *Start* menu or select the **Autodesk Inventor** icon on the desktop to start *Autodesk Inventor*. The *Autodesk Inventor* main window will appear on the screen.

2. Select the **New File** icon with a single click of the left-mouse-button in the *Launch* toolbar as shown.

3. Select the **Metric** tab and in the *Template* list, then select **Standard(mm).iam** (*Standard Inventor Assembly Model* template file).

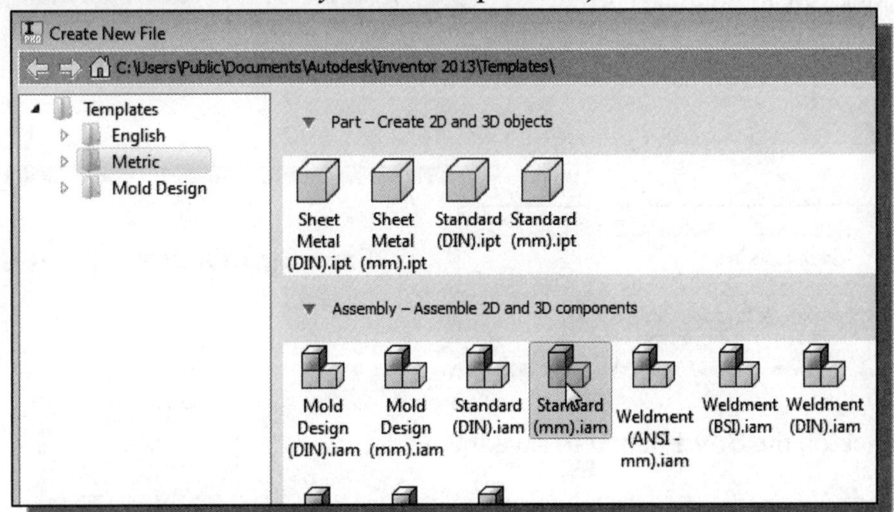

4. Click on the **Create** button in the *New File* dialog box to accept the selected settings and start a new assembly model.

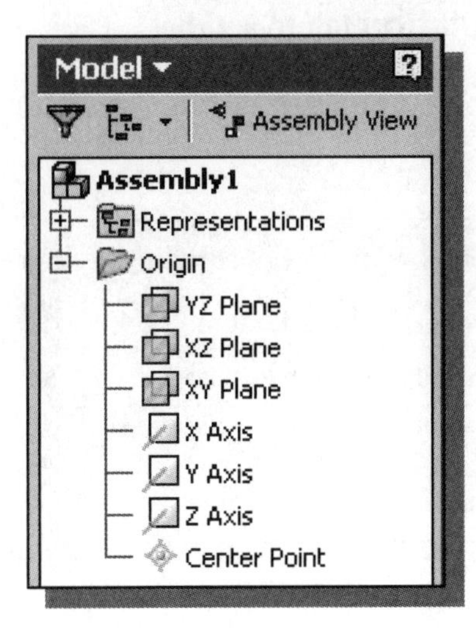

- In the *browser* window, **Assembly1** is displayed with a set of work planes, work axes and a work point. In most aspects, the usage of work planes, work axes and work point is very similar to that of the *Inventor Part Modeler*.

- Notice, in the *Ribbon* toolbar panels, several *component* options are available, such as **Place Component**, **Create Component** and **Place from Content Center**. As the names imply, we can use parts that have been created or create new parts within the *Inventor Assembly Modeler*.

The *Inventor* Spur Gear Generator

- In this section, we will illustrate the procedures to create gears with the **Spur Gear** component generator. Note that the nylon gears in the *Tamiya Tiger kit* are specifically designed for the *Tamiya kits*, and therefore the tooth profiles do not match with the gear profiles generated using the *Inventor Design Accelerator*.

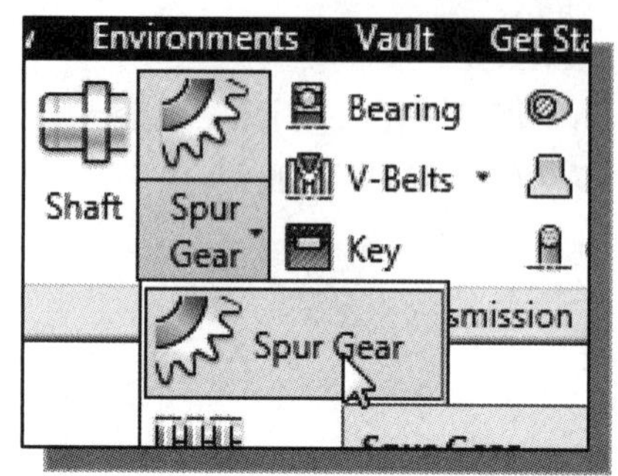

1. In the *Design* tab of the Ribbon toolbar area, select the **Spur Gear** command by left-mouse-clicking the icon.

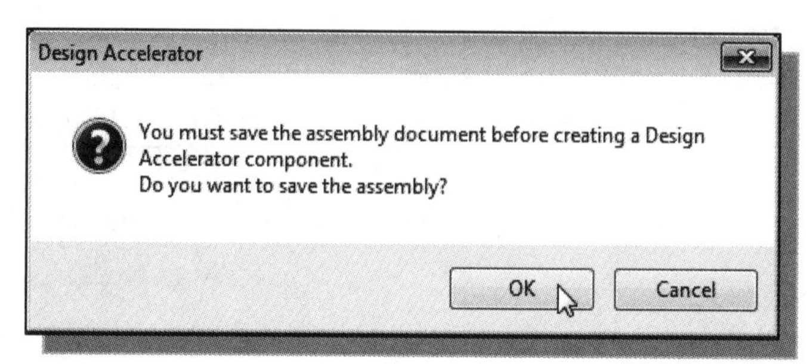

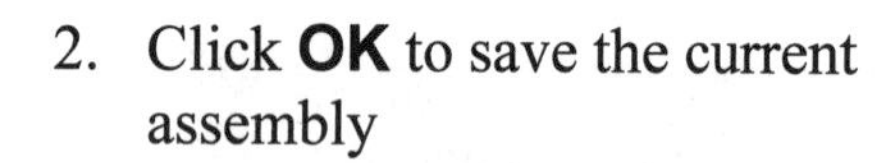

2. Click **OK** to save the current assembly

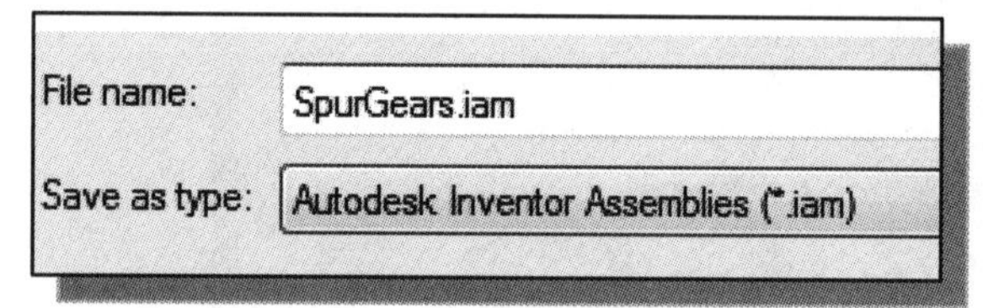

3. Enter **SpurGears** as the assembly model name.

4. Click on the **Save** button to save the model file.

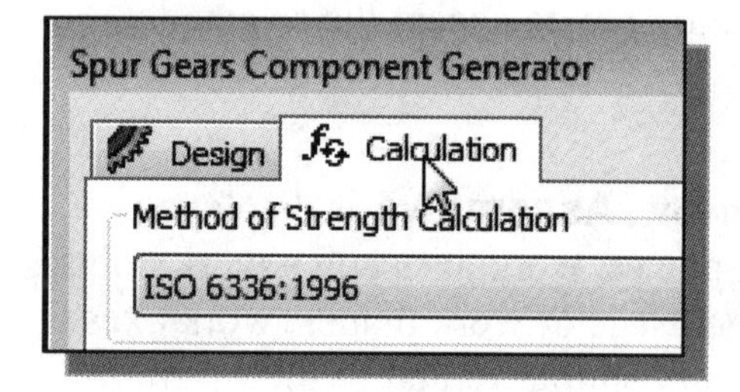

5. Click on the **Calculation** tab to set the associated power rating of the system.

6. Choose **Bach (Simple design)** as the method of strength calculation.

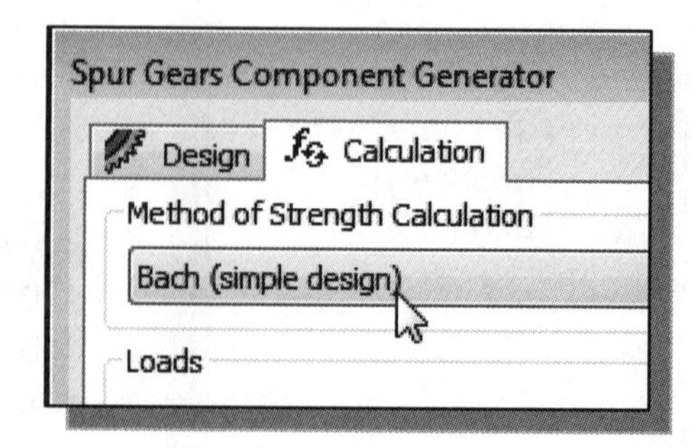

7. In the *Loads* section, set the variables using the following values: ***Power*** to **0.001 KW**, ***Speed*** to **500 rpm** and ***Efficiency*** to **0.90** as shown in the figure.

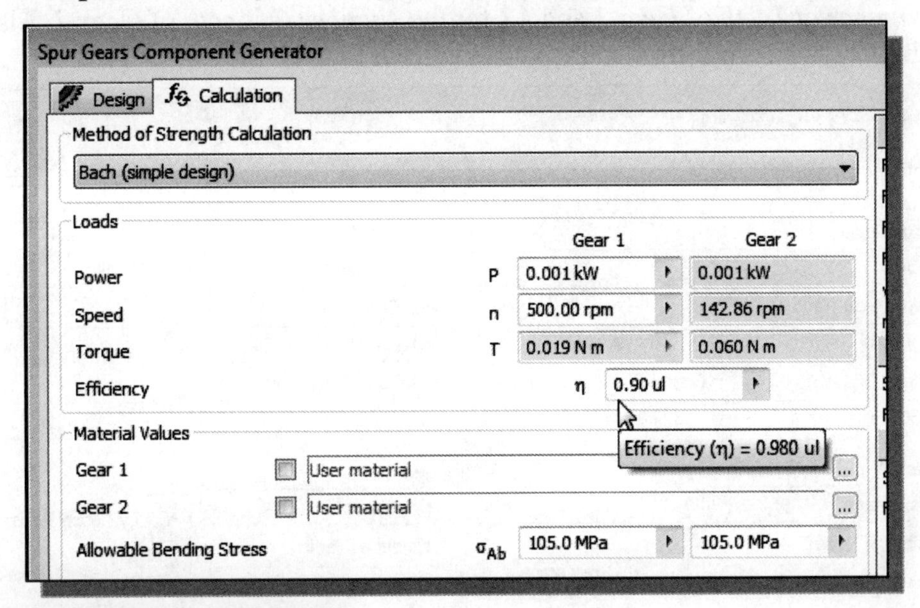

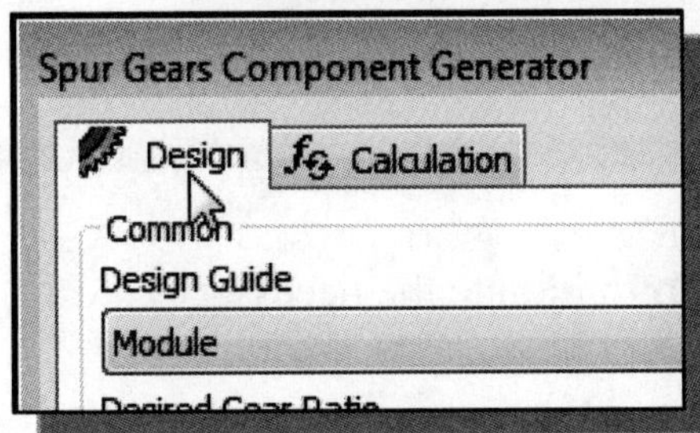

8. Click on the **Design** tab to switch back to adjust the settings of the related gears.

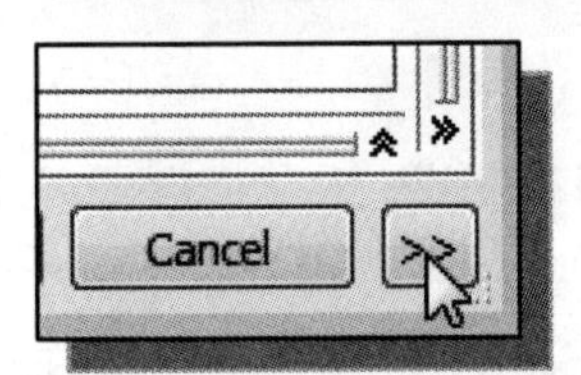

9. Click on the [>>] button to show the additional settings.

10. Set the *Input Type* to **Number of Teeth** and *Size Type* to **Module** and *Reaching Center Distance* to **Teeth Correction**. Also set the *Unit Tooth Sizes* to the values as shown.

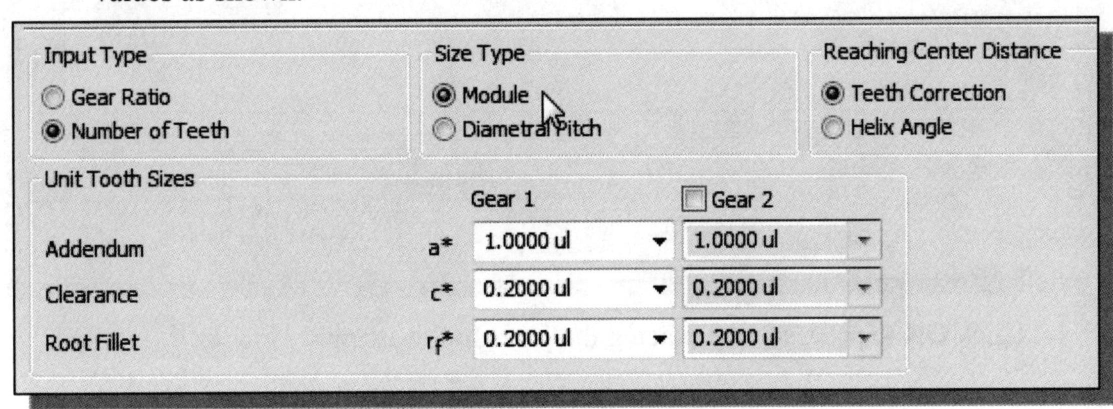

11. Set the *Design Guide* to **Module**, *Center Distance* to **13.5 mm** and *Pressure Angle* to **22.5 degrees**. Set the gear option to **Component** and enter **12** for the *Number of Teeth* of *Gear 1* and **42** for the *Number Of Teeth* of *Gear 2*. Also set the *Facewidth* to **5 mm** and *Unit Correction* to **0.1**.

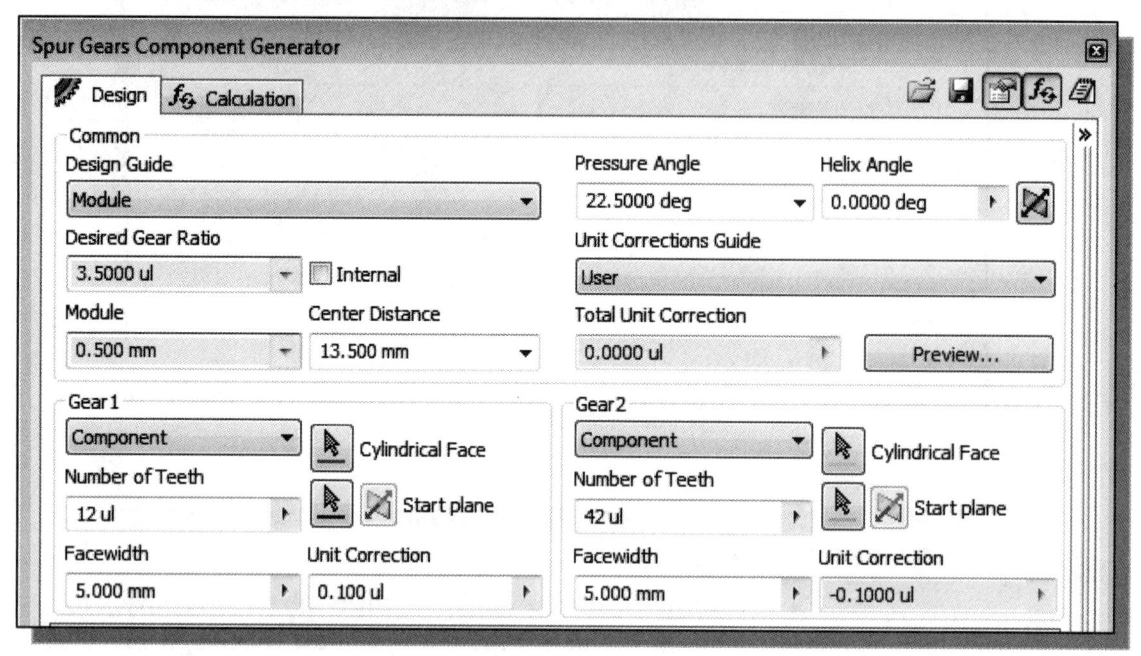

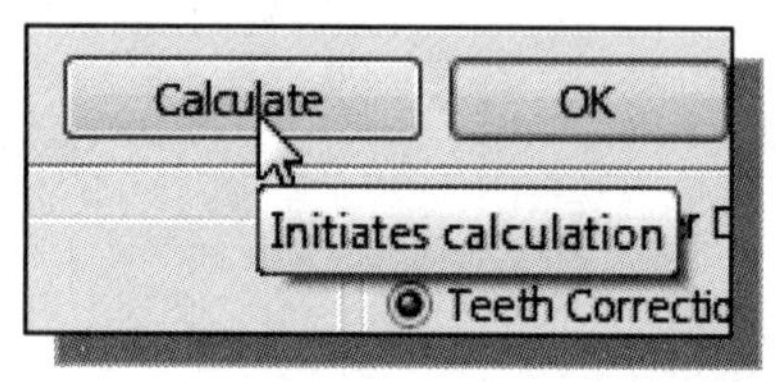

12. Click **Calculation** to initiate the necessary calculations to determine the gear profiles.

13. The results of the calculations are listed in the dialog box as shown. In this case, *Inventor* indicated that the gears will need to be adjusted for proper operation.

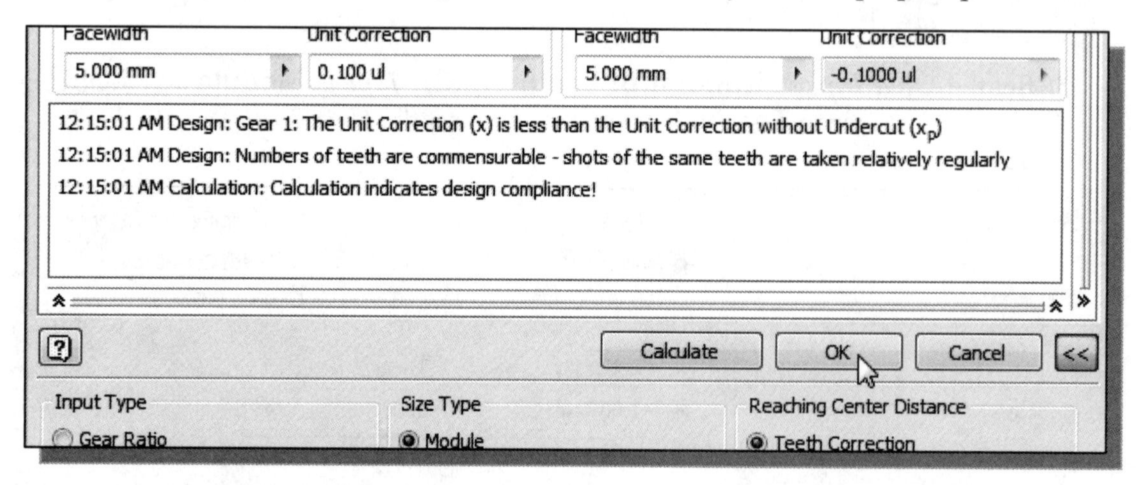

14. Click **OK** to proceed with placing the gears in the assembly model.

15. Click **OK** to accept the default naming of the files for the new gears.

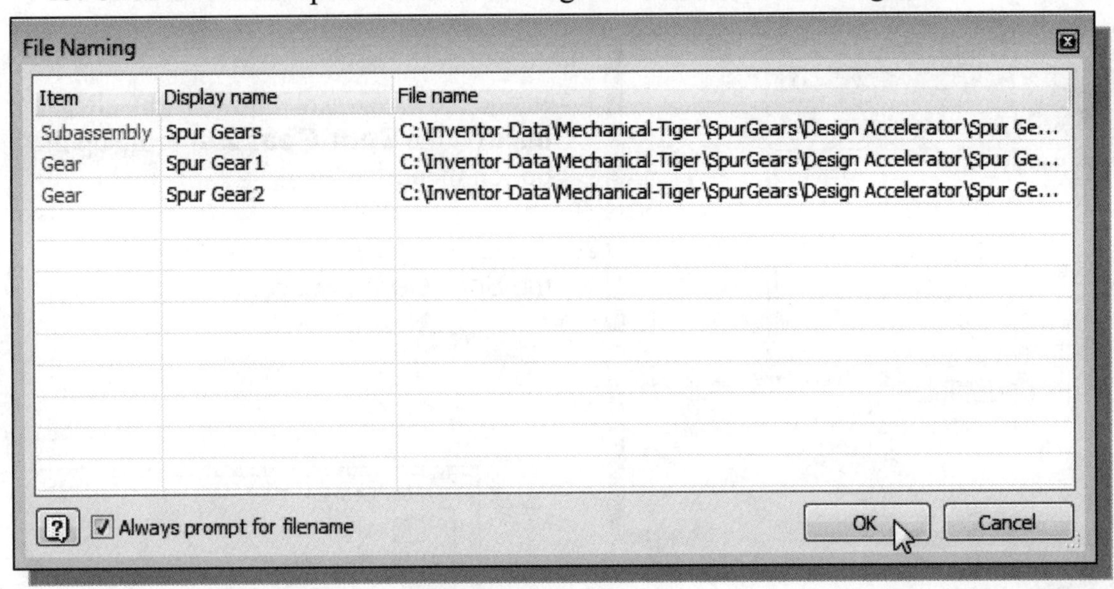

16. Place the gears in the assembly model by clicking in an empty location inside the graphics window.

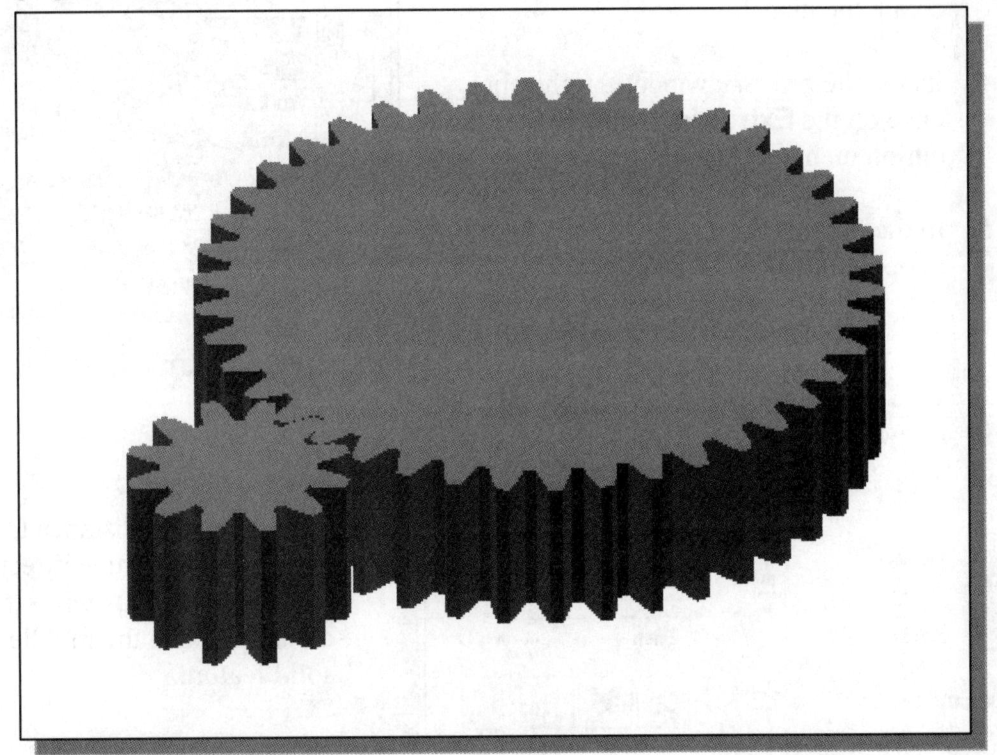

- The *Autodesk Inventor Design Accelerator* provides a convenient way to create a gear set. Note the creation of true gear tooth profiles can be very time consuming by traditional methods. The created gears can also be edited and modified just as a regular *Inventor* part. In the next sections, we will modify these two gears to create the parts needed for the *Tamiya Tiger kit*.

Modify the Generated Gears

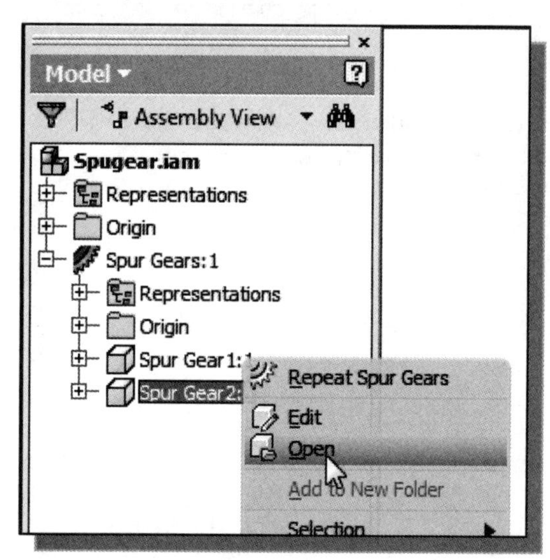

1. Inside the *browser* window, **right-click once** on the **Spur Gear 2** to bring up the option menu.

2. In the option menu, choose **Open** to open the *Spur Gear 2* model.

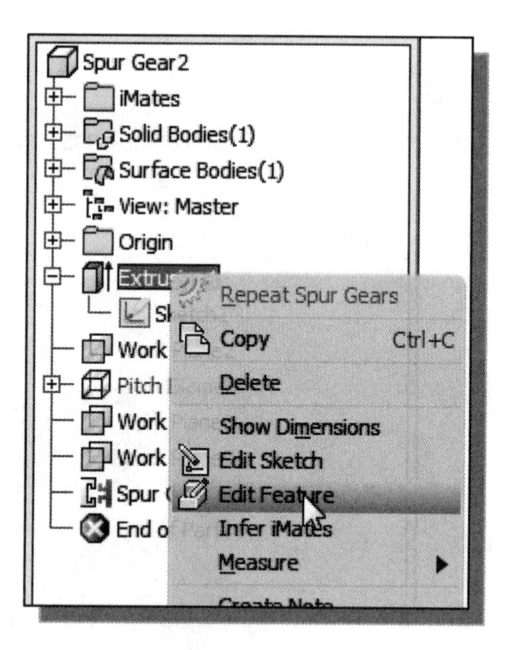

3. On your own, examine the *Spur Gear 2* model. Note the different features used to create the model.

4. Inside the *browser* window, right-click once on the **Extrusion1** to bring up the option menu.

5. In the option menu, choose **Edit Feature** to open the selected feature.

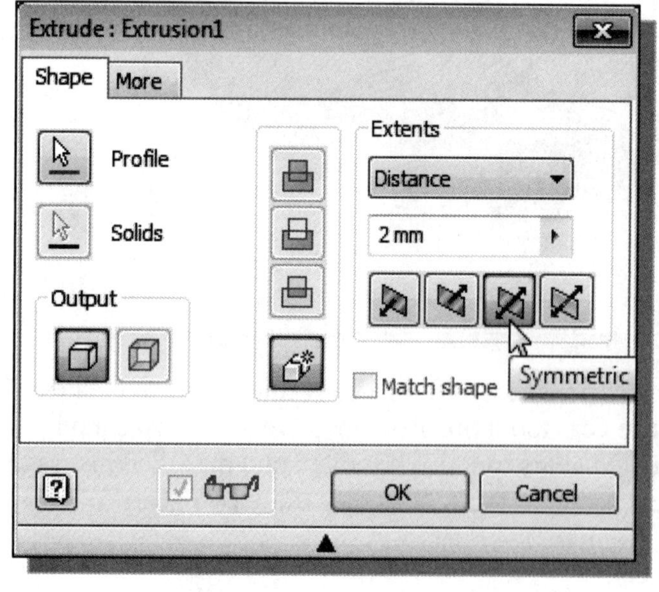

6. Set the extrude distance to **2 mm** and the extrude direction to **Symmetric**. This will set the datum plane in the middle of the solid feature.

7. Click **OK** to accept the settings and end the Edit Feature command.

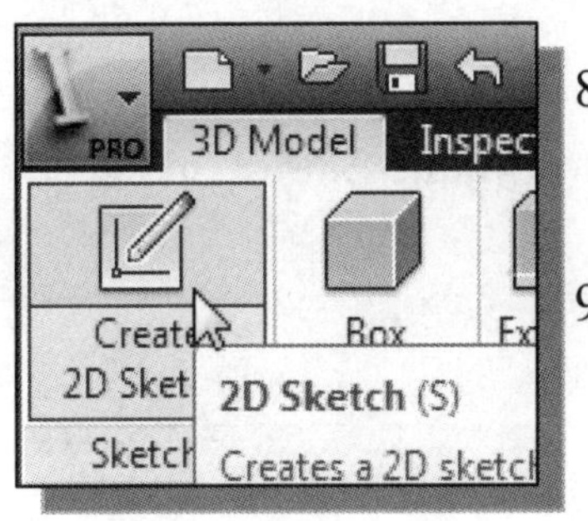

8. In the *Sketch* panel, select the **Create 2D Sketch** command by left-mouse-clicking once on the icon.

9. In the graphics window, select the **top plane** as the sketching plane.

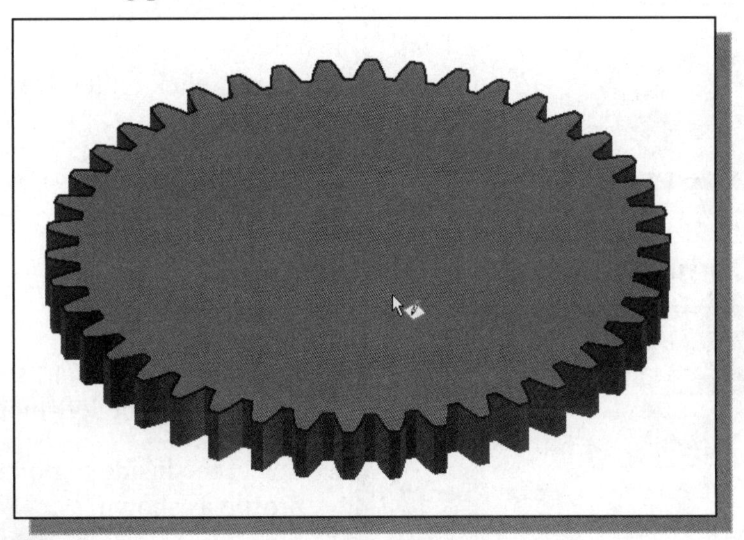

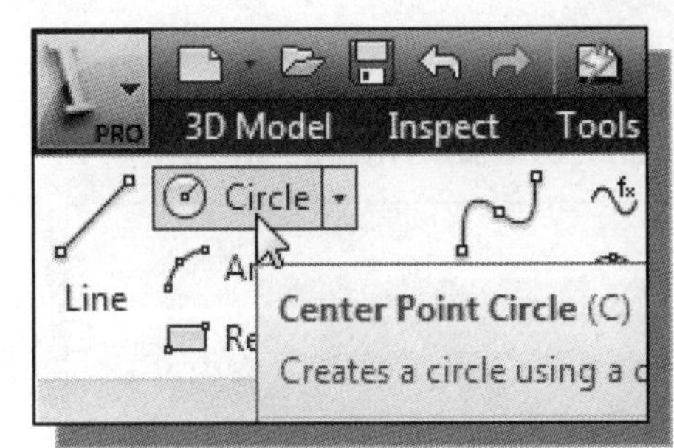

10. Select the **Center point circle** command by clicking once with the left-mouse-button on the icon in the *Draw* toolbar.

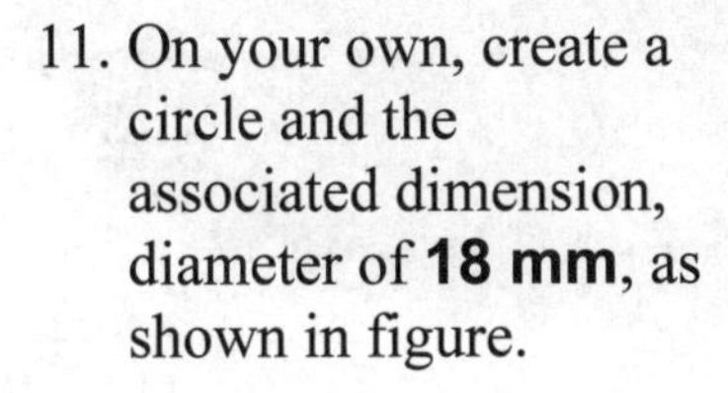

11. On your own, create a circle and the associated dimension, diameter of **18 mm**, as shown in figure.

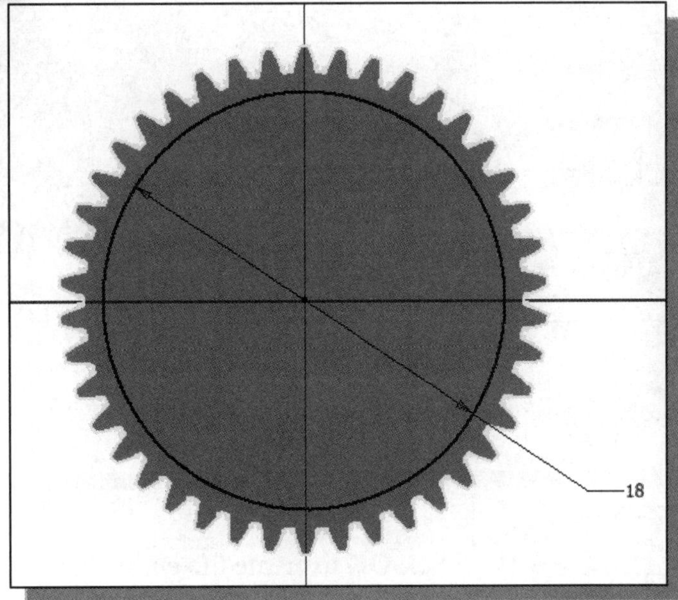

12. Select **Finish Sketch** in the *Ribbon* toolbar to exit the **Sketch** mode.

13. In the *View Cube* area, click on the **Home** icon to reset the display to the isometric view.

Complete the Solid Feature

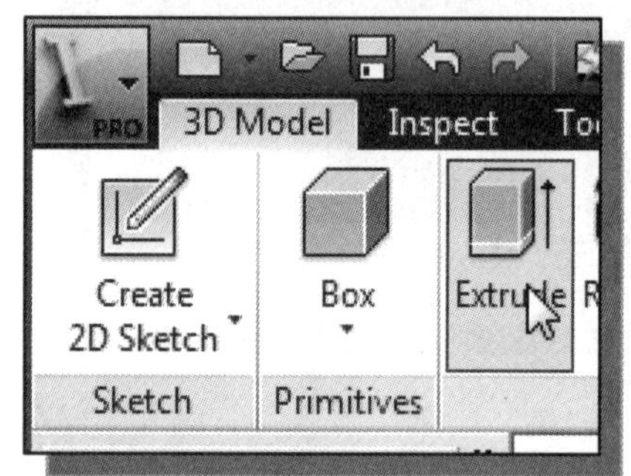

1. In the *Create* toolbar, select the **Extrude** command by releasing the left-mouse-button on the icon.
2. Select the inside region of the 2D sketch to create a profile as shown.

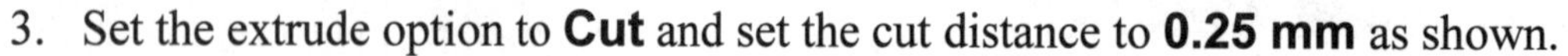

3. Set the extrude option to **Cut** and set the cut distance to **0.25 mm** as shown.

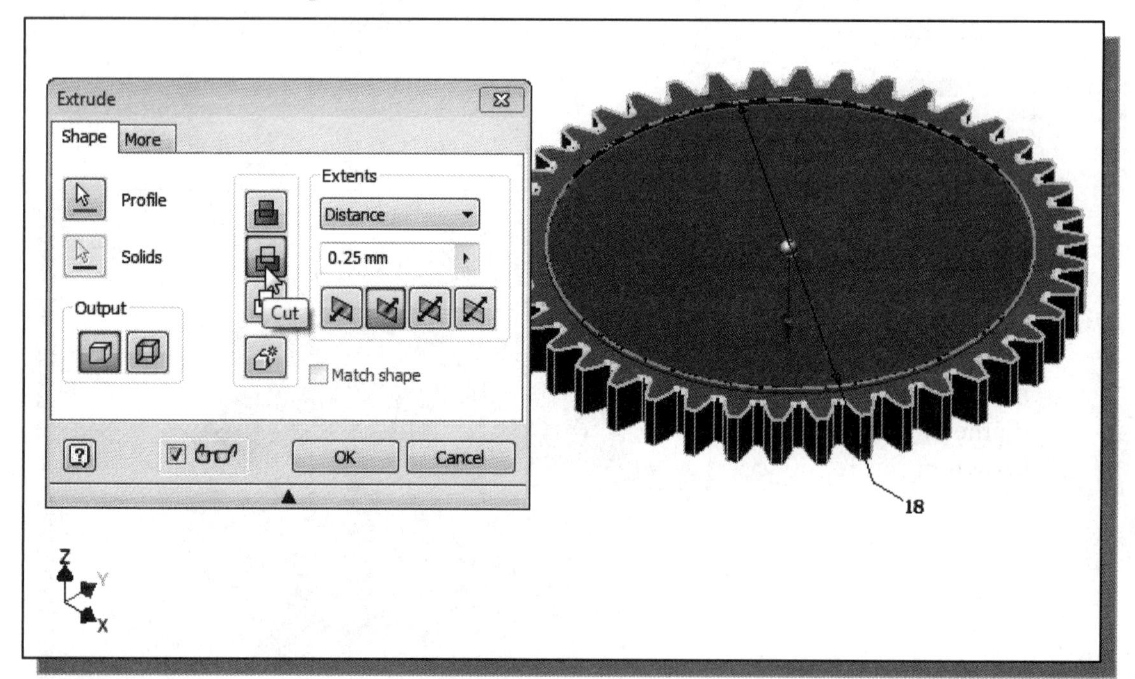

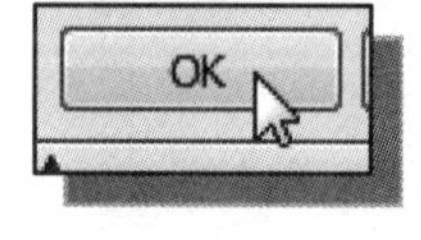

4. Click **OK** to create the cut feature as shown.

Create a Mirrored Feature

❖ The **Mirror** command can be used to create duplicates of existing features, with respect to a mirror image plane.

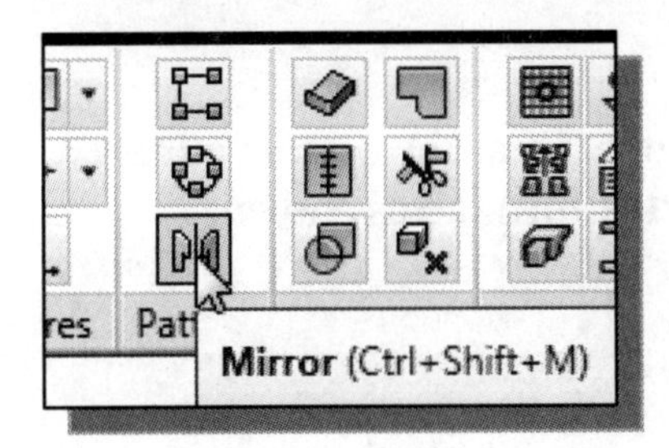

1. In the *Pattern* panel, select the **Mirror Image** command as shown.

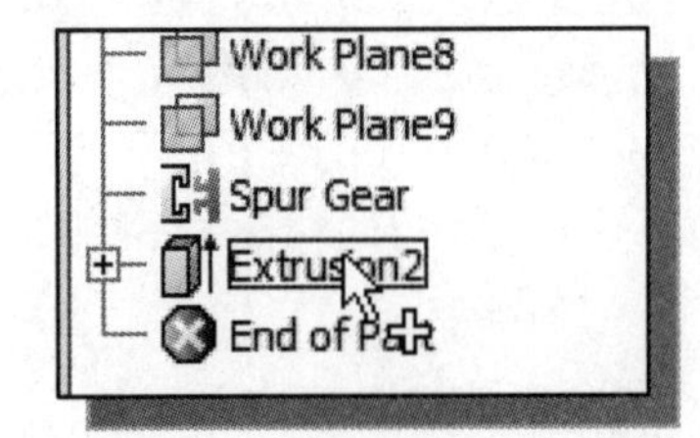

2. In the *browser* window, select the **cut extrusion** we just created.

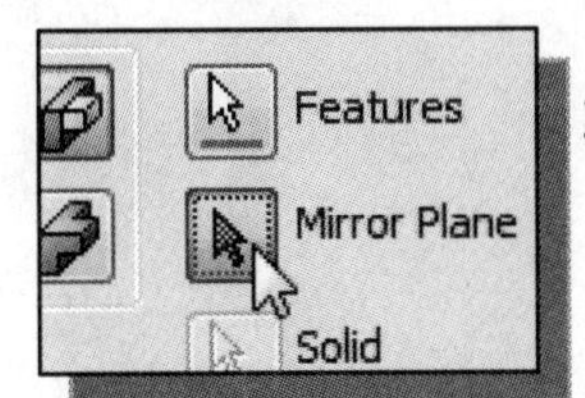

3. Click on the **Mirror Plane** icon, in the *Mirror* dialog box as shown.

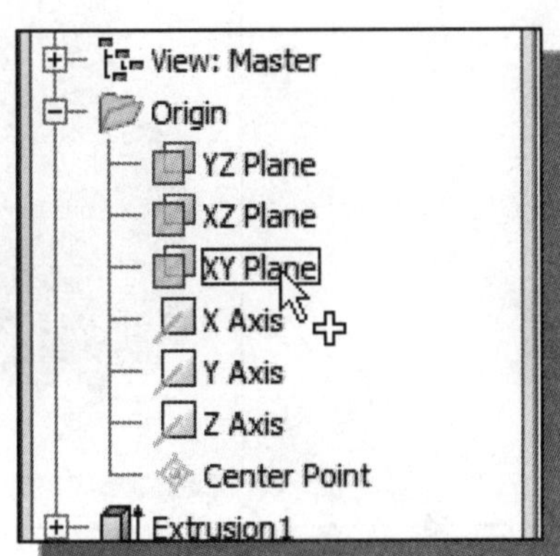

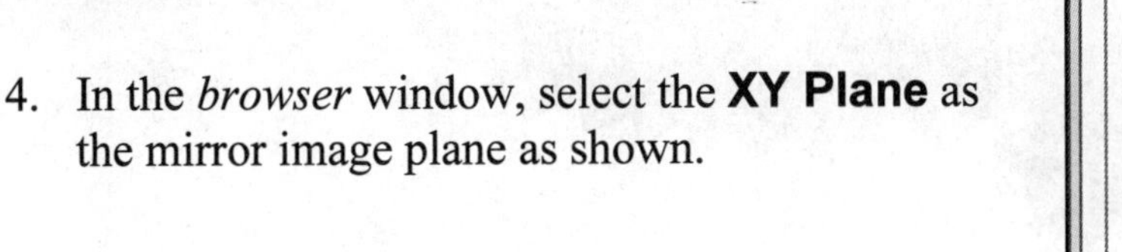

4. In the *browser* window, select the **XY Plane** as the mirror image plane as shown.

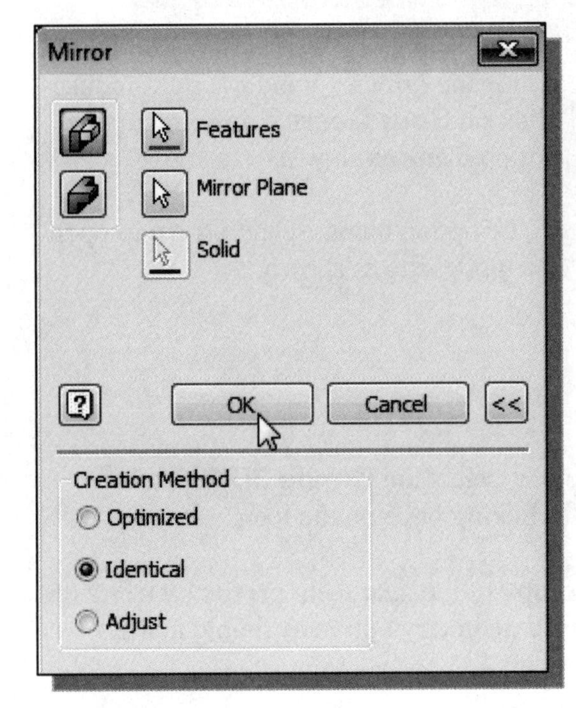

5. In the *Mirror* dialog box, click on the **[<<]** button to view the additional settings as shown.

❖ Note that three options are available. The mirrored duplicate can be created as identical to the original feature, or allow some settings to be adjusted.

6. Click **OK** to create the *Mirrored* feature.

Import the Profile of the Pinion Gear

- For the next feature, we will import the profile of the gear teeth of the pinion gear from the generated gear assembly.

1. In the graphics window, near the bottom edge, select the **SpurGears.iam** tab as shown.

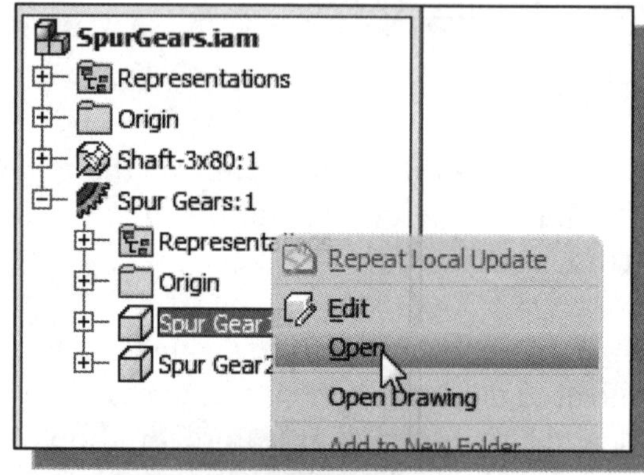

2. Inside the *browser* window, right-mouse-click on **Spur Gears:1** to bring up the option menu as shown.

3. In the option menu, select **Open** to open the model file as shown.

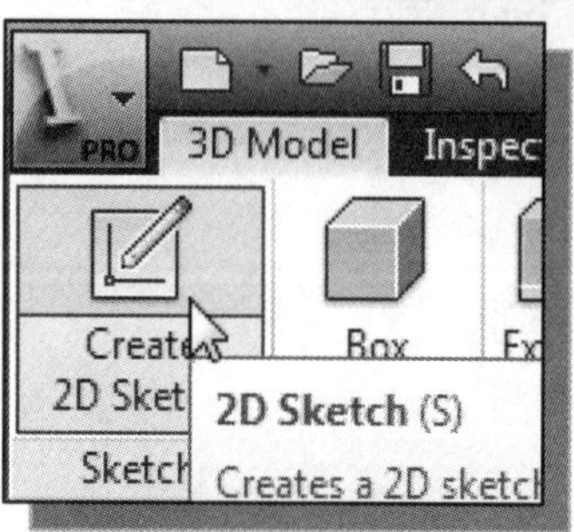

4. In the *Sketch* toolbar, select the **Create 2D Sketch** command by left-clicking once on the icon.

- We will create a copy of the gear teeth profile by using the automatic projected geometry option available in the Sketch mode.

5. Select the **top plane** of the pinion gear as the sketching plane.

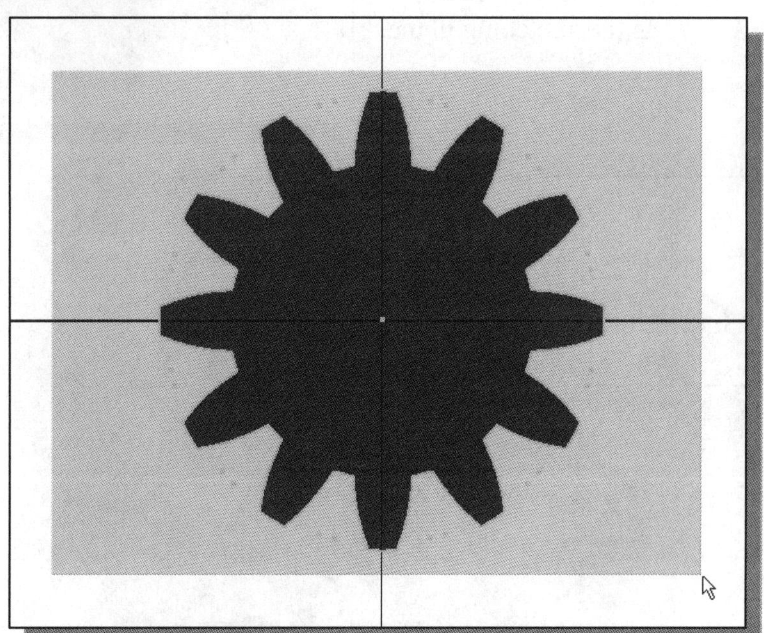

6. In the graphics window, use the left-mouse-button to enclose the pinion gear inside a ***selection window*** as shown.

- By default, *Inventor* Sketch mode will automatically project all of the geometry located on the sketching plane.

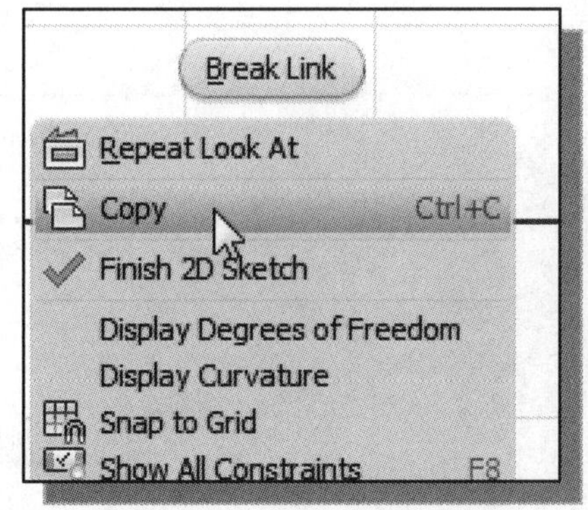

7. Inside the graphics window, right-mouse-click once to bring up the option menu as shown.

8. Select **Copy** as shown. Note that we are using the copy function of the system clipboard.

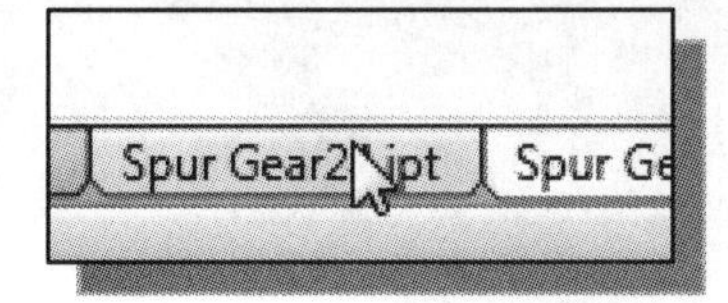

9. In the graphics window, near the bottom edge, select the **SpurGear21.ipt** tab as shown.

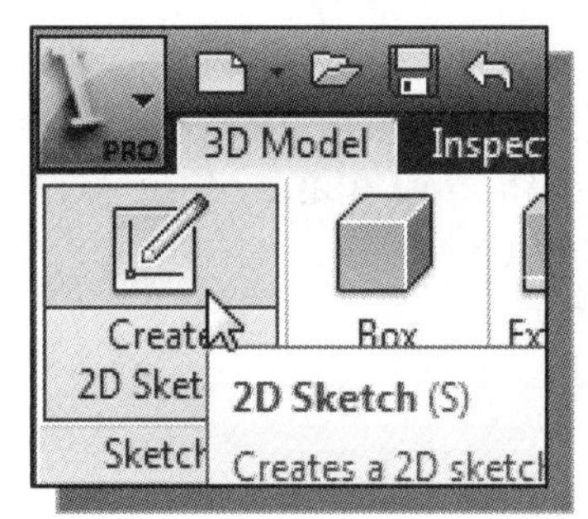

10. In the *Sketch* toolbar, select the **Create 2D Sketch** command by left-clicking once on the icon.

11. Select the **top plane** of the last cut feature as the sketching plane.

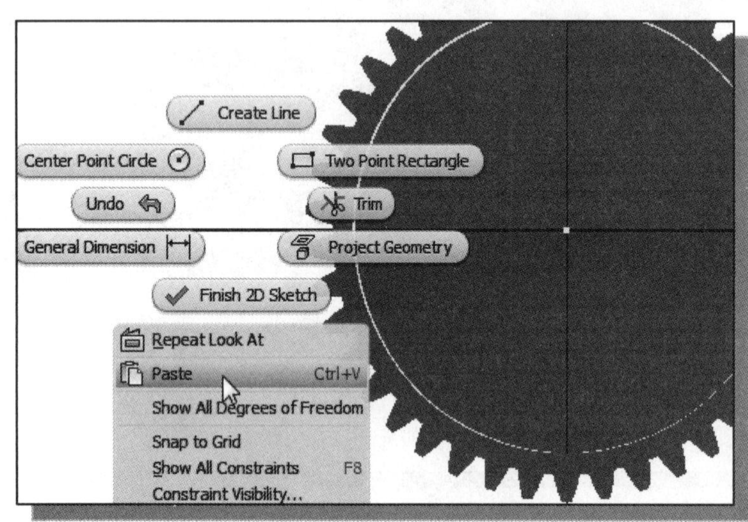

12. Inside the *browser* window, right-mouse-click to bring up the option menu as shown.

13. In the option menu, select **Paste** to paste the profile of the pinion gear as shown.

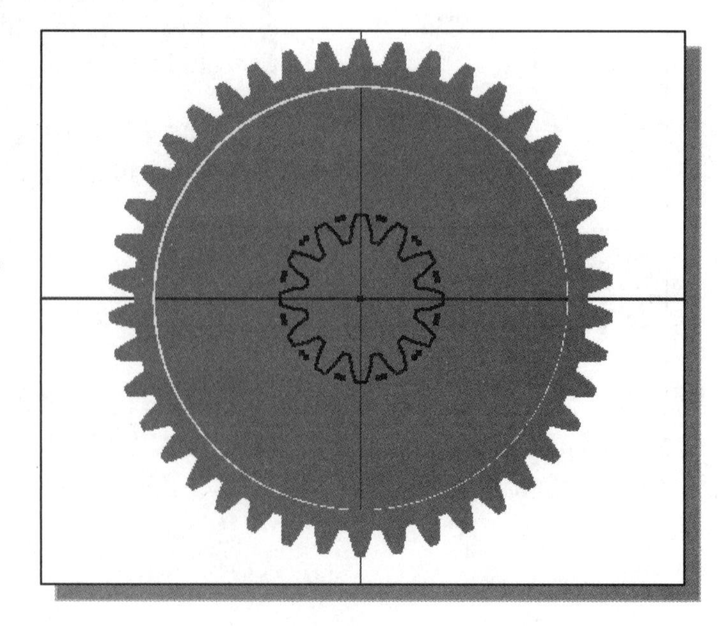

- Note that the orientation and location of the pasted pinion gear profile is aligned to the *center point* of the coordinate system.

14. Select **Finish Sketch** in the *Ribbon* toolbar to exit the **Sketch** mode.

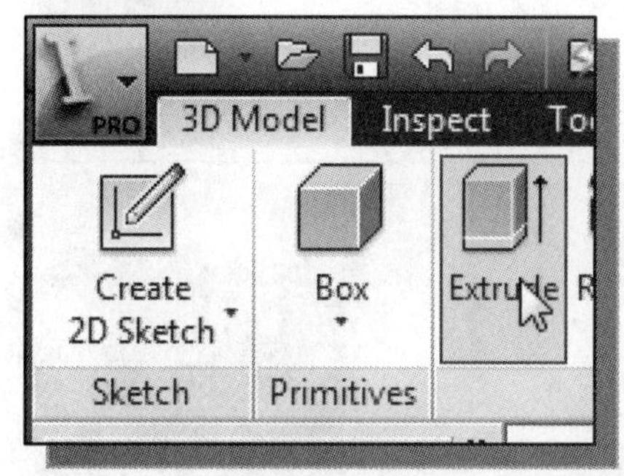

15. In the *Create* toolbar, select the **Extrude** command by clicking the left-mouse-button on the icon.

16. Select the inside region of the 2D sketch to create a profile as shown.

17. Set the extrude distance to **4.75 mm** and create the feature as shown.

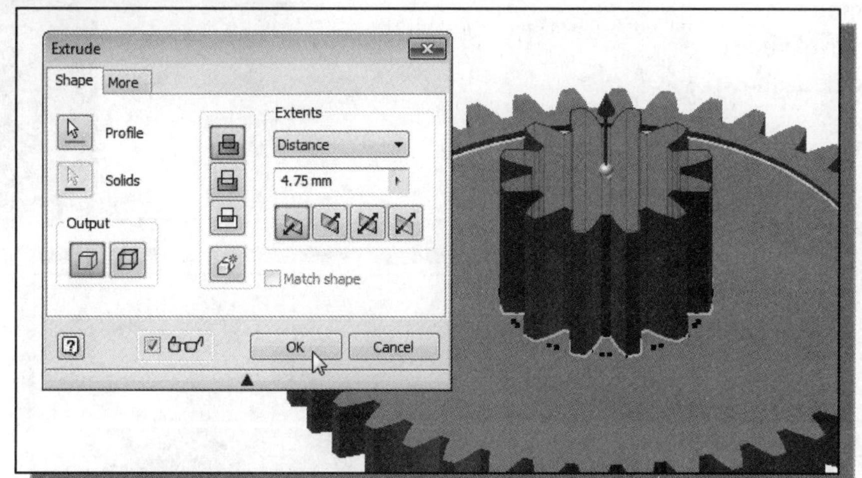

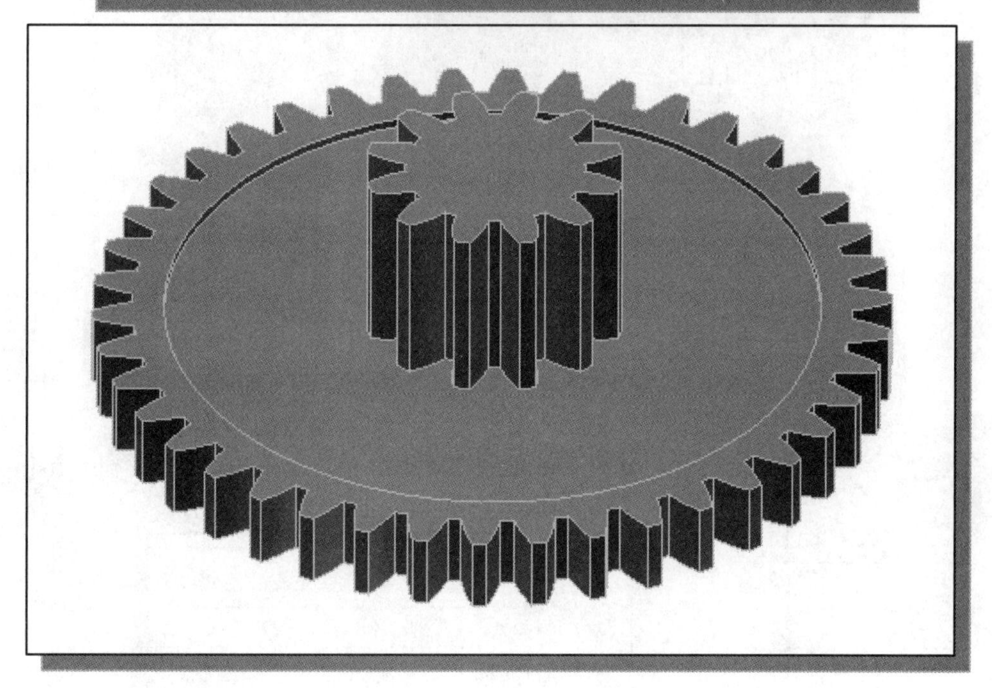

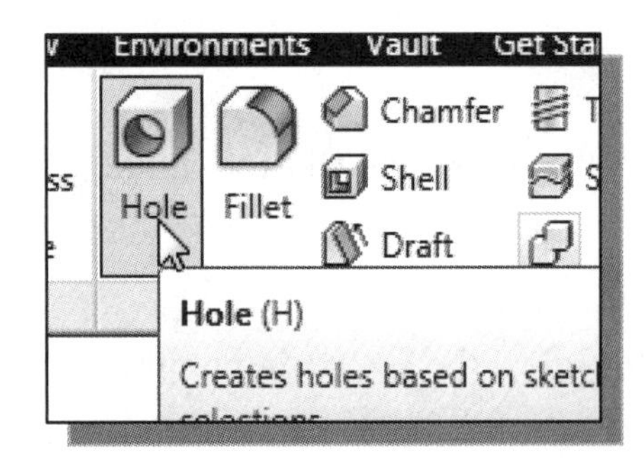

18. In the *Modify* toolbar, select the **Hole** command by clicking the left-mouse-button on the icon.

19. On your own, create a concentric hole feature, diameter **3mm**, aligned to the center of the solid model as shown.

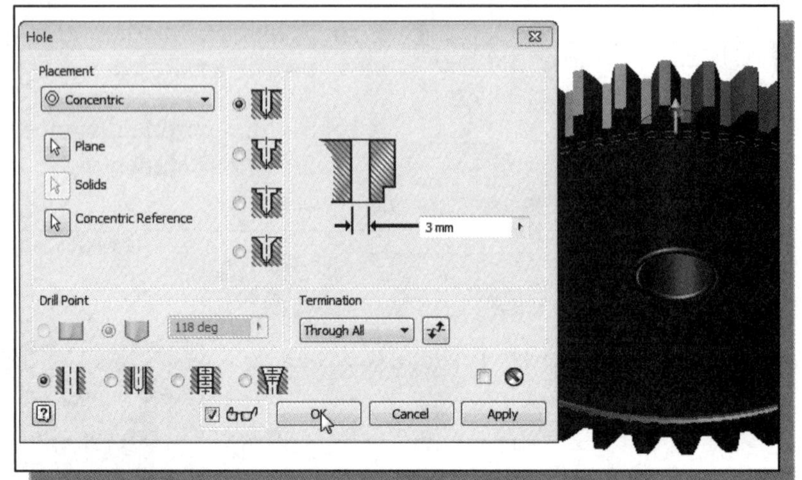

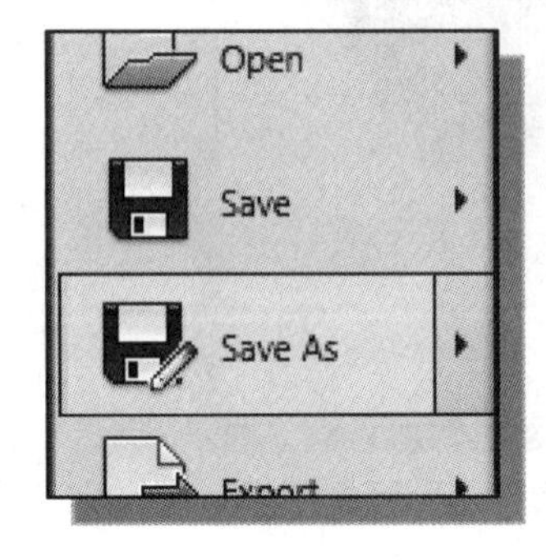

20. On your own, save the current model as a new part using **G2-Spur Gear** as the part file name.

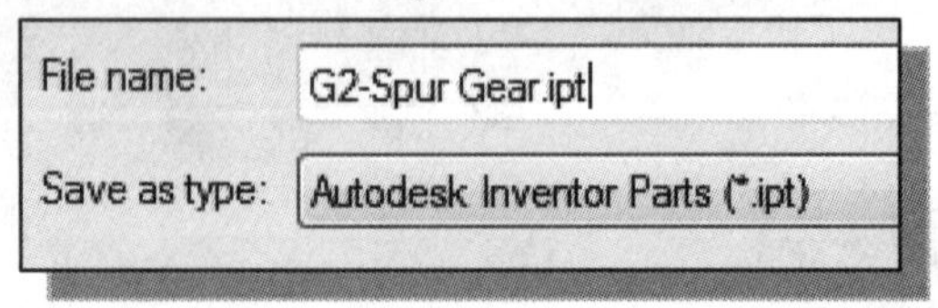

Create the *G3-Spur Gear* Part

❖ Next we will modify the current **Spur Gear 21** part to create the *G3-Spur Gear* part.

1. Switch to the **Spur Gear 21** part by left-mouse-clicking on the associated tab as shown.

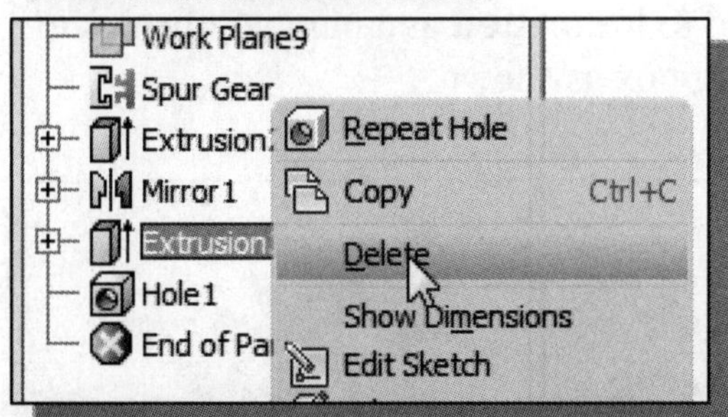

2. Inside the *browser* window, right-mouse-click on **Extrusion3** to bring up the option menu as shown.

3. In the option menu, select **Delete** to remove the extrusion feature as shown.

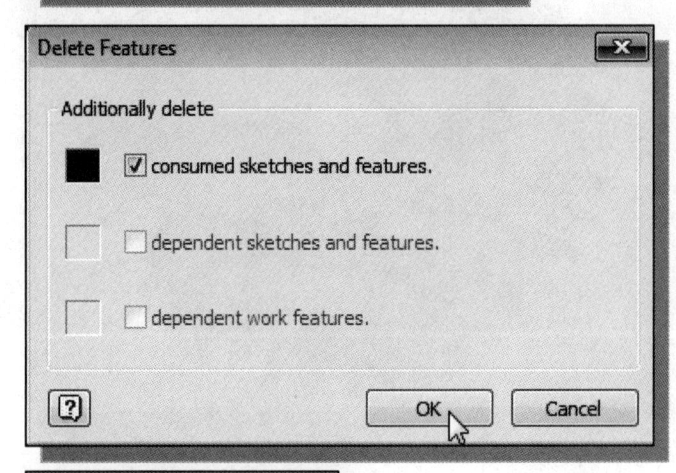

4. Inside the *Delete Features* dialog box, confirm the **delete associated sketches** option is turned ***ON***.

5. Click **OK** to accept the settings and delete the selected feature.

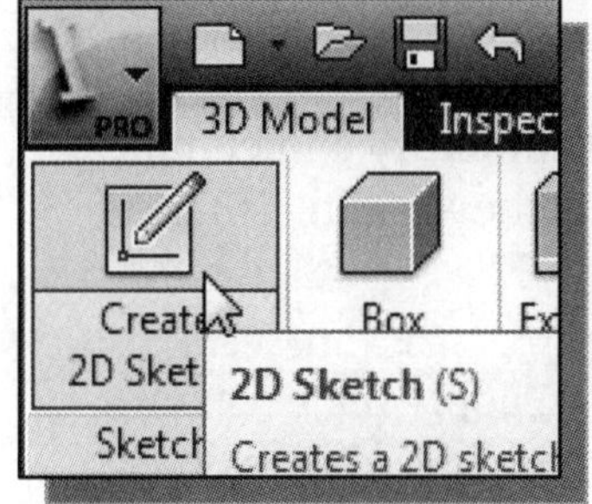

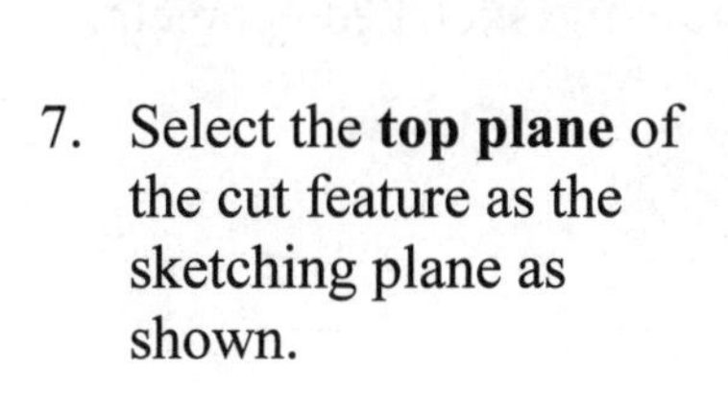

6. In the *Sketch* toolbar, select the **Create 2D Sketch** command by left-clicking once on the icon.

7. Select the **top plane** of the cut feature as the sketching plane as shown.

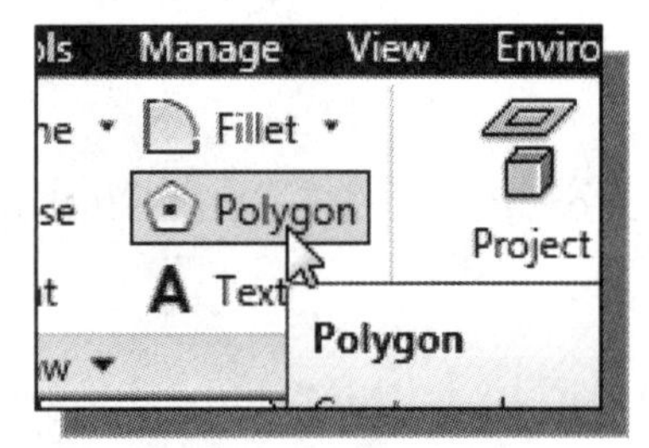

8. In the *Draw* toolbar panel, select the **Polygon** command by left-mouse-clicking once on the icon.

9. Set the polygon option to **Inscribed** and number of sides to **6** in the *Polygon* dialog box as shown.

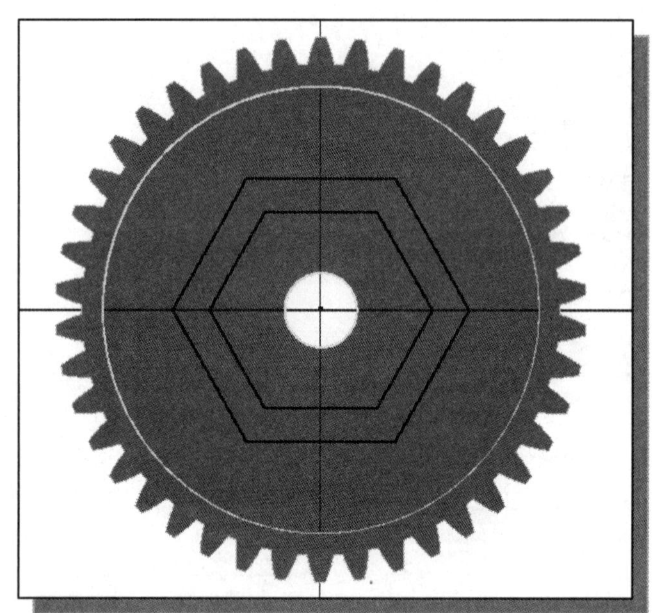

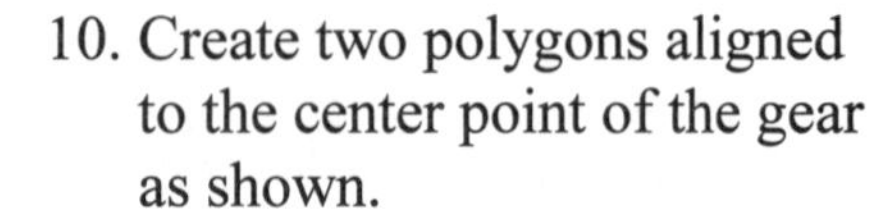

10. Create two polygons aligned to the center point of the gear as shown.

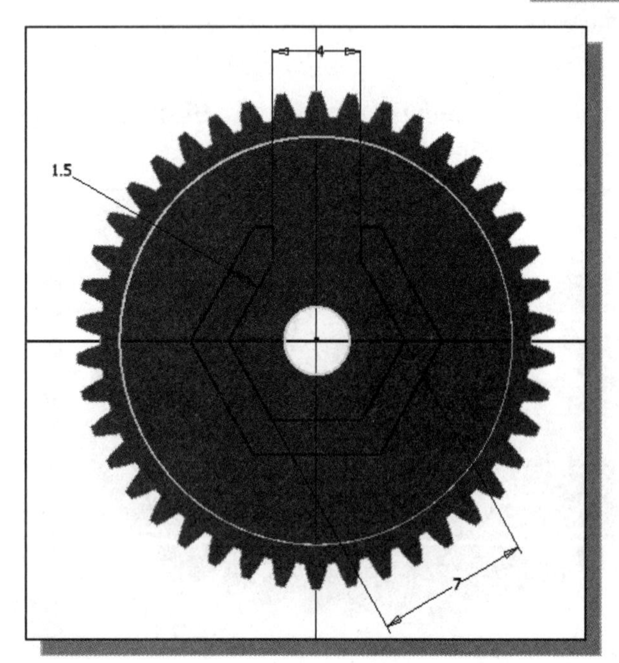

11. On your own, modify and complete the sketch as shown in the figure.

12. On your own, apply the necessary constraints and three dimensions, **7 mm**, **4 mm**, and **1.5 mm** to fully constrain the sketched geometry.

13. Select **Finish Sketch** in the *Ribbon* toolbar to exit the **Sketch** mode.

14. In the *Create* toolbar, select the **Extrude** command by releasing the left-mouse-button on the icon.
15. Select the inside region of the 2D sketch to create a profile as shown.

16. Set the extrude distance to **2.75 mm** and create the feature as shown.

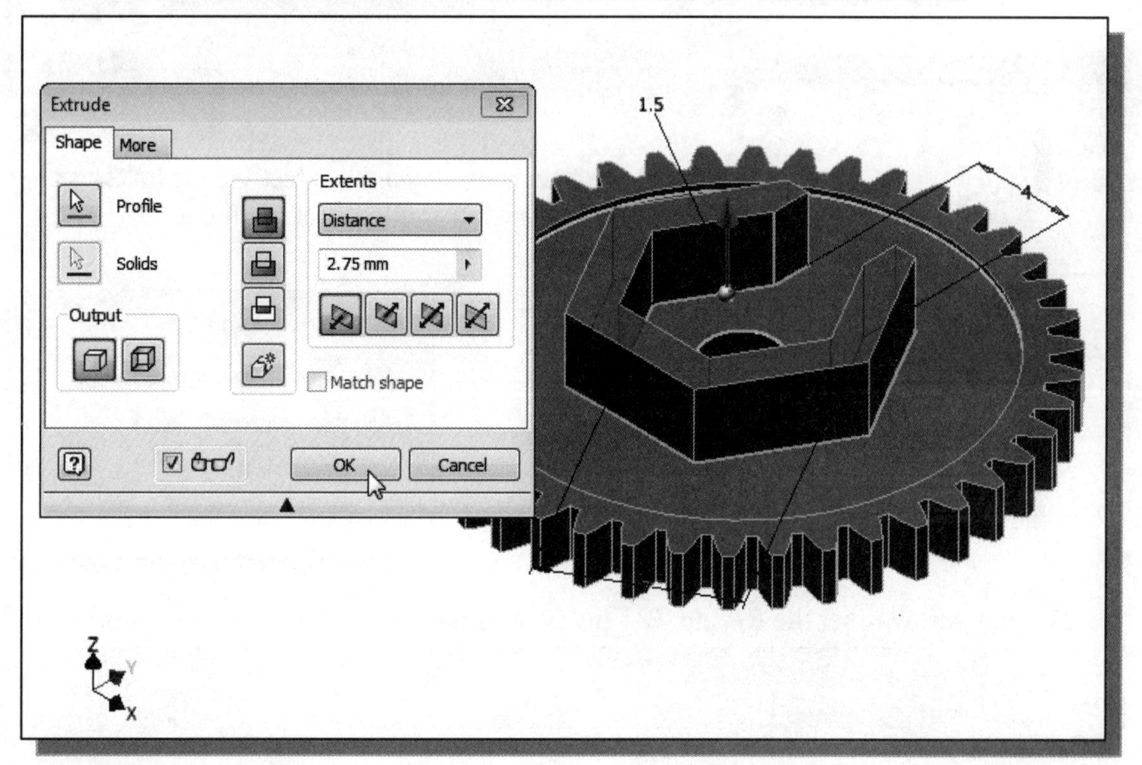

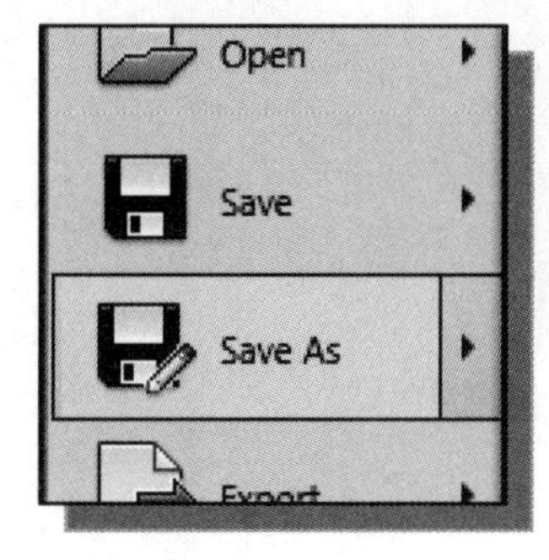

17. On your own, save the current model as a new part using **G3-Spur Gear** as the part file name.

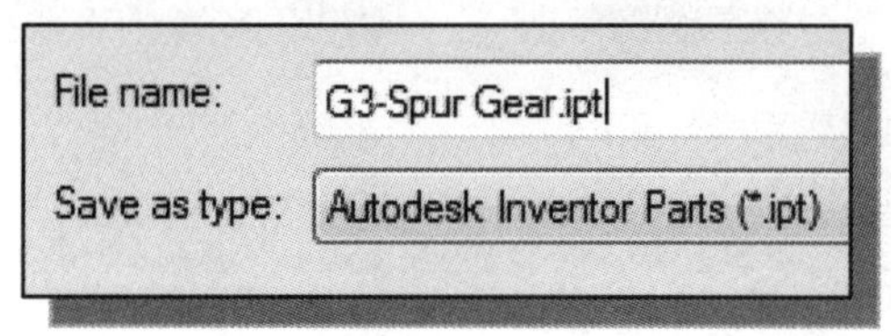

Create another Spur Gear Set

❖ For the two other gears, we will first create another set of spur gears by using the *Design Accelerator*.

1. In the graphics window, near the bottom edge, select the **SpurGears.iam** tab as shown.

2. In the *Ribbon* panel, select the **Design** tab by left-mouse-clicking the tab.

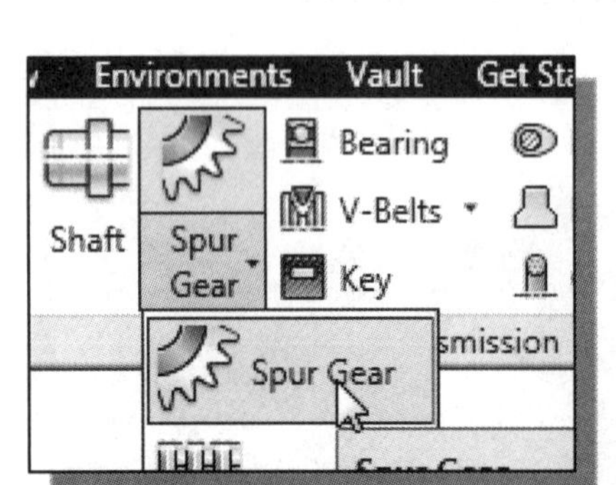

3. In the *Transmisson* panel, select the **Spur Gear** command by left-mouse-clicking the icon.

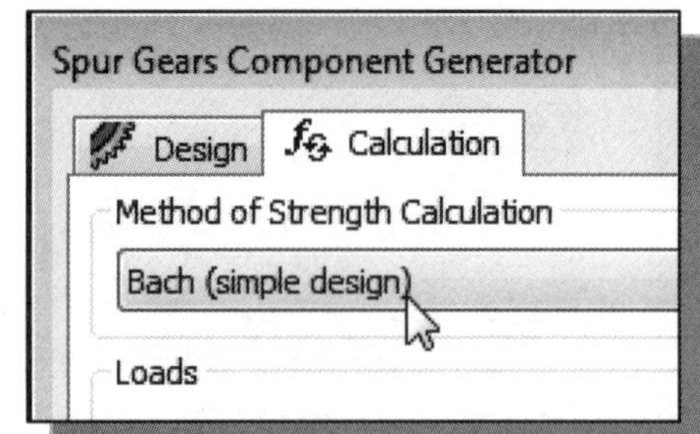

4. Click on the **Calculation** tab to set the associated power rating of the system. Choose **Bach (Simple design)** as the *Method of Strength Calculation.*

5. On your own, set the ***8T*** and ***72T*** gears information as shown in the figure.

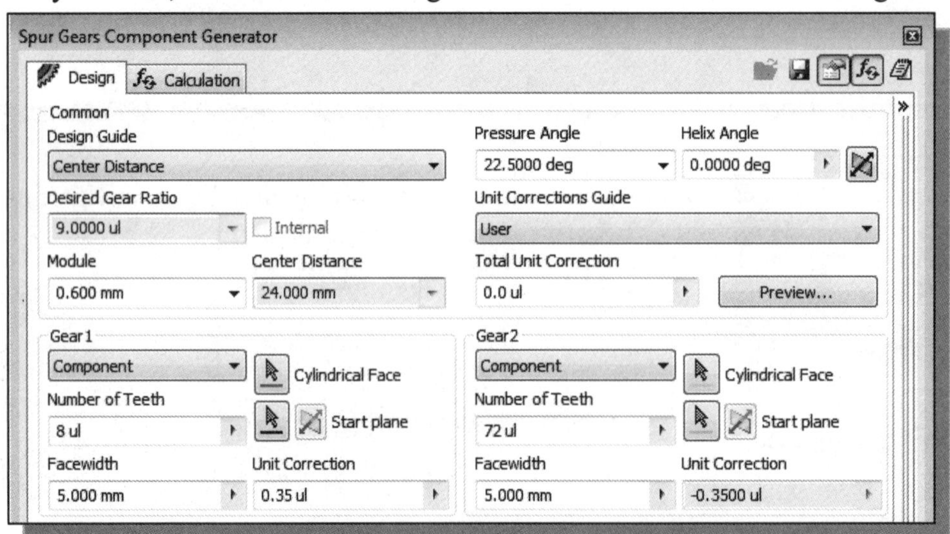

6. The results of the calculations are listed in the dialog box as shown. In this case, *Inventor* indicated that the gears will need to be adjusted for proper operation.

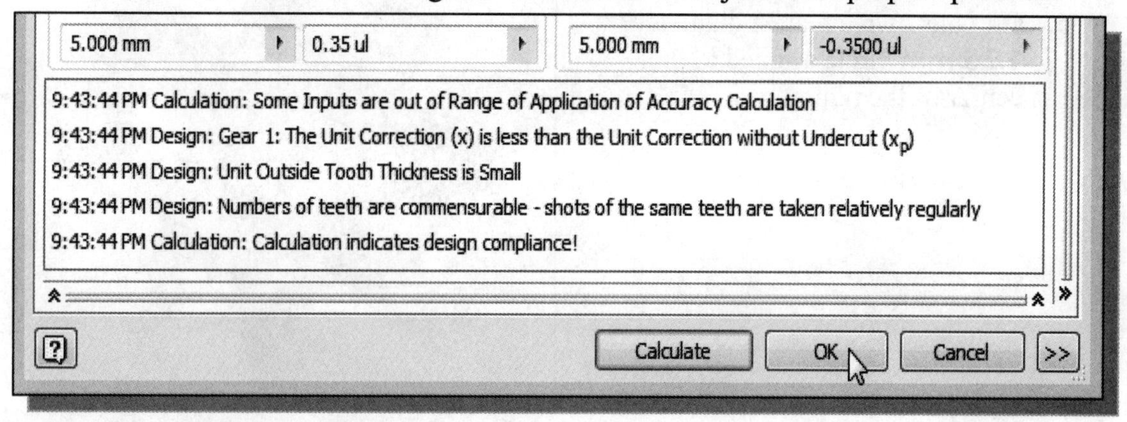

7. Click **OK** to accept the settings. On your own, place a copy of the generated gears in the assembly model.

Create the G0-Pinion Part

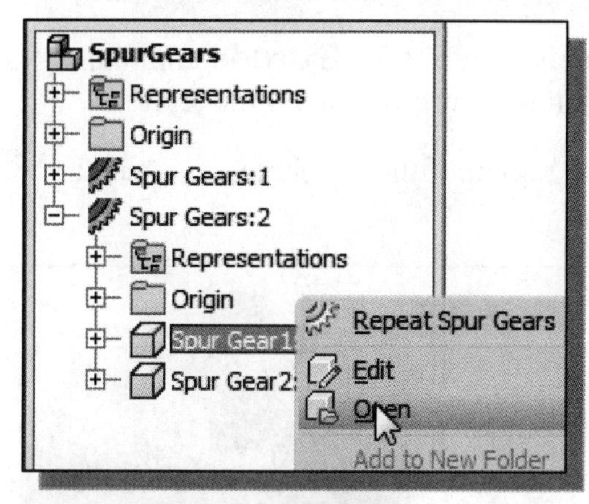

1. Inside the *browser* window, right-mouse-click on the **pinion gear**, on the second set of gears, to bring up the option menu as shown.

2. In the option menu, select **Open** to open the selected model as shown in the figure.

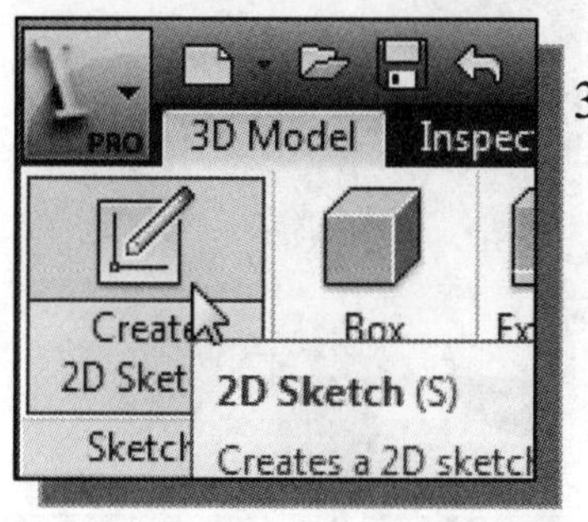

3. In the *Sketch* toolbar, select the **Create 2D Sketch** command by left-clicking once on the icon.

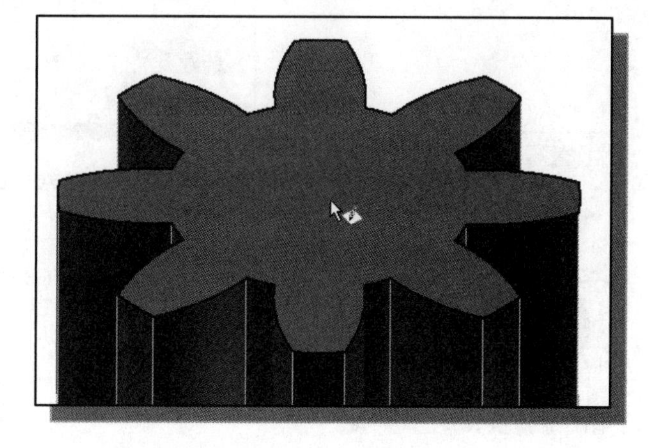

4. Select the **top plane** of the pinion gear model as the sketching plane.

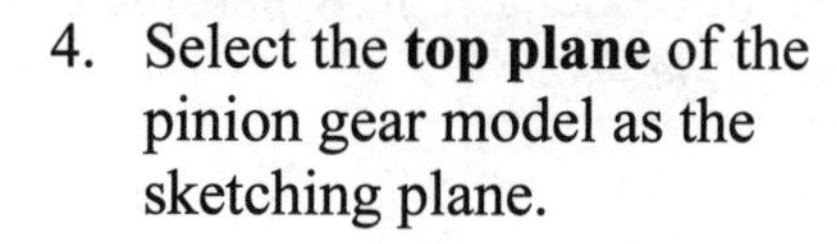

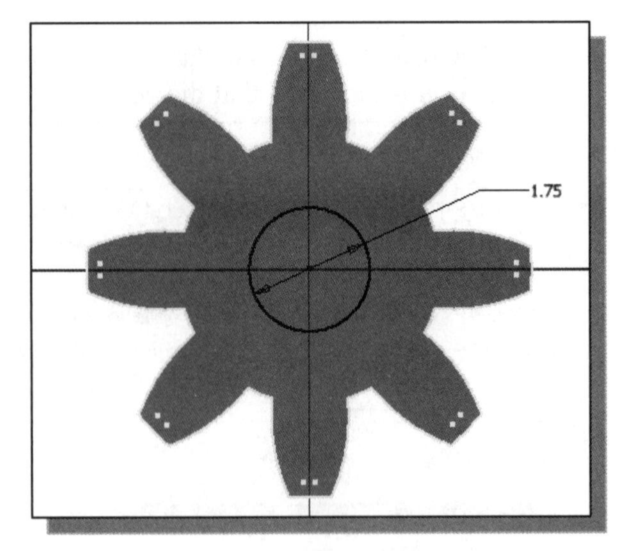

5. On your own, create a diameter **1.75 mm** circle aligned to the center of the pinion gear.

6. Select **Finish Sketch** in the *Ribbon* toolbar to exit the Sketch mode.

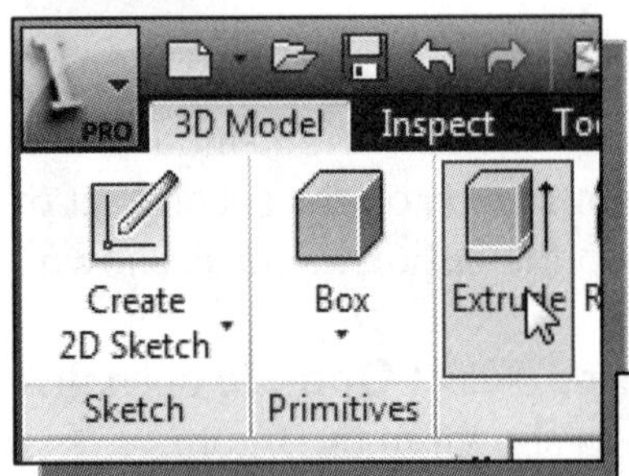

7. In the *Create* toolbar, select the **Extrude** command by releasing the left-mouse-button on the icon.

8. Select the inside region of the 2D sketch to create a profile as shown.

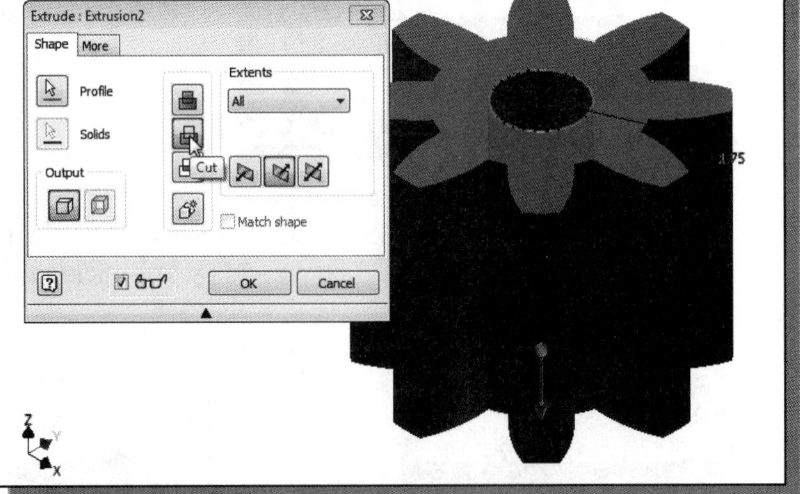

9. Set the extrude extents to **All** and extrude option to **Cut** as shown.

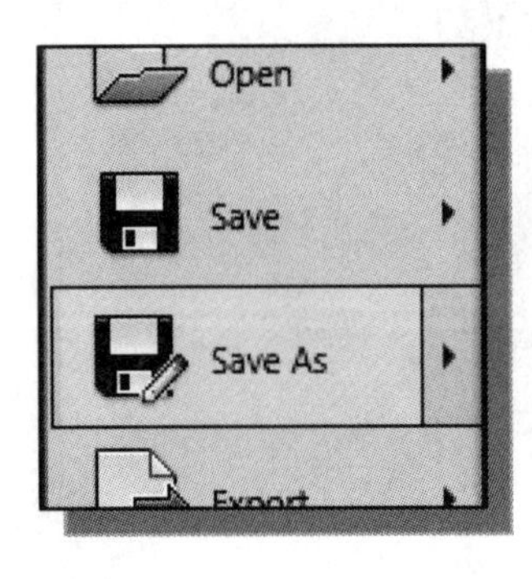

10. On your own, save the current model as a new part using **G0-Pinion** as the part file name.

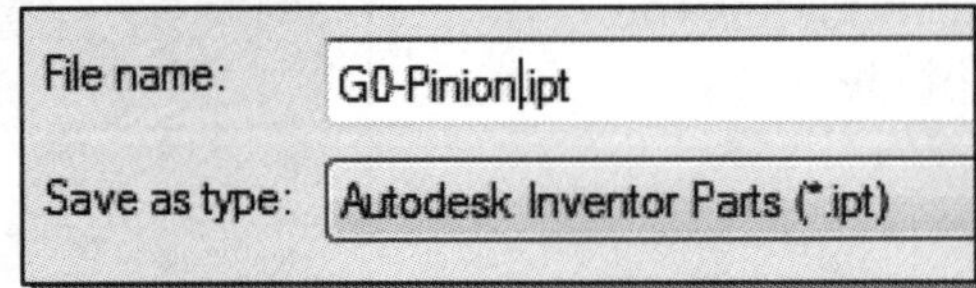

Start a New Part File

❖ The last gear we need for the *Tamiya Tiger kit* is a non-standard compound gear. We will create this gear using several options, with a portion of it using the **Revolve** *Pattern* commands.

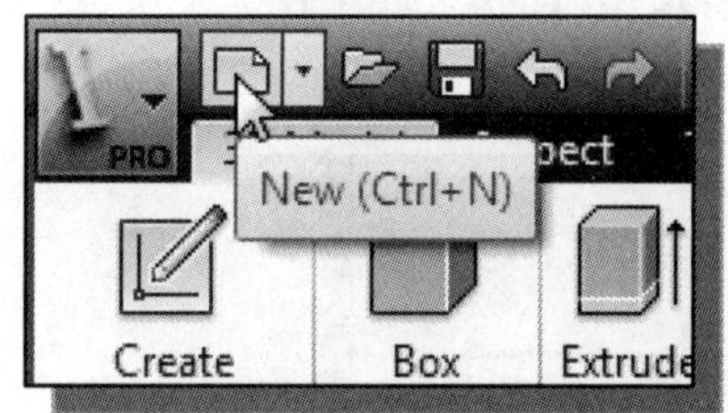

1. In the *Quick Access* menu, select **New** to start a new file.

2. In the *New File* dialog box, select the **Standard (mm).ipt** template by left-mouse-clicking once on the icon.

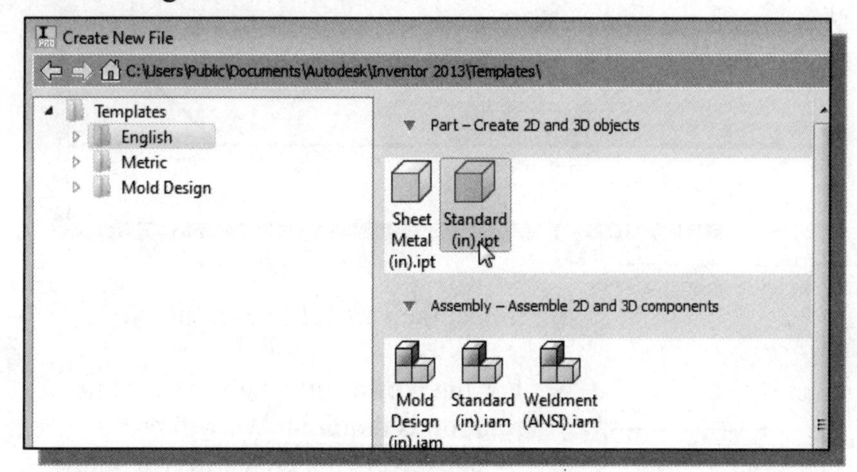

3. Click **Create** to accept the selection and start a new file.

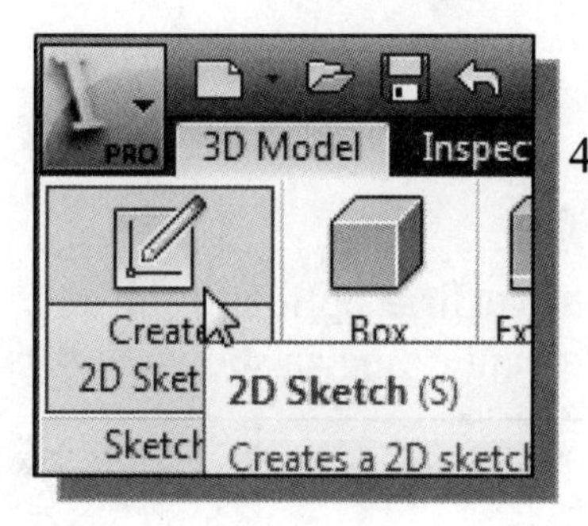

4. In the *Sketch* toolbar, select the **Create 2D Sketch** command by left-clicking once on the icon.

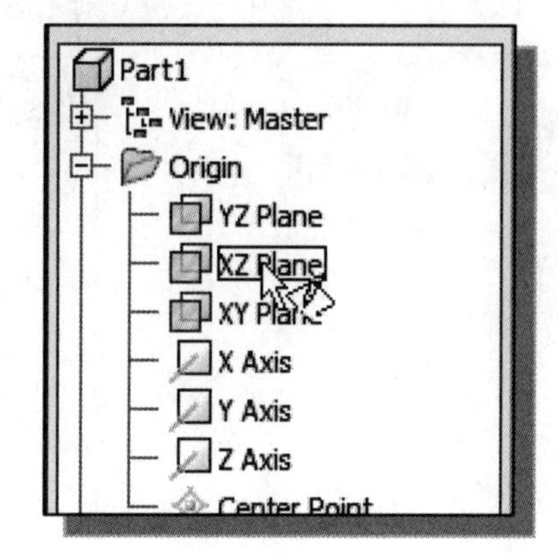

5. Inside the *browser* window, select the **XZ Plane** as the sketching plane as shown.

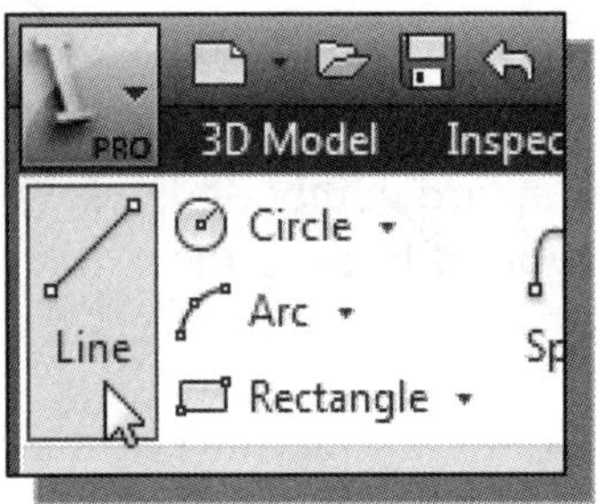

6. Select the **Line** command by clicking once with the left-mouse-button on the icon in the *Draw* toolbar.

7. Create **a closed region** sketch to the right side of the center point; also create **a vertical line** aligned to the center point of the coordinate system as shown.

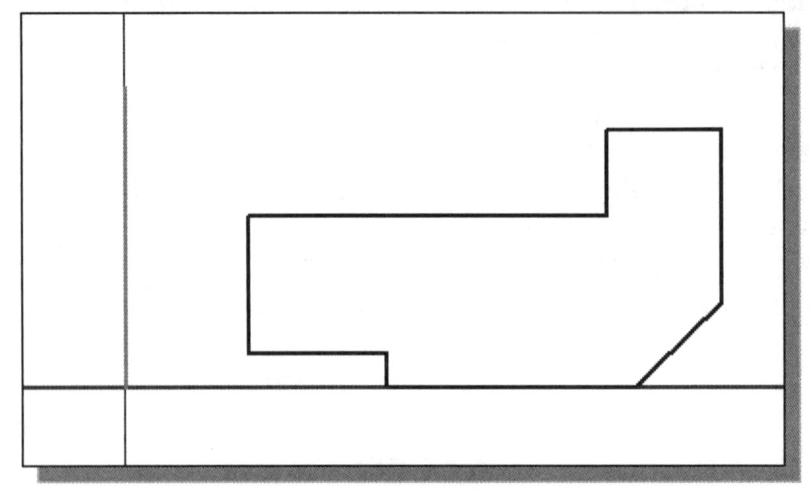

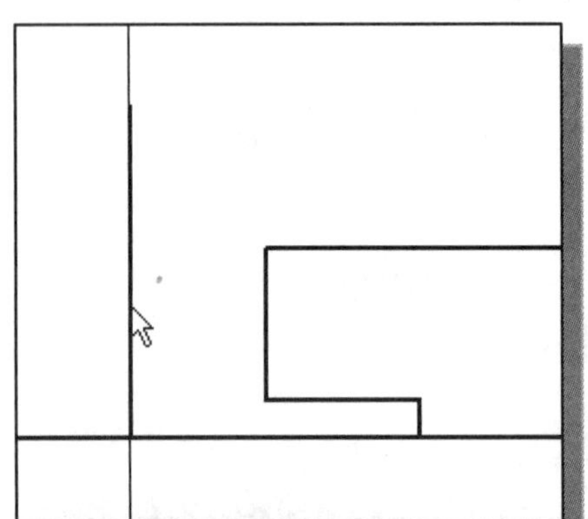

8. Select the vertical line as shown.

❖ For the **Revolve** command, a center axis of rotation is required. We will create a center line by converting the created vertical line.

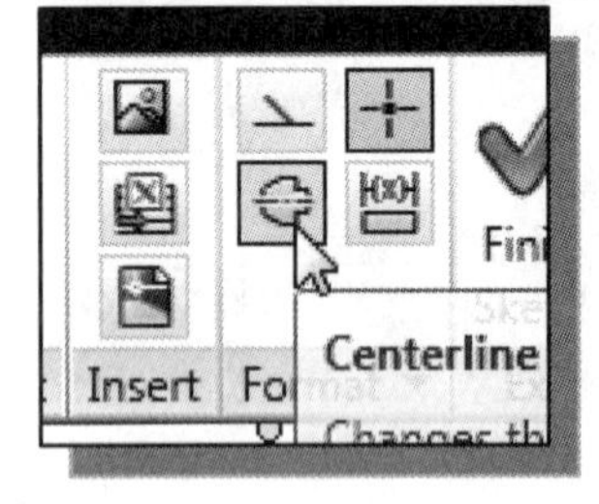

9. In the *Ribbon* toolbar, select the **Centerline** option to convert the created vertical line to a center line as shown.

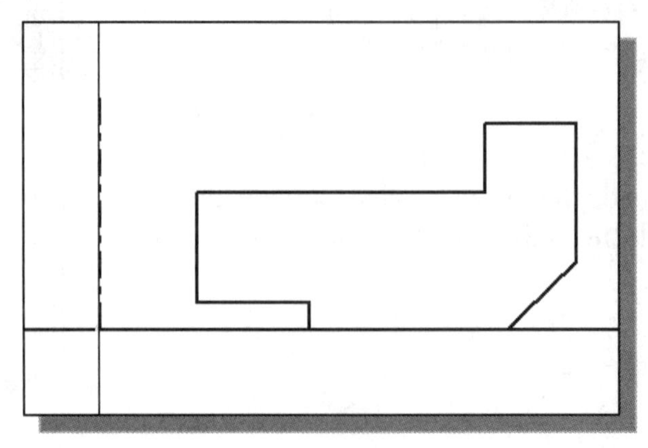

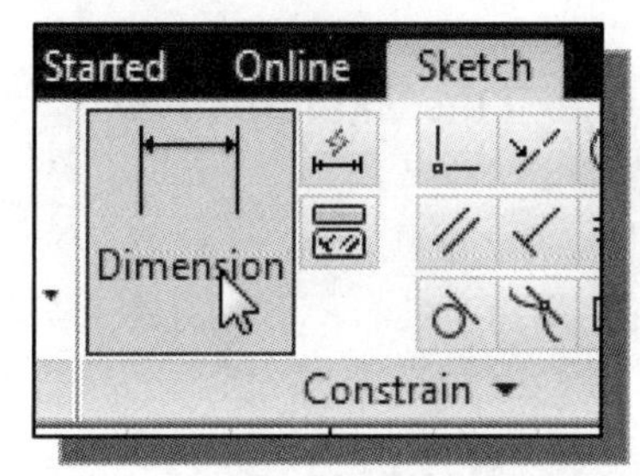

10. Select the **Smart Dimension** command in the *Constrain* panel.

11. Pick the **center line** as the first entity to dimension as shown in the figure.

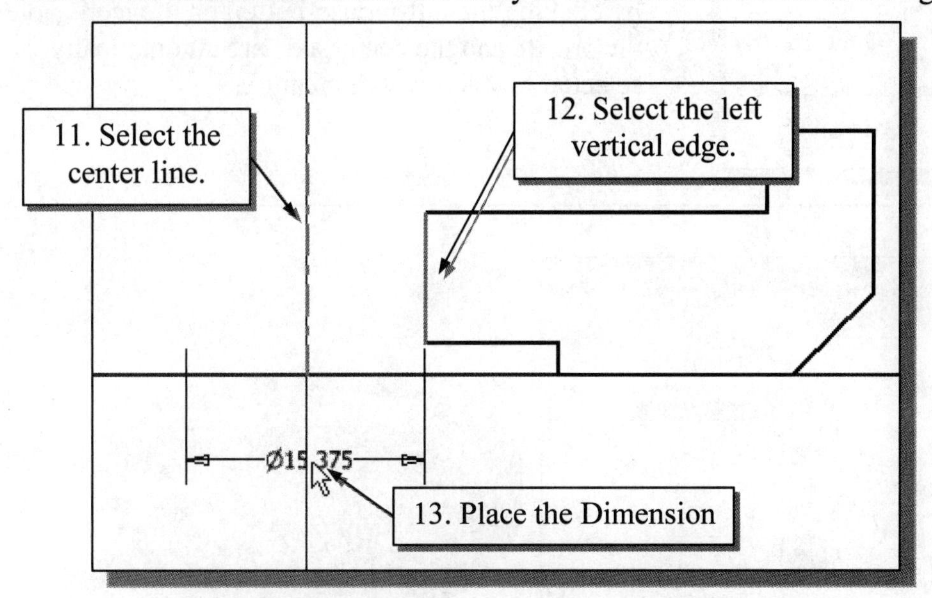

12. Select the **left vertical line** as the second object to dimension.

13. Place the dimension text below the center point of the coordinate system.

- To create a dimension that will account for the symmetrical nature of the design, pick the axis of symmetry, pick the entity, and then place the dimension. (The **Linear Dimension** option is also available through the right-mouse-click option menu.)

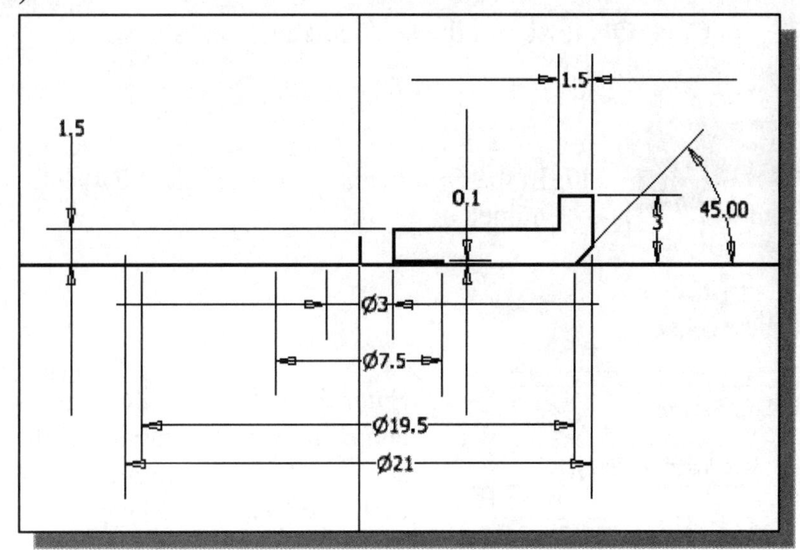

14. Select **Finish Sketch** in the *Ribbon* toolbar to exit the **Sketch** mode.

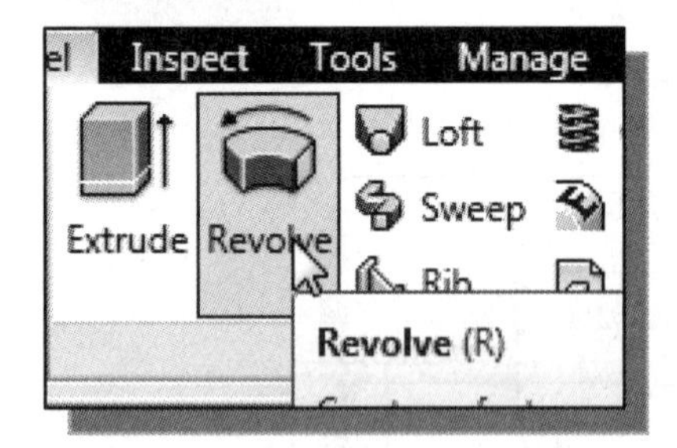

15. In the *Create* toolbar, select the **Revolve** command by clicking the left-mouse-button on the icon. Note the profile and the center axis are automatically selected for the revolved feature.

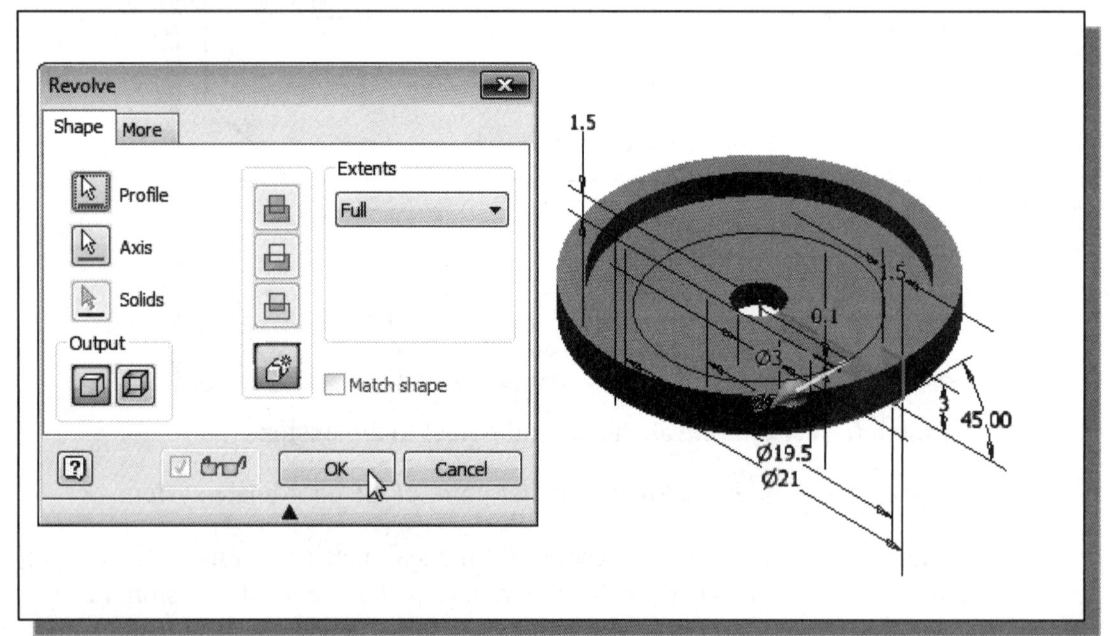

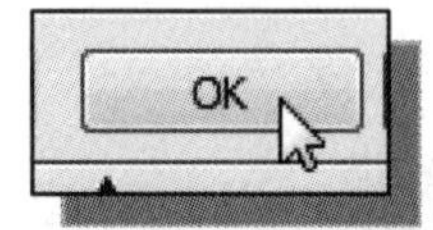

16. Click **OK** to accept the selection and start a new file.

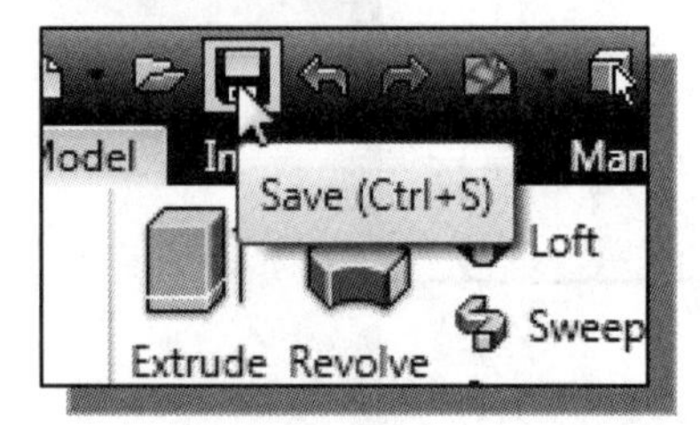

17. In the *Quick Access* menu, select **Save** to save the model.

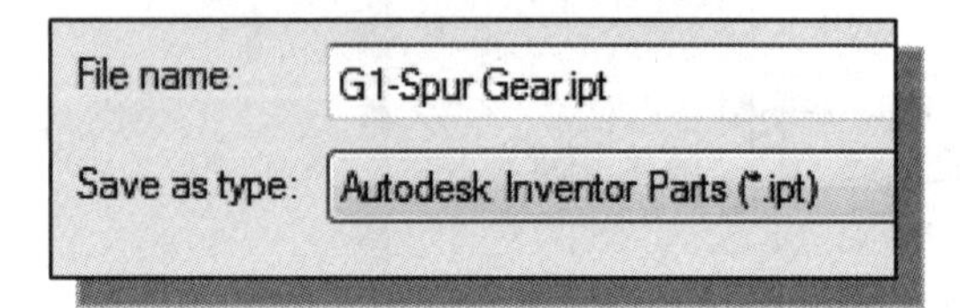

18. Enter the **G1-Spur Gear** as the part file name and save the model. Note that additional modifications will be performed in the next sections.

Export/Import the Generated Gear Profile

- We will create the *teeth* by importing the tooth profile from the gear generated by the *Design Accelerator*.

1. Switch to the **SpurGears** assembly by clicking on the associated tab.

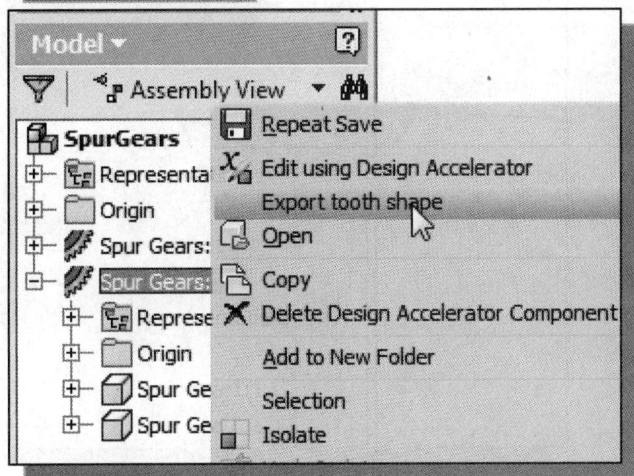

2. Inside the *browser* window, right-mouse-click on the second set of **SpurGears** to bring up the option menu as shown.

3. In the option menu, select **Export tooth shape** to proceed to export the tooth profile.

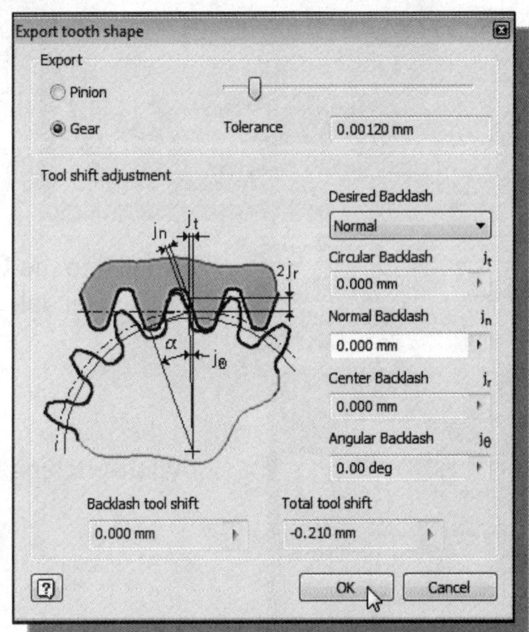

4. In the *Export tooth shape* dialog box, select the **Gear profile** to export.

5. Click **OK** to exit the command.

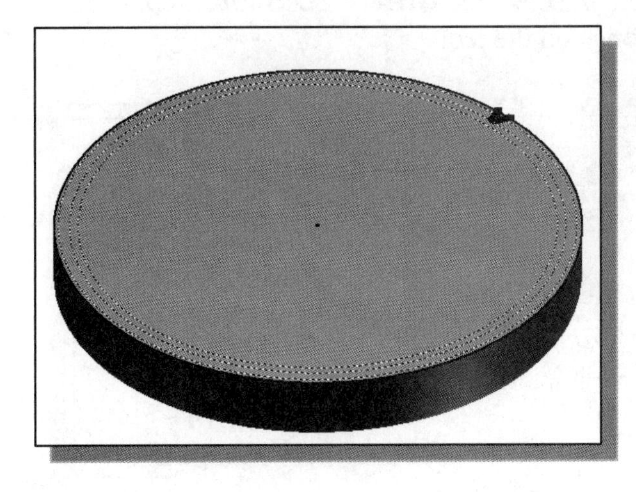

- Note the tooth profile with of the selected gear is displayed in a separate window.

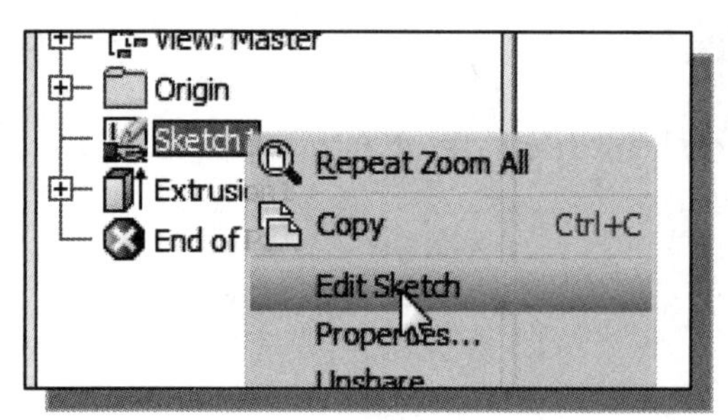

6. Inside the *browser* window, right-mouse-click on the **Sketch1** to bring up the *option menu* and select **Edit Sketch** as shown.

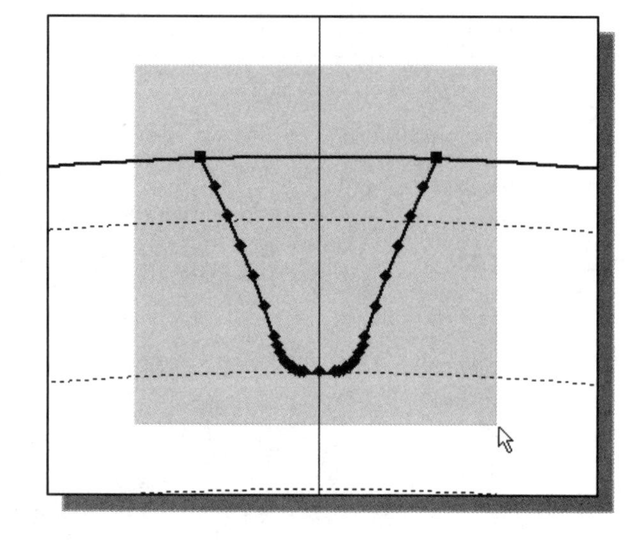

7. In the graphics window, use the left-mouse-button to enclose the tooth profile inside a **selection window** as shown.

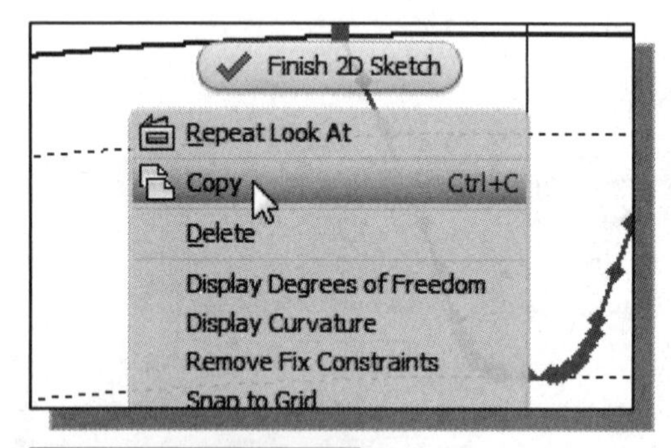

8. In the graphics window, right-mouse-click on bring up the option m*enu* and select **Copy** as shown.

9. Switch to the **G1-Spur Gear** part by clicking on the associated tab.

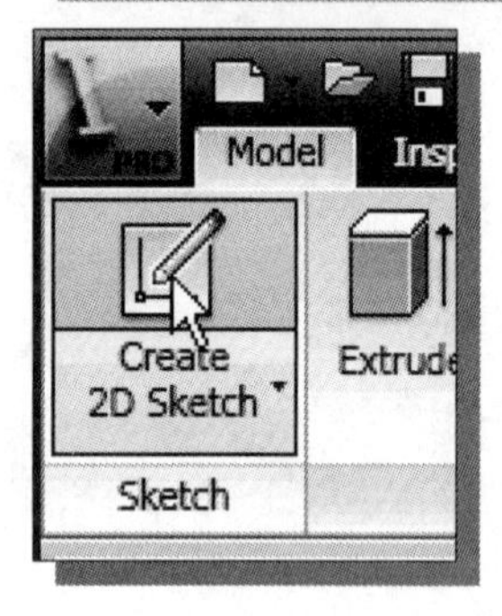

10. In the *Sketch* toolbar, select the **Create 2D Sketch** command by left-clicking once on the icon.

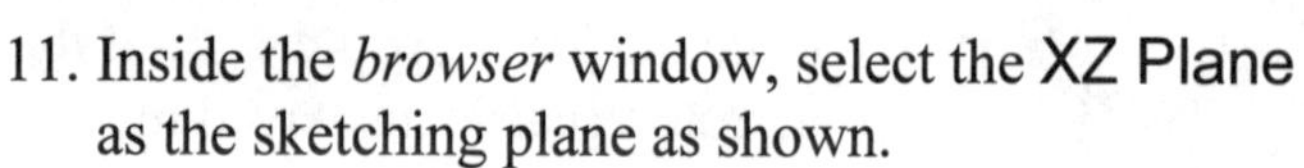

11. Inside the *browser* window, select the XZ Plane as the sketching plane as shown.

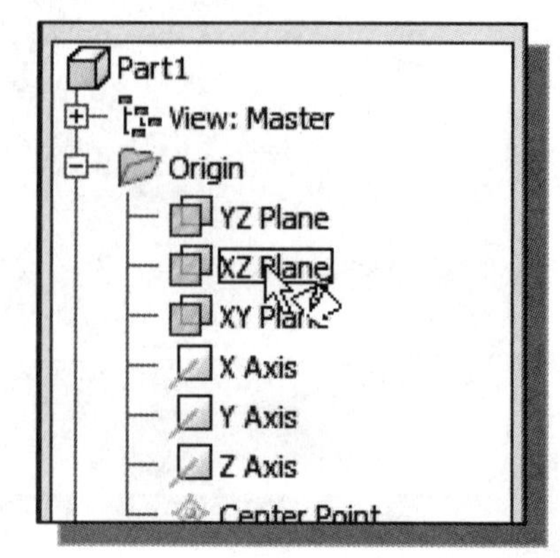

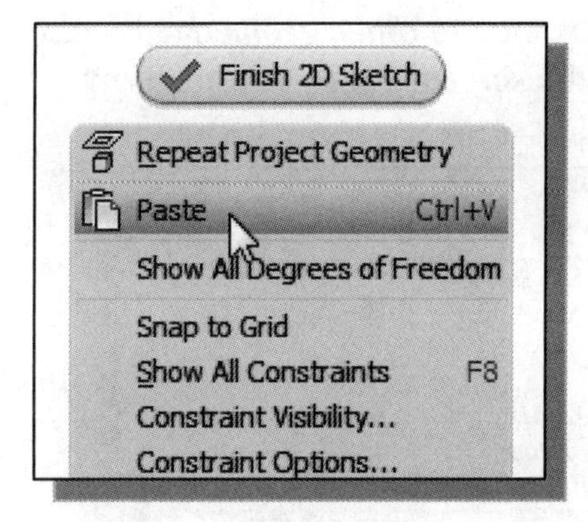

12. In the graphics window, right-mouse-click to bring up the option menu and select **Paste** as shown.

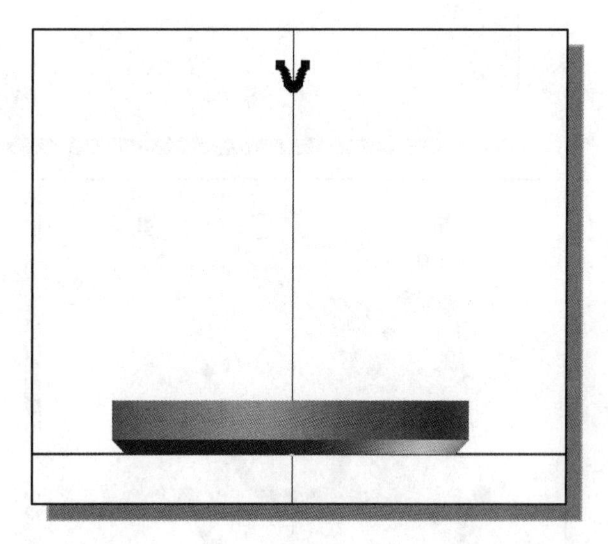

- Note the tooth profile is now imported into the current sketch. The imported geometry is placed above the solid model.

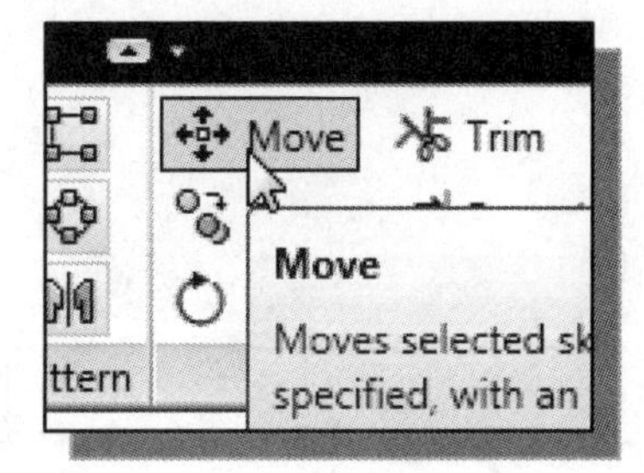

13. In the *Modify* toolbar, select the **Move** command by left-clicking once on the icon.

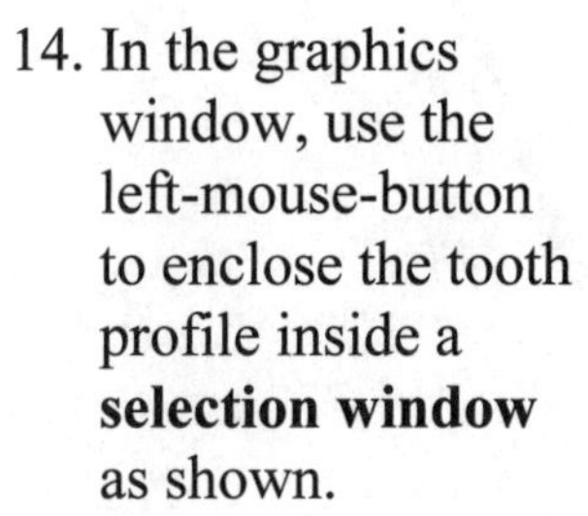

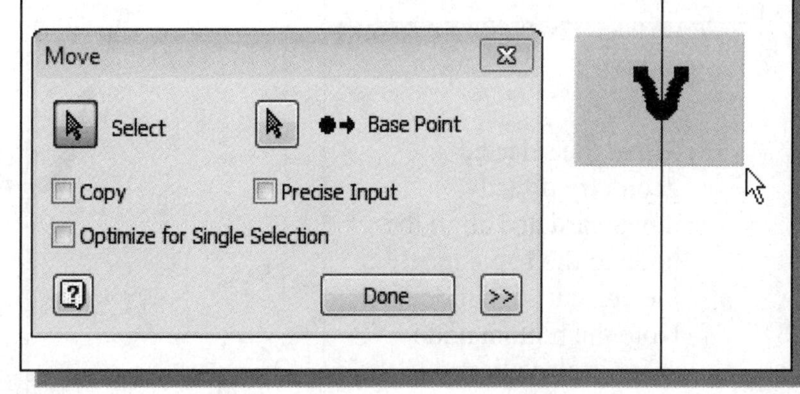

14. In the graphics window, use the left-mouse-button to enclose the tooth profile inside a **selection window** as shown.

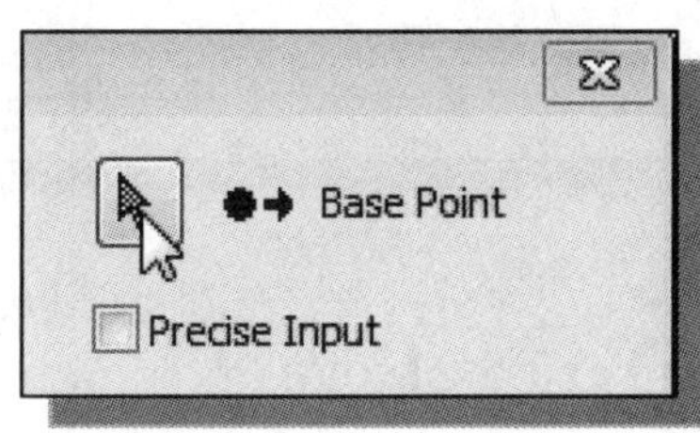

15. In the *Move* dialog box, click on the **Base Point** icon as shown.

16. The message "*The geometry being edited is constrained to other geometry. Would you like those constraints removed?*" appears on the screen. Click **Yes** to allow adjustments to the selected geometry.

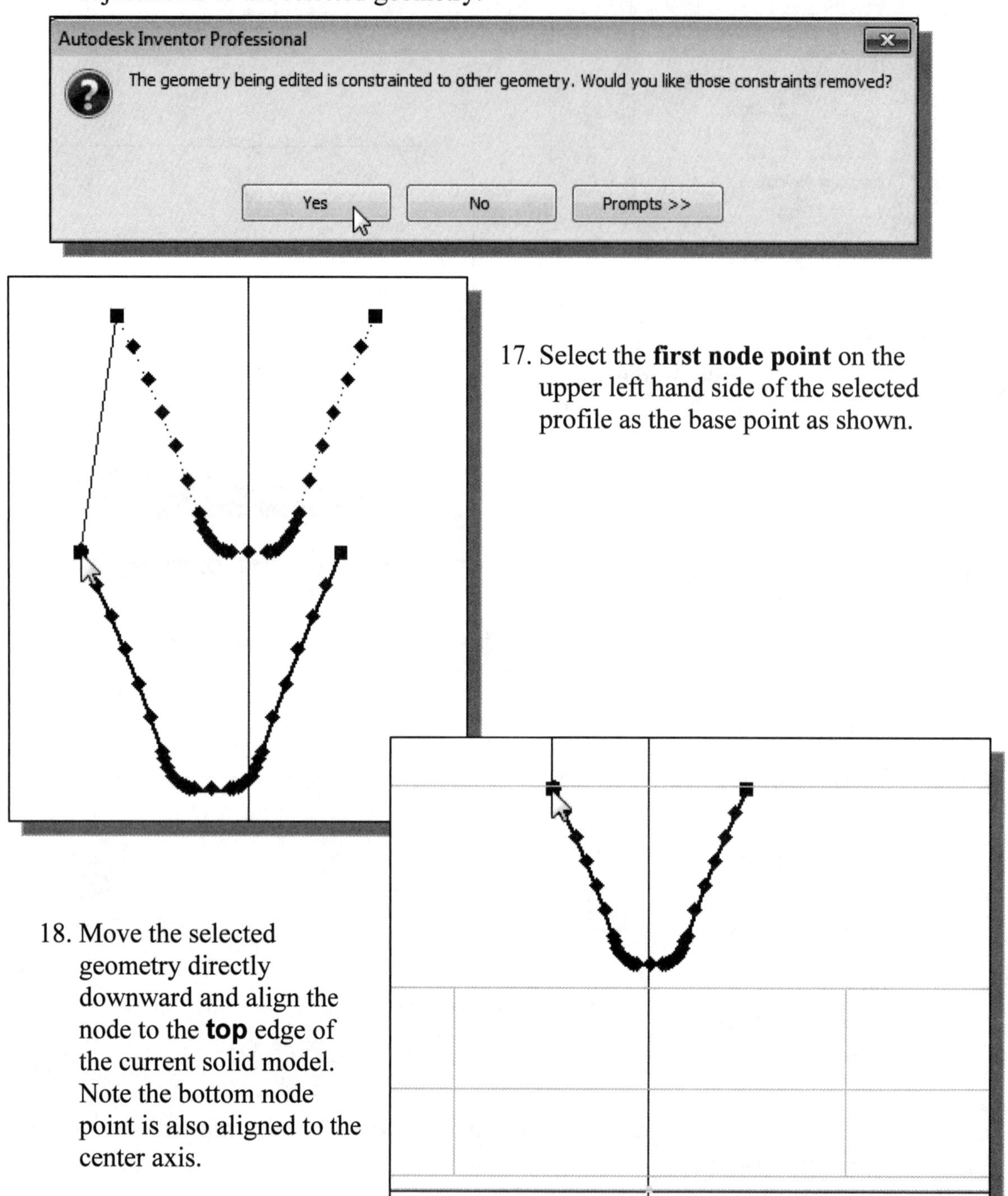

17. Select the **first node point** on the upper left hand side of the selected profile as the base point as shown.

18. Move the selected geometry directly downward and align the node to the **top** edge of the current solid model. Note the bottom node point is also aligned to the center axis.

19. Click **Done** to accept the change and exit the Move command.

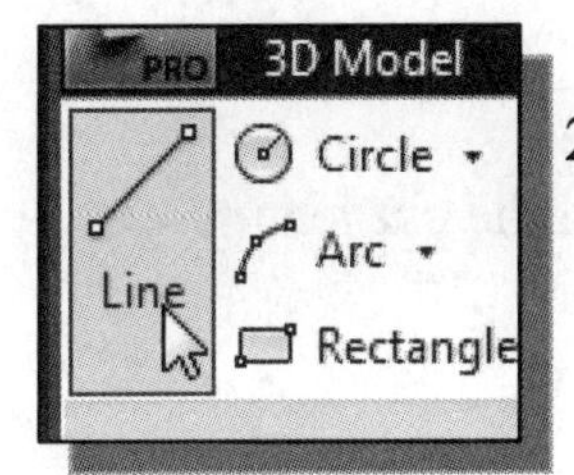

20. Select the **Line** command by clicking once with the left-mouse-button on the icon in the *Draw* toolbar.

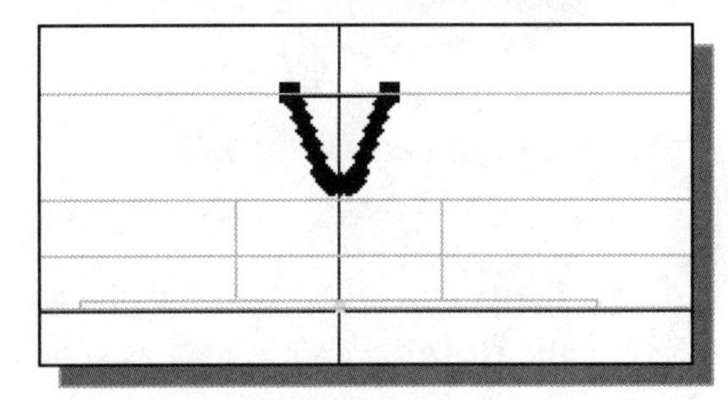

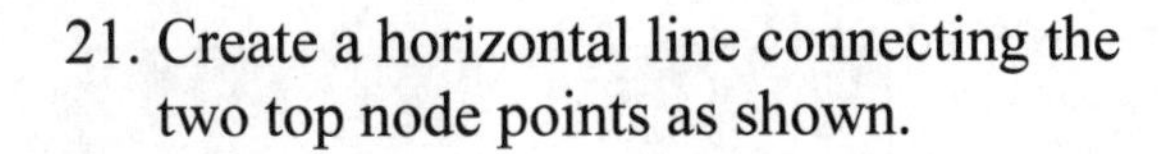

21. Create a horizontal line connecting the two top node points as shown.

22. Select **Finish Sketch** in the *Ribbon* toolbar to exit the **Sketch** mode.

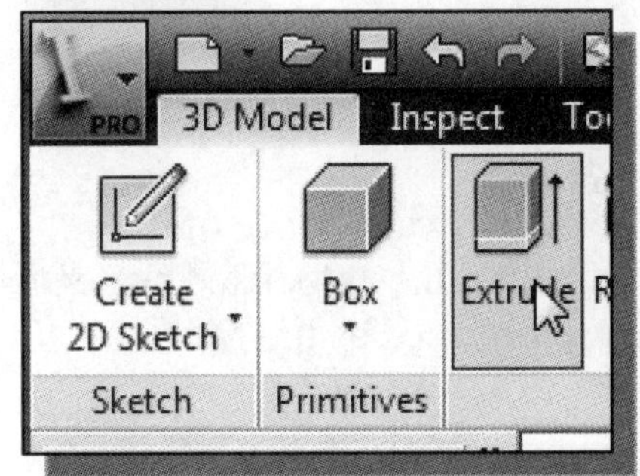

23. In the *Create* toolbar, select the **Extrude** command by clicking the left-mouse-button on the icon.

24. Select the inside region of the 2D sketch profile as shown.

25. On your own, complete the **Cut feature** as shown.

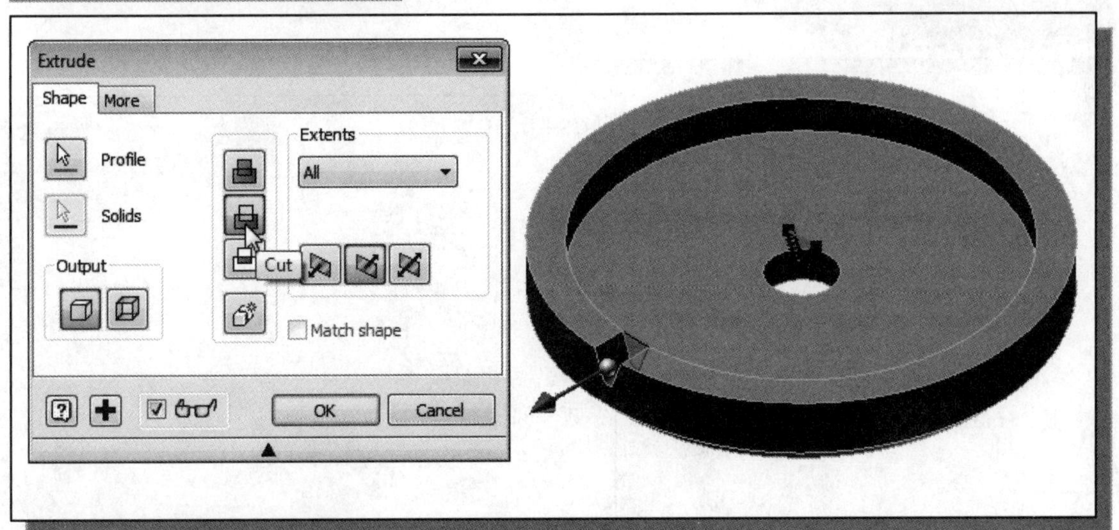

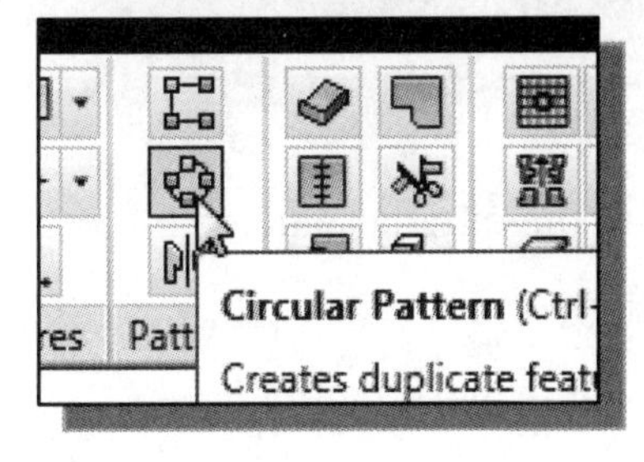

26. In the *Pattern* toolbar, select the **Circular Pattern** command by clicking the left-mouse-button on the icon.

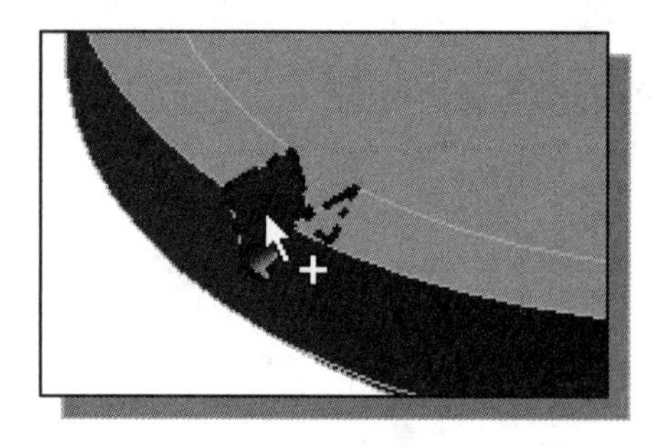

27. In the graphics window, select the cut feature we just created.

28. In the *Circular Pattern* dialog box, click on the **Rotation Axis** icon as shown.

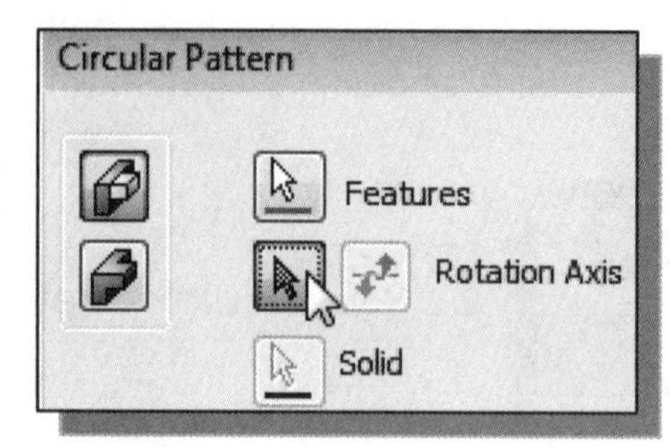

29. Select the cylindrical surface of the revolved feature; the associated center axis is therefore used as the rotation axis.

30. In the *Circular Pattern* dialog box, enter **40** for the number of pattern features and **360 degrees** in the *Placement* section as shown.

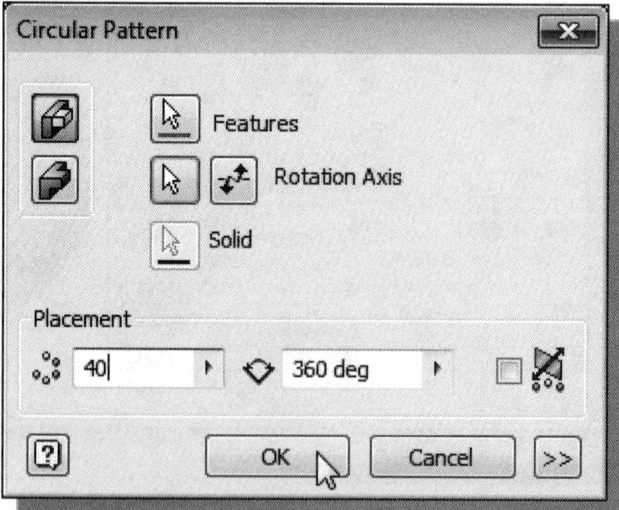

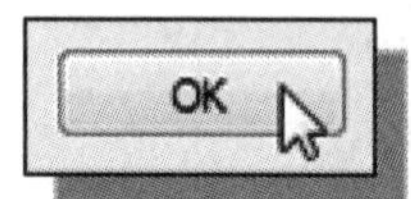

31. Click **OK** to accept the settings and create the circular pattern.

- The **Circular Pattern** command can be used to quickly create duplicates about a center axis of rotation.

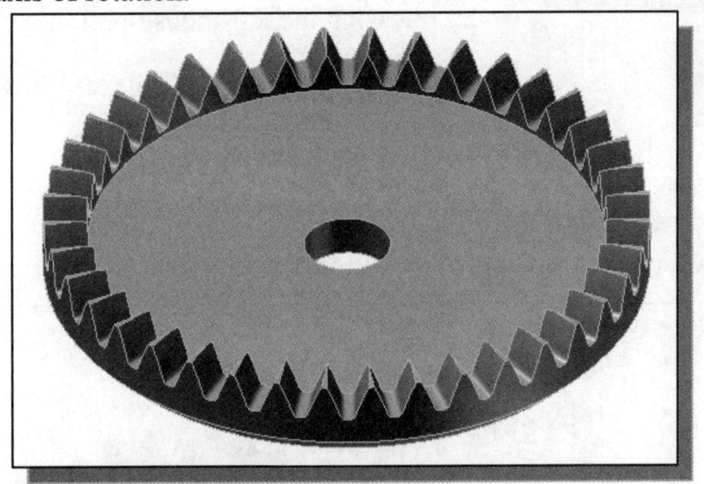

32. Switch to the ***SpurGears*** assembly by clicking on the associated tab.

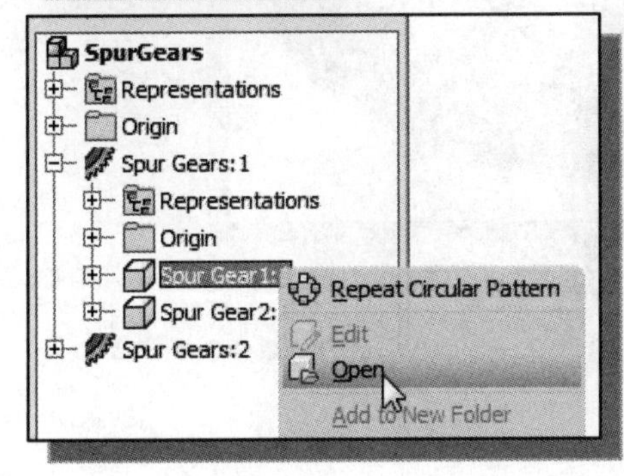

33. Inside the *browser* window, right-mouse-click on the ***Pinion Gear*** of the first set of gears to bring up the option menu as shown.

34. In the option menu, select **Open** to open the selected model as shown in the figure.

35. On your own, copy and paste the profile of the *Pinion Gear* into a new 2D sketch on the top face of the ***G1-Spur Gear*** as shown.

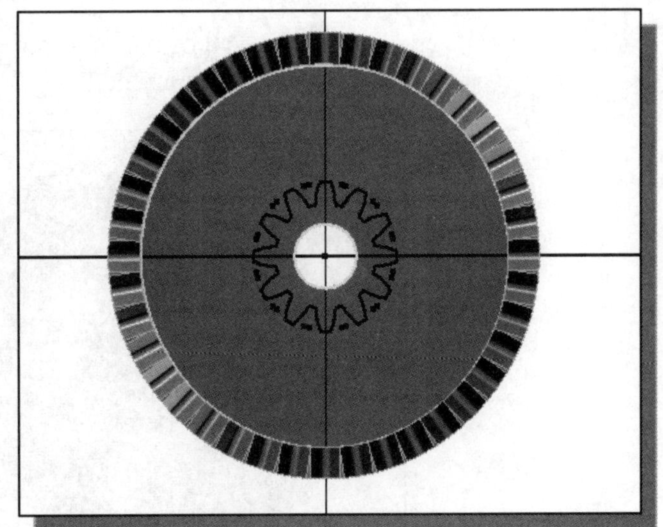

36. Select **Finish Sketch** in the *Ribbon* toolbar to exit the Sketch mode.

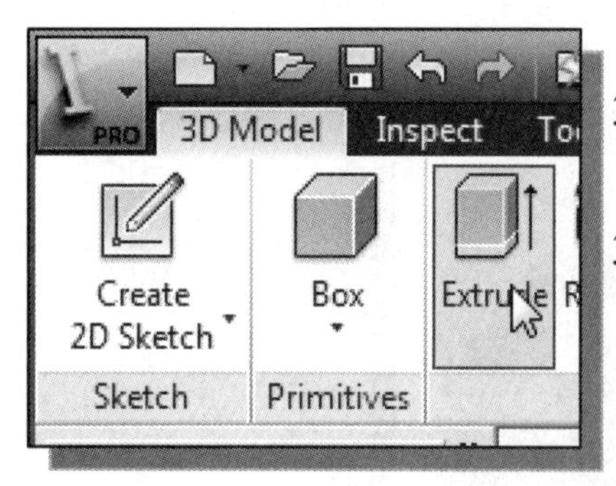

37. In the *Create* toolbar, select the **Extrude** command by releasing the left-mouse-button on the icon.

38. Select the inside region of the 2D sketch to create a profile as shown.

39. On your own, create the **6.5mm** extrusion feature as shown.

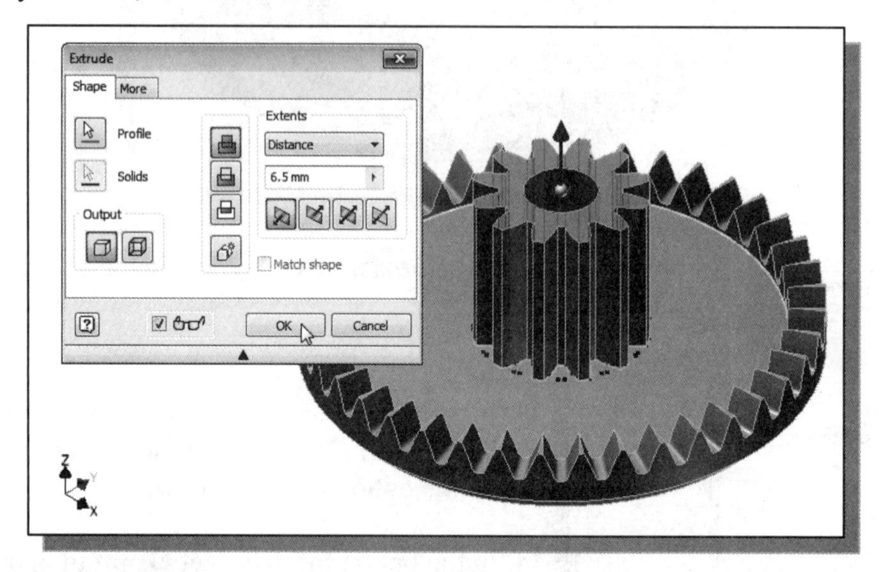

40. In the *Quick Access* menu, select **Save** to save the ***G1-Spur Gear*** model.

- All of the required gears for the *Tamiya Tiger* kit have been created.

Inventor Content Center

In this lesson, we will examine some of the procedures that are available in *Autodesk Inventor* to reuse existing 2D data and 3D parts. In *Autodesk Inventor*, standard parts commonly used in industry are available and custom parts can also be created through the *Inventor* component generators.

In *Autodesk Inventor*, we have the option of using the standard parts library through what is known as the ***Content Center***. The *Content Center* consists of multiple libraries of standard parts that have been created based on industry standards. Significant amounts of time can be saved by using these parts. Note that we can also create and publish our libraries so that others can reuse our parts. The *Content Center* has two modes for two distinct roles: **Consumer** and **Editor**. In the **Consumer** mode, we access the *Content Center* libraries and use the parts as a consumer. In the **Editor** mode, we can define the different categories, and also define the iterations for the parts the consumer can select and use.

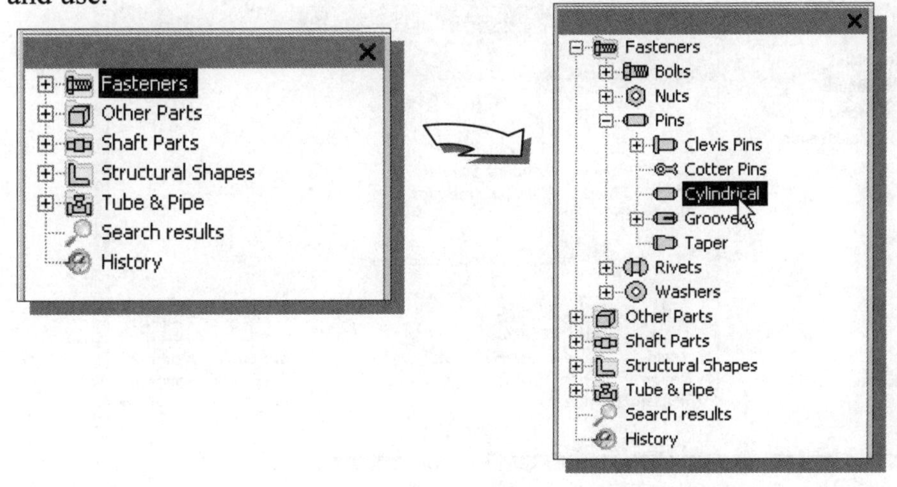

Besides using the standard part libraries through the ***Content Center***, other types of design components can also be created through the *Autodesk Inventor* ***Design Accelerator Panel***. A variety of design components can be generated through the different tools available in the *Design Accelerator Panel*. Note that both the ***Content Center*** and the ***Design Accelerator Panel*** are accessible only through the **Assembly Modeling** module.

In this chapter, the procedures of accessing the *Content Center* and the *Design Accelerator Panel* are illustrated. For the *Tamiya Tiger* design, custom parts will be created from both the standard library and the *Design Accelerator*. Note that the nylon gears in the *Tamiya Tiger* design are specifically designed for the *Tamiya kits*, and therefore the tooth profiles do not match with the gears generated in *Autodesk Inventor*.

Start another *Autodesk Inventor* Assembly Model

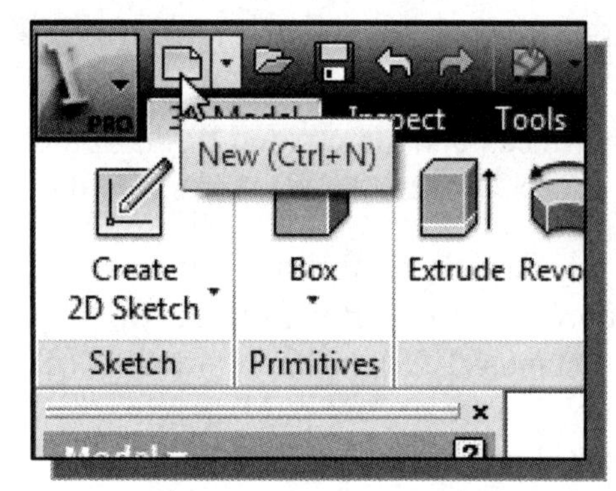

1. Select the **New File** icon with a single click of the left-mouse-button in the *Launch* toolbar as shown.

2. Select the **Metric** tab and in the *Template* list, then select **Standard(mm).iam** (*Standard Inventor Assembly Model* template file).

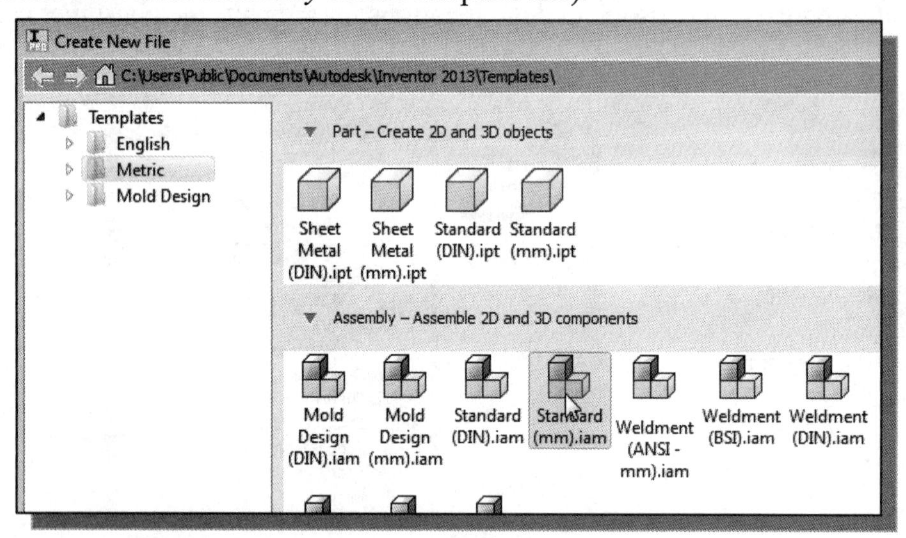

3. Click on the **Create** button in the *New File* dialog box to accept the selected settings.

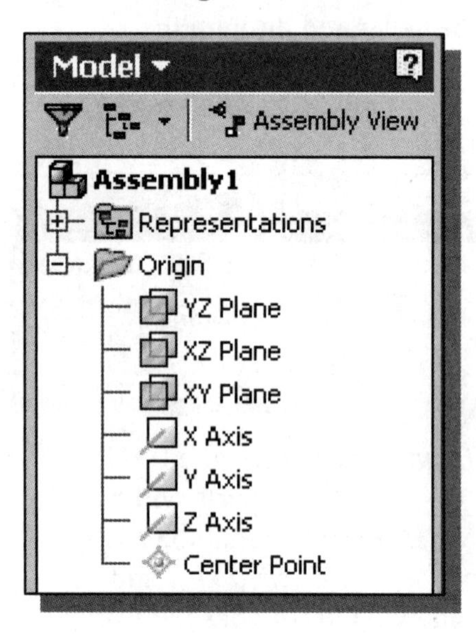

- In the *browser* window, **Assembly1** is displayed with a set of work planes, work axes and a work point. In most aspects, the usage of work planes, work axes and work point is very similar to that of the *Inventor Part Modeler*.

- Notice, in the *Ribbon* toolbar panels, several *component* options are available, such as **Place Component**, **Create Component** and **Place from Content Center**. As the names imply, we can use parts that have been created or create new parts within the *Inventor Assembly Modeler*.

Using the Content Center

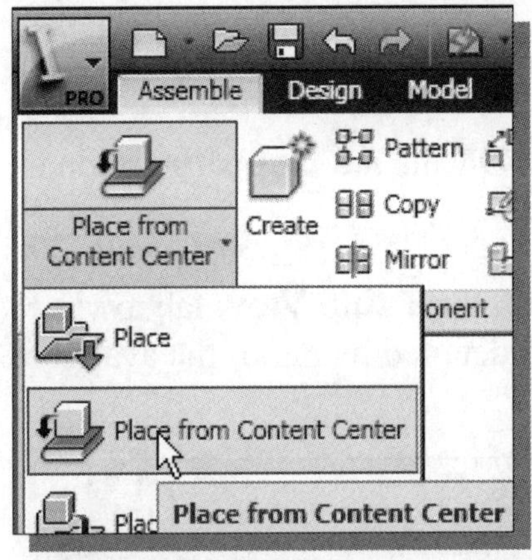

1. In the *Assemble* panel select the **Place from Content Center** command by left-mouse-clicking the icon.

2. Select the **Fasteners** category and then **Pins** as shown.

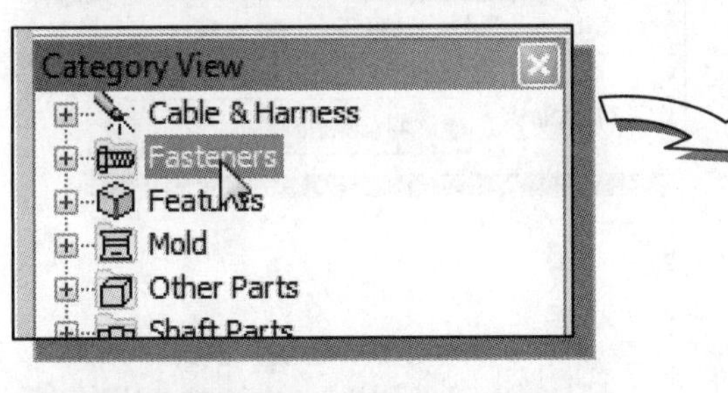

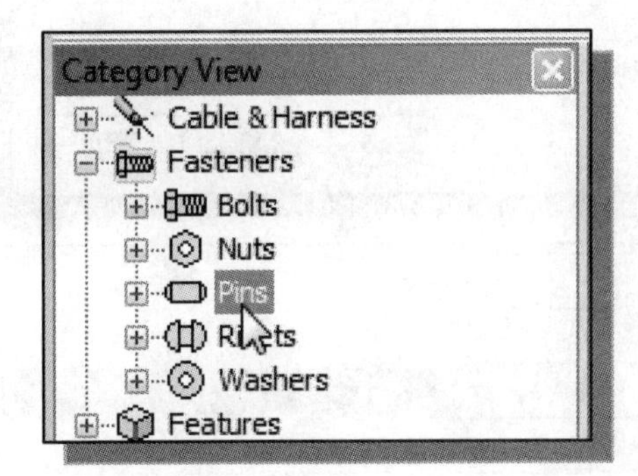

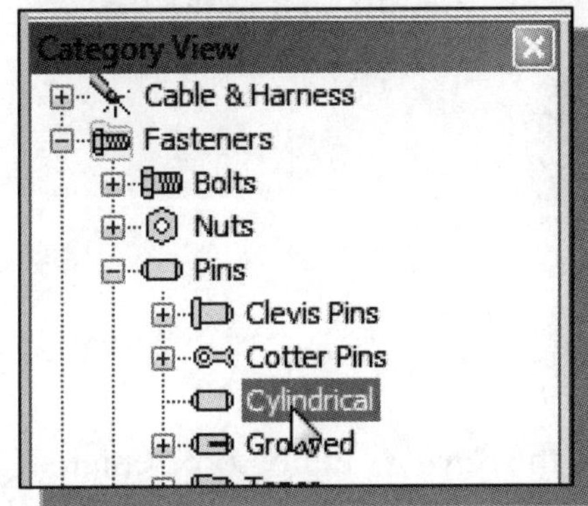

3. Choose **Cylindrical** under **Pins** as shown in the figure.

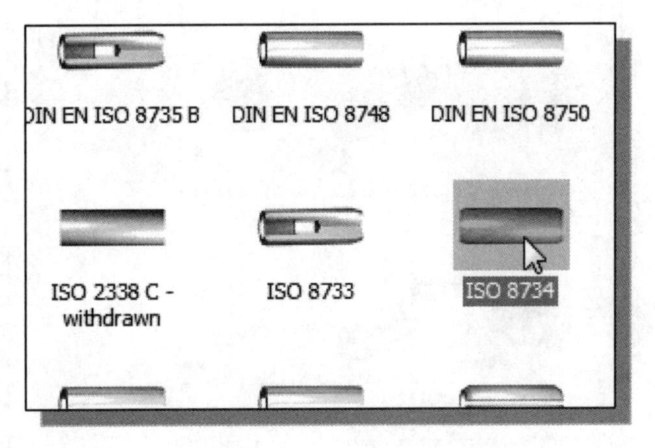

4. Select **ISO 8734** as shown in the figure.

5. Click **OK** to enter the selection dialog box.

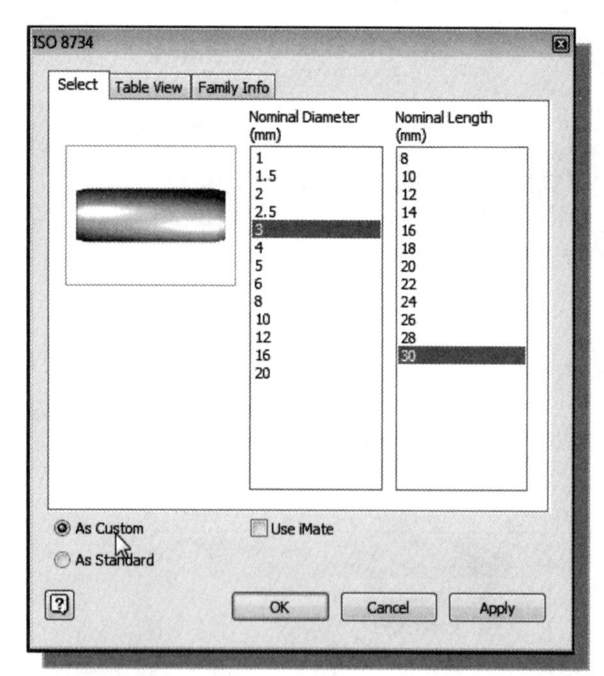

6. Set the designation diameter to **3** mm.

7. Set the nominal length to **30** mm.

8. Switch *ON* the **As Custom** option as shown.

9. Click on the **Table View** tab to view a more detailed listing of the available pins.

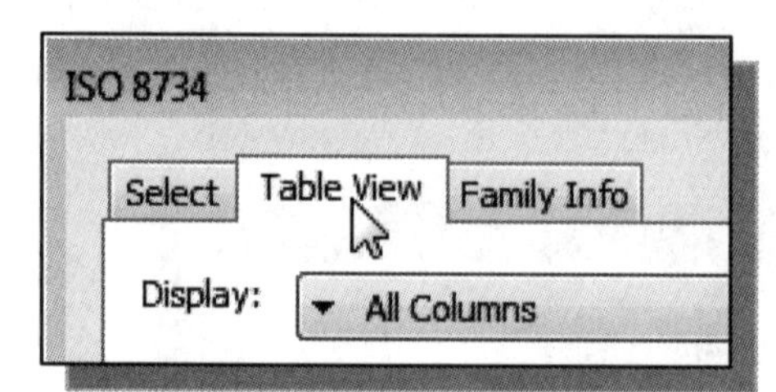

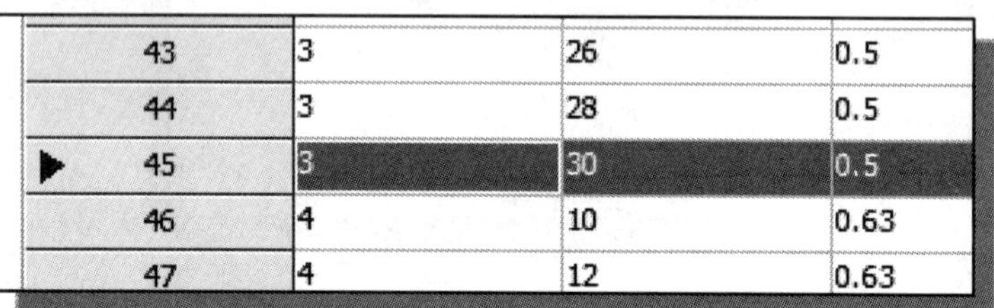

43	3	26	0.5
44	3	28	0.5
▶ 45	3	30	0.5
46	4	10	0.63
47	4	12	0.63

➢ Note the selected pin is highlighted in the list.

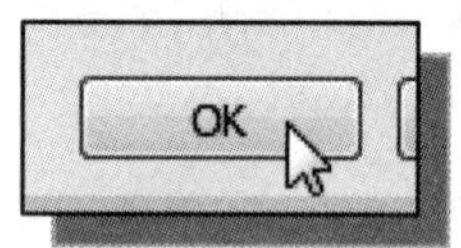

10. Click **OK** to accept the selection.

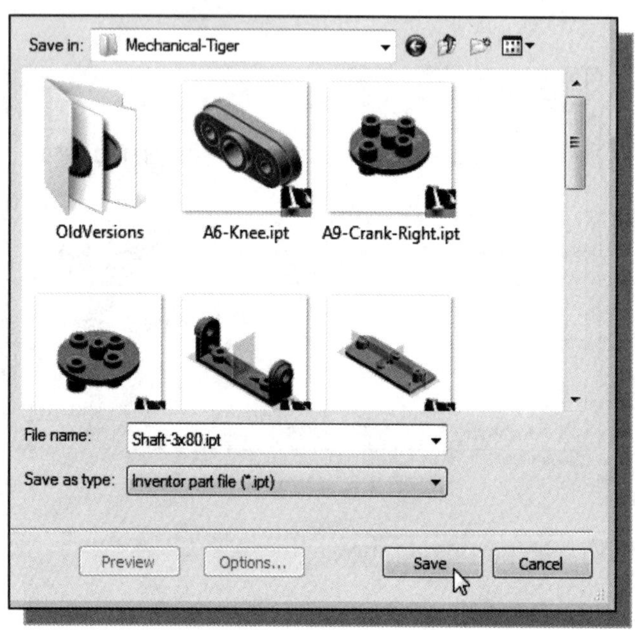

11. In the *Save As* dialog box, enter **Shaft-3x80.ipt** as the model filename as shown.

12. Click **Save** to save the model.

- Note that by saving the part as a custom file, this will allow the adjustment of the length of the shaft to the required *80 mm*.

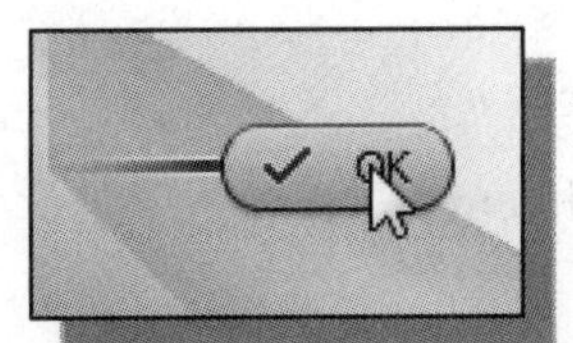

13. Inside the graphics window, right-mouse-click once to bring up the option menu and select **OK** to end the **Place Component** command.

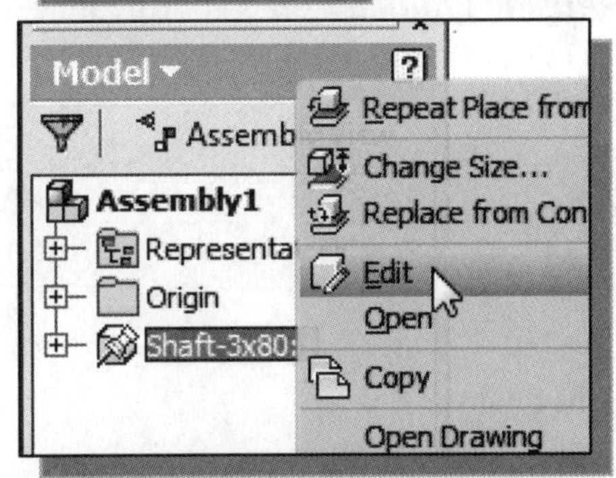

14. Inside the *browser* window, right-mouse-click once on the **Shaft-3x80** part to bring up the option menu and select **Edit** to modify the part.

15. Click on the **Manage** tab to switch to the *Manage* toolbar.

16. Click the **Parameters** icon to bring up the *Parameters* dialog box.

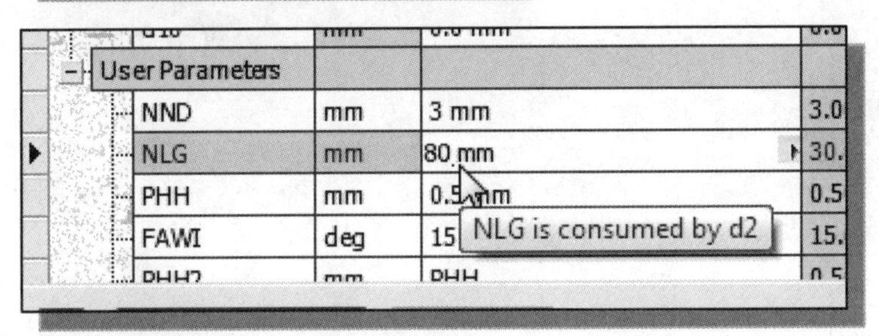

17. Edit the value of the *NLG* (Nominal Length) variable to **80 mm** as shown.

18. Click **Done** to accept the adjustment and exit the parameters dialog box.

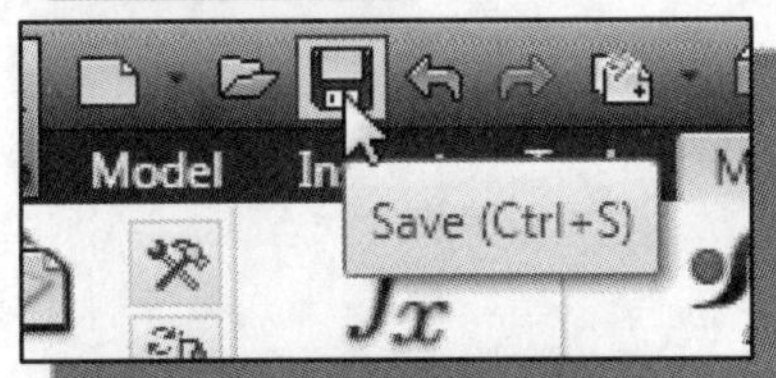

19. Click **Save** to save the changes to the model.

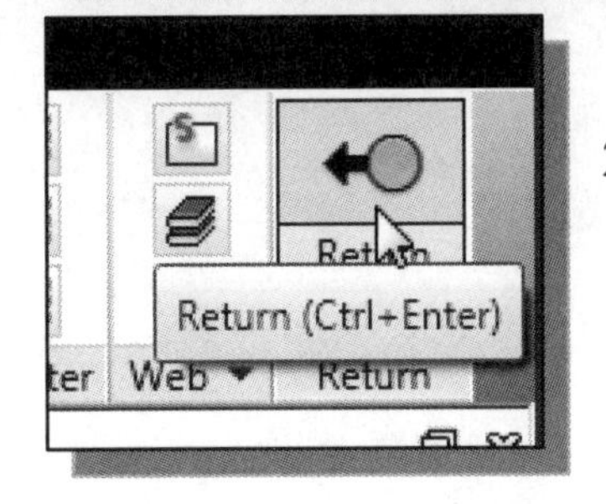

20. In the *Ribbon* toolbar area, click **Return** to end the editing of the *Shaft Model* and return to the assembly model.

Questions:

1. List and explain the differences of three types of commonly used gears.

2. What is the difference between ***pitch diameter*** and ***outside diameter*** on a spur gear?

3. How do we access the ***Autodesk Inventor Spur Gear Generator***?

4. What is the definition of ***Gear Ratio*** in a *Gear Train*?

5. What is included in the ***Autodesk Content Center***?

6. Why is it important to identify symmetrical features in designs?

7. When and why should we use the ***Pattern*** option?

8. How do we create a ***linear diameter dimension*** for a revolved feature?

9. How do we export and reuse the ***tooth profile*** of a *spur gear* created by the ***Autodesk Inventor Spur Gear Generator***?

Exercises:

1. **Shaft Support** (Dimensions are in inches.)

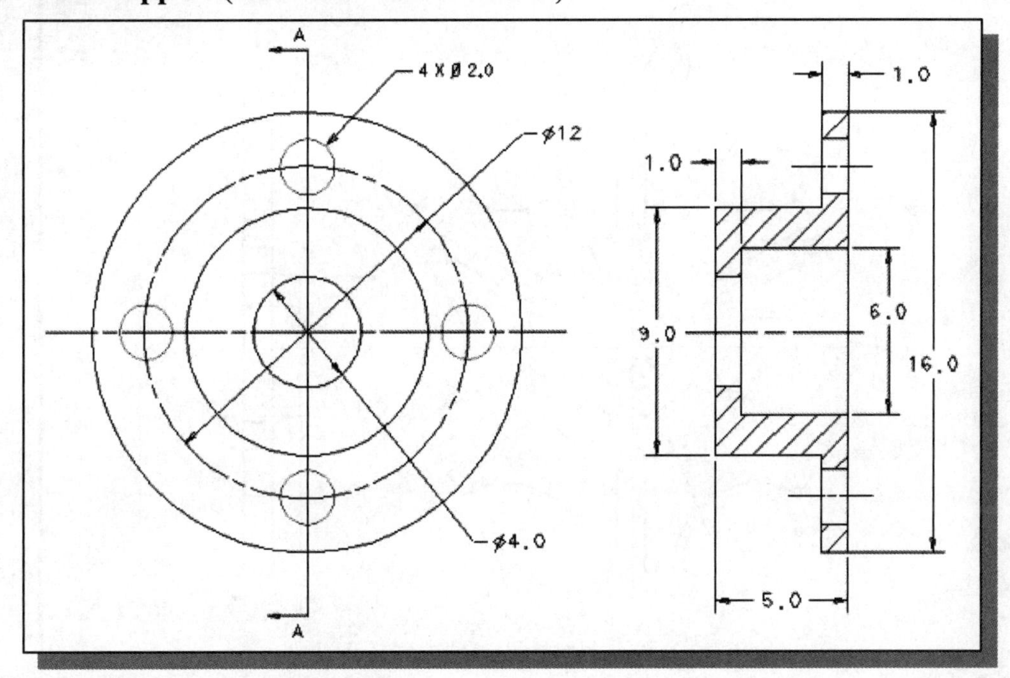

2. **Switch Base** (Dimensions are in inches.)

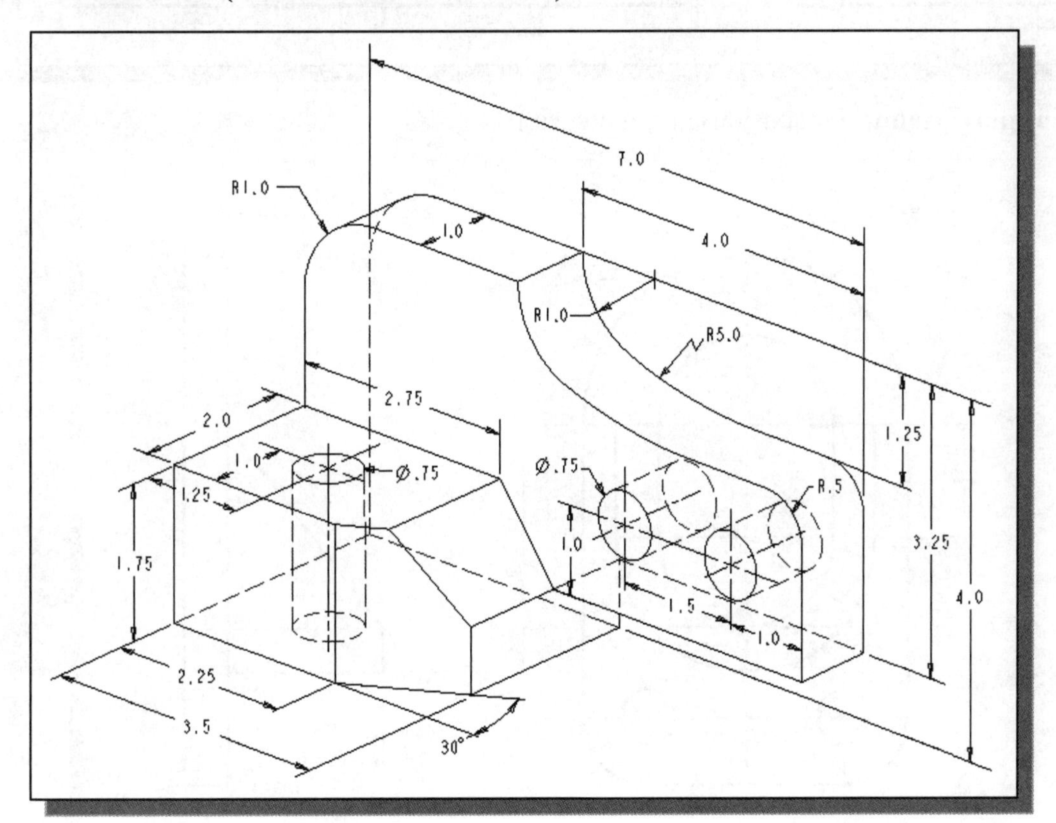

3. **Geneva Wheel** (Dimensions are in inches.)

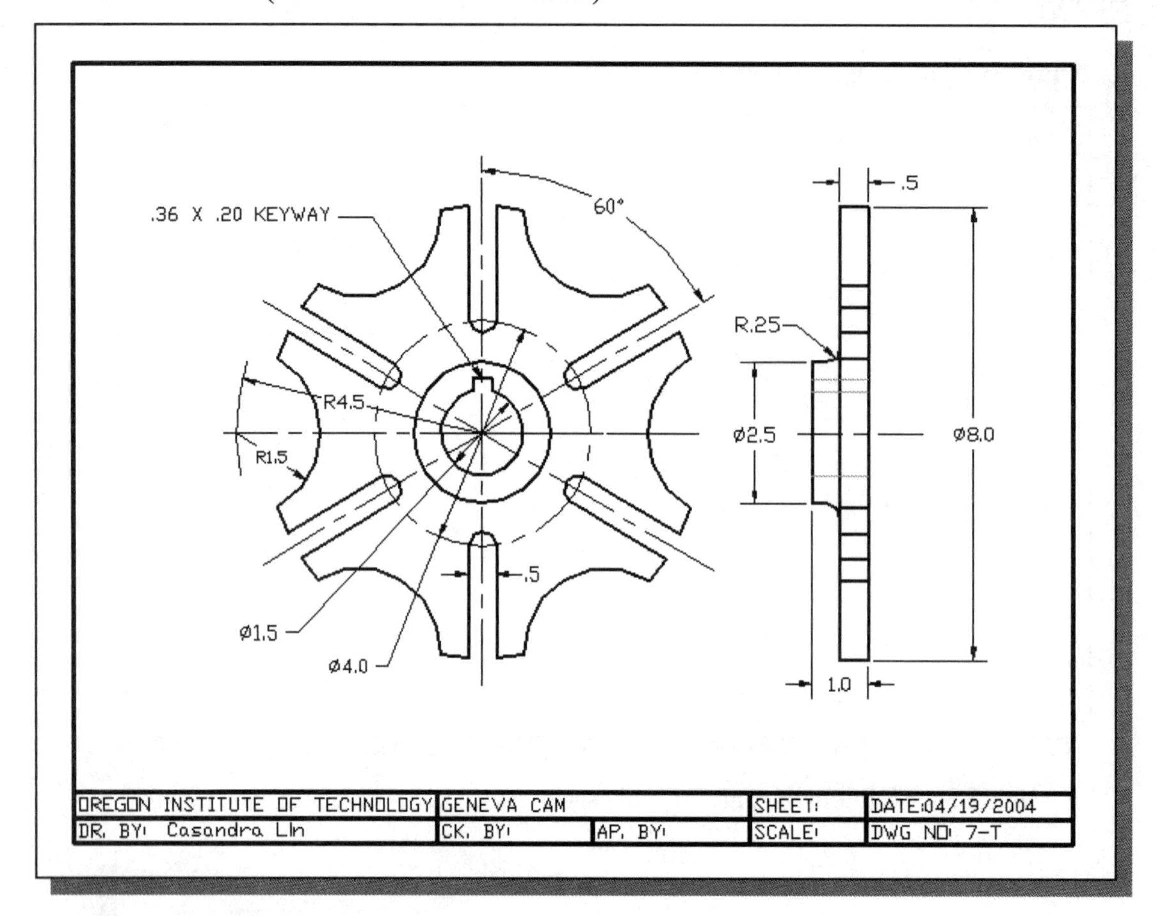

4. **Support Mount** (Dimensions are in inches.)

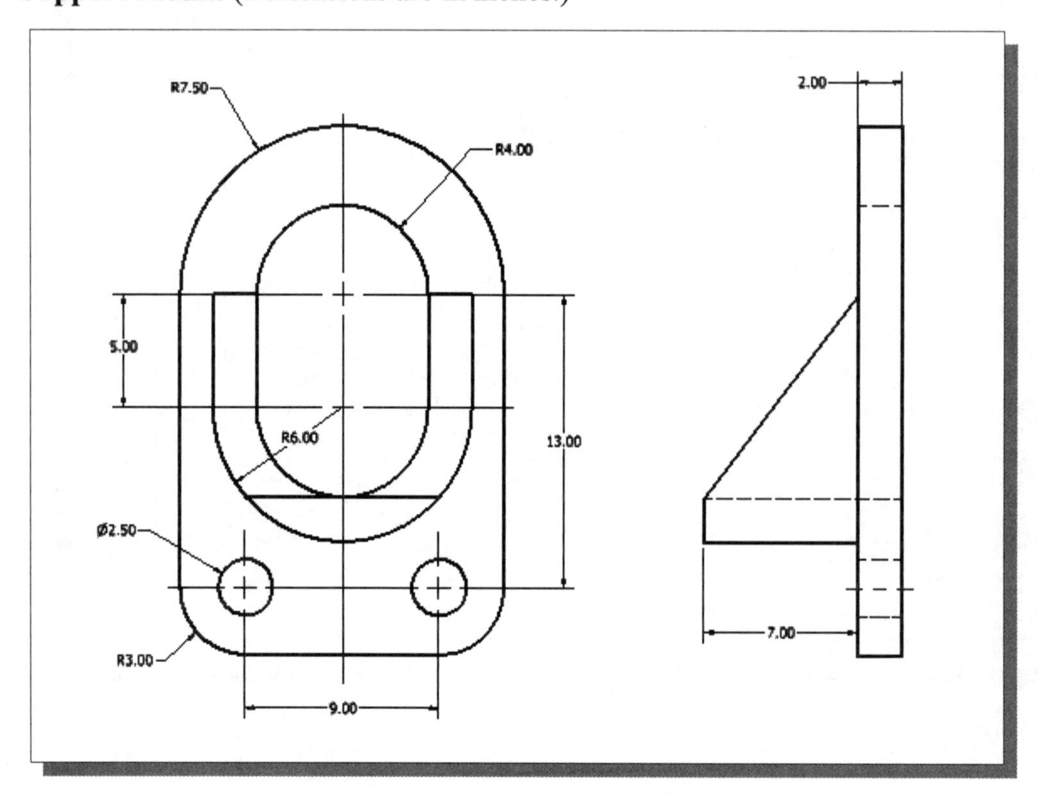

Chapter 9
Advanced 3D Construction Tools

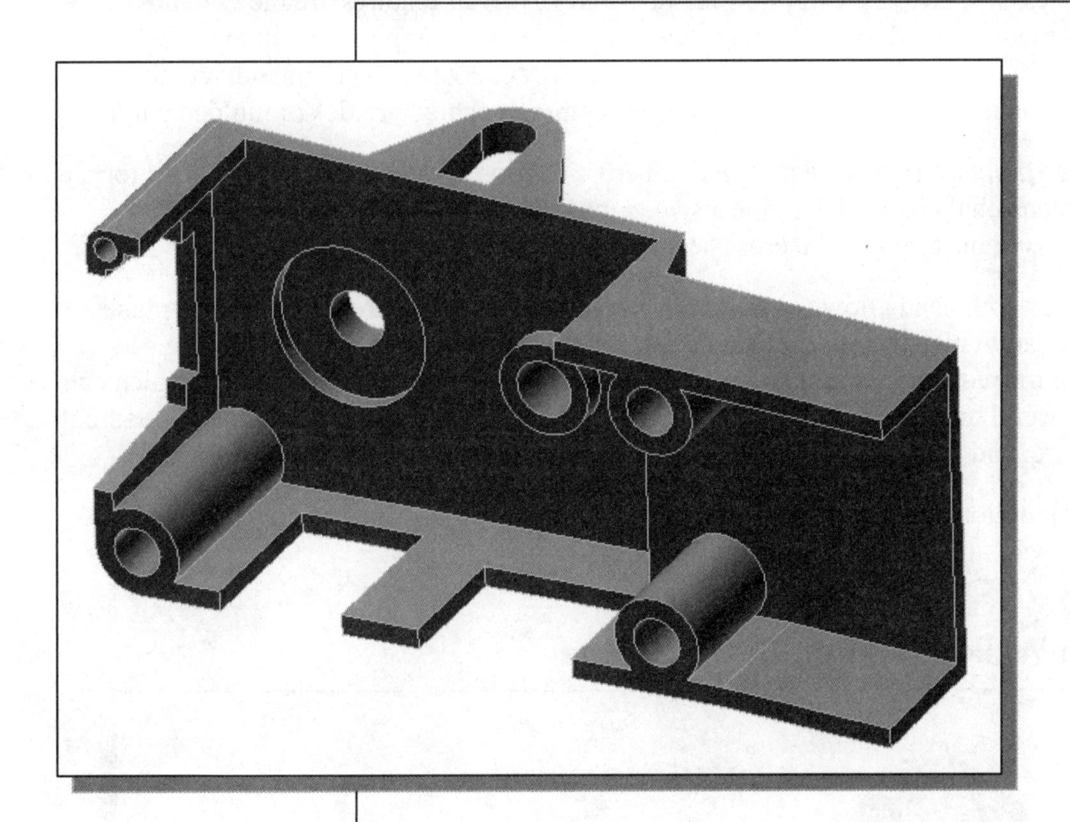

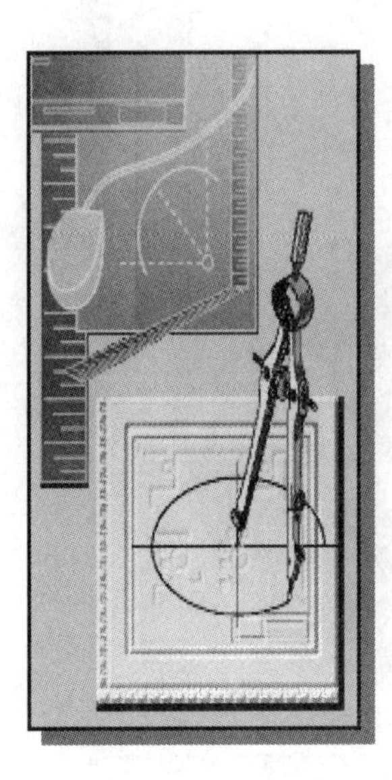

Learning Objectives

- **Understand the Concepts Behind the Different 3D Construction Tools**
- **Set up Multiple Work Planes**
- **Create Swept Features**
- **Use the Shell Command**
- **Create 3D Rounds & Fillets**

Introduction

Autodesk Inventor provides an assortment of three-dimensional construction tools to make the creation of solid models easier and more efficient. As demonstrated in the previous lessons, creating **extruded** features and **revolved** features are the two most common methods used to create 3D models. In this next example, we will examine the procedures for using the **Sweep** command, the **Mirror Feature** command, and the **Shell** command. These types of features are common characteristics of molded parts.

The **Sweep** option is defined as moving a cross-section through a path in space to form a three-dimensional object. To define a sweep in *Autodesk Inventor*, we define two sections: the trajectory and the cross-section.

The **Mirror** command allows us to create mirror images of features. A reference plane, such as a datum plane or an existing surface, is required to use the command. We can create a mirrored feature while maintaining the original parametric definitions, which can be quite useful in creating symmetrical features. For example, we can create one quadrant of a feature, and then mirror it twice to create a solid with four identical quadrants

The **Shell** option is defined as hollowing out the inside of a solid, leaving a shell of specified wall thickness.

A Thin-Walled Design: *Battery Case*

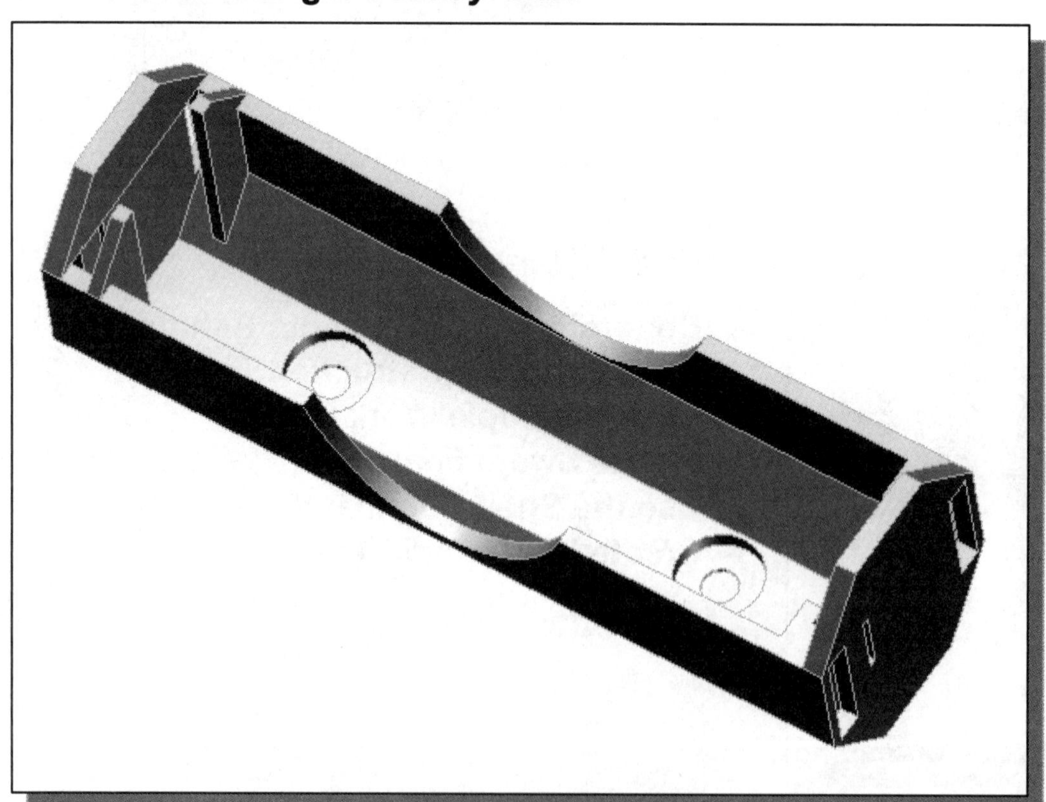

Modeling Strategy

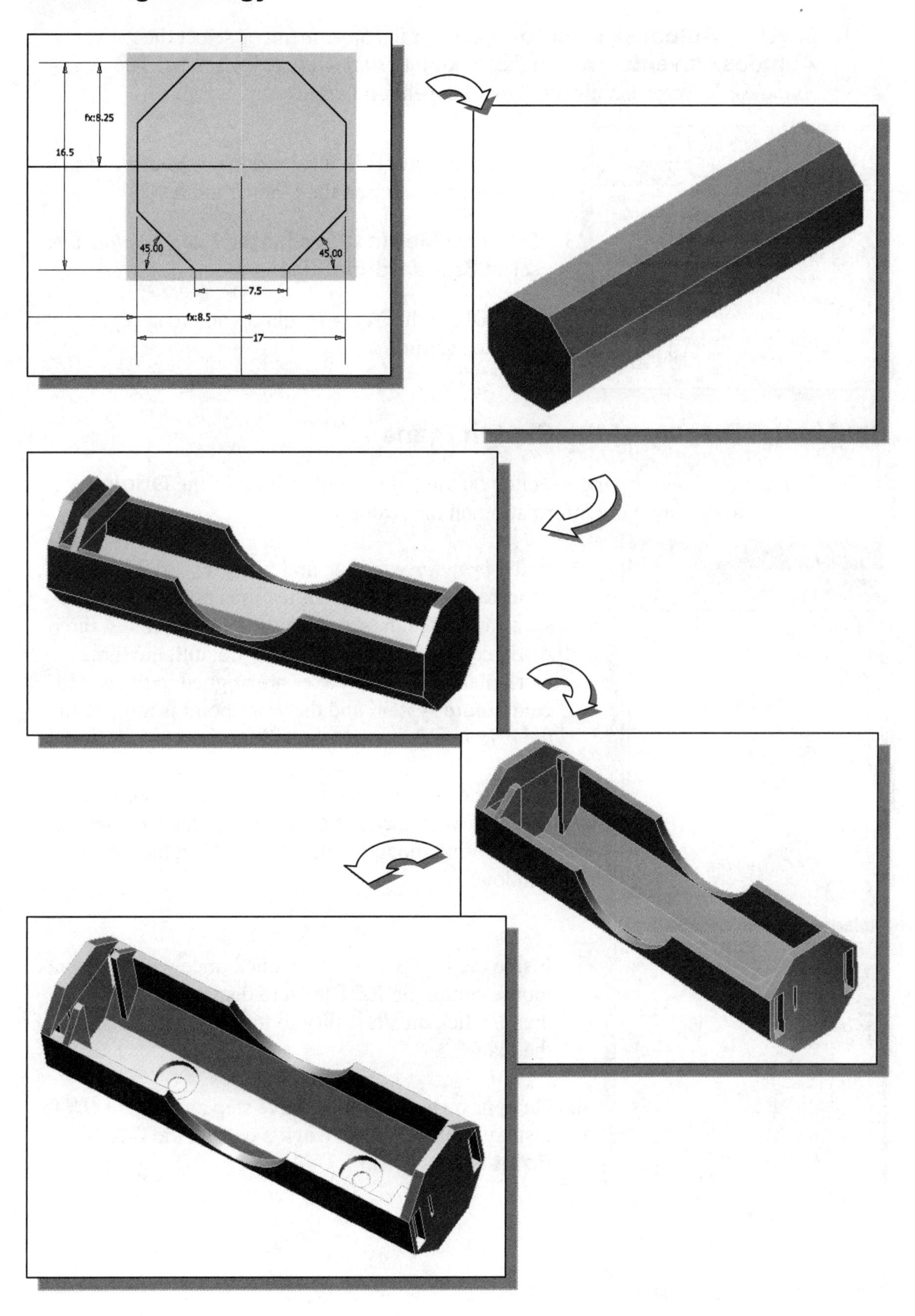

Starting *Autodesk Inventor*

1. Select the **Autodesk Inventor** option on the *Start* menu or select the **Autodesk Inventor** icon on the desktop to start *Autodesk Inventor*. The *Autodesk Inventor* main window will appear on the screen.

2. Select the **New File** icon with a single click of the left-mouse-button in the *Launch* toolbar.

3. Select the **Metric** tab, and in the *File Template* area select **Standard(mm).ipt**.

4. Pick **OK** in the *New File* dialog box to accept the selected settings.

Set Up the Display of the Sketch Plane

1. In the *part browser* window, click on the [+] symbol in front of the **Origin** feature to display more information on the feature.

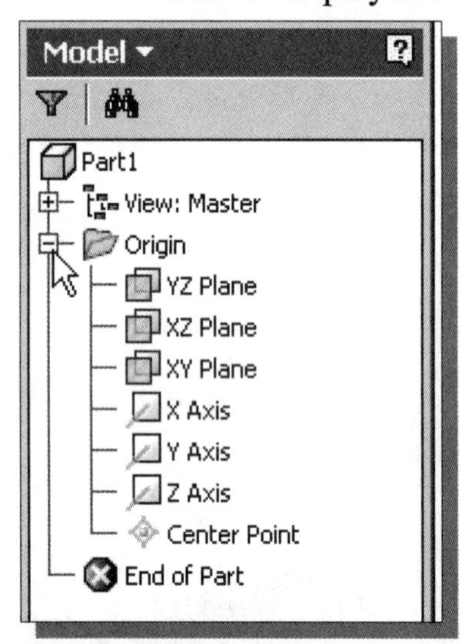

❖ In the *browser* window, notice a new part name appeared with seven work features established. The seven work features include three work planes, three work axes and a work point. By default, the three work planes and work axes are aligned to the **world coordinate system** and the work point is aligned to the *origin* of the **world coordinate system.**

2. Inside the *browser* window, move the cursor on top of the third work plane, **XZ Plane**. Notice a rectangle, representing the work plane, appears in the graphics window.

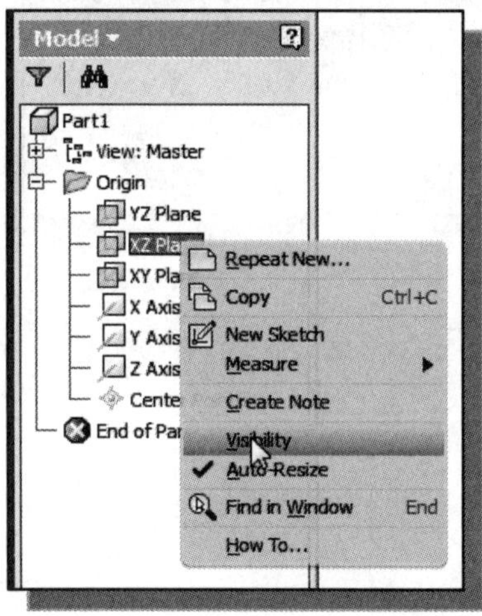

3. Inside the *browser* window, click once with the right-mouse-button on **XZ Plane** to display the option menu. Click on **Visibility** to toggle *ON* the display of the plane.

4. On your own, repeat the above step and toggle ***ON*** the display of the **X** and **Z work axes** and the **Center Point** on the screen.

Create the Base Feature

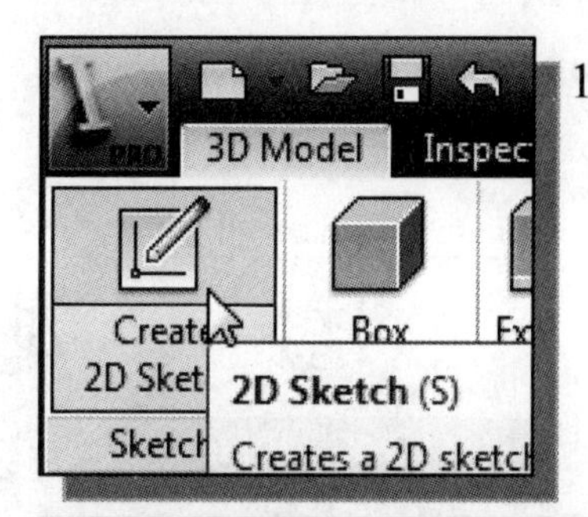

1. In the *Sketch* toolbar select the **Create 2D Sketch** command by left-clicking once on the icon.

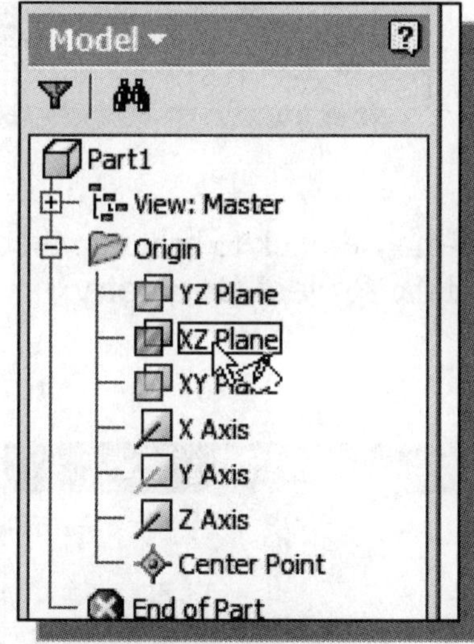

2. In the *Status Bar* area, the message: "*Select face, work plane, sketch or sketch geometry.*" is displayed. Select the **XZ Plane**, by left-clicking once on any edge of the XZ Plane in the graphics window or in the *browser* window as shown.

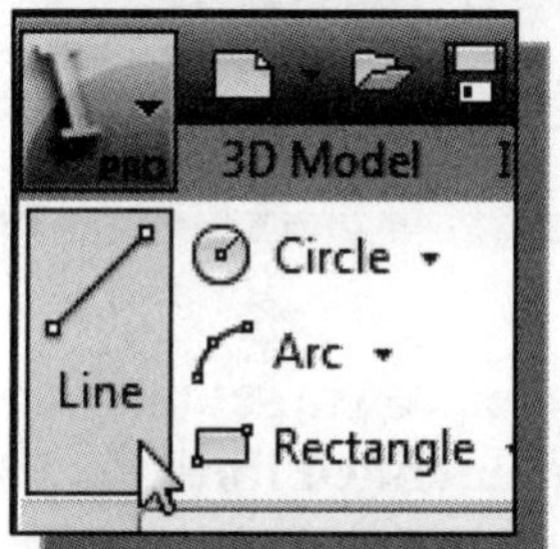

3. Move the graphics cursor to the **Line** icon in the *Draw* toolbar. A *Help-tip box* appears next to the cursor and a brief description of the command is displayed at the bottom of the drawing screen: "*Creates Straight line segments and tangent arcs.*"

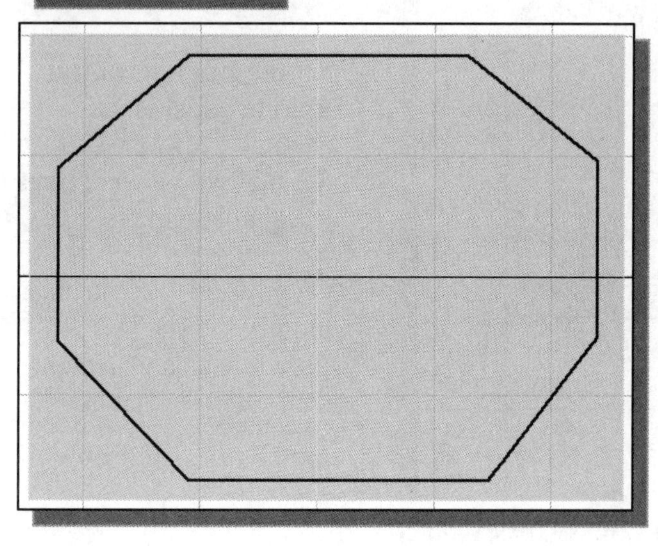

4. Create eight connected line segments of arbitrary size, with the center near the projected center point as shown.

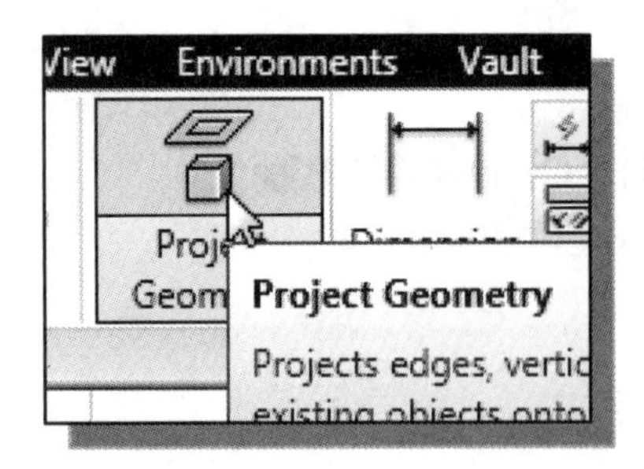

5. Click on the **Project Geometry** icon in the *Draw* panel.

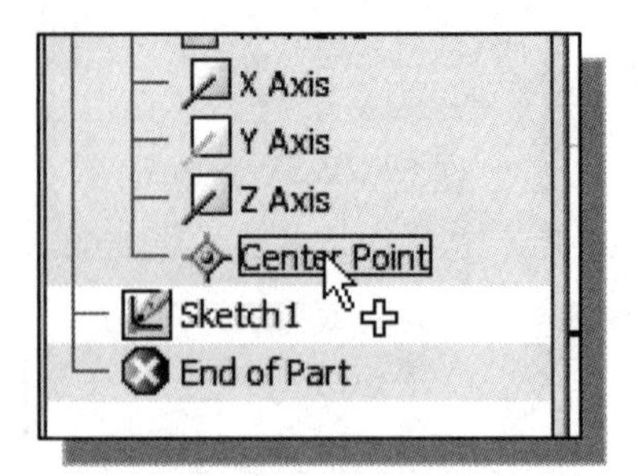

6. Select the **Center Point** in the *browser* window as shown.

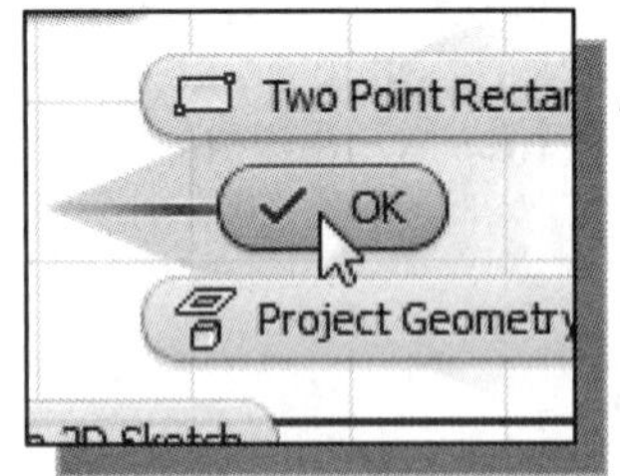

7. Inside the graphics window, right-mouse-click to bring up the option menu and select **OK** to end the Project Geometry command.

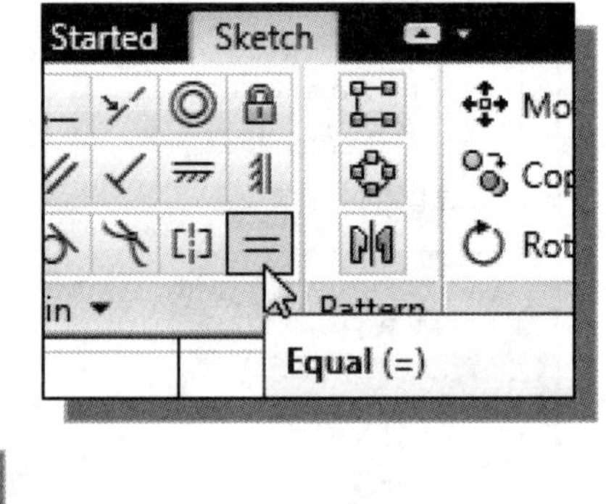

8. Click on the **Equal Constraint** icon in the *Constrain* panel.

9. On your own, set the **four inclined lines** to equal length.

10. Set the **two horizontal lines** to equal length.

11. Set the **two vertical lines** to equal length.

12. On your own, create and modify the **six dimensions** and center the sketch as shown.

13. In the *Ribbon* toolbar, click once with the left-mouse-button on **Finish Sketch** to end the **Sketch** option.

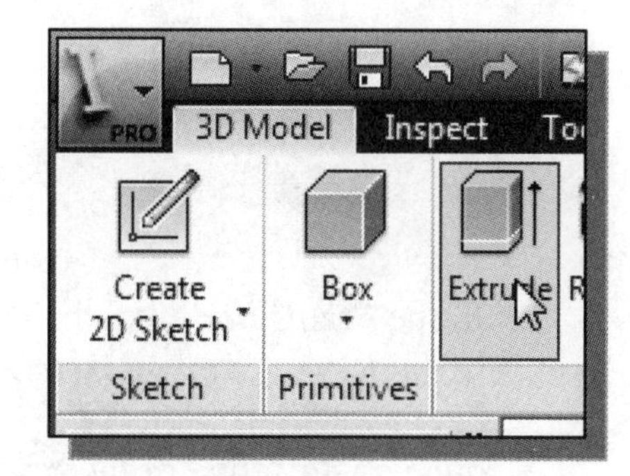

14. In the **Model** tab (the tab that is located in the ***Ribbon***) select the **Extrude** command by clicking with the left-mouse-button on the icon.

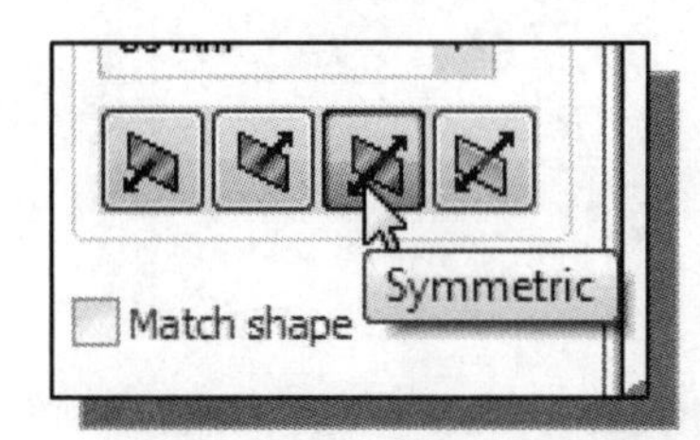

15. Set the extrude direction to **Symmetric** as shown.

16. In the *Extrude* popup window, enter **60mm** as the extrusion distance. Notice that the sketch region is automatically selected as the extrusion profile.

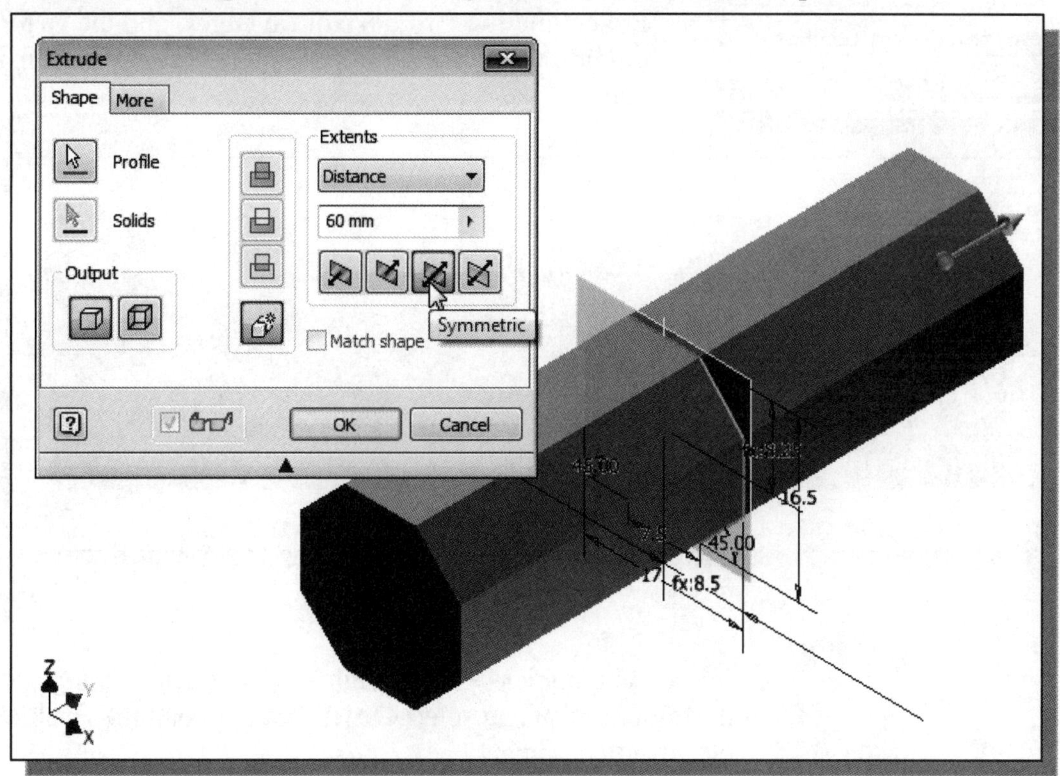

17. Click **OK** to create the extruded feature.

Create a Cut Feature

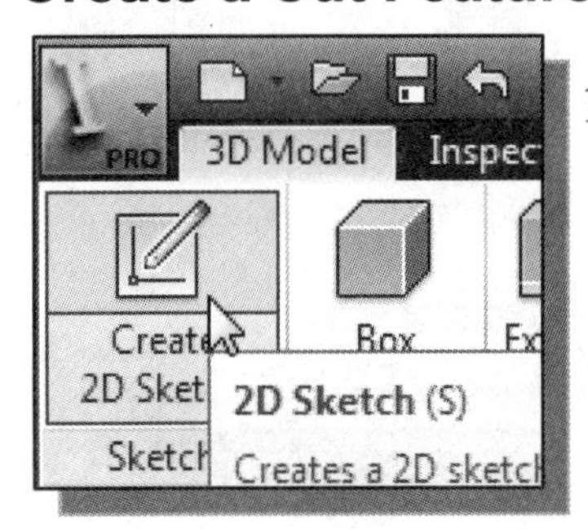

1. In the *Sketch* toolbar select the **Create 2D Sketch** command by left-clicking once on the icon.

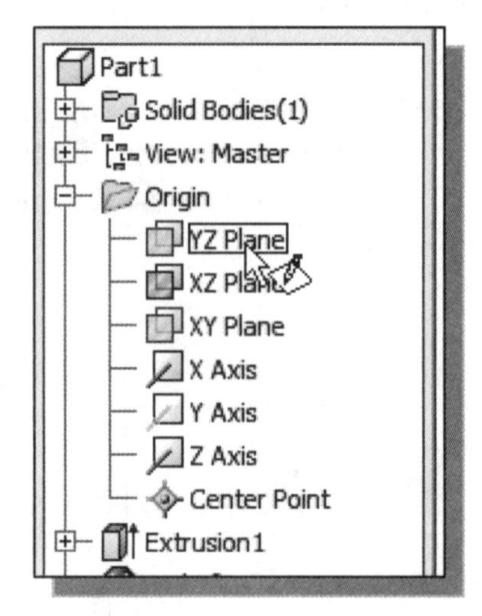

2. In the *Status Bar* area, the message: "*Select face, workplane, sketch or sketch geometry.*" is displayed. Select the **YZ Plane**, by left-clicking once on any edge of the YZ Plane in the graphics window or in the *browser* window as shown.

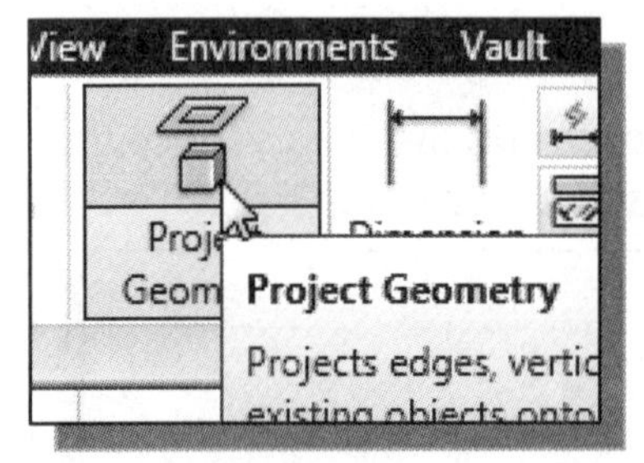

3. Click on the **Project Geometry** icon in the *Draw* panel.

4. Project the **top two horizontal edges**, and the **two outside vertical edges** of the model to the sketching plane.

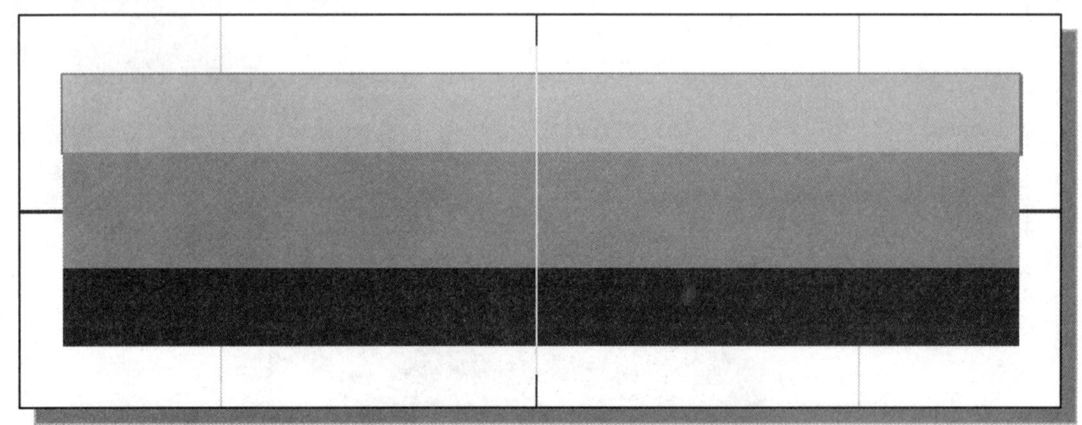

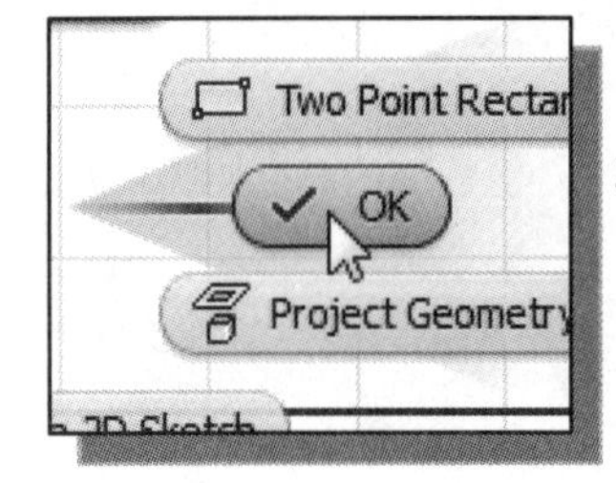

5. Inside the graphics window, right-mouse-click to bring up the option menu and select **Done [Esc]** to end the Project Geometry command.

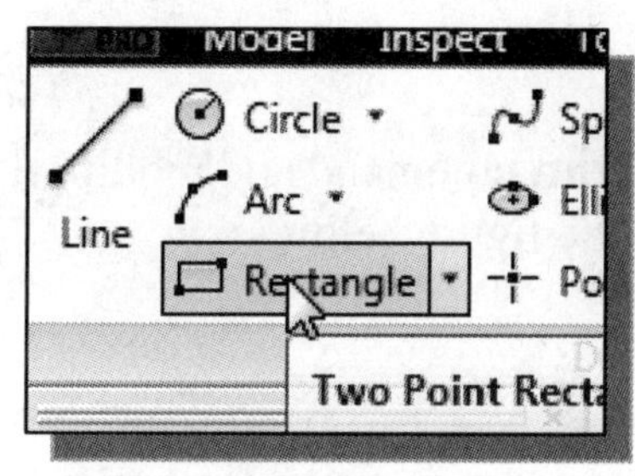

6. Click on the **Rectangle** icon in the *Draw* panel.

7. Create a rectangle aligned to the two horizontal lines projected to the sketching plane as shown.

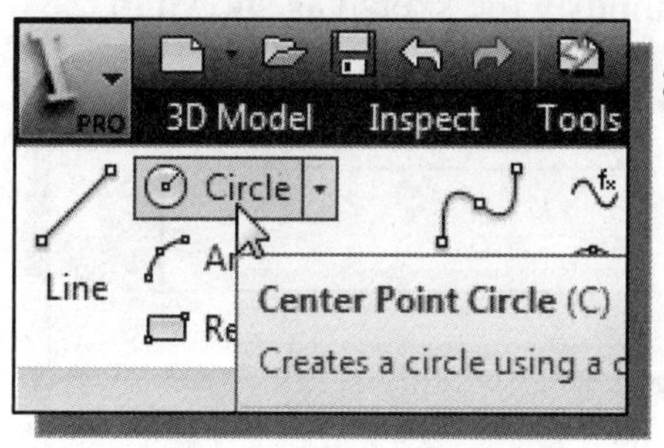

8. Select the **Circle** command by clicking once with the left-mouse-button on the icon in the *Draw* toolbar.

9. On your own, create a circle with the center aligned to the top horizontal projected line as shown.

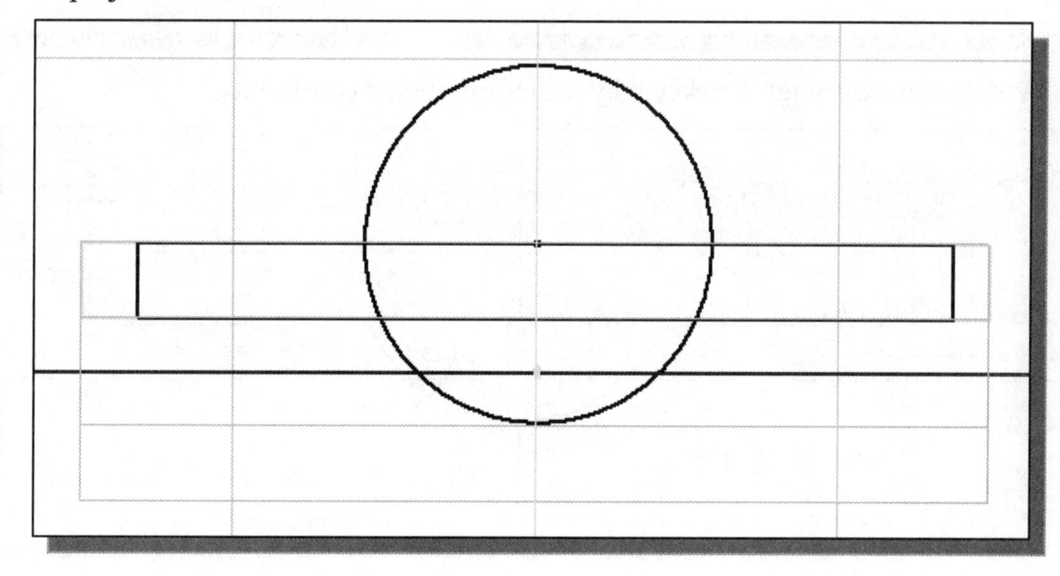

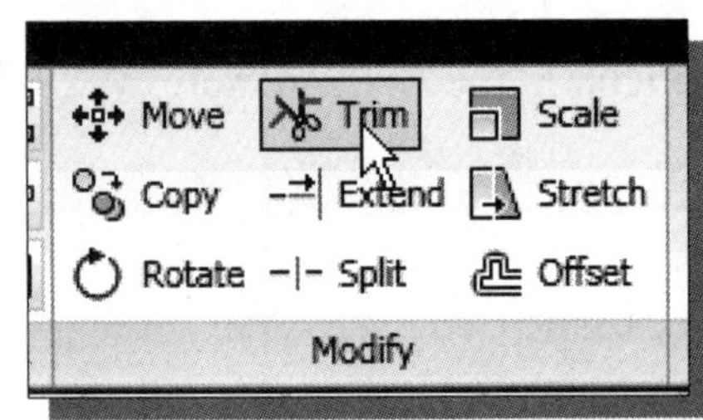

10. On your own, use the **Trim** command and modify the sketch as shown in the figure below.

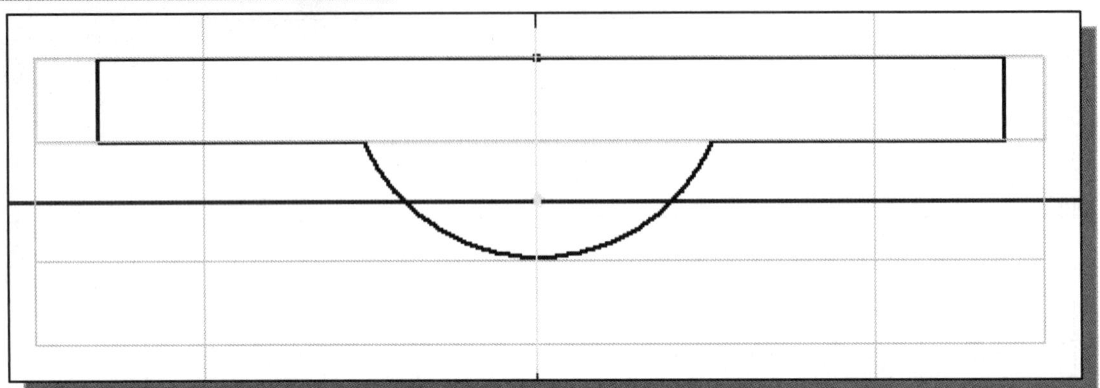

11. On your own, use the **Line** command and also modify the sketch as shown in the figure below.

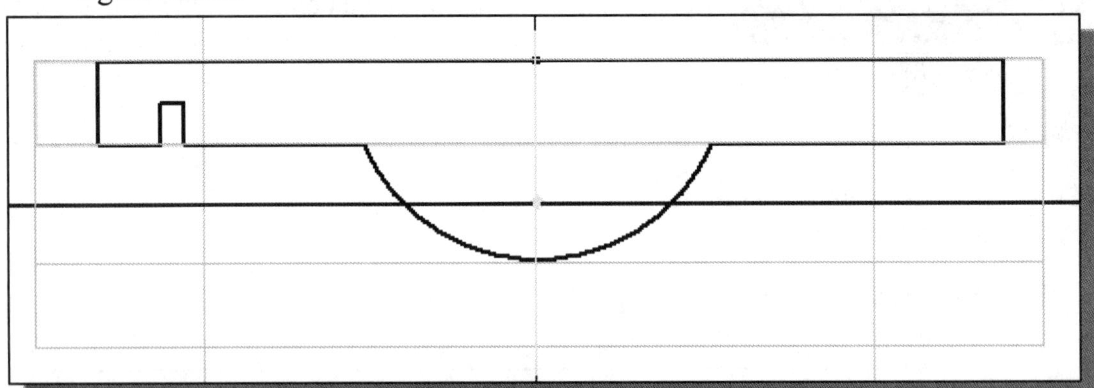

12. On your own, align the sketch by adding additional constraints.

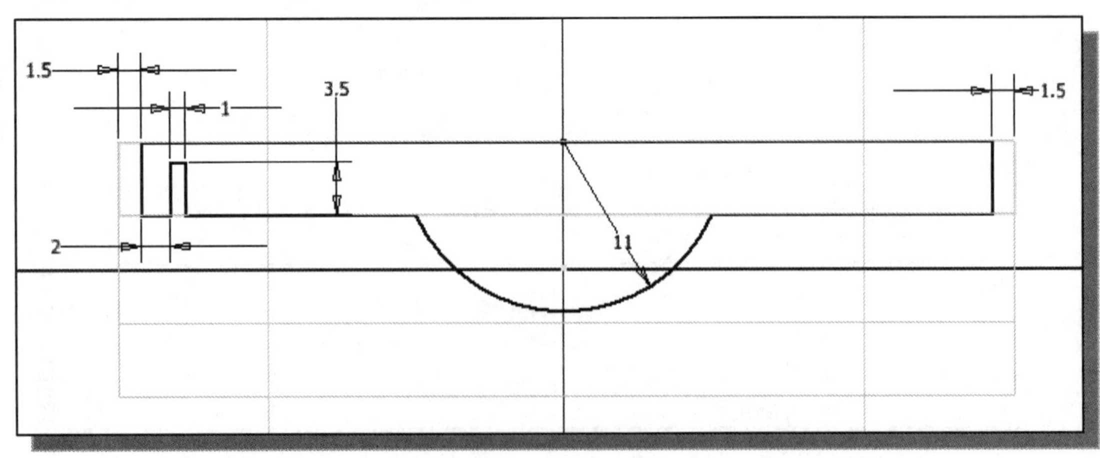

13. On your own, also create and modify the dimensions as shown.

14. In the *Ribbon* toolbar, click once with the left-mouse-button on **Finish Sketch** to end the Sketch option.

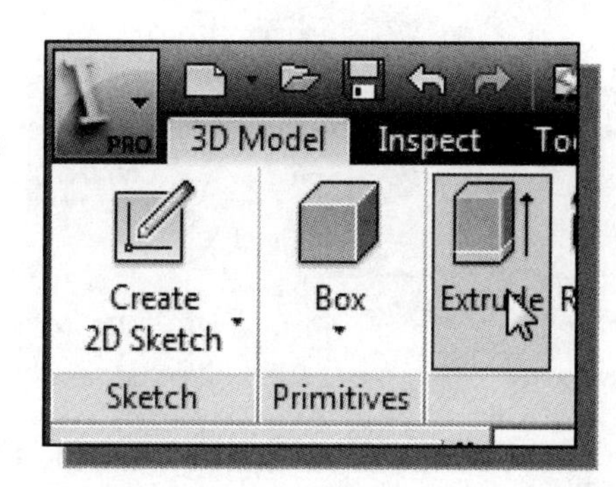

15. In the Model tab (the tab that is located in the ***Ribbon***) select the **Extrude** command by clicking with the left-mouse-button on the icon.

16. Click inside the sketched region to select as the extrusion profile.

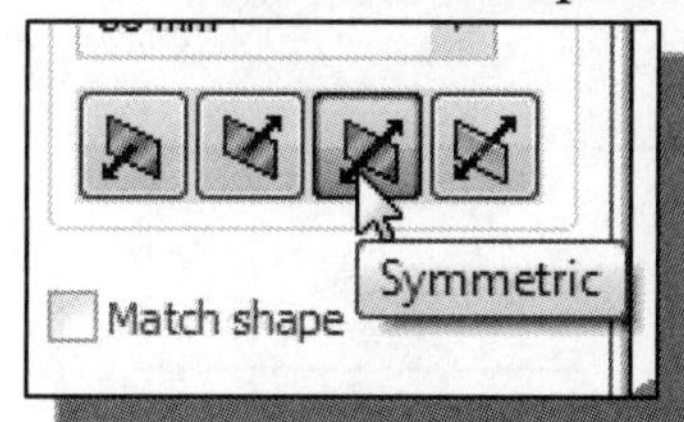

17. Set the extrude direction to **Symmetric** as shown.

18. In the *Extrude* dialog box, set the extrude option to **Cut** and enter **20mm** as the extrusion distance.

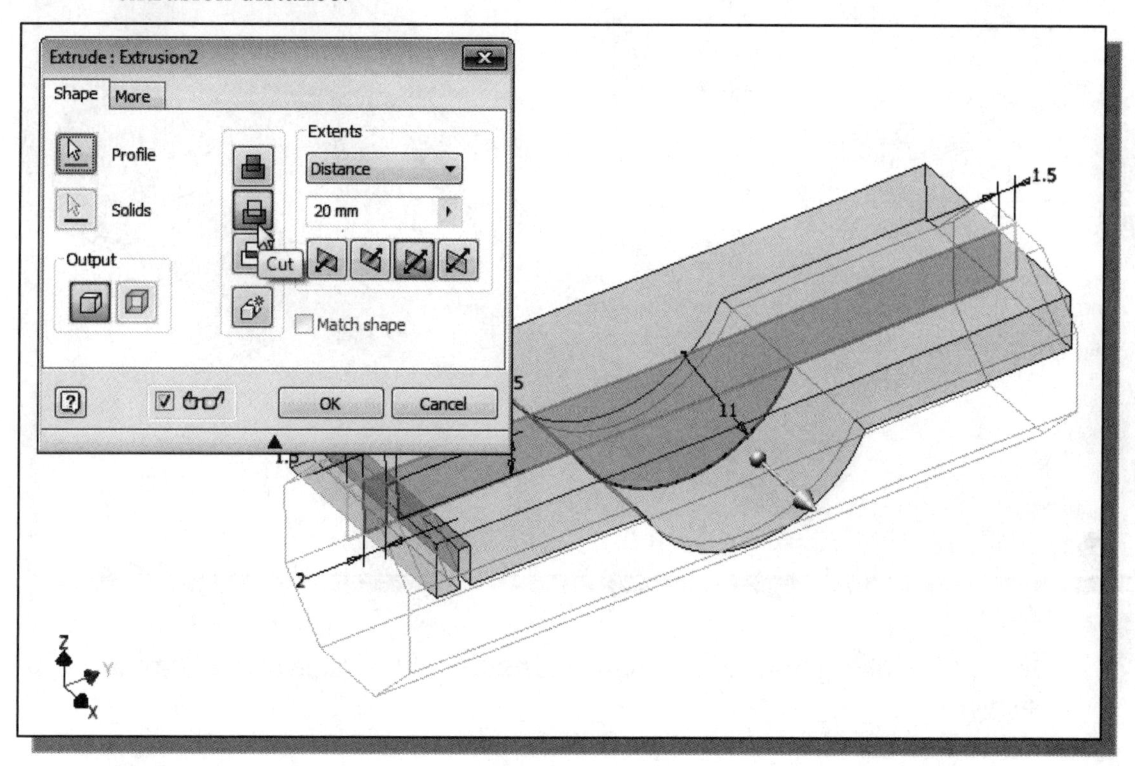

19. Click **OK** to create the extruded feature.

Create a Shell Feature

- The **Shell** command can be used to hollow out the inside of a solid, leaving a shell of specified wall thickness.

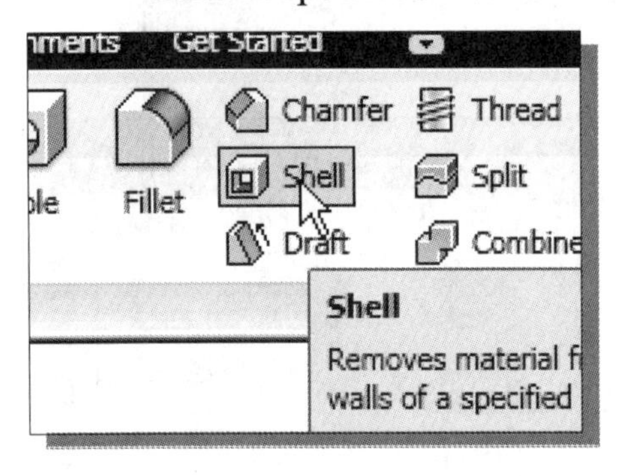

1. In the *Modify* toolbar, select the **Shell** command by left-clicking once on the icon.

2. On your own, use the **ViewCube** or the **3D Rotation** quick-key [**F4**] to display the back faces of the model as shown below.

3. In the *Shell* dialog box, the **Remove Faces** option is activated. Select the four faces as shown below.

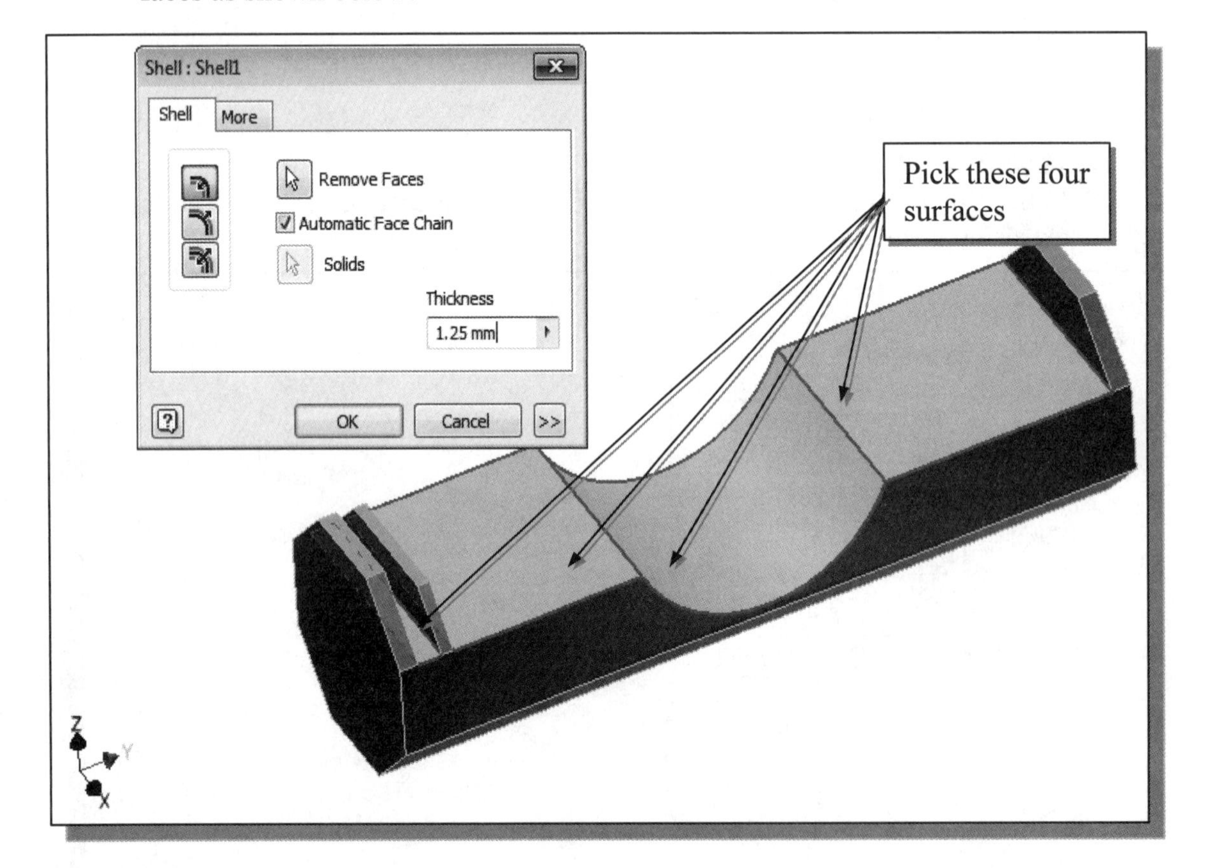

4. In the *Shell* dialog box, set the option to **Inside** with a value of **1.25mm** as shown.

5. In the *Shell* dialog box, click on the **OK** button to accept the settings and create the shell feature.

- Note the **Shell** command hollowed out the inside of the solid model.

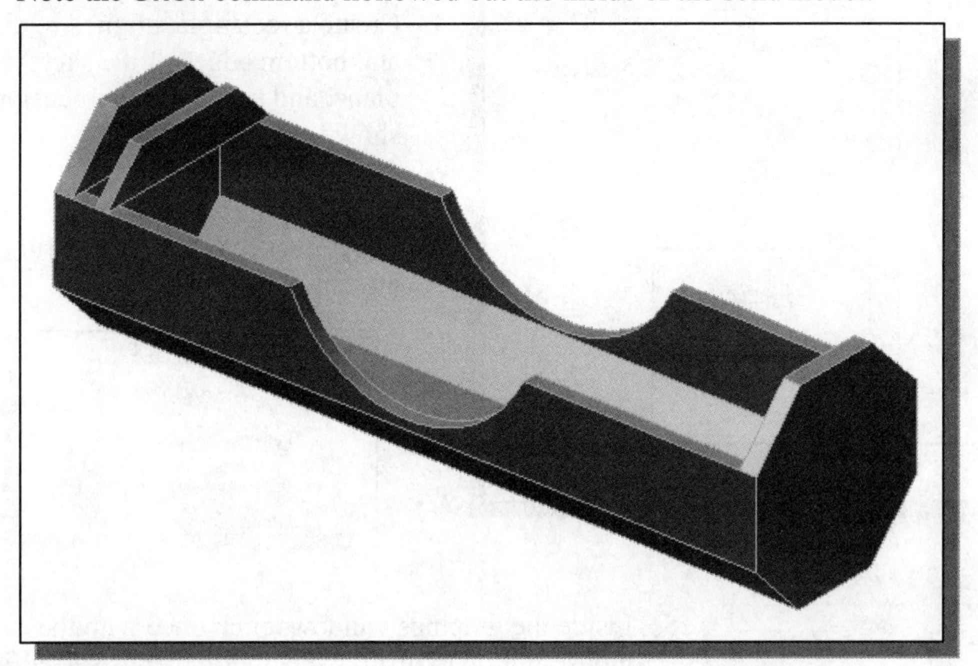

Create a Cut Feature

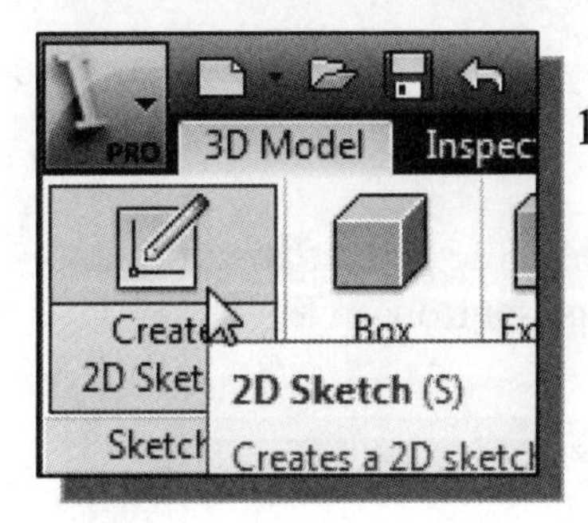

1. In the *Sketch* toolbar select the **Create 2D Sketch** command by left-clicking once on the icon.

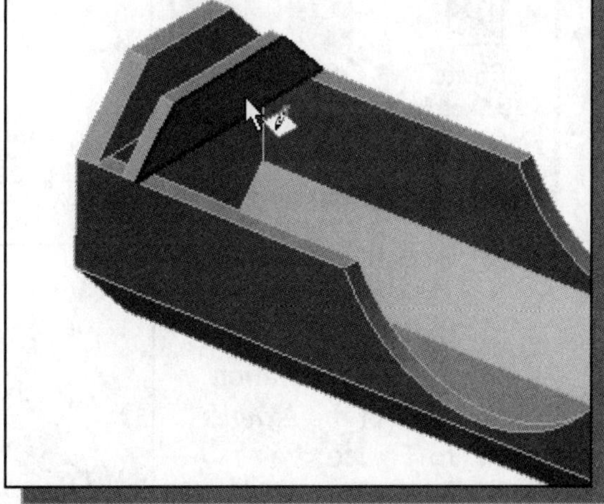

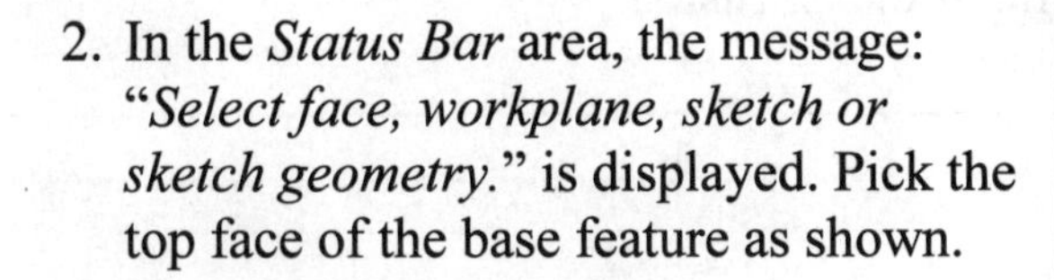

2. In the *Status Bar* area, the message: "*Select face, workplane, sketch or sketch geometry.*" is displayed. Pick the top face of the base feature as shown.

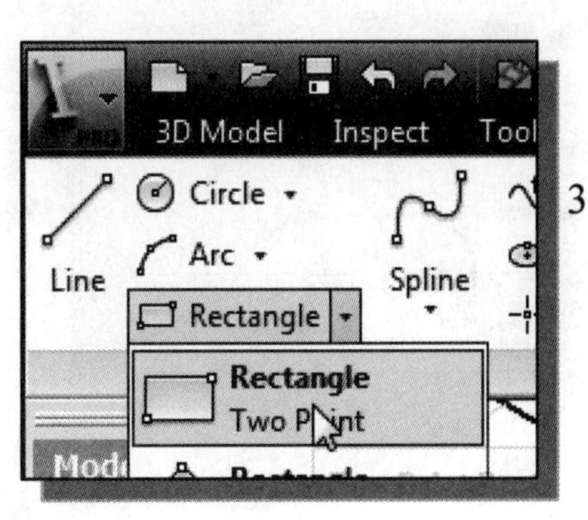

3. Select the **Two Point Rectangle** command by clicking once with the left-mouse-button on the icon in the *Draw* panel.

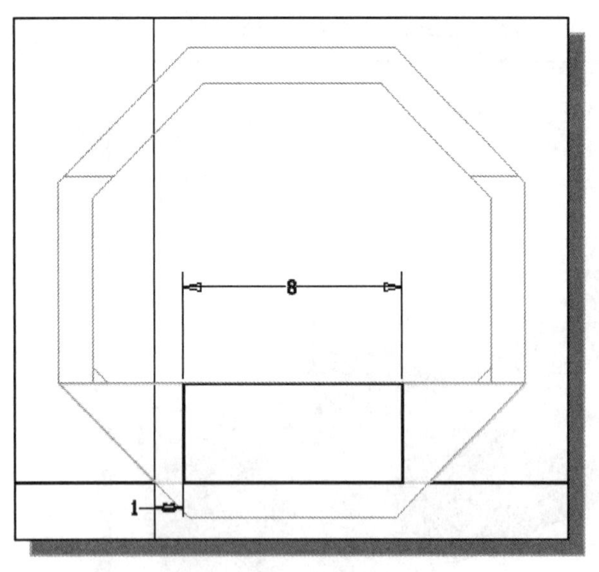

4. Create a rectangle, aligned to the top and bottom edges of the sketching plane, and modify the dimensions as shown.

- Hint: Use the **Select Other** option to select the necessary geometric entities.

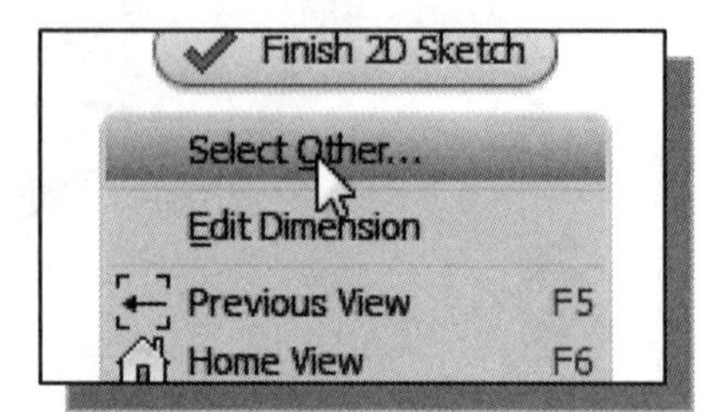

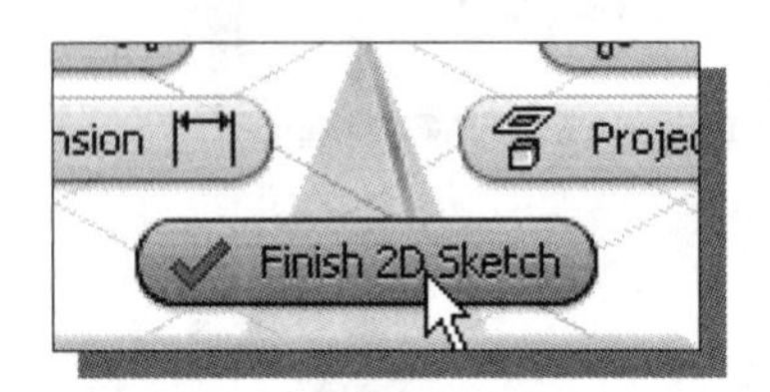

5. Inside the graphics window, click once with the right-mouse-button to display the option menu. Select **Finish 2D Sketch** in the popup menu to end the Sketch option.

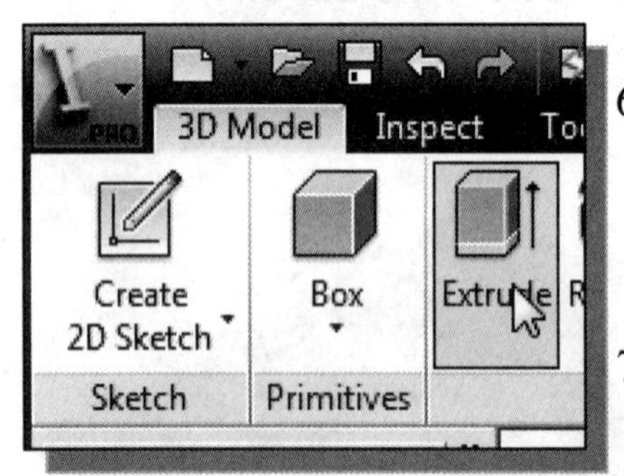

6. In the *Create* toolbar, select the **Extrude** command by releasing the left-mouse-button on the icon.

7. Select the inside region of the rectangle as the profile of the extrusion.

8. Inside the *Extrude* dialog box, select the **Cut** operation and set the *Extents* to **To Next** as shown.

9. In the *Extrude* dialog box, click on the **OK** button to proceed with creating the cut feature.

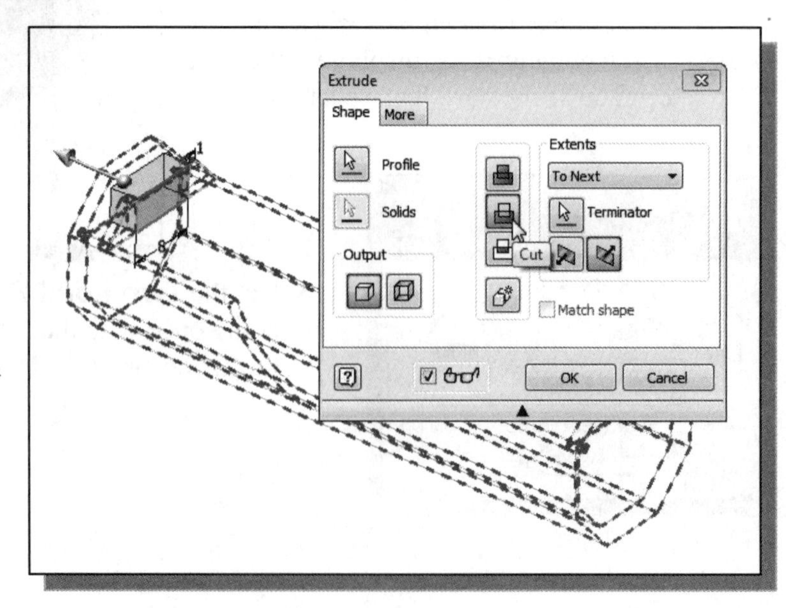

Create another Extruded Feature

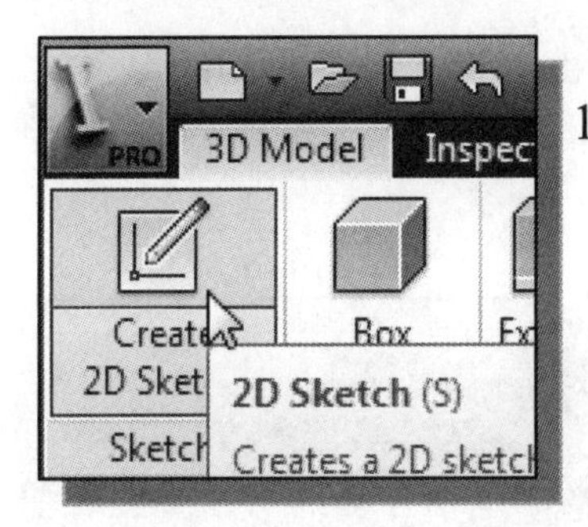

1. In the *Sketch* toolbar select the **Create 2D Sketch** command by left-clicking once on the icon.

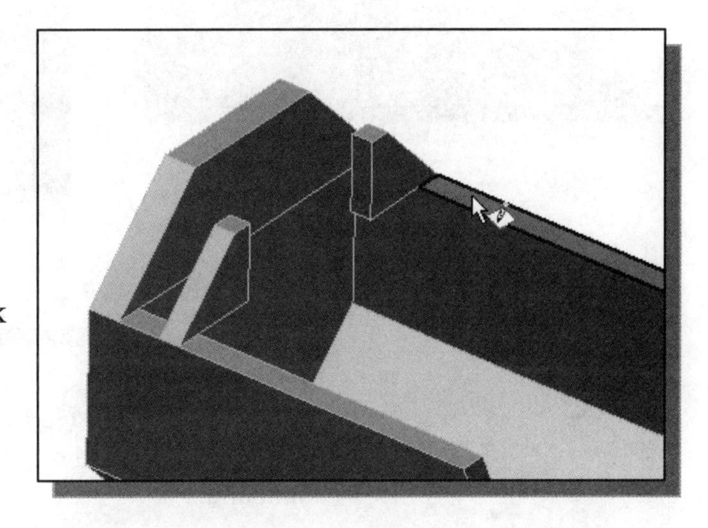

2. In the *Status Bar* area, the message: "*Select face, workplane, sketch or sketch geometry.*" is displayed. Pick the top face of the solid model as shown.

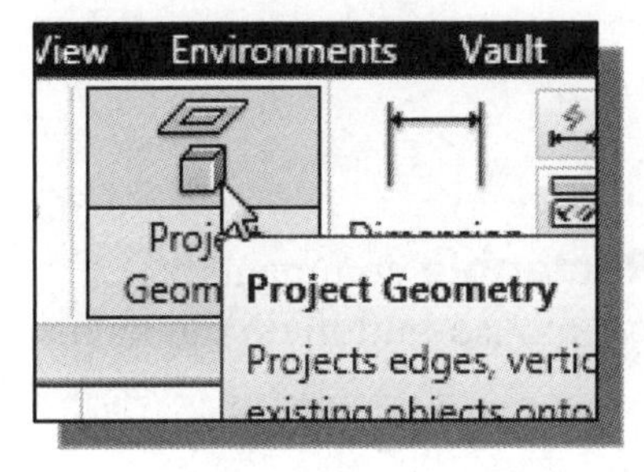

3. Click on the **Project Geometry** icon in the *Draw* panel.

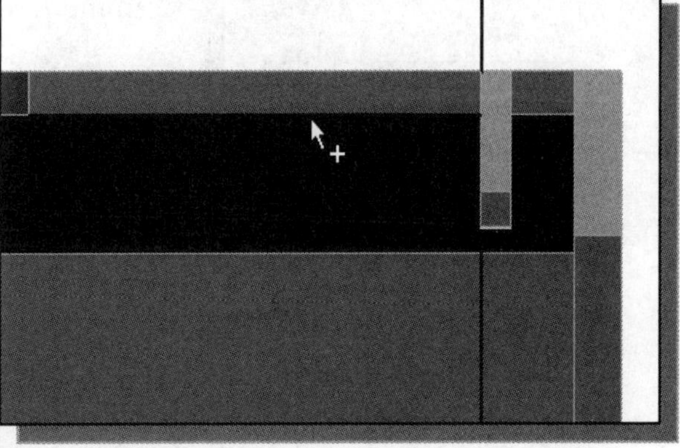

4. Select the **inside edges** near the upper part of the model as shown.

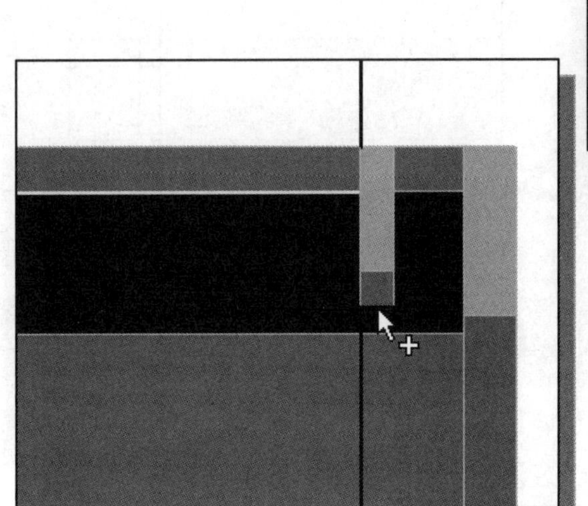

5. On your own, select the two corresponding edges in the lower part of the model.

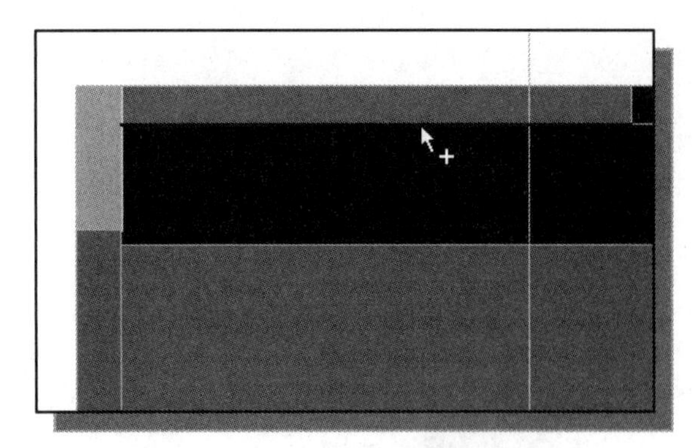

6. Select the inside edges on the other end of the model as shown.

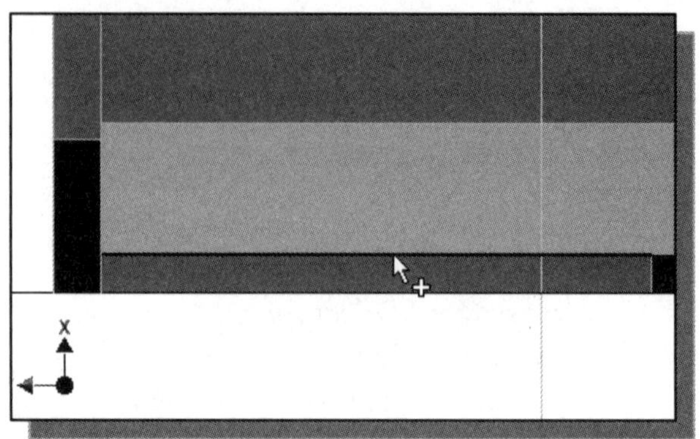

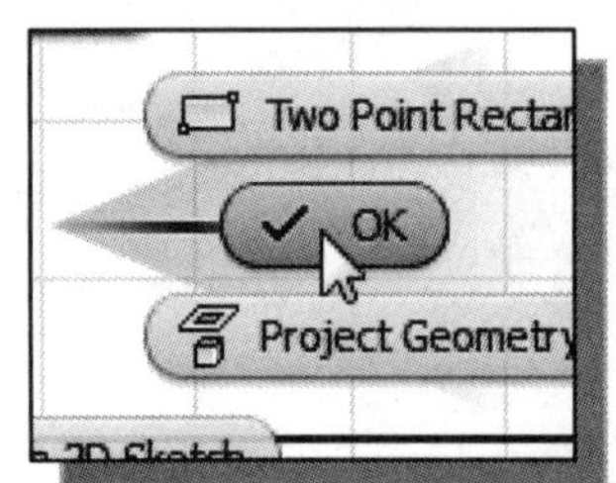

7. Inside the graphics window, **right-mouse-click** to bring up the option menu and select **OK** to end the Project Geometry command.

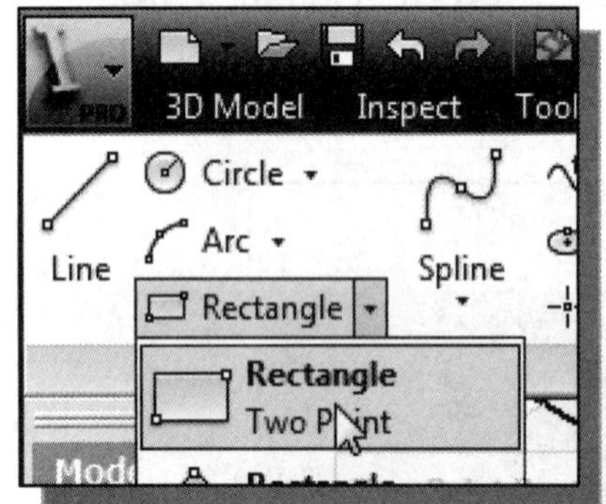

8. Select the **Two Point Rectangle** command by clicking once with the left-mouse-button on the icon in the *Draw* panel.

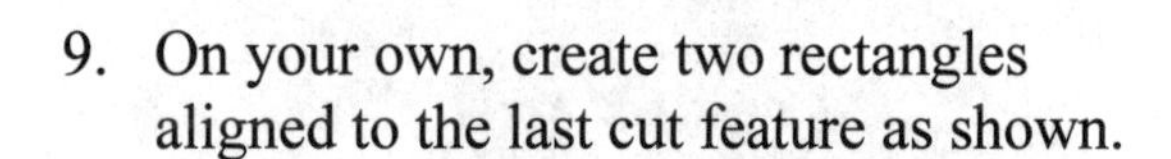

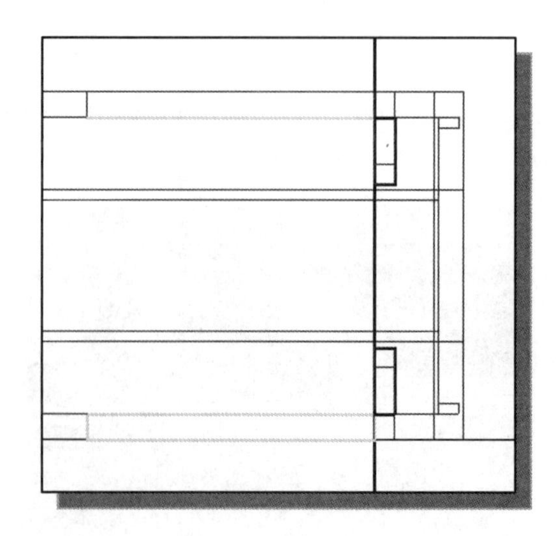

9. On your own, create two rectangles aligned to the last cut feature as shown.

- Note the two rectangles are fully constrained as shown near the bottom of the window.

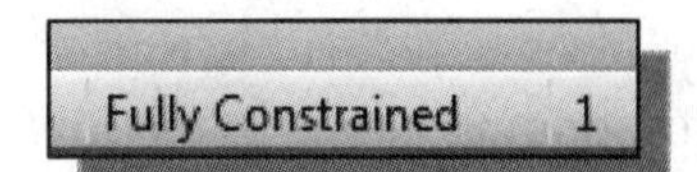

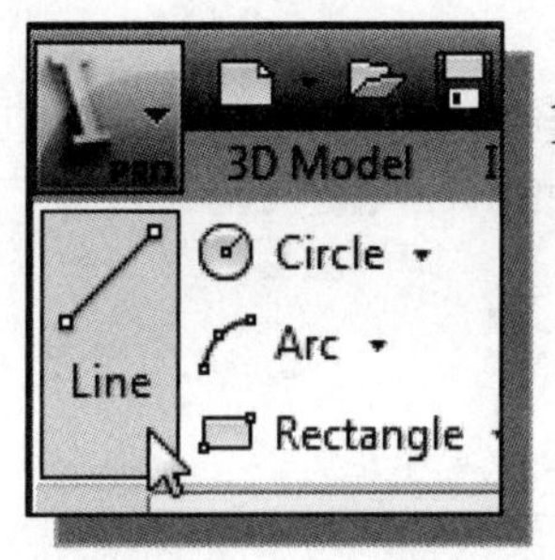

10. Move the graphics cursor to the **Line** icon in the *Sketch* toolbar. A *Help-tip box* appears next to the cursor and a brief description of the command is displayed at the bottom of the drawing screen: "*Creates Straight line segments and tangent arcs.*"

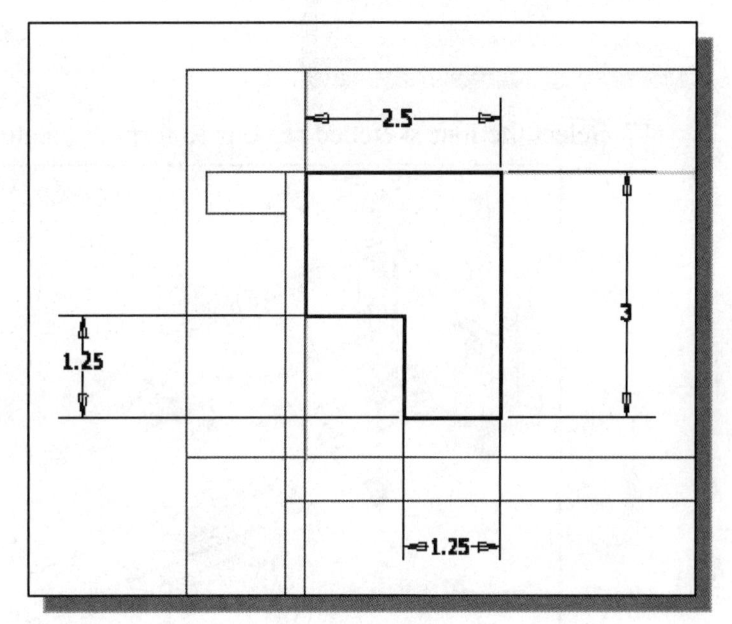

11. Create six connected line segments, with the corners aligned to the projected geometry as shown.

12. On your own, create and modify the **four dimensions** as shown.

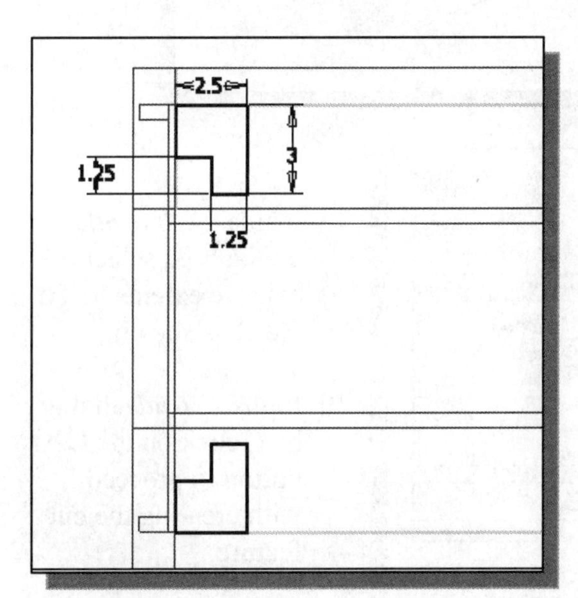

13. On your own, create another closed region on the lower part of the model as shown.

14. On your own, apply proper constraints, such as the **Equal** constraint, to fully constrain the line segments.

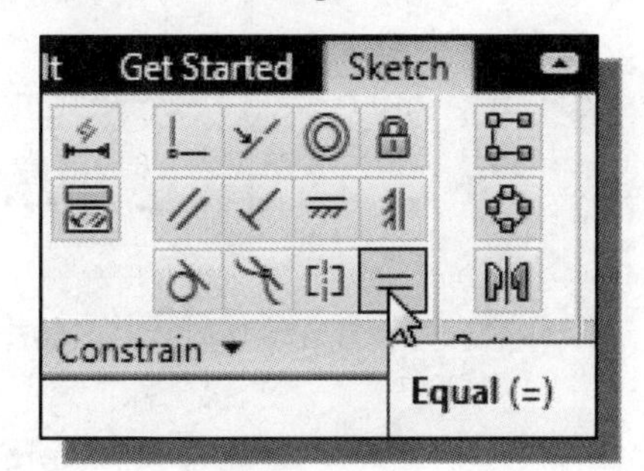

15. In the *Ribbon* toolbar, click once with the left-mouse-button on **Finish Sketch** to end the **Sketch** option.

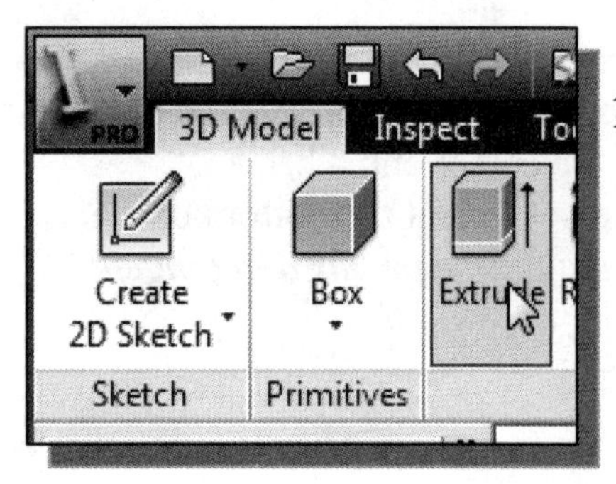

16. In the *Create* toolbar, select the **Extrude** command by releasing the left-mouse-button on the icon.

17. Select the four sketched regions to form the feature profile of the extrusion.

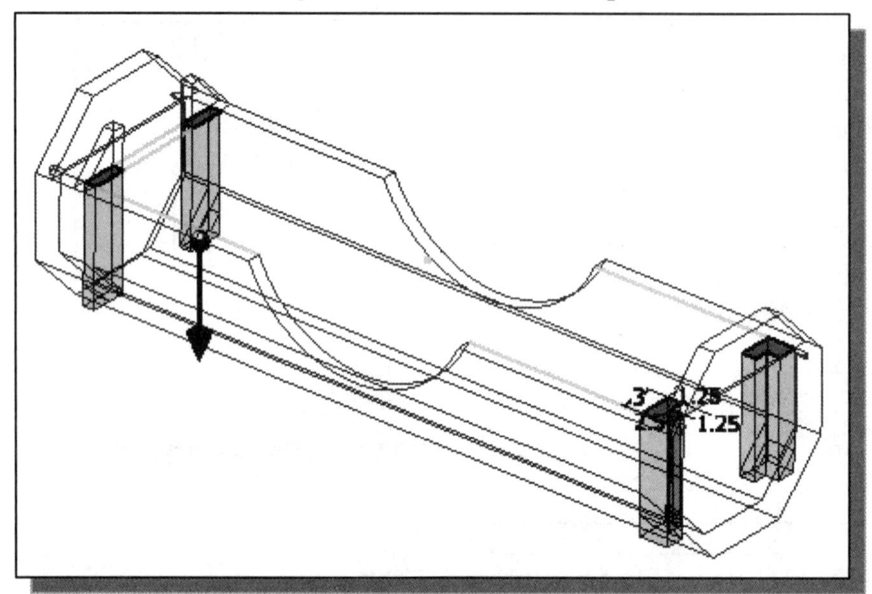

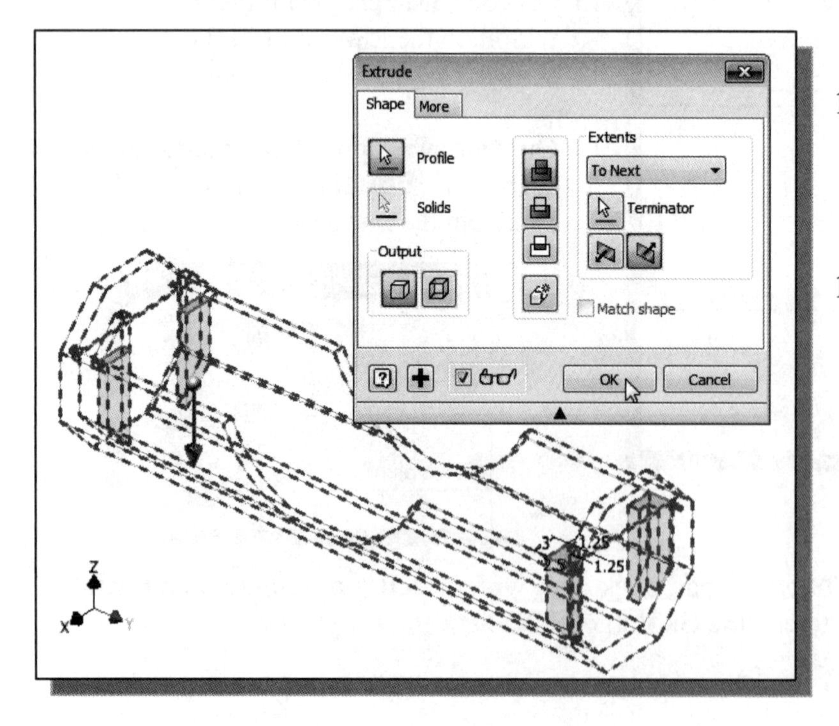

18. Inside the *Extrude* dialog box, select the extrude extents to **To Next** as shown.

19. In the *Extrude* dialog box, click on the **OK** button to proceed with creating the cut feature.

Create another Cut Feature

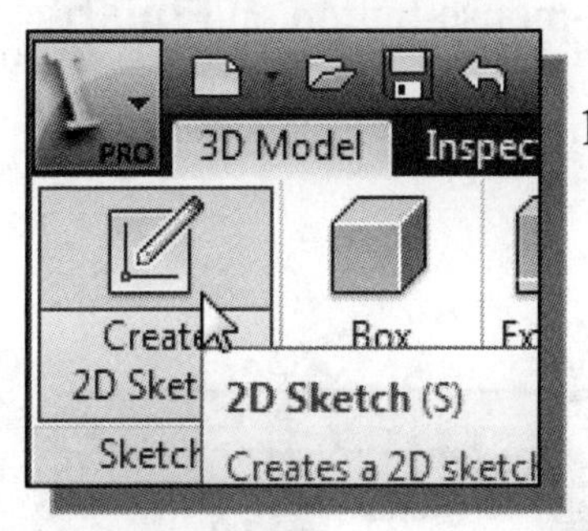

1. In the *Sketch* toolbar select the **Create 2D Sketch** command by left-clicking once on the icon.

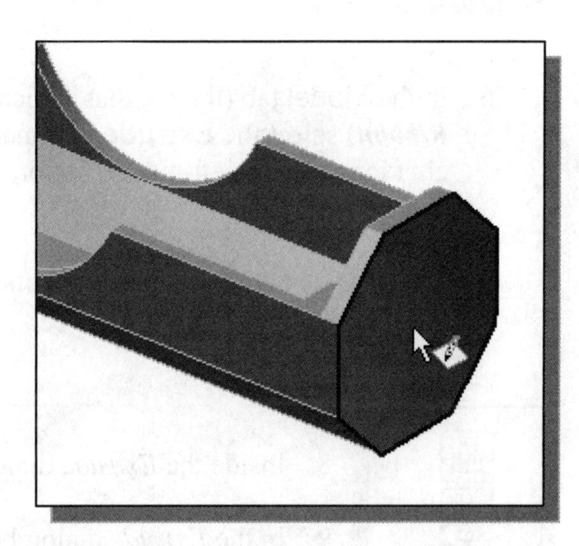

2. In the *Status Bar* area, the message: "*Select face, workplane, sketch or sketch geometry.*" is displayed. Pick the side face of the solid model as shown.

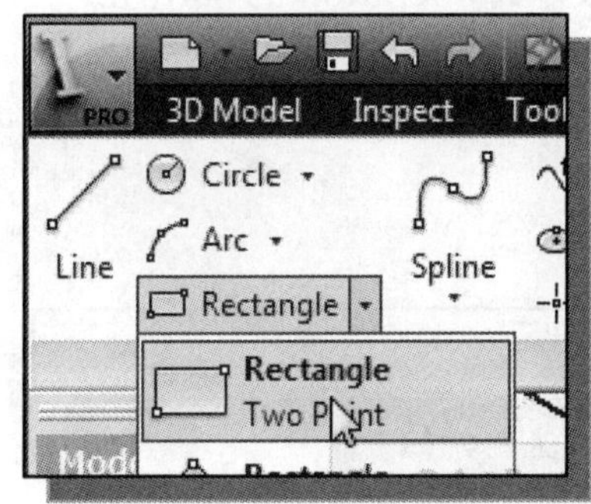

3. Select the **Two Point Rectangle** command by clicking once with the left-mouse-button on the icon in the *Draw* panel.

4. On your own, create three rectangles aligned to the center point as shown.

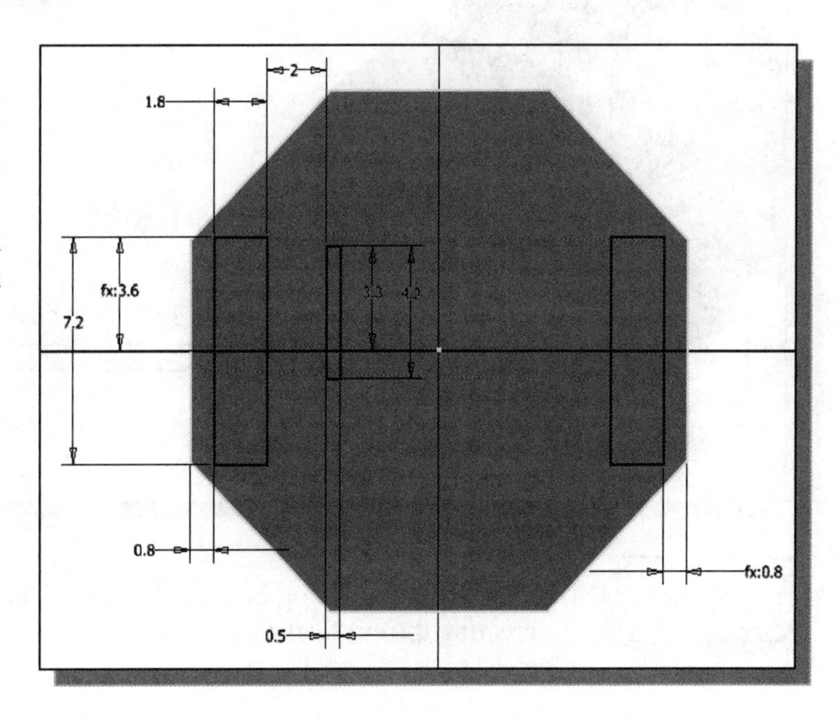

5. In the *Ribbon* toolbar, click once with the left-mouse-button on **Finish Sketch** to end the **Sketch** option.

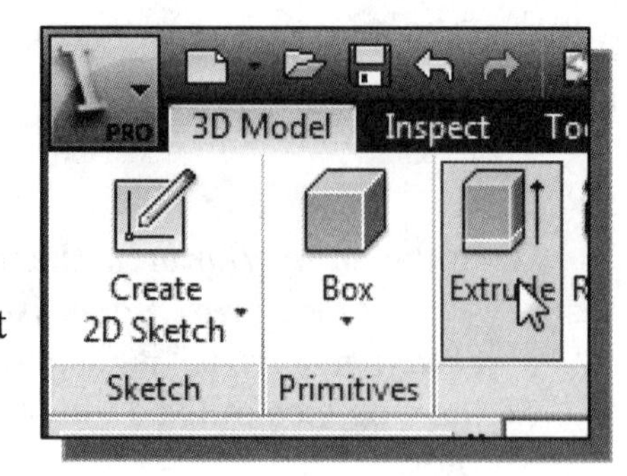

6. In the **Model** tab (the tab that is located in the ***Ribbon***) select the **Extrude** command by clicking with the left-mouse-button on the icon.

7. Click inside the three sketched rectangles to set them as the extrusion profile.

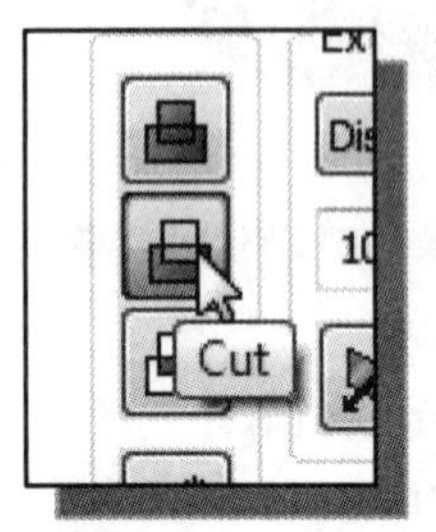

8. Inside the *Extrude* dialog box, select the **Cut** operation as shown.

9. In the *Extrude* dialog box, set the extents distance to **1.5mm** as shown.

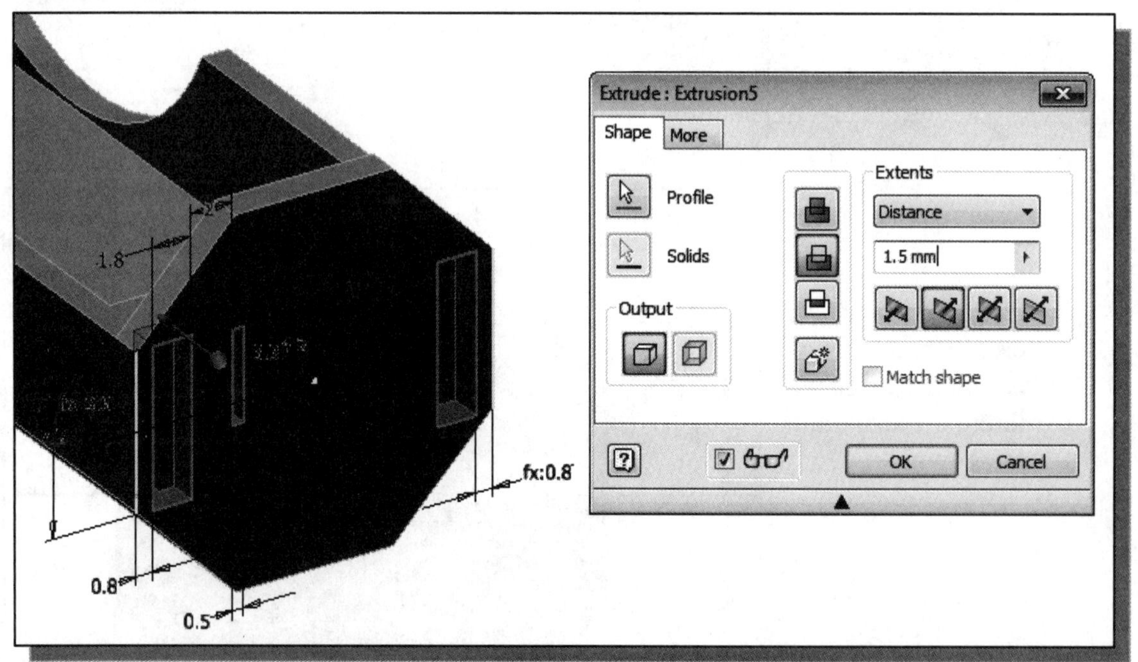

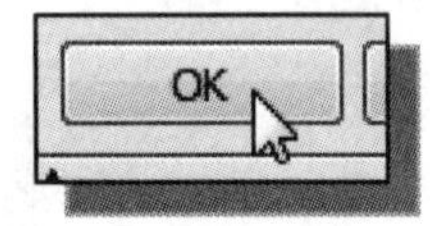

10. In the *Extrude* dialog box, click on the **OK** button to proceed with creating the cut feature.

Mirrored Features

- In *Autodesk Inventor*, features can be mirrored to create and maintain complex symmetrical features. We can mirror a feature about a work plane or a specified surface. We can create a mirrored feature while maintaining the original parametric definitions, which can be quite useful in creating symmetrical features.

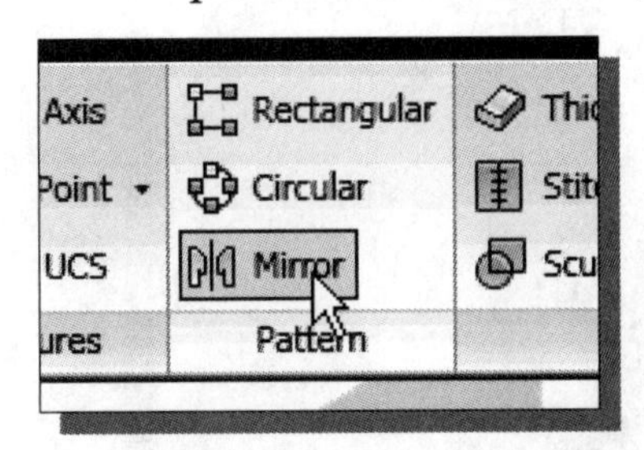

1. In the *Pattern* panel, select the **Mirror Feature** command by releasing the left-mouse-button on the icon.

2. In the *Mirror Pattern* dialog box, the **Features** button is activated. *Autodesk Inventor* expects us to select features to be mirrored. In the *browser* window, select the last cut feature.

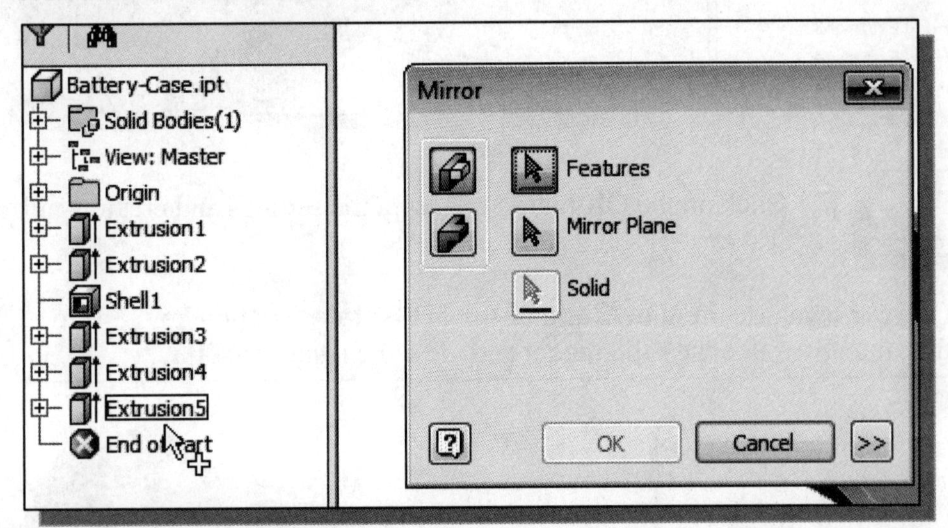

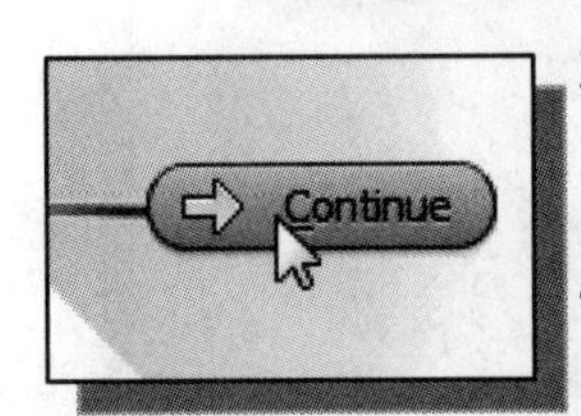

3. Inside the graphics window, right-mouse-click to bring up the option menu.

4. Select **Continue** in the option list to proceede with the Mirror Feature command.

5. In the *Mirror Pattern* dialog box, the **Mirror Plane** button is activated. *Autodesk Inventor* expects us to select a planar surface about which to mirror. In the prompt area, the message "*Select plane to mirror about*" is displayed.

6. Select the surface to the right, as shown, as the planar surface about which to mirror.

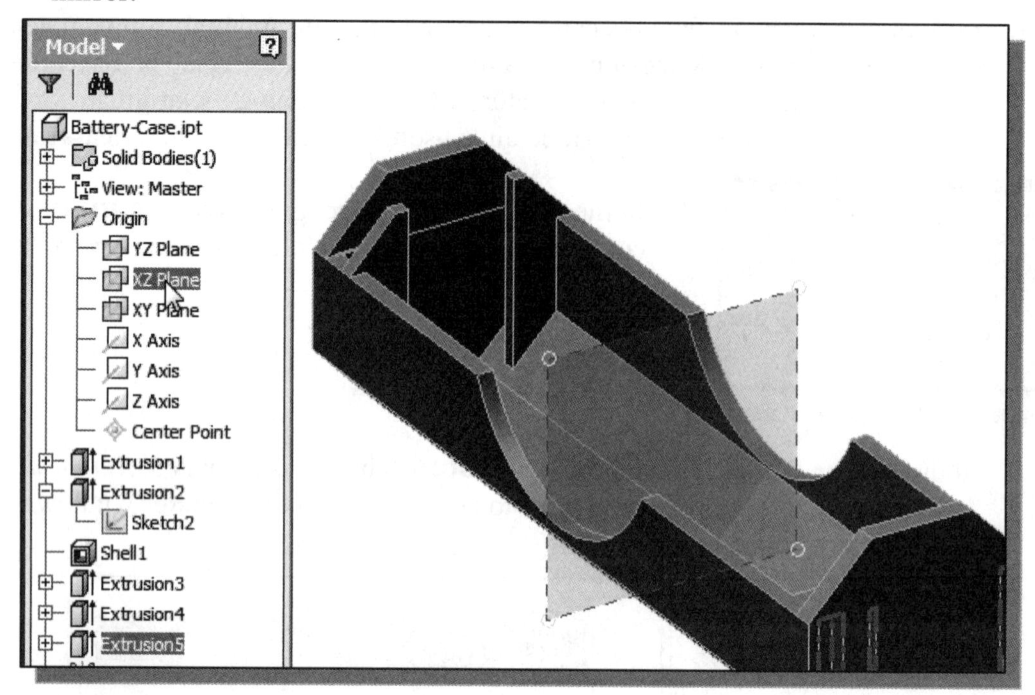

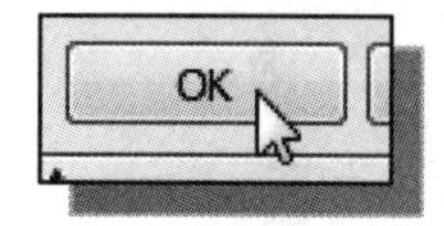

7. Click on the **OK** button to accept the settings and create a mirrored feature.

8. On your own, use the ViewCube or the 3D-Rotate function key [**F4**] to dynamically rotate the solid model and view the resulting solid.

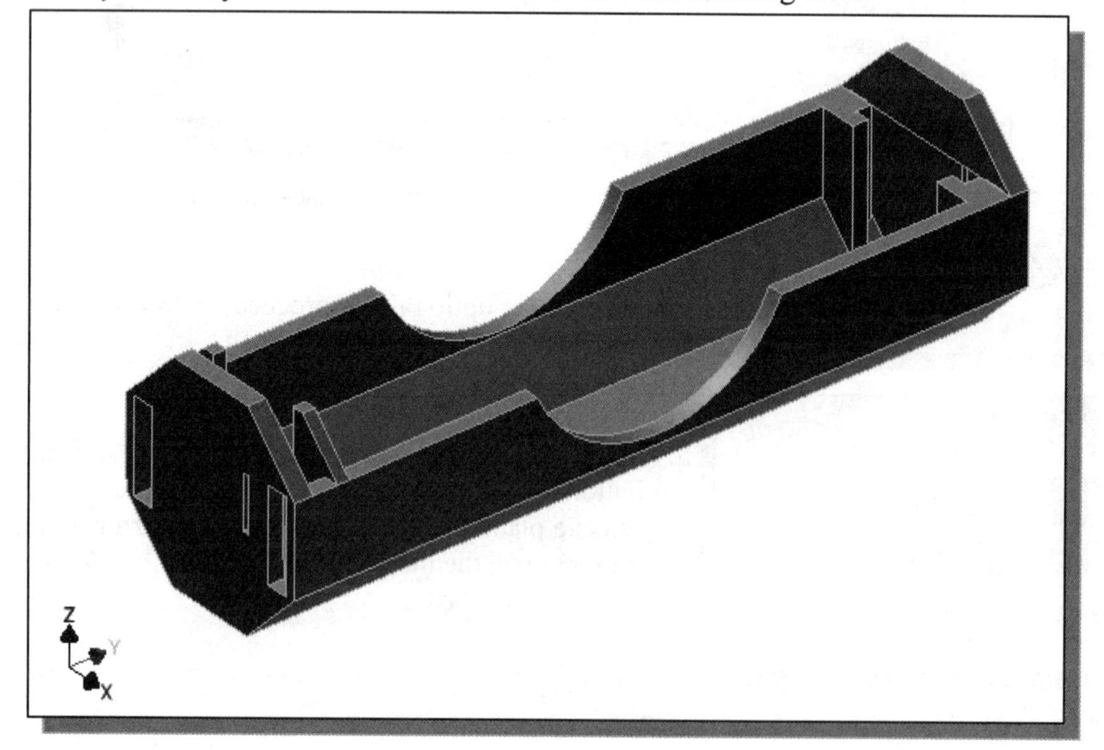

Create another Cut Feature

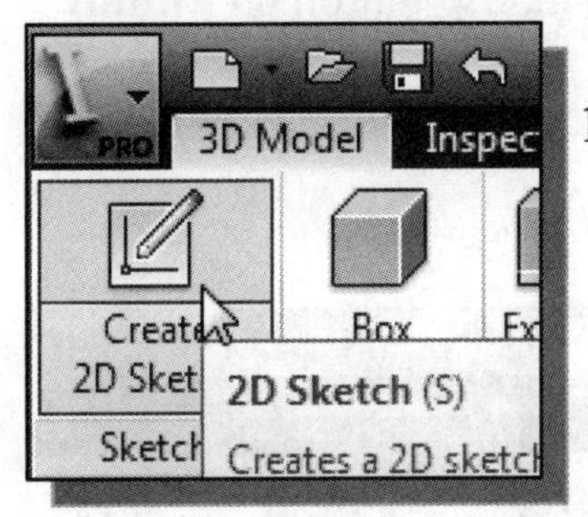

1. In the *Sketch* toolbar select the **Create 2D Sketch** command by left-clicking once on the icon.

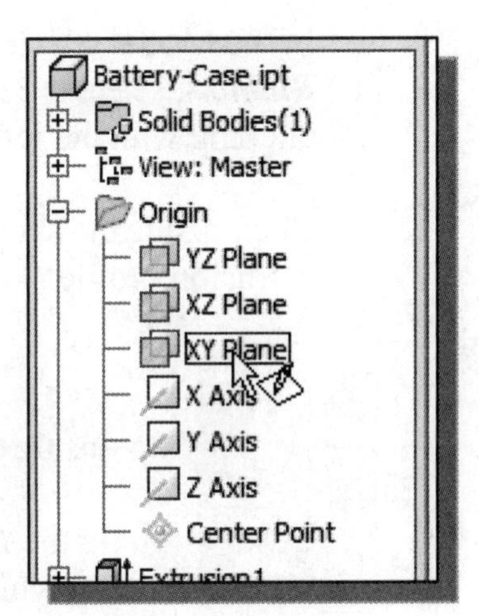

2. In the *Status Bar* area, the message: "*Select face, workplane, sketch or sketch geometry.*" is displayed. Select the **XY Plane**, by left-clicking once on any edges of the XY Plane in the graphics window or in the *browser* window as shown.

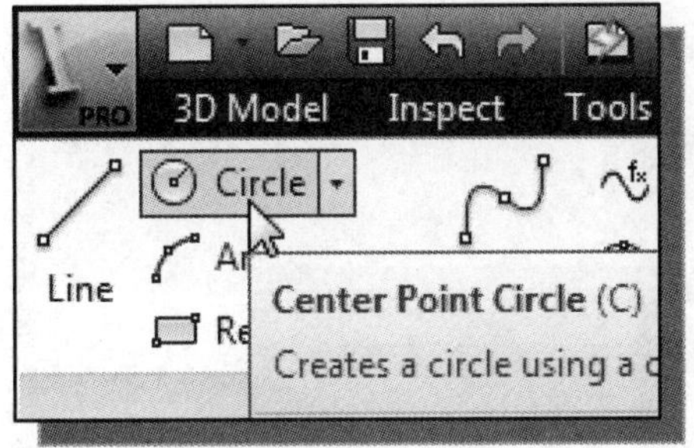

3. Select the **Center Point Circle** command by clicking once with the left-mouse-button on the icon in the *Draw* panel.

4. On your own, create two circles aligned to the center point as shown.

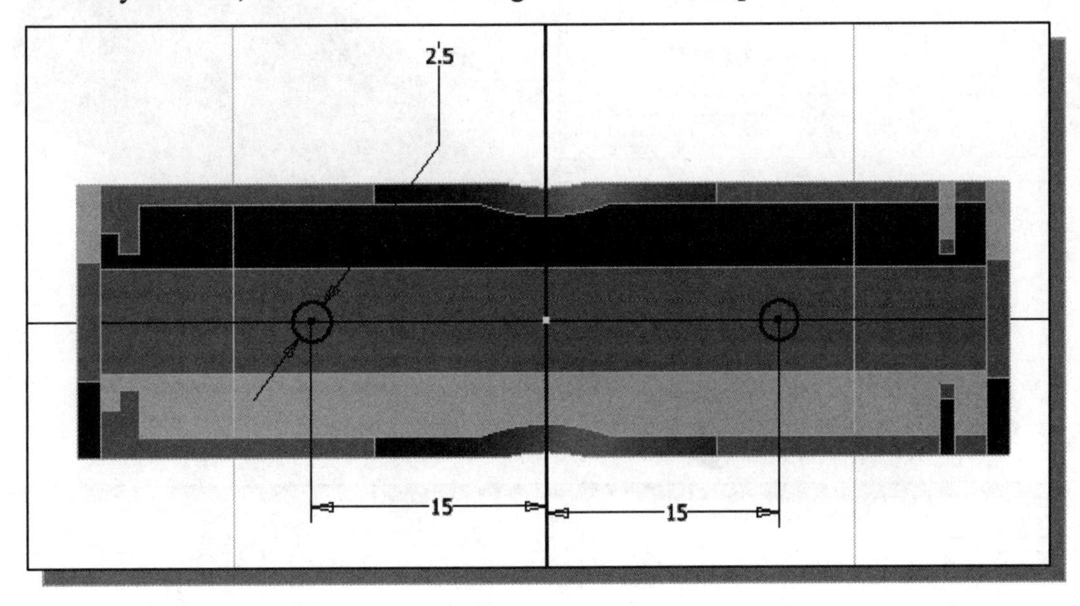

5. In the *Ribbon* toolbar, click once with the left-mouse-button on **Finish Sketch** to end the Sketch option.

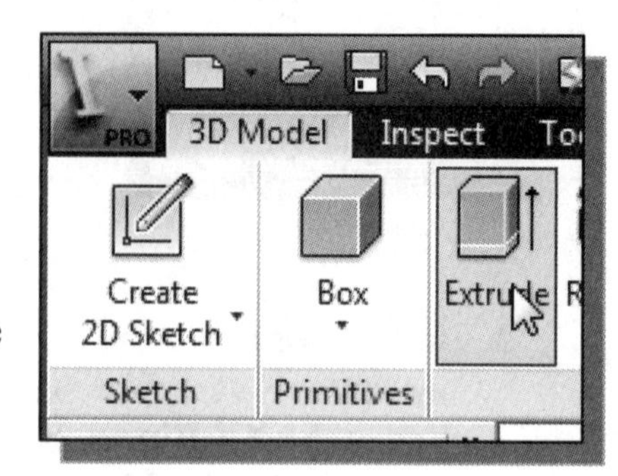

6. In the Model tab (the tab that is located in the ***Ribbon***), select the **Extrude** command by clicking with the left-mouse-button on the icon.

7. Click inside the sketched two circles to set as the extrusion profile.

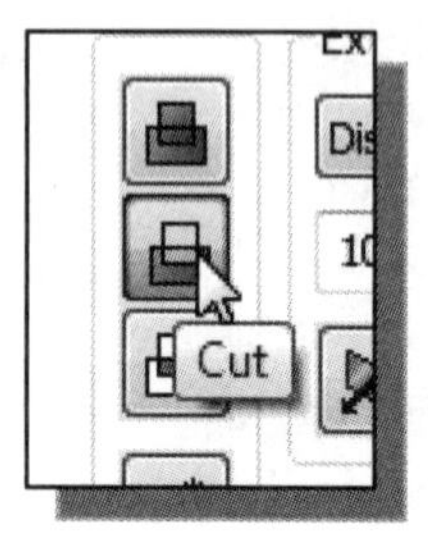

8. Inside the *Extrude* dialog box, select the **Cut** operation as shown.

9. In the *Extrude* dialog box, set the extents distance to **Thru All** as shown.

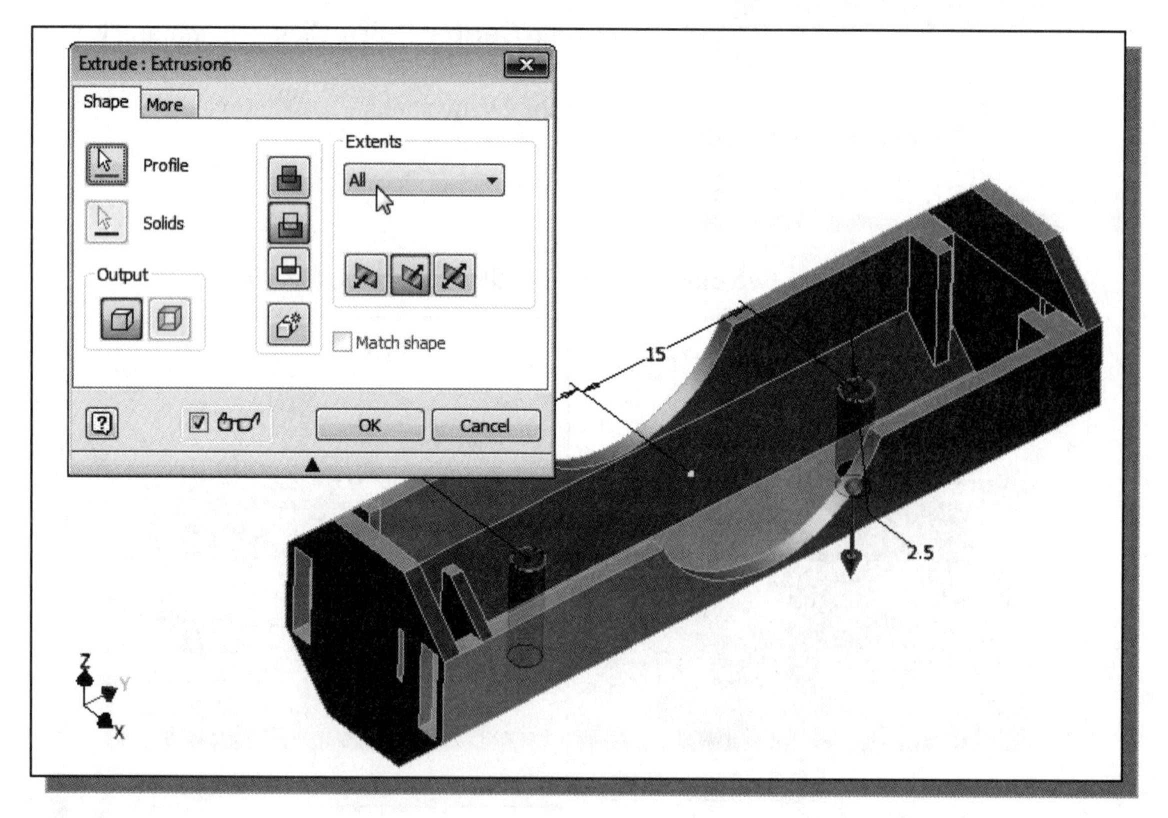

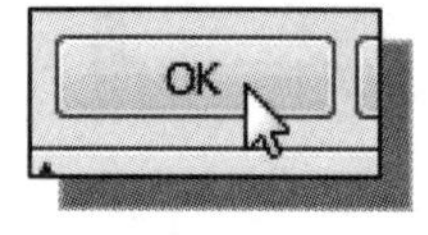

10. In the *Extrude* dialog box, click on the **OK** button to proceed with creating the cut feature.

Create the Last Feature

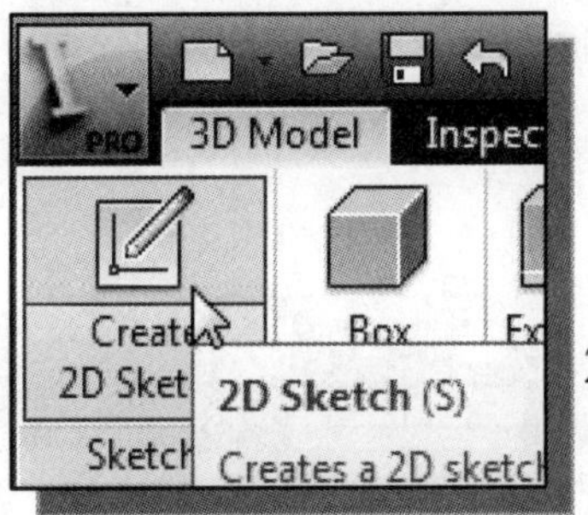

1. In the *Sketch* toolbar select the **Create 2D Sketch** command by left-clicking once on the icon.

2. In the *Status Bar* area, the message: "*Select face, workplane, sketch or sketch geometry.*" is displayed. Pick the inside face of the solid model as shown.

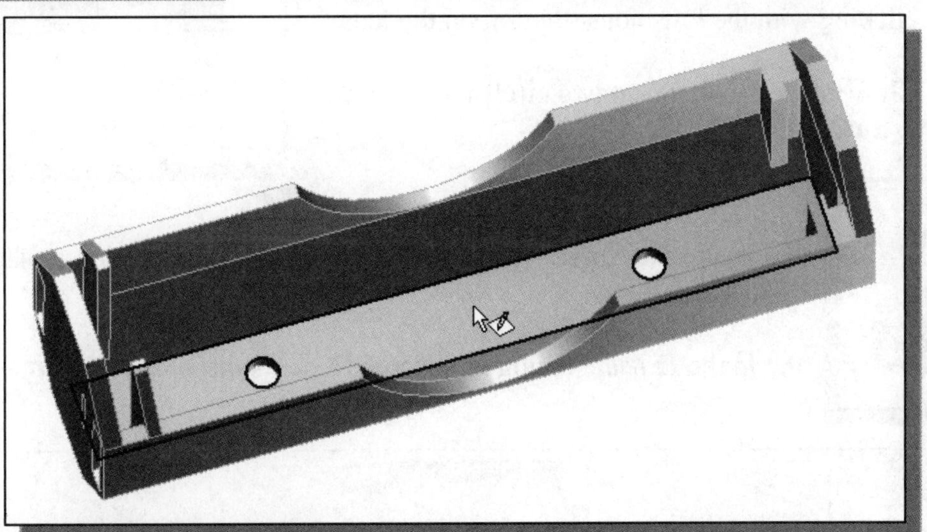

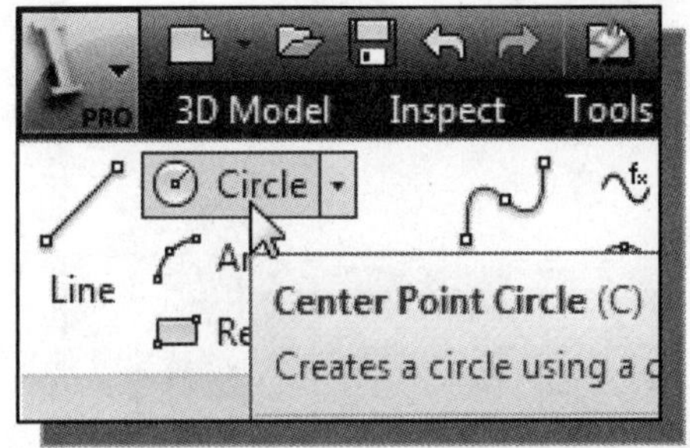

3. Select the **Center Point Circle** command by clicking once with the left-mouse-button on the icon in the *Draw* panel.

4. On your own, create two circles aligned to the center points as shown.

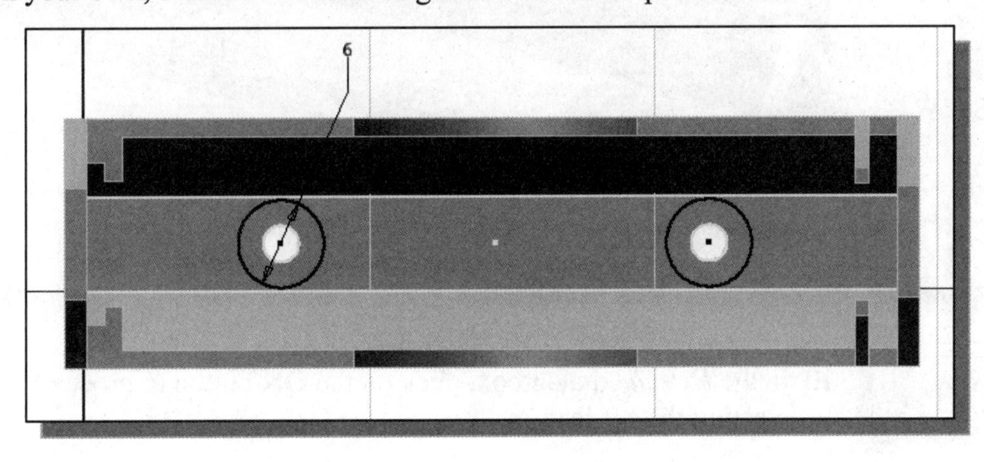

5. In the *Ribbon* toolbar, click once with the left-mouse-button on **Finish Sketch** to end the Sketch option.

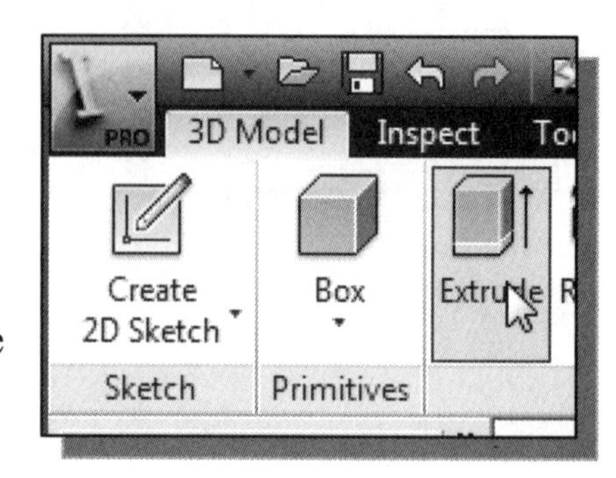

6. In the Model tab (the tab that is located in the ***Ribbon***) select the **Extrude** command by clicking with the left-mouse-button on the icon.

7. Click inside the sketched two circles to set as the extrusion profile.

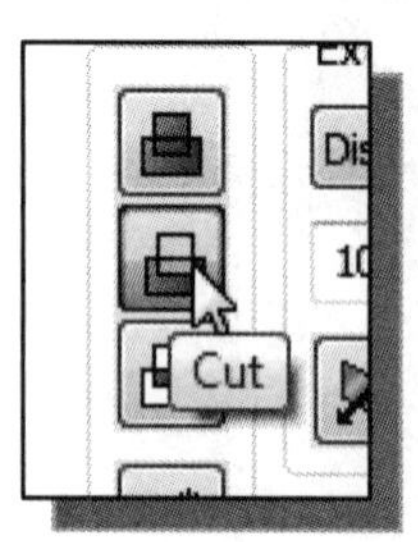

8. Inside the *Extrude* dialog box, select the **Cut** operation as shown.

9. In the *Extrude* dialog box, set the extents distance to **1mm** as shown.

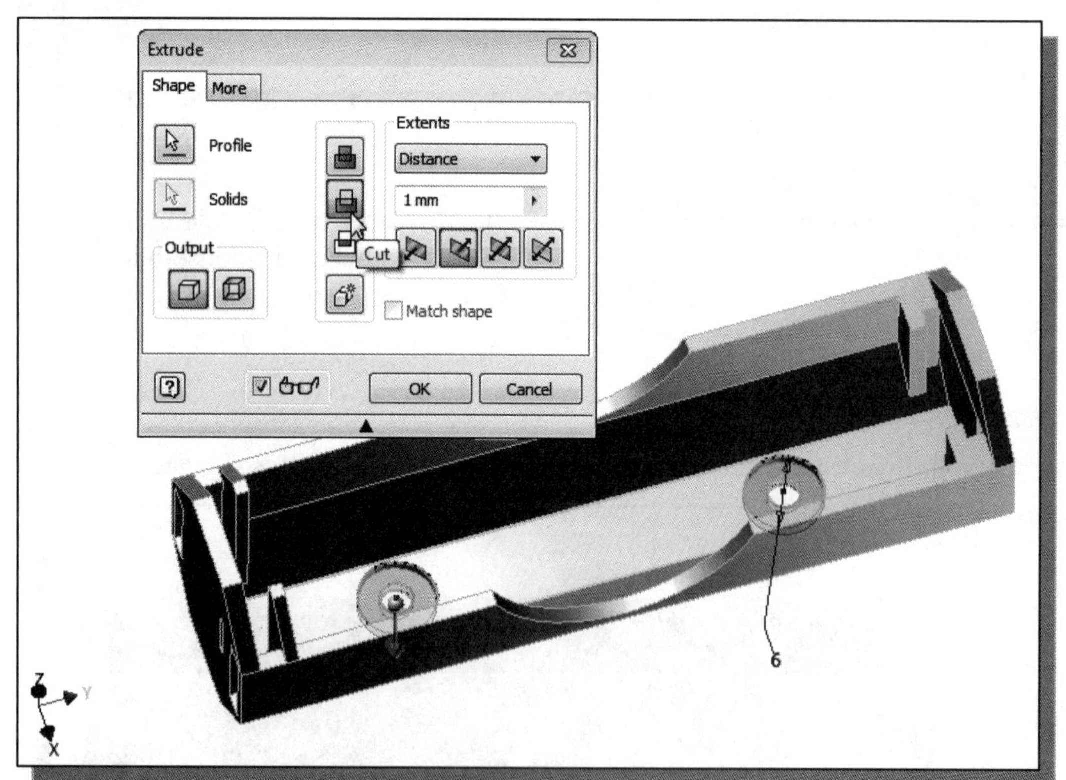

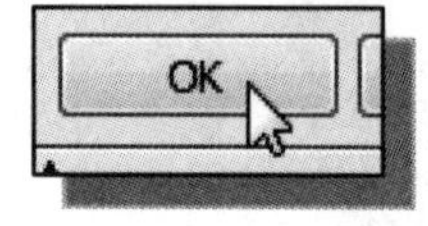

10. In the *Extrude* dialog box, click on the **OK** button to proceed with creating the cut feature.

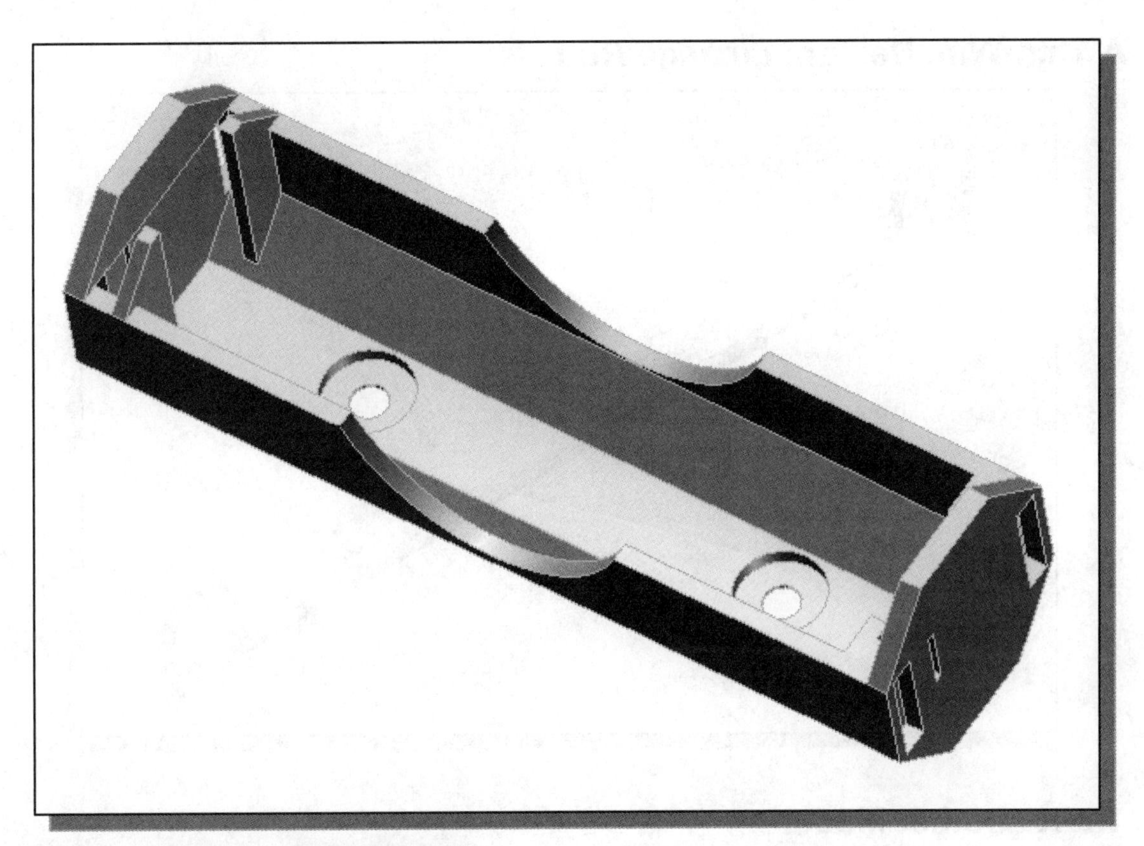

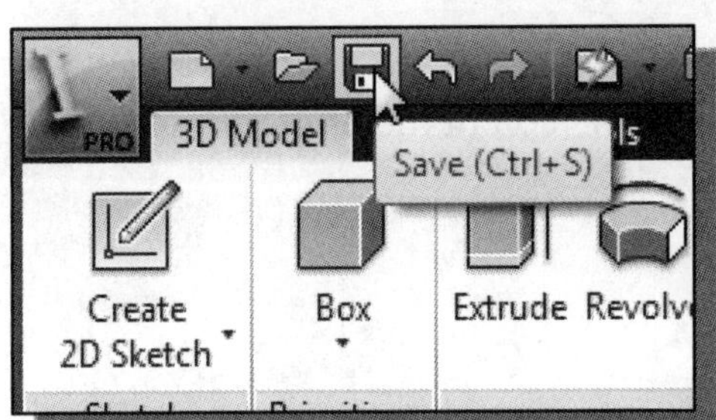

11. Select **Save** in the *Quick Access* toolbar, or you can also use the "**Ctrl-S**" combination (hold down the "Ctrl" key and hit the "S" key once) to save the part.

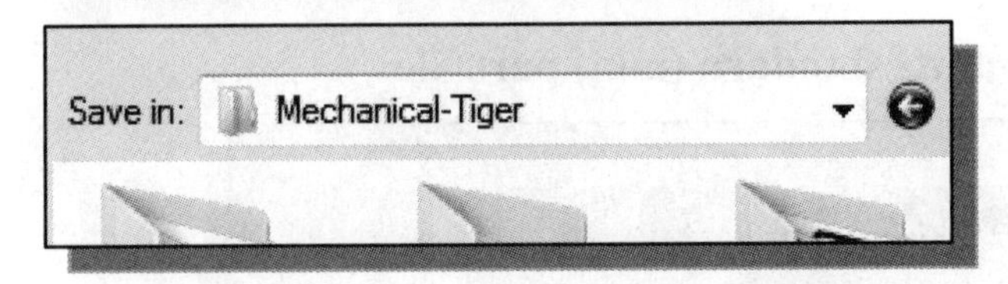

12. Confirm the *folder* is set to the Tamiya kit folder, **Mechanical-Tiger**, as shown.

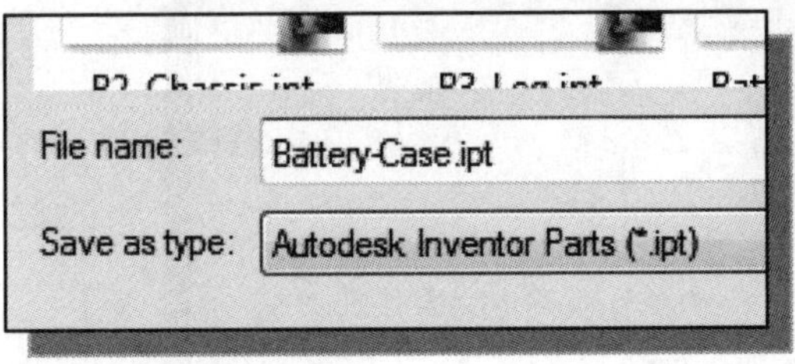

13. In the *File name* editor box, enter **Battery-Case.ipt** as the file name.

14. Click on the **Save** button to save the file.

A Thin-Wire Design: *Linkage Rod*

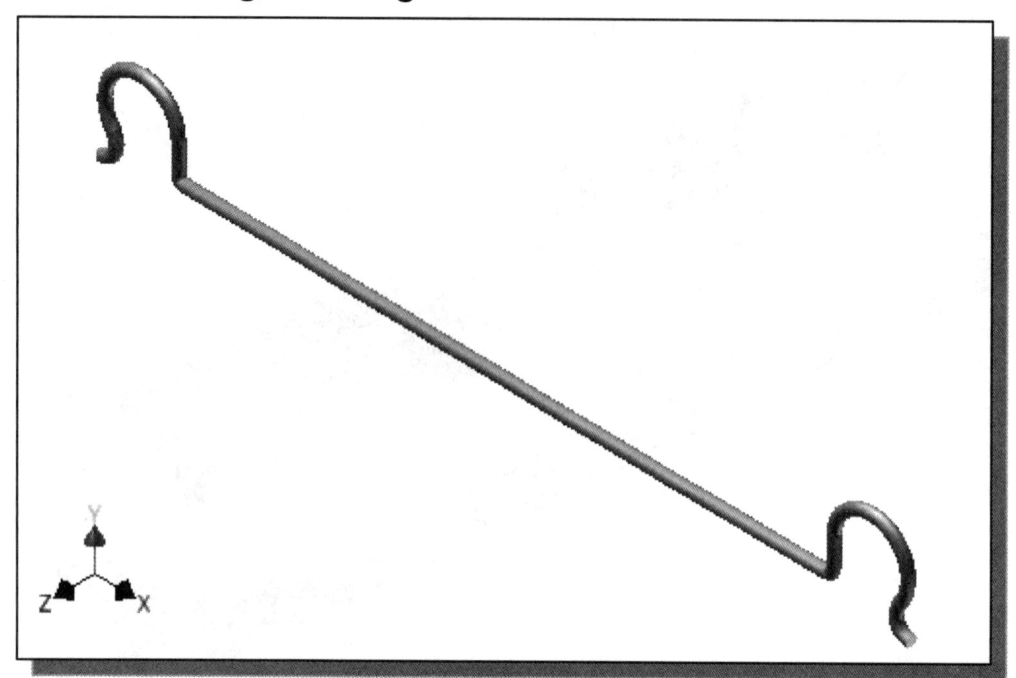

Start another Model

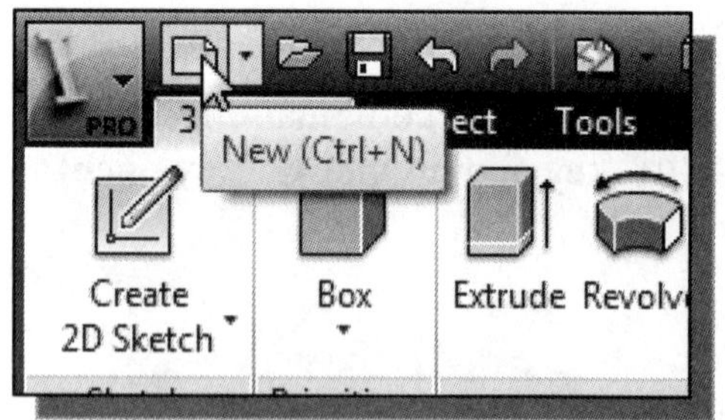

1. Click on the **New** icon in the *Standard* toolbar.

2. On your own, start a new **Metric units standard (mm) part** file.

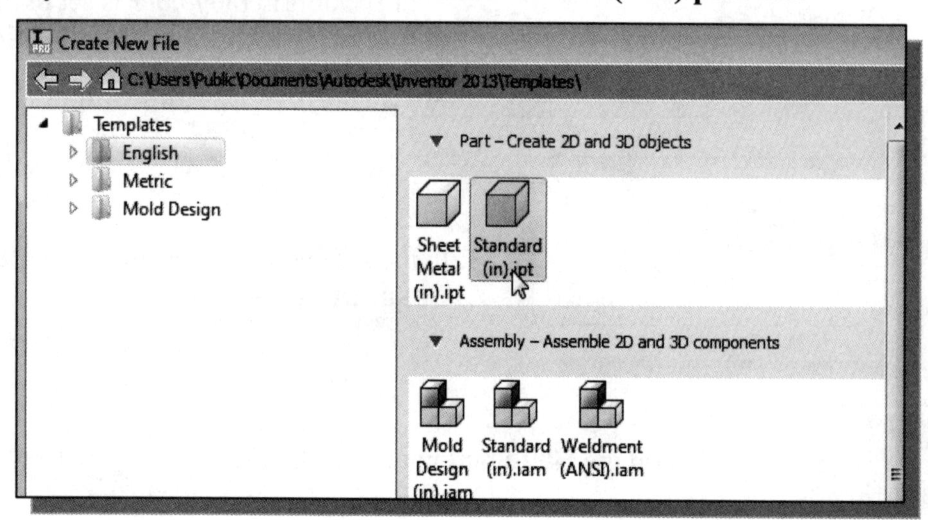

Create a Swept Feature

❖ The **Sweep** operation is defined as moving a planar section through a planar (2D) or 3D path in space to form a three-dimensional solid object. The path can be an open curve or a closed loop, but must be on an intersecting plane with the profile. The **Extrusion** operation, which we have used in the previous lessons, is a specific type of sweep. The **Extrusion** operation is also known as a ***linear sweep*** operation, in which the sweep control path is always a line perpendicular to the two-dimensional section. Linear sweeps of unchanging shape result in what are generally called ***prismatic solids*** which means solids with a constant cross-section from end to end. In *Autodesk Inventor*, we create a ***swept feature*** by defining a path and then a 2D sketch of a cross section. The sketched profile is then swept along the planar path. The Sweep operation is used for objects that have uniform shapes along a trajectory.

♦ Define a Sweep Path

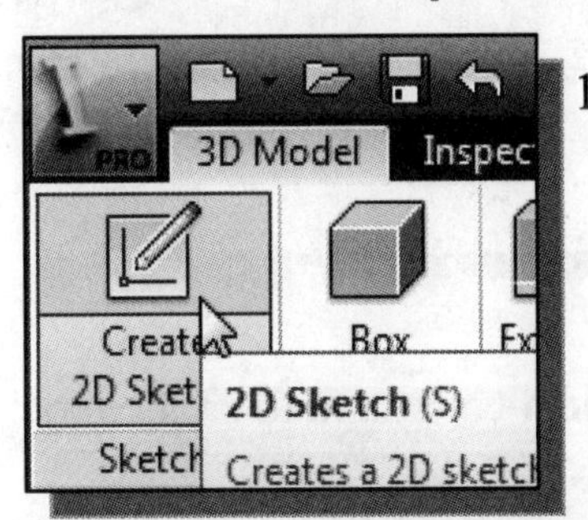

1. In the *Sketch* toolbar select the **Create 2D Sketch** command by left-clicking once on the icon.

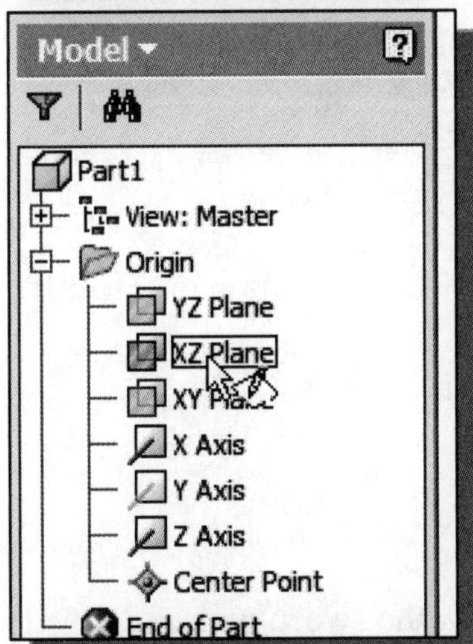

2. Select the **XZ Plane**, by left-clicking once on the XZ Plane, in the *browser* window.

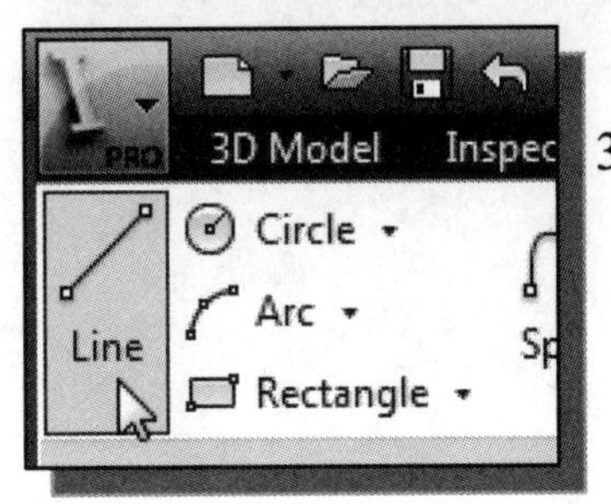

3. Move the graphics cursor to the **Line** icon in the *Draw* toolbar. A *Help-tip box* appears next to the cursor and a brief description of the command is displayed at the bottom of the drawing screen: "*Creates Straight line segments and tangent arcs.*"

4. On your own, create the **sweep path** with two line segments and 3 arcs as shown. Note the right endpoint of the horizontal line is aligned to the **Center Point** of the coordinate system. Also, the center of the left arc is also aligned horizontally to the **Center Point**.

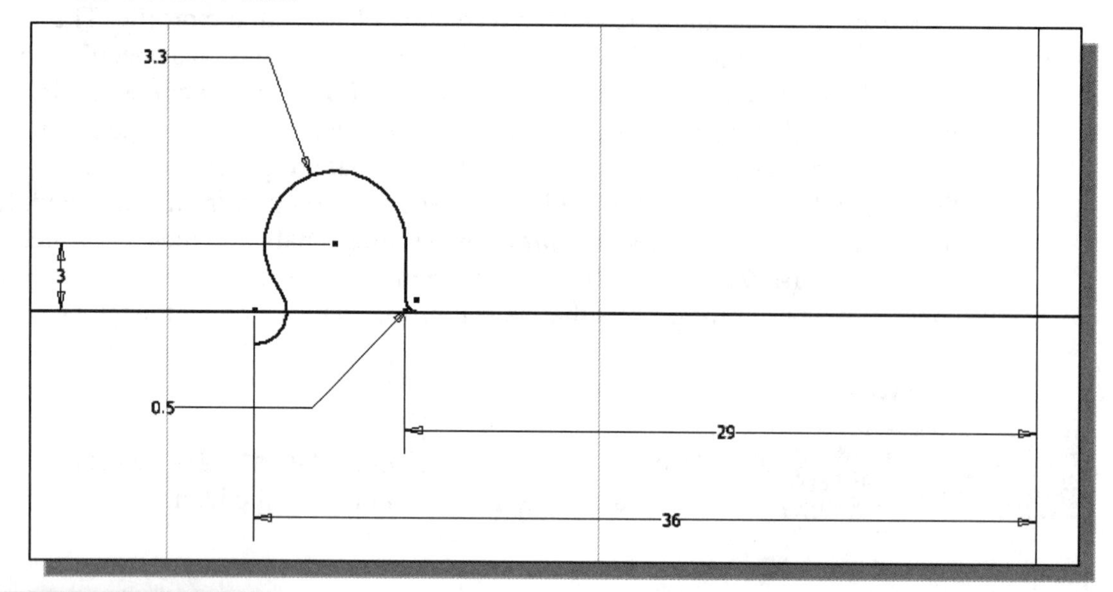

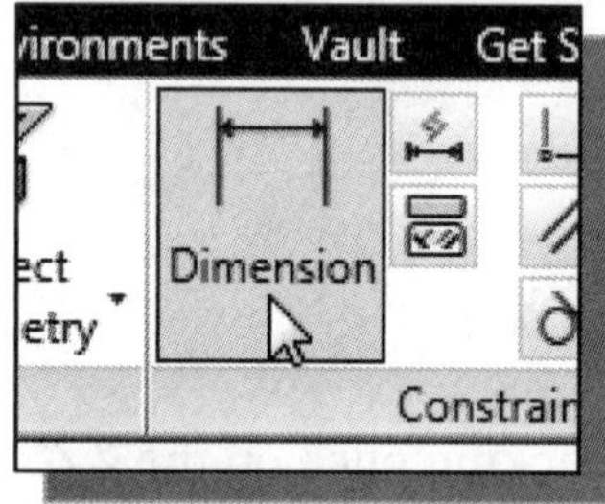

5. On your own, use the **Dimension** command to create the necessary dimensions as shown.

6. In the *Ribbon* toolbar, click once with the left-mouse-button on **Finish Sketch** to end the **Sketch** option.

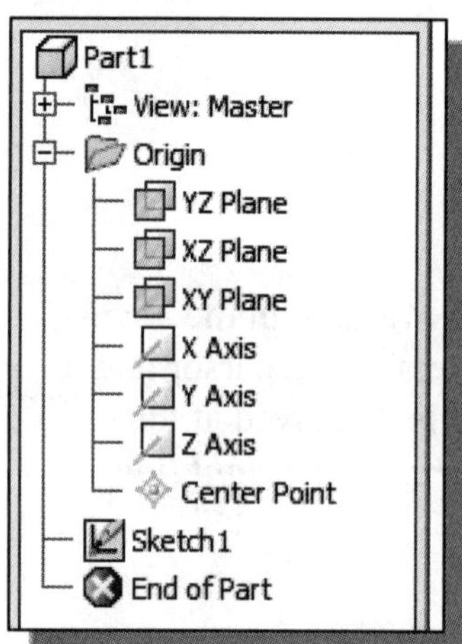

- Notice in the *browser* window, the sweep path is defined as the first 2D sketch.

♦ Define the Sweep Section

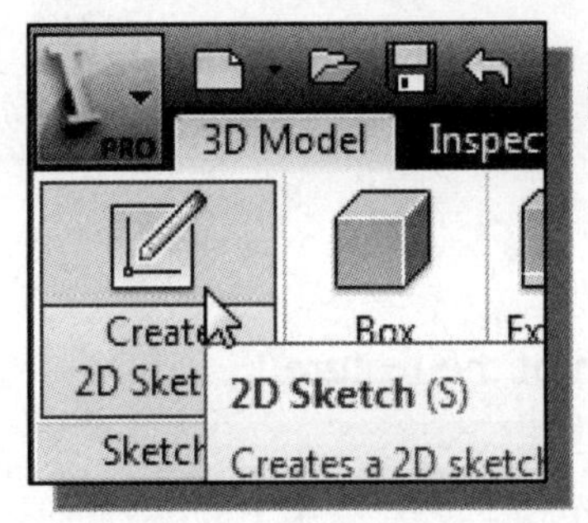

1. In the *Sketch* toolbar select the **Create 2D Sketch** command by left-clicking once on the icon.

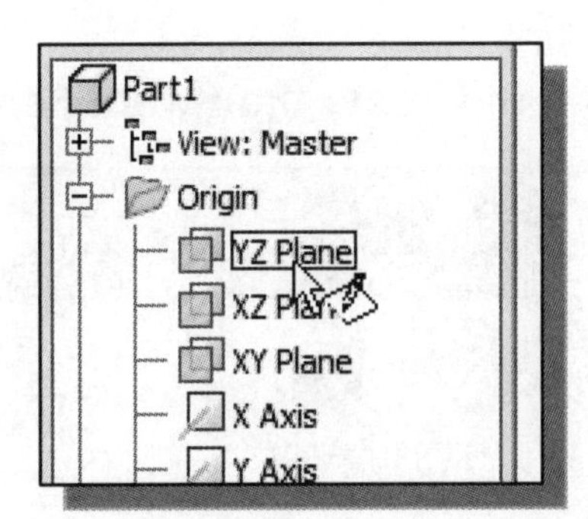

2. In the *Status Bar* area, the message: "*Select face, workplane, sketch or sketch geometry*" is displayed. Select the **YZ Plane**, by left-clicking once on the YZ Plane, in the *browser* window.

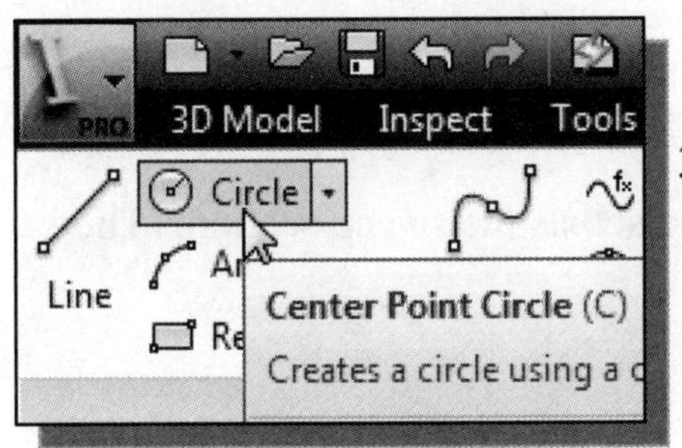

3. Select the **Center Point Circle** command by clicking once with the left-mouse-button on the icon in the *Draw* panel.

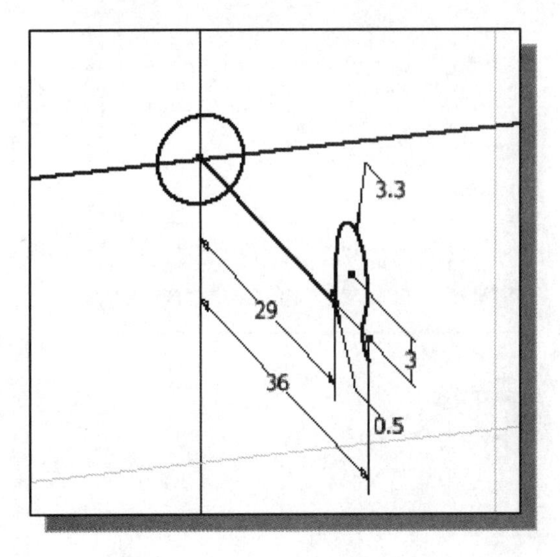

4. Create a circle that is aligned to the Center Point of the coordinate system as shown.

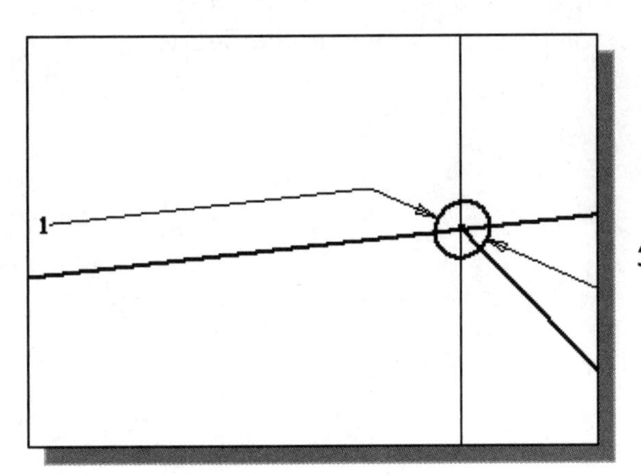

5. On your own, create the size dimension, **1mm**, to adjust the circle as shown.

6. In the *Ribbon* toolbar, click once with the left-mouse-button on **Finish Sketch** to end the **Sketch** option.

- The created 2D sketch will be used as the sweep section of the feature.

♦ Create the Swept Feature

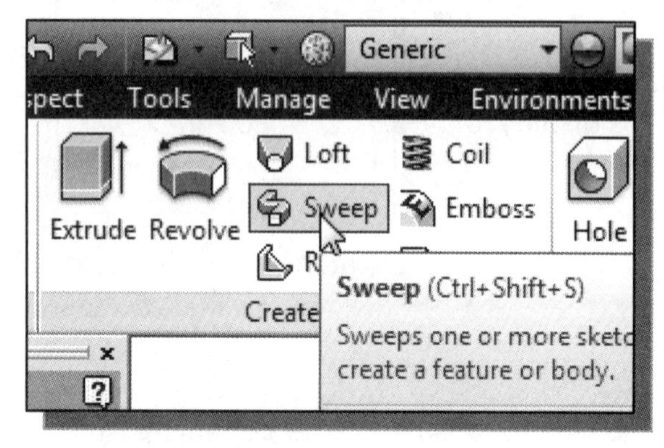

1. In the *Create* toolbar, select the **Sweep** command by left-clicking once on the icon.

2. Notice the circle is automatically selected to be used as the sweep section. (The *sweep section* must be a closed region.)

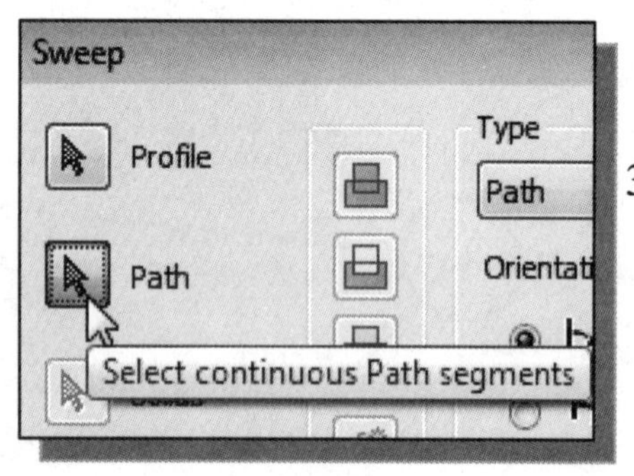

3. Notice the **Path** option is activated once the profile has been defined.

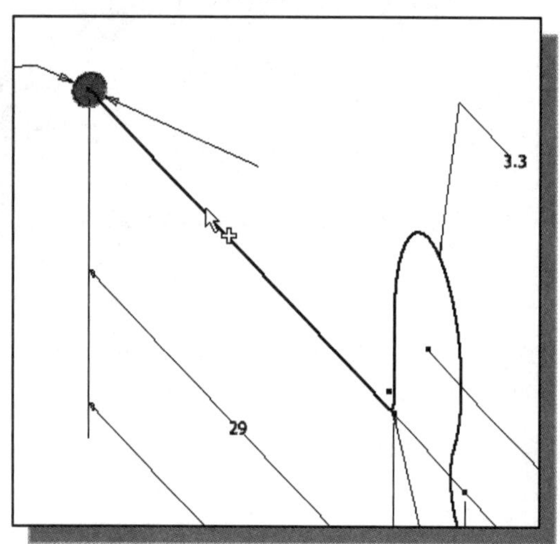

4. Click on the open curve we created as the sweep path as shown in the figure.

5. In the *Sweep* dialog box, confirm the settings are as shown.

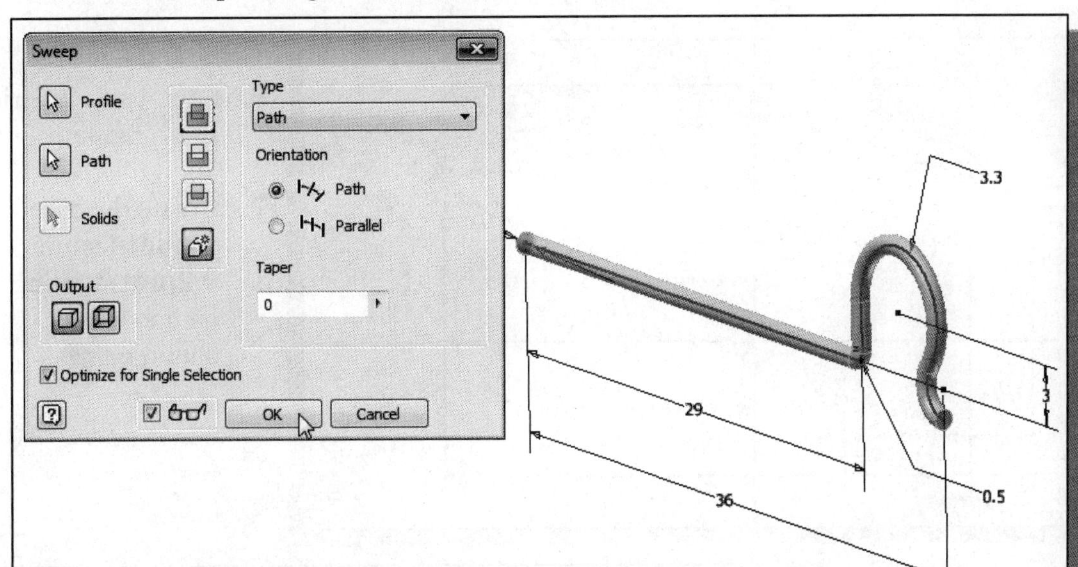

6. Click the **OK** button to accept the settings and create the *swept* feature.

Create a Mirrored Feature

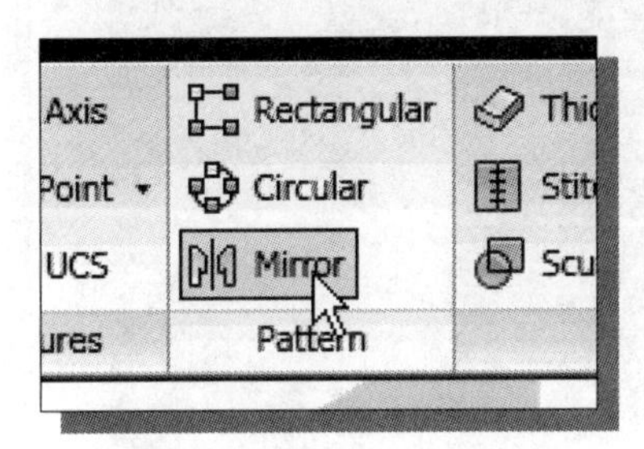

1. In the *Pattern* panel, select the **Mirror Feature** command by releasing the left-mouse-button on the icon.

2. On your own, complete the model as shown.

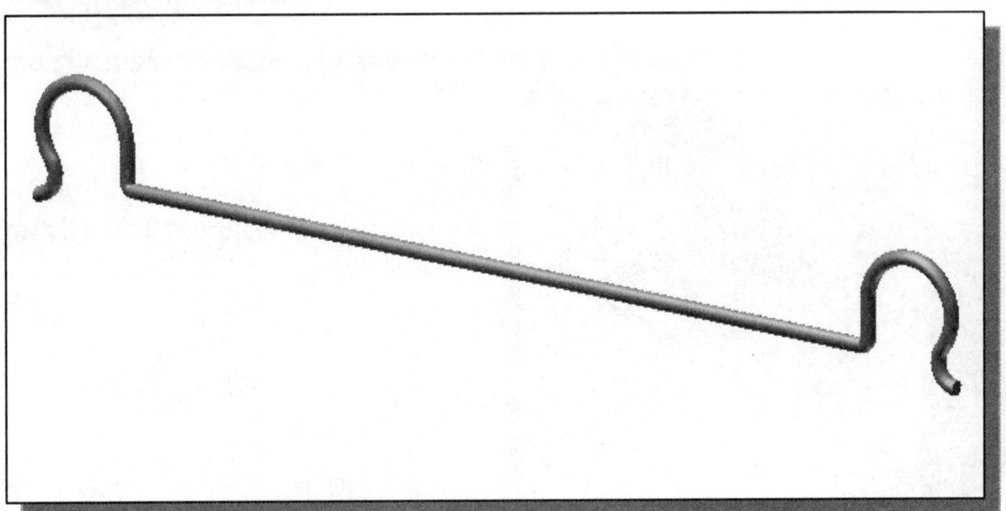

3. Save the model as **Linkage-Rod.ipt**.

The *Gear Box Right* Part

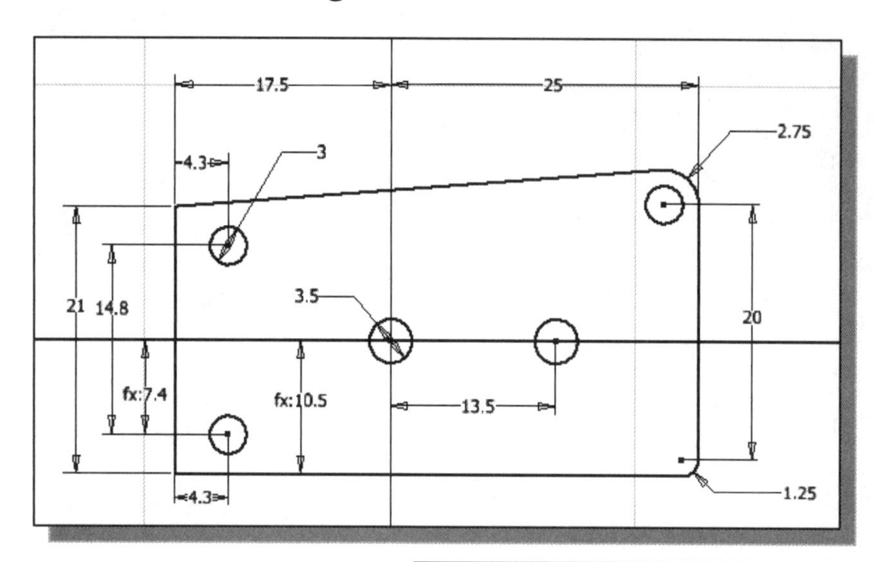

1. Start a new Metric (mm) model.

2. Create the base extrude feature, **9.2mm**, starting on one of the datum planes.

3. Use the **Shell** command, with shell thickness **1.25 mm**, and remove the back face and the shorter side face as shown.

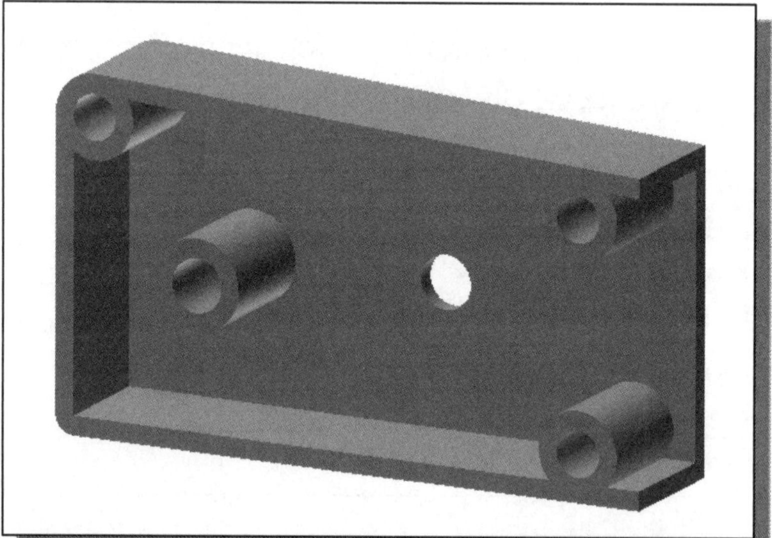

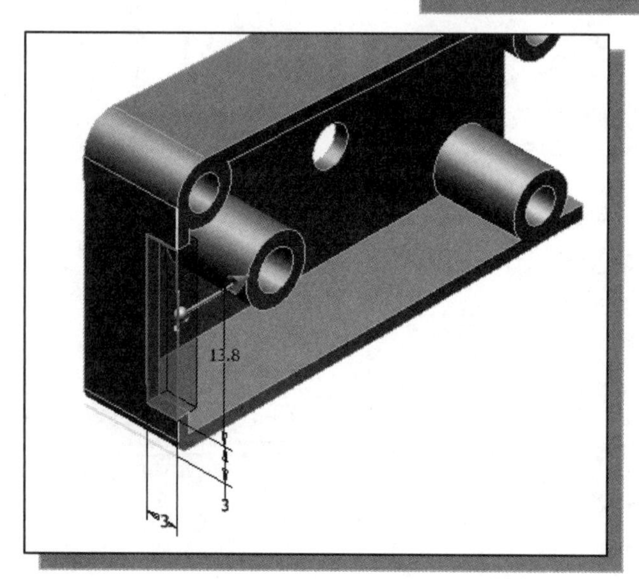

4. Create a cut feature, **13.8x3.0mm**, as shown.

5. Shorten the center support by **5 mm**.

6. Create an undercut feature, diameter **10 mm** and depth **0.5 mm**, as shown.

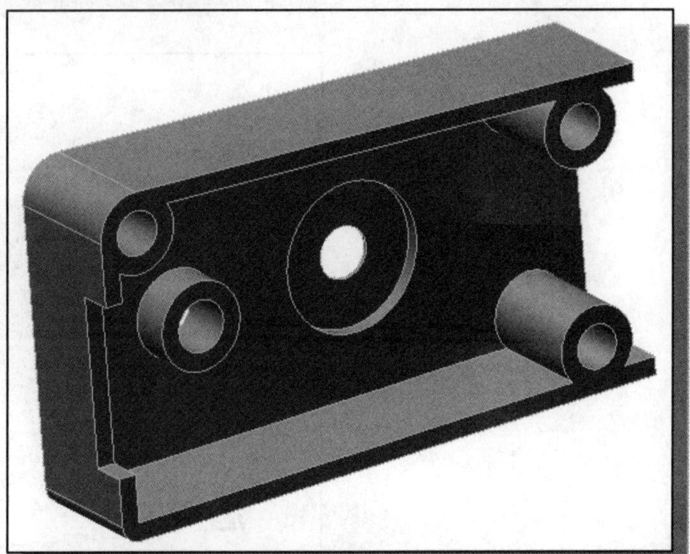

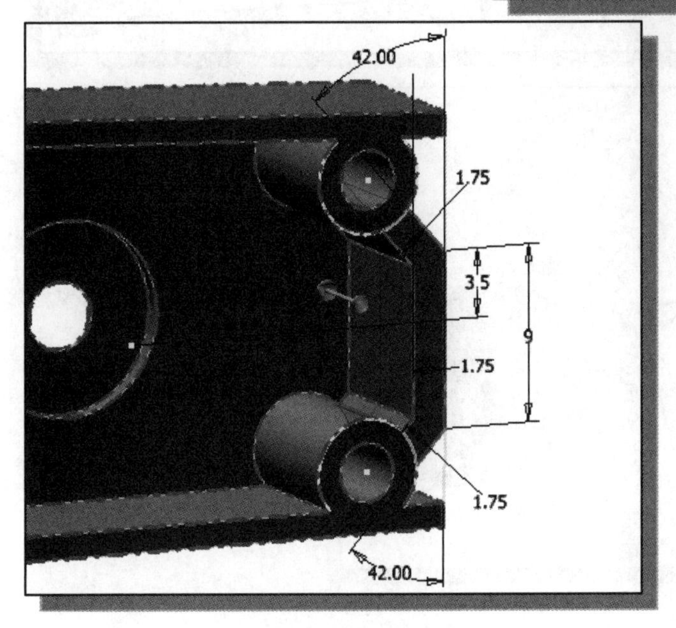

7. Create the additional support wall as shown.

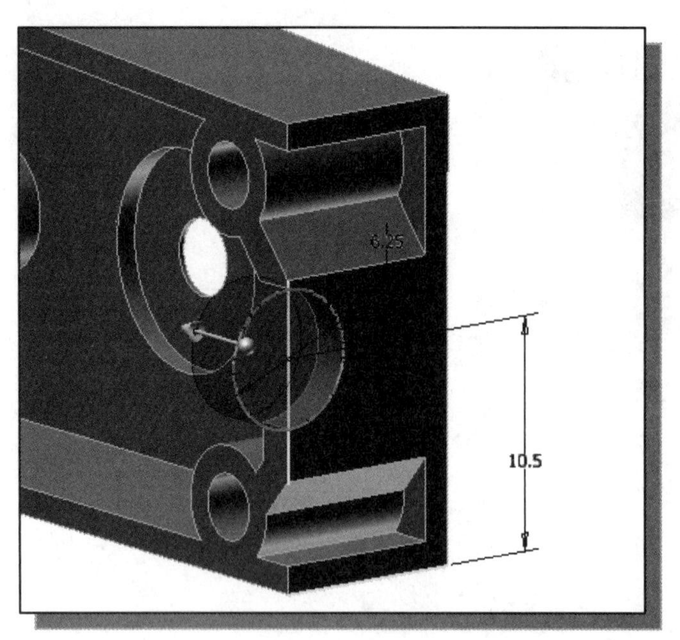

8. Create a cut feature, diameter **6.25 mm** and depth **3 mm**, as shown.

9. Create another cut feature **27 mm** long, as shown.

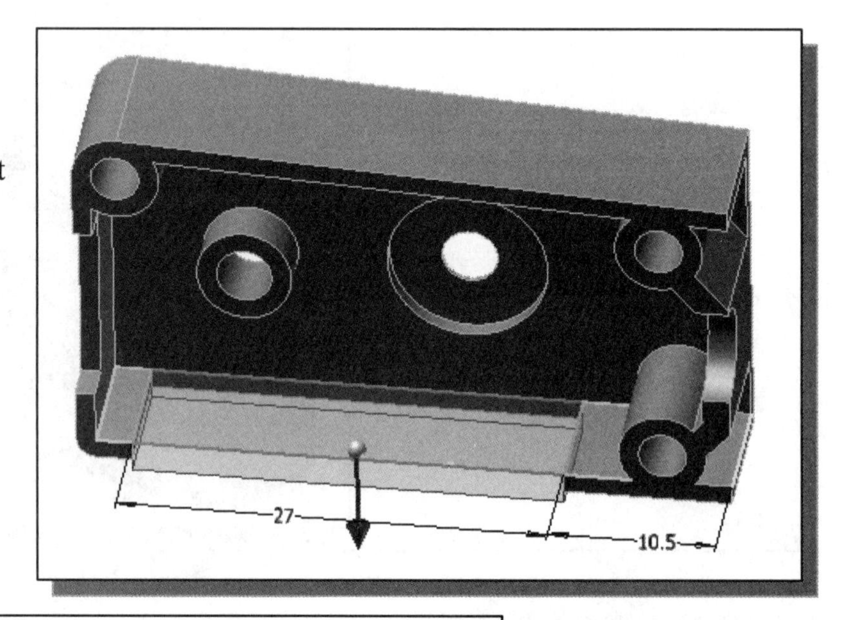

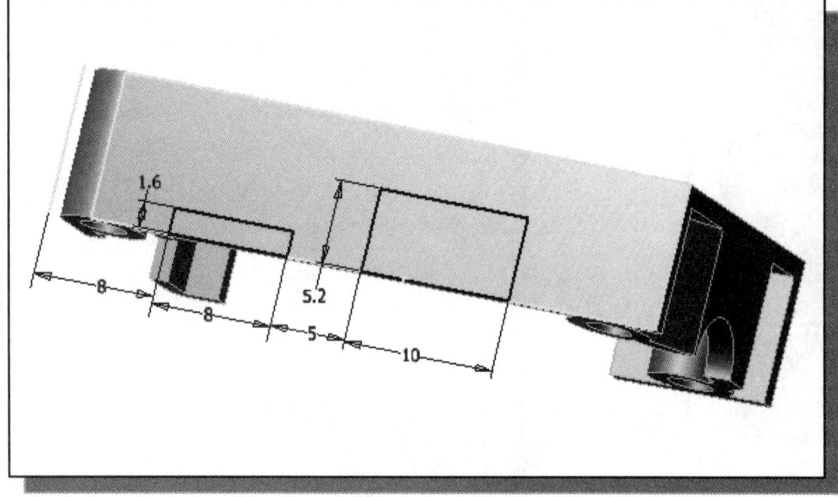

10. Create another cut feature on the opposite side of the model.

11. Create another circular extruded feature, diameter **7.3 mm** and depth **2 mm**, as shown.

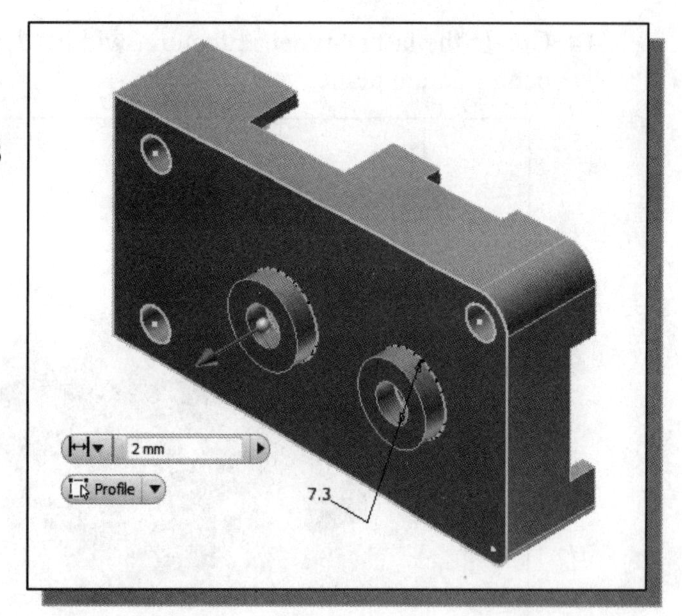

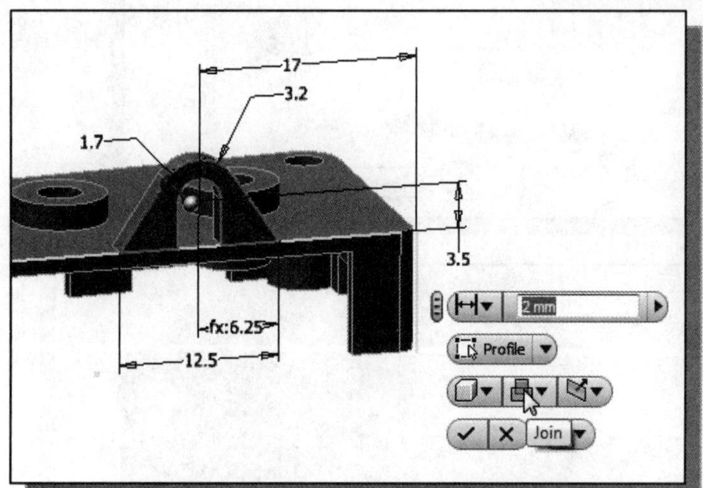

12. Add the side feature on the top surface as shown.

13. On the rounded corner, add another circular extruded feature, diameter **1.5 mm** and depth **11 mm**, as shown.

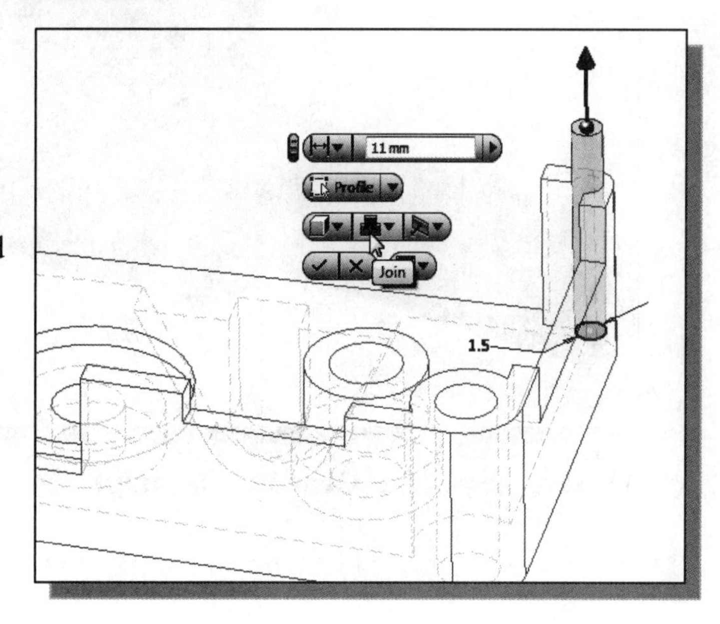

14. Create the last symmetric feature, width **11.5mm**, with the sketch aligned to the center datum plane.

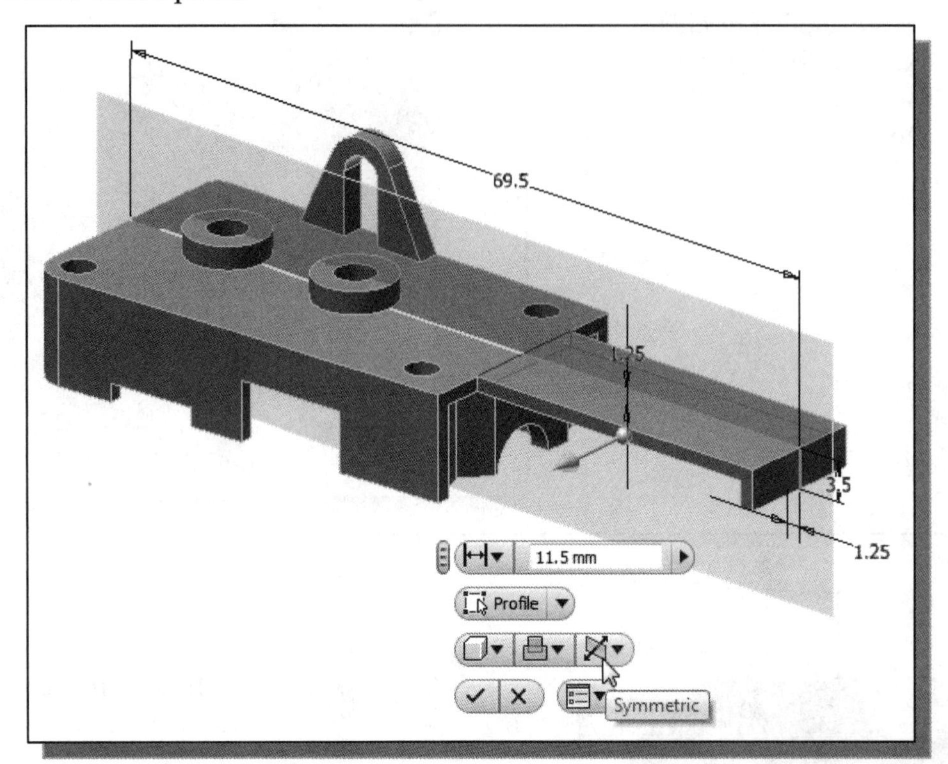

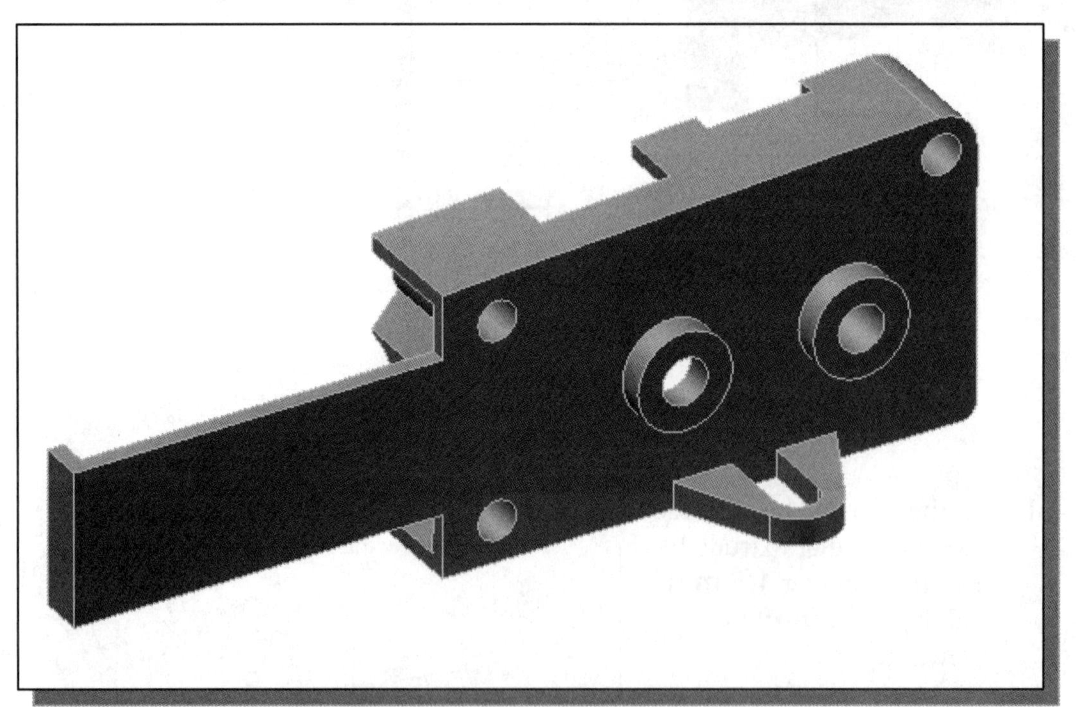

15. Save the model as **GearBox-Right.ipt**.

The *Gear Box Left* Part

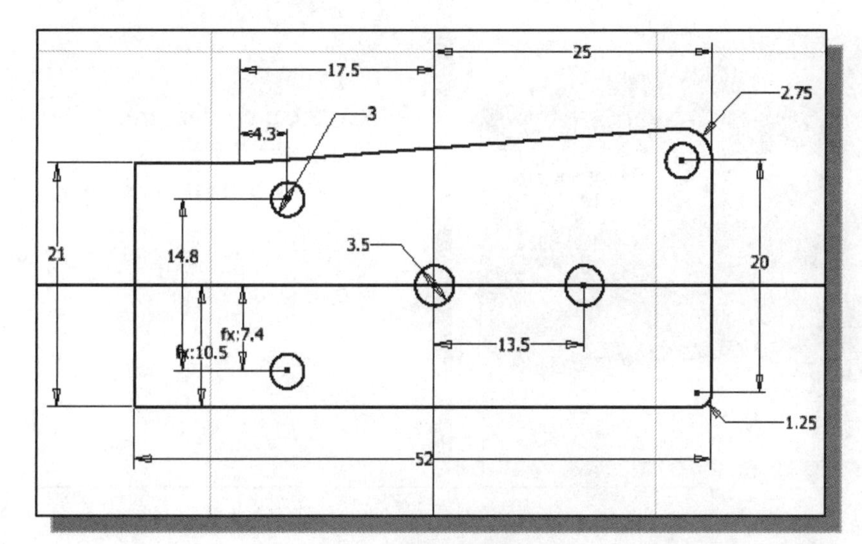

1. Start a new Metric (mm) model.

2. Create the base extruded feature, depth **14.0 mm**, starting on one of the datum plane.

3. Create a cut feature, **20.5 mm x 4.5 mm** and **75 degrees**.

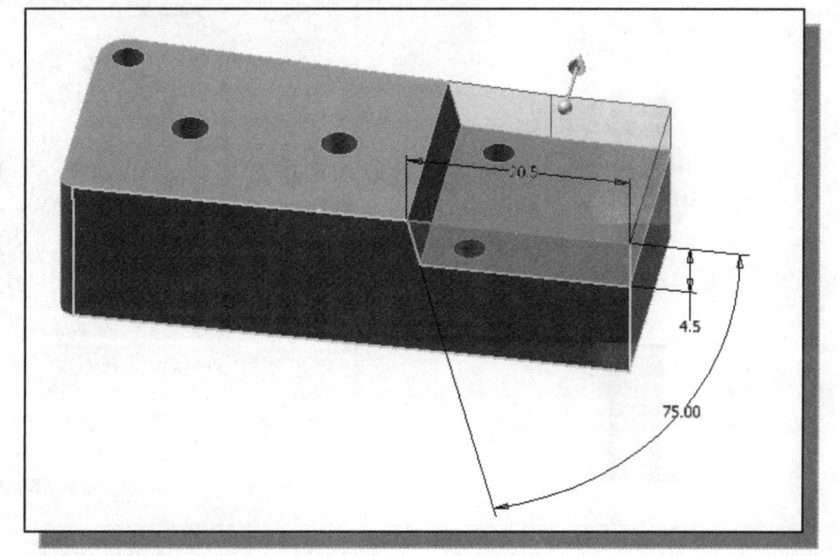

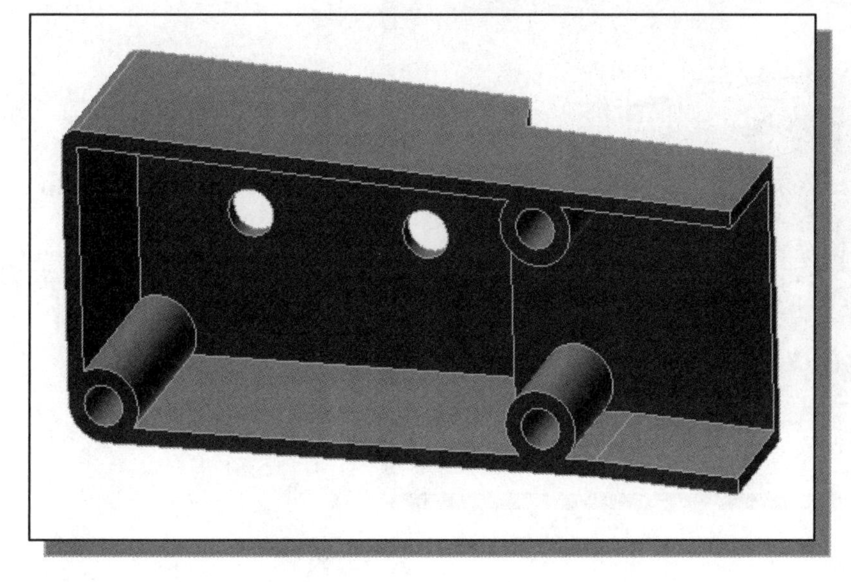

4. Use the **Shell** command, with shell thickness **1.25 mm**, and remove the back face and the shorter side face as shown.

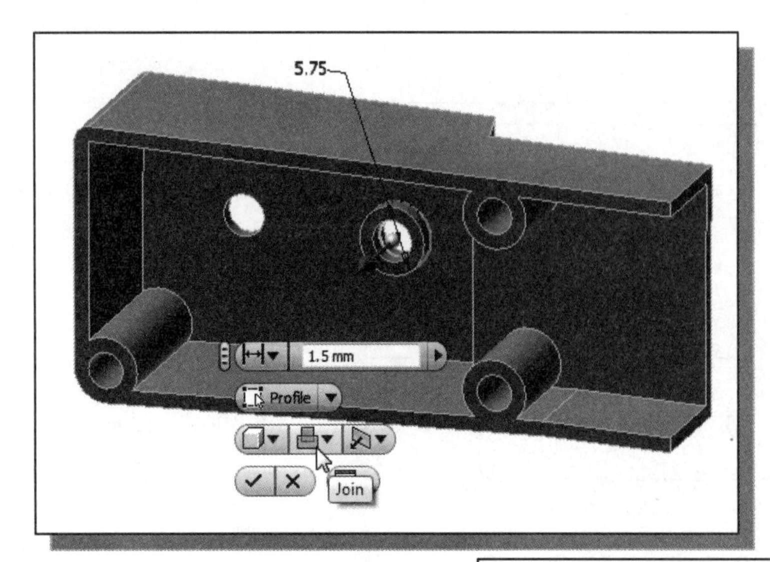

5. Create a cut feature, diameter **5.75 mm** and depth **1.5 mm**, as shown.

6. Create a cut feature on the side surface with the dimensions as shown.

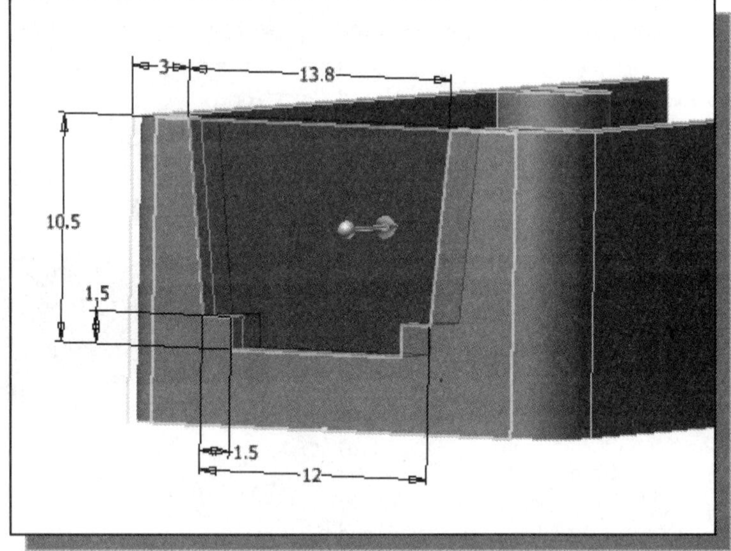

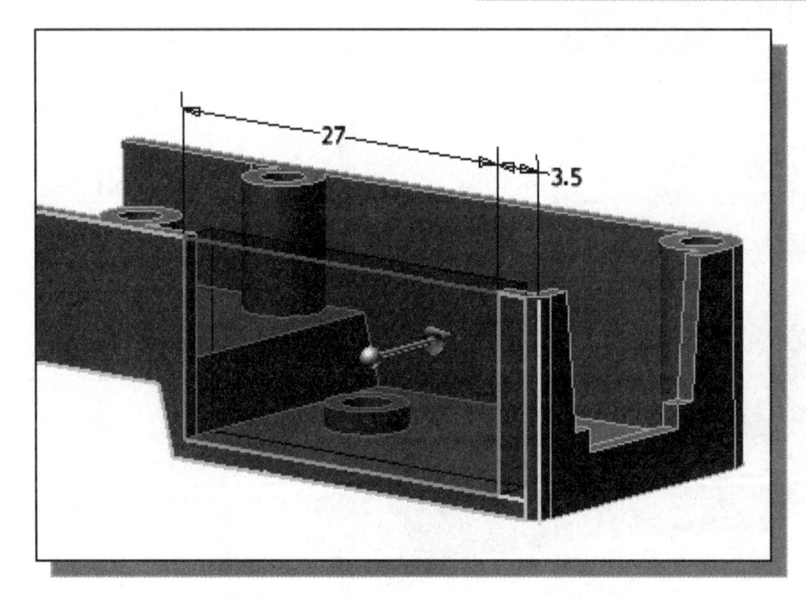

7. Create another cut feature on the adjacent surface, with the bottom edge aligned to the inside edge of the model as shown.

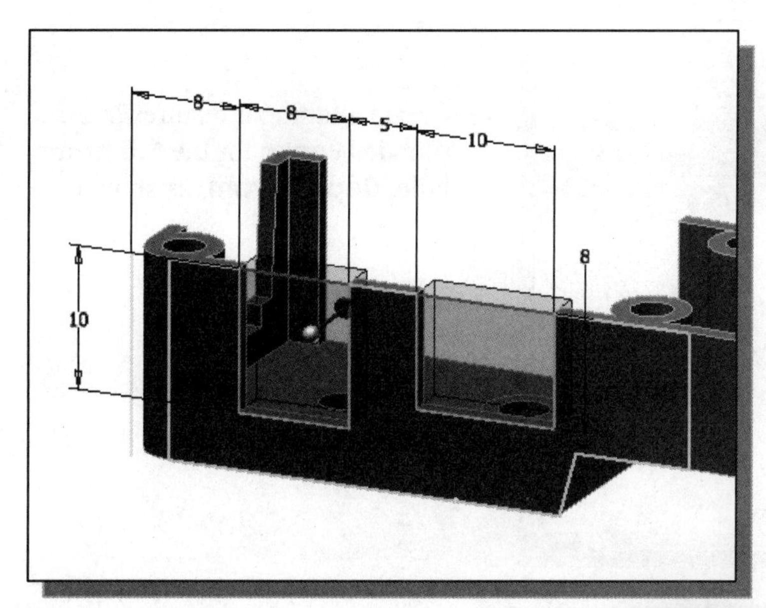

8. Create another cut feature on the opposite surface as shown.

9. Create another circular extruded feature, diameter **7.3 mm** and depth **2 mm**, as shown.

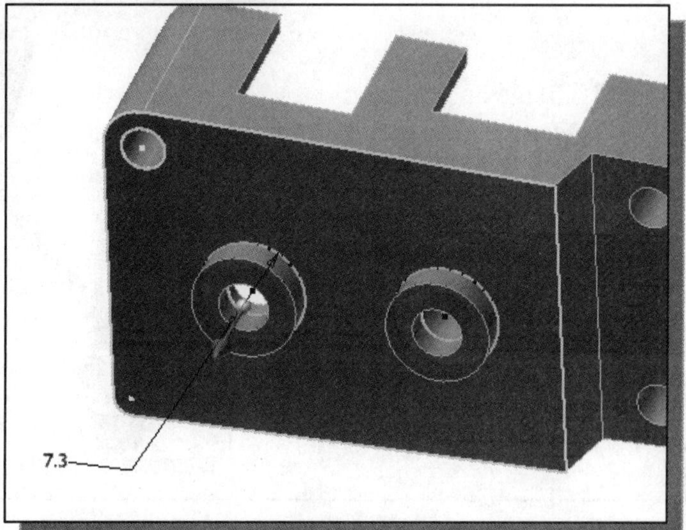

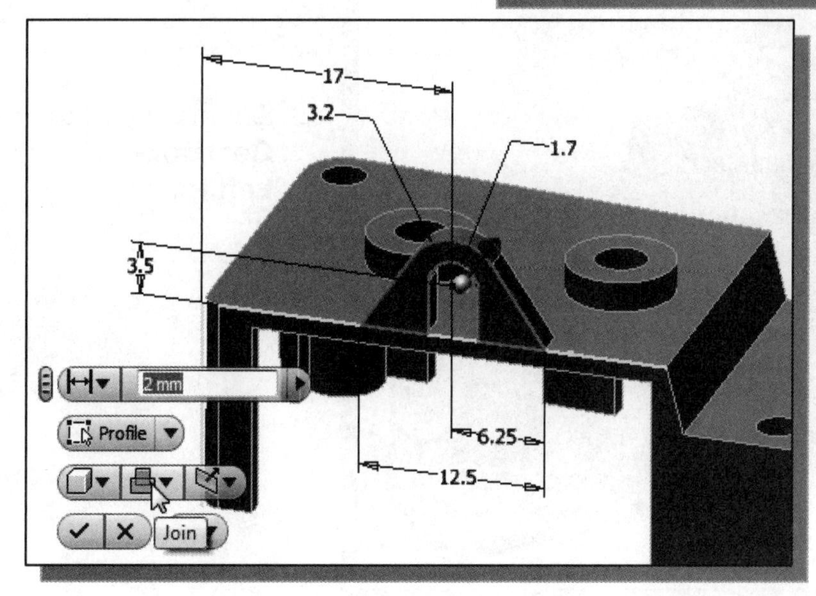

10. Add the side feature on the top surface as shown.

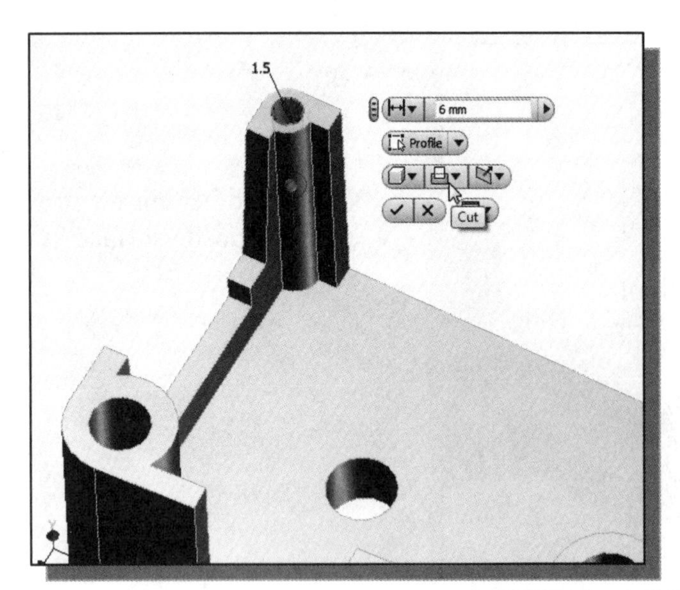

11. Create additional features at the rounded corner for the **1.5 mm** hole, **depth 6 mm**, as shown.

12. Create an undercut feature, diameter **10 mm** and depth **1 mm**, as shown.

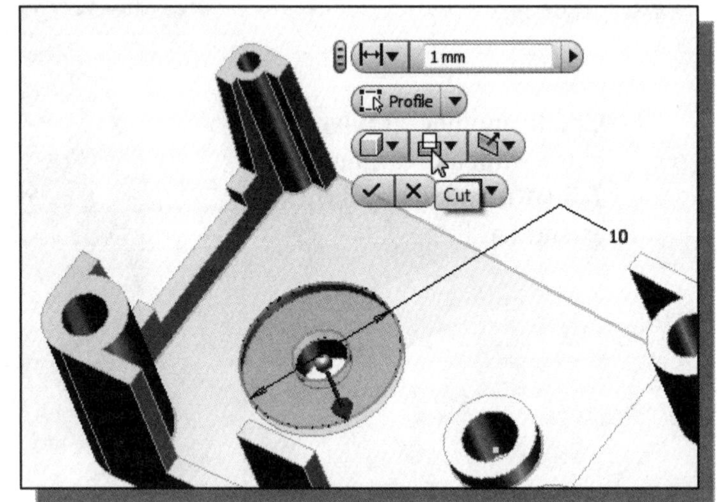

13. Save the model as **GearBox-Left.ipt**.

Questions:

1. Keeping the *History Tree* in mind, what is the difference between *cut with a pattern* and *cut each one individually*?

2. What is the difference between **Sweep** and **Extrude**?

3. What are the advantages and disadvantages of creating fillets using the **3D Fillets** command and creating fillets in the 2D profiles?

4. Describe the steps used to create the *Shell* feature in the lesson.

5. How do we modify the *Shell* parameters after the model is built?

6. Describe the elements required in creating a *Swept* feature.

7. Create sketches showing the steps you plan to use to create the model shown on the next page:

Exercises:

1. **Motor Housing** (Dimensions are in inches.)

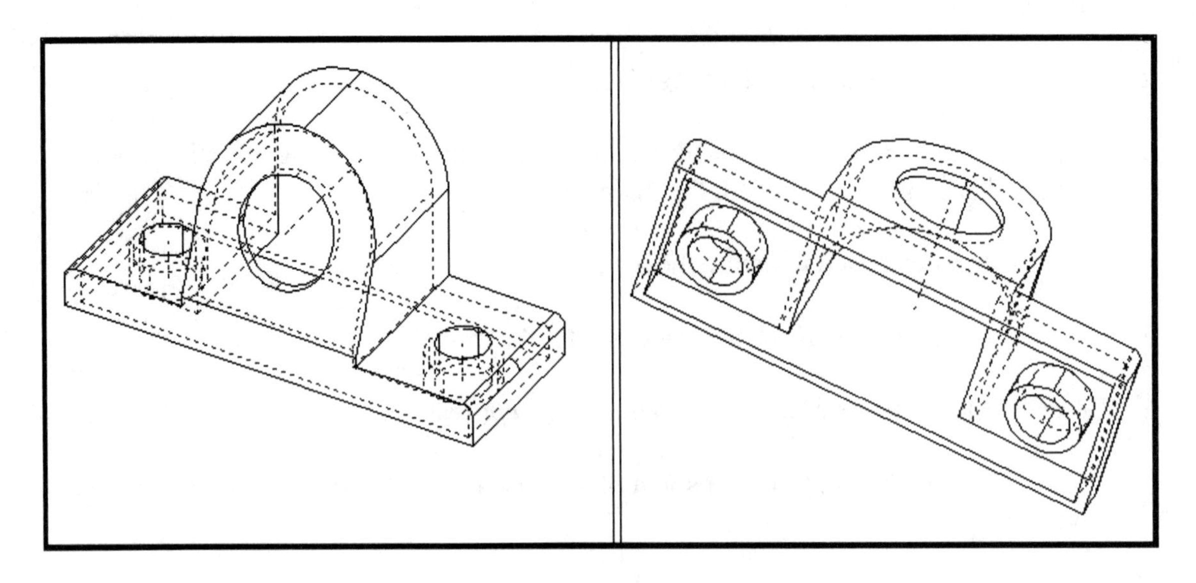

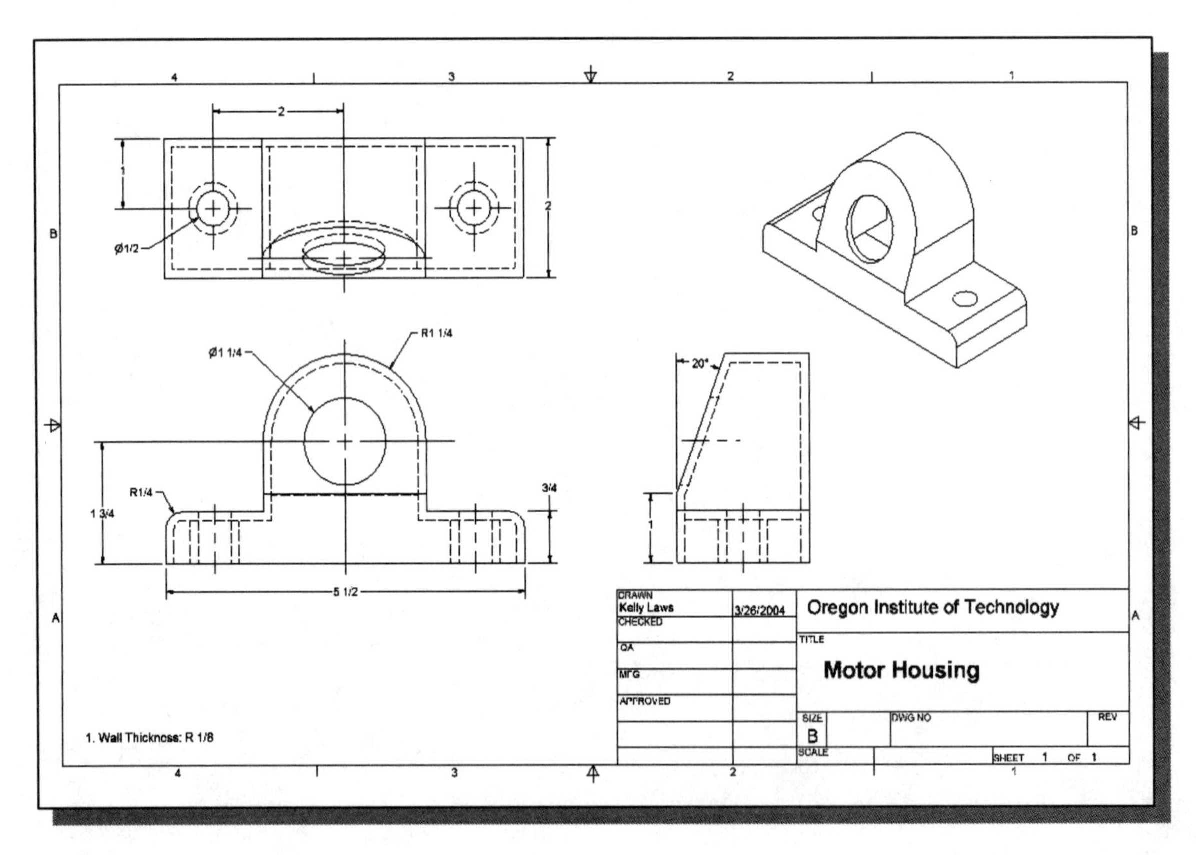

2. **Guide Base** (Dimensions are in mm.)

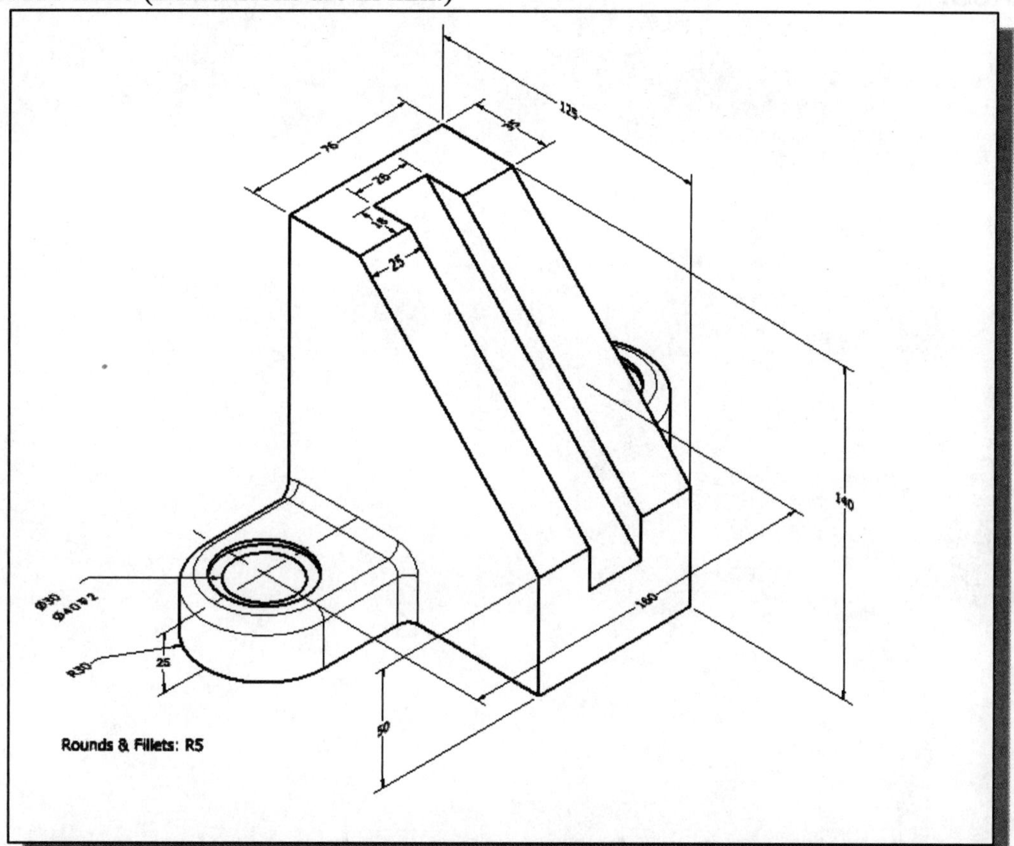

3. **Piston Cap** (Dimensions are in inches)

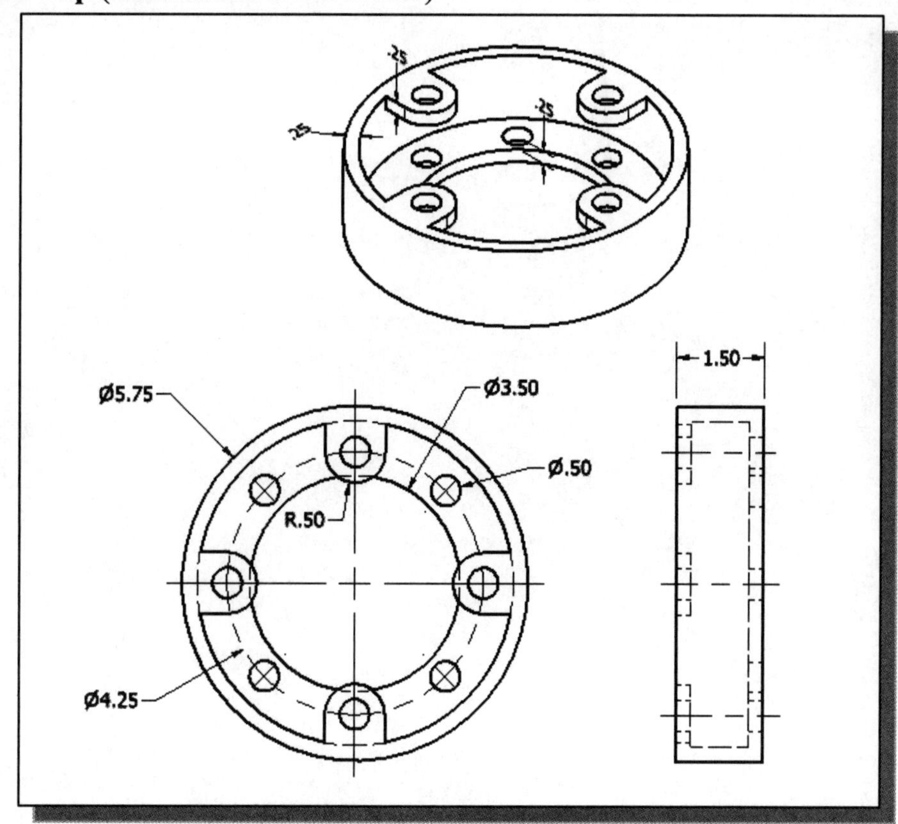

Notes:

Chapter 10
Planar Linkage Analysis Using GeoGebra

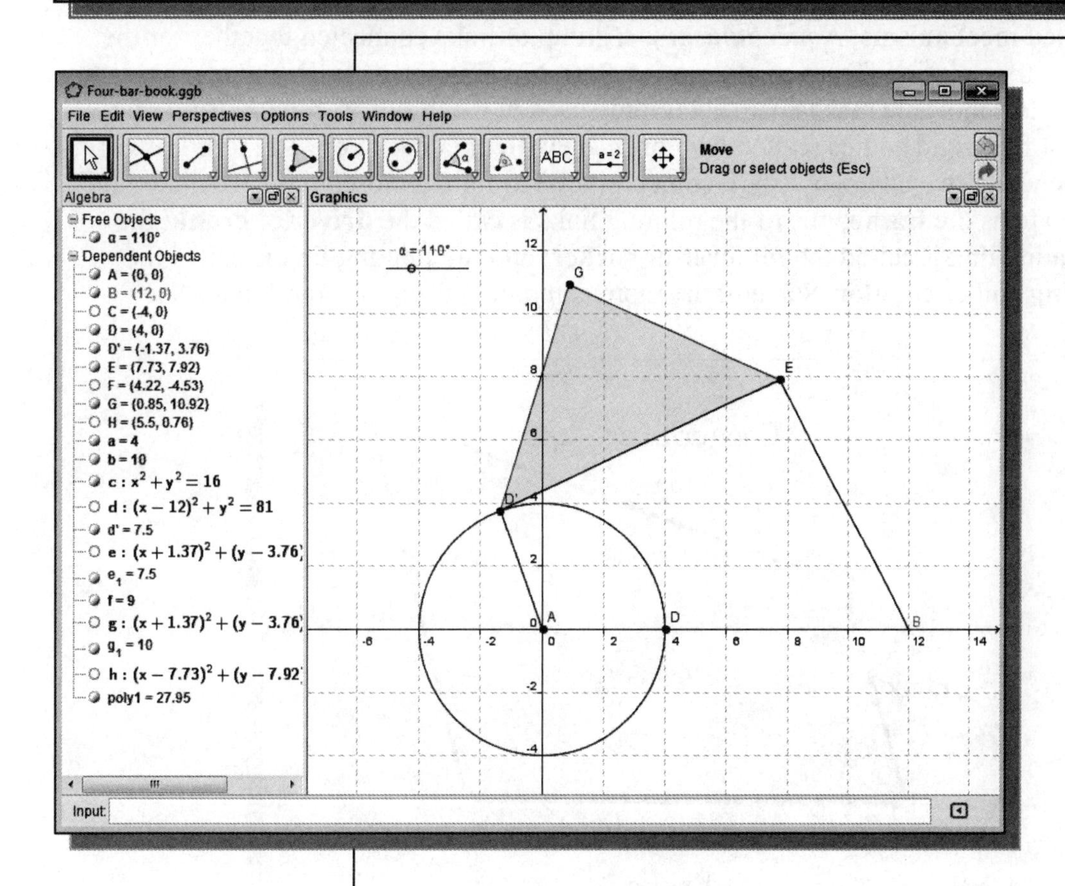

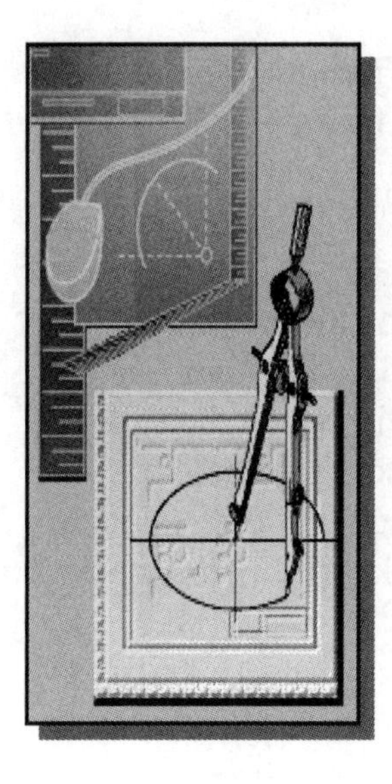

Learning Objectives

- Understand the Basic Type of Planar Four-Bar Linkage
- Learn the Basic Geometric Tools Available in GeoGebra
- Use GeoGebra to Construct Planar Four-Bar Linkage
- Use GeoGebra to Create Animations of Four-Bar Linkages

Introduction to Four-Bar Linkage

A machine is a system or device consisting of fixed and moving parts that modifies mechanical energy to do work. Assemblies within a machine that control movement are often called **mechanisms.** A ***mechanism*** is a group of links connected together for the purpose of transmitting forces or motions. A **four-bar linkage** or **four-bar mechanism** is the simplest movable mechanism commonly used in machines. A ***four-bar linkage*** consists of four rigid bodies (called *bars* or *links*) with one link fixed. The four links are each attached to two others by single joints (pivots) to form a closed loop. The fixed link is referred to as the **frame**; one of the rotating links is called the **driver** or **crank**; the other rotating link is called the **follower** or **rocker**; and the floating link is called the **connecting rod** or **coupler**. Some of the more commonly used four-bar linkages are listed below.

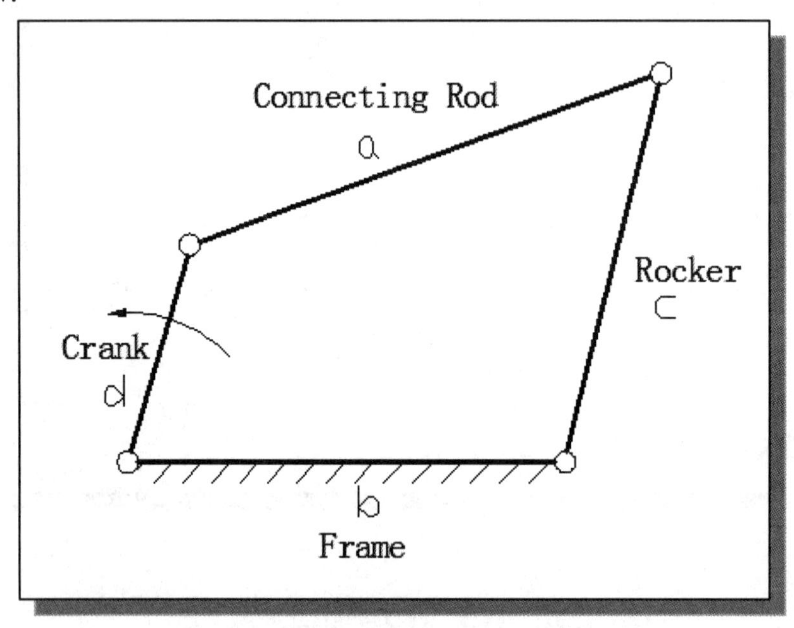

- Depending upon the arrangement and lengths of the links, different types of motions can be generated. Different mechanisms can also be formed by fixing different links of the same chain.

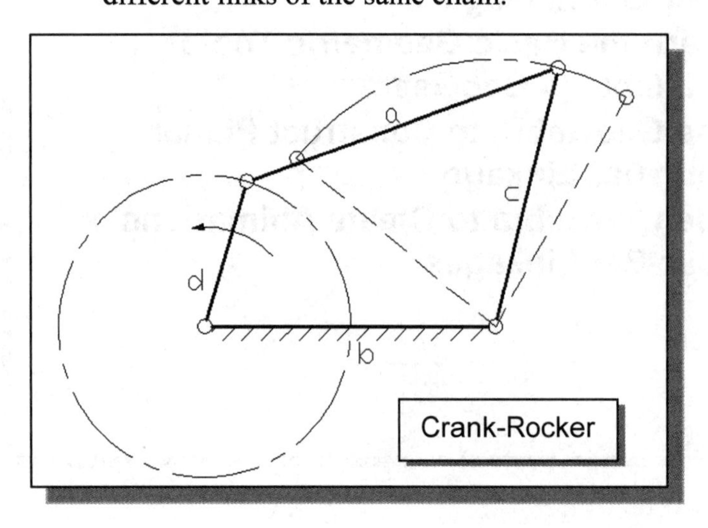

➢ The **Crank-Rocker** mechanism is a four-bar linkage in which the shorter link makes a complete revolution and the opposite link rocks (oscillates) back and forth.

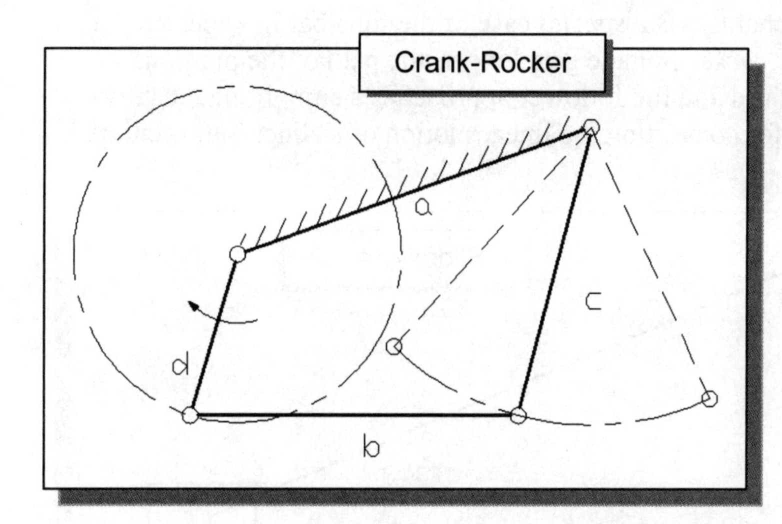

- Note that by fixing the opposite link of the same mechanism, another **Crank-Rocker** mechanism is formed.

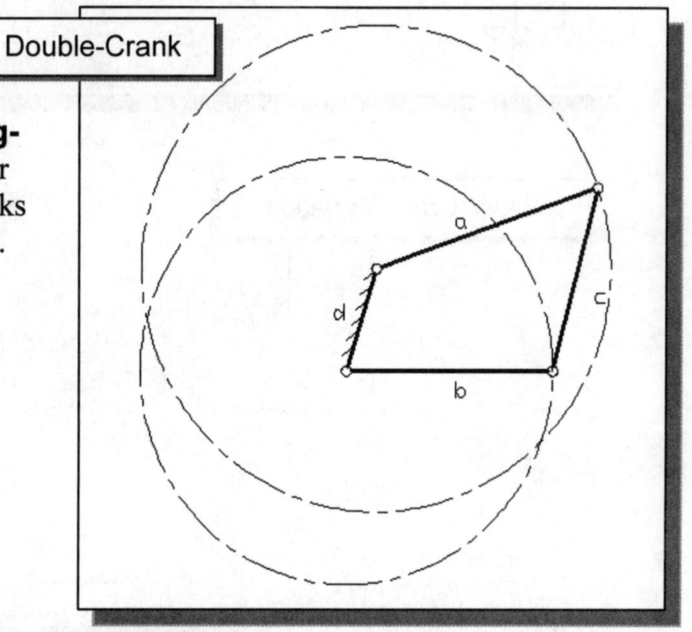

➤ The **Double-Crank** or **Drag-Link** mechanism is a four-bar linkage with two opposite links making complete revolutions.

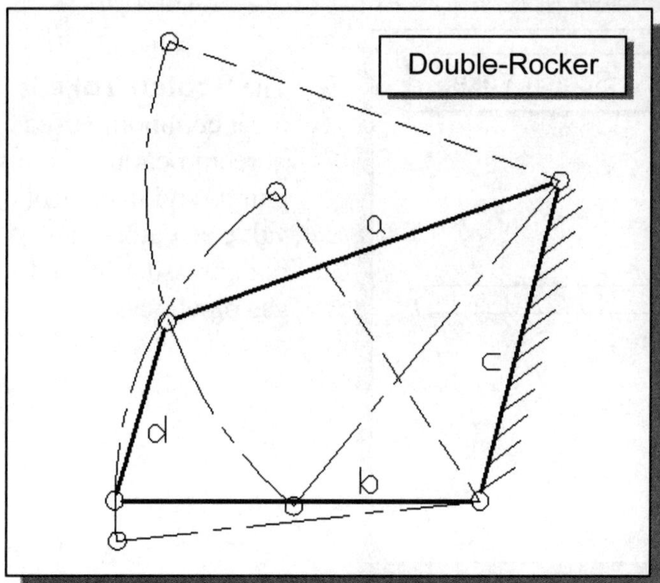

➤ The **Double-Rocker** mechanism is a four-bar linkage in which the crank and follower rock (oscillate) back and forth; none of the links can make full revolution.

- Note the four different mechanisms above are formed by fixing different links of the same links.

➢ The **Slider-Crank** mechanism is a special case of the four-bar linkage. As the follower link of a *crank-rocker* linkage gets longer, the path of the pin joint between the connecting rod and the follower approaches a straight line. **Slider-Crank** is a mechanism for converting the linear motion of a slider into rotational motion or vice-versa.

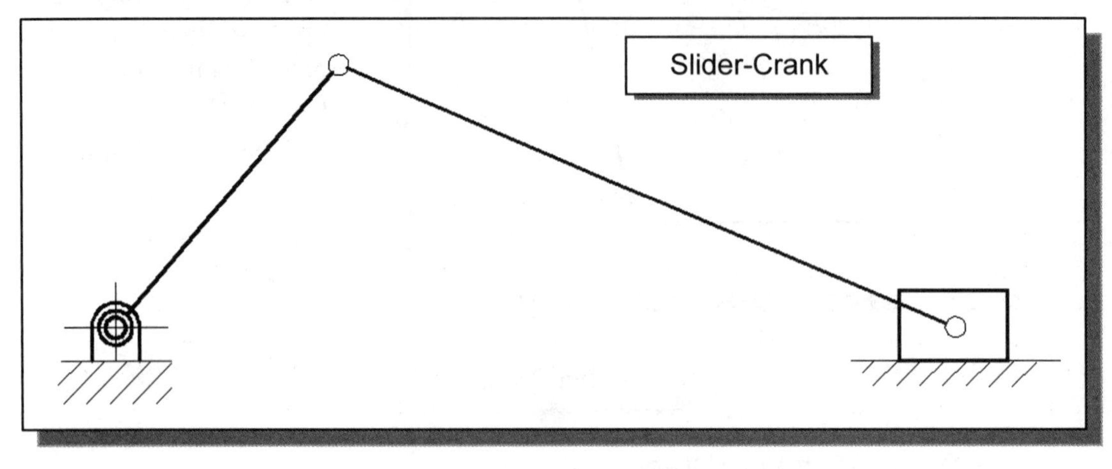

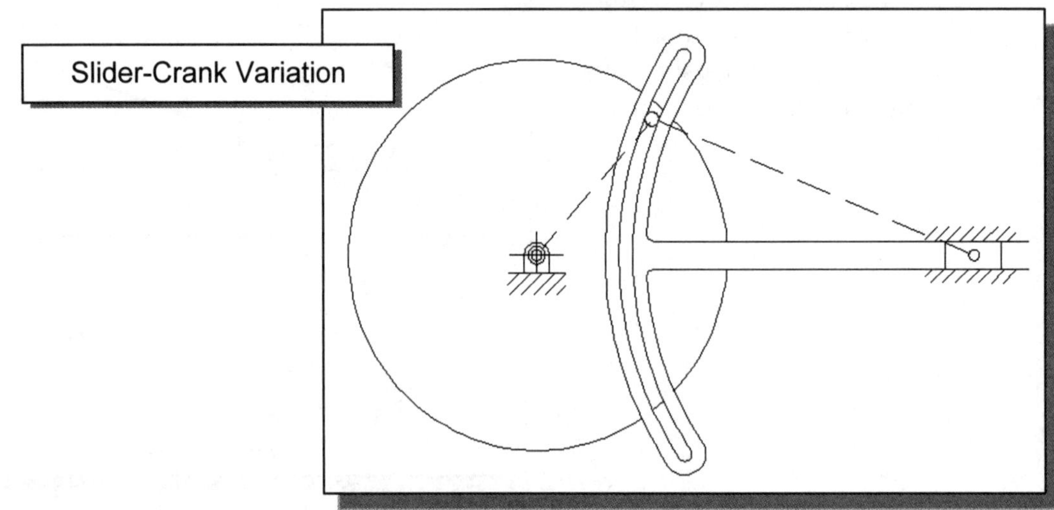

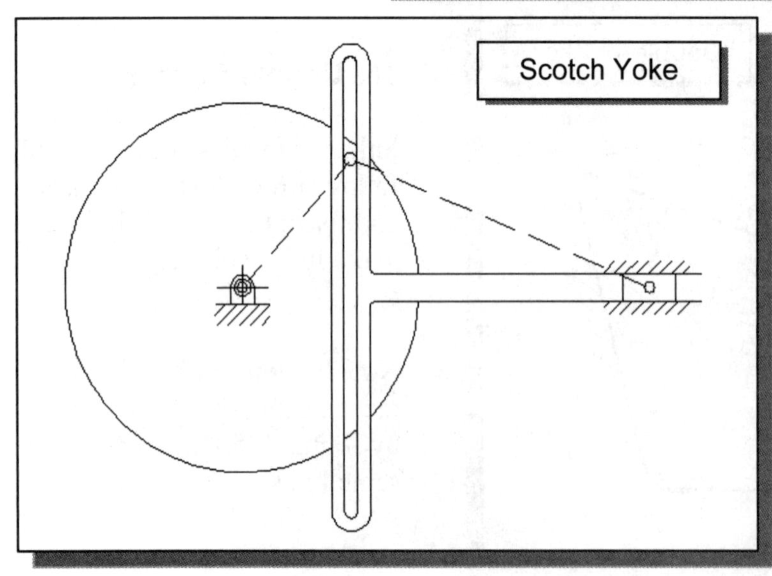

➢ The **Scotch Yoke** is most commonly used in reciprocating piston pumps and in control valve actuators in high pressure oil and gas pipelines.

Introduction to GeoGebra

GeoGebra is an award winning **interactive dynamic geometry software** that joins geometry, algebra and calculus. ***Interactive dynamic geometry software*** is a type of computer program that allows the creation and then manipulation of geometric constructions. In most *interactive geometry software*, constructions can be made with points, vectors, segments, lines, polygons, conic sections, inequalities, implicit polynomials and functions.

The development of *GeoGebra* was started by Prof. Markus Hohenwarter for mathematic education. Prof. Markus Hohenwarter started the project in 2001 at the *University of Salzburg*, continuing it at *Florida Atlantic University* (2006–2008), *Florida State University* (2008–2009), and now at the *University of Linz* together with the help of open-source developers and translators all over the world.

In *GeoGebra*, geometric entities can be entered and modified directly on screen, or through the *Input Bar*. *GeoGebra* has the ability to use variables for numbers, vectors and points; derivatives and integrals of functions can also be evaluated.

The main webpage of *GeoGebra* is at http://www.geogebra.org/cms/en:

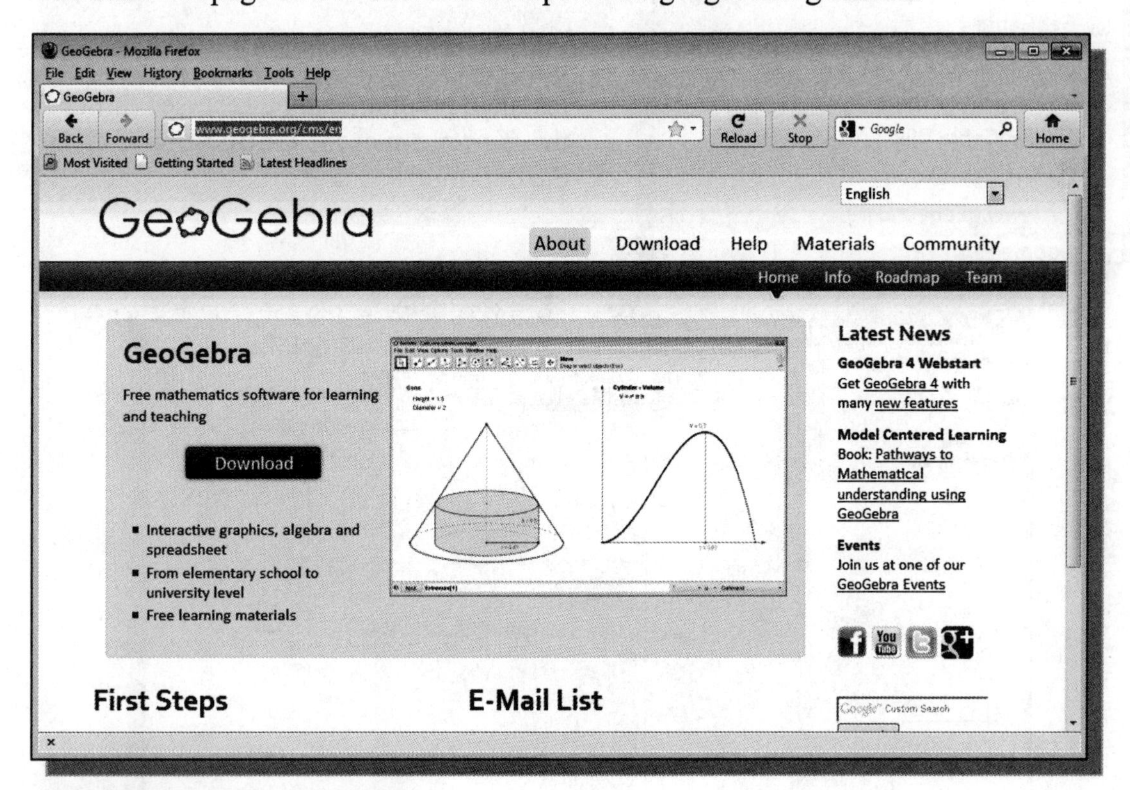

- *GeoGebra* is an open source free software, written in Java and thus available for multiple platforms, including *Windows*, *Mac* and *Linux* systems.

- Two versions of *GeoGebra* are available for download. *GeoGebra* is the full version and *GeoGebraPrim* has the same capabilities as the full version but uses a more simplified user interface. Files created in *GeoGebra* can be loaded in *GeoGebraPrim* and vice versa.

- Note the installation of the ***Java Platform*** is required as *GeoGebra* is written in Java.

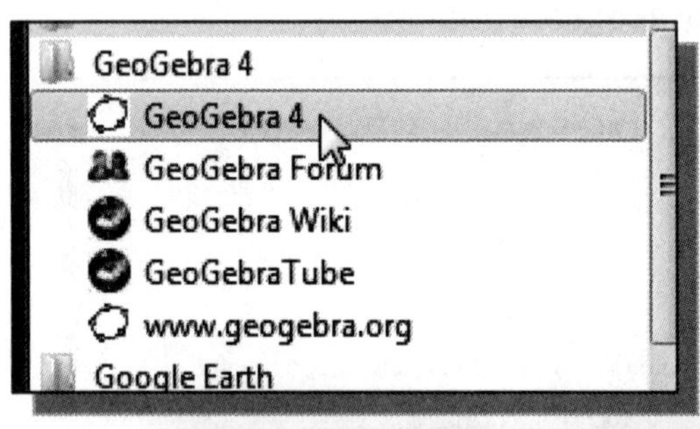

- For most versions of *GeoGebra*, the installed program can be accessed through the *Start* menu or the **GeoGebra** icon on the desktop.

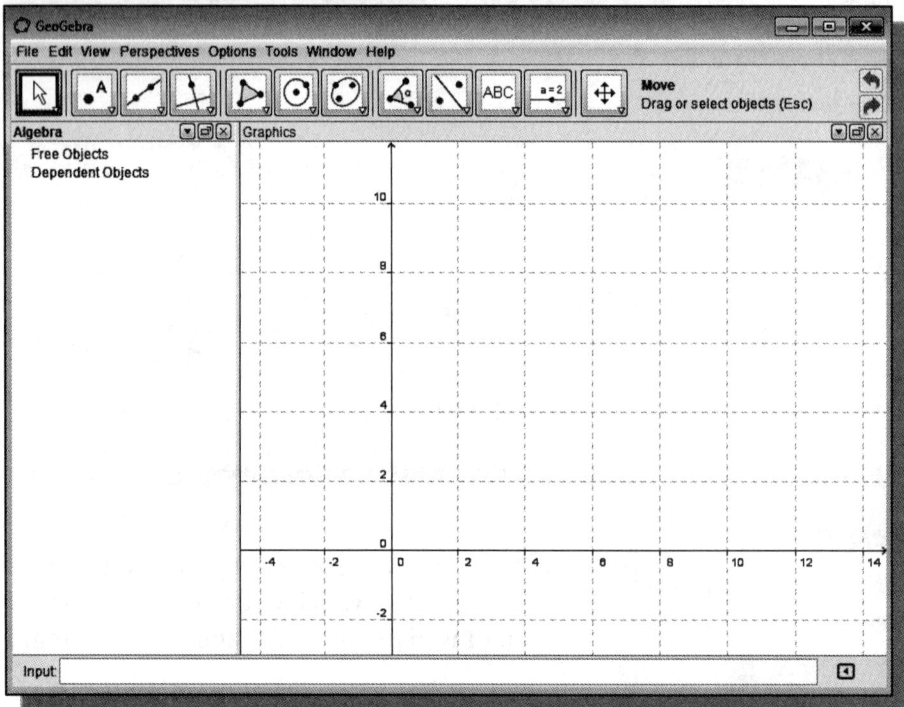

1. Select the **GeoGebra** option on the *Start* menu or select the **GeoGebra** icon on the desktop to start *GeoGebra*. The *GeoGebra* main window will appear on the screen.

2. In the **View** pull-down menu, switch on the **Grid** display by clicking on the icon with the left-mouse-button as shown.

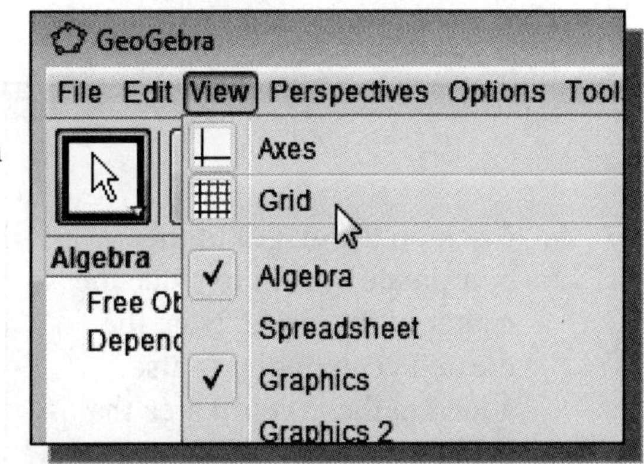

- The *GeoGebra* screen layout contains the ***pull-down*** menus, the ***Standard*** toolbar, the ***Algebra*** area, the ***graphics*** area, and the ***Input Bar*** at the bottom of the screen.

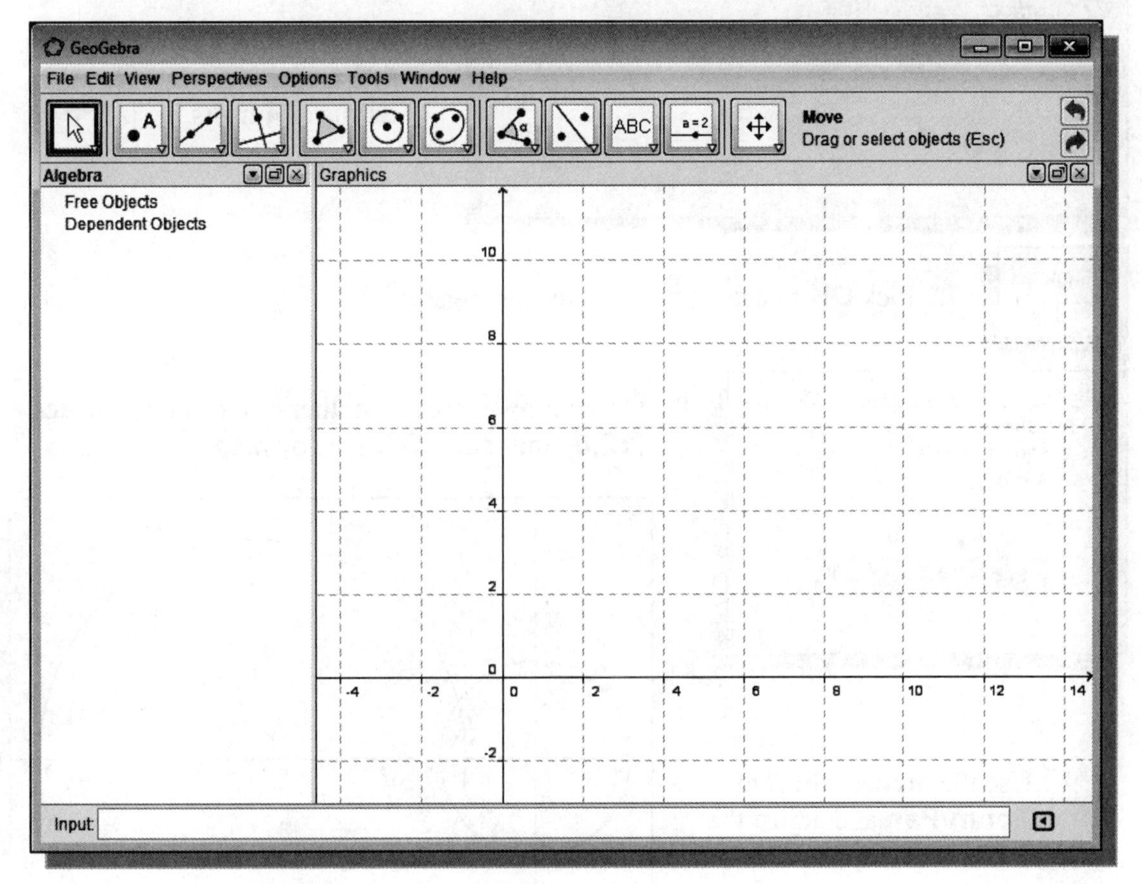

3. On your own, adjust the display of the *GeoGebra* main screen so that the *Algebra* area and the *Input Bar* are available as shown.

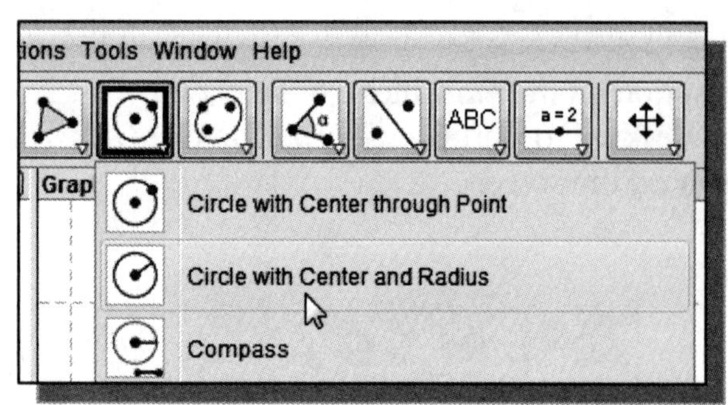

4. Select the **Circle with Center and Radius** icon in the *Standard* toolbar area.

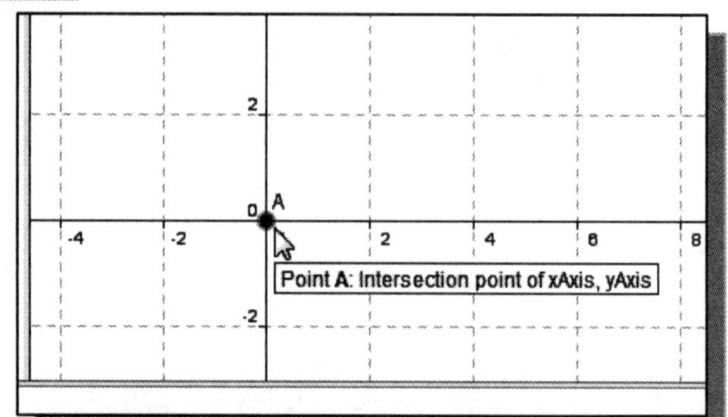

5. Click on the **origin** of the coordinate system to place the center of the circle. Note the created center point is also added in the *Algebra* area that is toward the left.

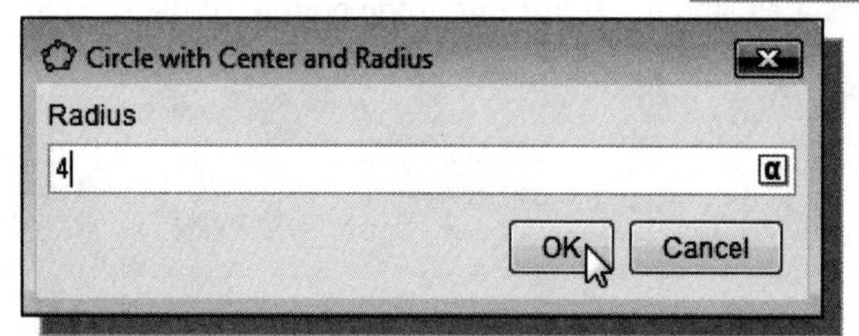

6. In the *Radius* input box, enter **4** to create a circle with a radius of **4 units** in size.

7. Pick **OK** to accept the selected settings.

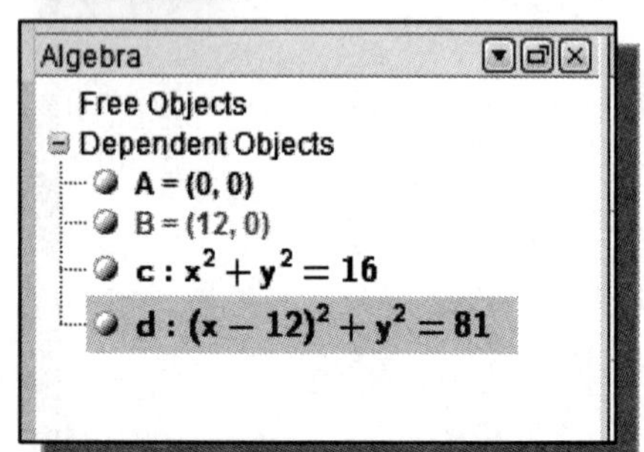

8. On your own, create another circle at coordinates **(12,0)** and radius **9.0** as shown.

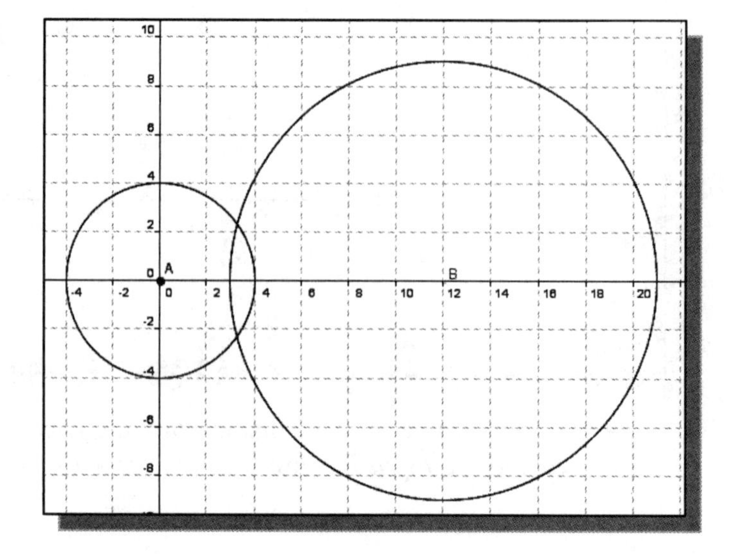

9. Use the **mouse wheel** to **Zoom/Pan** and adjust the current display.

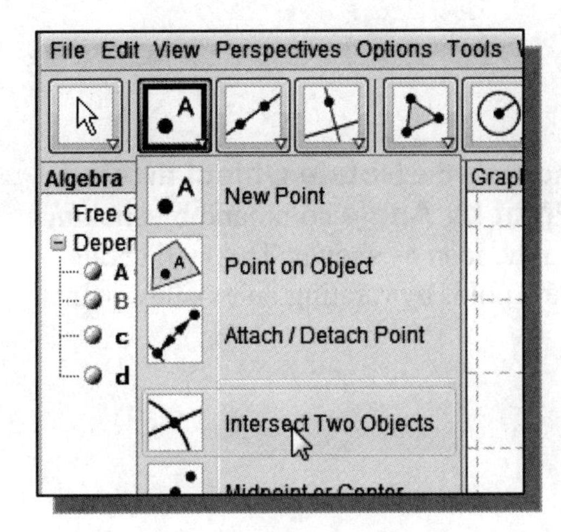

10. Activate the **Intersect Two Objects** command by clicking on the icon as shown. This will create points at the intersections of two selected objects.

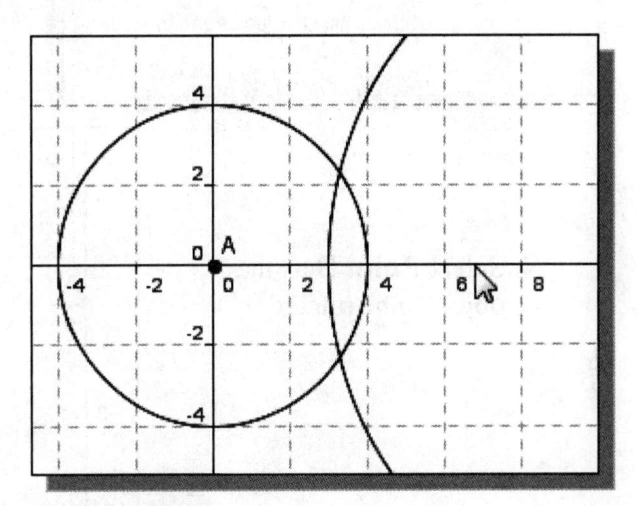

11. Select the **X Axis** as the first entity to define the intersection points.

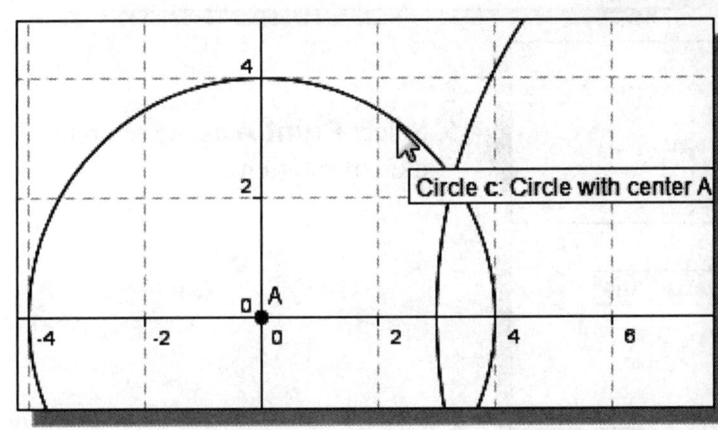

12. Select the **smaller circle** as the second entity to define the intersection points.

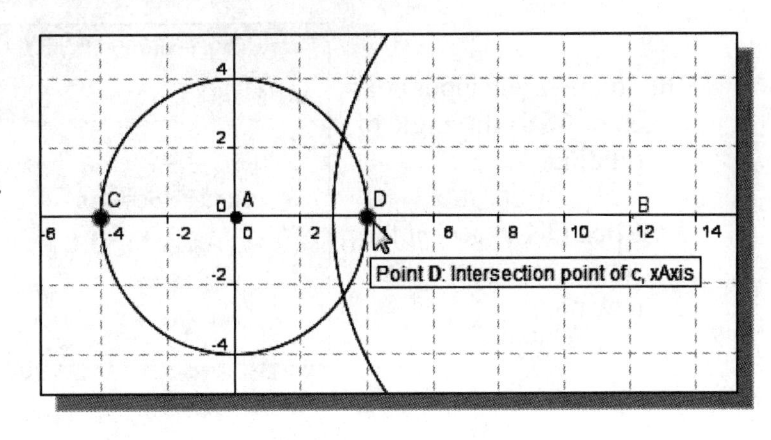

- Note two intersection points are created, **Point C** and **Point D**.

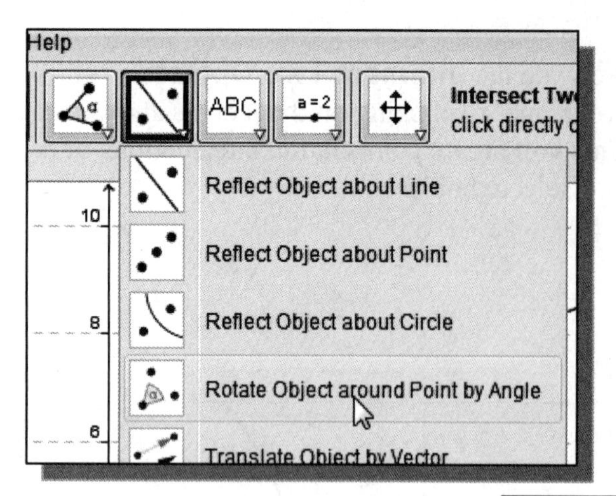

13. Activate the **Rotate Object around Point by Angle** command by clicking on the icon as shown. This will create a new point by rotating an existing point.

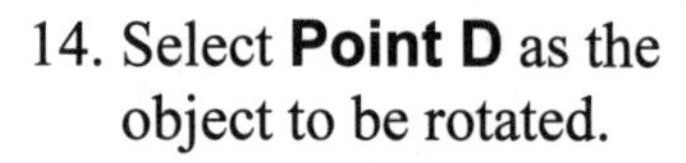

14. Select **Point D** as the object to be rotated.

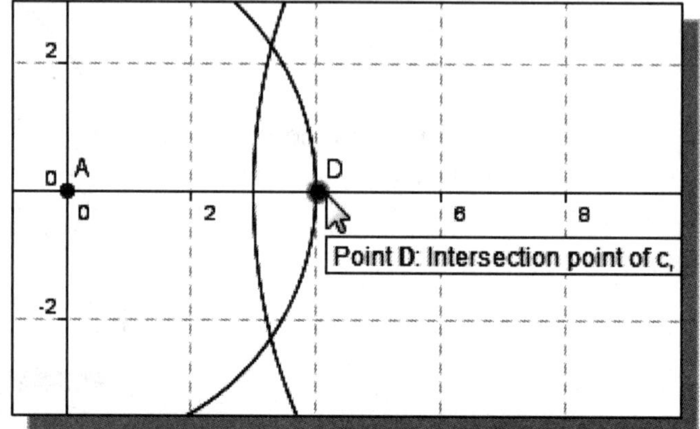

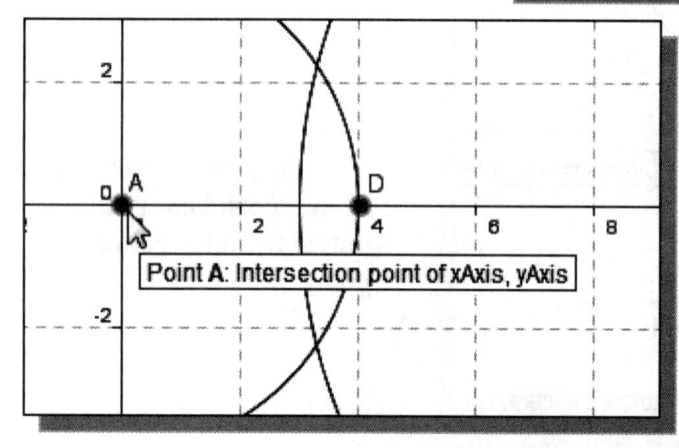

15. Select **Point A** as the reference axis of rotation.

16. In the *Angle* input box, enter **45** as the angle of rotation.

17. Click **OK** to accept the setting and create the new point.

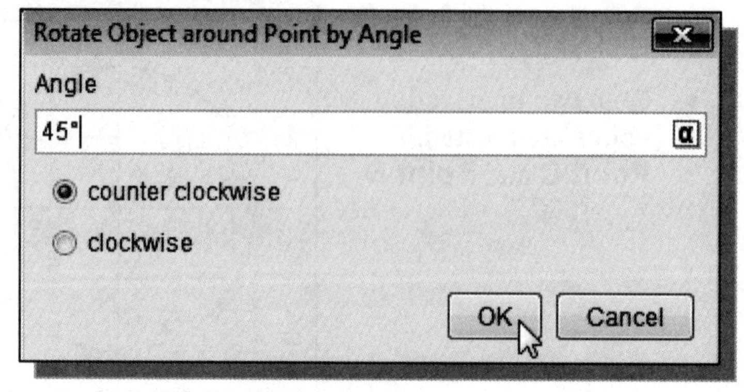

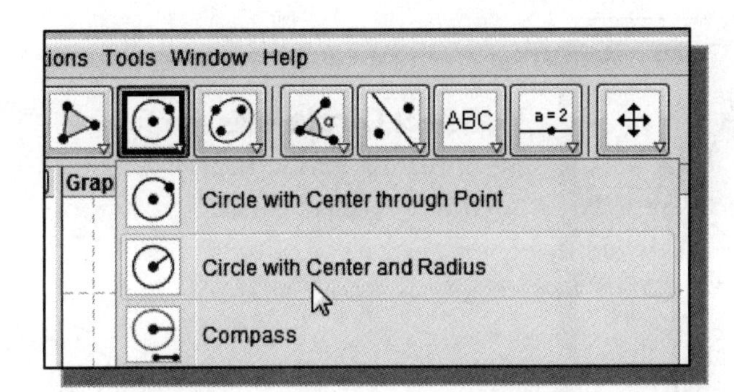

18. Select the **Circle with Center and Radius** icon in the *Standard* toolbar area.

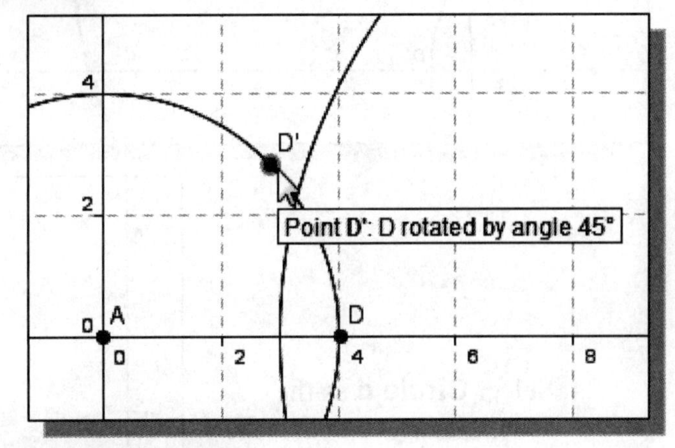

19. Select **Point D'** as the center of the new circle. Note the created center point is also added in the *Algebra* area that is toward the left.

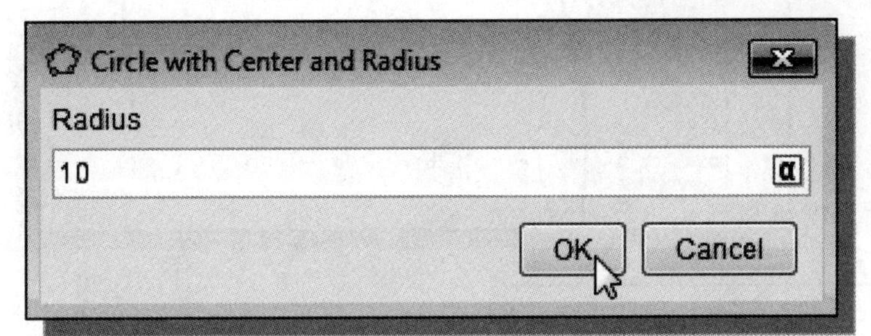

20. In the *Radius* input box, enter **10** to create a circle with a radius of **10 units** in size.

21. Click **OK** to accept the selected settings.

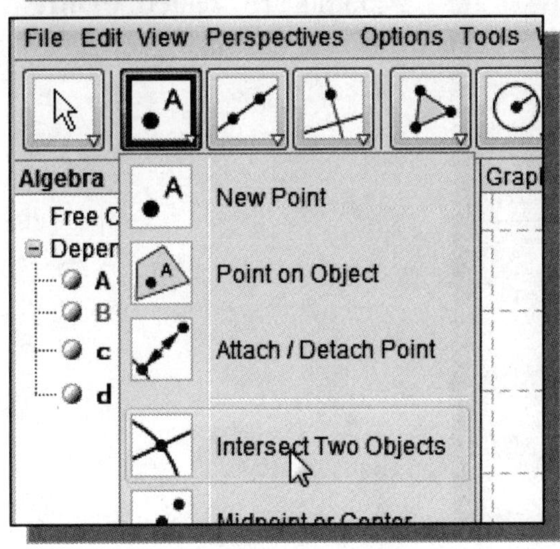

22. Activate the **Intersect Two Objects** command by clicking on the icon as shown. This will create points at the intersections of two selected objects.

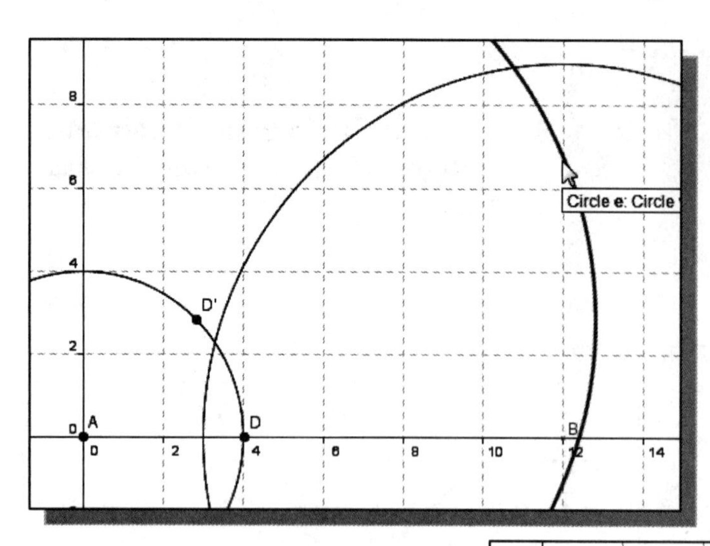

23. Select **Circle e** as the first entity to define the intersection points.

24. Select **Circle d** as the second entity to define the intersection points.

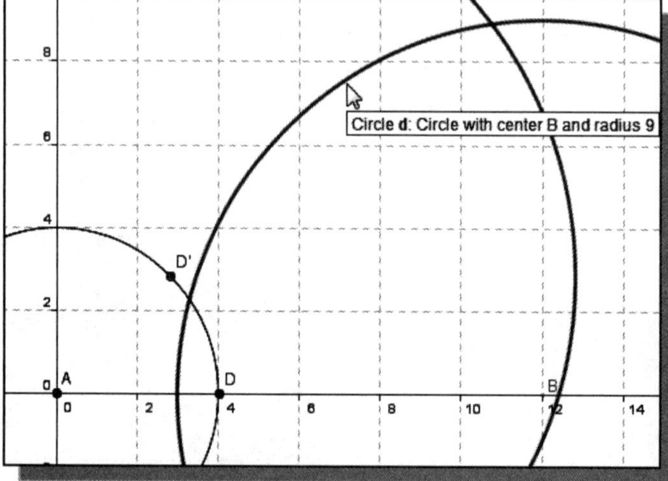

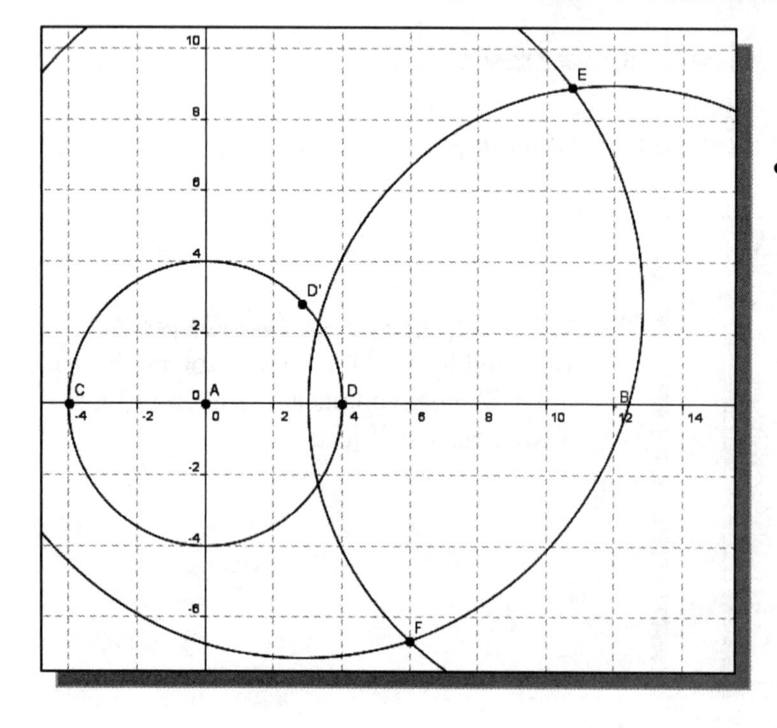

- Note two intersection points are created, **Point E** and **Point F**.

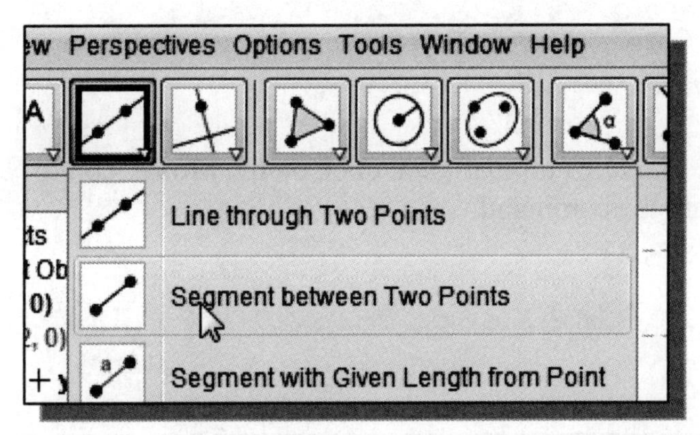

25. Activate the **Segment between Two Points** command by clicking on the icon as shown. This will create a line by selecting two endpoints.

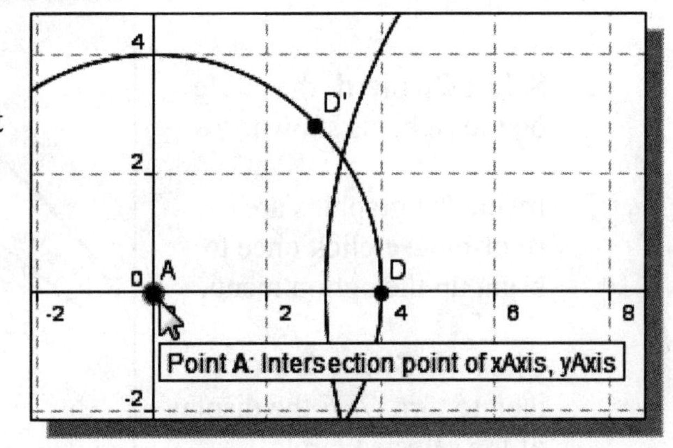

26. Click on **Point A** to place the first point of the line.

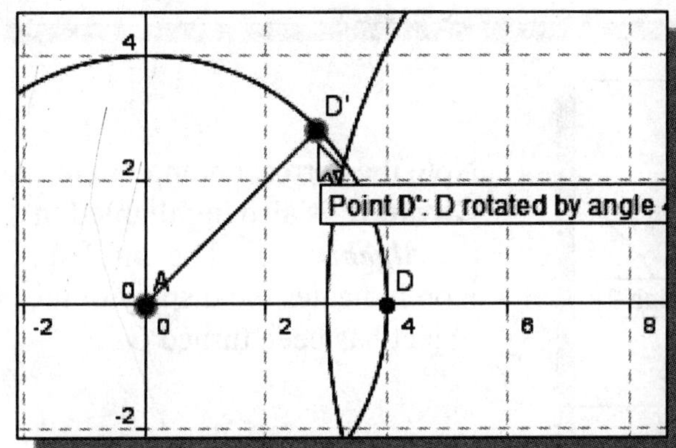

27. Click on **Point D'** to create a line as shown.

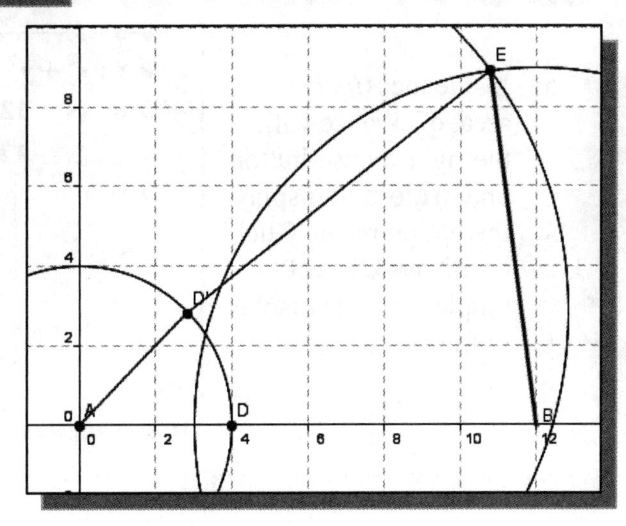

28. On your own, create two additional line segments, **Line DE** and **Line EB** as shown.

Hide the Display of Objects

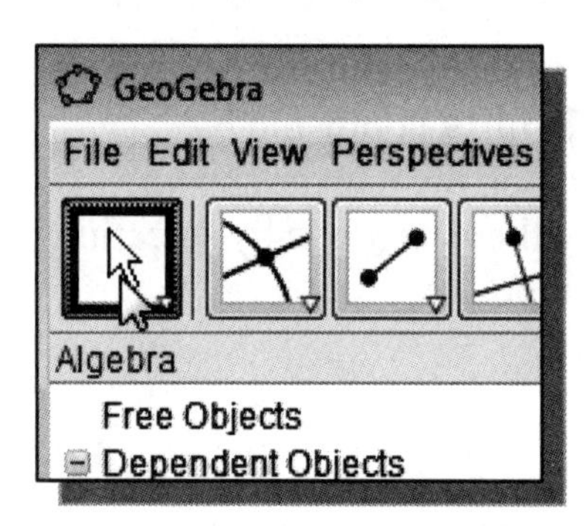

1. In the *Standard* toolbar area, click on the **Move** icon to activate the command.

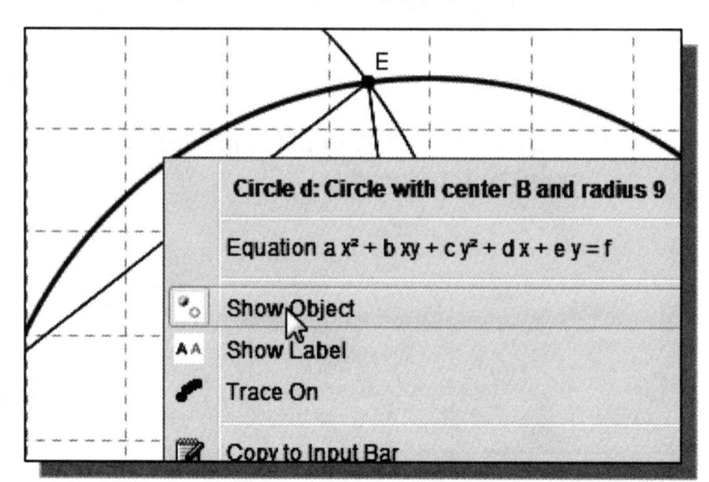

2. Select **Circle d**, the circle on the right, as shown.
3. Inside the graphics area, right-mouse-click once to bring up the option menu.
4. Click the **Show Object** icon to turn *OFF* the display of the selected circle.

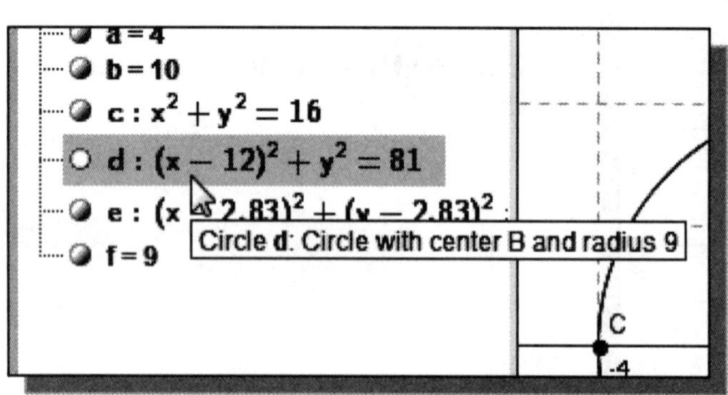

- Note the corresponding circle, **Circle d**, is also highlighted in the ***Algebra*** area. The unfilled icon indicates the display of the object has been turned *OFF*.

5. Inside the ***Algebra*** area, click once with the right-mouse-button on **Circle e** to display the option menu. Click on **Show Object** to toggle *OFF* the display of the circle.

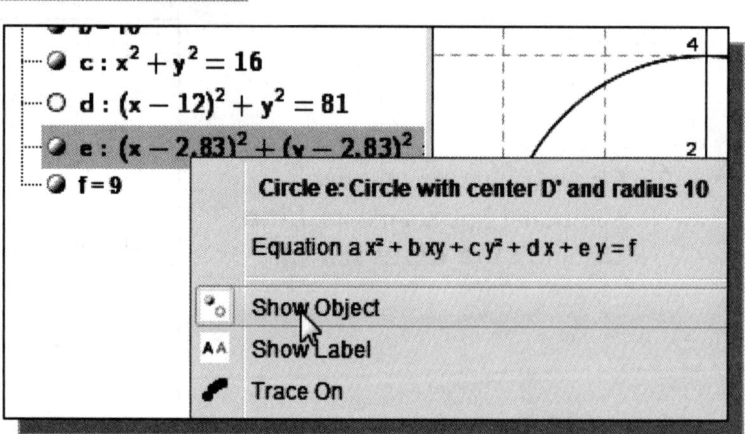

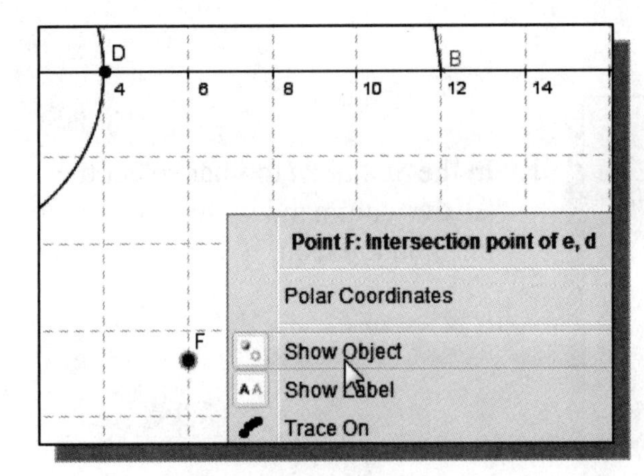

6. Select **Point F**, the point below the line segments, as shown.

7. On your own, turn *OFF* the display of the object through the option menu as shown.

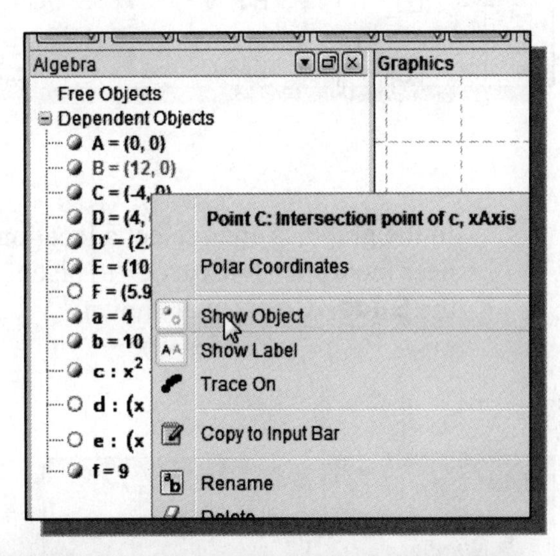

8. In the ***Algebra*** area, click once with the right-mouse-button on **Point C** to display the option menu. Click on **Show Object** to toggle *OFF* the display of the selected item.

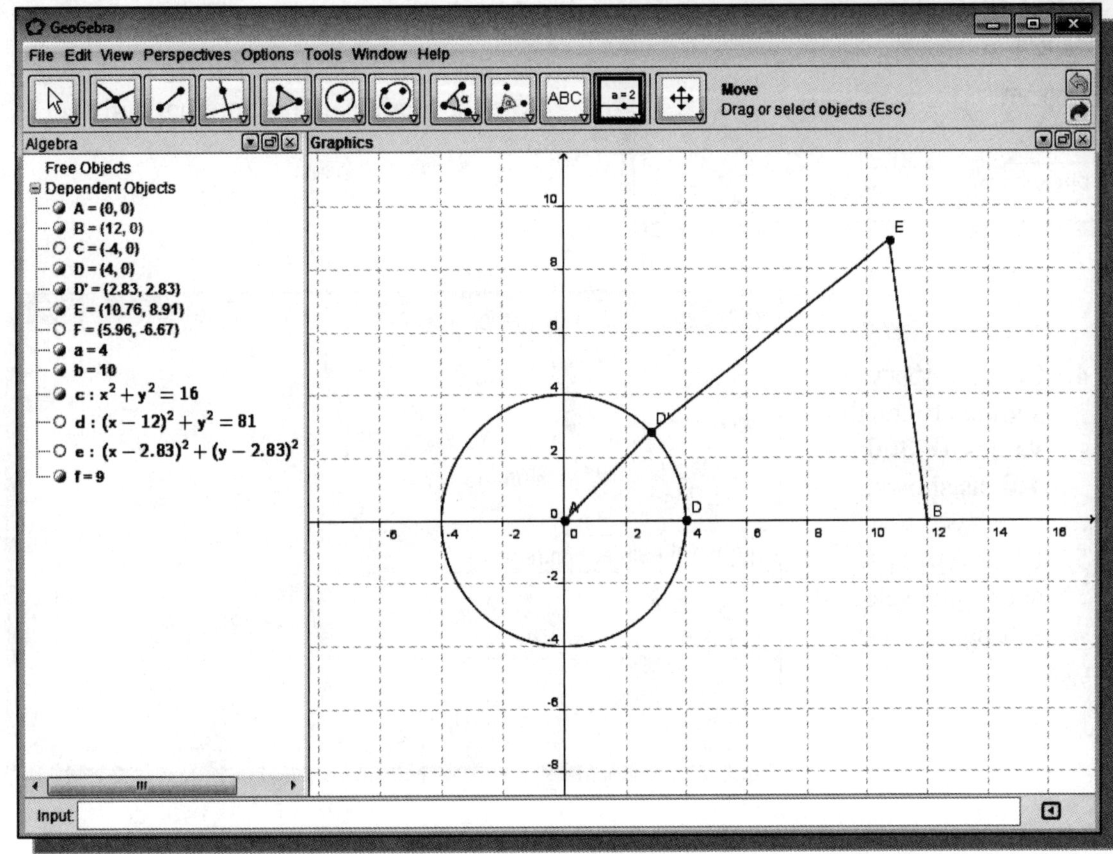

Adding a Slider Control

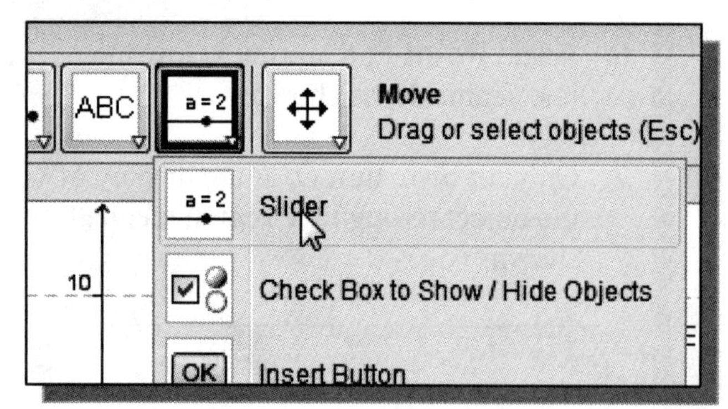

1. In the *Standard* toolbar select the **Slider** command by left-clicking once on the icon.

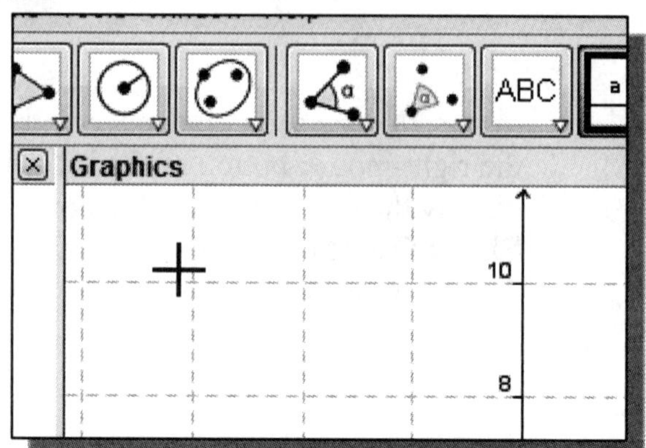

2. In the graphics area, select a location near the **upper left corner** to place the **Slider Control** as shown.

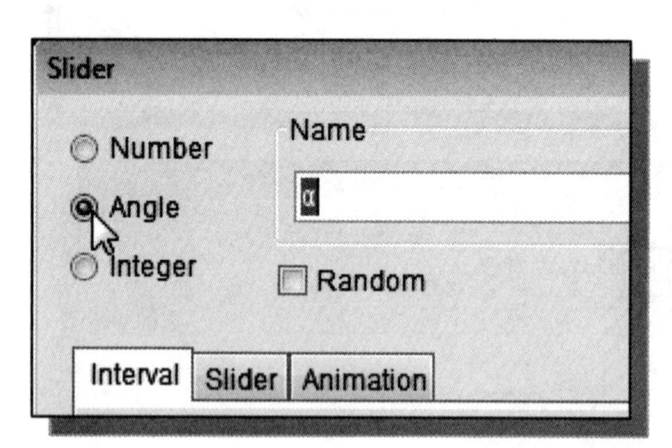

3. Set the *Slider* option to **Angle** as shown. Note the angle name is automatically set to **α**.

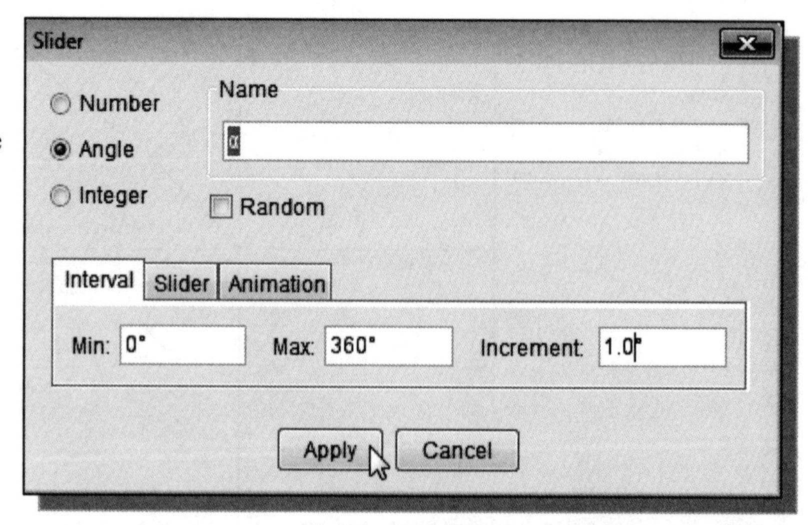

4. Set the **Interval** settings to the three values, **0**, **360** and **1.0**, as shown.

5. Click **Apply** to accept the selected settings

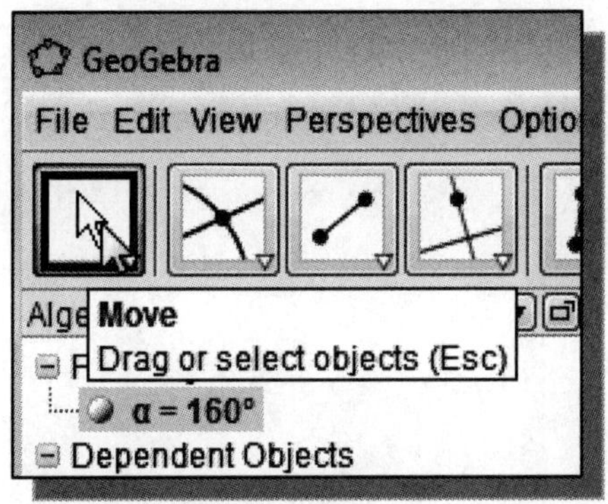

6. In the *Standard* toolbar area, click on the **Move** icon to activate the command.

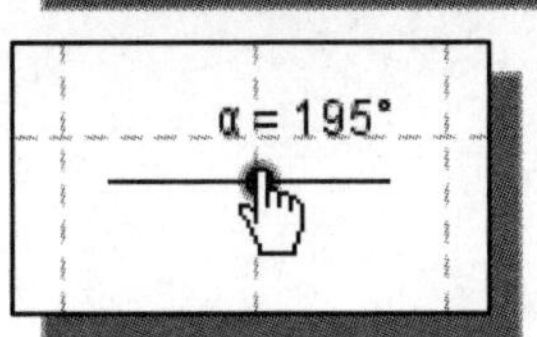

7. Drag the **handle** of the **Slider Control**, and notice **Angle α** is adjusted.

8. In the ***Algebra*** area, double click with the left-mouse-button on **Point D'** to enter the **Object Redefine** mode.

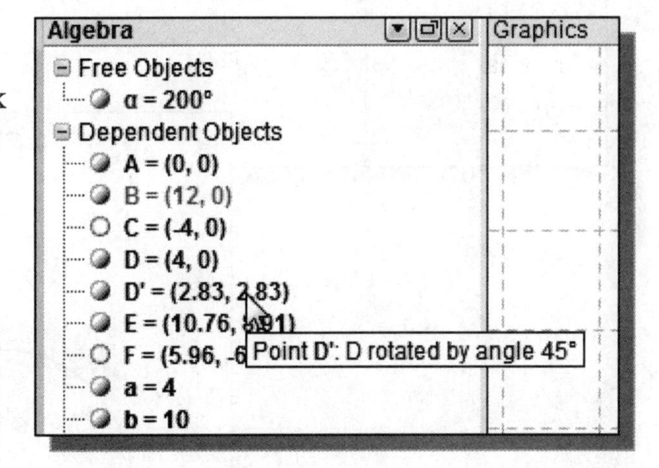

9. Modify the **angle variable**, the second variable in the edit box, to **α** as shown.

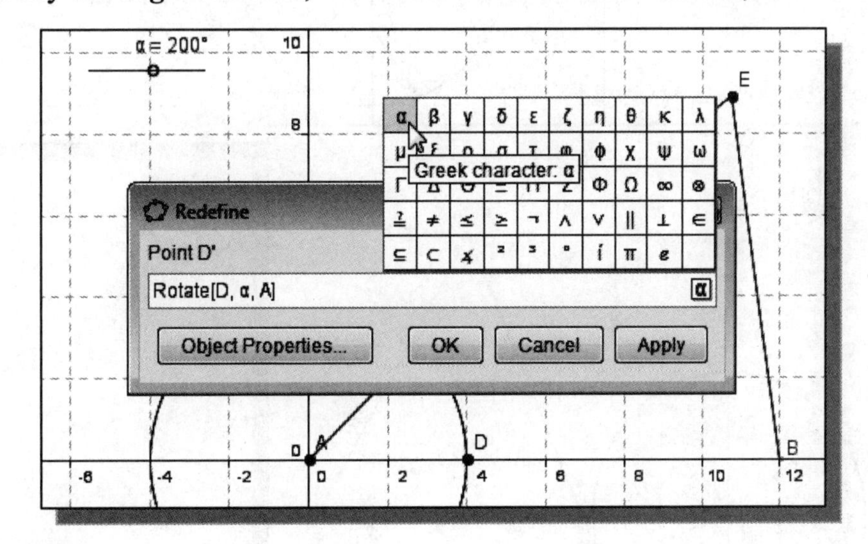

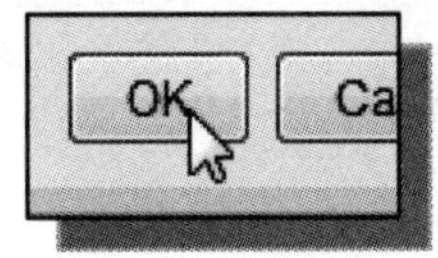

10. Click **OK** to accept the setting and exit the Redefine command.

11. Drag the **handle** of the **Slider Control**, and notice the positions of the links of the four-bar linkage are adjusted accordingly.

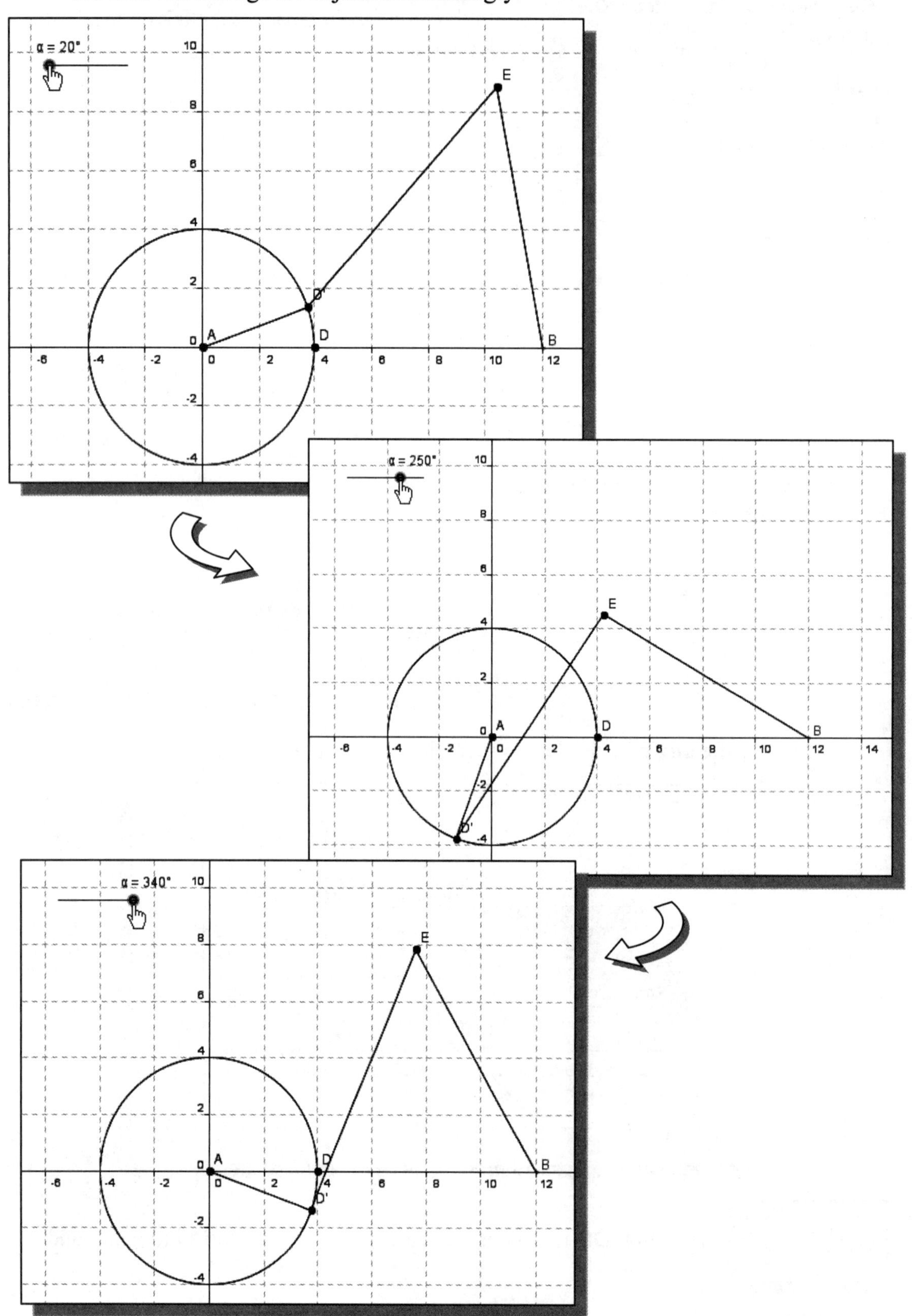

Using the Animate Option

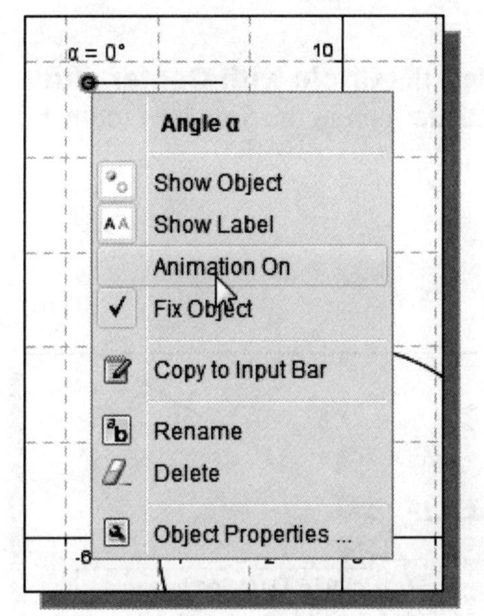

1. Inside the graphics area, click once with the right-mouse-button on **Slider Control** to display the option menu. Click on **Animation On** to toggle *ON* the animation of the Slider Control.

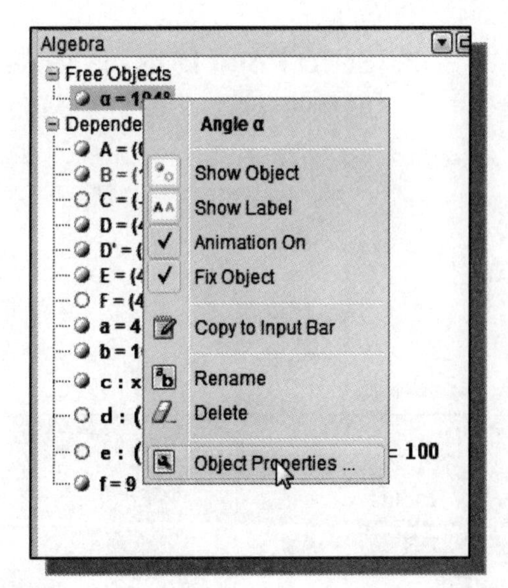

2. Inside the ***Algebra*** area, click once with the right-mouse-button on **Angle** to display the option menu. Click on **Object Properties** to activate the command.

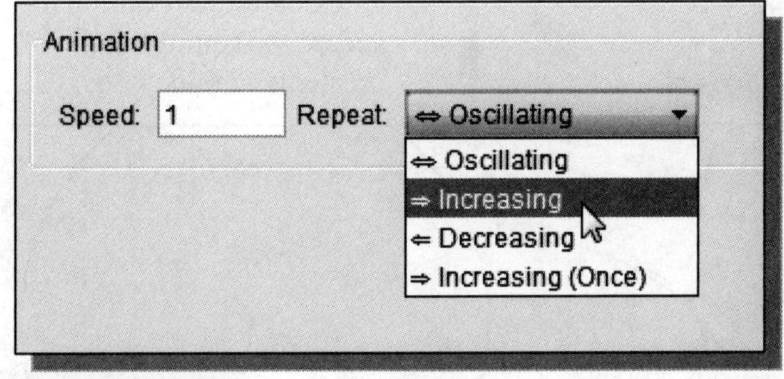

3. Set the animation repeat option to **Increasing** as shown.

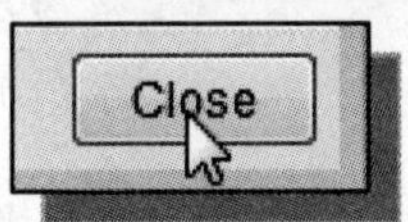

4. Click **Close** to accept the setting and exit the *object properties* command.

5. On your own, examine the animation of the four-bar linkage; turn ***OFF*** the animation option before proceeding to the next section.

Tracking the Path of a Point on the Coupler

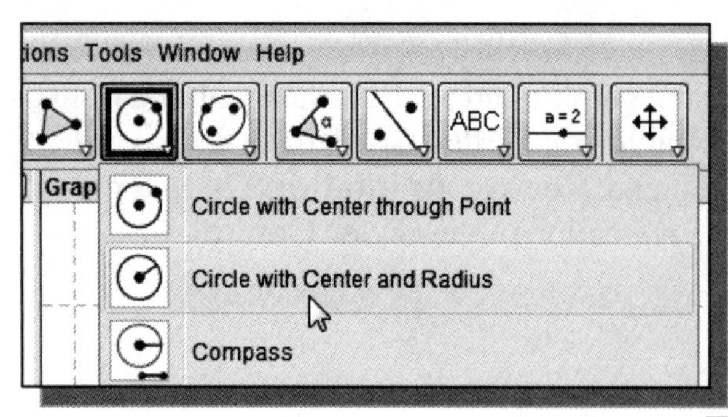

1. Select the **Circle with Center and Radius** icon in the *Standard* toolbar area.

2. Select **Point D'** as the center of the new circle.

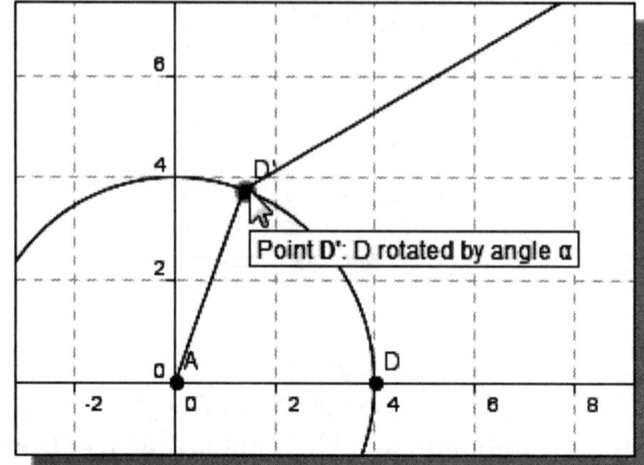

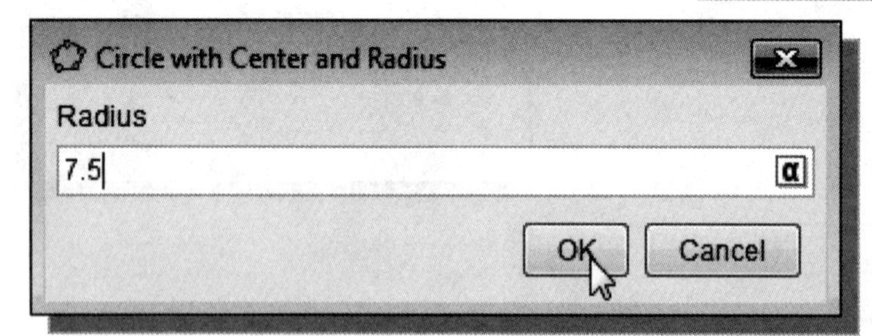

3. In the *Radius* input box, enter **10** as the radius.

4. Click on the **OK** button to accept the settings and create the circle.

5. On your own, create another **radius 7.5 circle** centered at **Point E** as shown.

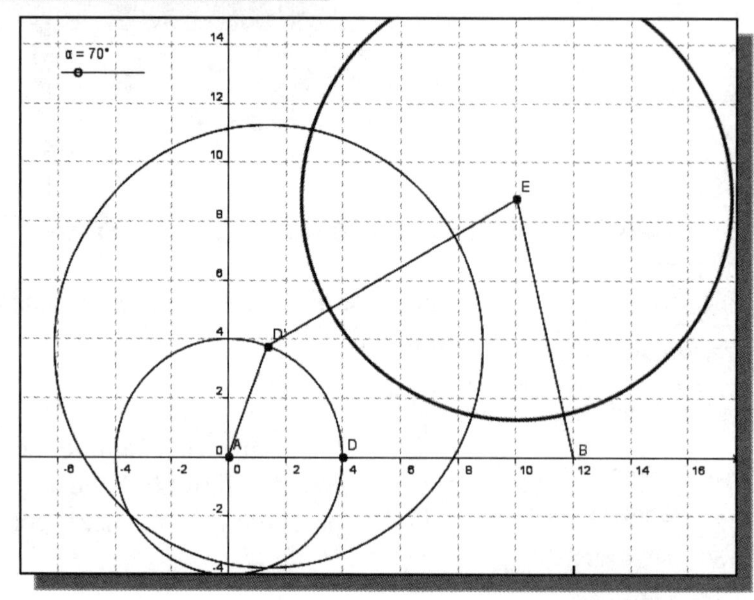

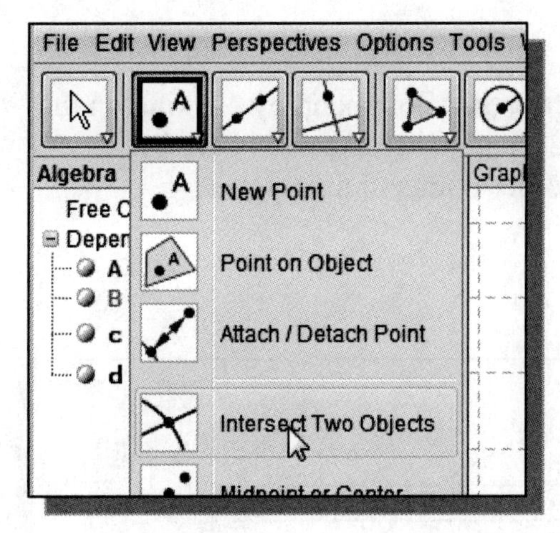

6. Activate the **Intersect Two Objects** command by clicking on the icon as shown. This will create points at the intersections of two selected objects.

7. Select **Circle g** as the first entity to define the intersection points.

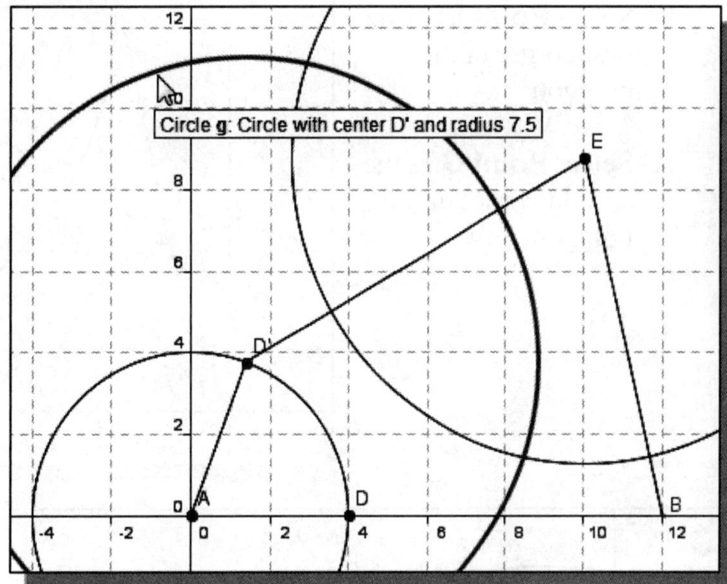

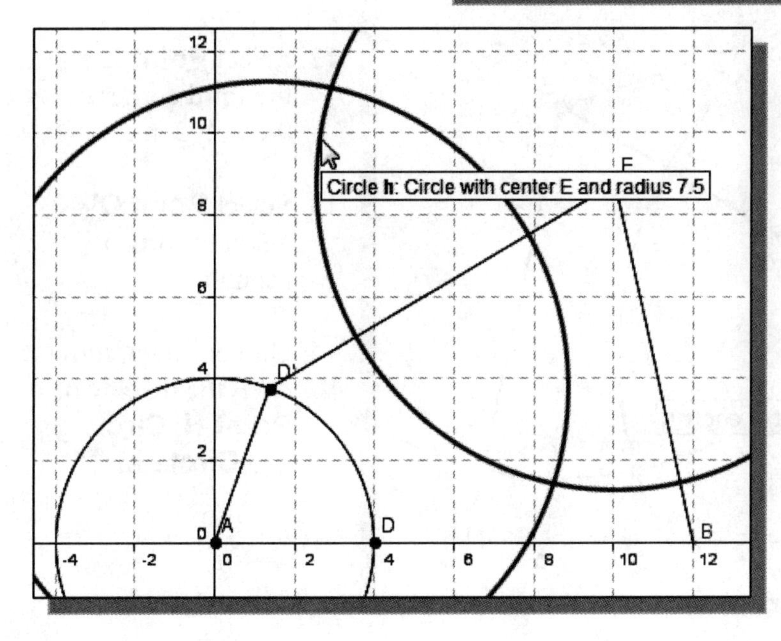

8. Select **Circle h** as the second entity to define the intersection points.

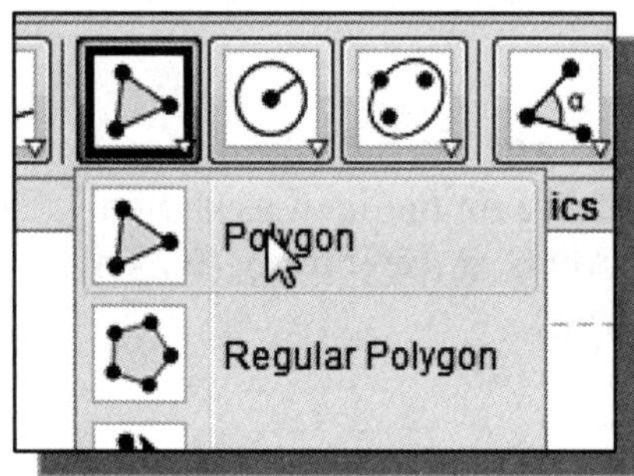

9. Activate the **Polygon** command by clicking on the icon as shown. This will create a polygon by defining the corner points of a polygon.

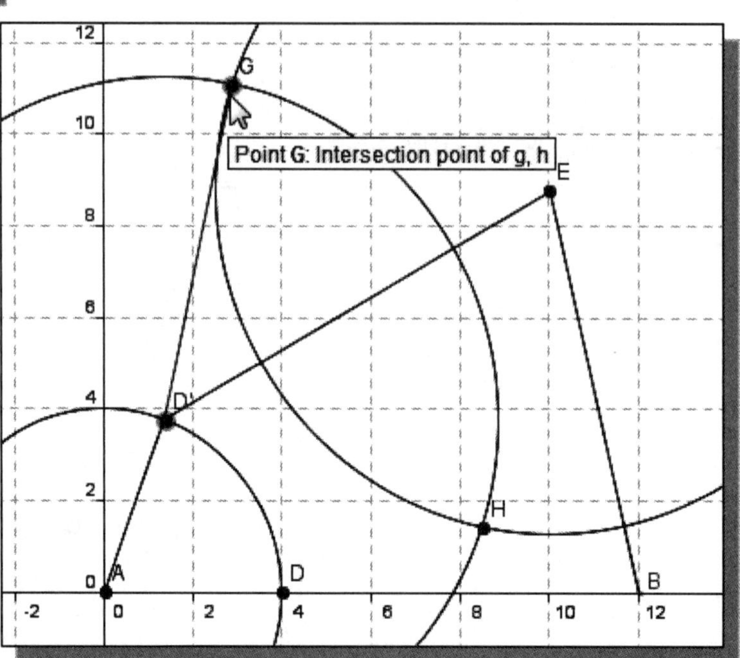

10. Select **Point D'** as the first corner of the polygon.

11. Select **Point G** as the second corner of the polygon.

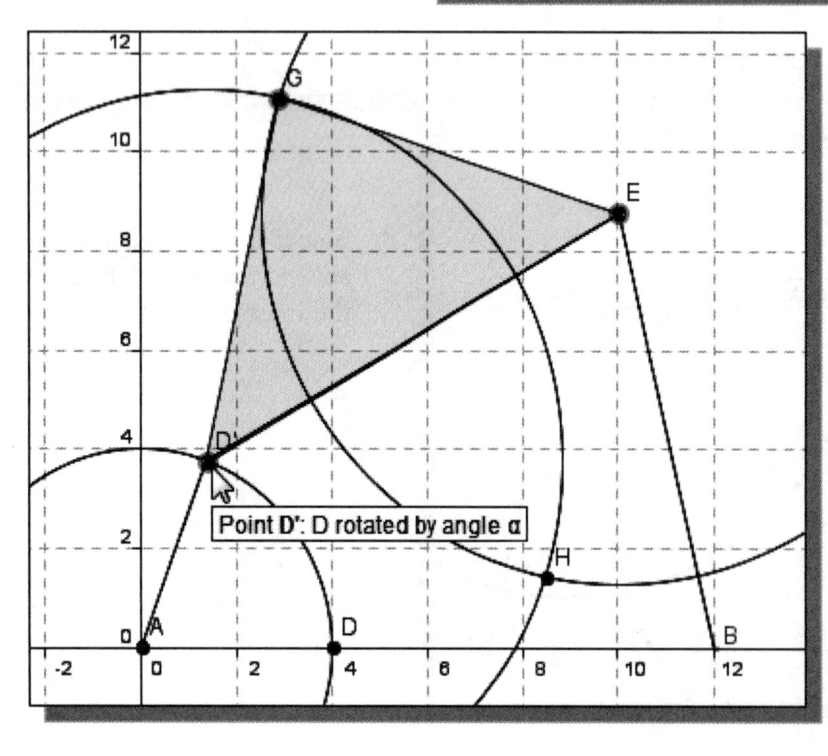

12. Select **Point E** as the third corner of the polygon.

13. Select **Point D'** again to form a triangle.

14. On your own, turn *OFF* the display of **Point H**, **Circle g** and **Circle h**.

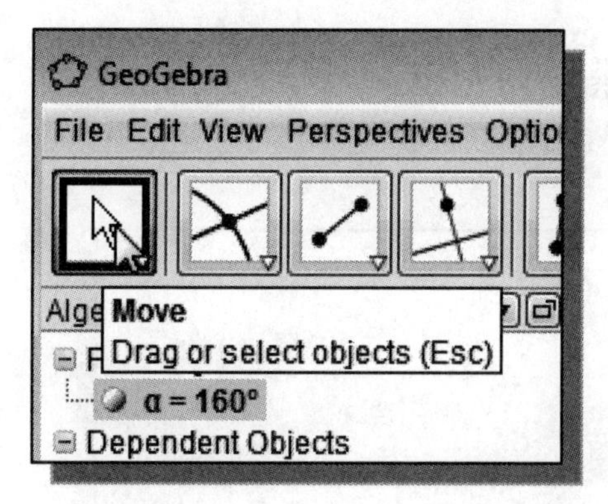

15. In the *Standard* toolbar area, click on the **Move** icon to activate the command.

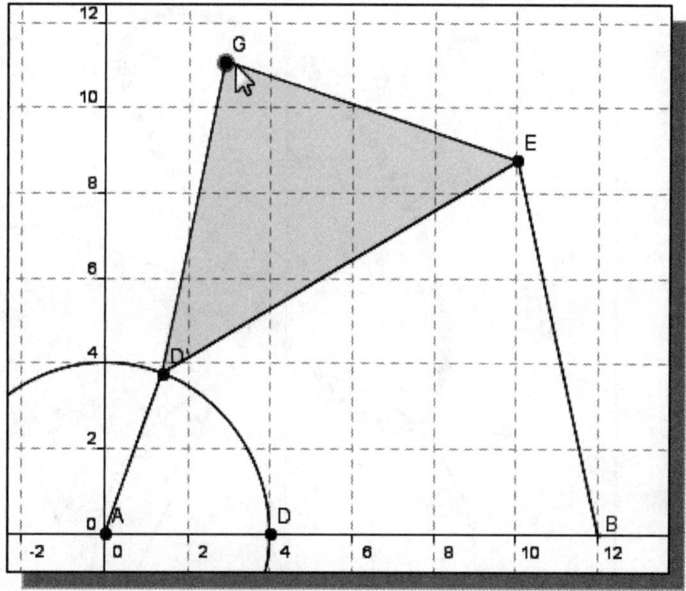

16. In the graphics area, select **Point G** as shown.

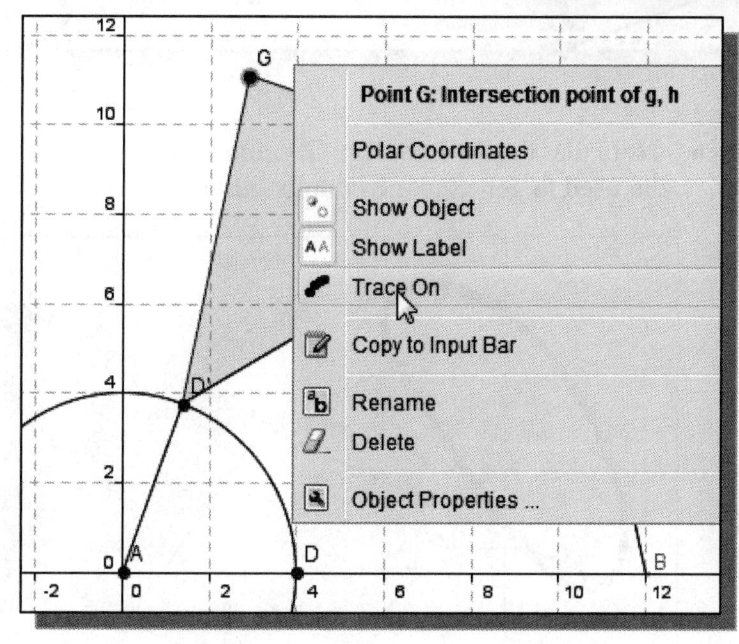

17. Inside the graphics area, click once with the right-mouse-button to display the option menu. Click on **Trace On** to activate the command.

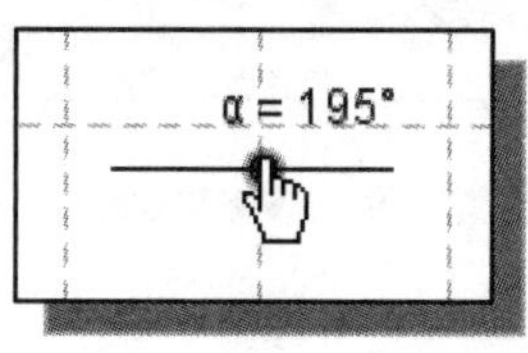

18. Drag the **handle** of the **Slider Control**, and notice **Angle α** is adjusted.

* The locus of point G, also known as a **coupler curve**, is generated as **Angle α** is adjusted. Note that by varying the lengths of the links in the mechanism, different coupler curves can be generated. The *interactive dynamic geometry* software can be used to aid linkage design and analysis.

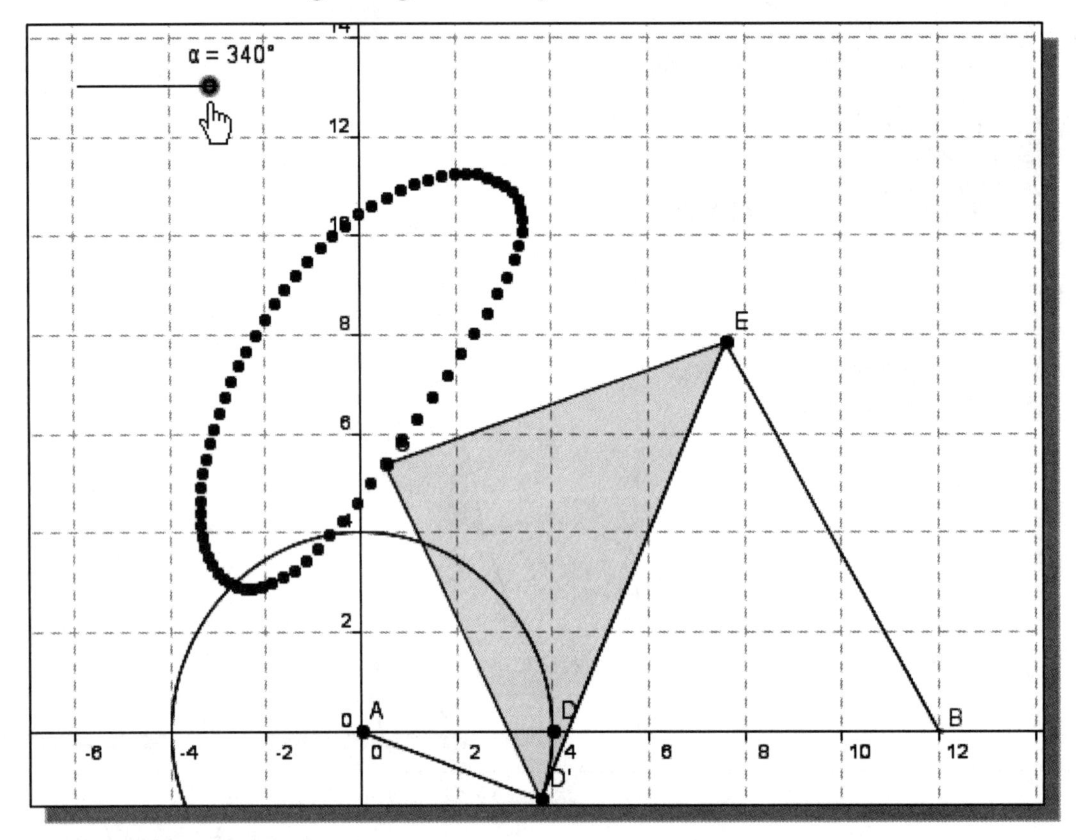

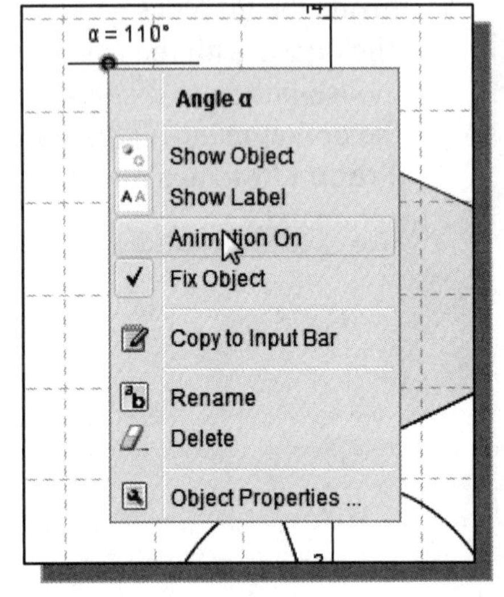

* Note that the **Animation On** command can also be used to generate the coupler curve.

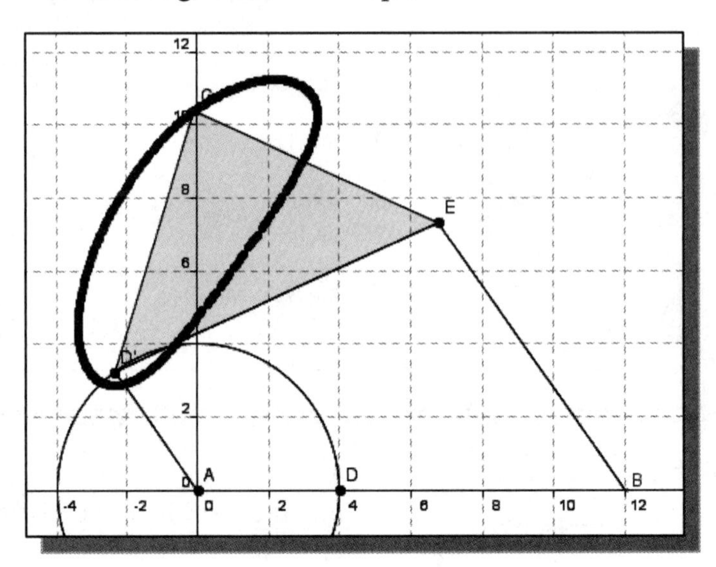

Exercises:

1. **Crank-Slider Mechanism**

AB = 2.5, BC = 5

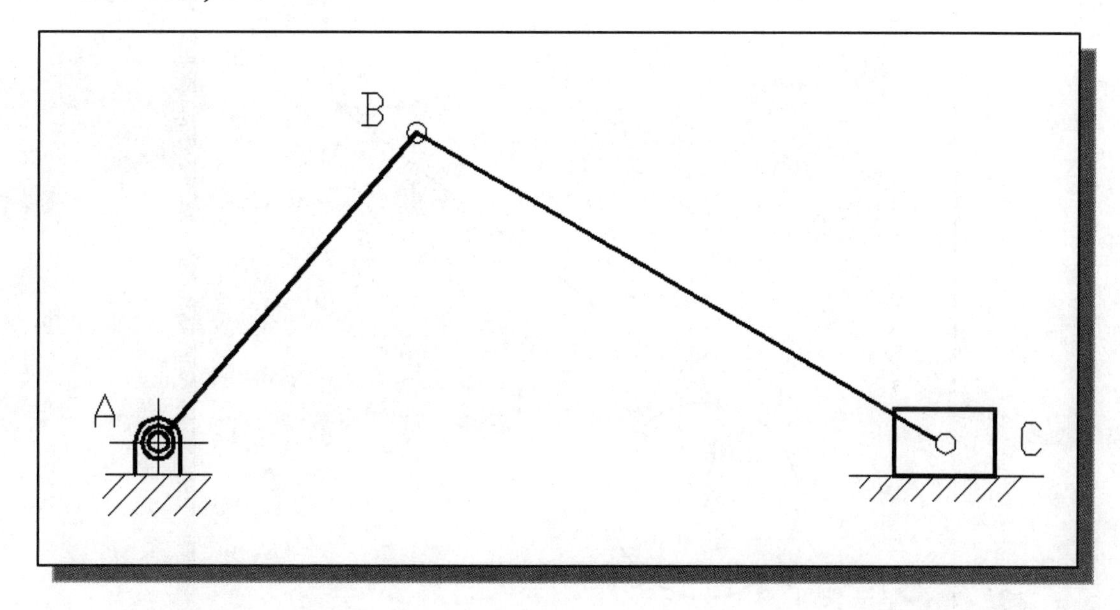

2. **Watt Straight Line Mechanism**

AD''=BE, ED''=1/2 AD'', AB = 2 BE, G is at the midpoint of ED''.
A and B are fixed points.

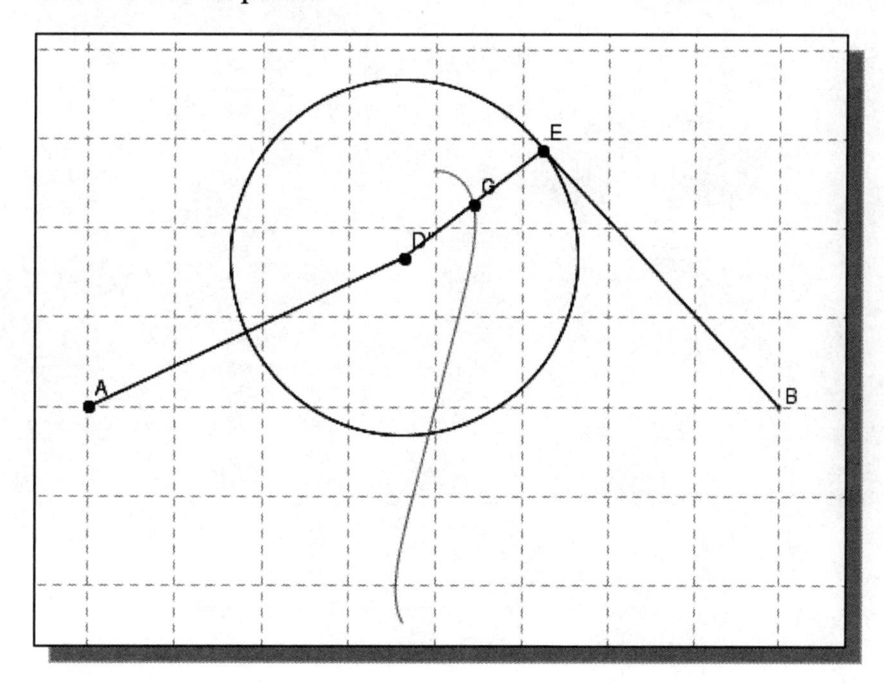

3. **Hoekens Straight Line Mechanism**

BE = EC' = EH = 2.5 AC', AB = 2 AC'
A and B are fixed points and Link C'-E-H is one rigid link.

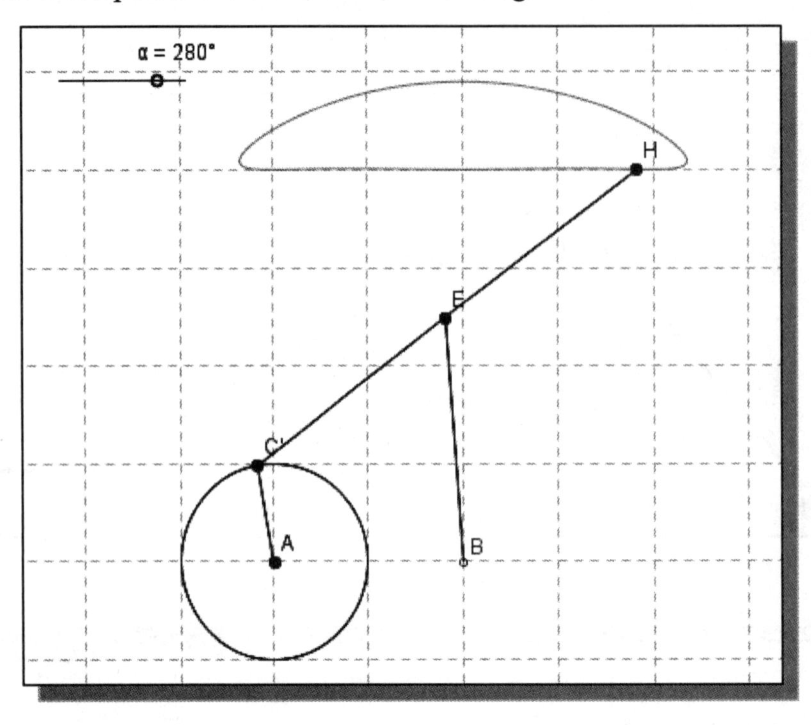

Lesson 11
Design Makes the Difference

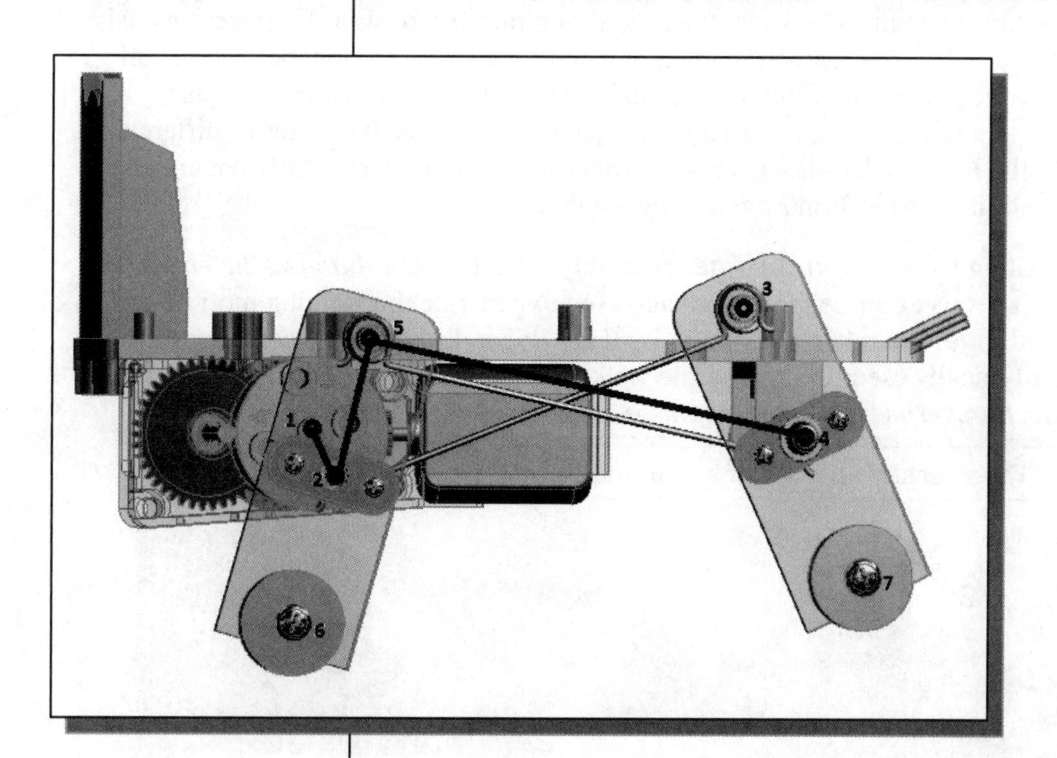

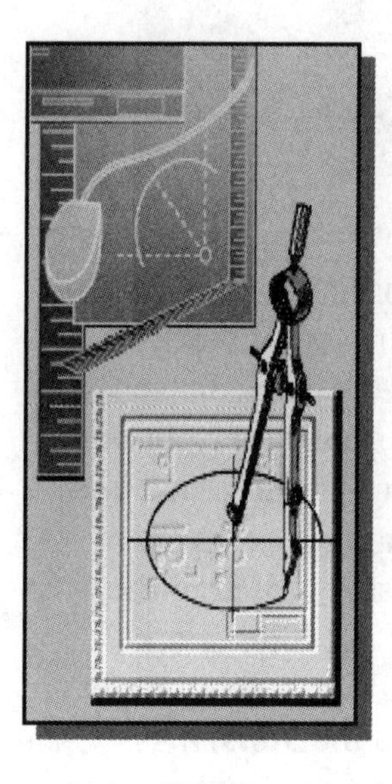

Learning Objectives

- Identify the Compound Mechanism used in the Mechanical Tiger
- Create and Construct the Leg Linkage using GeoGebra
- Generate the Foot Path of the Mechanical Tiger Design
- Examine the Jansen Mechanism and the Klann Mechanism.

Engineering Analysis – How does this work?

Have you ever wondered how a certain machine worked and how the machine can work better? Possessing this type of engineering curiosity may indicate you are an engineer at heart. The modeling of the *Mechanical Tiger* design in the previous chapters simply provides a starting point for further studying of **engineering design**. Engineering design is the ability to create and transform ideas and concepts into a product definition that meets the desired objective. Engineering design also involves activities, such as dissecting and analyzing, to identify the advantages and/or disadvantages of different designs. In the last two decades, Computer Aided Engineering has greatly enhanced engineers' abilities to perform *Engineering Analyses*.

Prior to creating the *Mechanical Tiger* assembly models in the *Autodesk Inventor* 3D environment, let us examine the leg linkage used to generate the walking motion. The *Mechanical Tiger* is a relatively simple *Walking Robot* design. More complex linkage designs are typically used for larger scale walking robots; two such mechanisms are the ***Jansen Mechanism*** and ***Klann Mechanism***.

Theo Jansen and one of his Strand Beasts

Theo Jansen is a Dutch artist and an engineer. He builds large artworks which resemble skeletons of animals that are able to walk using *wind power*. His animated artworks, also known as ***kinetic sculptures***, are a fusion of art and engineering. One of Theo Jansen's most inspirational quotes is: ***"The walls between art and engineering exist only in our minds."*** In the recent year, Theo Jansen has also incorporated artificial intelligence into his creations, so that the *kinetic sculptures* can avoid the ocean by changing course, or anchor themselves when strong wind is detected. His works were first revealed to the public in 2006, after he had worked on them for 16 years. His *kinetic sculptures* are relatively light-weight, with a fairly well-balanced mechanism assembly; and not much power is needed to drive them. The main eight-bar linkage, known as the ***Jansen Mechanism***, was developed by Theo Jansen himself. The mechanism development was done through the use of computers using a genetic algorithm.

The ***Walking Beast***, built by *Moltensteelman*, is a mechanical creature that walks on eight legs. The creature weighs approximately 6 1/2 tons (13,000 lbs), and stands 11′ tall, 8′ 4″ wide and 24′ long. It has a step height of 41 inches and a stride of 5 feet. The machine is powered by a 454 cubic inch Chevy V-8 engine connected to a modified TH400 transmission and two Klune reduction gear boxes coupled to a modified Rockwell 2 1/2 ton military differential that supplies power to the crank shafts of the legs. The leg linkage used is a six-bar mechanism designed by ***Joe Klann***. The *Klann Mechanism* was developed by Joe Klann in 1994 as an expansion of *Burmester* curves, which are used to develop four-bar double-rocker linkages. Since the shape of the *Klann Mechanism* resembles the leg of a spider, many of the walking robots built using this mechanism are known as the **mechanical spiders**.

In this chapter, we will first examine the leg mechanism of the *Mechanical Tiger*. We will also examine the ***Jansen Mechanism*** and ***Klann Mechanism***. It should be noted that a majority of the mechanisms used in mechanical design are 2D planar mechanisms. In most cases, it is more practical to perform 2D analyses prior to the more time consuming 3D analyses.

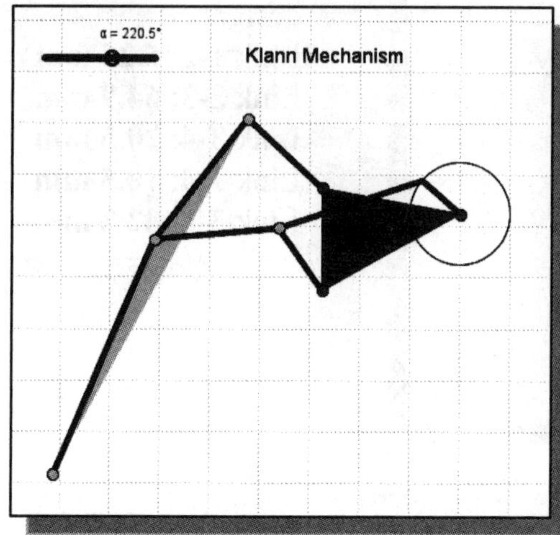

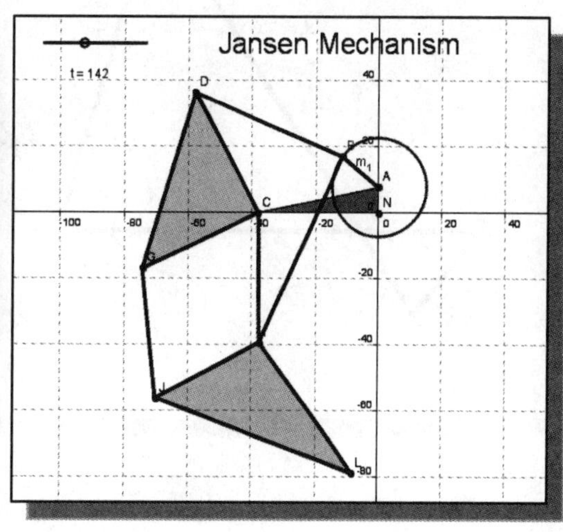

Identify the Six-bar Linkage of the *Mechanical Tiger*

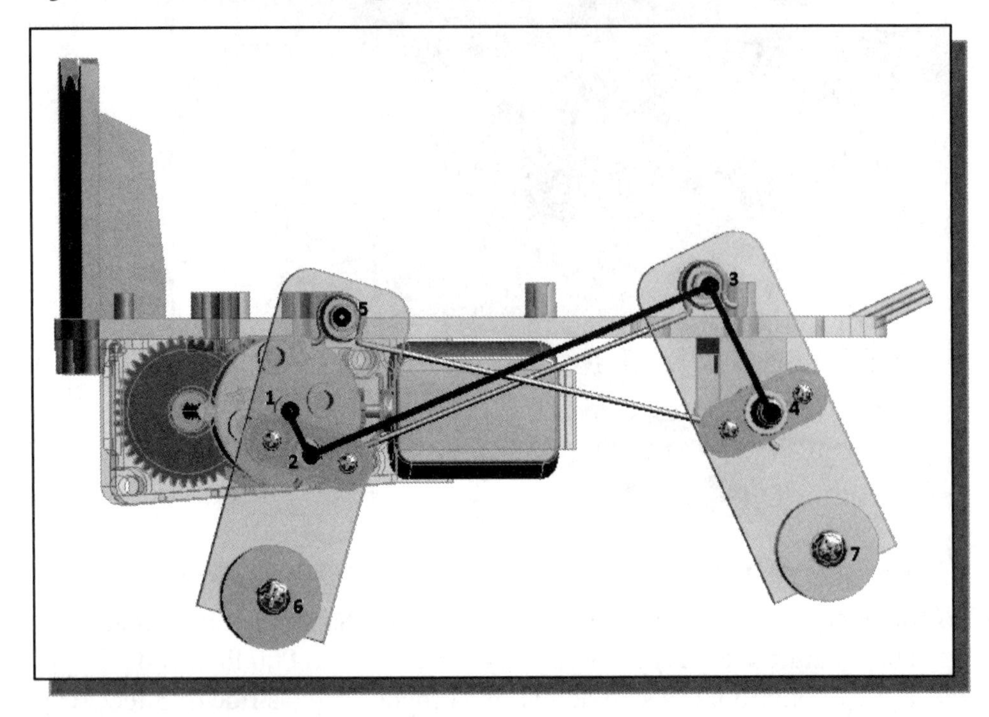

The leg linkage of the *Mechanical Tiger* has six-links and the design consists of a compound mechanism, two combined four-bar mechanisms. The first four-bar mechanism is formed by Points 1-2-3-4, where link 1-2 (from Point 1 to Point 2) is on the rotating A9-crank part; Point 1 is aligned to the axis of the rotating shaft. Point 2 is a moving point on the Crank part. Link 1-2 is therefore the **crank**, or **driver** link. Point 4 is aligned to the rear axle. Since Point 1 and Point 4 are fixed to the chassis, link 1-4 is a fixed link, also known as the **frame**. Link 3-4 is the **follower** link, which rocks back and forth about Point 4. Link 2-3 is the connecting rod. These four links form a four-bar ***Crank-Rocker*** mechanism, as discussed in the previous chapter.

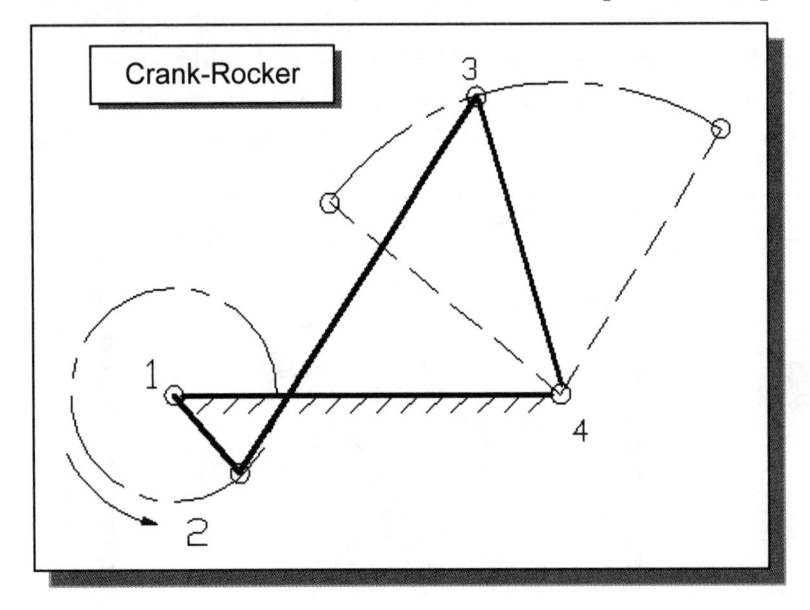

Length of the links used by the *Mechanical Tiger* design:
Link 1-2: **5.25 mm**
Link 2-3: **64.9 mm**
Link 3-4: **20.5 mm**
Link 1-4: **70.5 mm**
Link 3-7: **42.2 mm**

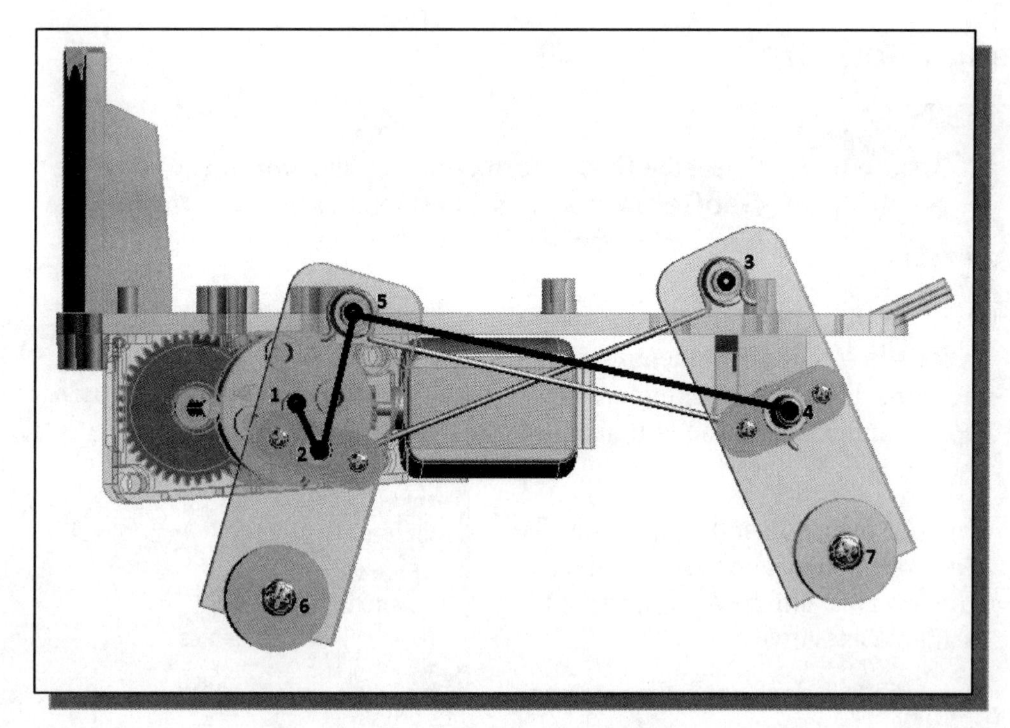

The second four-bar mechanism is formed by Points 1-2-5-4, where link 1-2 is still the **crank**, or **driver**. Both Point 1 and Point 4 are aligned to the front and rear axles; link 1-4 is the **frame**. Link 5-4 is the **follower** link, which rocks back and forth about Point 4. Link 2-5 is the **connecting rod**. These four links also form a four-bar ***Crank-Rocker*** mechanism, similar to the first four-bar mechanism. Note that by using the same driver and frame for both mechanisms, the front legs and back legs are synchronized. The two sets of connecting rod and follower are generating the necessary rocking motion for the walking motion. For the feet locations of the design, Points 6 and 7, they are driven by the rocker link 3-4 in the first mechanism and the connecting rod 2-5 of the second mechanism. Point 7 is part of the link 3-4 and Point 6 is part of the link 2-5. The combination of the two sets of four-bar mechanisms is well designed to generate the walking motion. By cleverly attaching the second set of linkages on the other side, all four legs are synchronized to generate the desired motion.

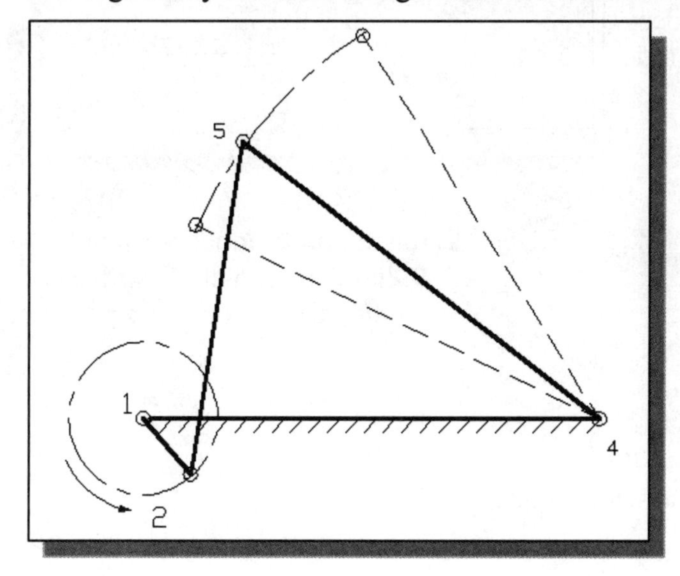

Length of the links used by the *Mechanical Tiger* design:
Link 1-2: **5.25 mm**
Link 2-5: **64.9 mm**
Link 4-5: **20.5 mm**
Link 1-4: **70.5 mm**
Link 5-6: **42.2 mm**

Starting *GeoGebra*

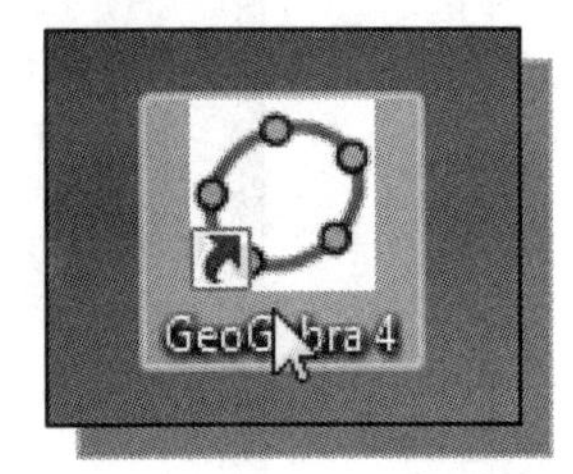

1. Select the **GeoGebra** option on the *Start* menu or select the **GeoGebra** icon on the desktop to start *the software*. The *GeoGebra* main window will appear on the screen.

2. In the **View** pull-down menu, switch on the **Grid** display by clicking on the icon with the left-mouse-button as shown.

3. On your own, adjust the display of the *GeoGebra* main screen so that the *Algebra* area and the *Input Bar* are available as shown.

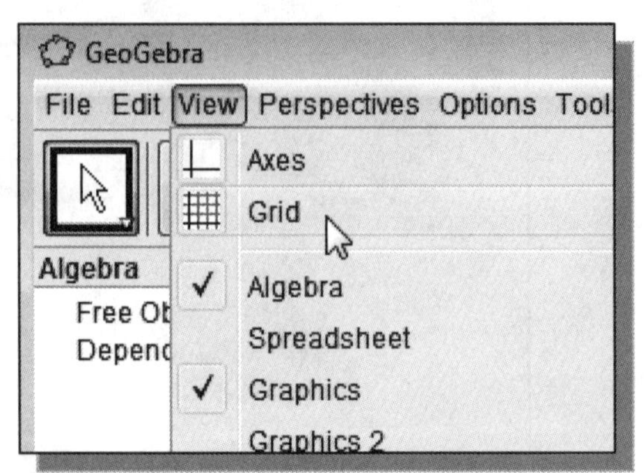

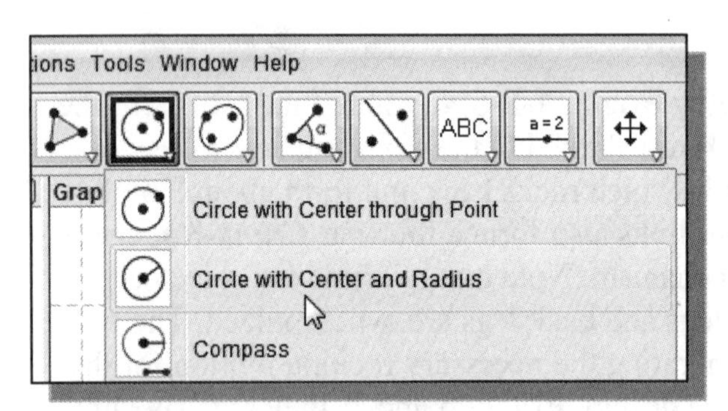

4. Select the **Circle with Center and Radius** icon in the *Standard* toolbar area.

5. Click on the **origin** of the coordinate system to place the center of the circle. Note the created center point is also added in the *Algebra* area that is toward the left.

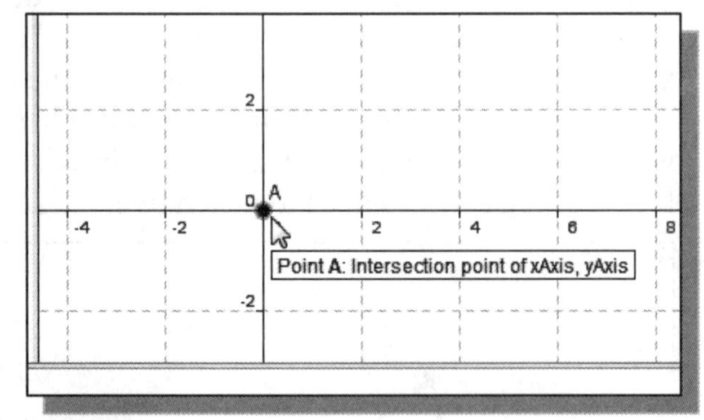

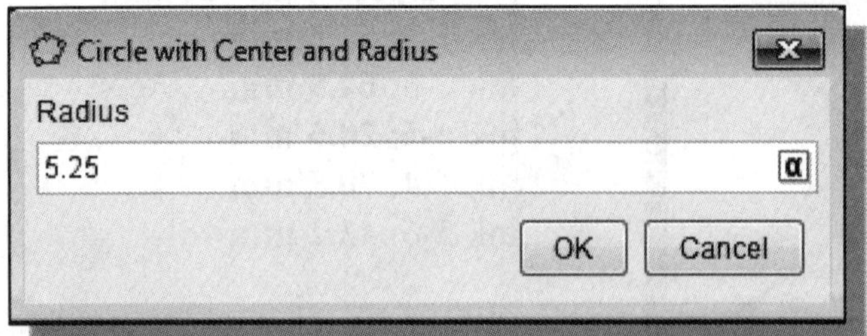

6. In the *Radius* input box, enter **5.25** to create a circle with a radius of **5.25 mm** in size.

7. Pick **OK** to accept the selected settings.

8. In the *Input bar* area, enter **B = (70.5,0)** to create Point B at the specified coordinates.

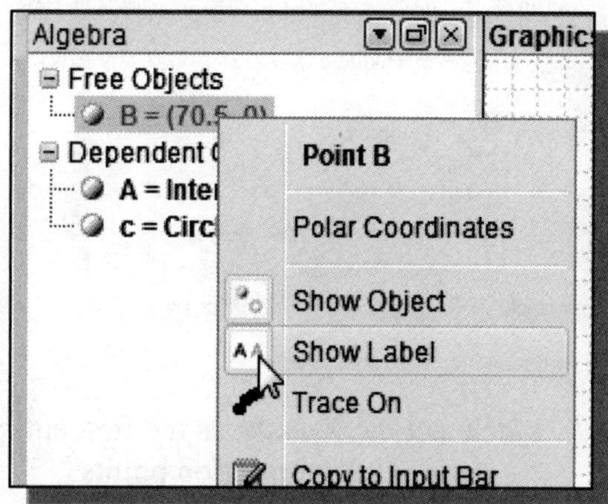

9. In the *Algebra* area, right-mouse-click once on Point B and switch *ON* the display of the label as shown.

10. On your own, use the mouse-wheel to adjust the display of the graphics area so that Point B is visible.

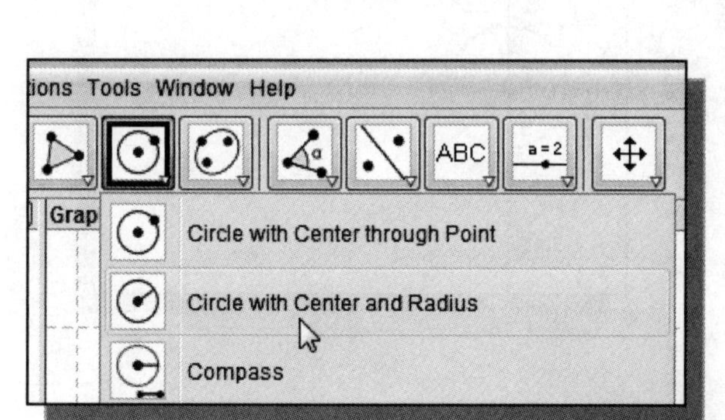

11. Select the **Circle with Center and Radius** icon in the *Standard* toolbar area.

12. Click on **Point B** to place the center of the circle.

13. In the *Radius* input box, enter **20.5** to create a circle with a radius of **20.5 mm** in size.

14. Pick **OK** to accept the selected settings.

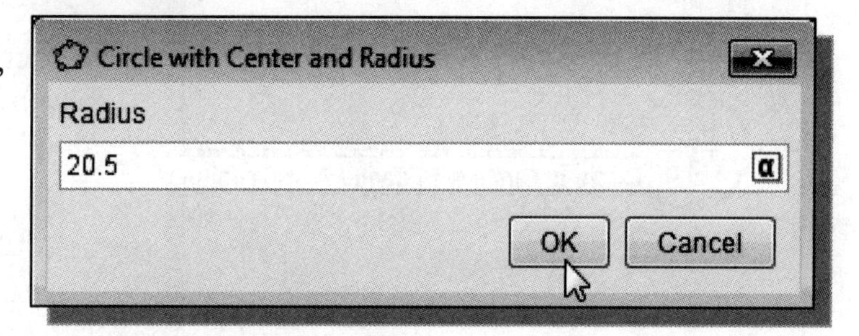

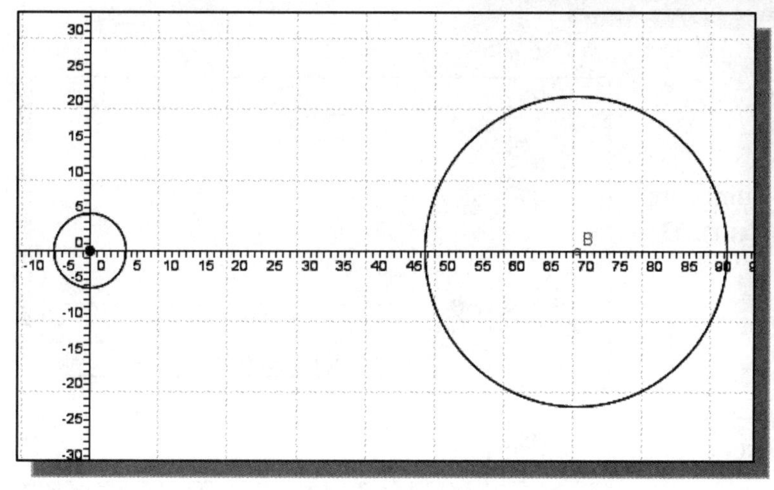

15. Use the **mouse wheel** to Zoom/Pan and adjust the current display.

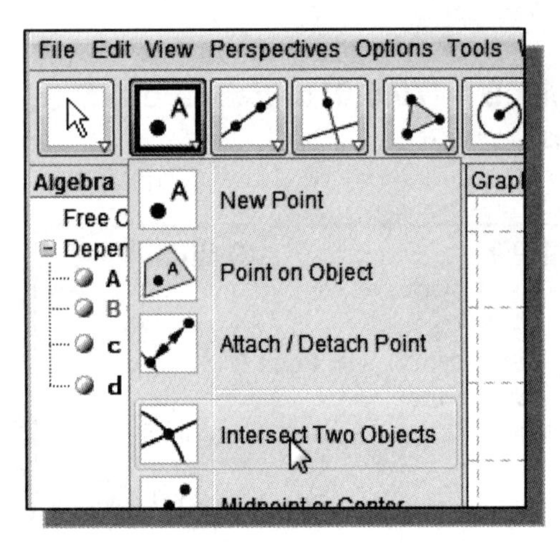

16. Activate the **Intersect Two Objects** command by clicking on the icon as shown. This will create points at the intersections of two selected objects.

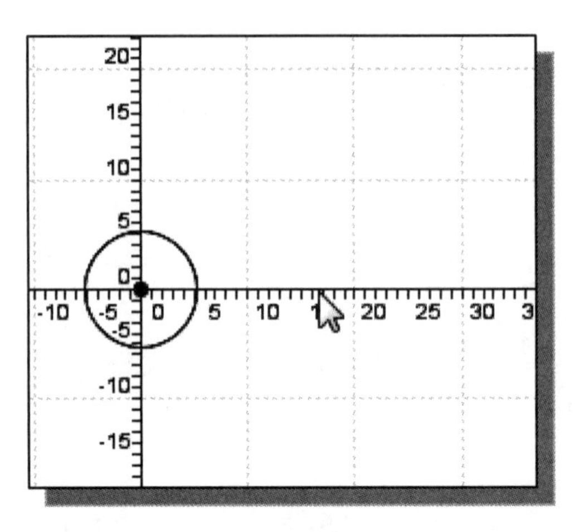

17. Select the **X-Axis** as the first entity to define the intersection points.

18. Select the **smaller circle** as the second entity to define the intersection points.

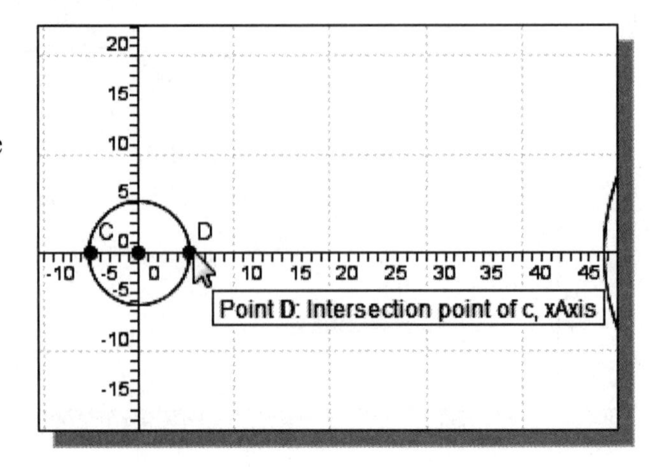

- Note two intersection points are created, **Point C** and **Point D**.

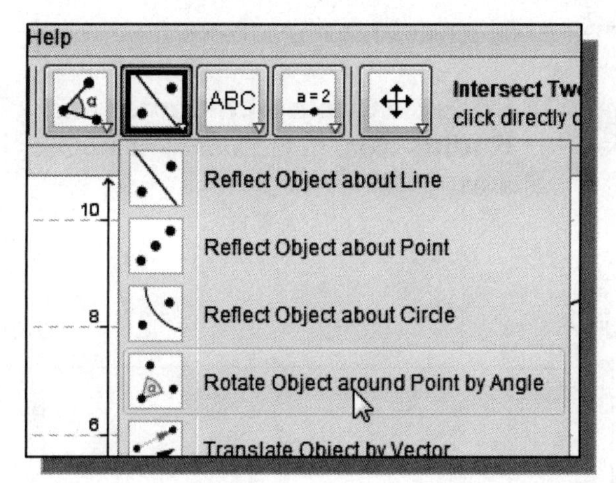

19. Activate the **Rotate Object around Point by Angle** command by clicking on the icon as shown. This will create a new point by rotating an existing point.

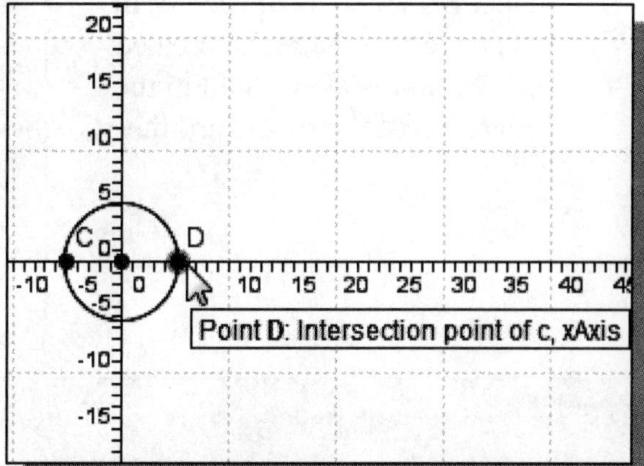

20. Select **Point D** as the object to be rotated.

21. Select **Point A** as the reference axis of rotation.

22. In the *Angle* input box, enter **45** as the angle of rotation.

23. Click **OK** to accept the setting and create the new point.

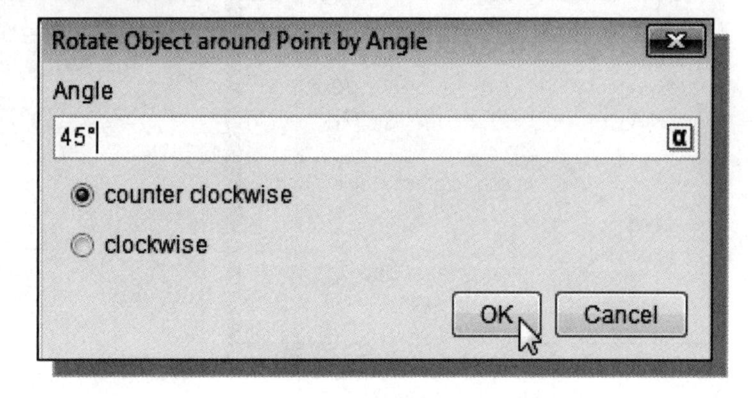

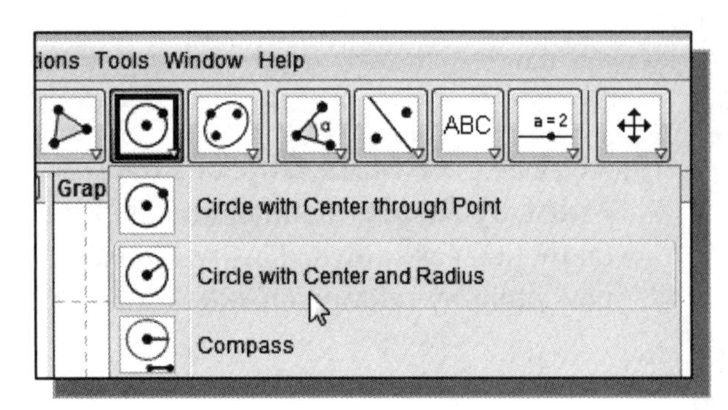

24. Select the **Circle with Center and Radius** icon, in the *Standard toolbar* area.

25. Select **Point D'** as the center of the new circle. Note the created center point is also added in the *Algebra* area that is toward the left.

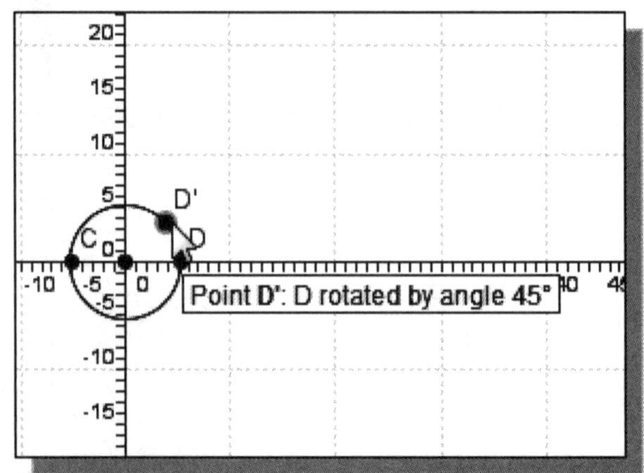

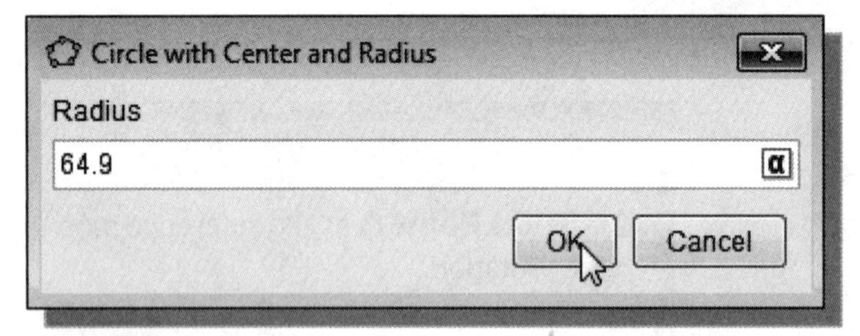

26. In the *Radius* input box, enter **64.9** to create a circle with a radius of **64.9 mm** in size.

27. Click **OK** to accept the selected settings.

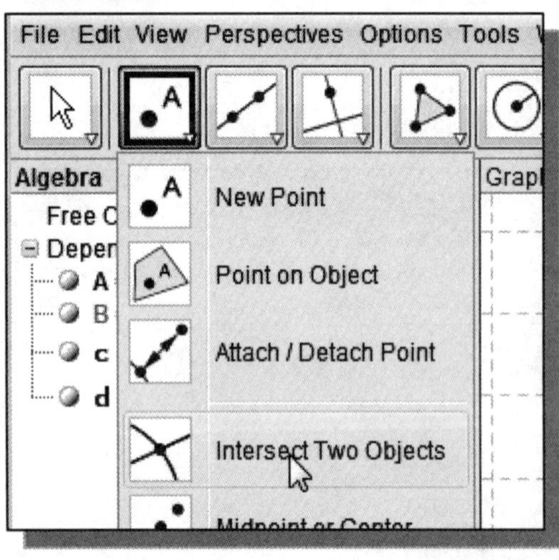

28. Activate the **Intersect Two Objects** command by clicking on the icon as shown. This will create points at the intersections of two selected objects.

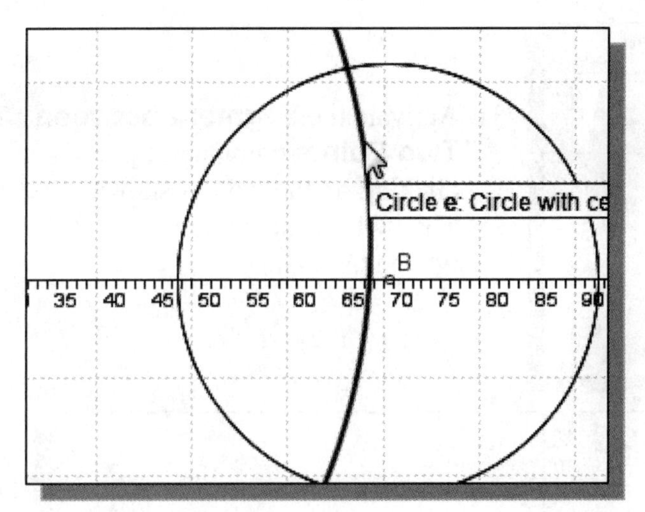

29. Select **Circle e** as the first entity to define the intersection points.

30. Select **Circle d** as the second entity to define the intersection points.

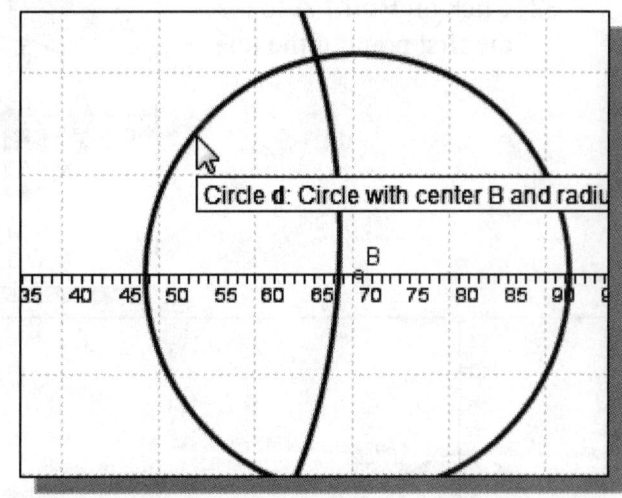

- Note two intersection points are created, **Point E** and **Point F**.

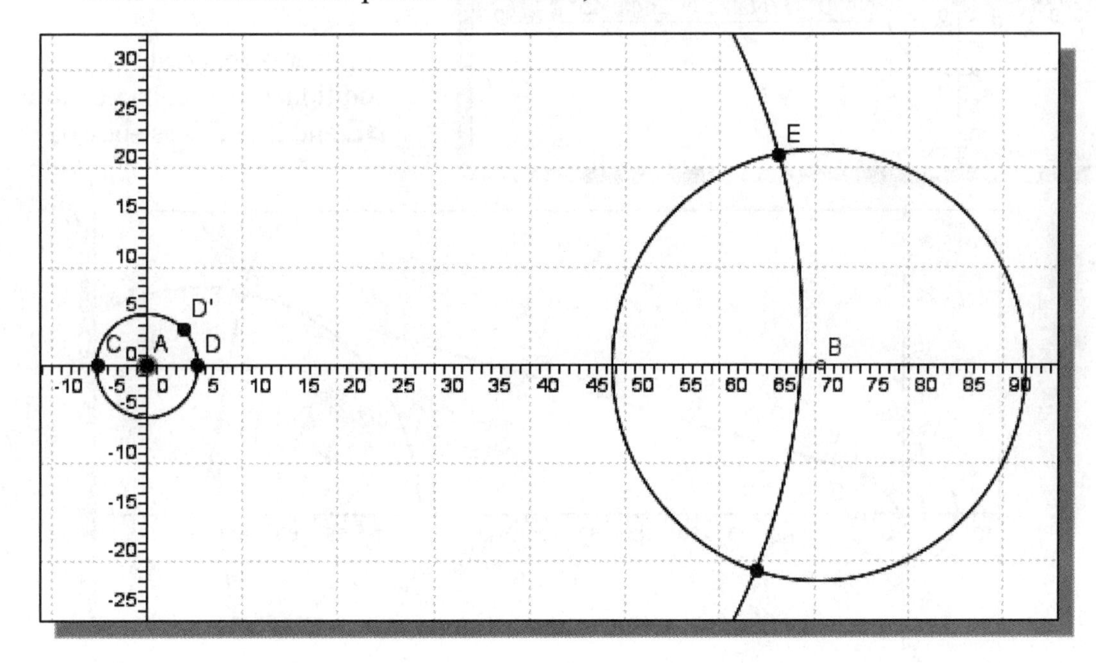

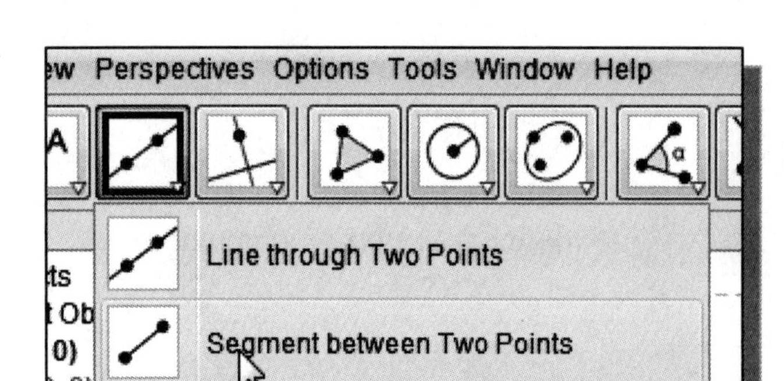

31. Activate the **Segment between Two Points** command by clicking on the icon as shown. This will create a line by selecting two endpoints.

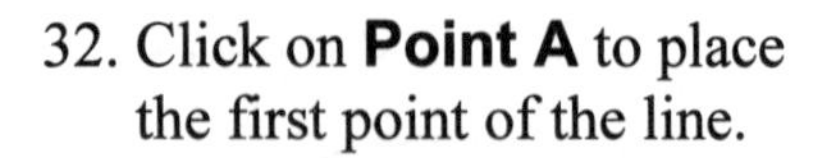

32. Click on **Point A** to place the first point of the line.

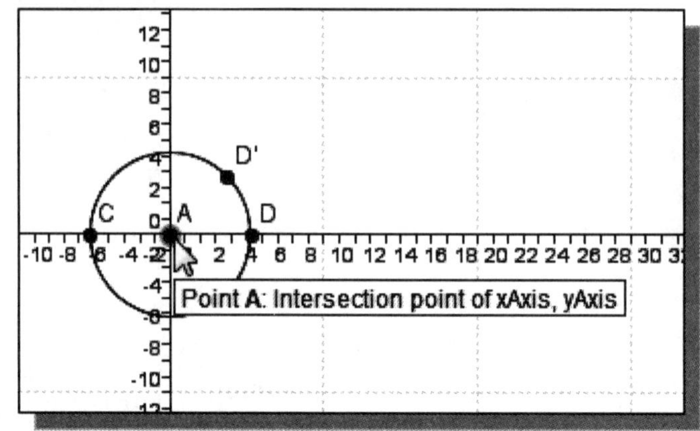

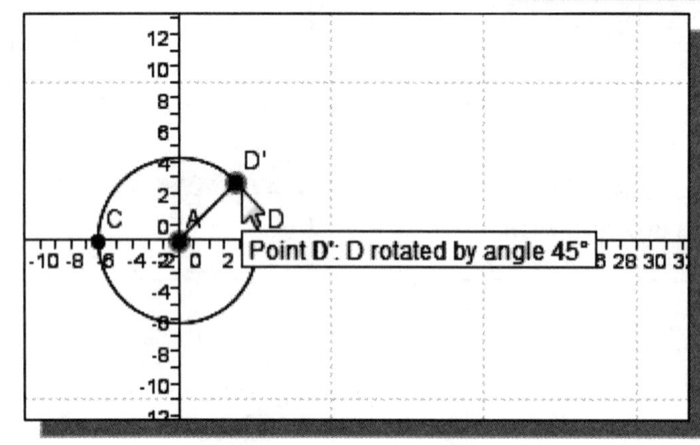

33. Click on **Point D'** to create a line as shown.

34. On your own, create two additional line segments, **line DE** and **line EB** as shown.

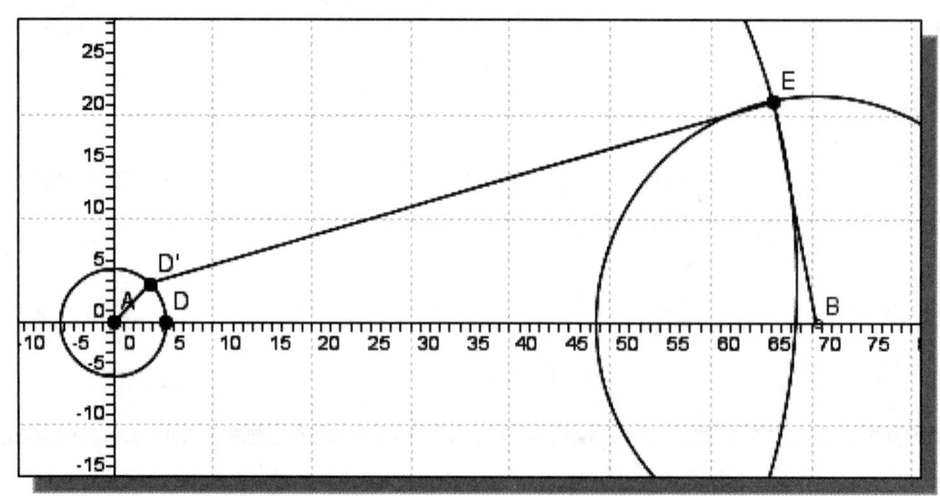

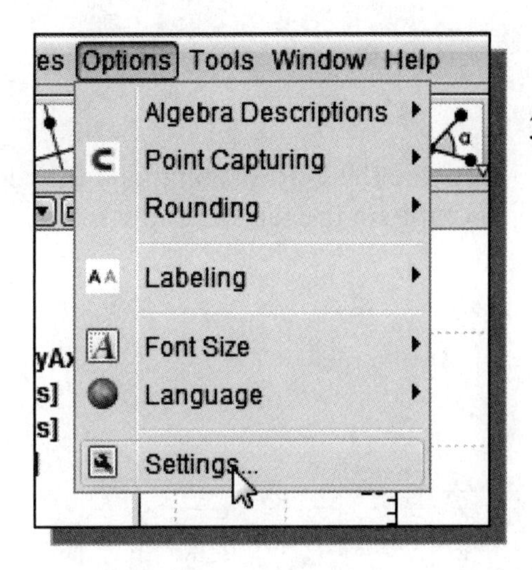

35. In the pull-down menu, select **Options → Settings** to access the *Settings* dialog box.

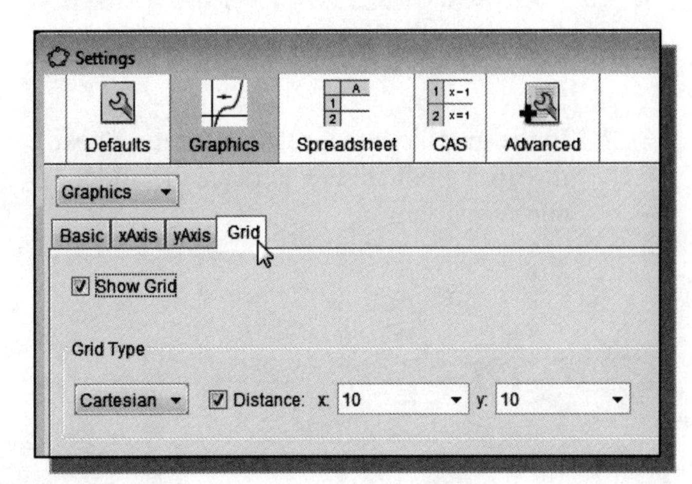

36. Click on the **Graphics** icon and select the **Grid** tab.

37. On your own, adjust the settings of the **Grid** display as shown.

38. Click **Save Settings** to store the adjusted settings.

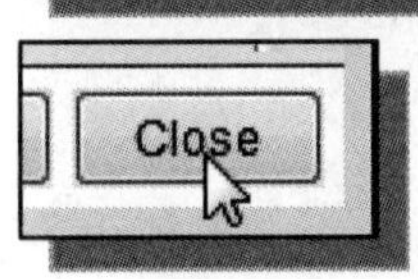

39. Click **Close** to exit the ***Settings*** dialog box.

40. On your own, hide the extra geometry so that the display is as shown in the figure.

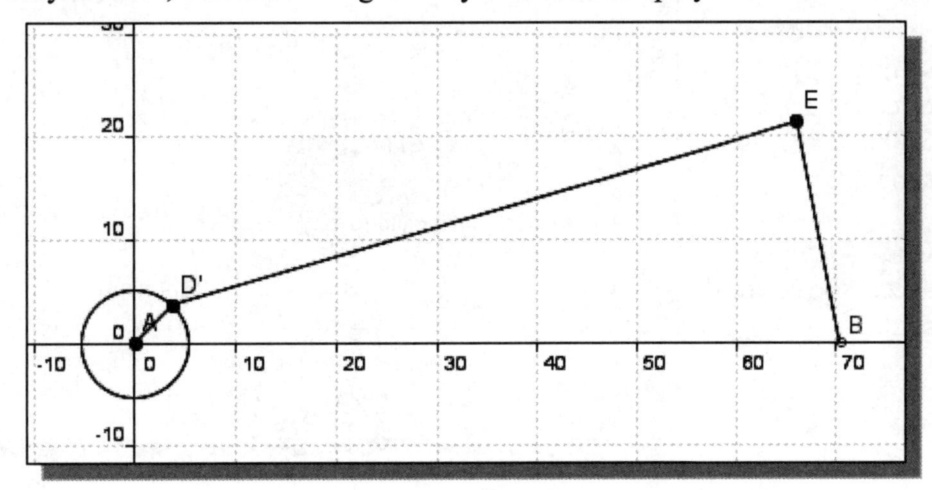

Adding a Slider Control

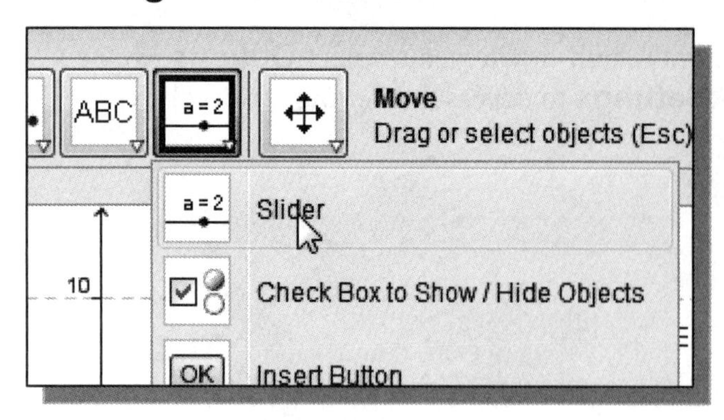

1. Activate the **Slider** command by clicking on the icon as shown.

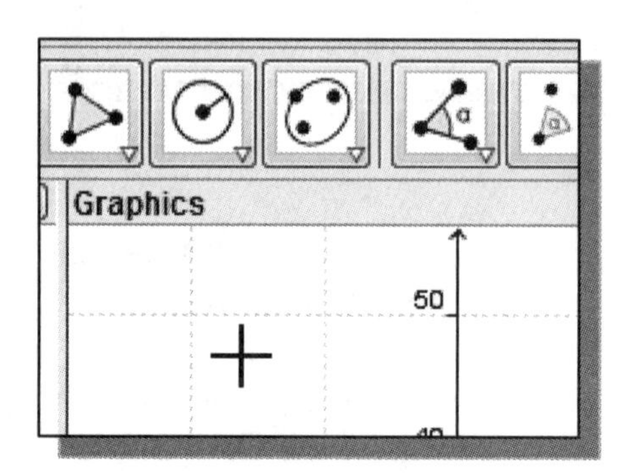

2. In the graphics area, select a location near the **upper left corner** to place the slider control as shown.

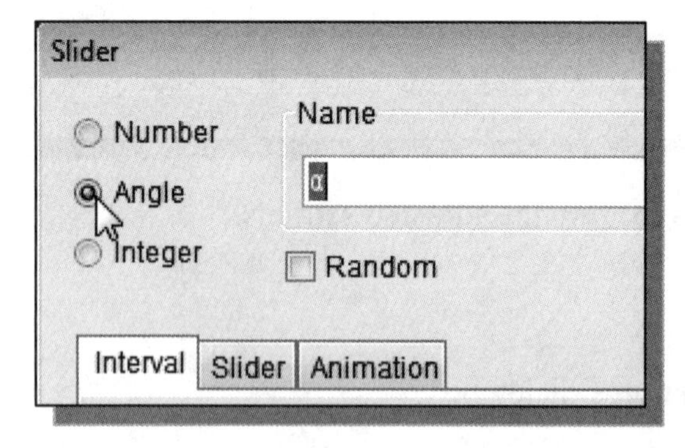

3. Set the slider option to **Angle** as shown. Note the angle name is automatically set to **α** when the angle option is selected.

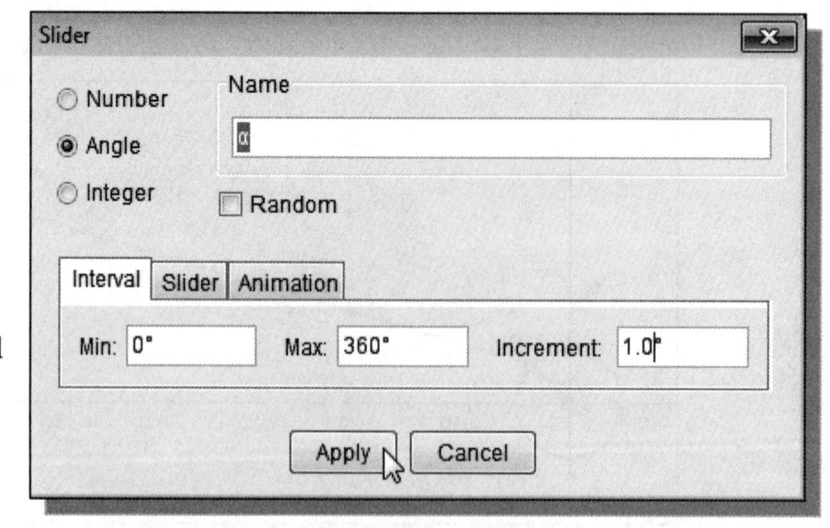

4. Set the **Interval** settings to the three values, **0**, **360** and **1.0**, as shown.

5. Click **Apply** to accept the selected settings.

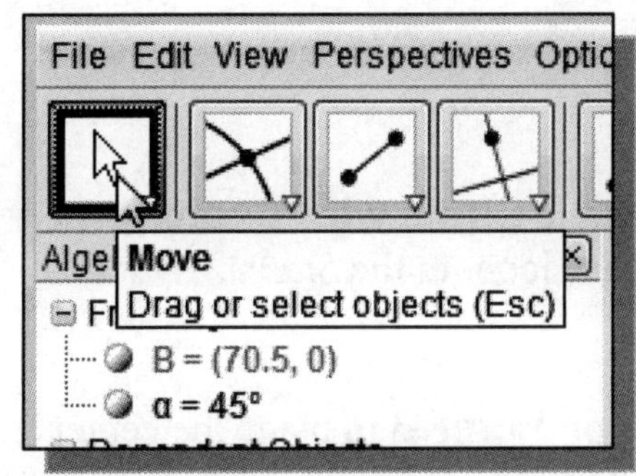

6. In the *Standard* toolbar area, click on the **Move** icon to activate the command.

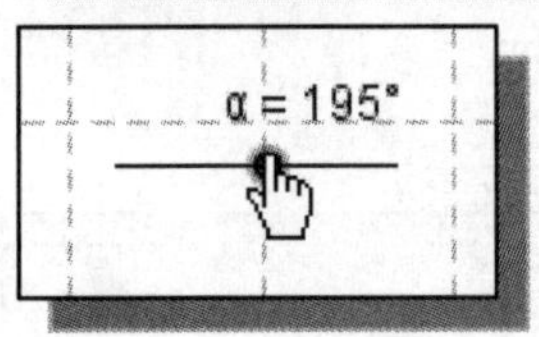

7. Drag the **handle** of the slider control, and notice **Angle α** is adjusted.

8. In the *Algebra* area, double click with the left-mouse-button on **Point D'** to enter the **Object Redefine** mode.

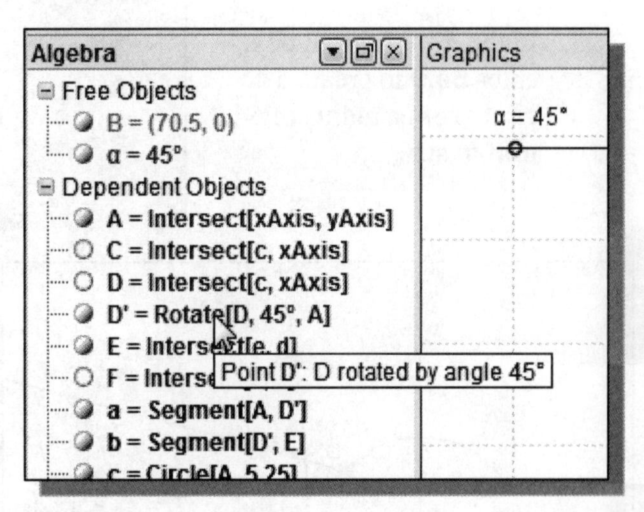

9. Modify the **angle variable**, the second variable in the edit box, to **α** as shown.

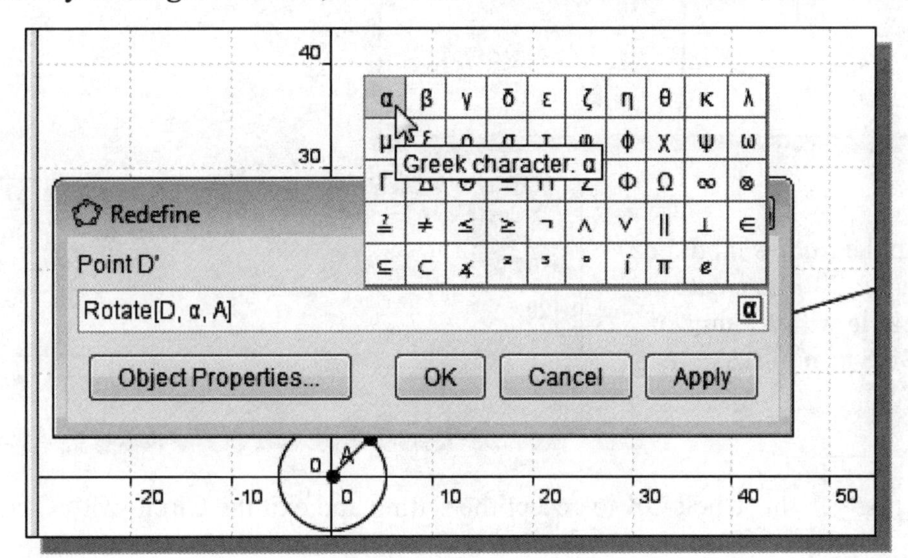

10. Click **OK** to accept the setting and exit the **Redefine** command.

Create the Second Four-bar Mechanism

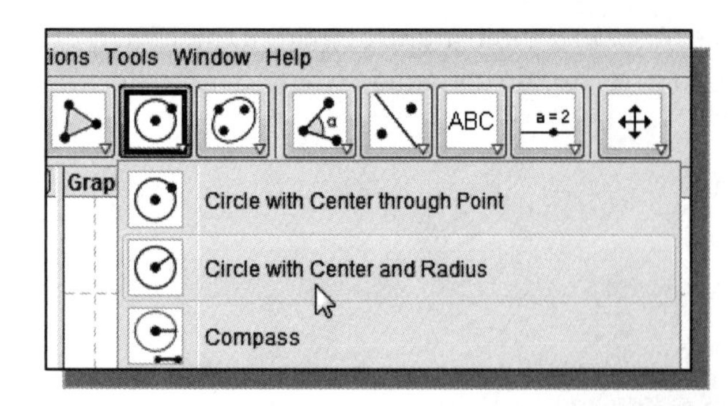

1. Select the **Circle with Center and Radius** icon, in the *Standard* toolbar area.

2. Click on **Point B** to place the center of the circle.

3. In the *Radius* input box, enter **64.9** to create a circle with a radius of **64.9 mm** in size.

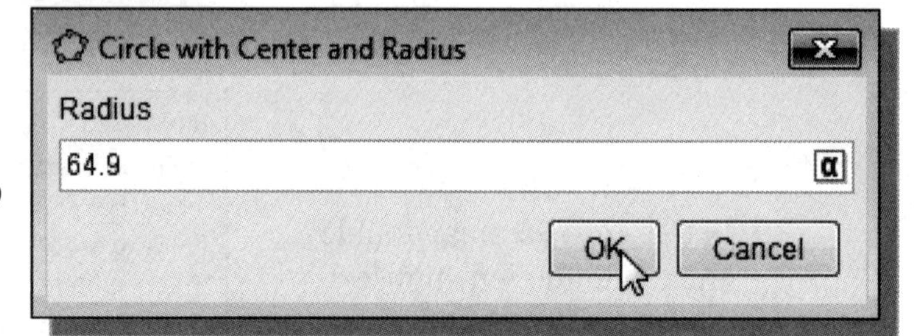

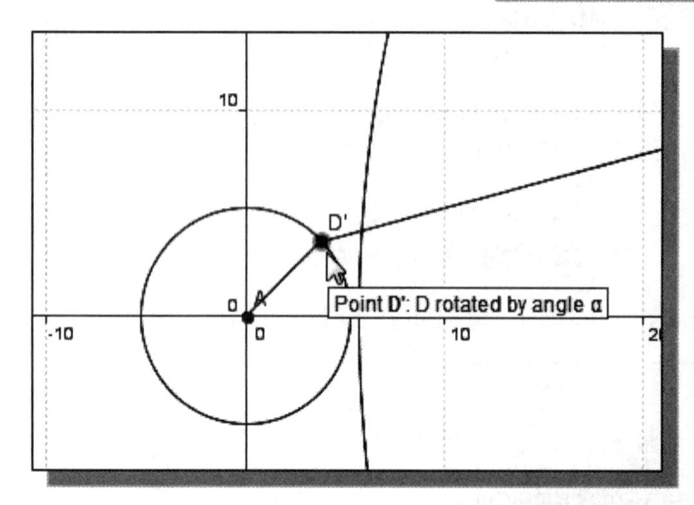

4. Next, click on **Point D'** to place the center of the circle.

5. In the *Radius* input box, enter **20.5** to create a circle with a radius of **20.5 mm** in size.

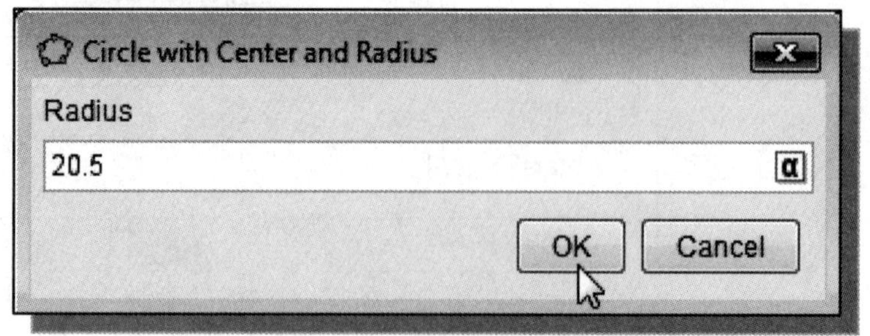

6. Click **OK** to accept the setting and exit the **Circle with Center and Radius** command.

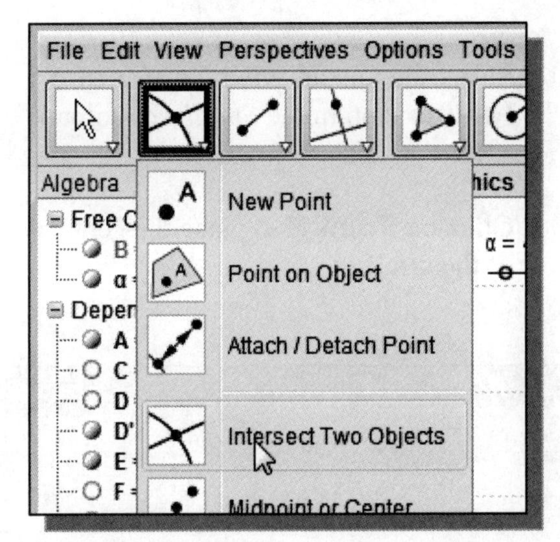

7. Activate the **Intersect Two Objects** command by clicking on the icon as shown.

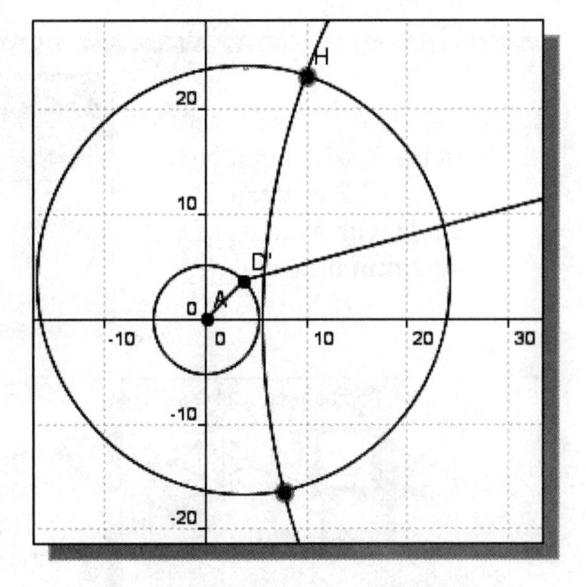

8. Select the two circles we just created to define the intersection points.

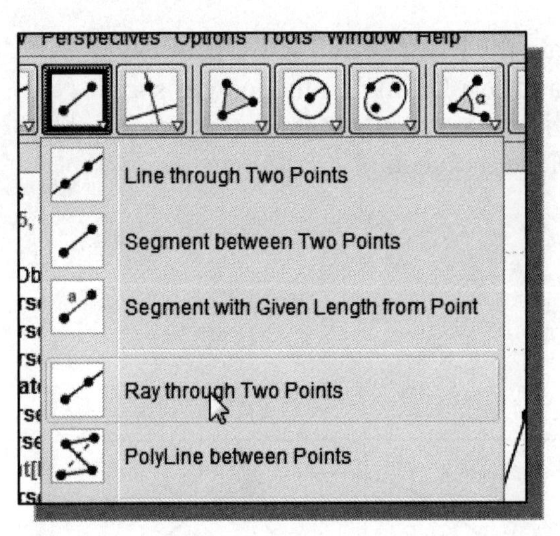

9. Activate the **Ray through Two Points** command by clicking on the icon as shown. This command will create a ray of line through two selected points.

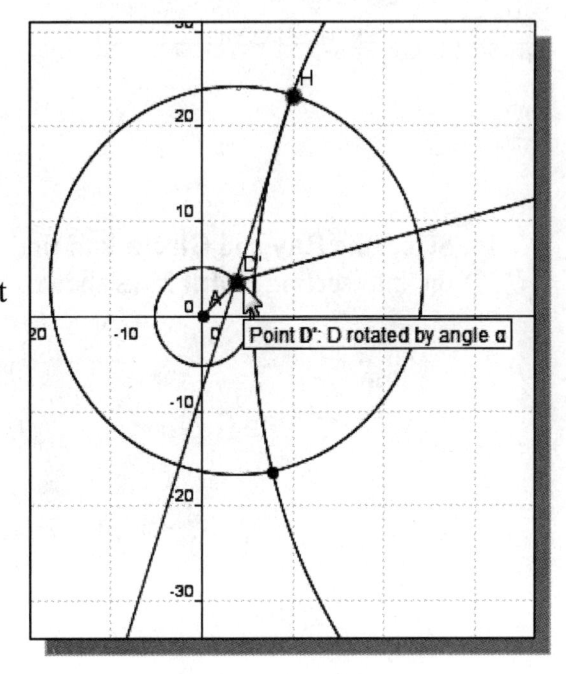

10. Create a ray by selecting **Point H** first and then pick **Point D'** as shown.

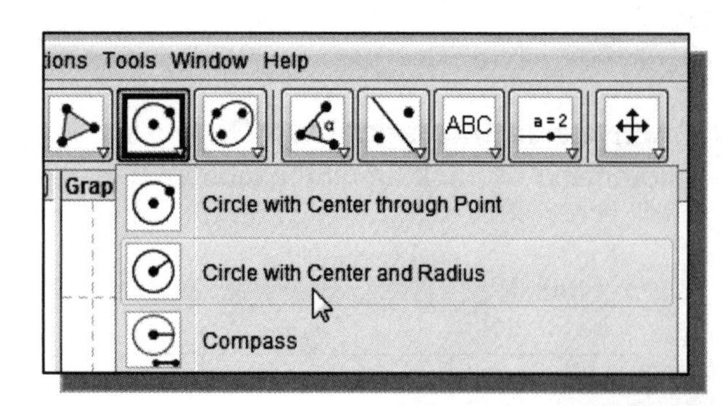

11. Select the **Circle with Center and Radius** icon in the *Standard* toolbar area.

12. Click on **Point H** to place the center of the circle.

13. In the *Radius* input box, enter **42.2** to create a circle with a radius of **42.2 mm** in size.

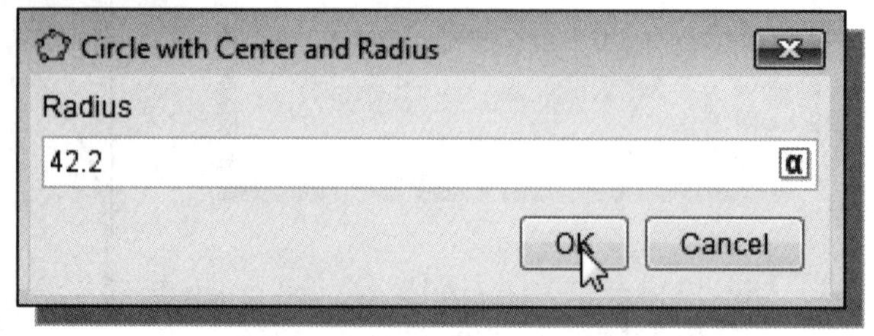

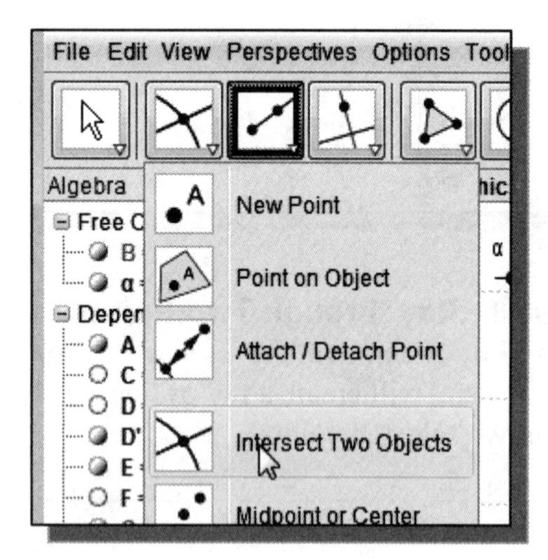

14. Activate the **Intersect Two Objects** command by clicking on the icon as shown. This will create points at the intersections of two selected objects.

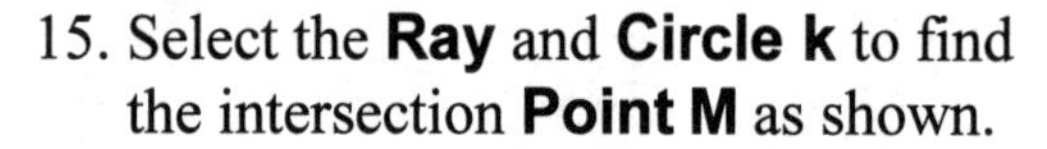

15. Select the **Ray** and **Circle k** to find the intersection **Point M** as shown.

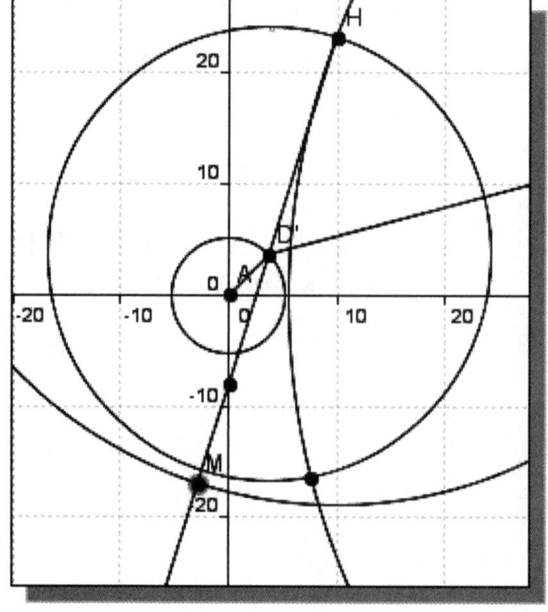

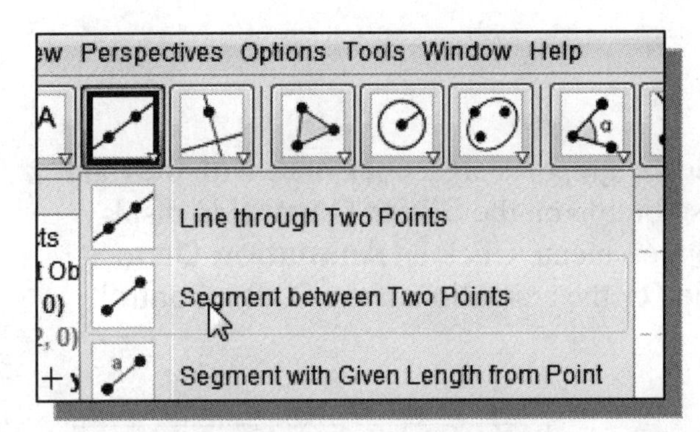

16. Activate the **Segment between Two Points** command by clicking on the icon as shown.

17. On your own, create **line KH** and **line BH** as shown.

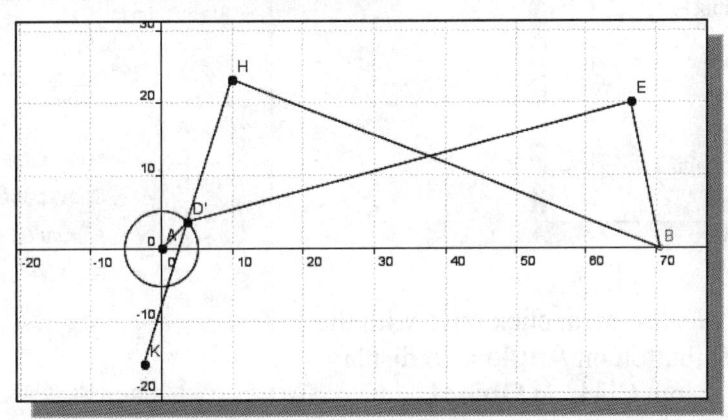

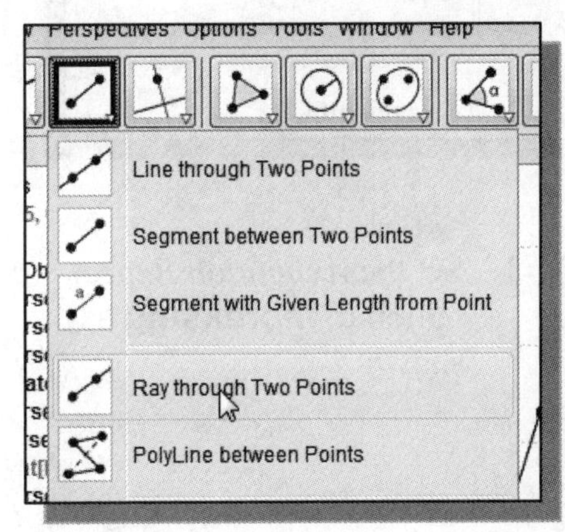

18. On your own, repeat the above steps and create a **Ray** through **Point E** and **Point B**.

19. Also create a circle with the center point aligned to **Point E** and a radius of **42.2 mm**.

20. Locate the intersection between the *ray* and the *circle*.

21. On your own, complete the geometry construction as shown.

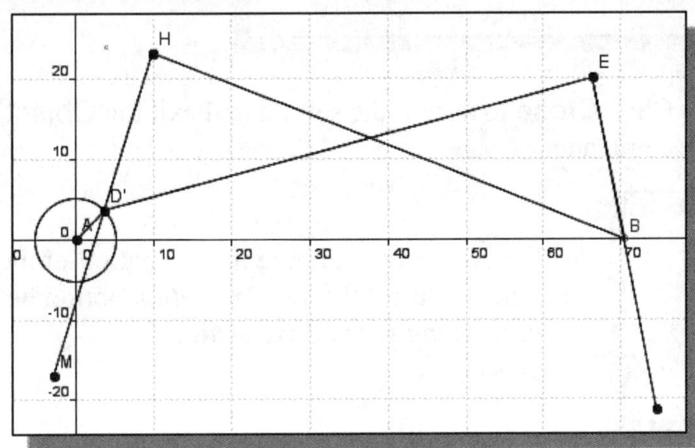

Using the Animate Option

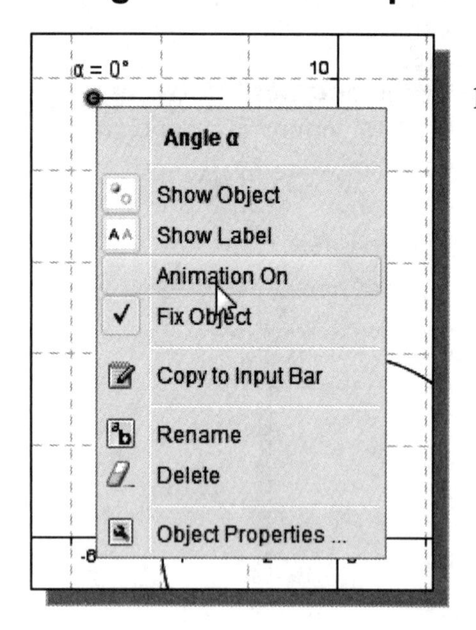

1. Inside the graphics area, click once with the right-mouse-button on the **Slider Control** to display the option menu. Click on **Animation On** to toggle *ON* the animation of the Slider Control.

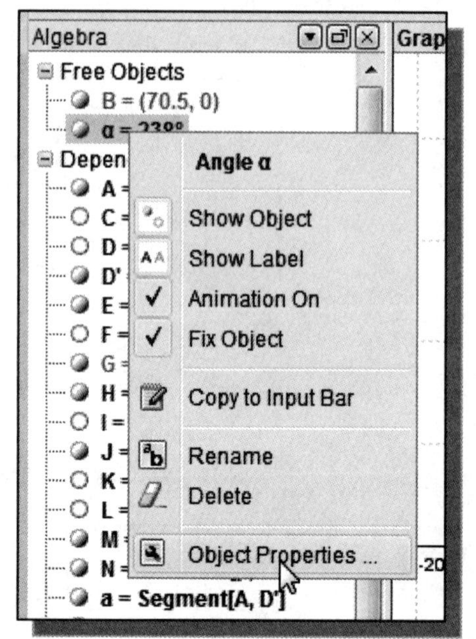

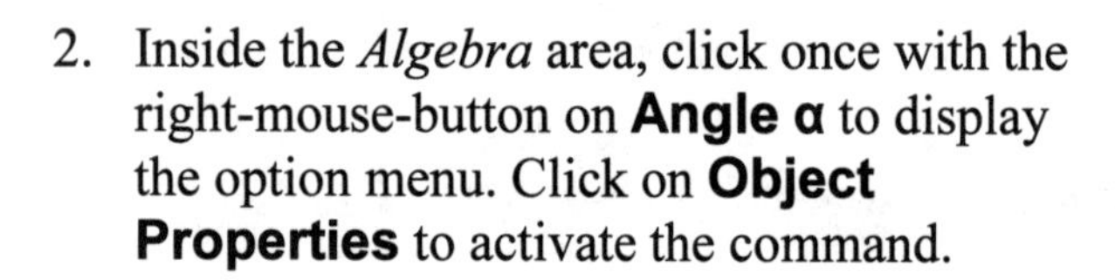

2. Inside the *Algebra* area, click once with the right-mouse-button on **Angle α** to display the option menu. Click on **Object Properties** to activate the command.

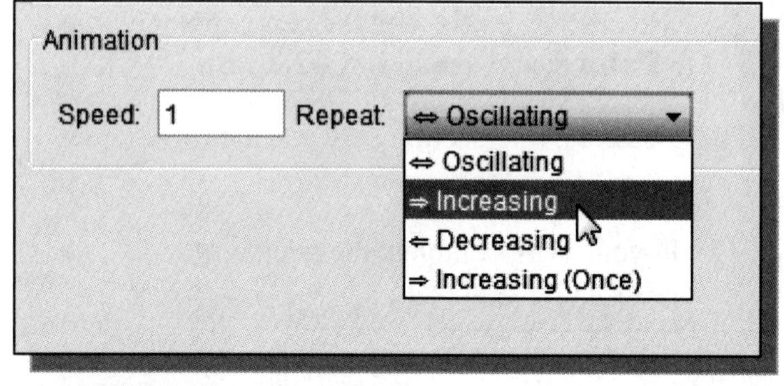

3. Set the *Animation Repeat* option to **Increasing** as shown.

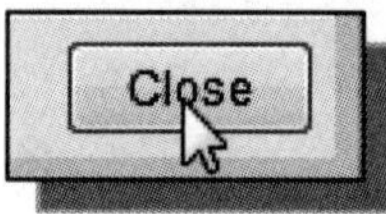

4. Click **Close** to accept the setting and exit the Object Properties command.

5. On your own, examine the animation of the four-bar linkage; turn *OFF* the *Animation* option before proceeding to the next section.

Tracking the Paths of the Feet

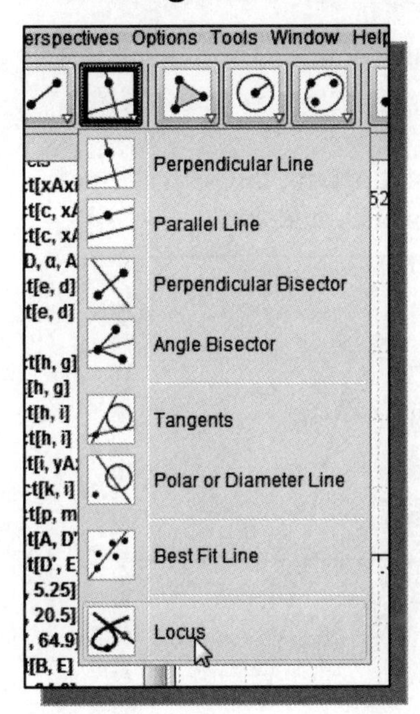

1. Select the **Locus** command in the *Standard* toolbar area.

2. Select the point representing the front foot, **Point M**, as shown.

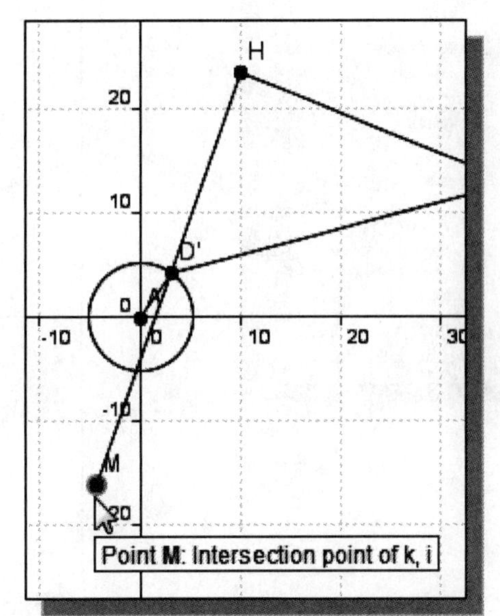

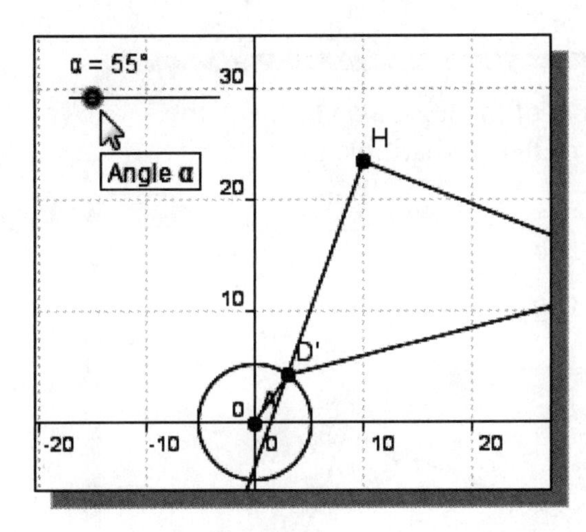

3. Next select the **Slider** as the increment source.

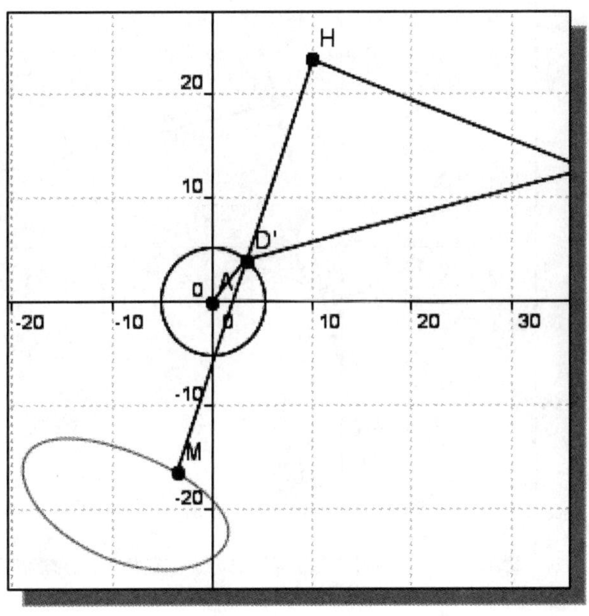

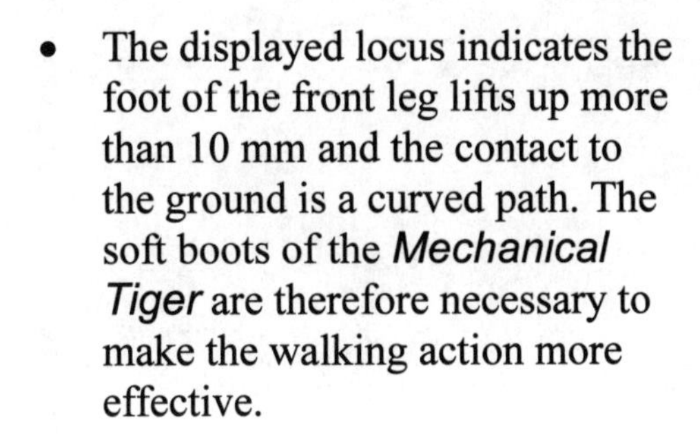

- The displayed locus indicates the foot of the front leg lifts up more than 10 mm and the contact to the ground is a curved path. The soft boots of the *Mechanical Tiger* are therefore necessary to make the walking action more effective.

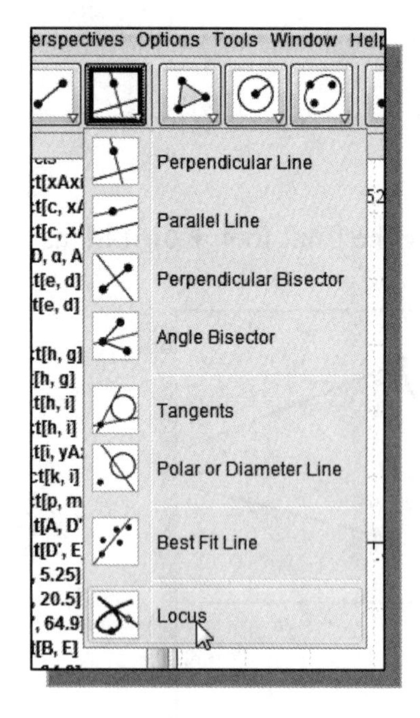

4. On your own, use the **Locus** command and display the path of the rear foot.

   - Note the rear foot does not lift up; the path is an arc. The rear leg only swings back and forth.

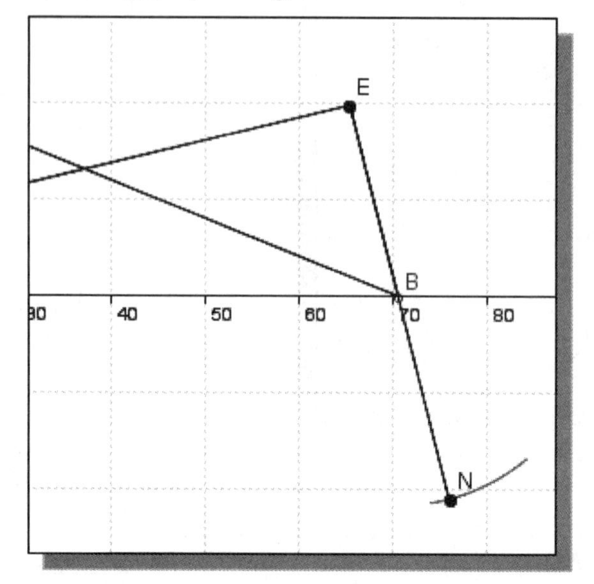

5. On your own, show the loci of the top joints of the legs; also turn *ON* the *Animation* option to observe the behavior of the six-bar linkage.

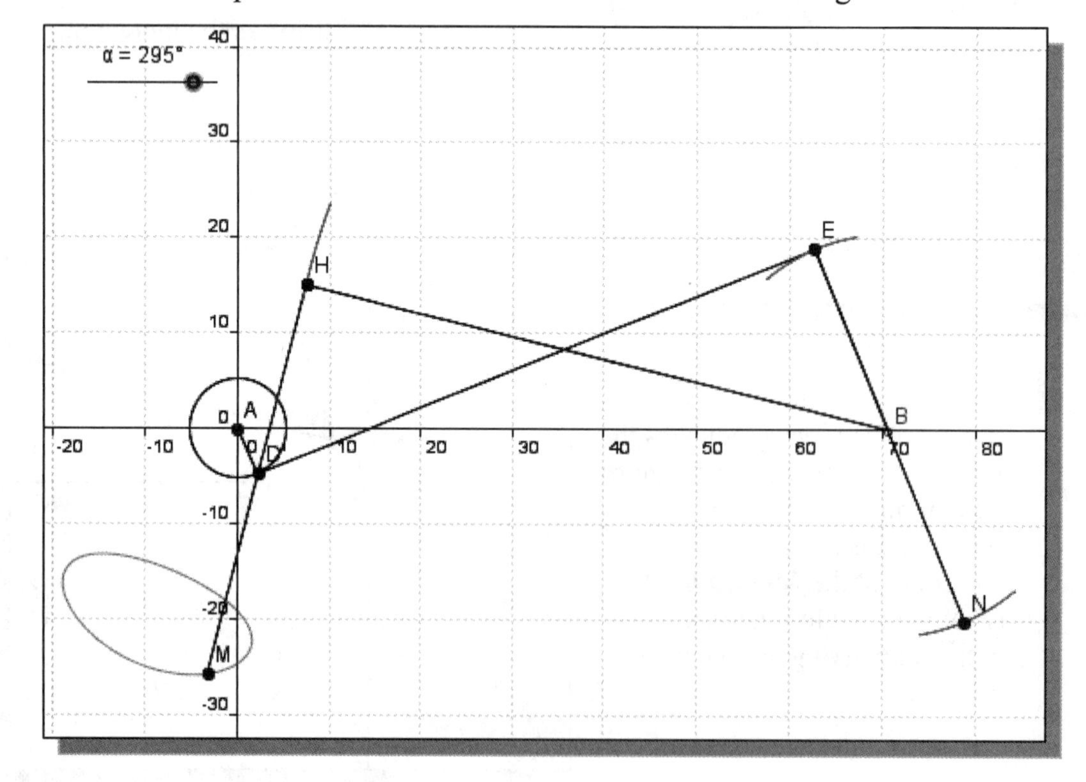

Adjusting the Crank Length

- Mounting the front leg to the different holes on the crank will result in a different stride length and lift distance.

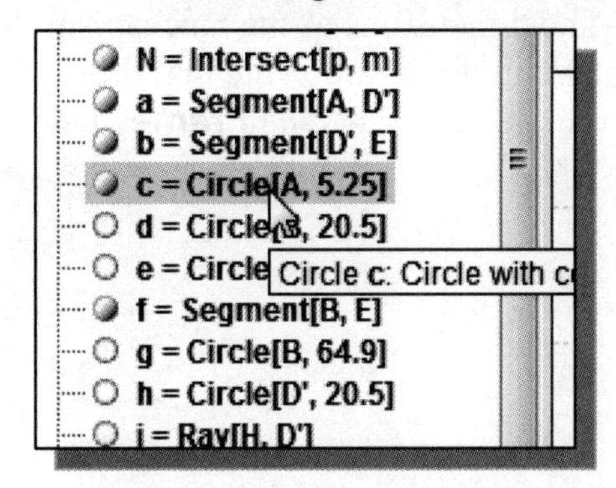

1. In the *Algebra* area, double click with the left-mouse-button on **Circle c** to enter the **Object Redefine** mode.

2. Enter **8.25** as the new radius value.

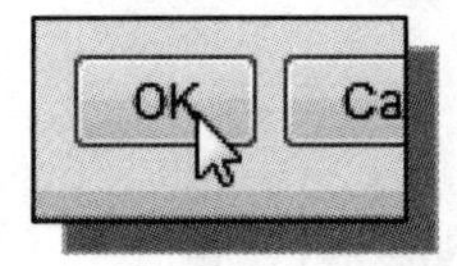

3. Click **OK** to accept the setting and exit the Redefine command.

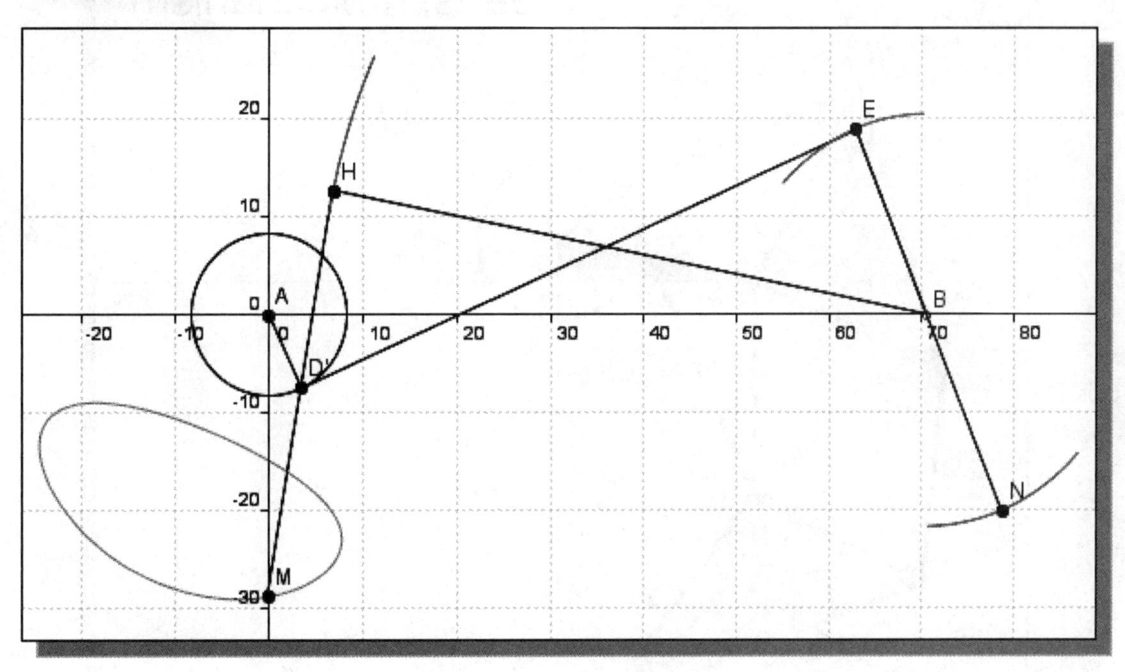

- Note the locus of the front foot has been updated; both the height and length have almost doubled in size.

The Jansen Mechanism

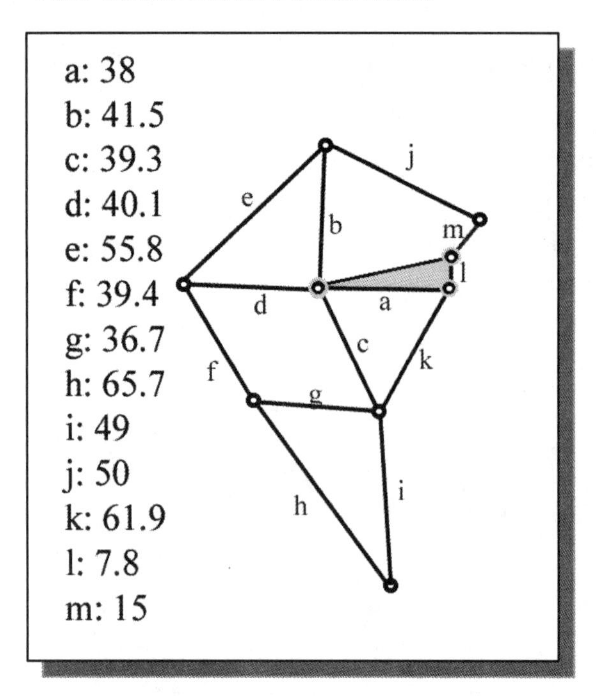

The ***Jansen Mechanism*** is an eight-bar linkage. The dimensions of the mechanism are as shown in the figure. Note that the shaded triangle represents the **Frame**, and **m** is the **Crank**.

- The locus of the foot shows the design grants very straight contact path to the ground and also has good lift height.

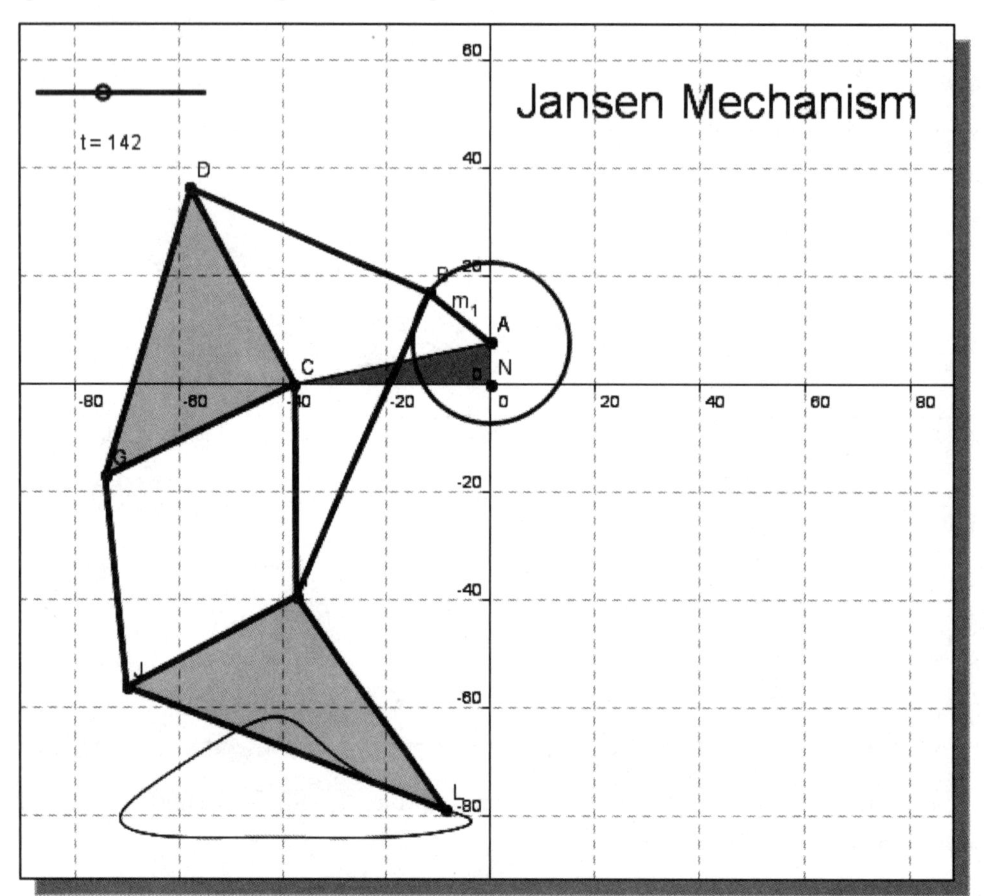

The Klann Mechanism

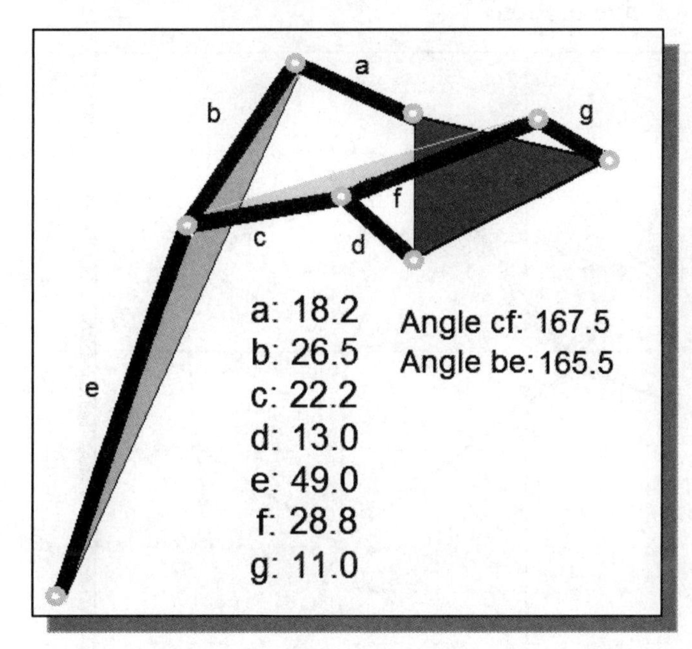

The ***Klann Mechanism*** is a six-bar linkage; the dimensions of the mechanism are as shown in the figure. Note that the dark shaded triangle represents the **Frame**, the light shaded triangles indicate the two rigid links, and **g** is the **Crank**.

- The locus of the foot shows the design also grants relatively straight contact path to the ground and good height of the foot lift.

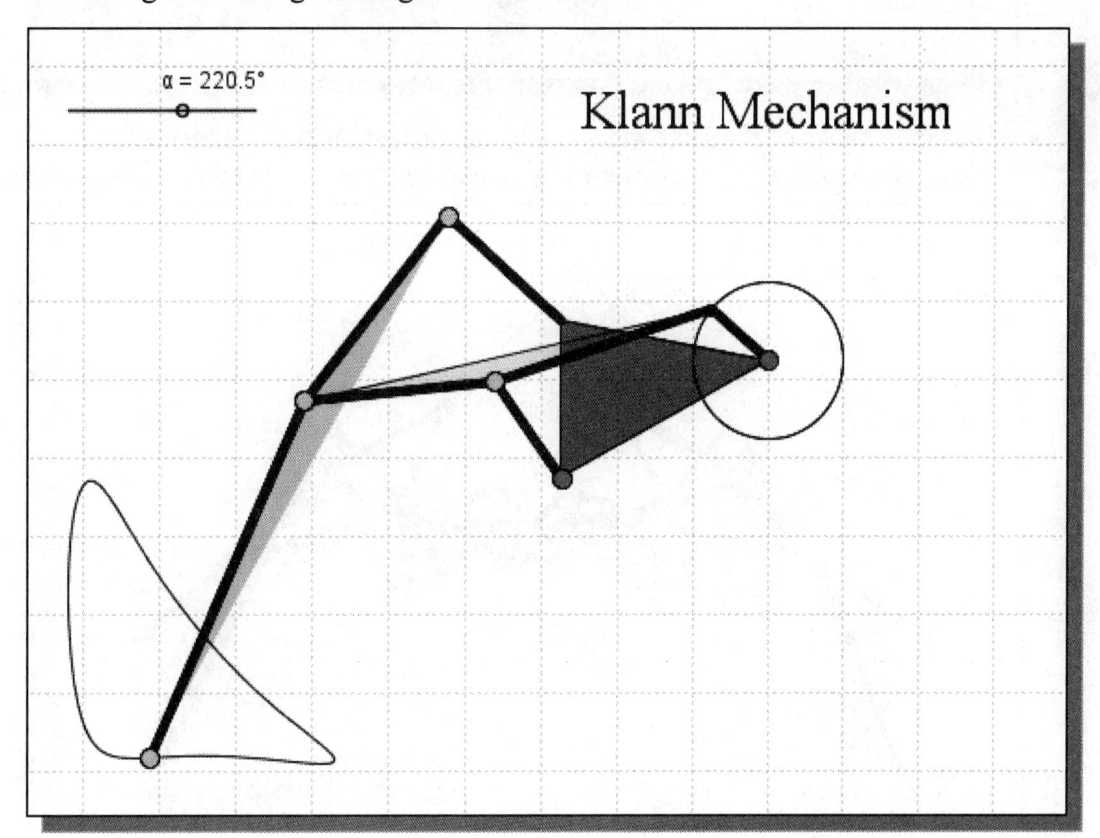

- Adding a second leg on the other side of the two mechanisms will show the configuration provides good walking motion.

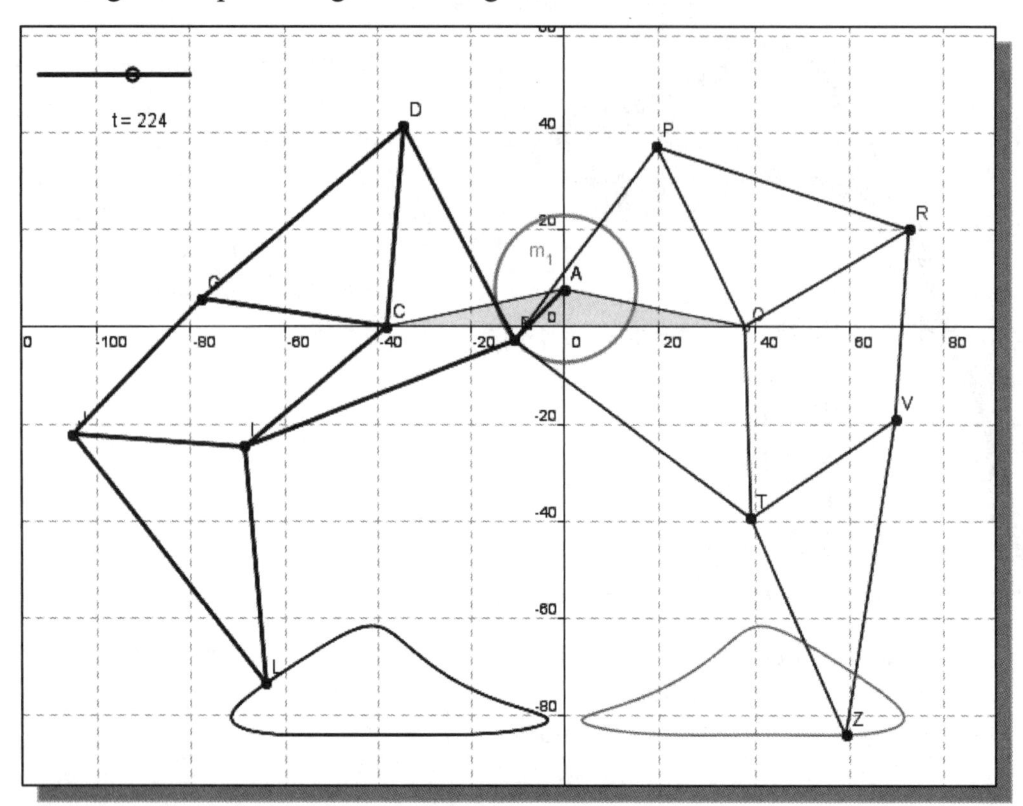

- Multiple copies of the same mechanism can provide better sychroization of the leg motions, also the center of gravity will always be at the center of the design.

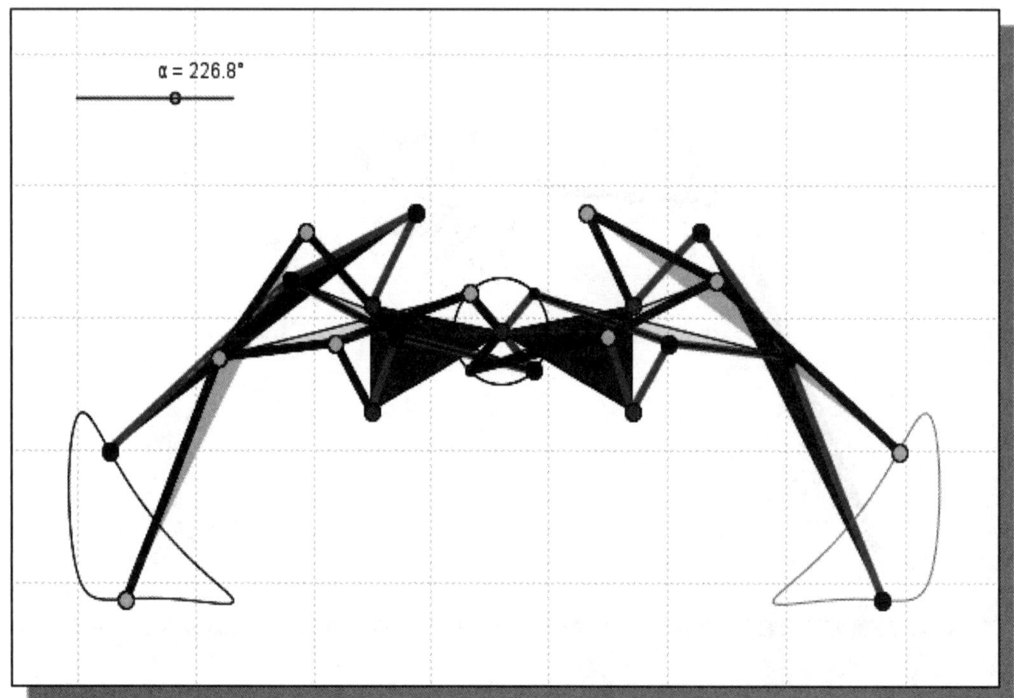

Exercises: (Design the following Mechanisms with *GeoGebra*)

1. **Peaucellier Straight Line Mechanism**

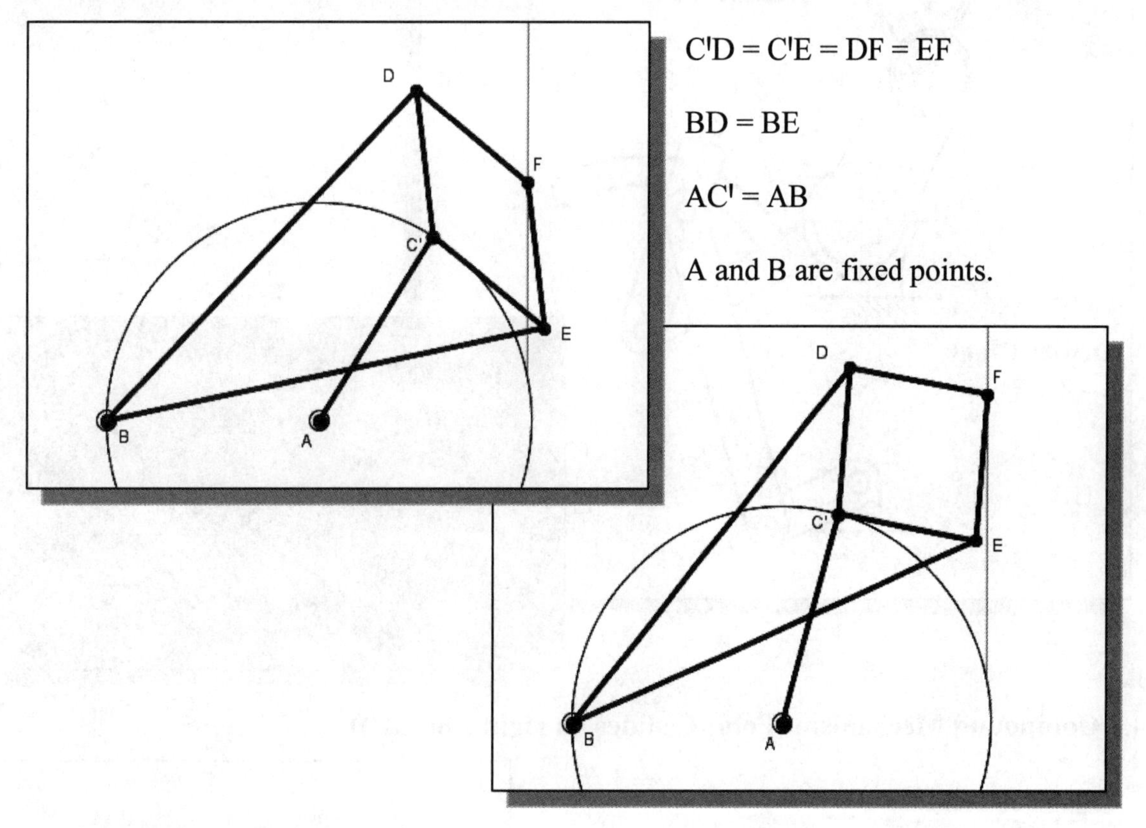

C'D = C'E = DF = EF

BD = BE

AC' = AB

A and B are fixed points.

2. **Compound Mechanism**

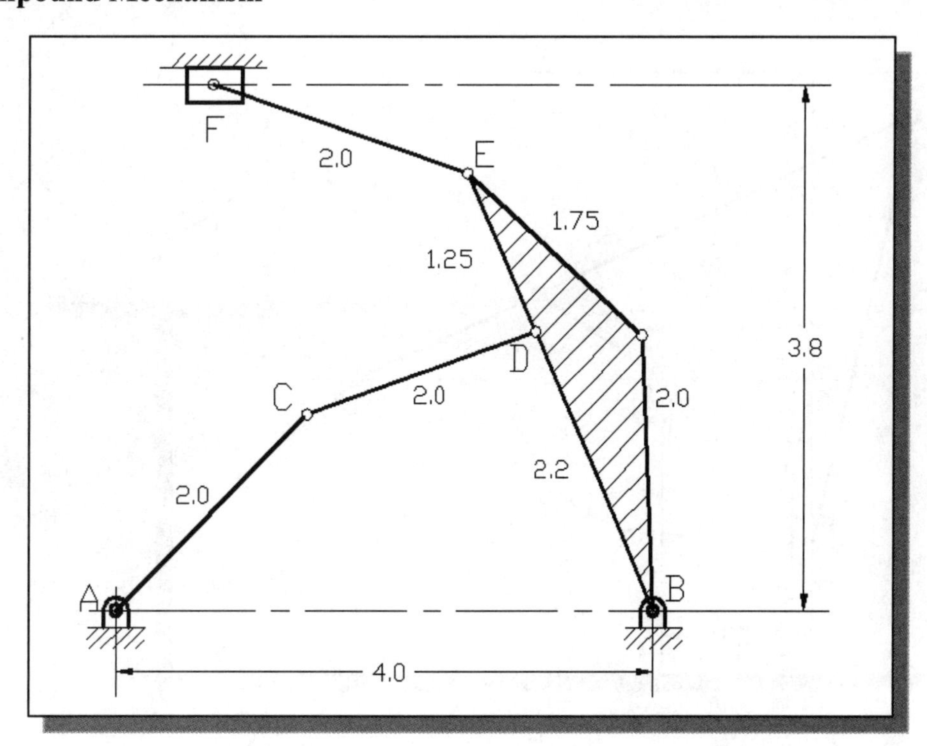

3. **Oscillating Sprinkler Mechanism**

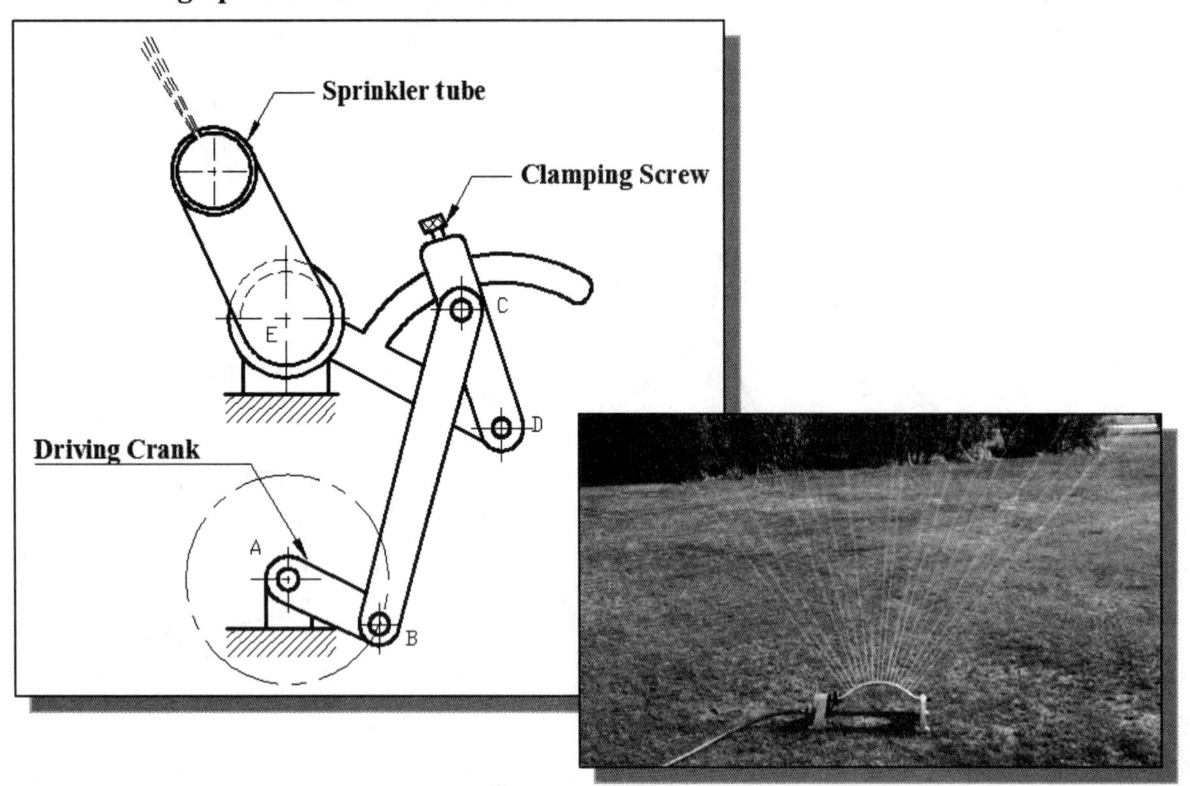

4. **Compound Mechanism (Point C slides on rigid Link BD)**

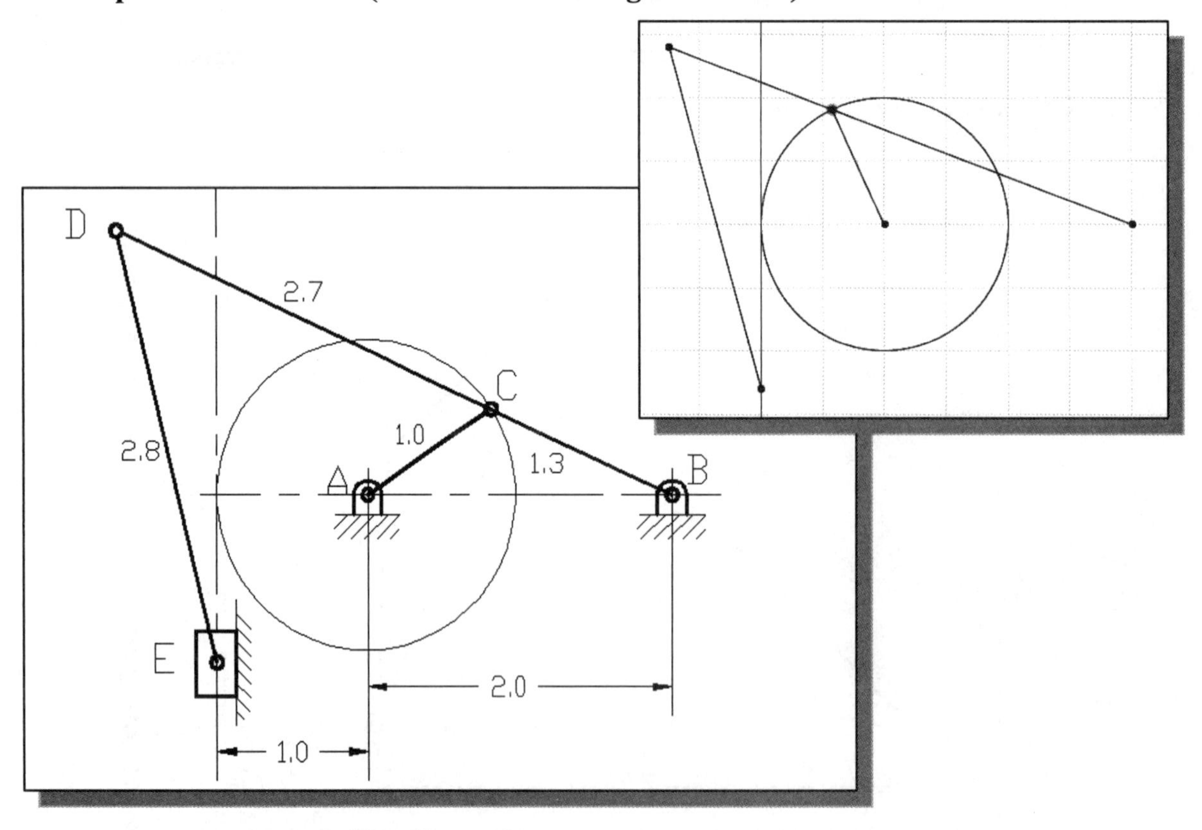

Chapter 12
Assembly Modeling and Motion Analysis

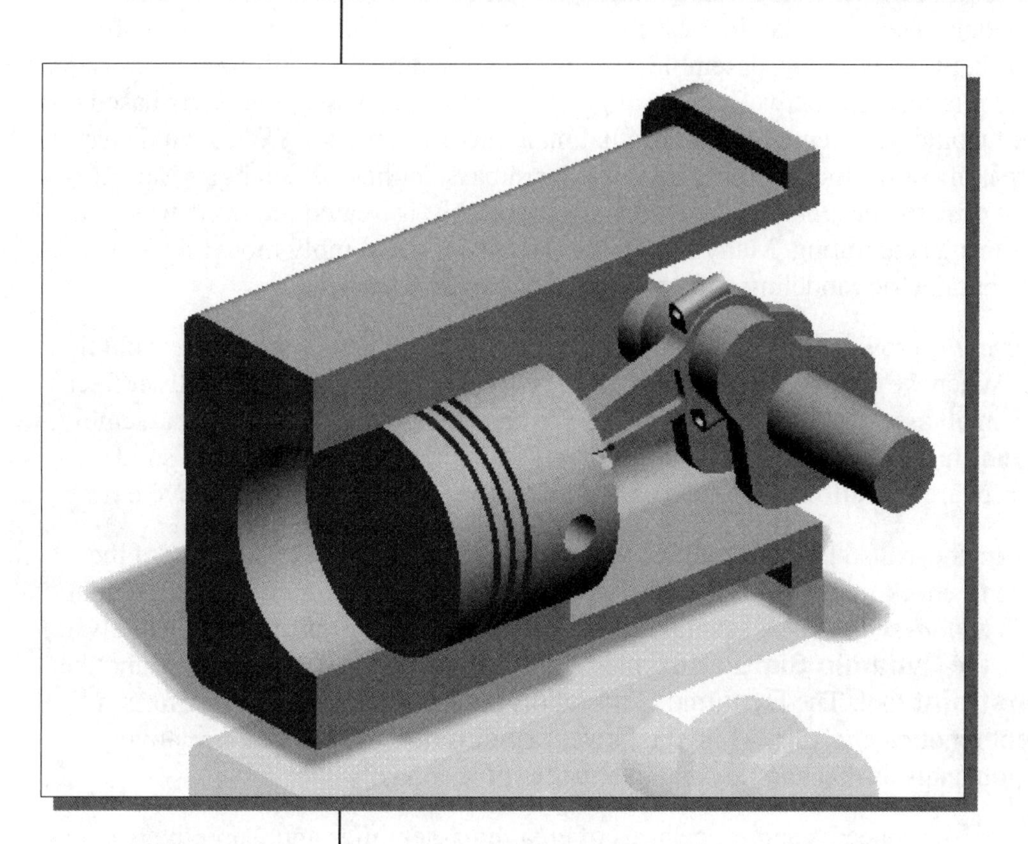

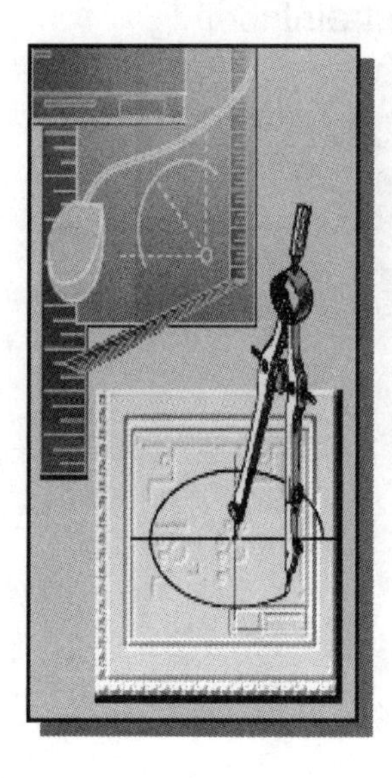

Learning Objectives

- ♦ Understand the Assembly Modeling Methodology
- ♦ Understand and Utilize Assembly Constraints
- ♦ Understand the Autodesk Inventor DOF Display
- ♦ Utilize the Autodesk Inventor Drive Constraint Option
- ♦ Record Animation Movies

Introduction

In the previous lessons, we have gone over the fundamentals of creating basic parts and drawings. In this lesson, we will examine the assembly modeling functionality of *Autodesk Inventor*. We will start with a demonstration on how to create and modify assembly models. The main task in creating an assembly is establishing the assembly relationships between parts. To assemble parts into an assembly, we will need to consider the assembly relationships between parts. It is a good practice to assemble parts based on the way they would be assembled in the actual manufacturing process. We should also consider breaking down the assembly into smaller subassemblies, which helps the management of parts. In *Autodesk Inventor*, a subassembly is treated the same way as a single part during assembling. Many parallels exist between assembly modeling and part modeling in parametric modeling software such as *Autodesk Inventor*.

Autodesk Inventor provides full associative functionality in all design modules, including assemblies. When we change a part model, *Autodesk Inventor* will automatically reflect the changes in all assemblies that use the part. We can also modify a part in an assembly. **Bi-directional full associative functionality** is the main feature of parametric solid modeling software that allows us to increase productivity by reducing design cycle time.

Motion analysis can also be performed to visually confirm the proper assembly of the designs, also to check for any interference between mating parts and any other potential problems. In *Autodesk Inventor*, several options are available to perform motion analysis, for example, the **Dynamic Simulation** module, the **Inventor Studio** module and the **Drive Constraint** tool. The Dynamic Simulation module can be used to perform a fairly in-depth motion analysis, while the Drive Constraint tool provides a relatively simple motion analysis that can be done in a matter of seconds.

In this chapter, the concepts and procedures of creating assemblies and using the *Content Center* for standard parts are illustrated. The procedure for basic motion analysis of the *Mechanical Tiger* assembly is also illustrated using the Drive Constraint tool.

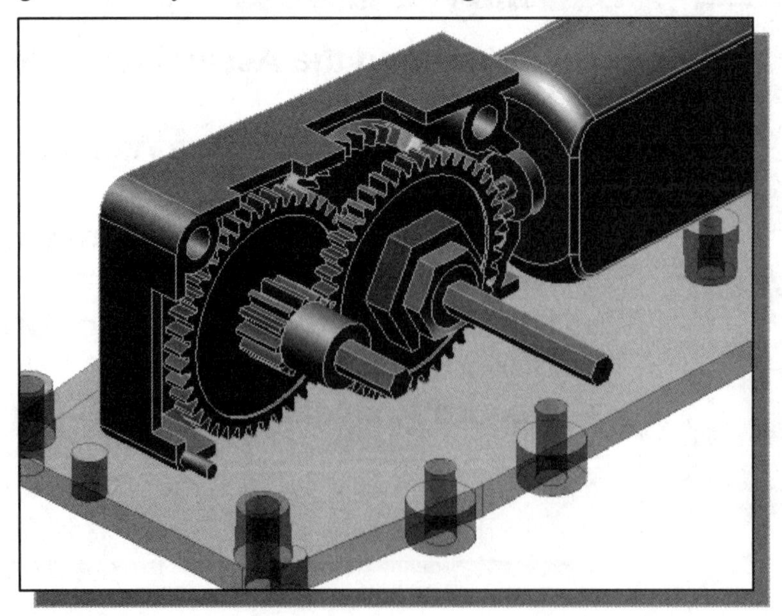

Assembly Modeling Methodology

The *Autodesk Inventor Assembly Modeler* provides tools and functions that allow us to create 3D parametric assembly models. An assembly model is a 3D model with any combination of multiple part models. *Parametric assembly constraints* can be used to control relationships between parts in an assembly model.

Autodesk Inventor can work with any of the assembly modeling methodologies:

The Bottom Up Approach

The first step in the *bottom up* assembly modeling approach is to create the individual parts. The parts are then pulled together into an assembly. This approach is typically used for smaller projects with very few team members.

The Top Down Approach

The first step in the *top down* assembly modeling approach is to create the assembly model of the project. Initially, individual parts are represented by names or symbols. The details of the individual parts are added as the project gets further along. This approach is typically used for larger projects or during the conceptual design stage. Members of the project team can then concentrate on the particular section of the project to which he/she is assigned.

The Middle Out Approach

The *middle out* assembly modeling approach is a mixture of the bottom-up and top-down methods. This type of assembly model is usually constructed with most of the parts already created, and additional parts are designed and created using the assembly for construction information. Some requirements are known and some standard components are used, but new designs must also be produced to meet specific objectives. This combined strategy is a very flexible approach for creating assembly models.

The different assembly modeling approaches described above can be used as guidelines to manage design projects. Keep in mind that we can start modeling our assembly using one approach and then switch to a different approach without any problems.

In this lesson, the *bottom up* assembly modeling approach is illustrated. All of the parts (components) required to form the assembly are created first. *Autodesk Inventor's* assembly modeling tools allow us to create complex assemblies by using components that are created in part files or are placed in assembly files. A component can be a subassembly or a single part, where features and parts can be modified at any time. The sketches and profiles used to build part features can be fully or partially constrained. Partially constrained features may be adaptive, which means the size or shape of the associated parts are adjusted in an assembly when the parts are constrained to other parts. The basic concept and procedure of using the adaptive assembly approach is demonstrated in the tutorial.

The *Mechanical Tiger* Assembly

Additional Parts

- Four additional parts are also required for the assembly: (1) ***Hex Shaft***, (2) ***Short Spacer***, (3) ***M1-Spacer*** and (4) ***A3-Spacer.*** On your own, create the four parts as shown below; save the models as separate part files as shown.

(1) ***Hex Shaft (Flat to Flat 2.5mm x 27mm, Hex-Shaft.ipt)***

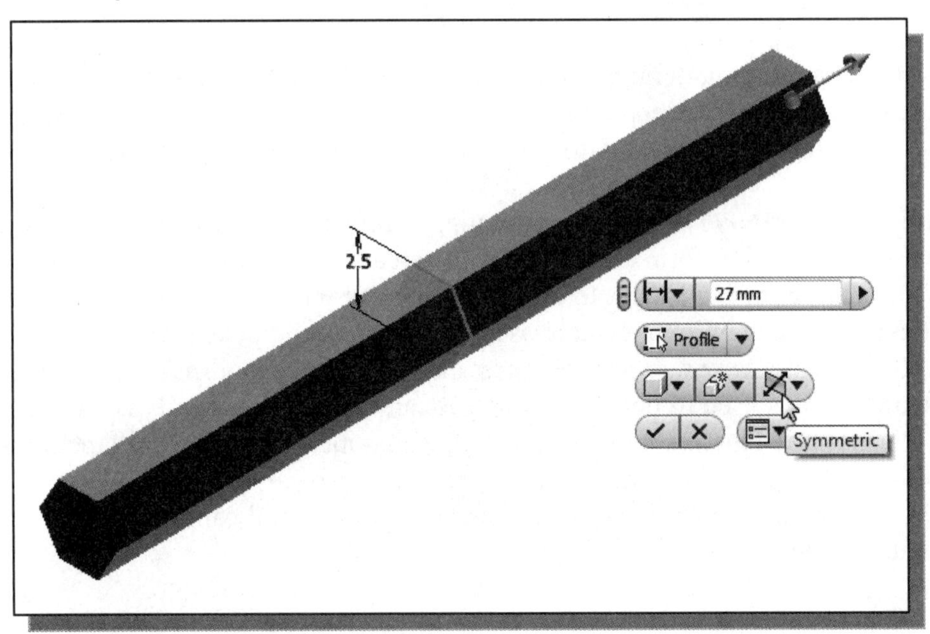

(2) ***Short Spacer (5.5mm x 3mm x 3.2mm, Spacer-Short.ipt)***

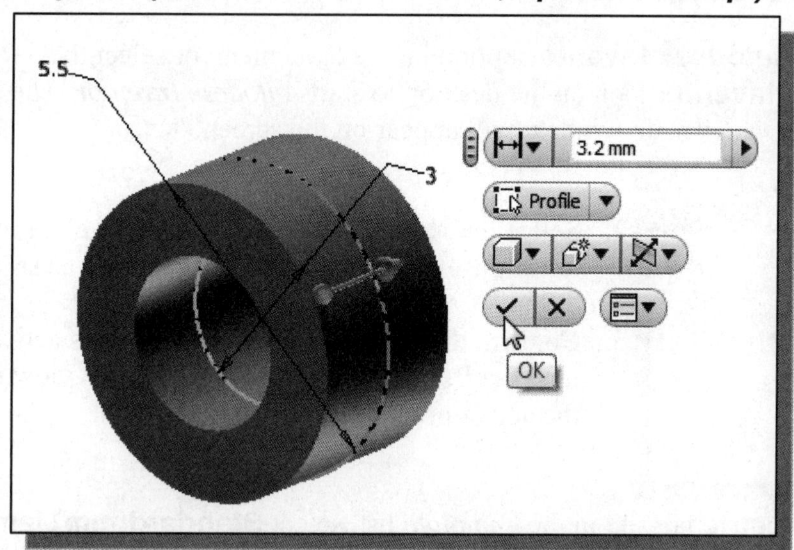

(3) ***M1-Spacer*** (***6mm x 3mm x 5mm, M1-Spacer.ipt***)

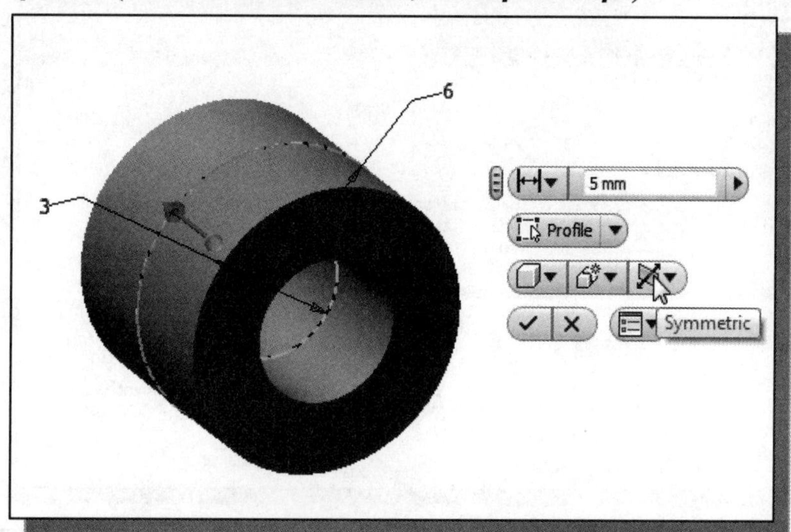

(4) ***A3-Spacer (6mm x 3mm x 7.5 mm, A3-Spacer.ipt)***

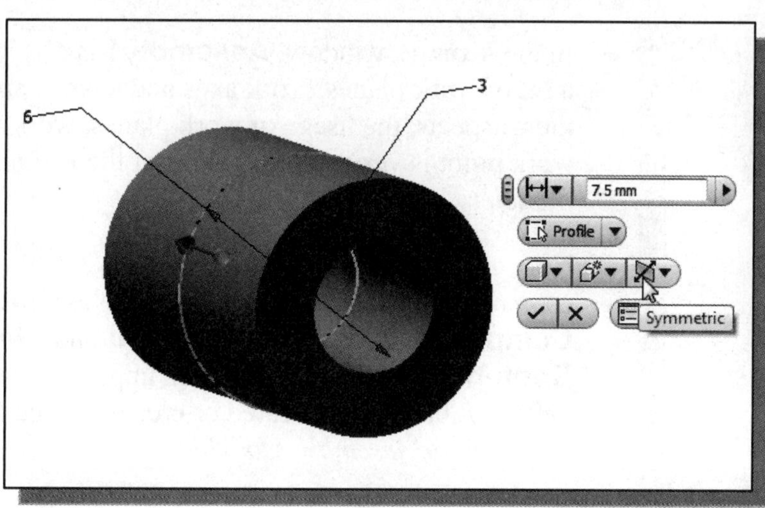

Starting *Autodesk Inventor*

1. Select the **Autodesk Inventor** option on the *Start* menu or select the **Autodesk Inventor** icon on the desktop to start *Autodesk Inventor*. The *Autodesk Inventor* main window will appear on the screen.

2. Select the **New File** icon with a single click of the left-mouse-button in the *Launch* toolbar as shown.

3. Confirm the ***Mechanical Tiger*** project is activated; note the **Projects** button is available to view/modify the active project.

4. Select the **Metric** tab and in the *Template* list, select **Standard(mm).iam** (*Standard Inventor Assembly Model* template file).

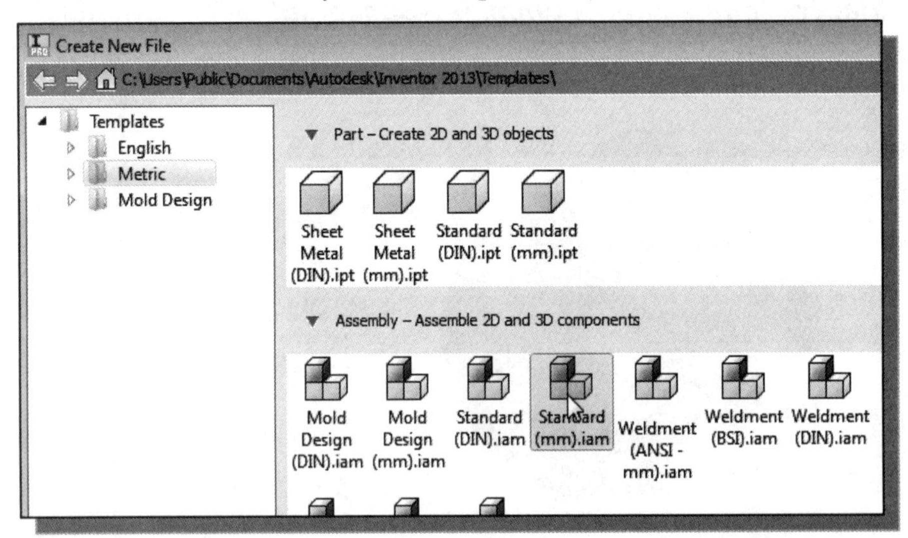

5. Click on the **Create** button in the *New File* dialog box to accept the selected settings.

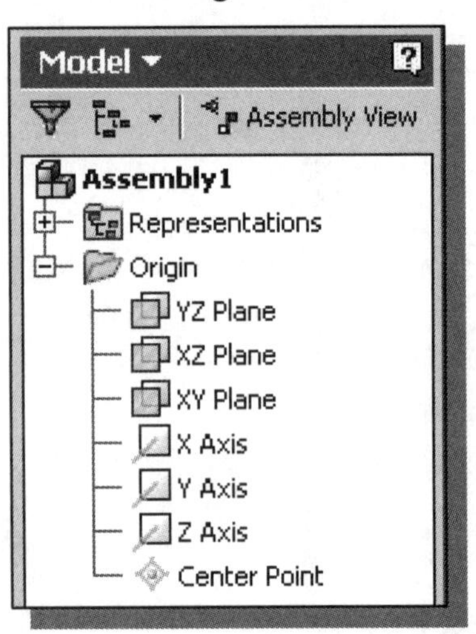

- In the *browser* window, **Assembly1** is displayed with a set of work planes, work axes and a work point. In most aspects, the usage of work planes, work axes and work point is very similar to that of the *Inventor Part Modeler*.

- Notice, in the *Ribbon* toolbar panels, several *component* options are available, such as **Place Component, Create Component** and **Place from Content Center**. As the names imply, we can use parts that have been created or create new parts within the *Inventor Assembly Modeler*.

Create the Leg Subassembly

- The first component placed in an assembly should be a fundamental part or subassembly. The first component in an assembly file sets the orientation of all subsequent parts and subassemblies. The origin of the first component is aligned to the origin of the assembly coordinates and the part is grounded (all degrees of freedom are removed). The rest of the assembly is built on the first component, the ***base component***. In most cases, this *base component* should be one that is **not likely to be removed** and **preferably a non-moving part** in the design. Note that there is no distinction in an assembly between components; the first component we place is usually considered as the *base component* because it is usually a fundamental component to which others are constrained. We can change the base component to a different base component by placing a new base component, specifying it as grounded, and then re-constraining any components placed earlier, including the first component. For our project, we will use the ***B3-Leg*** as the base component in the assembly.

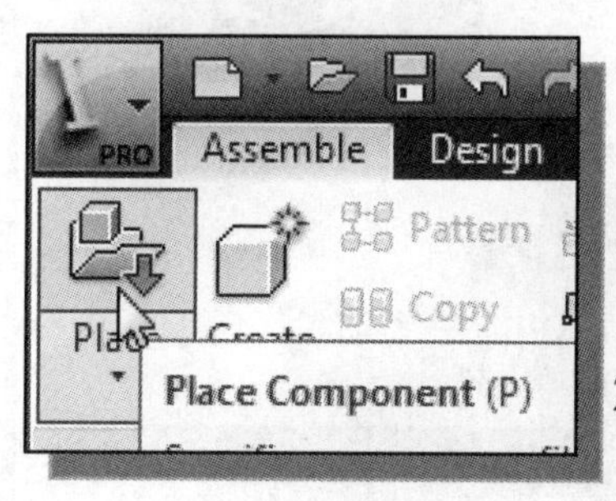

1. In the *Ribbon* toolbar panel, select the **Place Component** command by left-mouse-clicking the icon.

2. Select the ***B3-Leg*** (part file: ***B3-Leg.ipt***) in the list window.

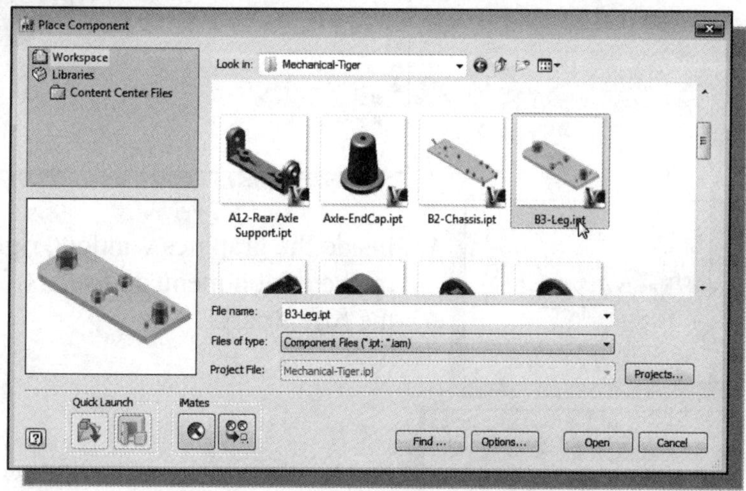

3. Click on the **Open** button to retrieve the model.

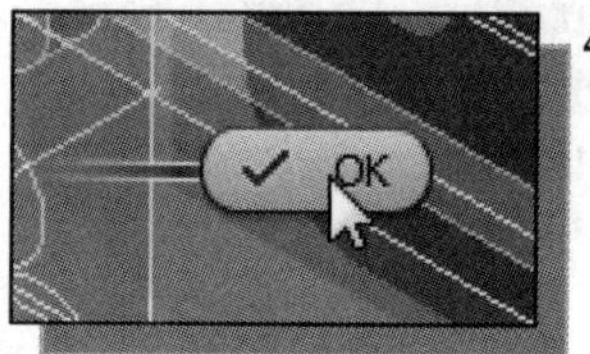

4. By default, the first component is automatically aligned to the origin of the assembly coordinates. We can also place multiple copies of the same component. Right-mouse-click once to bring up the option menu and select **OK** to end the placement of the *B3-Leg* part.

Placing the Second Component

➢ We will retrieve the *Knee* part as the second component of the assembly model.

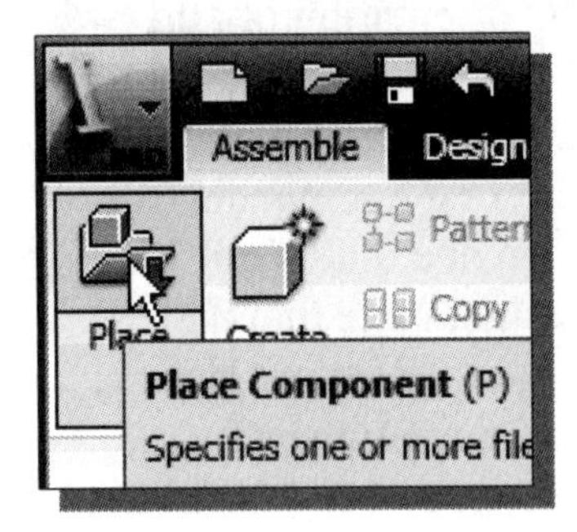

1. In the *Ribbon* toolbar panel, select the **Place Component** command by left-mouse-clicking the icon.

2. Select the ***Knee*** design (part file: ***A6-Knee.ipt***) in the list window. And click on the **Open** button to retrieve the model.

3. Place the *A6-Knee* part toward the upper right corner of the graphics window, as shown in the figure.

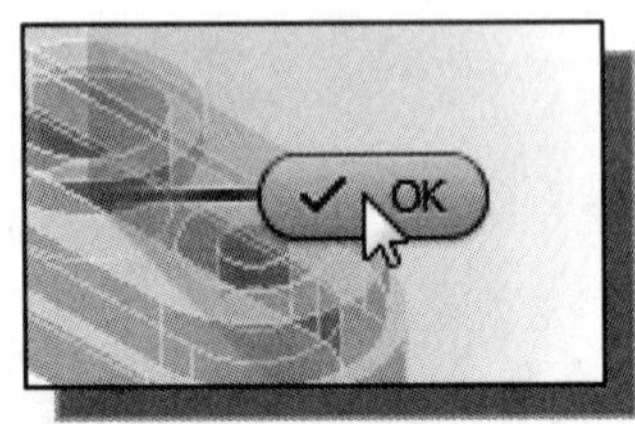

4. Inside the graphics window, right-mouse-click once to bring up the option menu and select **OK** to end the placement of the *Knee* part.

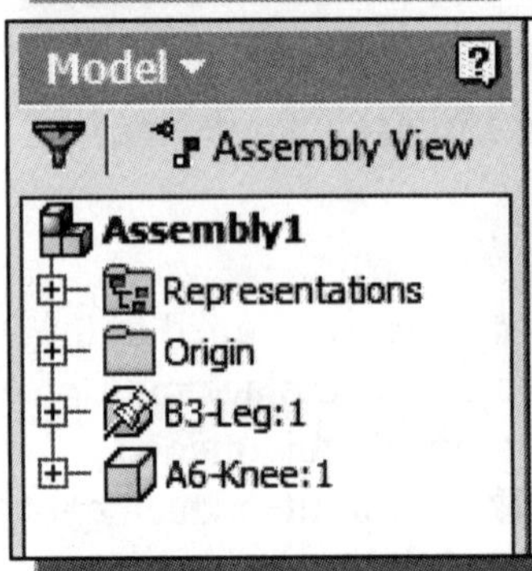

- Inside the *browser* window, the retrieved parts are listed in their corresponding order. The **Pin** icon in front of the *B3-Leg* filename signifies the part is grounded and all ***six degrees of freedom*** are restricted. The number behind the filename is used to identify the number of copies of the same component in the assembly model.

Degrees of Freedom and Constraints

- Each component in an assembly has six **degrees of freedom (DOF)**, or ways in which rigid 3D bodies can move: movement along the X, Y, and Z axes (translational freedom), plus rotation around the X, Y, and Z axes (rotational freedom). *Translational DOF*s allow the part to move in the direction of the specified vector. *Rotational DOF*s allow the part to turn about the specified axis.

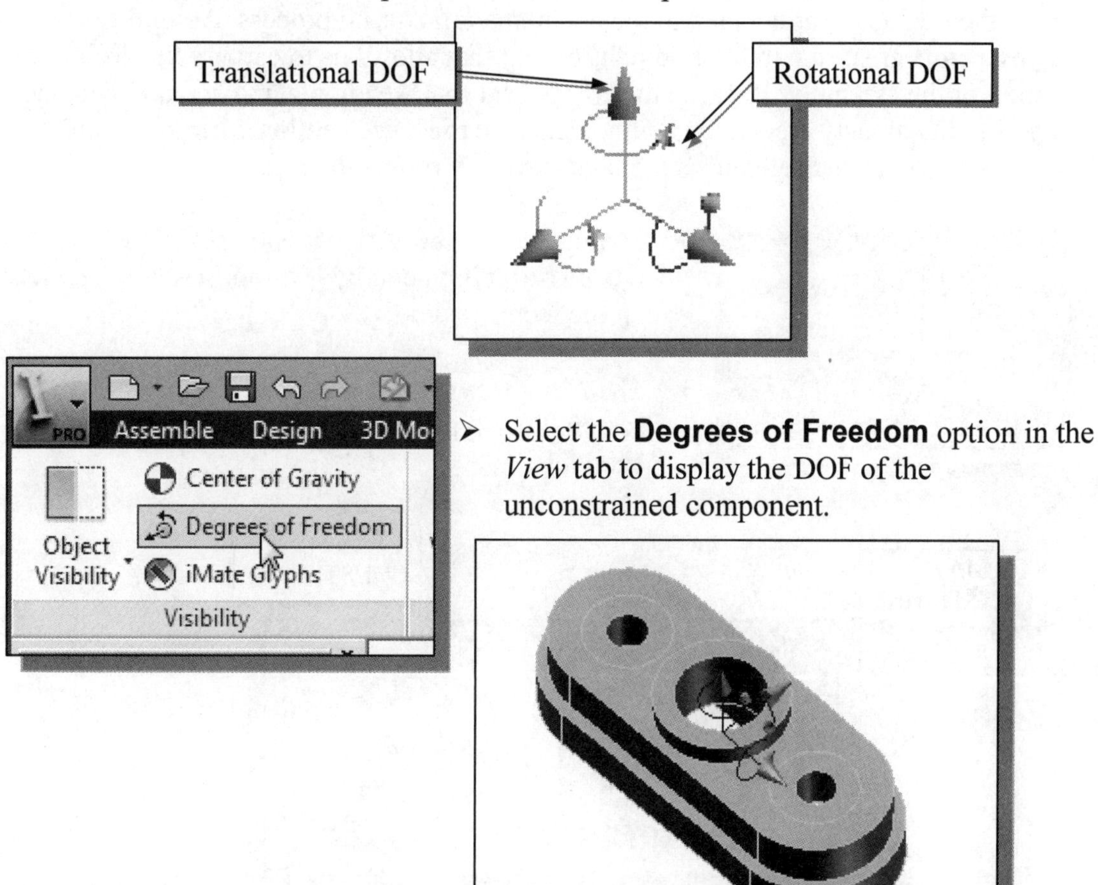

➢ Select the **Degrees of Freedom** option in the *View* tab to display the DOF of the unconstrained component.

- In *Autodesk Inventor*, the degrees-of-freedom symbol shows the remaining degrees of freedom (both translational and rotational) for one or more components of the active assembly. When a component is fully constrained in an assembly, the component cannot move in any direction. The position of the component is fixed relative to other assembly components. All of its degrees of freedom are removed. When we place an assembly constraint between two selected components, they are positioned relative to one another. Movement is still possible in the unconstrained directions.

➢ It is usually a good idea to fully constrain components so that their behavior is predictable as changes are made to the assembly. Leaving some degrees of freedom open can sometimes help retain design flexibility. As a general rule, we should use only enough constraints to ensure predictable assembly behavior and avoid unnecessary complexity.

Assembly Constraints

- We are now ready to assemble the components together. We will start by placing assembly constraints on the ***A6-Knee*** part and the ***B3-Leg***.

  To assemble components into an assembly, we need to establish the assembly relationships between components. It is a good practice to assemble components the way they would be assembled in the actual manufacturing process. **Assembly constraints** create a parent/child relationship that allows us to capture the design intent of the assembly. Because the component that we are placing actually becomes a child to the already assembled components, we must use caution when choosing constraint types and references to make sure they reflect the intent.

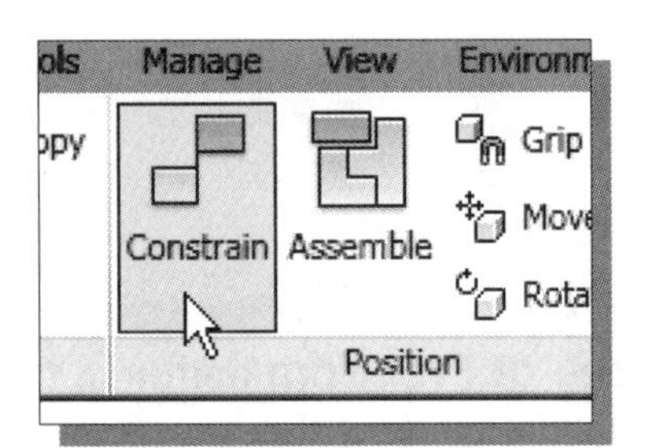

➢ Switch back to the *Assemble* tab; select the **Constrain** command by left-mouse-clicking once on the icon.

- The *Place Constraints* dialog box appears on the screen. Four types of assembly constraints are available.

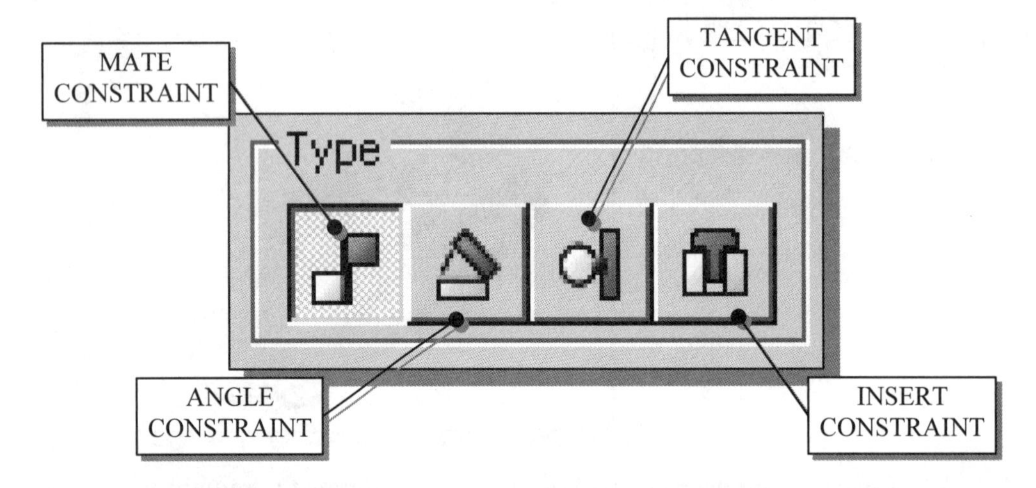

- Assembly models are created by applying proper *assembly constraints* to the individual components. The constraints are used to restrict the movement between parts. Constraints eliminate rigid body degrees of freedom (**DOF**). A 3D part has *six degrees of freedom* since the part can rotate and translate relative to the three coordinate axes. Each time we add a constraint between two parts; one or more DOF is eliminated. The movement of a fully constrained part is restricted in all directions. Four basic types of assembly constraints are available in *Autodesk Inventor*: Mate, Angle, Tangent and Insert. Each type of constraint removes different combinations of rigid body degrees of freedom. Note that it is possible to apply different constraints and achieve the same results.

- **Mate** – Constraint positions components face-to-face, or adjacent to one another, with faces flush. Removes one degree of linear translation and two degrees of angular rotation between planar surfaces. Selected surfaces point in opposite directions and can be **offset** by a specified distance. The Mate constraint positions selected faces normal to one another, with faces coincident.

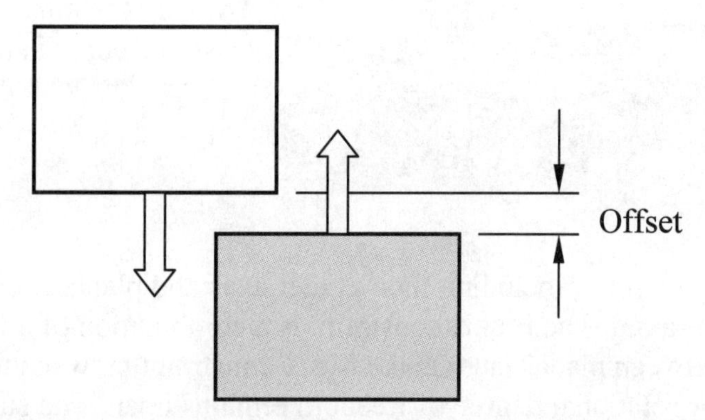

- **Flush** – Makes two planes coplanar with their faces aligned in the same direction. Selected surfaces point in the same direction and are offset by a specified distance. The Flush constraint aligns components adjacent to one another with faces flush and positions selected faces, curves, or points so that they are aligned with surface normals pointing in the same direction. (Note that the Flush constraint is listed as a selectable option in the Mate constraint.)

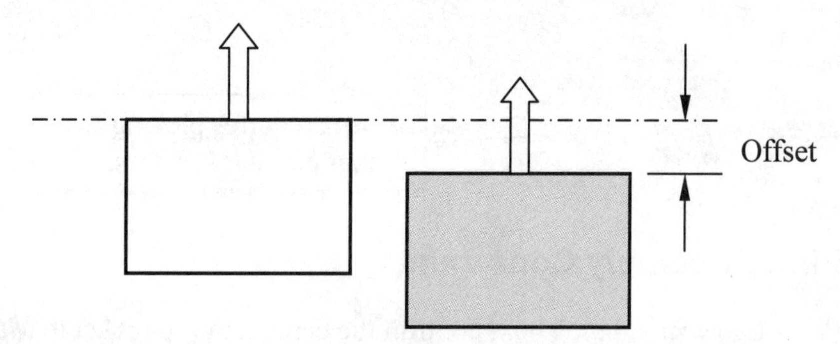

- **Angle** – Creates an angular assembly constraint between parts, subassemblies, or assemblies. Selected surfaces point in the direction specified by the angle.

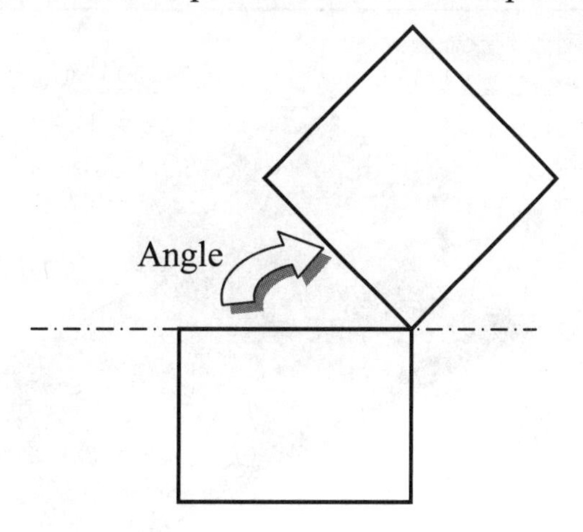

- **Tangent** – Aligns selected faces, planes, cylinders, spheres, and cones to contact at the point of tangency. Tangency may be on the inside or outside of a curve, depending on the selection of the direction of the surface normal. A **Tangent** constraint removes one degree of translational freedom.

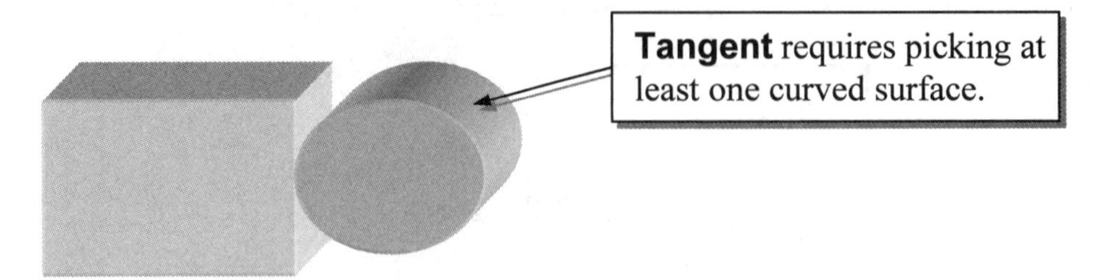

- **Insert** – Aligns two circles, including their center axes and planes. Selected circular surfaces become co-axial. The **Insert** constraint is a combination of a face-to-face **Mate** constraint between planar faces and a **Mate** constraint between the axes of the two components. A rotational degree of freedom remains open. The surfaces do not need to be full 360-degree circles. Selected surfaces can point in opposite directions or in the same direction and can be offset by a specified distance.

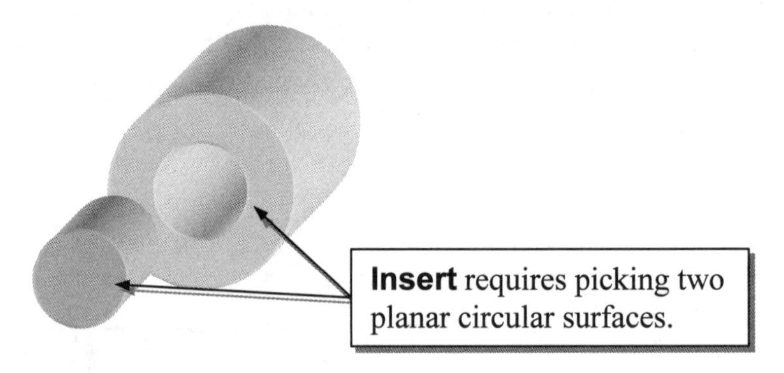

Apply the First Assembly Constraint

1. In the *Place Constraint* dialog box, confirm the constraint type is set to **Mate** constraint and select the **top horizontal surface** of the base part as the first part for the **Mate** alignment command.

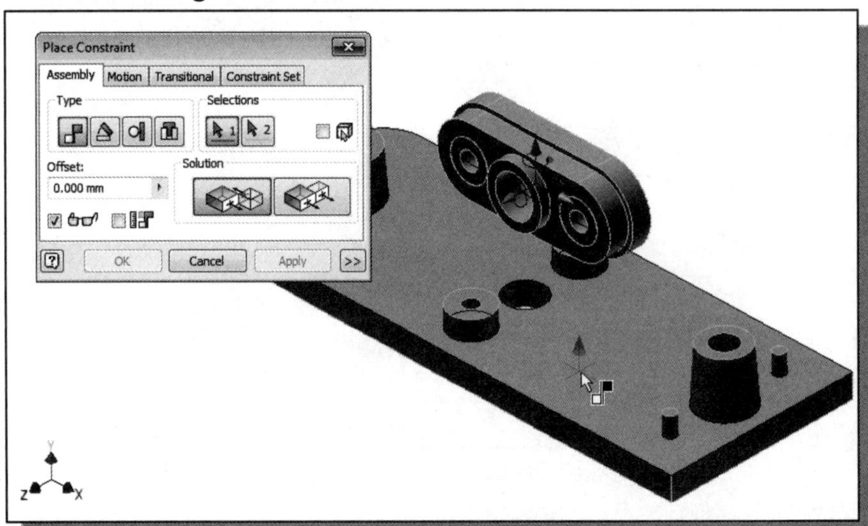

2. On your own, dynamically rotate the displayed model to view the bottom of the *Knee* part, as shown in the figure below.

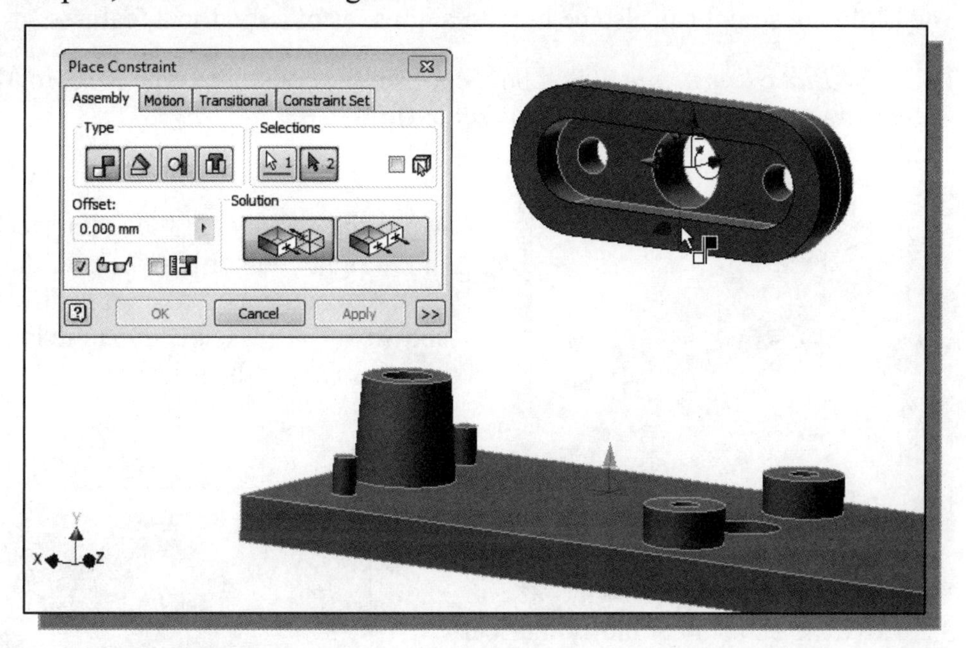

3. Click on the bottom face of the *Knee* part as the second part selection to apply the constraint. Note the direction normals shown in the figure; the Mate constraint requires the selection of opposite direction of surface normals.

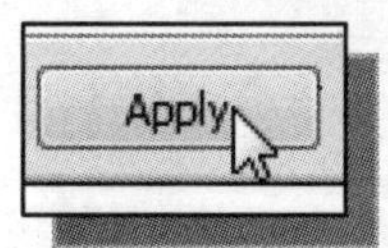

4. Click on the **Apply** button to accept the selection and apply the Mate constraint.

5. On your own, examine the model by using the dynamic Rotate command available in *Autodesk Inventor*.

❖ Notice the DOF symbol is adjusted automatically in the graphics window. The Mate constraint removes one degree of linear translation and two degrees of angular rotation between the selected planar surfaces. The *A6-Knee* part can still move along two axes and rotate about the third axis.

Apply a Second MATE Constraint

❖ The **Mate** constraint can also be used to align axes of cylindrical features.

1. In the *Place Constraint* dialog box, confirm the constraint type is set to **Mate** constraint and the *Offset* option is set to **0.00**.

2. Move the cursor near the cylindrical surface of the right counter bore hole of the *Knee* part. Select the axis when it is displayed as shown. (Hint: Use the dynamic **Rotation** option to assist the selection.)

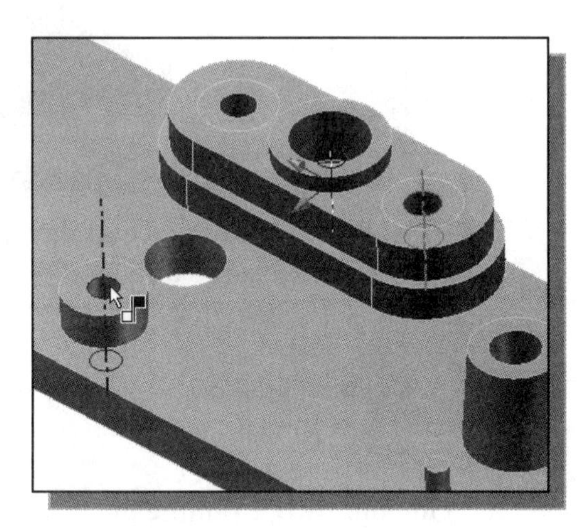

3. Move the cursor near the cylindrical surface of the small hole on the *B3-Leg* part. Select the axis when it is displayed as shown.

4. In the *Place Constraint* dialog box, click on the **Apply** button to accept the selection and apply the **Mate** constraint.

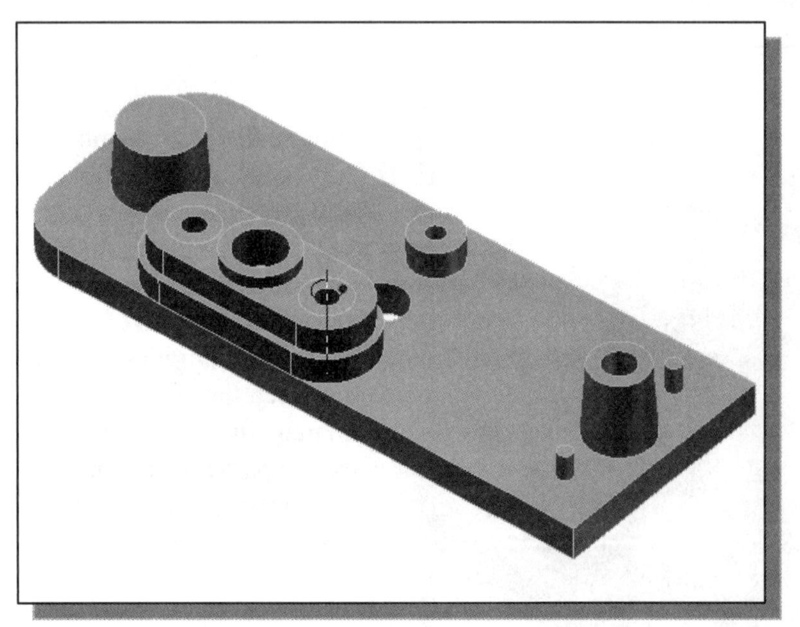

5. In the *Place Constraint* dialog box, click on the **Cancel** button to exit the **Place Constraint** command.

❖ The *A6-Knee* part appears to be placed in the length direction of the *Leg* part; the DOF symbol does indicate that the *Knee* part can still rotate about the displayed vertical axis.

Constrained Move

❖ To see how well a component is constrained, we can perform a ***constrained move***. A constrained move is done by dragging the component in the graphics window with the left-mouse-button. A constrained move will honor previously applied assembly constraints. That is, the selected component and parts constrained to the component move together in their constrained positions. A grounded component remains grounded during the move.

1. Inside the graphics window, move the cursor on top of the larger horizontal surface of the *knee* part as shown in the figure.

2. Press and hold down the left-mouse-button and drag the *A6-Knee* part downward.

❖ The *A6-Knee* part can still rotate about the displayed axis, while the applied constraints are maintained.

3. On your own, apply another **Mate** constraint to align the *A6-Knee* part as shown.

❖ Note the DOF symbol disappears, which indicates the assembly is fully constrained.

Placing the Third Component

➢ We will retrieve the *A8-Rod Pin* part as the third component of the assembly model.

1. In the *Assemble* panel (located in the *Ribbon* toolbar), select the **Place Component** command by left-mouse-clicking the icon.

2. Select the ***A8-Rod Pin*** design in the list window. And click on the **Open** button to retrieve the model.

3. Place the *A8-Rod Pin* part toward the upper right corner of the graphics window, as shown in the figure.

4. Inside the graphics window, right-mouse-click once to bring up the option menu and select **OK** to end the placement of the *A8-Rod Pin* part.

❖ Notice the *A8-Rod Pin* part was created using the Metric (mm) units set, while the *Leg* part was created using the English (inch) units set. Models created in *Autodesk Inventor* can use different units, for both size and location definitions. Any adjustments to dimensions can also be done using any units.

Apply an Insert Constraint

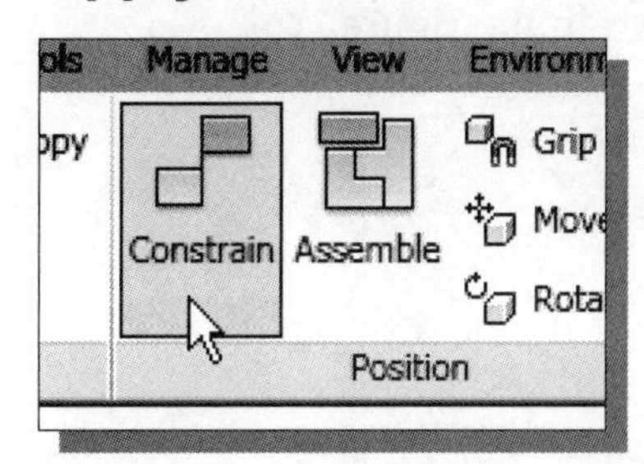

1. In the *Position* panel, select the **Constrain** command by left-mouse-clicking once on the icon.

2. In the *Place Constraint* dialog box, switch to the **Insert** constraint.

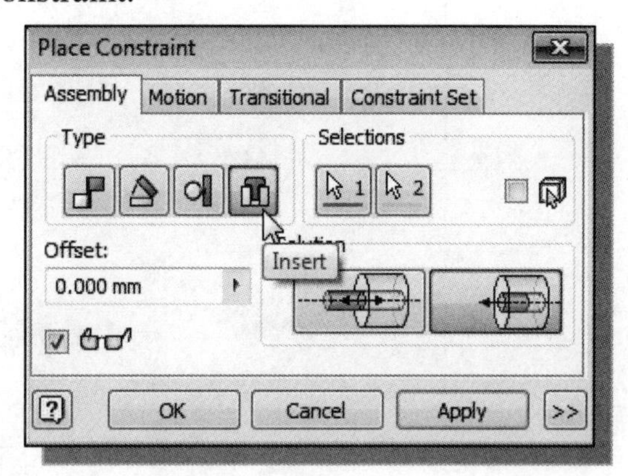

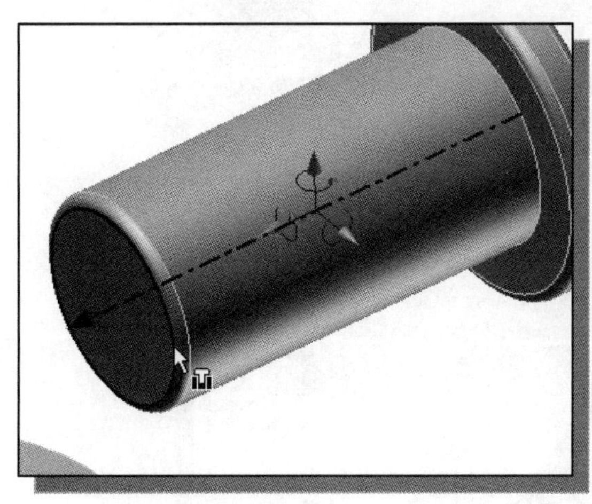

3. Select the end circle of the ***A8-Rod Pin*** part as the first object to apply the **Insert** constraint, as shown in the figure.

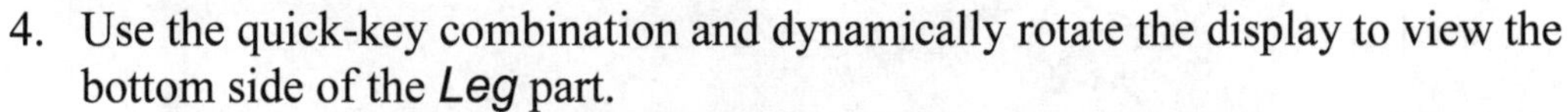

4. Use the quick-key combination and dynamically rotate the display to view the bottom side of the ***Leg*** part.

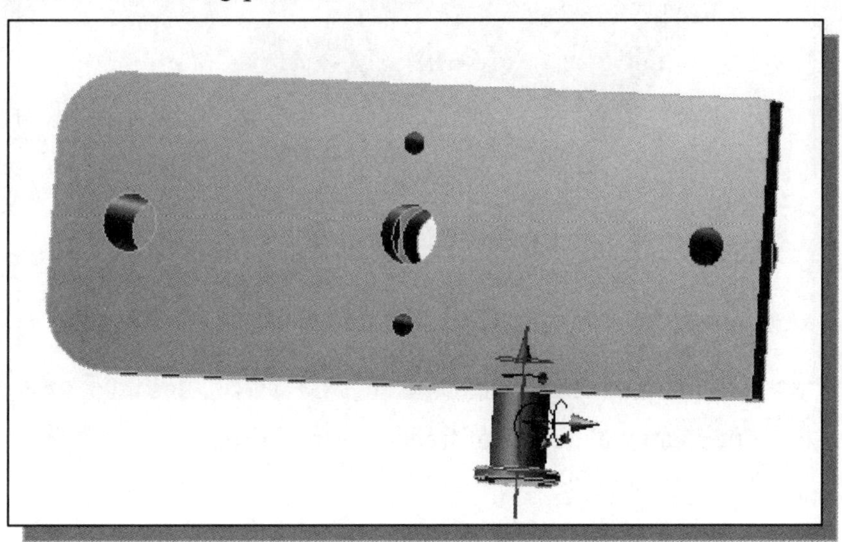

5. Select the **inside circle** at the end of the cylinder of the *B3-Leg* part as the second surface to apply the Insert constraint, as shown in the figure.

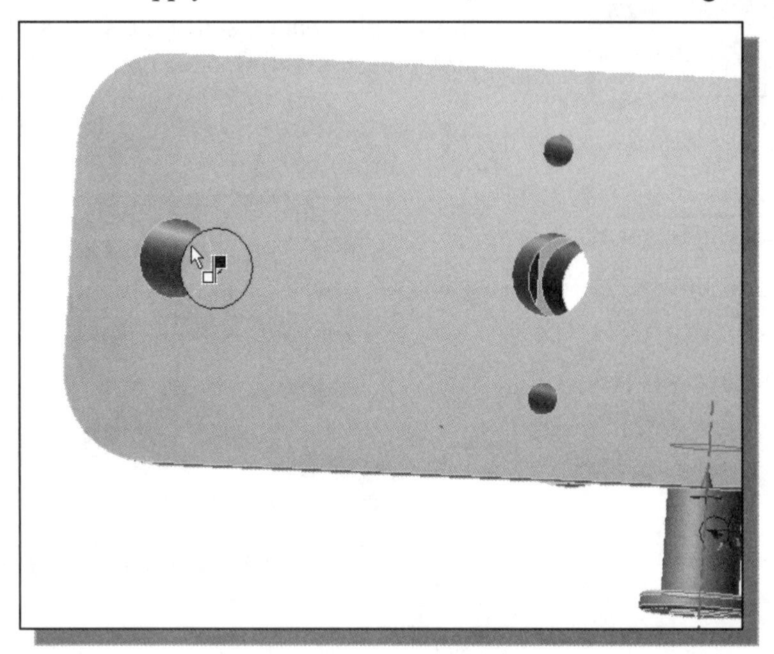

6. Click on the **Apply** button to accept the settings.

➢ Note that one rotational degree of freedom remains open; the *A8-Rod Pin* part can still freely rotate about the displayed DOF axis.

Apply a Flush Constraint

- Besides selecting the surfaces of solid models to apply constraints, we can also select the established work planes to apply the assembly constraints. This is an additional advantage of using the *BORN* technique in creating part models. For the *A6-Knee* part, we will apply another **Flush** constraint to two of the work planes and eliminate the last rotational DOF.

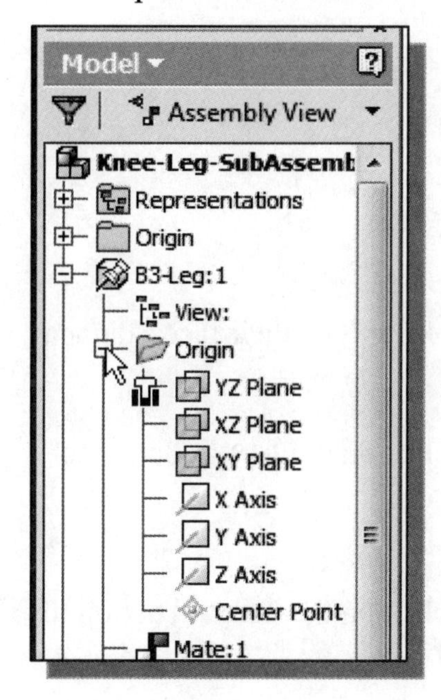

1. On your own, inside the *browser* window expand the **Origin** *work planes* of the *B3-Leg* and the *A6-Knee* parts.

2. In the *Place Constraint* dialog box, switch the constraint type to **Mate** constraint and the *Offset* option is set to **0.00**.

3. In the *Place Constraint* dialog box, switch the *Solution* option to **Flush** as shown.

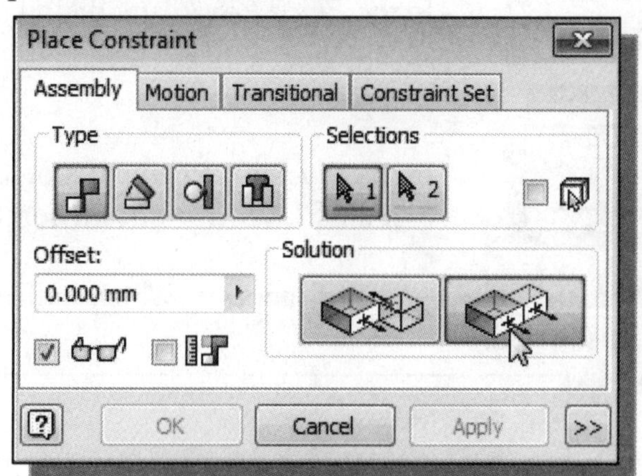

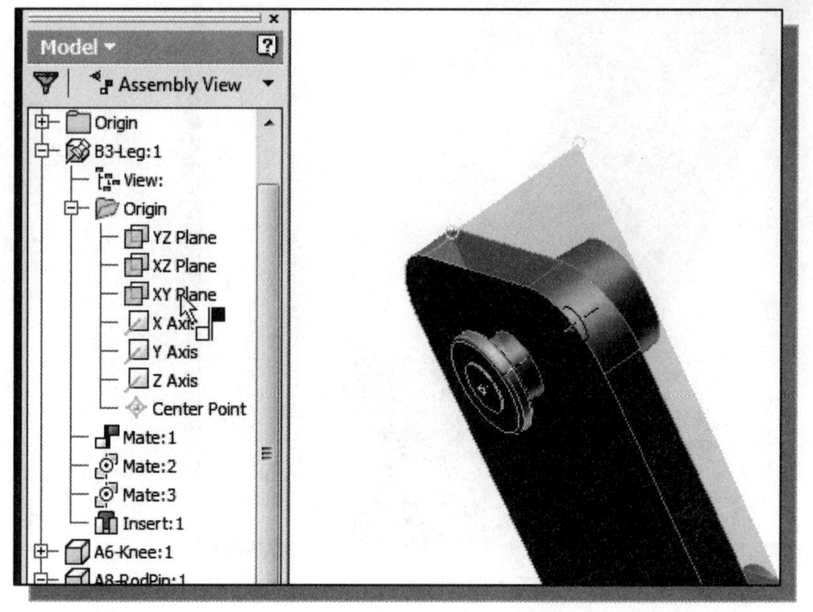

4. Select the **work plane** of the *B3-Leg* part aligned in the length direction as the first part for the **Flush** alignment command.

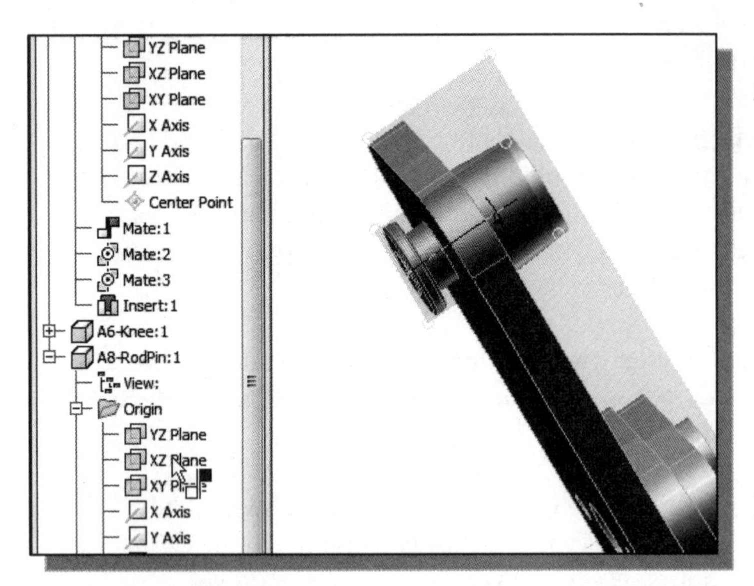

5. Select the corresponding **work plane** of the *A8-RodPin* part as the second part for the **Flush** alignment command.

❖ Note that the **Flush** constraint makes two planes coplanar with their faces aligned in the same direction.

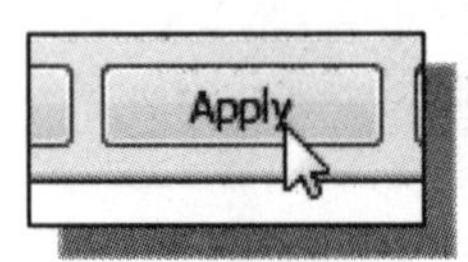

6. In the *Place Constraint* dialog box, click on the **Apply** button to accept the settings.

7. In the *Place Constraint* dialog box, click on the **Cancel** button to exit the Place Constraint command.

❖ Note the DOF symbol disappears, which indicates the assembly is fully constrained.

Edit Parts in the Assembly Mode

❖ We will place a copy of the *Boot* part in the assembly model. We will also illustrate the procedure to edit parts in the assembly mode.

1. In the *Assemble* panel (the toolbar is located in the *Ribbon* toolbar) select the **Place Component** command by left-mouse-clicking once on the icon.

2. Select the ***Boot*** design (part file: ***Boot.ipt***) in the list window. Then click on the **Open** button to retrieve the model.

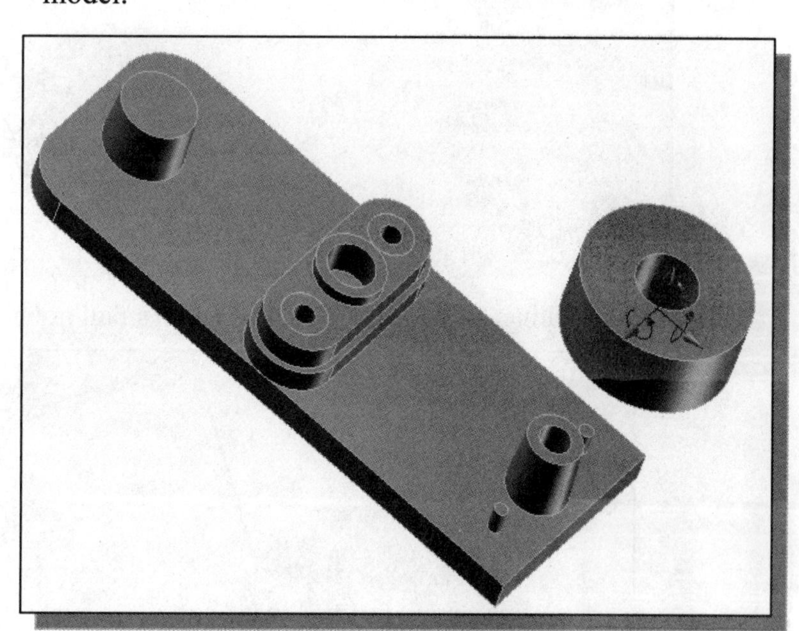

3. Place a copy of the **Boot** part on one side of the *A8-Rod Pin* by clicking once on the screen as shown in the figure.

4. Inside the graphics window, right-mouse-click once to bring up the option menu and select **Done** to end the Place Component command.

➢ Note that the two parts are created using different units. Let's adjust some of the dimensions so that they match.

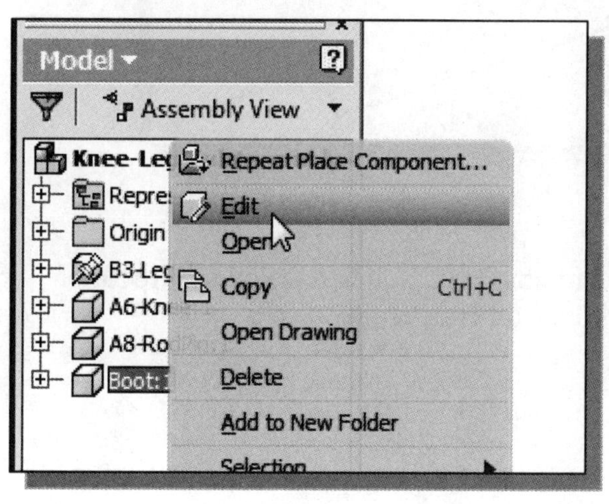

5. Inside the *browser* window, right-mouse-click once on top of the Boot part to bring up the option menu and select **Edit** to start the Edit Component mode.

6. Inside the *browser* window, right-mouse-click once on top of the last feature of the *Boot* part to bring up the option menu and select **Show dimensions**.

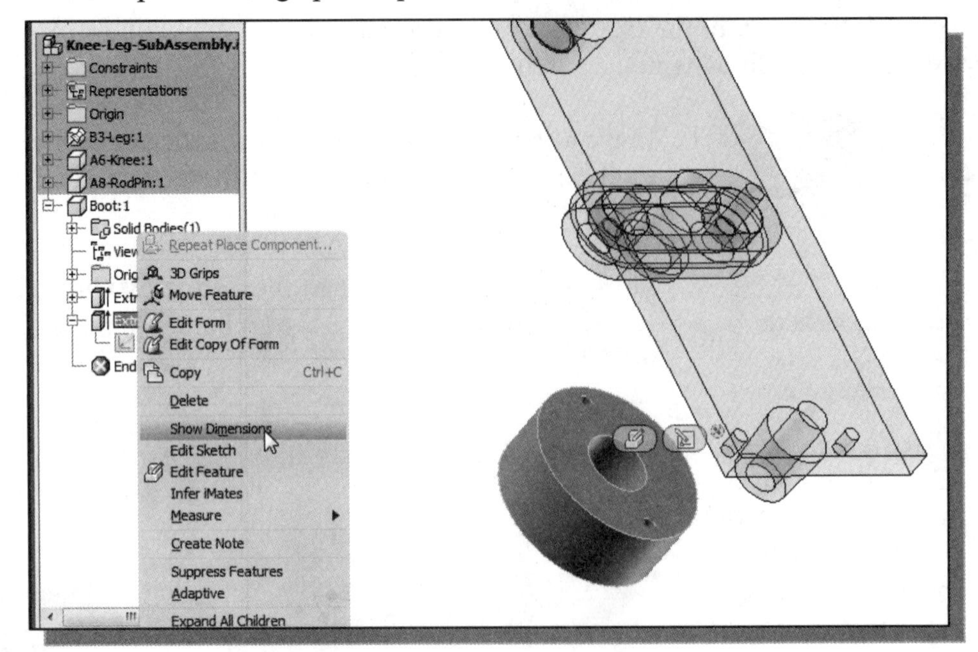

7. Note the dimensions associated to the two small holes of the *Boot* part.

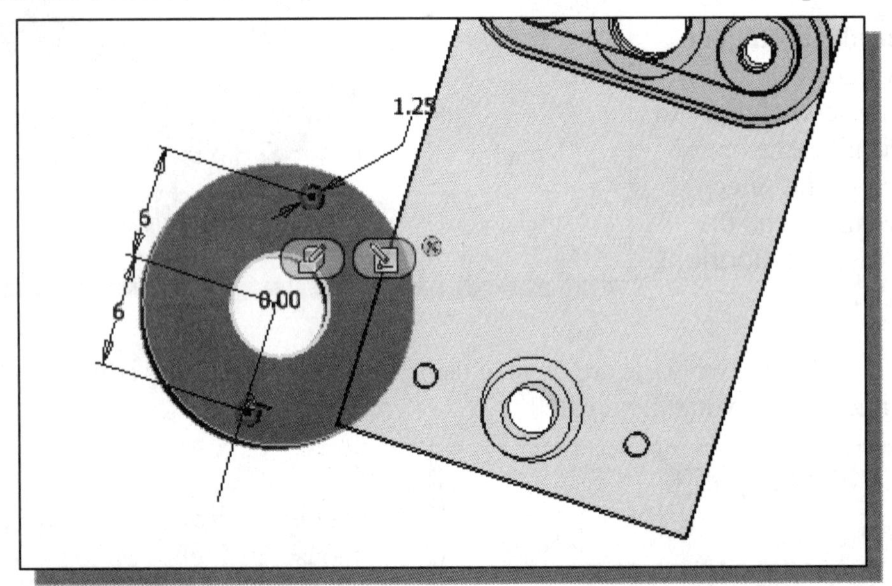

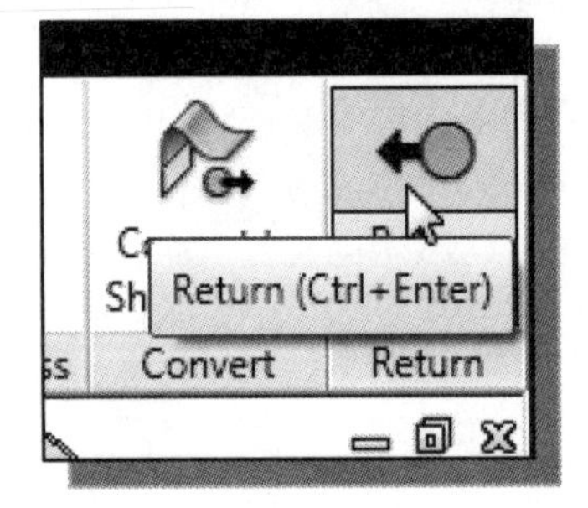

8. In the *Ribbon* toolbar panel, click once on the **Return** icon and exit the Edit Part mode.

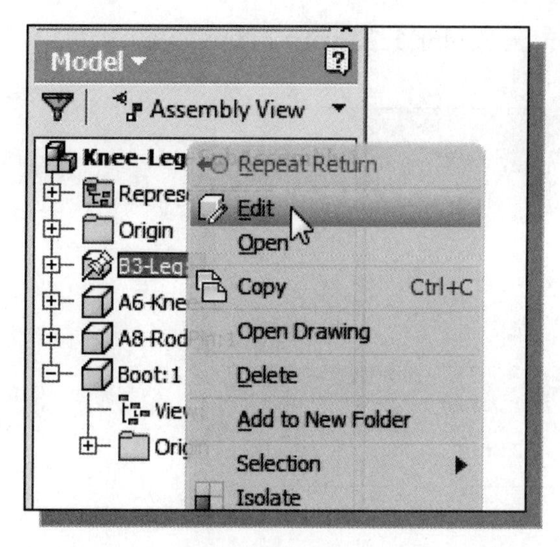

9. Inside the *browser* window, right-mouse-click once on top of the B3-Leg part to bring up the option menu and select **Edit** to start the Edit Component mode.

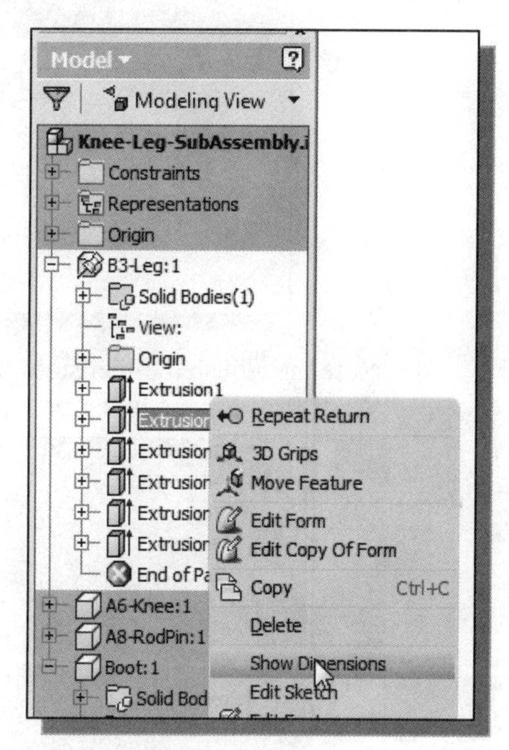

10. Inside the *browser* window, right-mouse-click once on top of the **Extrusion2** feature of the *B3-Leg* part to bring up the option menu and select **Show dimensions**.

11. Double-click on the *0.475 inch* dimension and enter **12 mm** in the *Edit Dimension* box as shown.

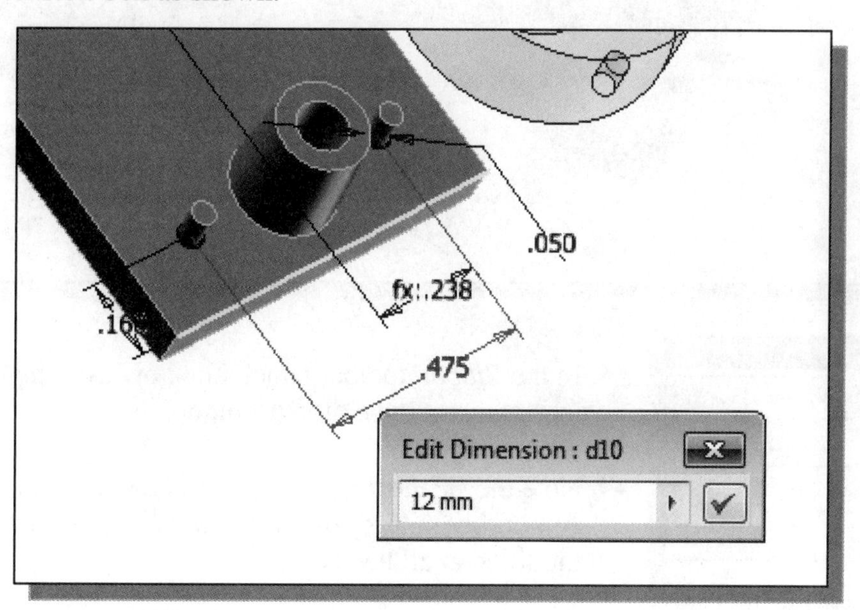

12. Double-click on the *0.05 inch* dimension and enter **1.23 mm** in the *Edit Dimension* box as shown.

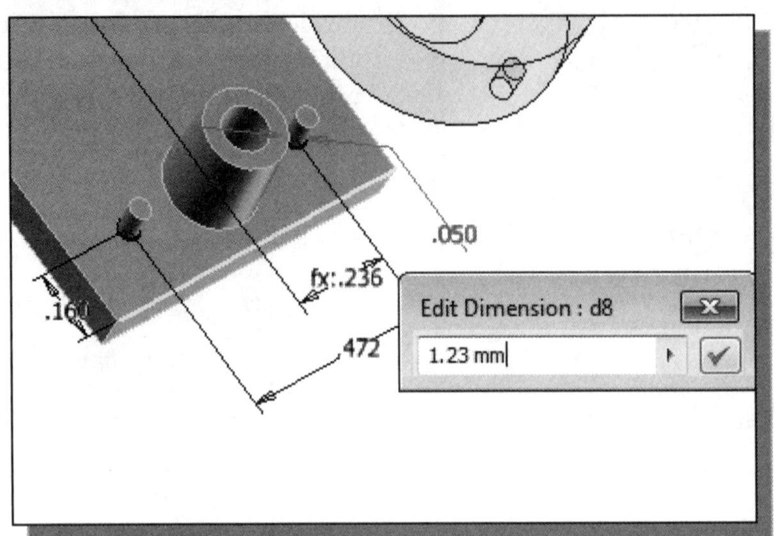

13. Note the adjusted dimensions are displayed in inches, the part units set established when the part was first created.

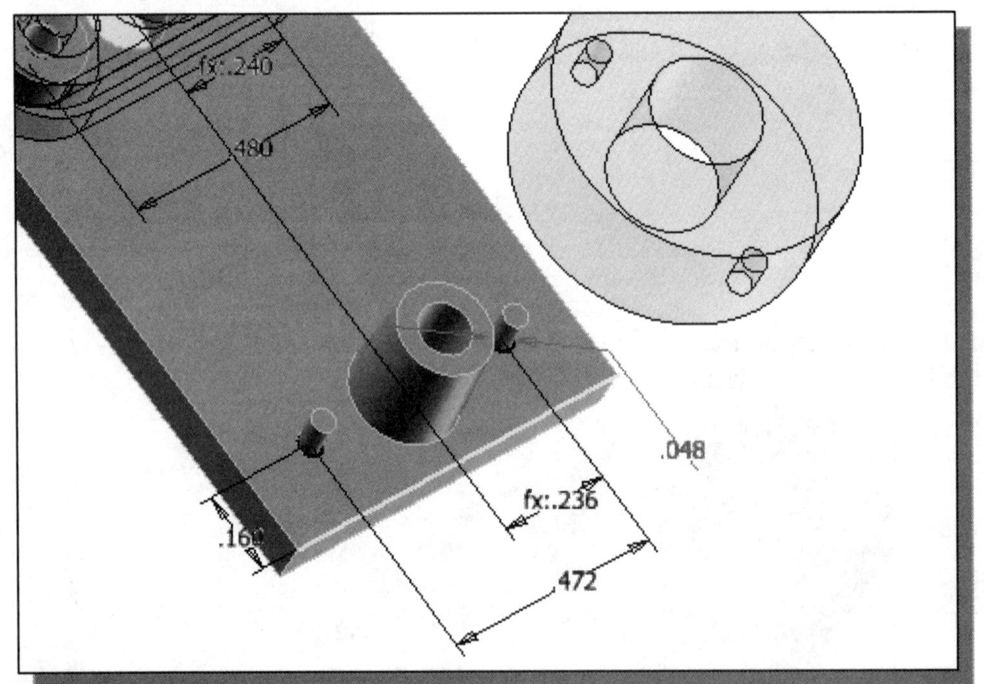

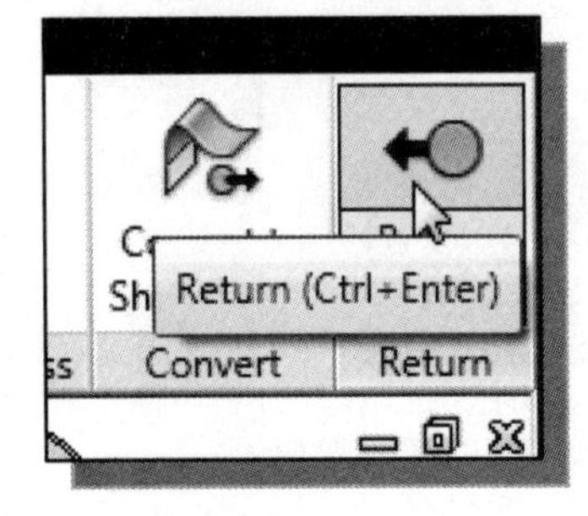

14. In the *Ribbon* toolbar panel, click once on the **Return** icon and exit the Edit Part mode.

- Note the *B3-Leg* part has been updated through the Assembly mode; *parametric modeling* allows parts to be modified at all levels.

Assemble the Boot Part

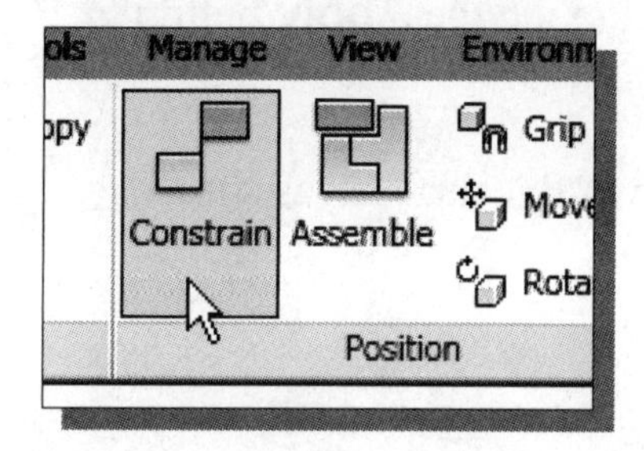

1. In the *Position* panel, select the **Constrain** command by left-mouse-clicking once on the icon.

2. In the *Place Constraint* dialog box, confirm the constraint type is set to **Mate** constraint.

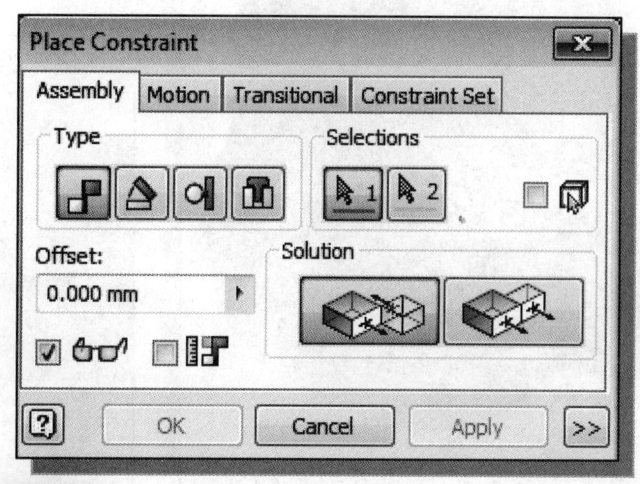

3. Select one of the small cylindrical surfaces of the *B3-Leg* part as the first object to apply the Mate constraint, as shown in the figure.

4. Select one of the small cylindrical surfaces of the *Boot* part as the second object to apply the Mate constraint as shown.

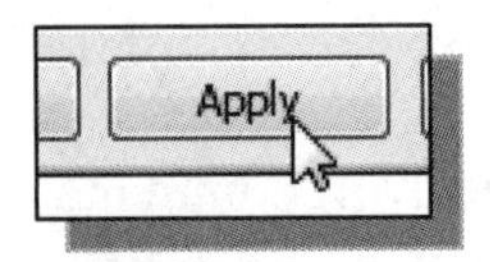

5. In the *Place Constraint* dialog box, click on the **Apply** button to accept the settings.

6. On your own, apply another **Mate** constraint to the two flat surfaces as shown.

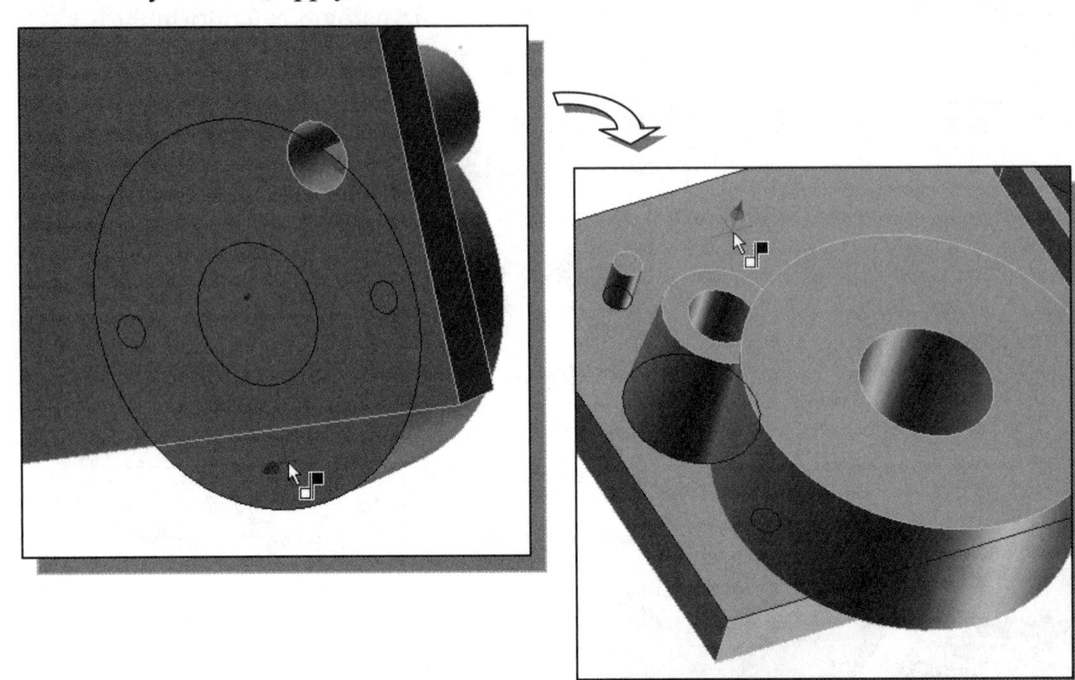

7. On your own, fully constrain the *Boot* part by using a **Mate** constraint to align the two associated ***center axes*** as shown.

Use the Content Center and Assemble Two Screws

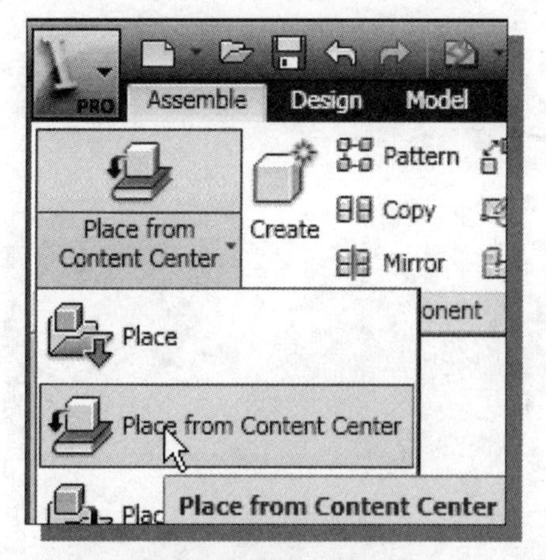

1. In the *Assemble* panel select the **Place from Content Center** command by left-mouse-clicking the icon.

2. Select the **Fasteners** category and then **Bolts → Round Head** as shown.

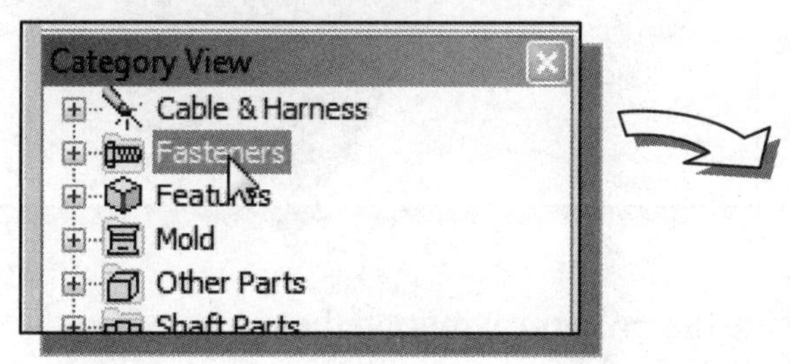

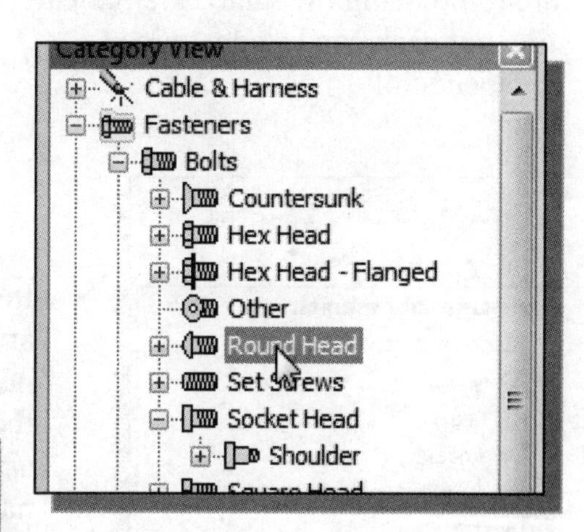

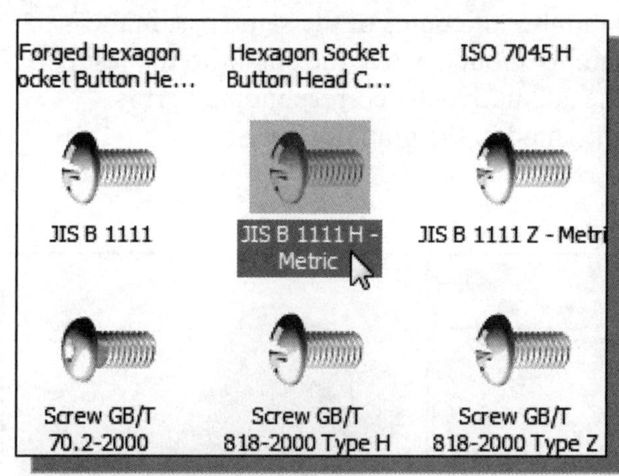

3. Select **JIS B 1111H-Metric** as shown in the figure.

4. Click **OK** to enter the selection dialog box.

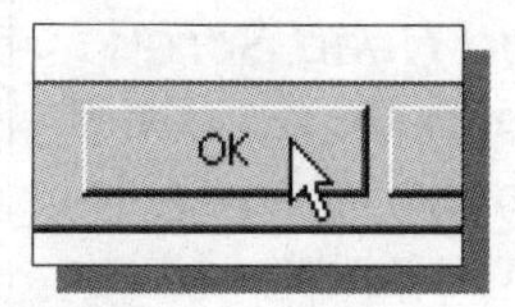

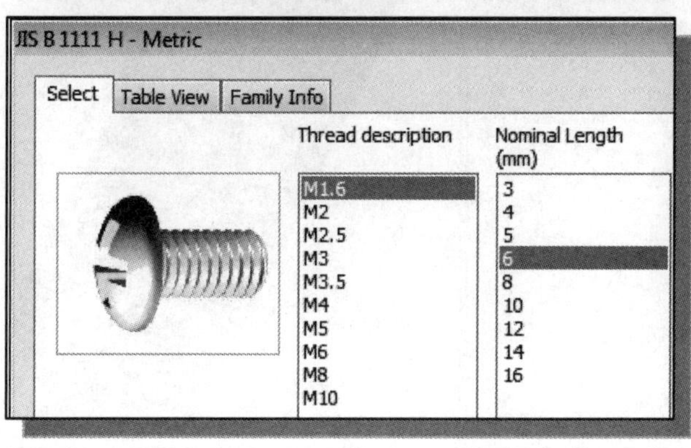

5. Click once inside the graphics window to accept the assembly model.

6. Set the thread type to **M1.6** and the nominal length to **6** mm.

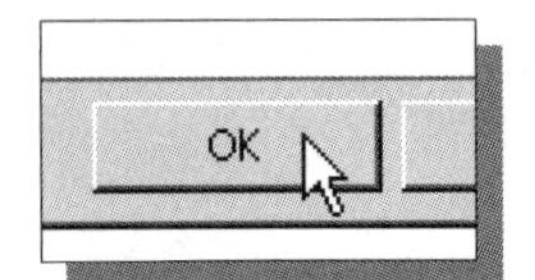

7. Click **OK** to accept the selection.

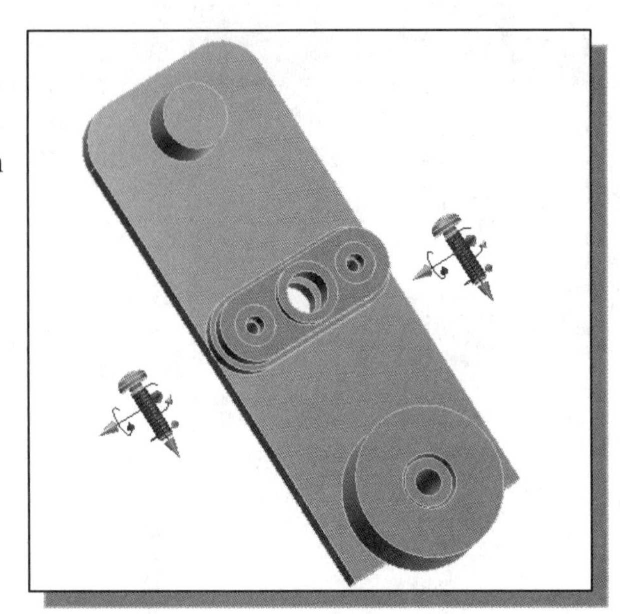

8. Place two copies of the ***M1.6x6 Round Head Screw*** part on both sides of the *B3-Leg* by clicking twice on the screen.

❖ Notice the DOF symbols displayed on the screen. Each *Round Head Screw* has six degrees of freedom. Both parts are referencing the same external part file, but each can be constrained independently.

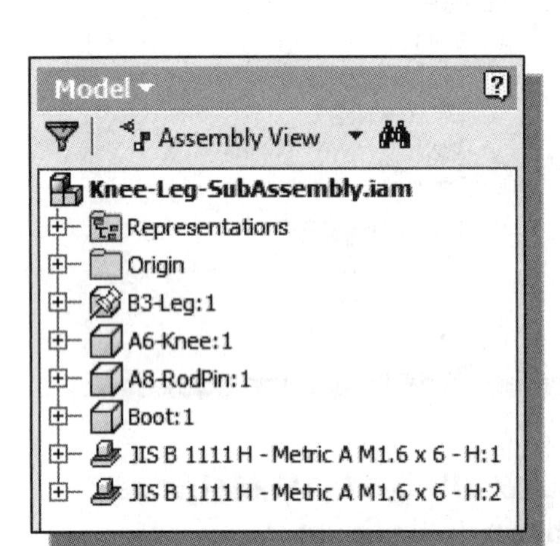

- Inside the *browser* window, the retrieved parts are listed in the order they were placed. The number behind the part name is used to identify the number of copies of the same part in the assembly model. Move the cursor to the last part name and notice the corresponding part is highlighted in the graphics window.

9. On your own, place a copy of the ***M3x8 Round Head Screw*** and apply the necessary constraints to assemble the three screws as shown.

10. In the *Quick Access* menu, select **Save** and save as the **Knee-Leg-SubAssembly** model.

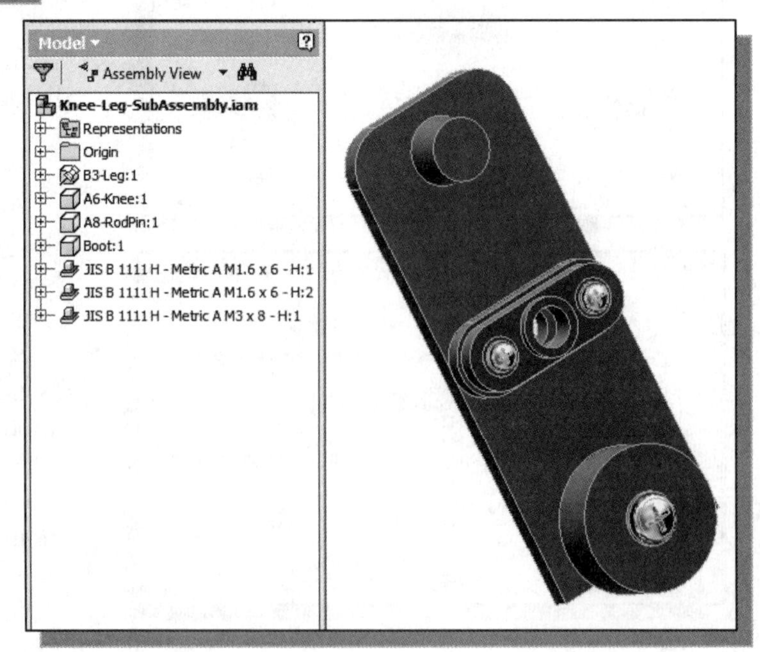

Start the *Main Assembly*

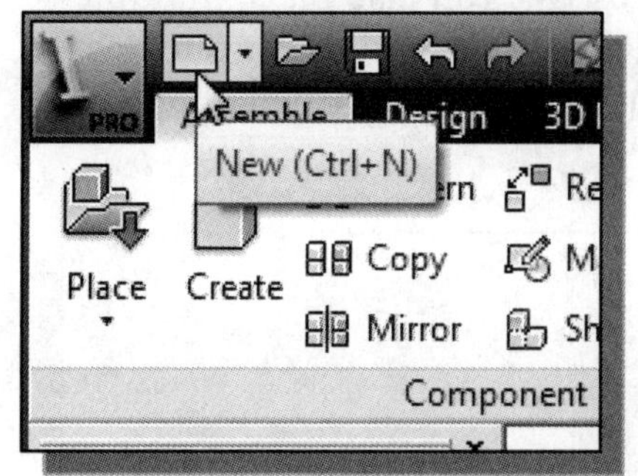

1. Select the **New File** icon with a single click of the left-mouse-button in the *Quick Access* toolbar as shown.

2. Select the **Metric** tab and in the *Template* list, select **Standard(mm).iam** (*Standard Inventor Assembly Model* template file).

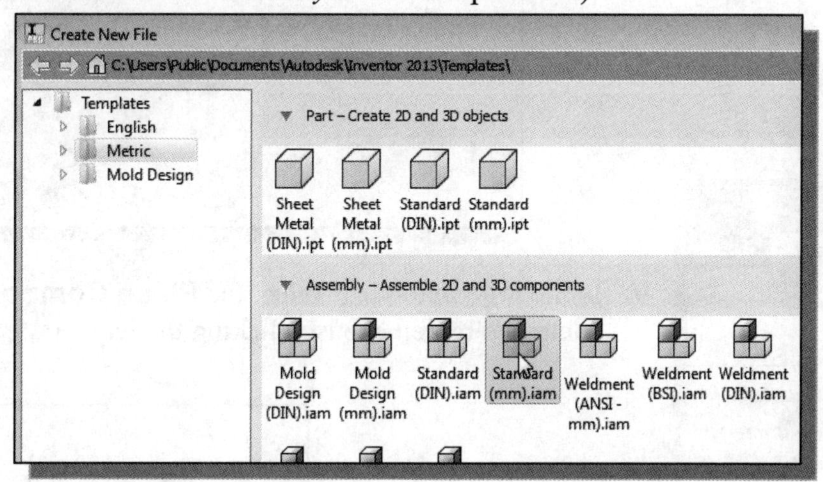

3. Click on the **Create** button in the *New File* dialog box to accept the selected settings.

4. In the *Ribbon* toolbar panel, select the **Place Component** command by left-mouse-clicking the icon.

5. Select the ***Chassis*** (part file: ***B2-Chassis.ipt***) in the list window.

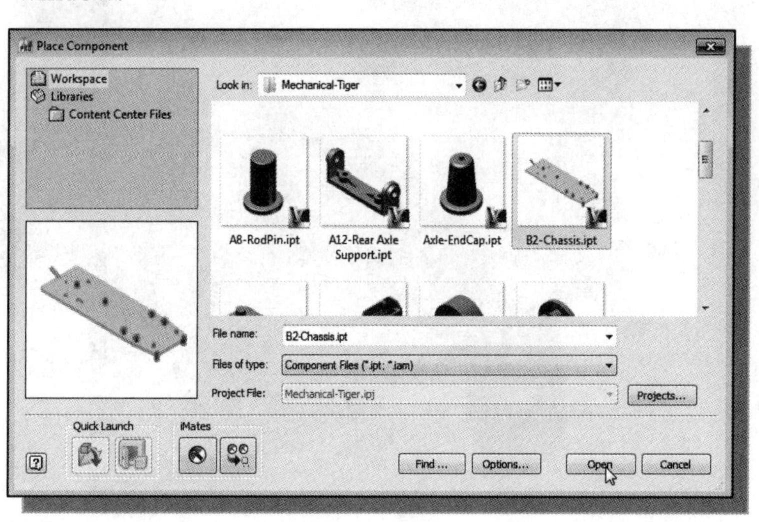

6. Click on the **Open** button to retrieve the model.

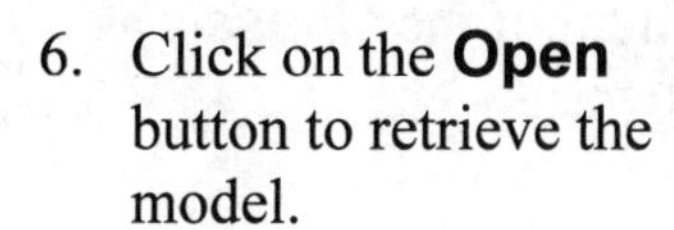

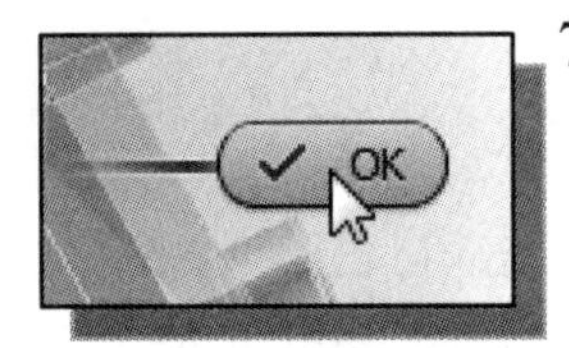

7. By default, the first component is automatically aligned to the origin of the assembly coordinates. We can also place multiple copies of the same component. Right-mouse-click once to bring up the option menu and select **OK** to end the placement of the ***B2-Chassis*** part.

8. On your own, rotate the display to the orientation as shown.

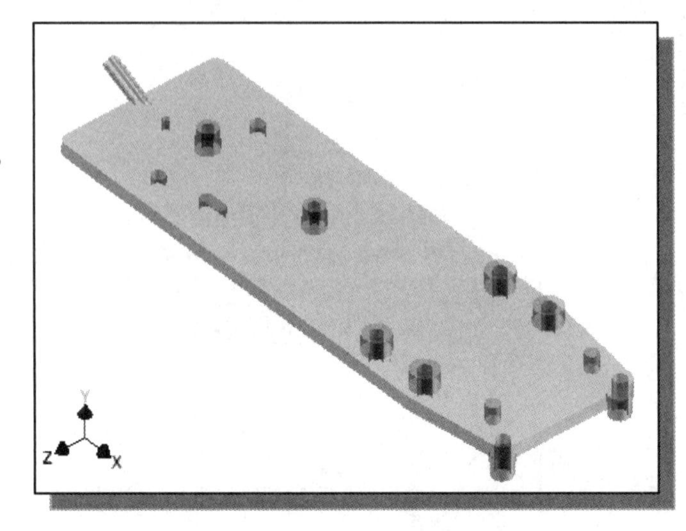

9. In the *Ribbon* toolbar, select the **Place Component** command by left-mouse-clicking the icon.

10. Select the ***Rear Axle Support*** (part file: ***A12- Rear Axle Support.ipt***) in the list window.

11. Place a copy of the selected part on the screen next to the chassis.

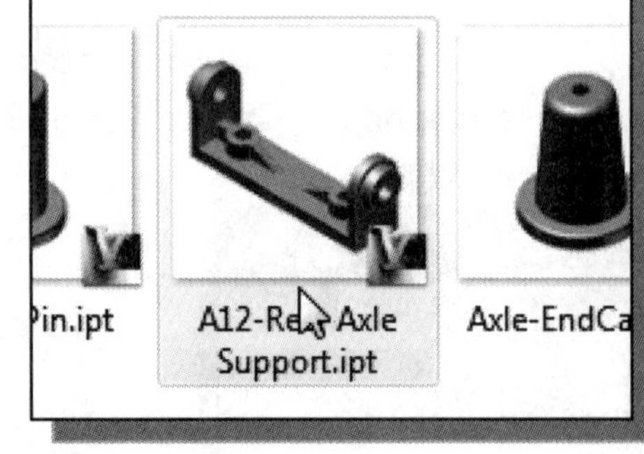

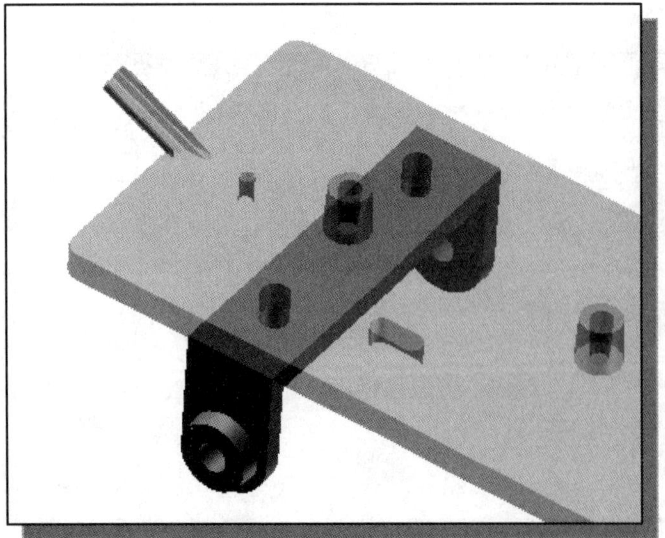

12. On your own, apply three **Mate** constraints to assemble the *Rear Axle Support* to the bottom of the *Chassis* as shown.

Assemble the Gear Box Right Part

1. In the *Ribbon* toolbar, select the **Place Component** command by left-mouse-clicking the icon.

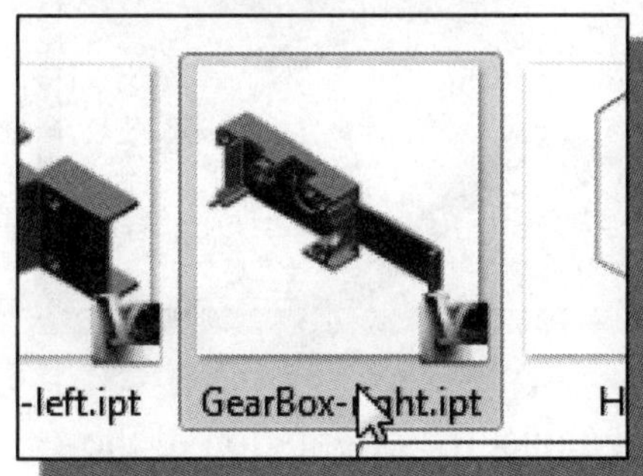

2. Select the ***GearBox-right*** in the list window.

3. Place a copy of the selected part on the screen next to the *Chassis*.

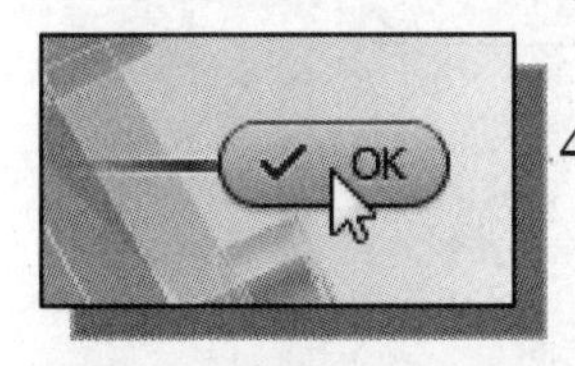

4. Right-mouse-click once to bring up the option menu and select **OK** to end the placement of the part.

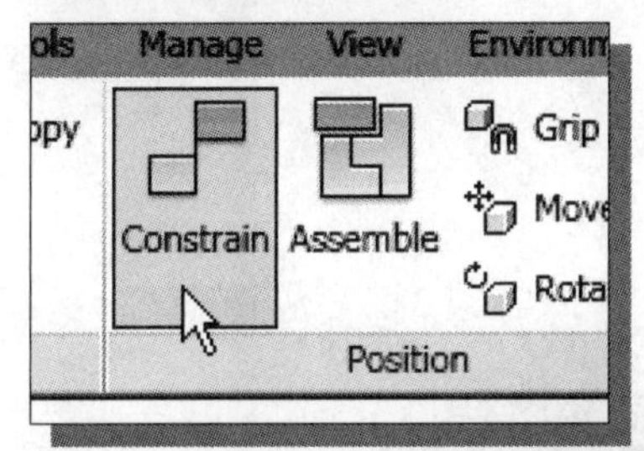

5. In the *Assembly* panel, select the **Constrain** command by left-mouse-clicking once on the icon.

6. On your own, apply a **Mate** constraint to assemble the *GearBox-right* part to the bottom of the *Chassis*.

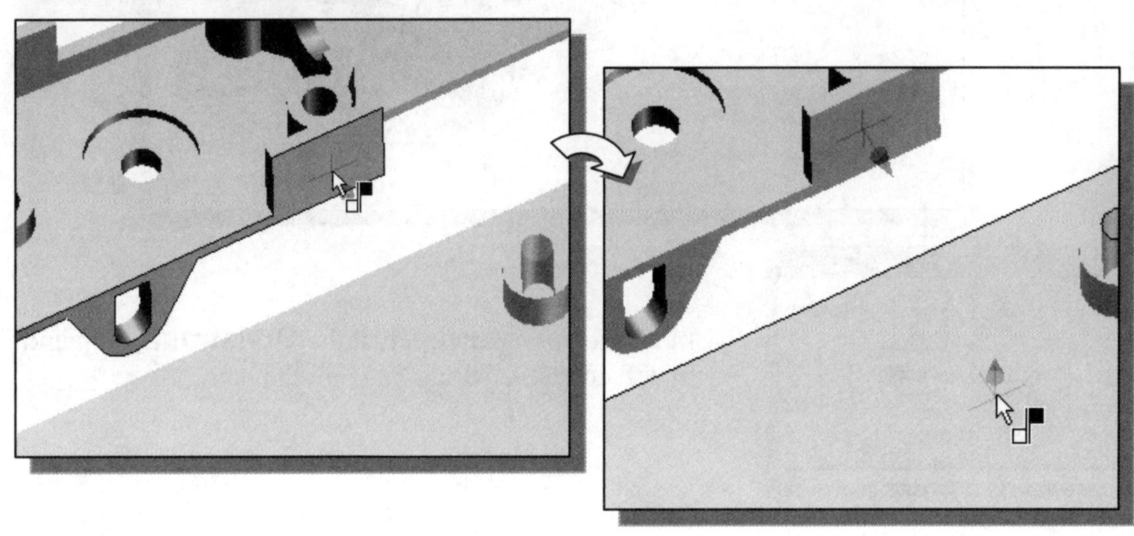

7. On your own, apply another **Mate** constraint to assemble the circular surface to the cylindrical surface of the *Chassis* as shown. (Note the figures below are viewing the bottom of the *Chassis*.)

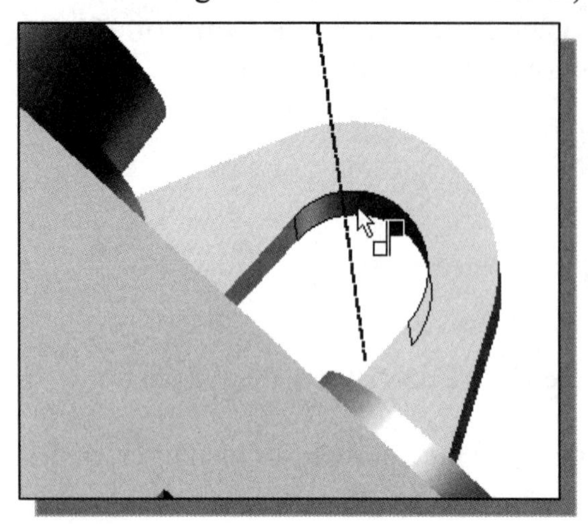

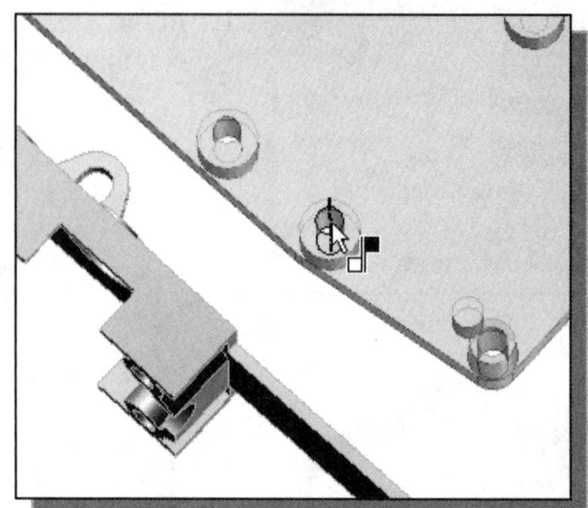

8. On your own, use the constrained move option and adjust the orientation of the *GearBox-right* part as shown.

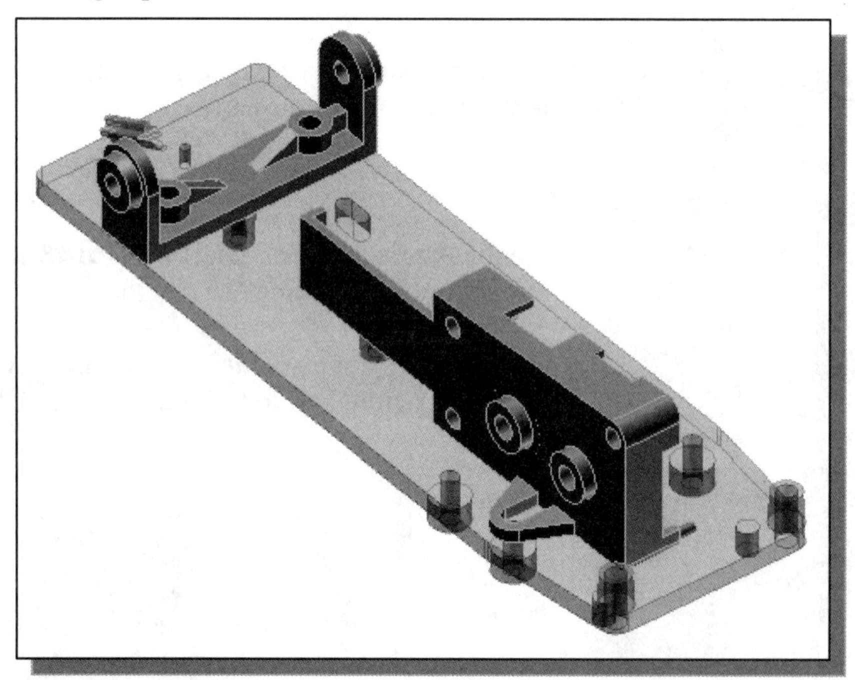

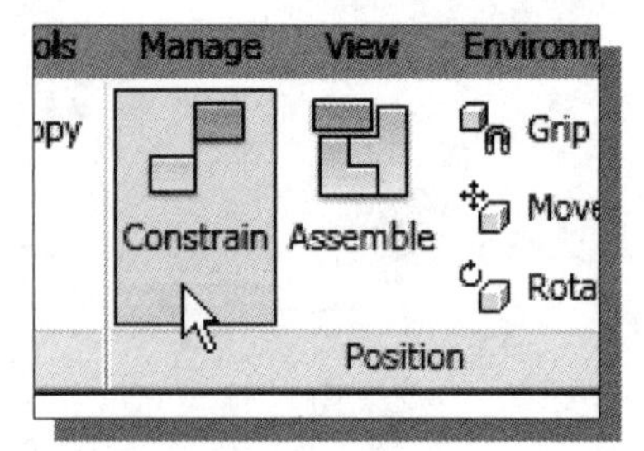

9. In the *Position* panel, select the **Constrain** command by left-mouse-clicking once on the icon.

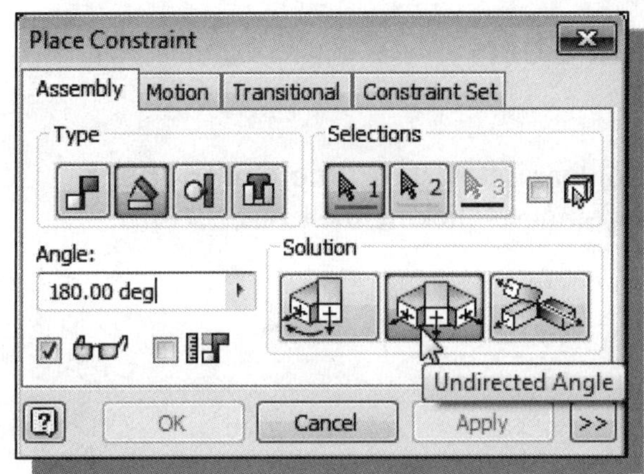

10. On your own, apply an **Angle** constraint with the **Undirected Angle** option to align the corresponding work planes. **XY Plane** of the chassis part and **XZ Plane** of the Gear-Box part, as shown. (Hint: Use either **0 degree** or **180 degrees** to create the alignment as shown.)

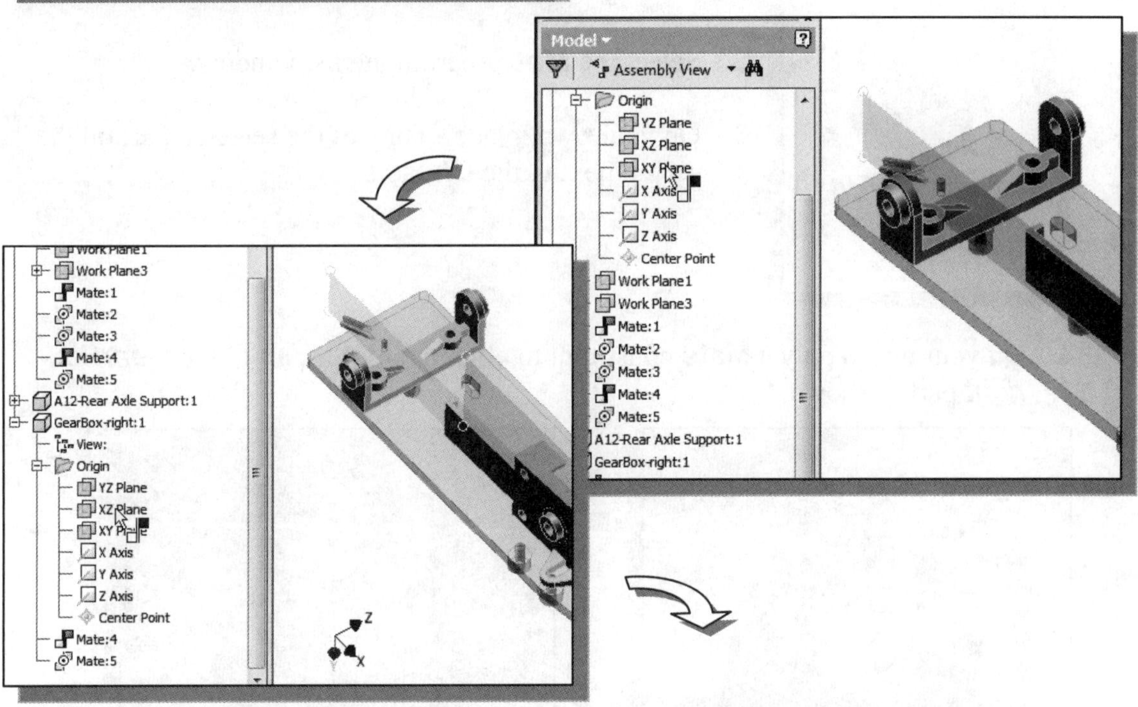

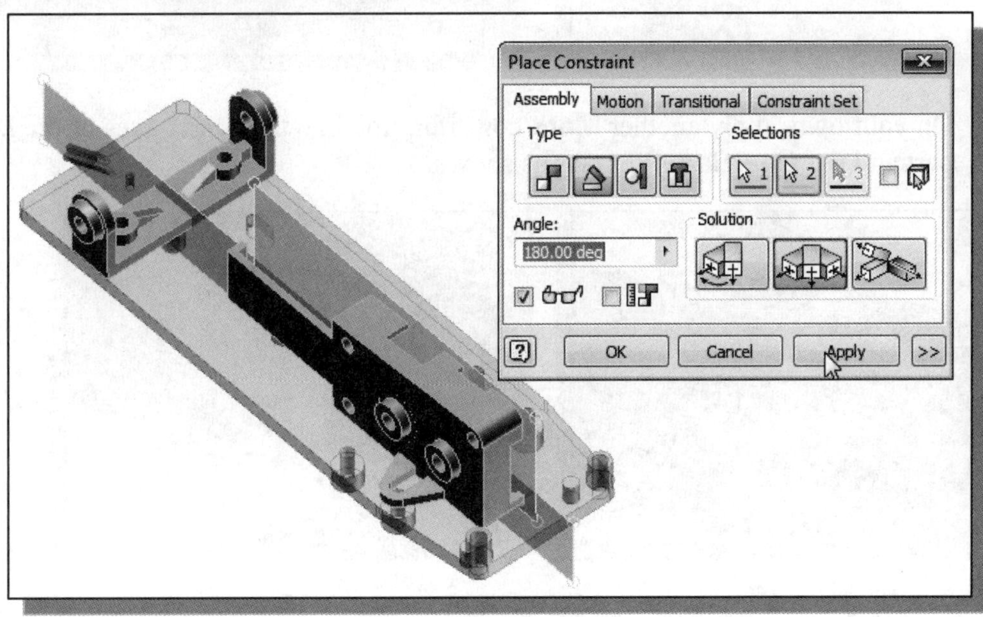

Assemble the Motor and the Pinion Gear

1. In the *Ribbon* toolbar, select the **Place Component** command by left-mouse-clicking the icon.

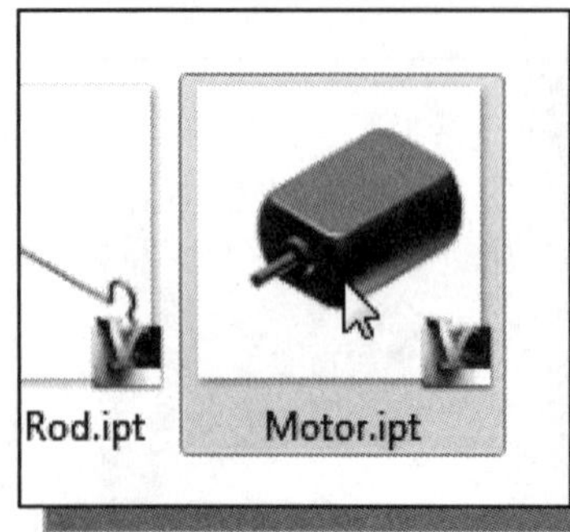

2. Select the ***Motor*** part in the list window.

3. On your own, place a copy of the selected part on the screen next to the *Chassis*.

4. On your own, apply a **Mate** constraint to align the *Motor* part to the *GearBox-right* part as shown.

5. On your own, apply another **Mate** constraint to align the center axes of the *Motor* part and the *GearBox-right* part as shown.

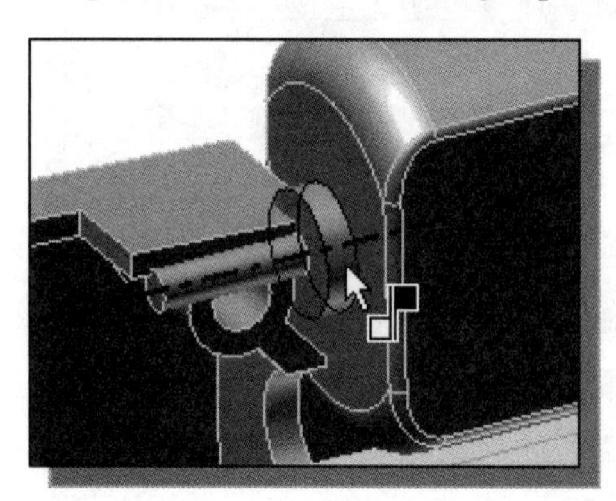

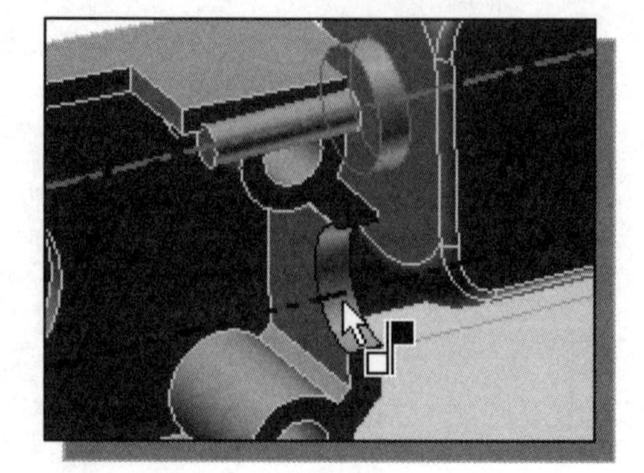

6. On your own, apply an **Undirected Angle** constraint to align the *Motor* part to the *GearBox-right* part as shown

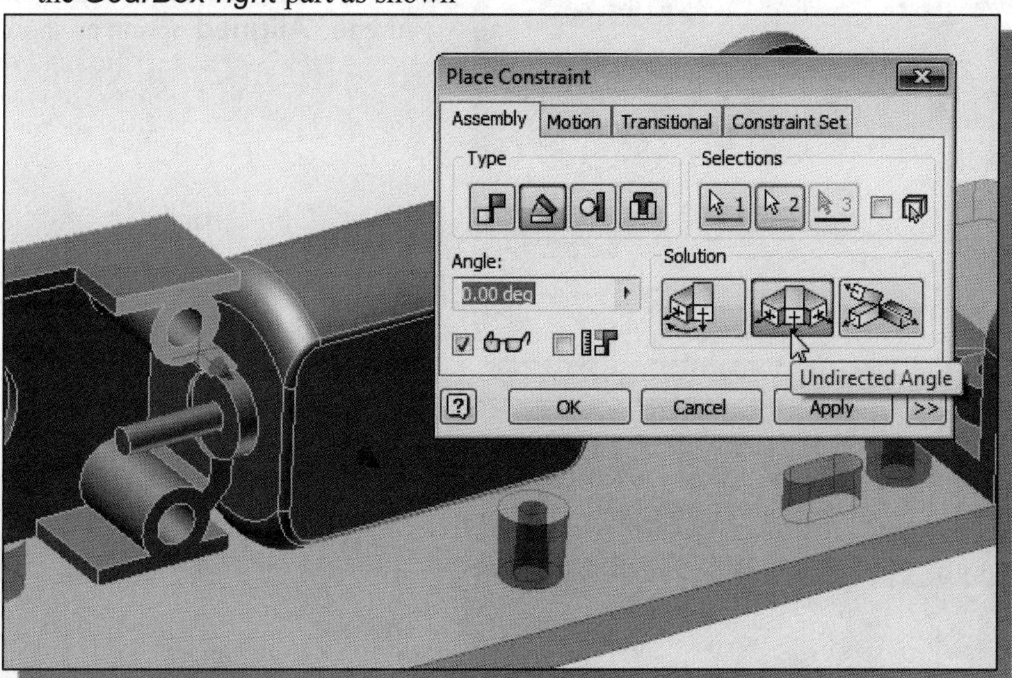

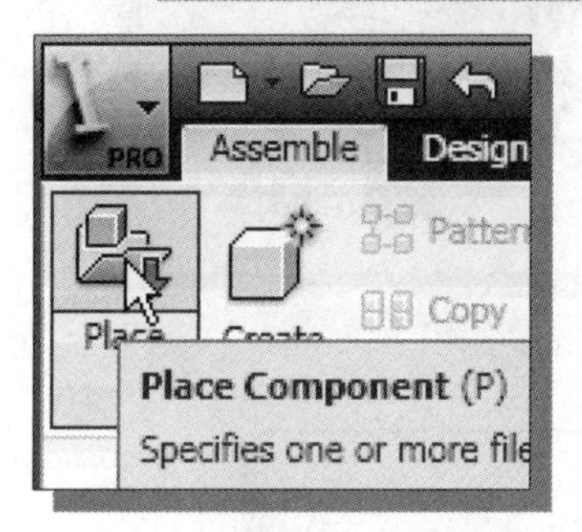

7. In the *Ribbon* toolbar, select the **Place Component** command by left-mouse-clicking the icon.

8. Select the ***G0-Pinion*** part in the list window.

9. On your own, place a copy of the selected part on the screen next to the *Chassis*.

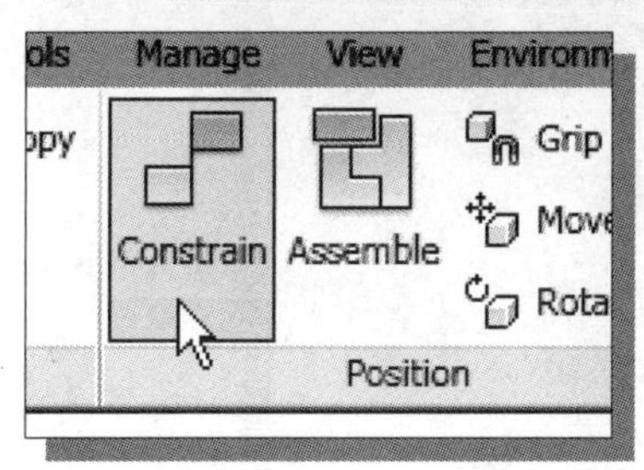

10. In the *Position* panel, select the **Constrain** command by left-mouse-clicking once on the icon.

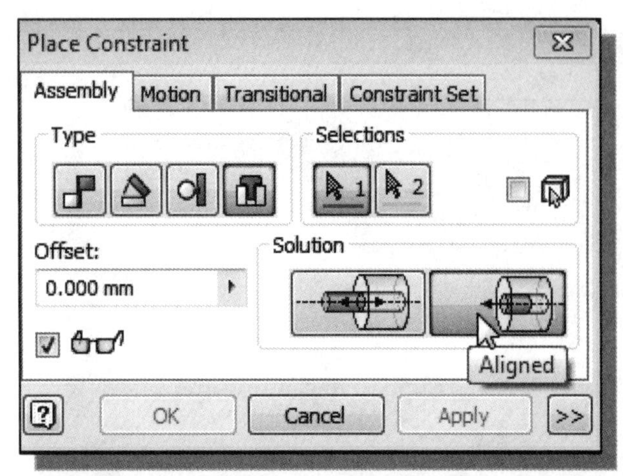

11. Choose the **Insert** constraint and use the **Aligned** option as shown.

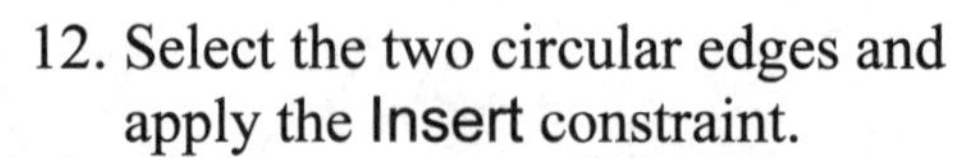

12. Select the two circular edges and apply the Insert constraint.

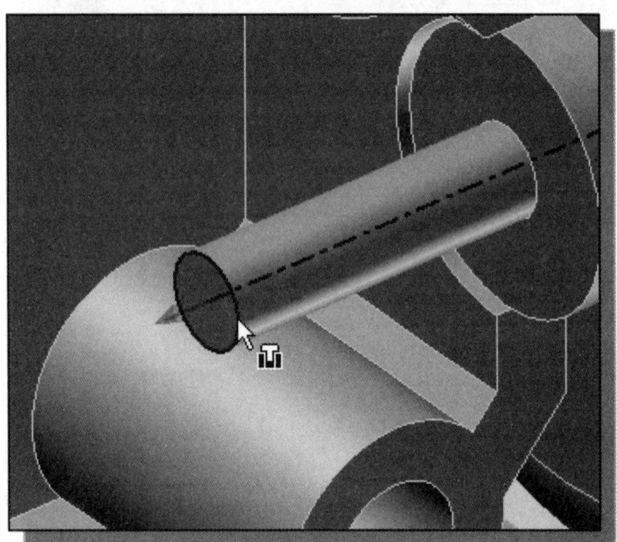

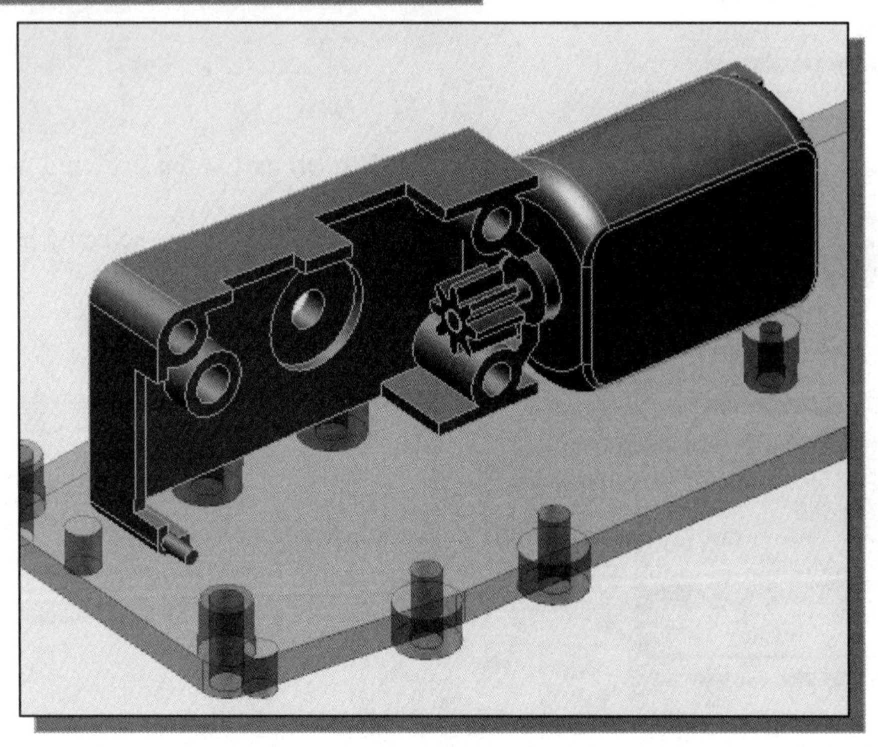

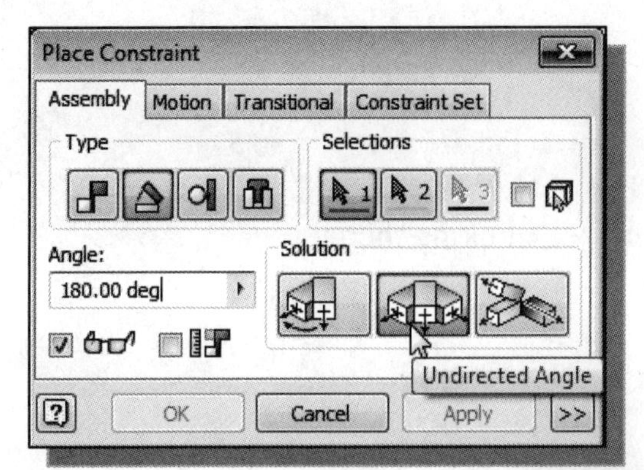

13. Choose the **Angular** constraint (Undirected Angle option and **180 degrees**).

14. Select the **YZ Plane** of the *Motor* part.

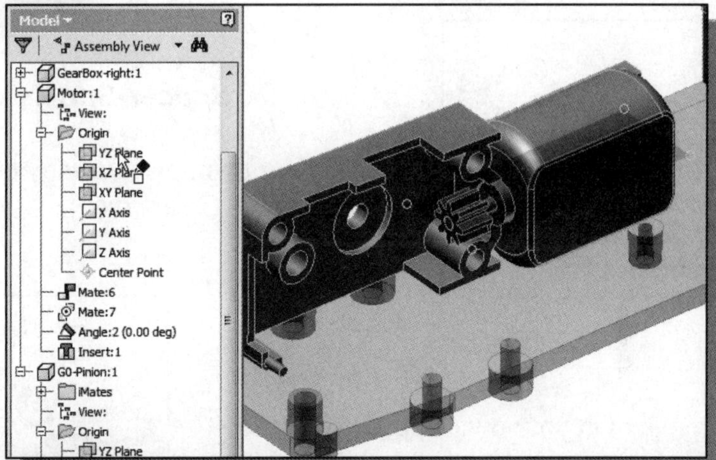

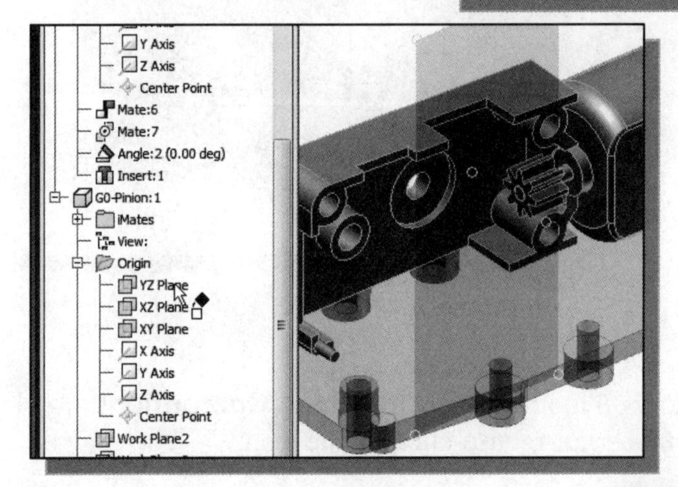

15. Select the **YZ plane** of the *G0-Pinion* part.

16. Set the *Angle* to **45 degrees** and click **Apply** to complete the assembly of the *G0-Pinion* part.

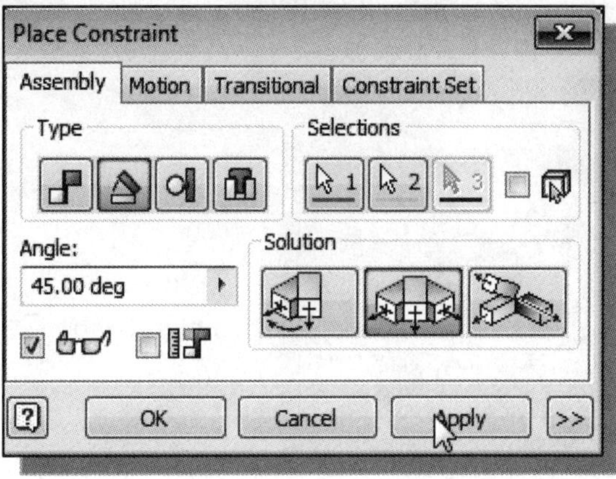

Assemble the G1 Gear

1. In the *Ribbon* toolbar, select the **Place Component** command by left-mouse-clicking the icon.

2. Select the ***Spacer-Short*** part in the list window.

3. On your own, place a copy of the selected part on the screen next to the *GearBox*.

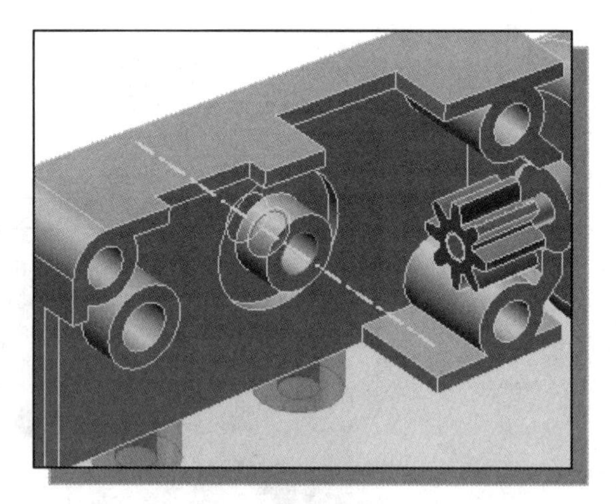

4. On your own, apply two **Mate** constraints to assemble the ***Spacer-Short*** as shown.

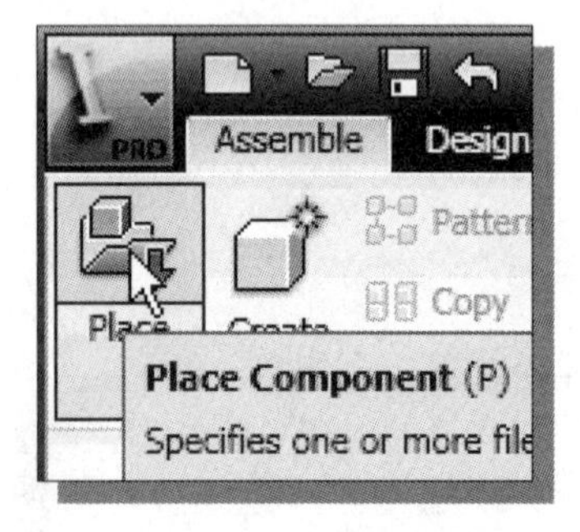

5. In the *Ribbon* toolbar, select the **Place Component** command by left-mouse-clicking the icon.

6. Select the ***G1-Spur Gear*** part in the list window.

7. On your own, place a copy of the selected part on the screen next to the *GearBox*.

8. On your own, apply two **Mate** constraints to assemble the *G1-Spur Gear* part next to the *Spacer-Short* part as shown.

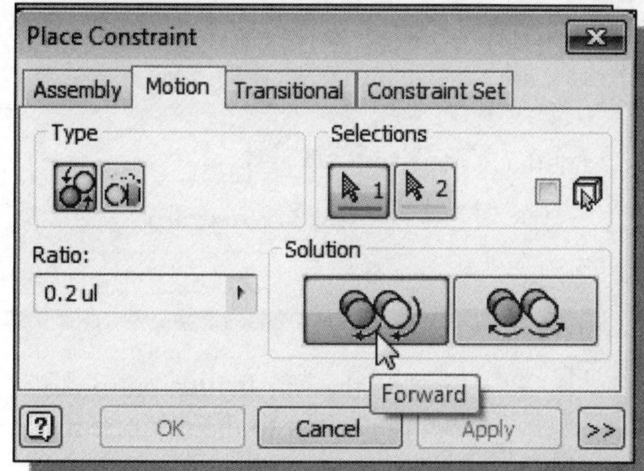

9. In the *Place Constraint* dialog box, select the **Motion** tab.

10. Enter **0.2** as the gear *Ratio*. (Why is the gear ratio 0.2?)

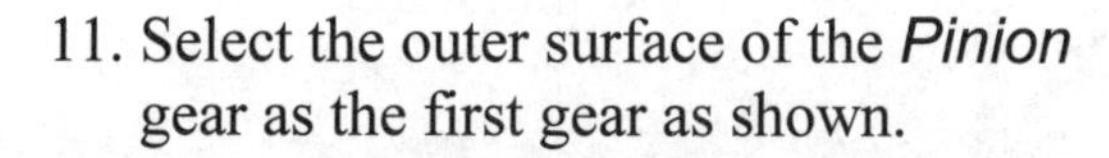

11. Select the outer surface of the *Pinion* gear as the first gear as shown.

12. Select the outer surface of the *G1-Spur Gear* as the second gear as shown.

13. Click **Apply** to accept the settings.

Animation with the *Inventor* Drive Constraint Tool

- *Autodesk Inventor's* **Drive Constraint** tool allows us to perform basic motion analysis by creating animations of assemblies with moving parts.

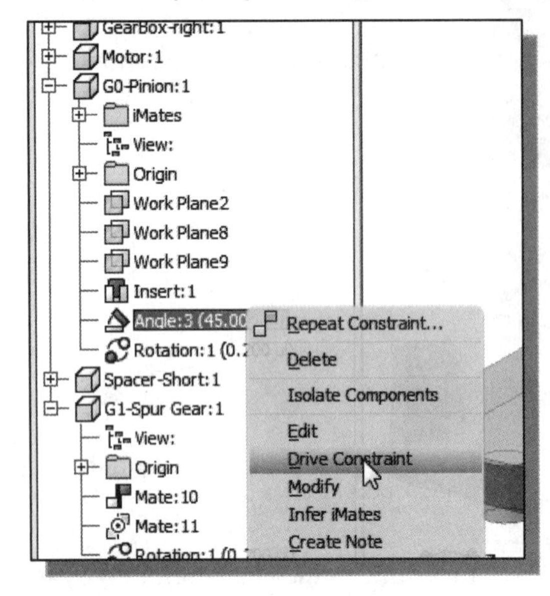

1. In the *Model History Tree* window, right-mouse-click on the **Angle** constraint (Pinion gear) to bring up the option menu and select **Drive Constraint** as shown.

2. In the *Drive Constraint* dialog box, enter **45.00** as the starting angle and **410.00** as the ending angle as shown.

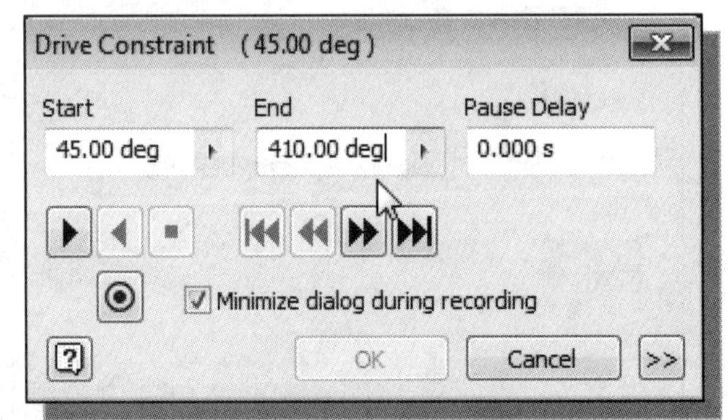

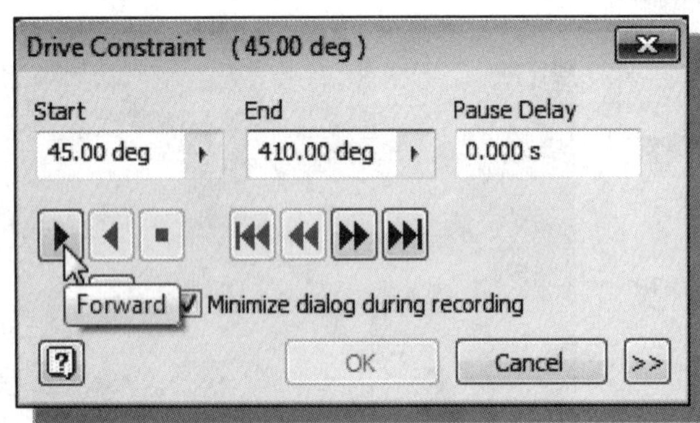

3. Click on the **Minimum** button to reset the driver to the minimum position.

4. Click on the **Forward** button to begin the animation.

5. Click on the **Reverse** button to animate in the reversed direction.

6. Click **Cancel** to exit the Drive Constraint command.

Assemble the G2 Gear and the G3 Gear

1. In the *Ribbon* toolbar, select the **Place Component** command by left-mouse-clicking the icon.

2. Select the ***Spacer-Short*** part in the list window.

3. On your own, place a copy of the selected part on the screen next to the *GearBox*.

4. On your own, apply two **Mate** constraints to assemble the second *Spacer-Short* as shown.

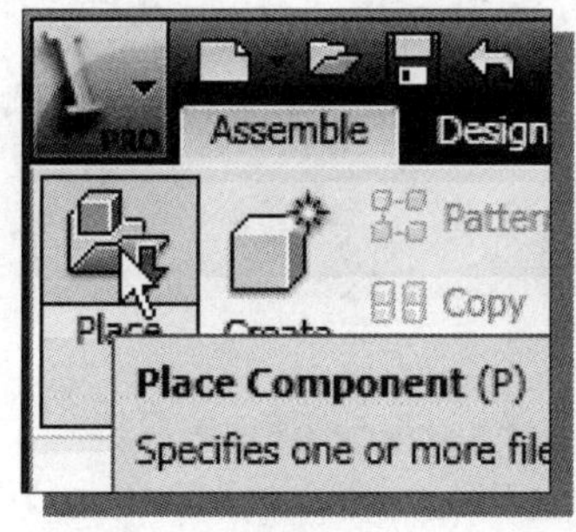

5. In the *Ribbon* toolbar, select the **Place Component** command by left-mouse-clicking the icon.

6. Select the ***G2-Spur Gear*** part in the list window.

7. On your own, place a copy of the selected part on the screen next to the *GearBox*.

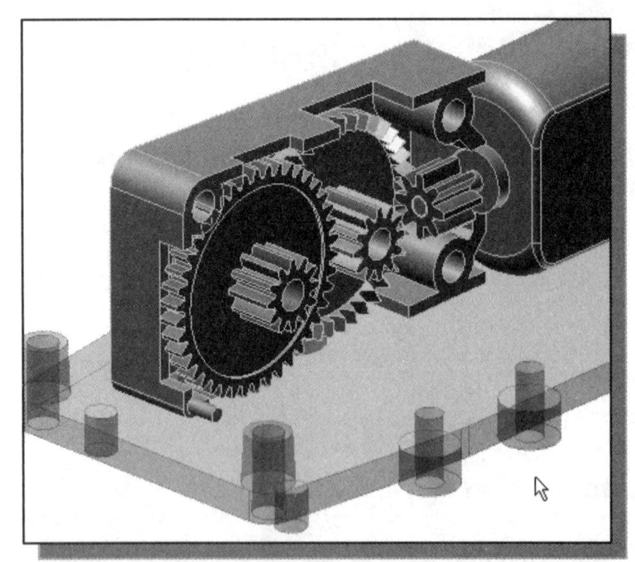

8. On your own, apply two **Mate** constraints to assemble the *G2-Spur Gear* part so that it is aligned to the *Spacer-Short* part as shown.

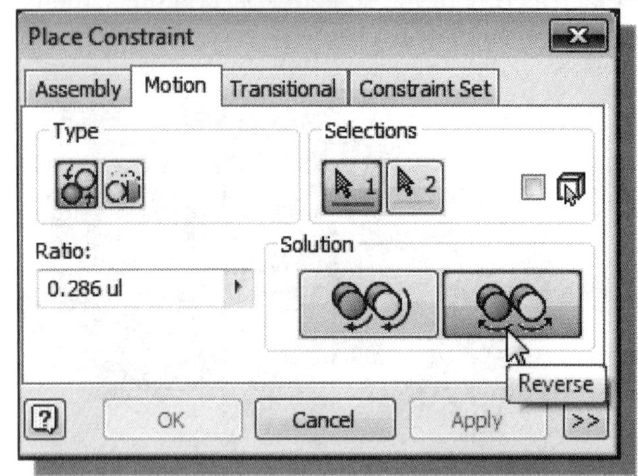

9. In the *Place Constraint* dialog box, select the **Motion** tab.

10. Enter **0.286** as the gear *Ratio*. (Why is the gear ratio 0.286?)

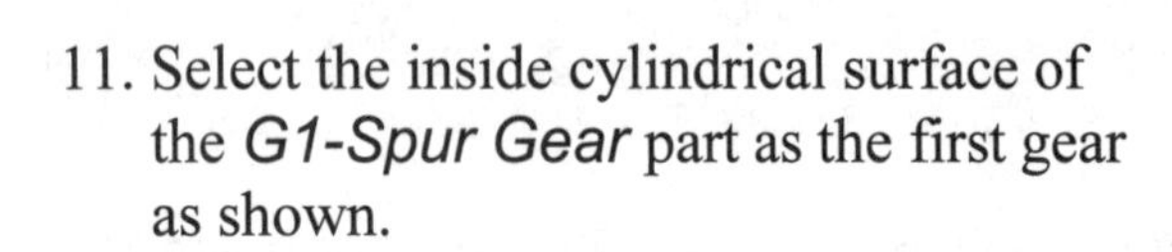

11. Select the inside cylindrical surface of the *G1-Spur Gear* part as the first gear as shown.

12. Select the outer surface of the *G2-Spur Gear* as the second gear as shown.

13. Click **Apply** to accept the settings.

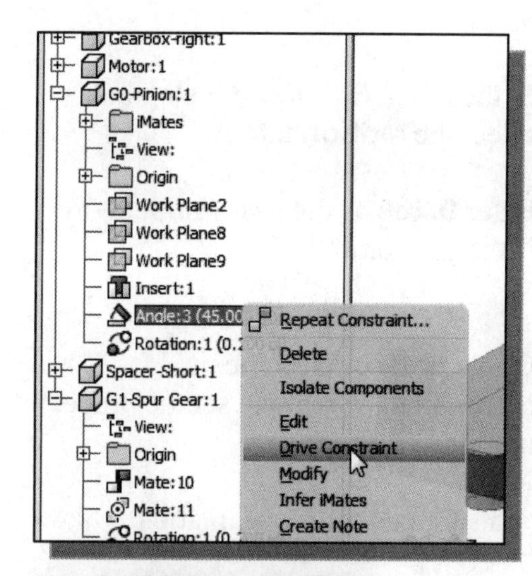

14. In the *Model History Tree* window, right-mouse-click on the **Angle** constraint (Pinion gear) to bring up the option menu and select **Drive Constraint** as shown.

15. On your own, confirm the gears are moving correctly with the Drive Constraint tool.

16. In the *Ribbon* toolbar, select the **Place Component** command by left-mouse-clicking the icon.

17. Select the ***G3-Spur Gear*** part in the list window.

18. On your own, place a copy of the selected part on the screen next to the *GearBox*.

19. On your own, apply two **Mate** constraints to assemble the *G3-Spur Gear* part so that it is aligned to the *G1-Spur Gear* part as shown.

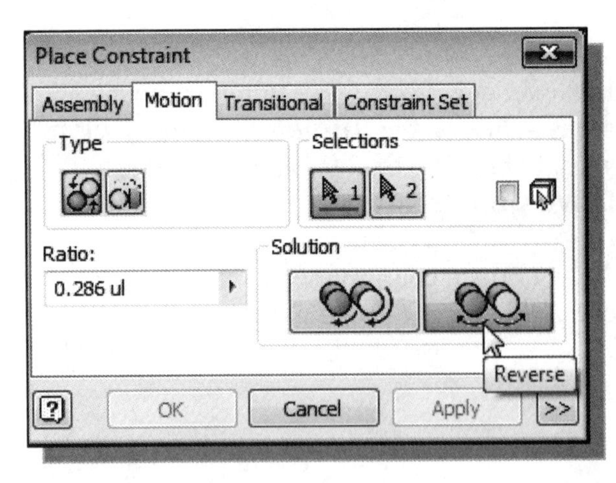

20. In the *Place Constraint* dialog box, select the **Motion** tab.

21. Enter **0.286** as the gear *Ratio*. (Why is the gear ratio 0.286?)

22. On your own, select the corresponding gears to set the proper motion constraint.

23. On your own, confirm the gears are properly constrained for the animation.

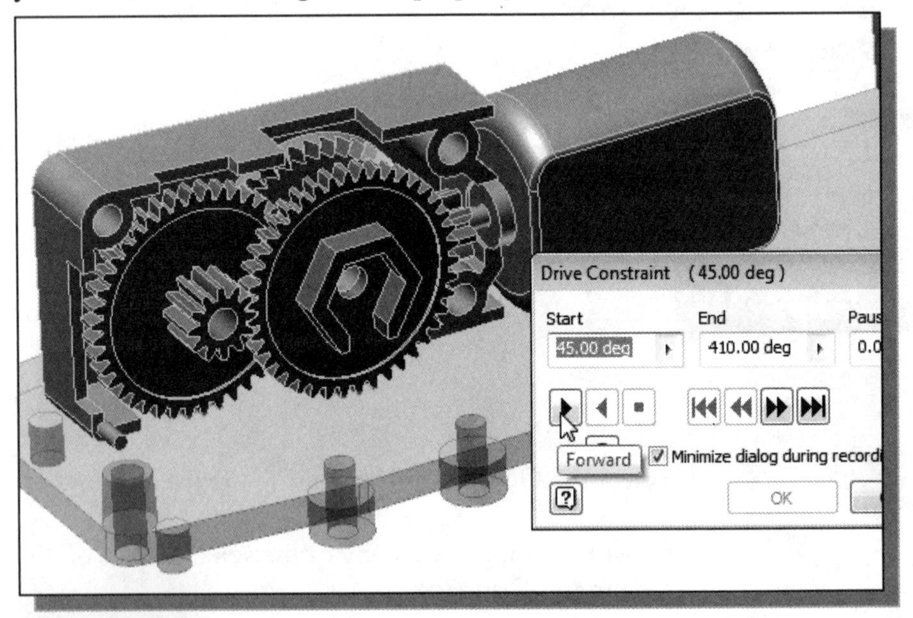

24. In the *Ribbon* toolbar, select the **Place Component** command by left-mouse-clicking the icon.

25. Select the ***Hex-Shaft with Collar*** part in the list window.

26. On your own, place a copy of the selected part on the screen next to the *GearBox*.

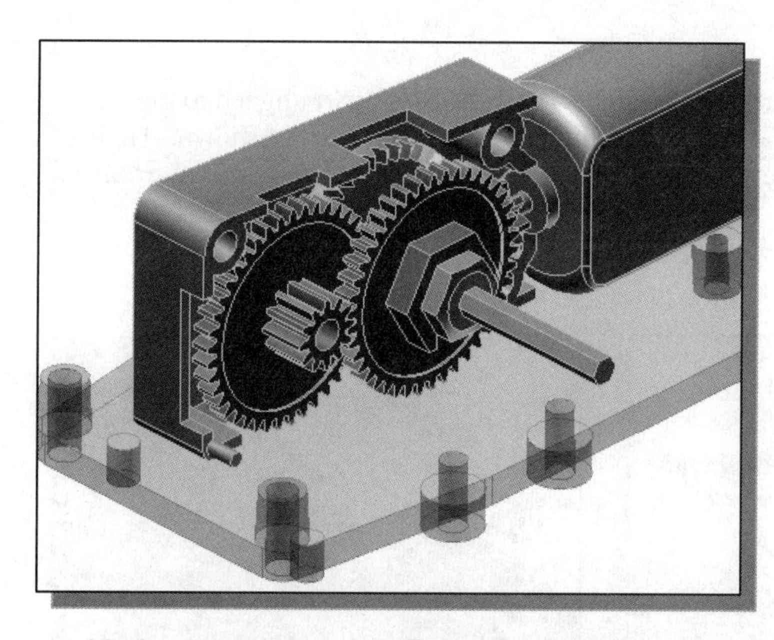

27. On your own, apply three constraints to assemble the ***Hex-Shaft-Collar*** part next to the ***G3-Spur Gear*** part as shown.

- The *Shaft* should be constrained so that it moves with the *G3-Spur Gear.*

28. On your own, use the **Drive Constraint** option to confirm that the animation of the gears is still accurate.

29. On your own, assemble the ***M1-Spacer*** part next to the ***G2-Spur Gear*** part using two **Mate** constraints as shown.

30. On your own, assemble the ***Hex-Shaft*** part aligned to the ***G2-Spur Gear*** part and the ***GearBox-right*** part as shown. (Hint: Use the work planes that pass through the center of the ***Shaft***.)

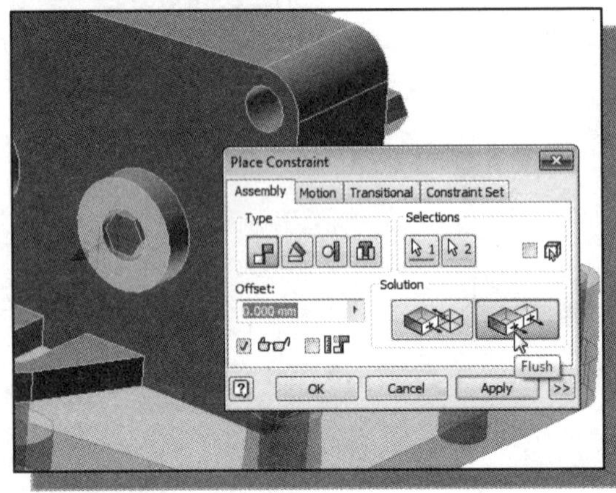

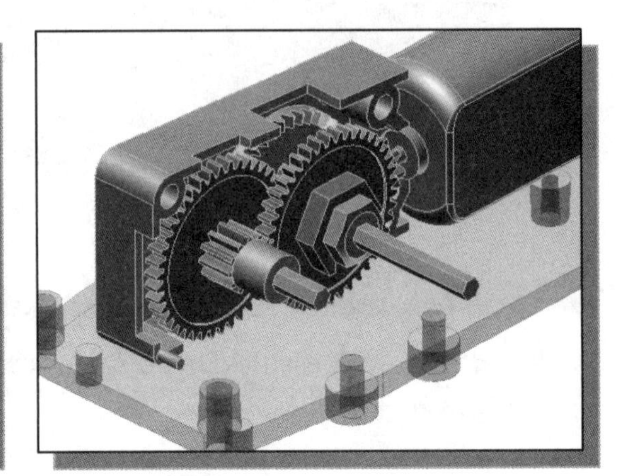

31. On your own, assemble the ***GearBox-left*** part as shown.

32. On your own, use the **Drive Constraint** command to confirm the animation of the gears is still accurate.

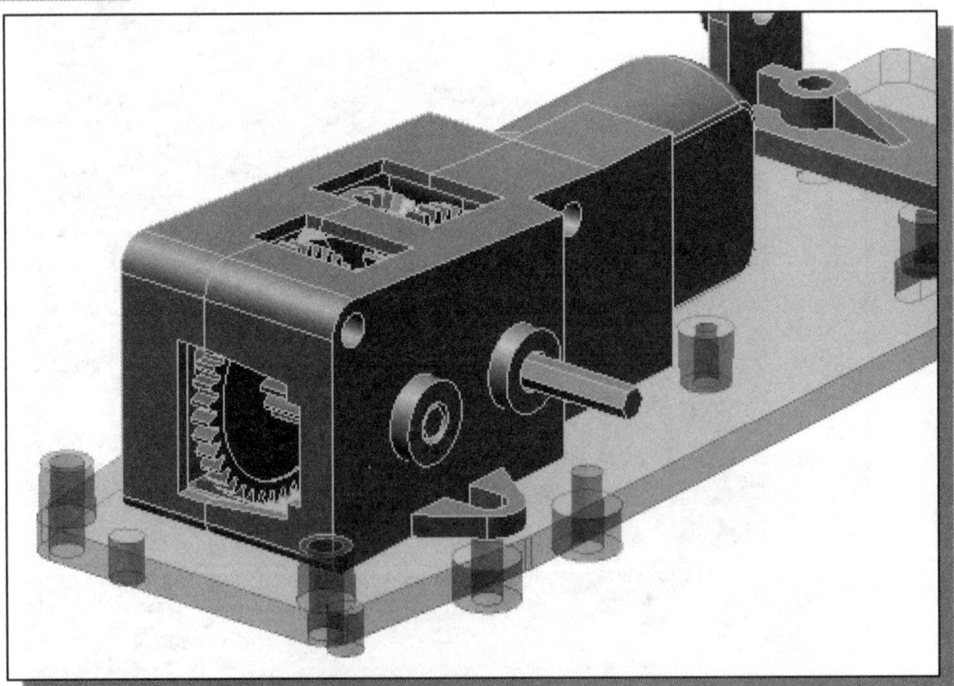

Assemble the Crank Parts

1. In the *Ribbon* toolbar, select the **Place Component** command by left-mouse-clicking the icon.

2. Select the ***A9-Crank-Right*** part in the list window.

3. On your own, place a copy of the selected part on the screen next to the *GearBox*.

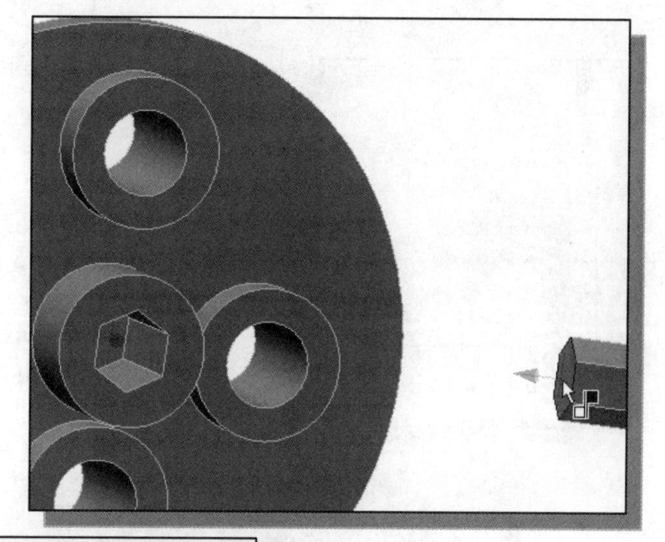

4. On your own, first apply a **Mate** constraint to align the end surface of the *Crank* to the *Hex-Shaft* part as shown

5. Apply a **Mate** constraint to align the center axis of the *Crank* to the *GearBox-right* part as shown.

6. Apply another **Mate** constraint to align the *Crank* by constraining the corresponding work plane to the *Hex-Shaft-Collar* part as shown.

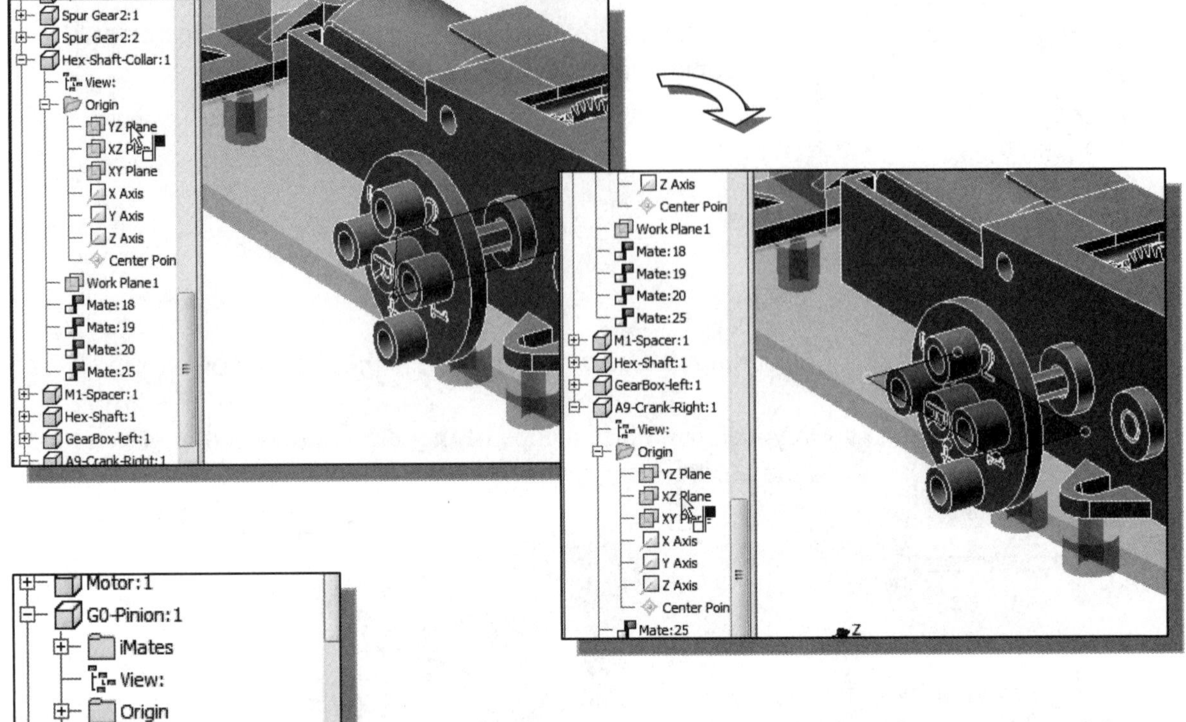

7. On your own, use the **Drive Constraint** option on *G0-Pinion* to confirm the animation is still accurate.

8. Set the *Repetitions* option to **5** in the extra option list to control the length of the animation as shown.

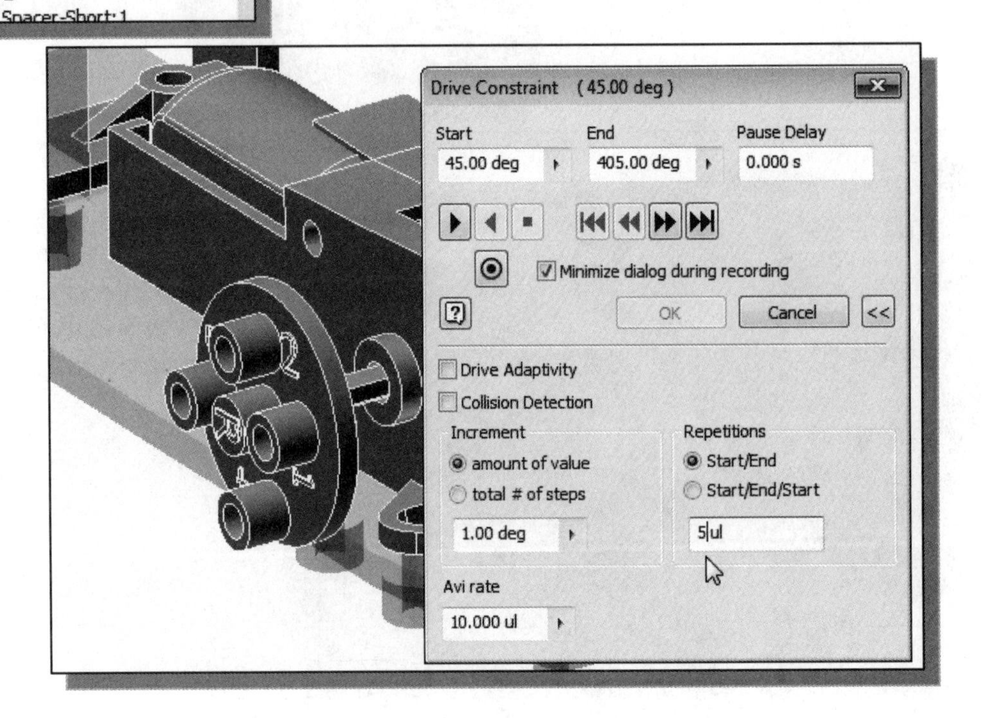

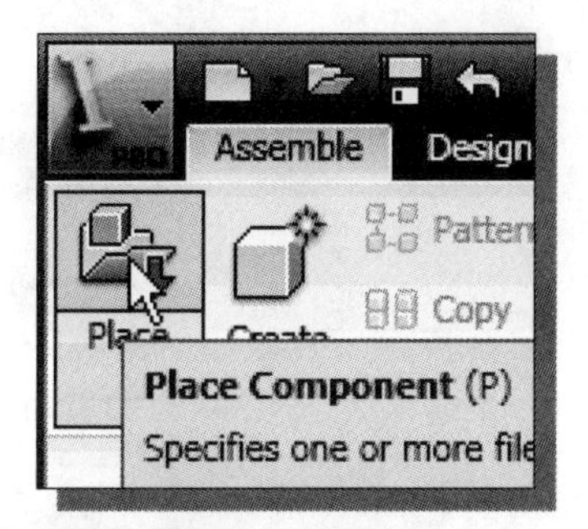

9. In the *Ribbon* toolbar, select the **Place Component** command by left-mouse-clicking the icon.

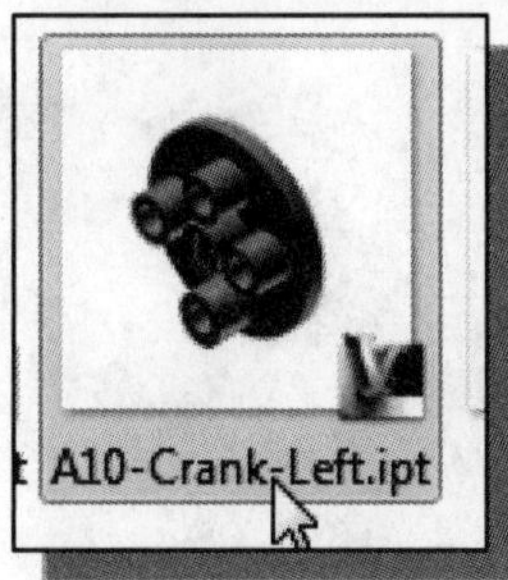

10. Select the ***A10-Crank-Left*** part in the list window.

11. On your own, place a copy of the selected part on the screen next to the *GearBox*.

12. On your own, first apply a **Mate** constraint to align the end surface of the *Crank* to the *Hex-Shaft* as shown

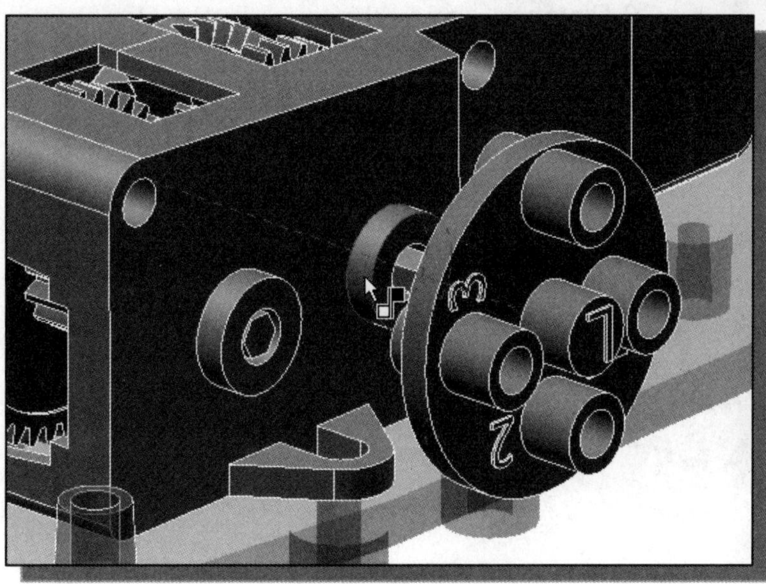

13. Apply a **Mate** constraint to align the center axis of the *Crank* to the *GearBox-right* part as shown.

14. Adjust the crank, using **drag and drop**, so that the ***hole No. 1*** is roughly aligned **180 degrees** to the *hole No. 1* on the *A9-Crank-Right* part.

15. Apply a **Flush** constraint to align the *Crank* by constraining the corresponding work plane to the *Hex-Shaft-Collar* part as shown.

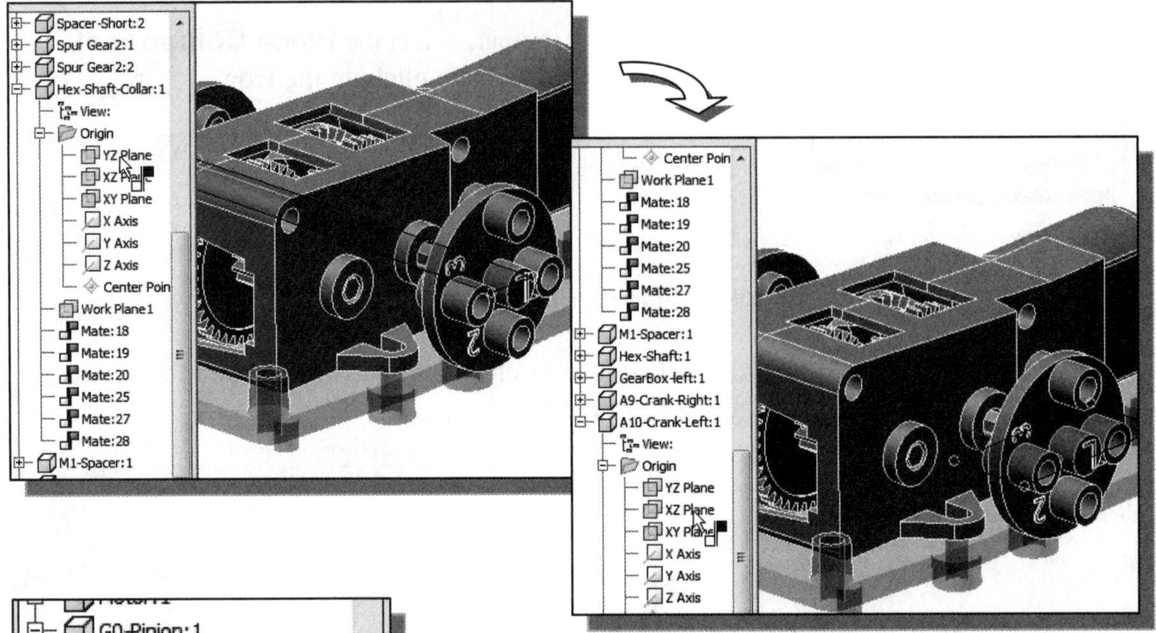

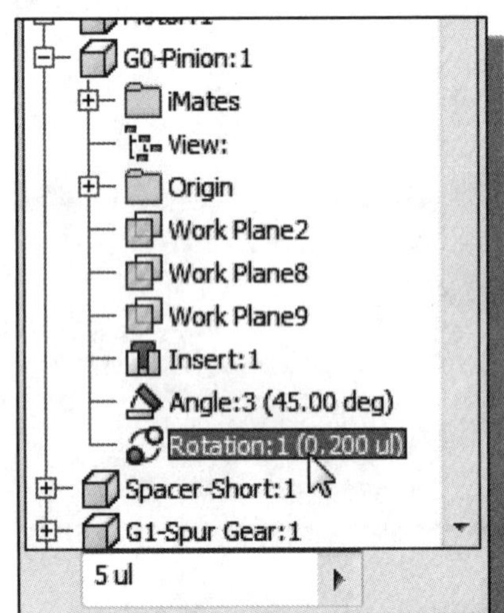

16. In the *browser* window, click on the **Rotation** option on *G0-Pinion* and adjust the gear ratio to **5** in the edit box located near the bottom of the screen as shown.

- By changing the gear ratio of the first pair of the gear train, the speed of the entire gear train is increased.

17. On your own, use the **Drive Constraint** option on *G0-Pinion* to confirm the animation is still correct.

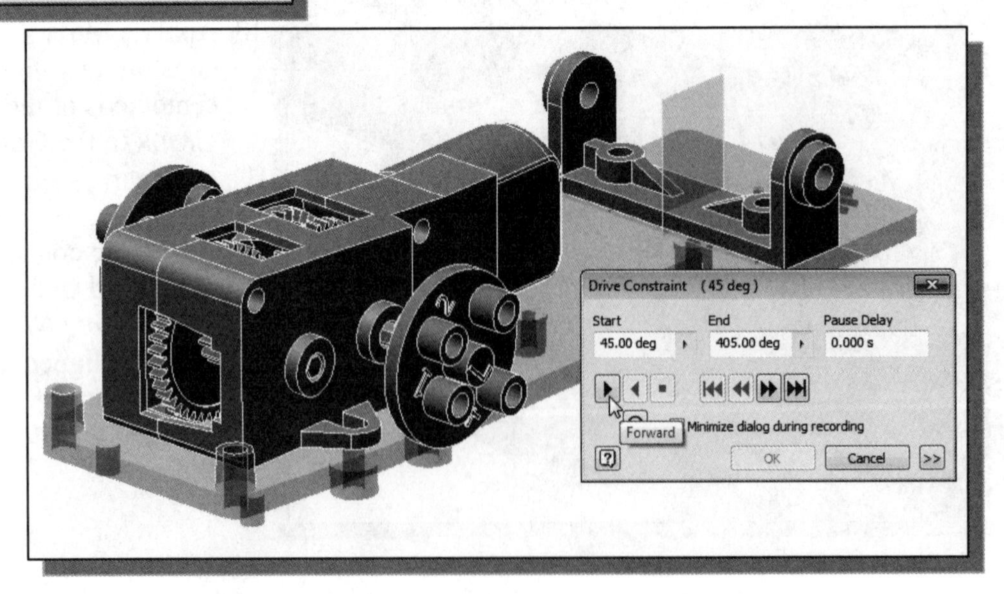

Assemble the Rear Shaft and Legs

1. In the *Ribbon* toolbar, select the **Place Component** command by left-mouse-clicking the icon.

2. Select the ***Shaft-3x80*** part in the list window.

3. On your own, place a copy of the selected part on the screen next to the *GearBox*.

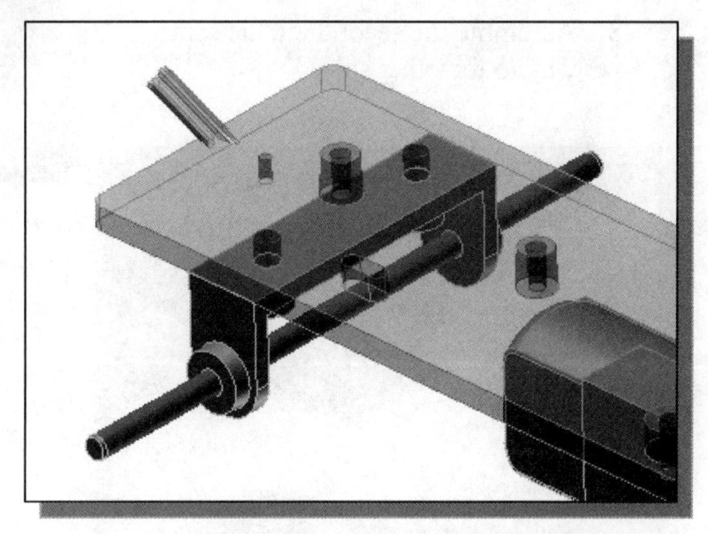

4. On your own, apply the necessary constraints to align the *Shaft* to the *Rear Axle Support* as shown

5. On your own, assemble two copies of the ***A3-Spacer*** part to the *Rear Axle Support* as shown.

6. On your own, place four copies of the ***Knee-Leg Sub-Assembly*** near the assembly model.

- In *Autodesk Inventor*, a subassembly is treated the same way as a single part during assembly.

7. Assemble one of the rear legs, so that the center axis is aligned to the *Shaft*. Also apply a **Mate** constraint with an offset distance of **2.0 mm** to the *A3-Spacer* as shown.

8. Assemble the second leg the same way to the other side.

9. Assemble the left front leg, with the center axis aligned to ***hole No. 1***, the center of the hole that is nearest to the center axis of the *Crank* model.

10. Also apply a **Mate** constraint with an offset distance of **2.0 mm** to the *A10-Crank* as shown.

11. Assemble the last leg the same way to the other side. (Hint: Also align the center axis to *hole No. 1*.)

12. On your own, use the **Constrained Move** option and adjust the orientation of the four legs as shown.

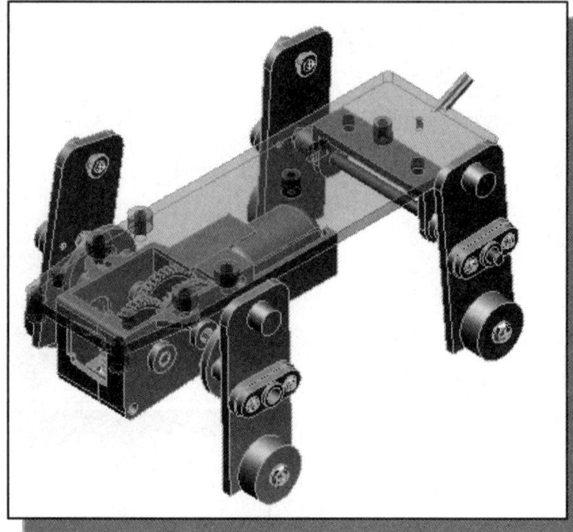

Assemble the Linkage-Rods

1. In the *Ribbon* toolbar, select the **Place Component** command by left-mouse-clicking the icon.

2. Select the ***Linkage-Rod*** part in the list window.

3. On your own, place four copies of the selected part on the screen next to the assembly model.

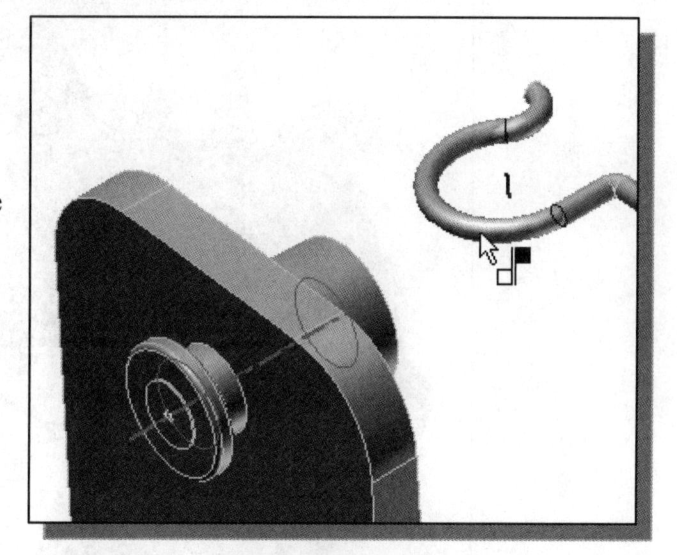

4. Apply a **Mate** constraint to align to the upper axis on one of the rear legs, so that the center axis on the *Linkage Rod* is aligned as shown.

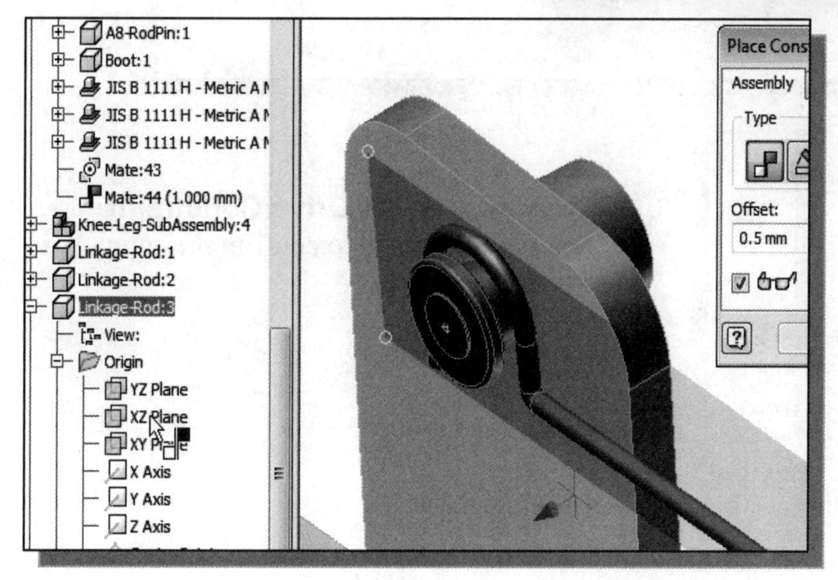

5. Apply a **Mate** constraint with an offset distance of **0.5 mm** to the **XZ Plane** to the inside surface of the leg as shown.

6. For the other end of the *Linkage Rod*, apply a **Mate** constraint to align the center axis on the *Linkage Rod* to the center axis where the front leg is attached to the *Crank* as shown.

7. On your own, repeat the above steps and assemble the other three *Linkage Rods*.

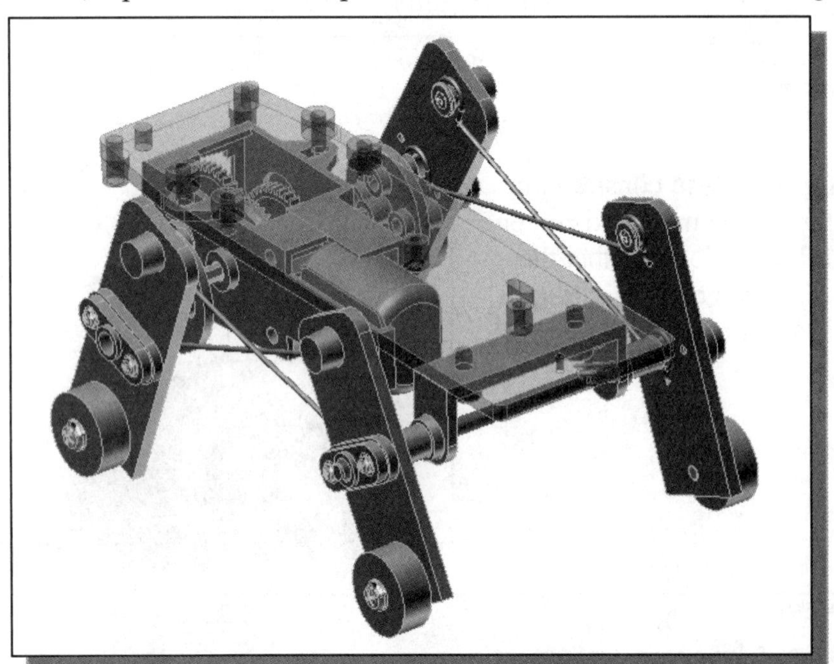

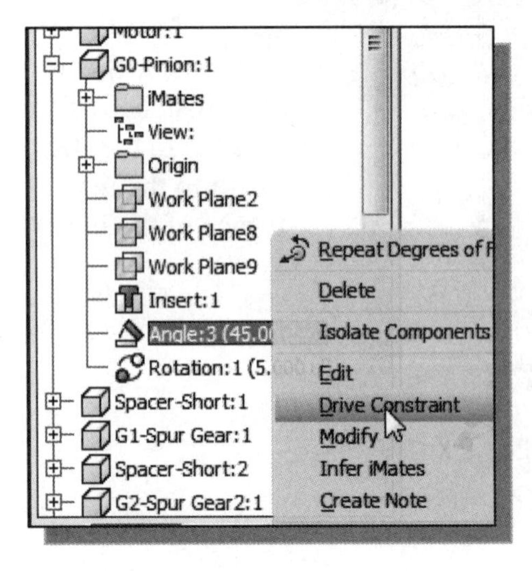

8. On your own, use the **Drive Constraint** option on *G0-Pinion* to confirm the animation is still correct.

Complete the Assembly Model

1. In the *Ribbon* toolbar, select the **Place Component** command by left-mouse-clicking the icon.

2. On your own, assemble the ***TigerHead*** to the assembly model as shown in the figure below.

3. On your own, assemble two copies of the ***Axle-End Cap*** parts and the ***Battery Case*** to the assembly model. (Hint: First adjust the two hole locations on the *Battery Case*.)

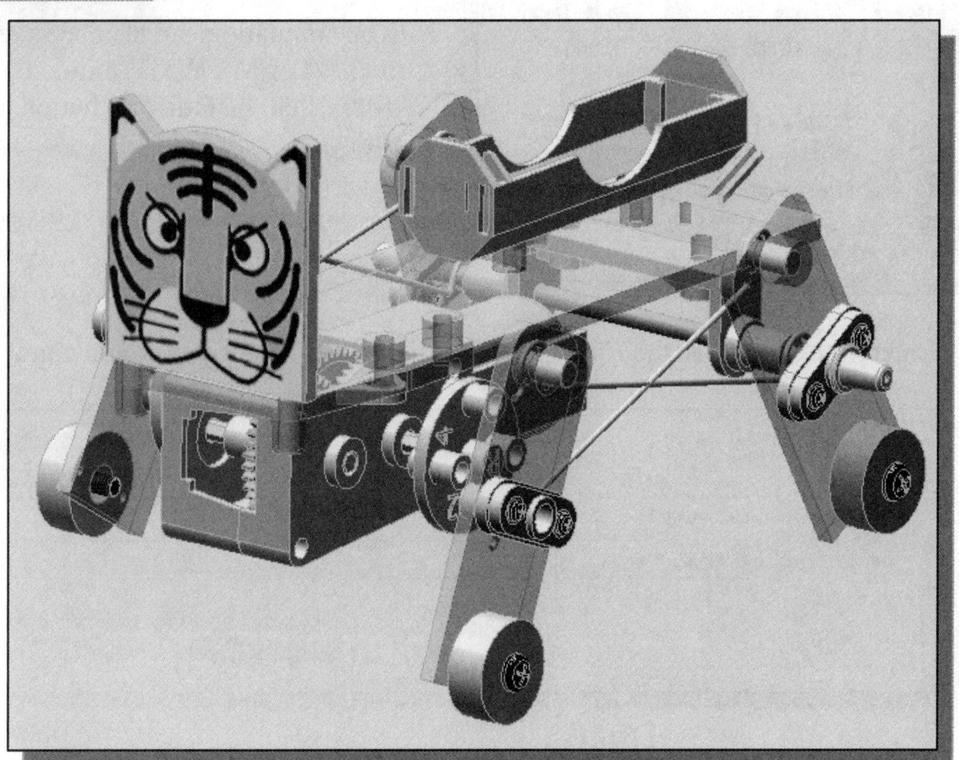

Record an Animation Movie

1. On your own, start the simulation and dynamically rotate the display while the simulation is running.

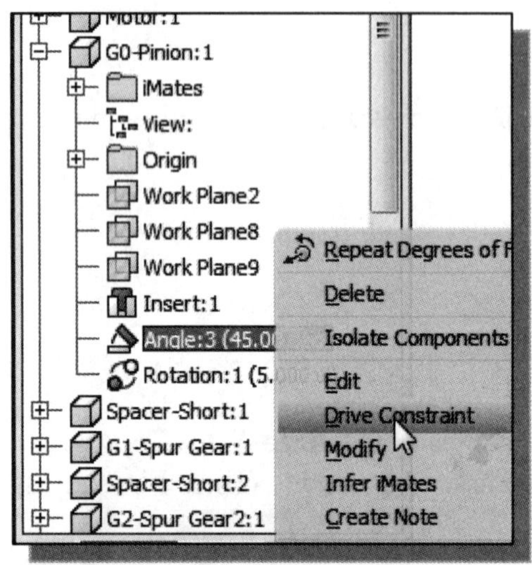

2. In the *browser* window, activate the **Drive Constraint** option by right-mouse-clicking on the **Angle** constraint of the *G0-Pinion* part as shown.

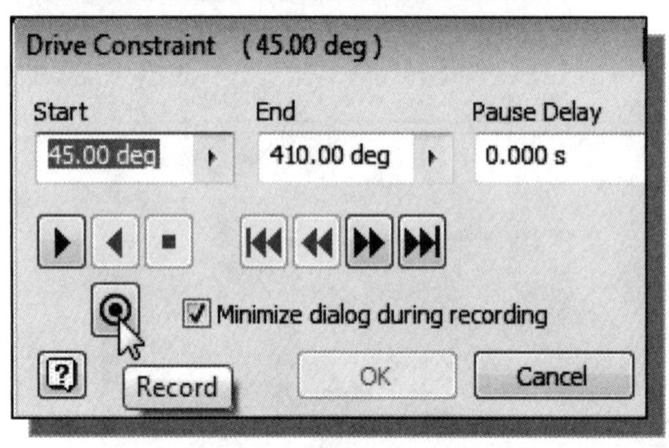

3. The simulation can also be saved as an AVI or MS WMV movie format file. Click the **Record** button as shown.

4. Enter **Tiger.wmv** as the *File name* and click **Save** to accept the file name.

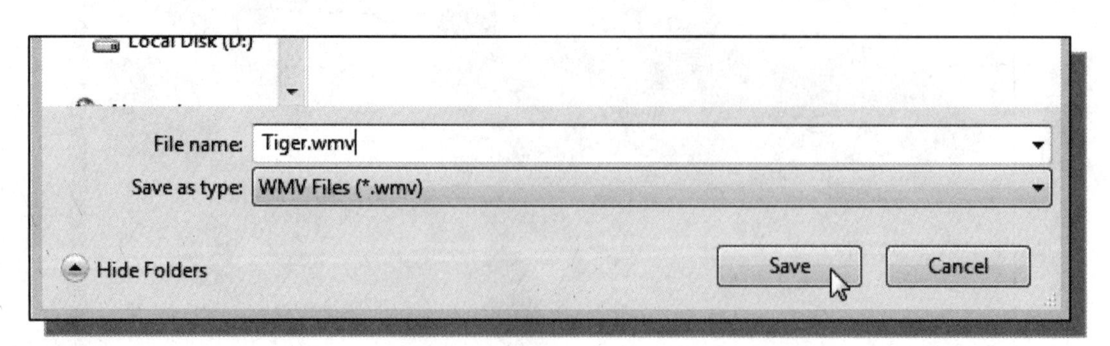

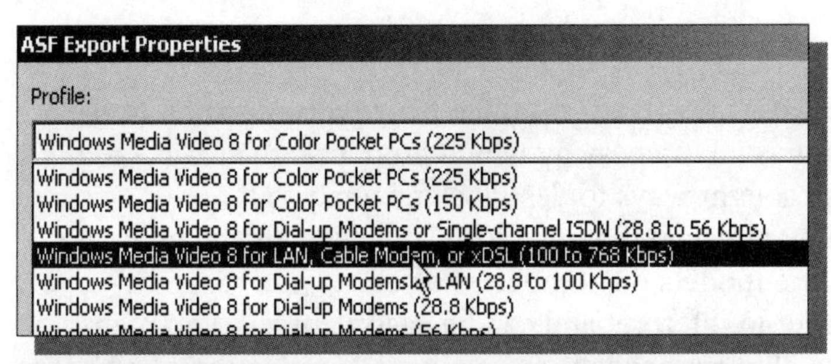

5. Select one of the profile settings to set the quality and speed of the file and click **OK** to proceed.

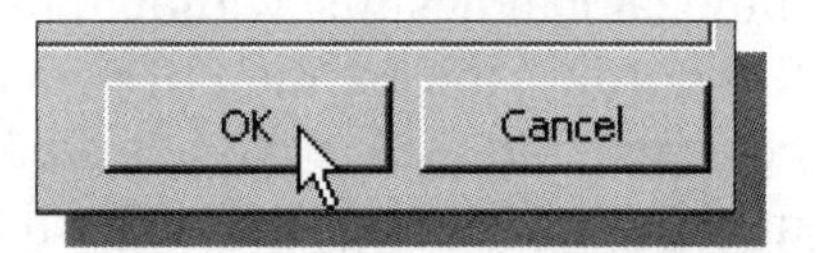

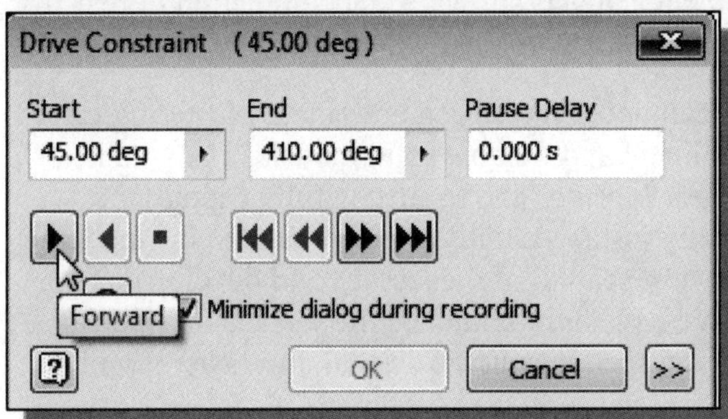

6. Click on the **Forward** button to begin the simulation, and also record the simulation.

❖ Note the *Minimize dialog during recording* option is turned *ON* by default.

7. On your own, view the recorded video file.

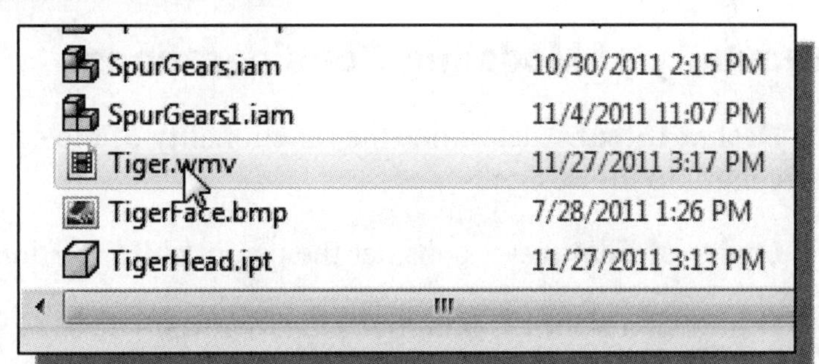

Conclusion

Engineering design includes all activities involved, from the original conception to the finished product. *Engineering design* is the process by which products are created and modified. For many years designers sought ways to describe and analyze three-dimensional designs without building physical models. With advancements in computer technology, the creation of parametric models on computers offers a wide range of benefits. Parametric models are easier to interpret and can be easily altered. Parametric models can be analyzed using finite element analysis software, and simulation of real-life loads can be applied to the models and the results graphically displayed.

Throughout this text, various modeling techniques have been presented. Mastering these techniques will enable you to create intelligent and flexible solid models. The goal is to make use of the tools provided by *Autodesk Inventor* and to successfully capture the **DESIGN INTENT** of the product. In many instances, only one approach to the modeling tasks was presented; you are encouraged to repeat all of the lessons and develop different ways of accomplishing the same tasks. We have only scratched the surface of *Autodesk Inventor's* functionality. The more time you spend using the system, the easier it will be to perform parametric modeling with *Autodesk Inventor*.

Summary of Modeling Considerations

- **Design Intent** – determine the functionality of the design; select features that are central to the design.
- **Order of Features** – consider the parent/child relationships necessary for all features.
- **Dimensional and Geometric Constraints** – the way in which the constraints are applied determines how the components are updated.
- **Relations** – consider the orientation and parametric relationships required between features and in an assembly.

Questions:

1. What is the purpose of using ***assembly constraints***?

2. List three of the commonly used ***assembly constraints***.

3. Describe the difference between the **Mate** constraint and the **Flush** constraint.

4. In an assembly, can we place more than one copy of a part? How is it done?

5. How should we determine the assembly order of different parts in an assembly model?

6. Can we perform a constrained move on fully constrained components?

7. Can *Autodesk Inventor* calculate the center of gravity of an assembly model? How do you activate this option?

8. Can assembly constraints be temporarily disabled in an assembly model? How?

9. Can *Autodesk Inventor* calculate the weight of an assembly model? How is this done?

Exercises: (Time: 60 minutes)

1. **Ratchet Mechanism**

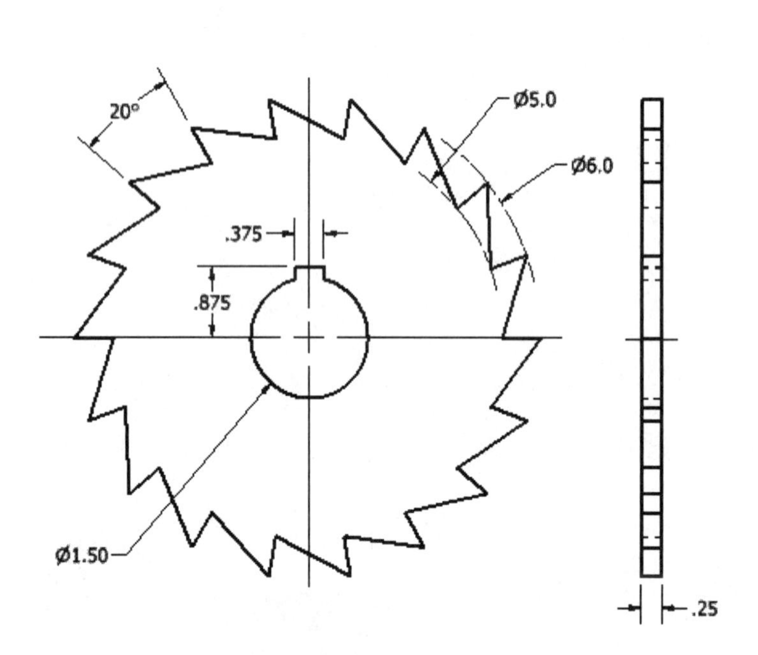

(1) Create the *Ratchet Plate*.
(2) Design the *Ratchet Pawl* part.
(3) Create an assembly model.
(4) Use the Contact Solver to perform a motion analysis.

2. **Quick Return Mechanism**

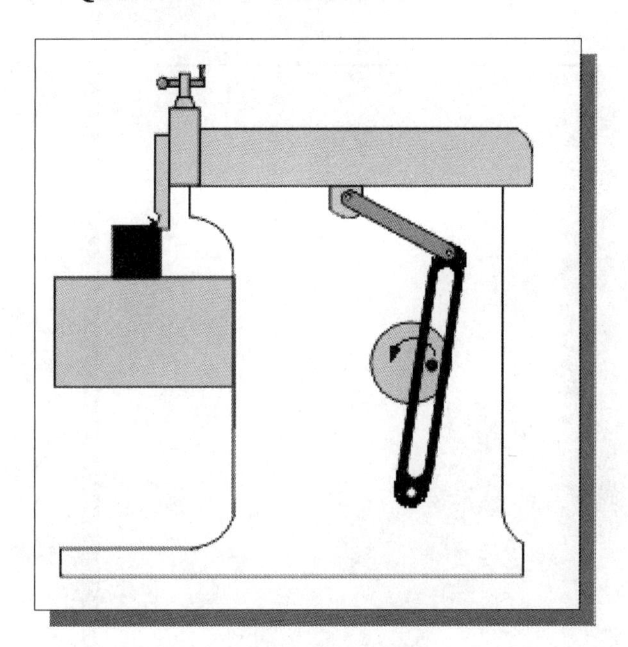

Design and create the necessary parts.

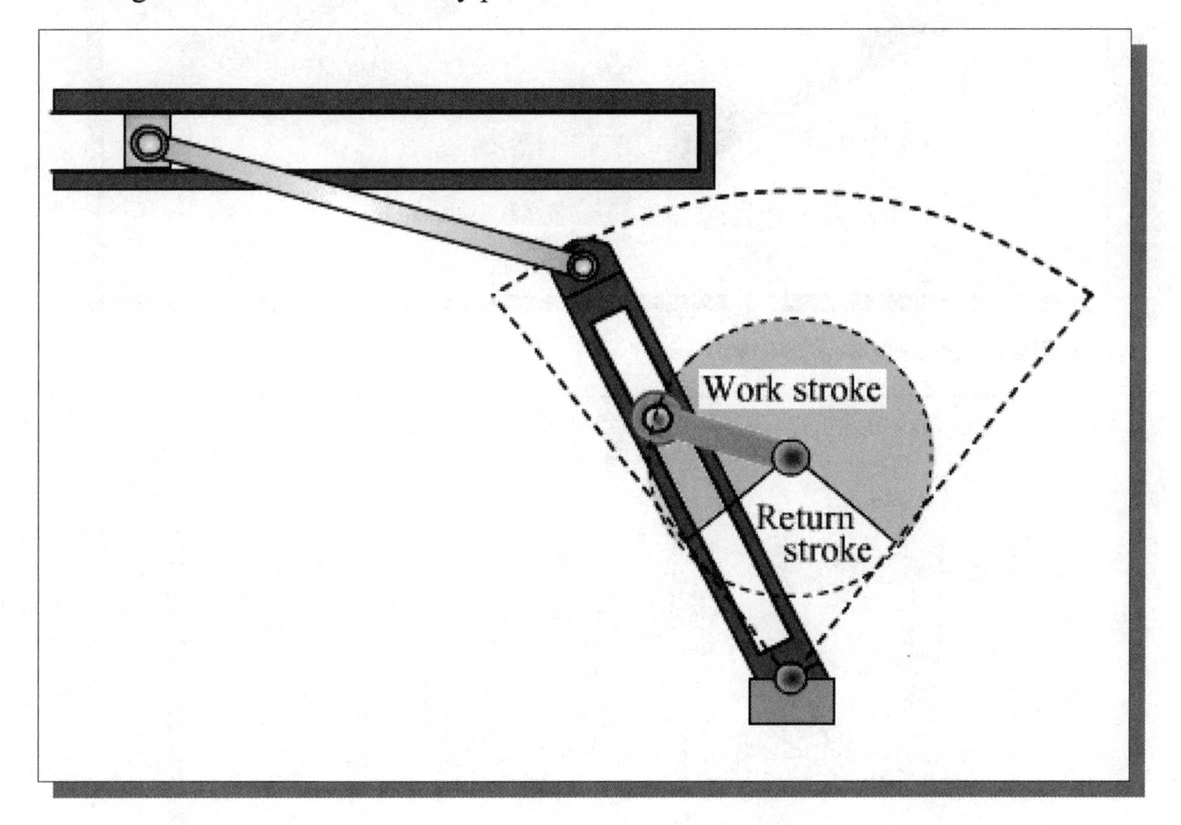

3. **Leveling Assembly** (Create a set of detail and assembly drawings. All dimensions are in mm.)

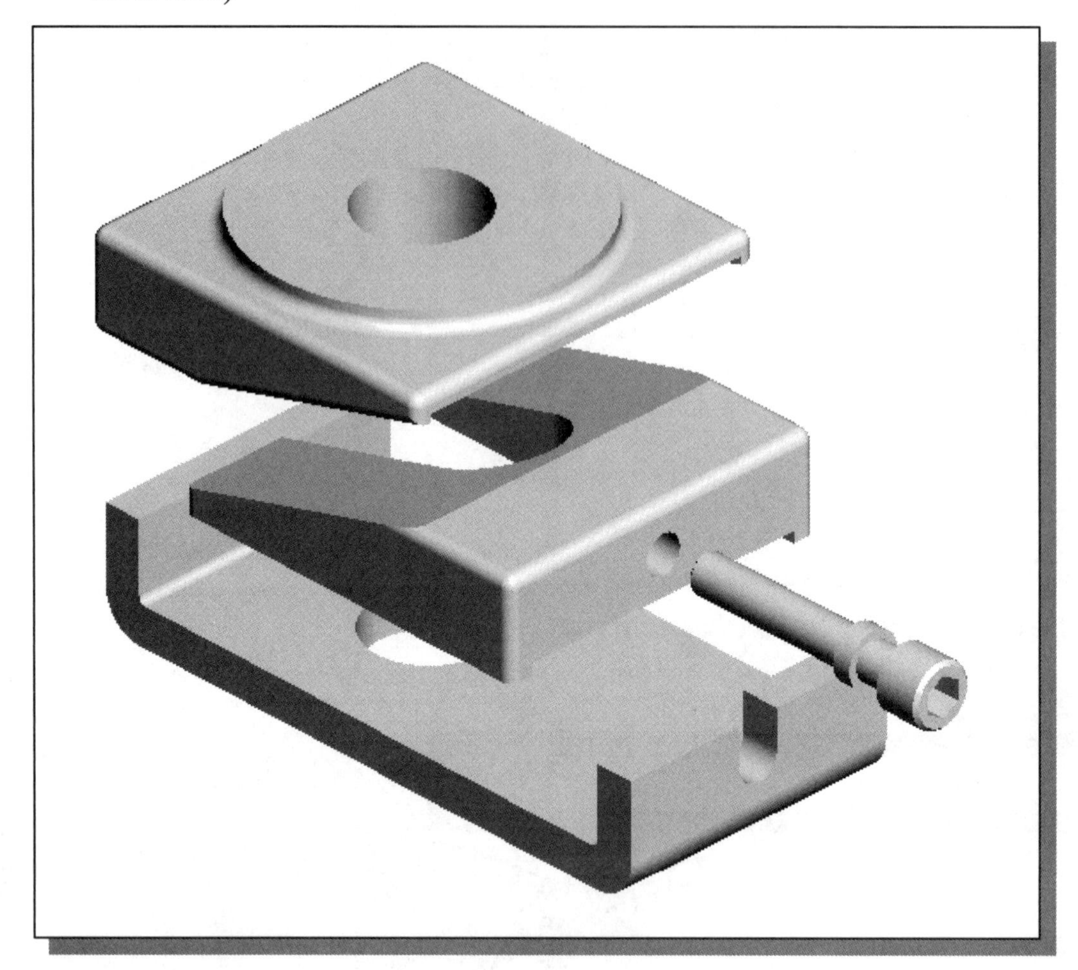

(a) **Base Plate**

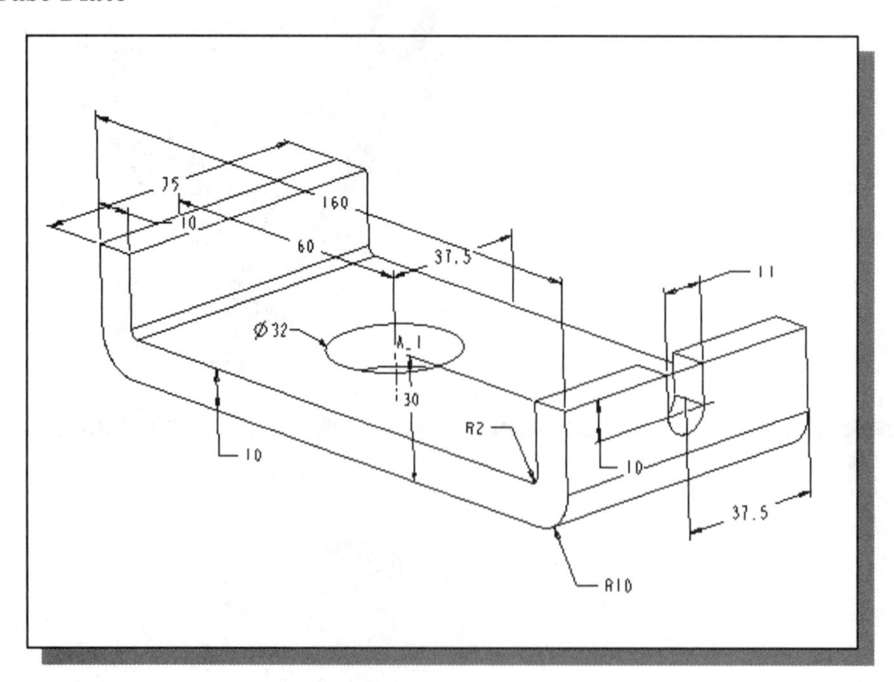

(b) **Sliding Block** (Rounds & Fillets: R3)

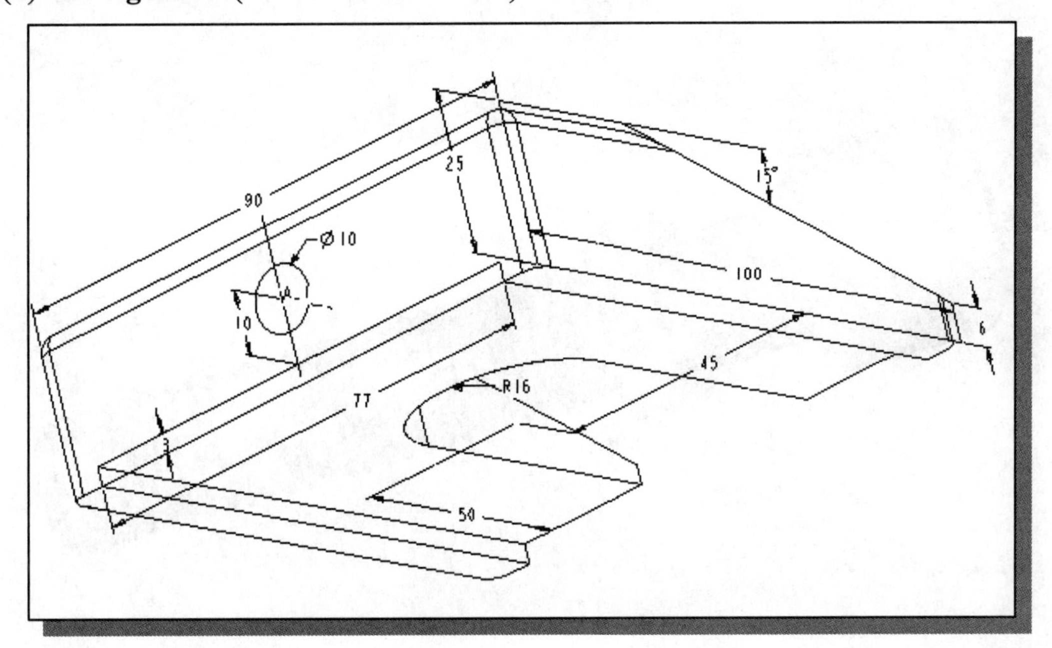

(c) **Lifting Block** (Rounds & Fillets: R3)

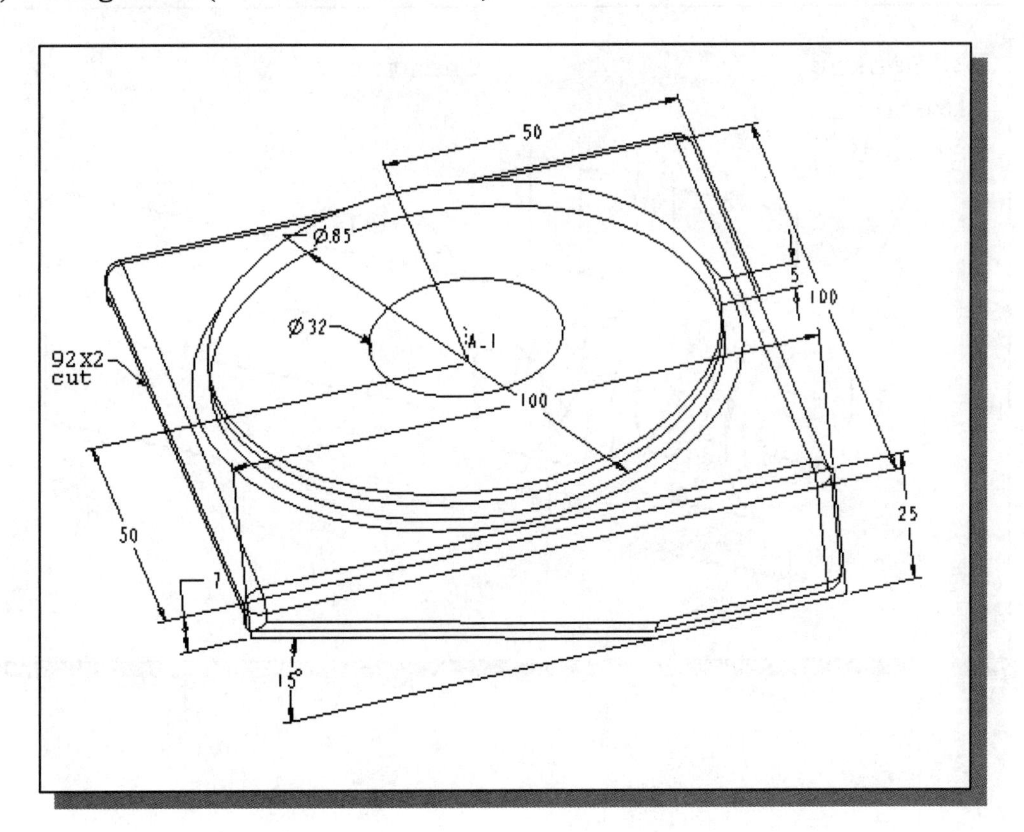

(d) **Adjusting Screw** (M10 X 1.5)

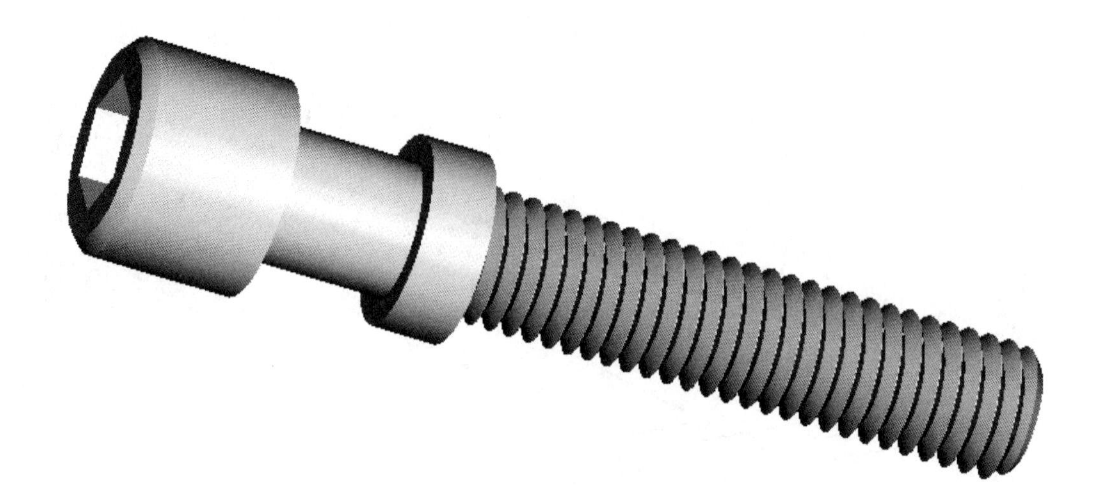

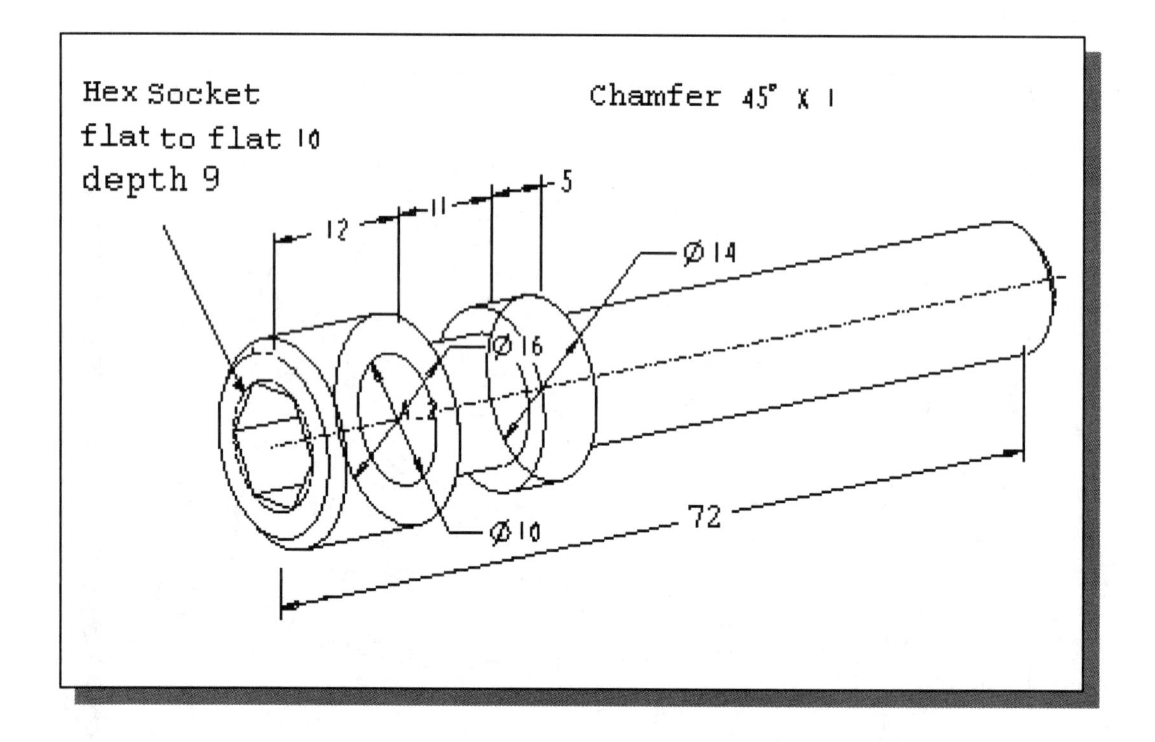

4. **Vise Assembly** (Create a set of detail and assembly drawings. All dimensions are in inches.)

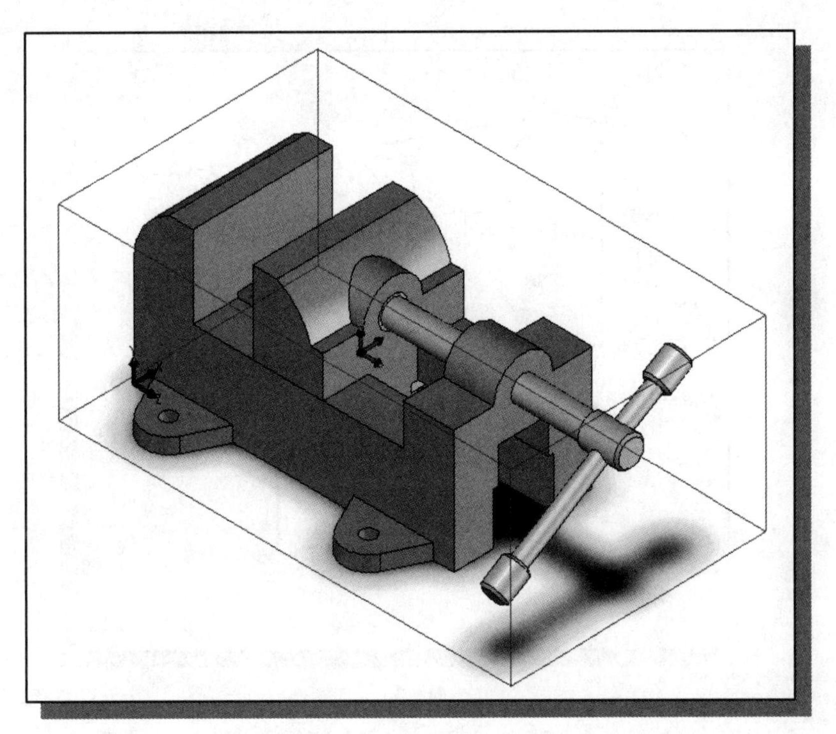

(a) **Base:** The 1.5 inch wide and 1.25 inch wide slots are cut through the entire base. Material: Gray Cast Iron.

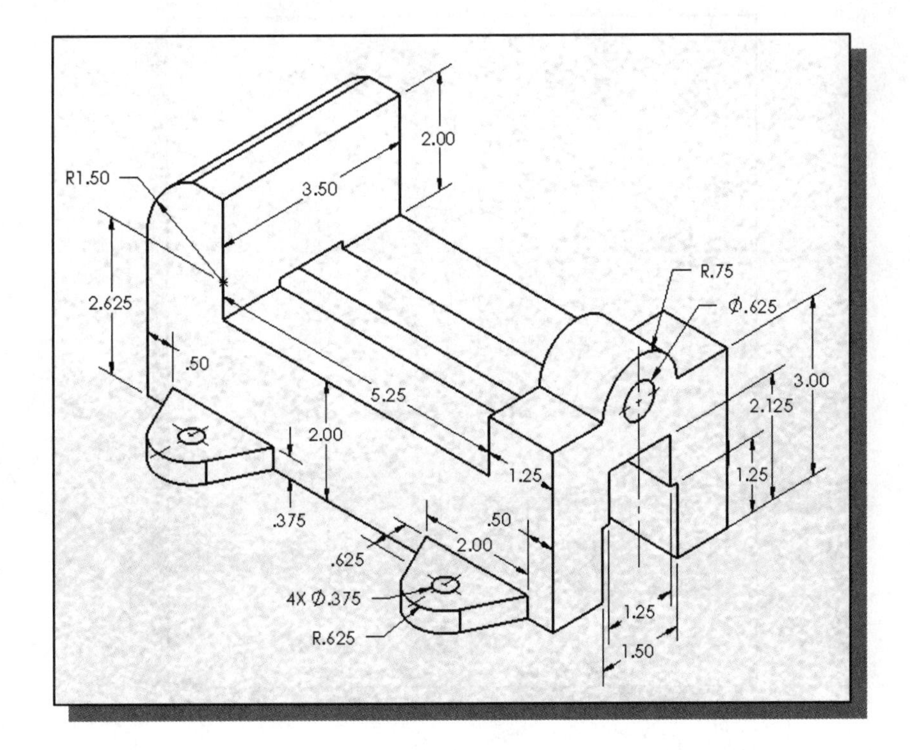

(b) **Jaw:** The shoulder of the jaw rests on the flat surface of the base and the jaw opening is set to 1.5 inches. Material: Gray Cast Iron.

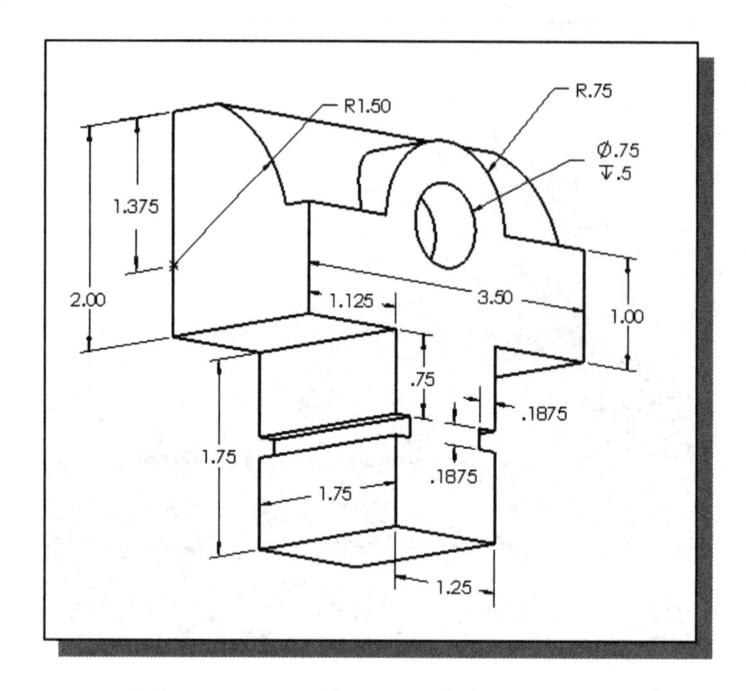

(c) **Key:** 0.1875 inch H x 0.375 inch W x 1.75 inch L. The keys fit into the slots on the jaw with the edge faces flush as shown in the sub-assembly to the right. Material: Alloy Steel.

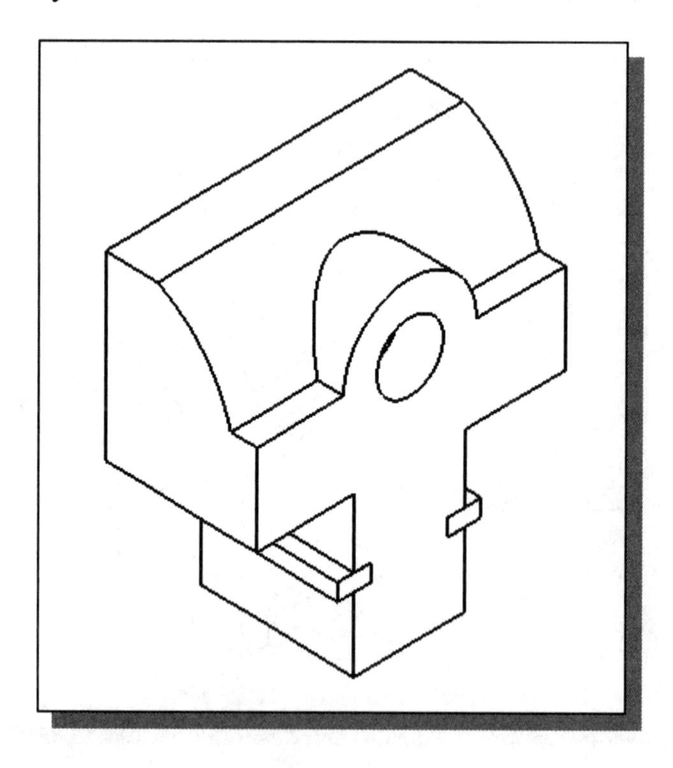

(d) **Screw:** There is one chamfered edge (0.0625 inch x 45º). The flat ∅ 0.75″ edge of the screw is flush with the corresponding recessed ∅ 0.75 face on the jaw. Material: Alloy Steel.

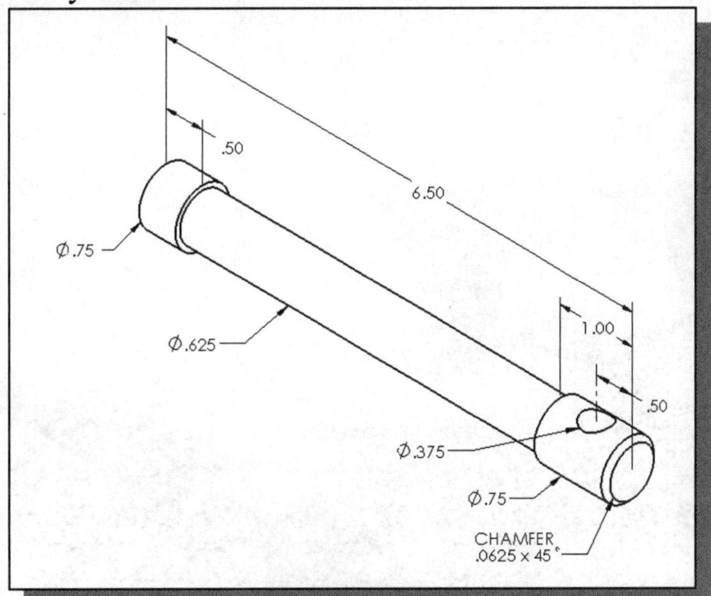

(e) **Handle Rod:** ∅ 0.375″ x 5.0″ L. The handle rod passes through the hole in the screw and is rotated to an angle of 30º with the horizontal as shown in the assembly view. The flat ∅ 0.375″ edges of the handle rod are flush with the corresponding recessed ∅ 0.735″ faces on the handle knobs. Material: Alloy Steel.

(f) **Handle Knob:** There are two chamfered edges (0.0625 inch x 45º).The handle knobs are attached to each end of the handle rod. The resulting overall length of the handle with knobs is 5.50″. The handle is aligned with the screw so that the outer edge of the upper knob is 2.0″ from the central axis of the screw. Material: Alloy Steel.

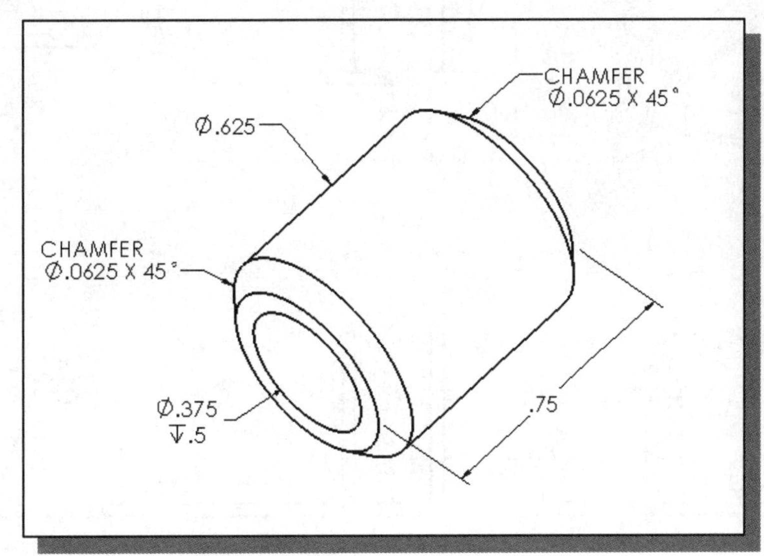

5. **Toggle-Clamp Assembly** (Time: 80 minutes)
(Create a set of detail and assembly drawings. All dimensions are in inches.)

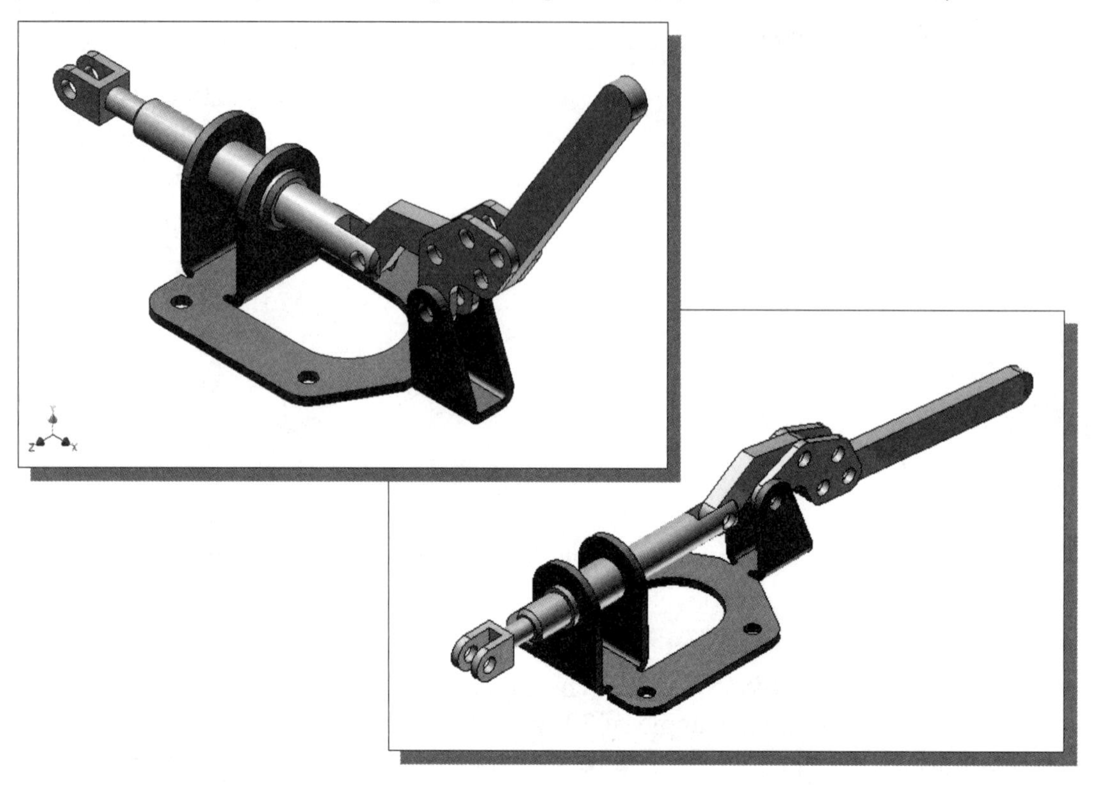

(a) **Sheet Metal Base**

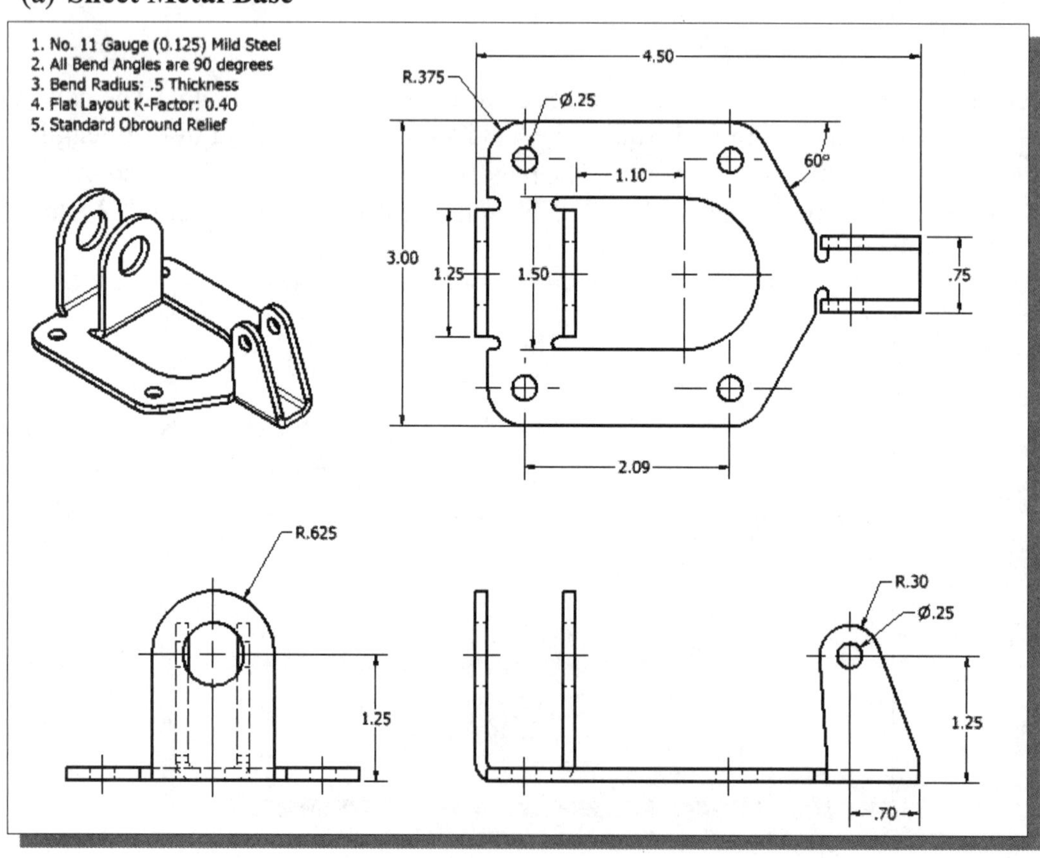

(b) **Connector**

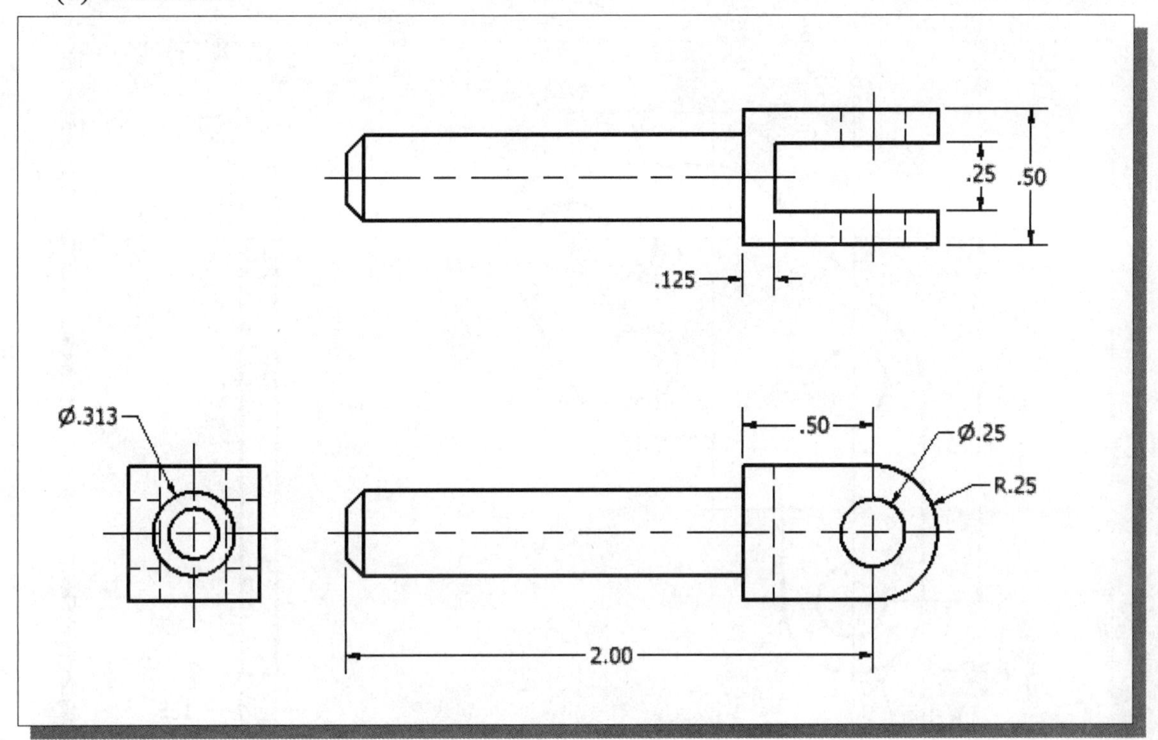

(c) **Handle**

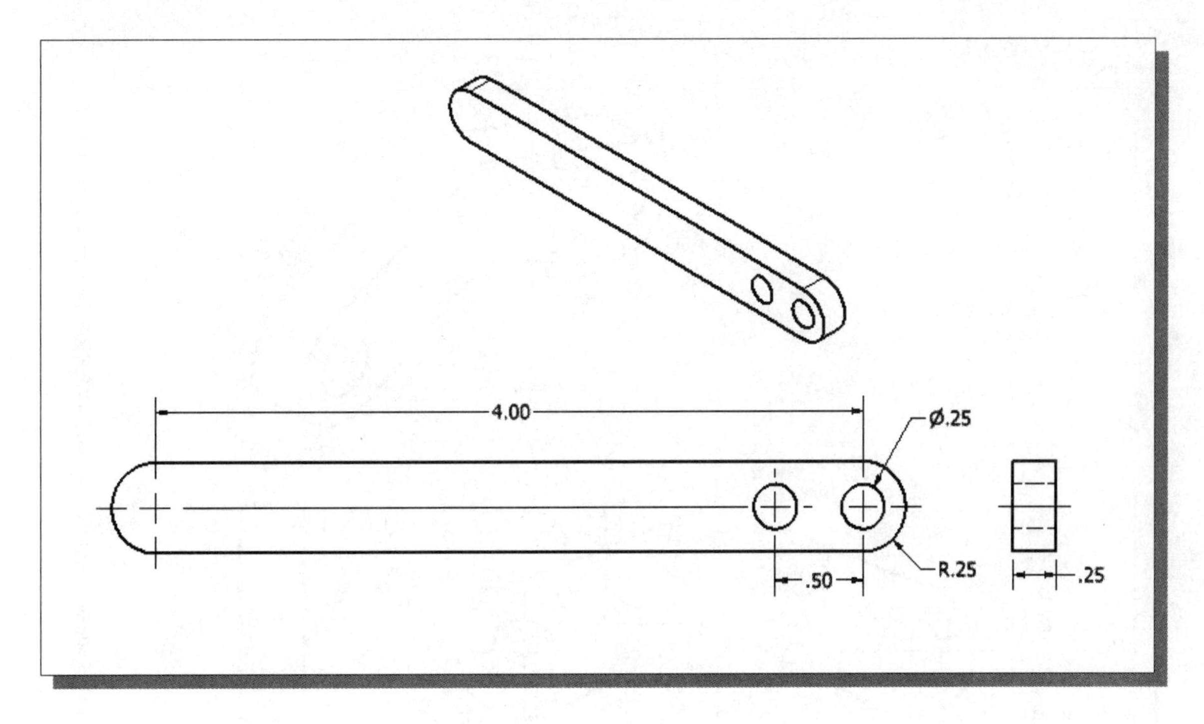

(d) **Joint Plate**

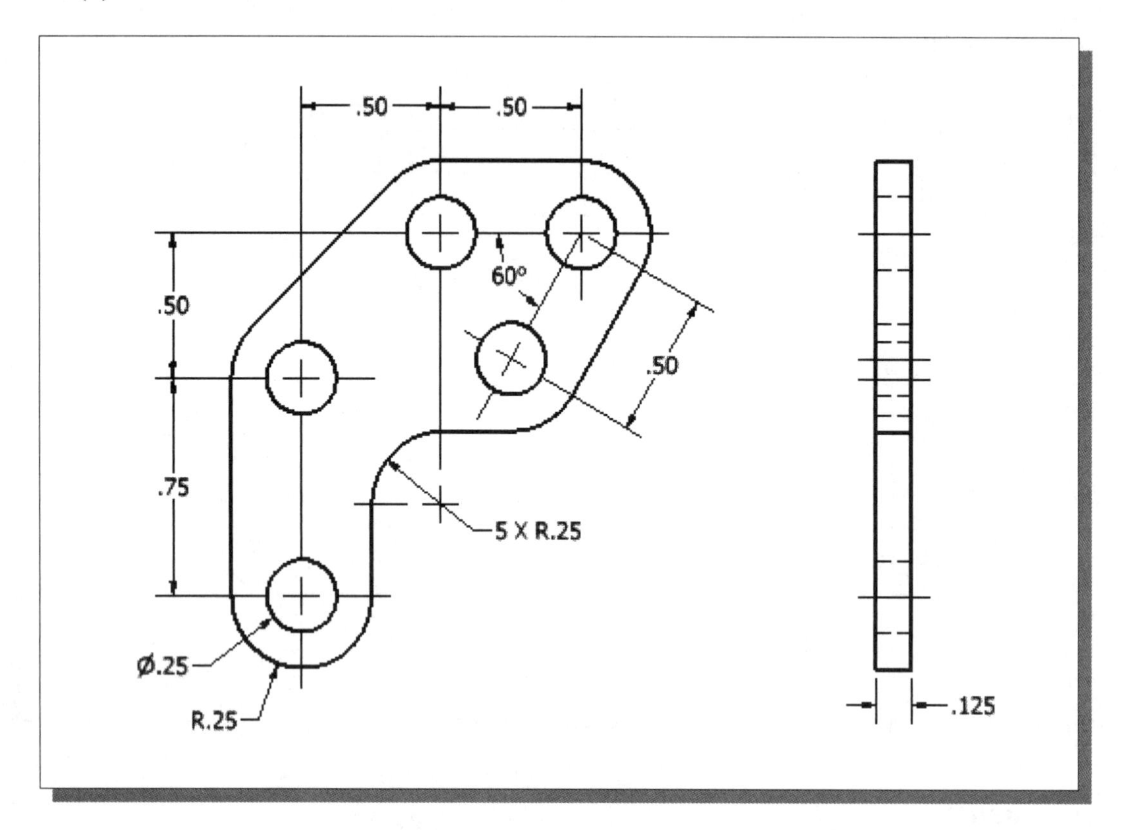

(e) **V-Link**

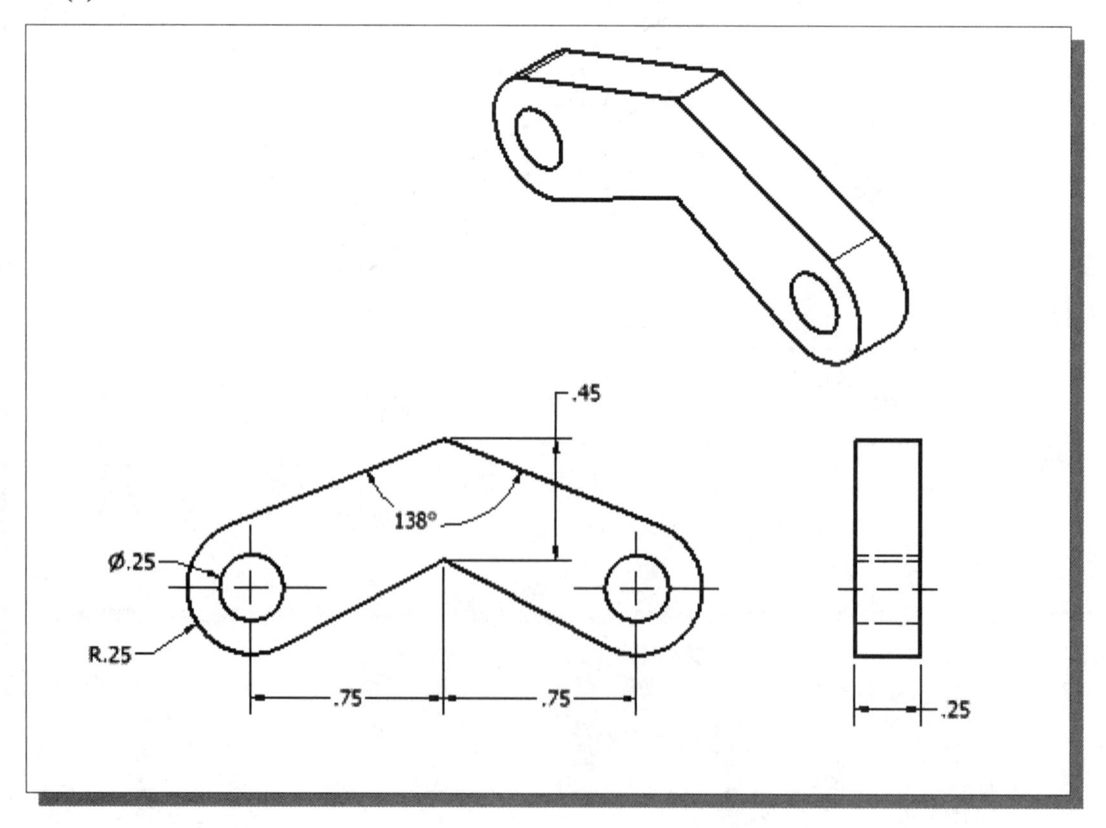

(f) **Rod**

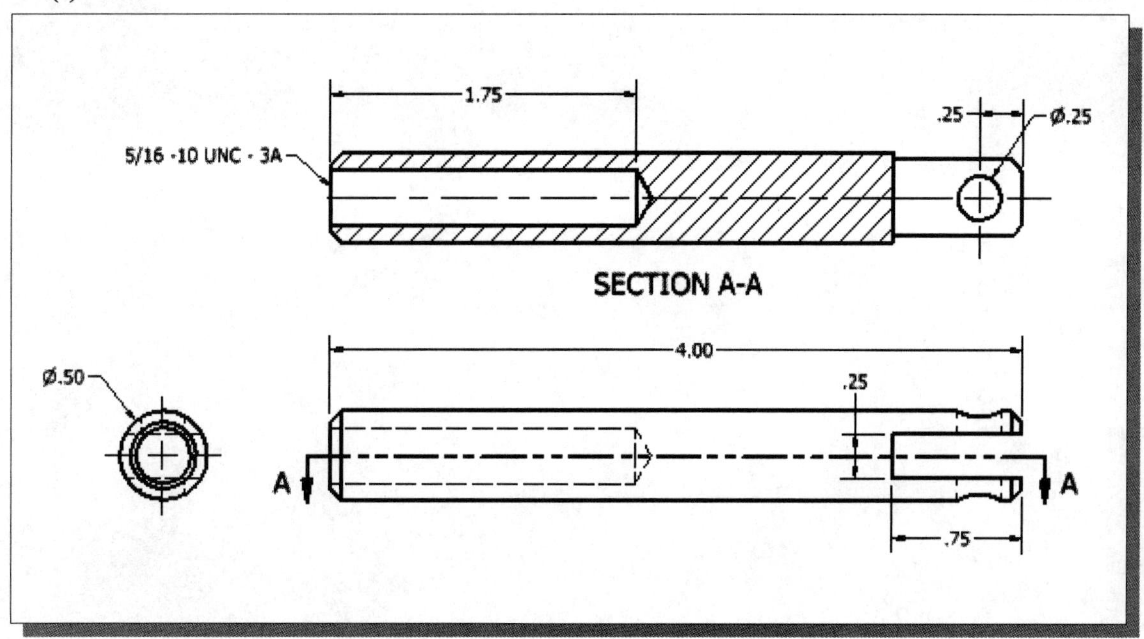

(g) **Bushing**

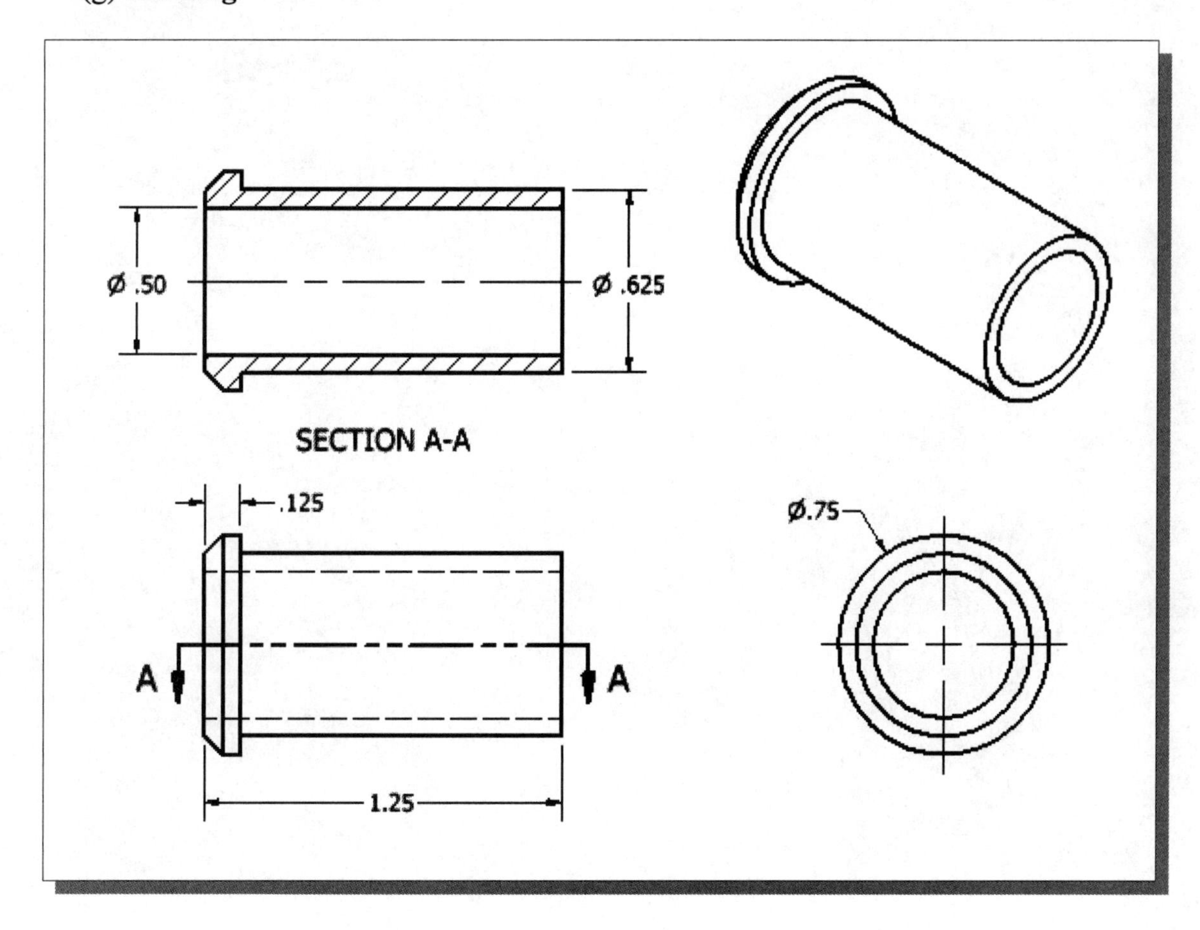

Notes:

INDEX

E

F

G

H

I

W

X

Y

Z